W9-CHO-031

Linear Circuits

General Information	1

Thermal Information	2

Operational Amplifiers	3

Voltage Comparators	4

Special Functions	5

Voltage Regulators	6

Data Acquisition	7

Appendix	A

Linear Circuits
Data Book

TEXAS
INSTRUMENTS

In Memoriam
Bryan E. Zimmerman
1929—1983

IMPORTANT NOTICE

Texas Instruments reserves the right to make changes at any time in order to improve design and to supply the best product possible.

Texas Instruments assumes no responsibility for infringement of patents or rights of others based on Texas Instruments applications assistance or product specifications, since TI does not possess full access to data concerning the use or applications of customer's products. TI also assumes no responsibility for customer product designs.

Information contained herein supersedes data published in The Linear Control Circuits Data Book, Second Edition, and the Voltage Regulator Data Book, SLVD001.

ISBN 0-89512-089-5
ISSN 0741-4226

INTRODUCTION

The rapid advance in high-tech digital processing creates new demands for microprocessor-compatible circuits that can sense, amplify, and convert analog signals or provide regulated power to a system. In this volume, Texas Instruments presents specifications and technical information on our broad line of integrated circuits designed for applications that involve analog signal conditioning. That product line includes:

- Operational amplifiers
- Voltage comparators
- Regulators
- Power supply monitors
- Switching-mode power supply circuits
- Hall-effect circuits
- Current mirrors
- Floppy-disk circuits for control, reading, or writing
- Timers
- A/D converters
- Video amplifiers
- Analog switches

These circuits span the recent rapid development of integrated circuit technology from classical bipolar through BIFET™ and BIDFET™ to TI's new LinCMOS™ processing that provides a step-function improvement in input impedance, power dissipation, and threshold stability. New surface-mount packages include both plastic and ceramic chip carriers and the small-outline packages that increase board density with little impact on power handling capability.

Ordering information and mechanical data are in the Appendix. Section 1 contains an alphanumeric index that lists page numbers for all the device types included, and each data sheet section provides a functional selection guide to the devices in that section.

While this volume offers design and specification data only for Linear components, complete technical data for any TI semiconductor product is available from:

Texas Instruments
Literature Response Center
P.O. Box 401560
Dallas, Texas 75240

Texas Instruments
Customer Response Center
(214) 995-6611

If you need sales assistance with any TI semiconductor products, please contact your nearest TI field sales office or TI authorized distributor. A listing can be found at the back of this data book.

General Information 1

Thermal Information 2

Operational Amplifiers 3

Voltage Comparators 4

Special Functions 5

Voltage Regulators 6

Data Acquisition 7

Appendix A

Contents

	Page
Section 1. General Information	1-1
Alphanumeric Index	1-4
Section 2. Thermal Information	2-1
Thermal Resistance of Integrated Circuits	2-3
Power Dissipation Derating Curves for Integrated Circuit Packages	2-4
Section 3. Operational Amplifiers	3-1
Selection Guide	3-3
Glossary of Terms and Definitions	3-13
Individual Data Sheets	3-17
Section 4. Voltage Comparators	4-1
Selection Guide	4-3
Glossary of Terms and Definitions	4-5
Individual Data Sheets	4-9
Section 5. Special Functions	5-1
Selection Guide	5-3
Individual Data Sheets	5-7
Section 6. Voltage Regulators	6-1
Selection Guide	6-3
Glossary of Terms and Definitions	6-8
Individual Data Sheets	6-11
Section 7. Data Acquisition	7-1
Selection Guide	7-3
Individual Data Sheets	7-5
Appendix A	A-1
Ordering Instructions	A-3
Mechanical Data	A-4

1

General Information

Contents

1

ADC0801	7-5	LM320-15	6-21
ADC0802	7-5	LM324	3-29
ADC0803	7-5	LM324A	3-29
ADC0804	7-11	LM330	6-27
ADC0805	7-11	LM337	6-17
ADC0808	7-17	LM339	4-25
ADC0809	7-17	LM339A	4-25
ADC0831	7-23	LM340-05	6-33
ADC0832	7-23	LM340-12	6-33
ADC0834	7-23	LM340-15	6-33
ADC0838	7-23	LM348	3-33
LM101A	3-17	LM350	6-41
LM106	4-9	LM358	3-36
LM107	3-21	LM358A	3-36
LM108	3-25	LM393	4-29
LM110	3-27	LM393A	4-29
LM111	4-15	LM2900	3-47
LM124	3-29	LM2901	4-25
LM139	4-25	LM2902	3-29
LM139A	4-25	LM2903	4-29
LM148	3-33	LM2904	3-36
LM158	3-36	LM2930-5	6-45
LM193	4-29	LM2930-8	6-45
LM201A	3-17	LM2931-5	6-51
LM206	4-9	LM3302	4-35
LM207	3-21	LM3900	3-47
LM208	3-25	MC1445	5-7
LM210	3-27	MC1458	3-53
LM211	4-15	MC1545	5-7
LM217	6-11	MC1558	3-53
LM218	3-43	MC3423	6-55
LM219	4-33	MC3469	5-9
LM224	3-29	MC3470	5-10
LM224A	3-29	MC3471	5-19
LM237	6-17	MC3303	3-57
LM239	4-25	MC3403	3-57
LM239A	4-25	MC3503	3-57
LM248	3-33	MC79L05	6-57
LM258	3-36	MC79L05A	6-57
LM258A	3-36	MC79L12	6-57
LM293	4-29	MC79L12A	6-57
LM293A	4-29	MC79L15	6-57
LM301A	3-17	MC79L15A	6-57
LM306	4-9	MC34060	6-61
LM307	3-21	MC35060	6-61
LM308	3-25	NE5532	3-63
LM310	3-27	NE5532A	3-63
LM311	4-15	NE5534	3-91
LM317	6-11	NE5534A	3-91
LM318	3-43	NE555	5-21
LM319	4-33	NE556	5-31
LM320-5	6-21	NE592	5-35
LM320-12	6-21	NE592A	5-35

General Information

1

General Information

OP-07C	3-67	TL061A	3-103
OP-07D	3-67	TL061B	3-103
OP-07E	3-67	TL062	3-103
OP-12A	3-71	TL062A	3-103
OP-12B	3-71	TL062B	3-103
OP-12C	3-71	TL064	3-103
OP-12E	3-71	TL064A	3-103
OP-12F	3-71	TL064B	3-103
OP-12G	3-71	TL066	3-113
OP-227E	3-77	TL066A	3-113
OP-227F	3-77	TL066B	3-113
OP-227G	3-77	TL068	3-123
RC4136	3-81	TL070	3-125
RC4193	6-67	TL070A	3-125
RC4558	3-87	TL071	3-125
RC4559	3-81	TL071A	3-125
RM4136	3-83	TL071B	3-125
RM4193	6-67	TL072	3-125
RM4558	3-87	TL072A	3-125
RV4136	3-83	TL072B	3-125
RV4558	3-87	TL074	3-125
SA555	5-21	TL074A	3-125
SA556	5-31	TL074B	3-125
SE555	5-21	TL075	3-125
SE555C	5-21	TL080	3-135
SE556	5-31	TL080A	3-135
SE556C	5-31	TL081	3-135
SE592	5-35	TL081A	3-135
SE5534	3-91	TL081B	3-135
SE5534A	3-91	TL082	3-135
SG1524	6-69	TL082A	3-135
SG1525A	6-81	TL082B	3-135
SG1527A	6-81	TL083	3-135
SG2524	6-69	TL083A	3-135
SG2525A	6-81	TL084	3-135
SG2527A	6-81	TL084A	3-135
SG3524	6-69	TL084B	3-135
SG3525A	6-81	TL085	3-135
SG3527A	6-81	TL087	3-145
SN28827	5-43	TL088	3-145
TL010	5-49	TL136	3-151
TL011	5-53	TL160	5-63
TL012	5-53	TL170	5-65
TL014	5-53	TL172	5-67
TL021	5-53	TL173	5-69
TL022	3-95	TL182	7-37
TL026	5-59	TL185	7-37
TL030	5-61	TL188	7-37
TL044	3-99	TL191	7-37
TL060	3-103	TL287	3-145
TL060A	3-103	TL288	3-145
TL060B	3-103	TL291	3-155
TL061	3-103	TL292	3-155

TEXAS INSTRUMENTS

POST OFFICE BOX 225012 • DALLAS, TEXAS 75265

1

General Information

TL294	3-155	TL1525A	6-153
TL317	6-91	TL1527A	6-153
TL321	3-157	TL2525A	6-153
TL322	3-161	TL2527A	6-153
TL331	4-37	TL3525A	6-153
TL430	6-95	TL3527A	6-153
TL431	6-99	TL7700	6-163
TL493	6-107	TL7702A	6-165
TL494	6-107	TL7705A	6-165
TL495	6-107	TL7709A	6-165
TL496	6-115	TL7712A	6-165
TL497A	6-119	TL7715A	6-165
TL499	6-124	TLC251	3-165
TL500	7-43	TLC251A	3-165
TL501	7-43	TLC251B	3-165
TL502	7-43	TLC252	3-175
TL503	7-43	TLC252A	3-175
TL505	7-57	TLC252B	3-175
TL506	4-39	TLC25L2	3-175
TL507	7-63	TLC25L2A	3-175
TL510	4-45	TLC25L2B	3-175
TL514	4-51	TLC25M2	3-175
TL520	7-67	TLC25M2A	3-175
TL521	7-67	TLC25M2B	3-175
TL522	7-67	TLC254	3-187
TL530	7-77	TLC254A	3-187
TL531	7-77	TLC254B	3-187
TL532	7-87	TLC25L4	3-187
TL533	7-87	TLC25L4A	3-187
TL580	6-125	TLC25L4B	3-187
TL592	5-73	TLC25M4	3-187
TL592A	5-73	TLC25M4A	3-187
TL592B	5-77	TLC25M4B	3-187
TL593	6-127	TLC261	3-199
TL594	6-127	TLC261A	3-199
TL595	6-127	TLC261B	3-199
TL601	7-95	TLC262	3-199
TL604	7-95	TLC262A	3-199
TL607	7-95	TLC262B	3-199
TL610	7-95	TLC264	3-199
TL710	4-59	TLC264A	3-199
TL712	4-63	TLC264B	3-199
TL721	4-65	TLC271	3-165
TL780-05	6-137	TLC271A	3-165
TL780-12	6-137	TLC271B	3-165
TL780-15	6-137	TLC272	3-175
TL783	6-141	TLC272A	3-175
TL810	4-67	TLC272B	3-175
TL811	4-73	TLC27L2	3-175
TL820	4-79	TLC27L2A	3-175
TL851	5-79	TLC27L2B	3-175
TL852	5-83	TLC27M2	3-175
TL1451	6-151	TLC27M2A	3-175

TEXAS
INSTRUMENTS
POST OFFICE BOX 225012 • DALLAS, TEXAS 75265

1

General Information

TLC27M2B	3-175	uA78L08	6-183
TLC274	3-187	uA78L08A	6-183
TLC274A	3-187	uA78L09	6-183
TLC274B	3-187	uA78L09A	6-183
TLC27L4	3-187	uA78L12	6-183
TLC27L4A	3-187	uA78L12A	6-183
TLC27L4B	3-187	uA78L15	6-183
TLC27M4	3-187	uA78L15A	6-183
TLC27M4A	3-187	uA78M05	6-189
TLC27M4B	3-187	uA78M06	6-189
TLC277	3-201	uA78M08	6-189
TLC27L7	3-201	uA78M10	6-189
TLC27M7	3-201	uA78M12	6-189
TLC372	4-83	uA78M15	6-189
TLC374	4-85	uA78M20	6-189
TLC532A	7-101	uA78M24	6-189
TLC533A	7-101	uA7905	6-201
TLC540	7-109	uA7906	6-201
TLC541	7-109	uA7908	6-201
TLC549	7-115	uA7912	6-201
TLC551	5-89	uA7915	6-201
TLC552	5-93	uA7918	6-201
TLC555	5-97	uA7924	6-201
TLC556	5-93	uA7952	6-201
TLC7126	7-119	uA79M05	6-207
uA702	3-211	uA79M06	6-207
uA709	3-215	uA79M08	6-207
uA709A	3-215	uA79M12	6-207
uA710	4-87	uA79M15	6-207
uA711	4-91	uA79M20	6-207
uA714C	3-219	uA79M24	6-207
uA714E	3-219	UC3846	6-217
uA714L	3-219	UC3847	6-217
uA723	6-169		
uA733	5-101		
uA741	3-223		
uA748	3-229		
uA2240	5-109		
uA7805	6-175		
uA7806	6-175		
uA7808	6-175		
uA7810	6-175		
uA7812	6-175		
uA7815	6-175		
uA7818	6-175		
uA7824	6-175		
uA7885	6-175		
uA78L02	6-183		
uA78L02A	6-183		
uA78L05	6-183		
uA78L05A	6-183		
uA78L06	6-183		
uA78L06A	6-183		

TEXAS
INSTRUMENTS
POST OFFICE BOX 225012 • DALLAS, TEXAS 75265

Linear Circuits

General Information	1
Thermal Information	2
Operational Amplifiers	3
Voltage Comparators	4
Special Functions	5
Voltage Regulators	6
Data Acquisition	7
Appendix	A

2

Thermal Information

THERMAL RESISTANCE

PACKAGE	PINS	JUNCTION-TO-CASE THERMAL RESISTANCE $R_{\theta JC}$(°C/W)	JUNCTION-TO-AMBIENT THERMAL RESISTANCE $R_{\theta JA}$(°C/W)
D plastic dual-in-line	8	51	172
	14, 16	33	131
FH ceramic chip carrier	20, 28	35	104
FK ceramic chip carrier	20	35	91
FN plastic chip carrier	20	37	114
J ceramic dual-in-line (glass-mounted chips)	14 thru 20	60	122
J ceramic dual-in-line[†] (alloy-mounted chips)	14 thru 20	29[†]	91[†]
JG ceramic dual-in-line (glass-mounted chips)	8	58	151
JG ceramic dual-in-line[†] (alloy-mounted chips)	8	26[†]	119[†]
LP plastic plug-in	3	40	160
LU plastic plug-in	3	40	178
N plastic dual-in-line	14 thru 20	72	143
	28	45	100
	40	40	100
P plastic dual-in-line	8	79	172
U ceramic flat	10, 14	55	185
W ceramic flat	14, 16	60	125

[†] In addition to those products so designated on their data sheets, all devices having a type number prefix of "SNC" or "SNM," or a suffix of "/883B" have alloy-mounted chips.

PLASTIC PACKAGES

These curves are for use with the continuous dissipation ratings specified on the individual data sheets. Those ratings apply up to the temperature at which the rated level intersects the appropriate derating curve or the maximum operating free-air temperature.

DISSIPATION DERATING CURVE

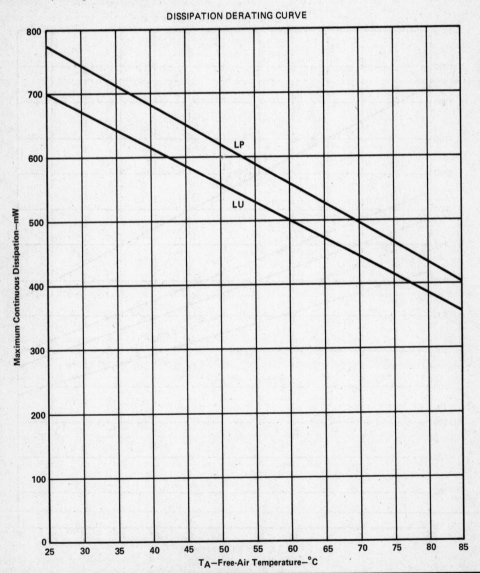

Thermal Information

2

PLASTIC PACKAGES (CONTINUED)

These curves are for use with the continuous dissipation ratings specified on the individual data sheets. Those ratings apply up to the temperature at which the rated level intersects the appropriate derating curve or the maximum operating free-air temperature.

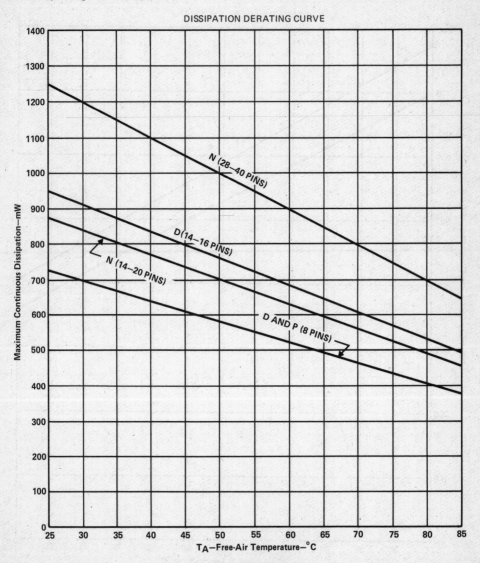

DISSIPATION DERATING CURVE

N (28—40 PINS)

D (14—16 PINS)

N (14—20 PINS)

D AND P (8 PINS)

Maximum Continuous Dissipation—mW

T_A—Free-Air Temperature—°C

TEXAS
INSTRUMENTS
POST OFFICE BOX 225012 • DALLAS, TEXAS 75265

FLAT PACKAGES

These curves are for use with the continuous dissipation ratings specified on the individual data sheets. Those ratings apply up to the temperature at which the rated level intersects the appropriate derating curve or the maximum operating free-air temperature.

DISSIPATION DERATING CURVE

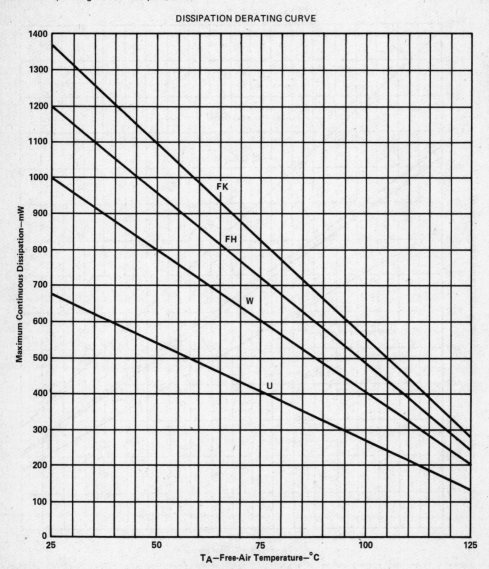

TEXAS INSTRUMENTS
POST OFFICE BOX 225012 • DALLAS, TEXAS 75265

CERAMIC DUAL-IN-LINE PACKAGES

These curves are for use with the continuous dissipation ratings specified on the individual data sheets. Those ratings apply up to the temperature at which the rated level intersects the appropriate derating curve or the maximum operating free-air temperature.

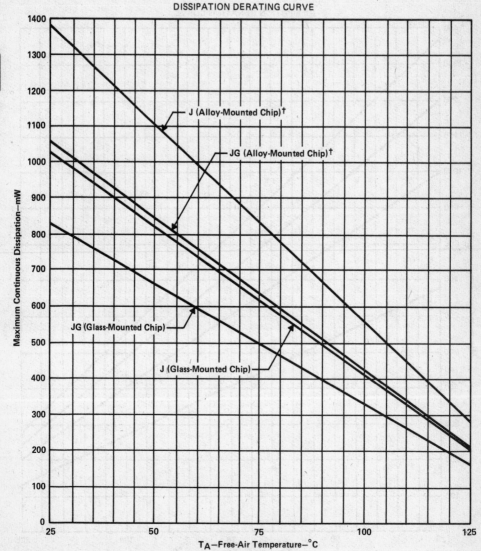

DISSIPATION DERATING CURVE

†In addition to those products so designated on their data sheets, all devices having a type number prefix of "SNC" or "SNM", or a suffix of "/883B" have alloy-mounted chips.

2

Thermal Information

TEXAS
INSTRUMENTS
POST OFFICE BOX 225012 • DALLAS, TEXAS 75265

General Information   1

Thermal Information   2

Operational Amplifiers   3

Voltage Comparators   4

Special Functions   5

Voltage Regulators   6

Data Acquisition   7

Appendix   A

3

Operational Amplifiers

OPERATIONAL AMPLIFIERS

noncompensated, single

military temperature range (values specified for T_A = 25°C)

DEVICE NUMBER	DESCRIPTION	B_1 (MHz) TYP	SR (V/μs) TYP	V_{IO} (mV) MAX	I_{IB} (nA) MAX	A_{VD} (V/mV) MIN	SUPPLY VOLTAGE (V) MIN	MAX	PACKAGES	PAGE
TL080M	BIFET, Low Power	3	13	6	0.2	25	±3.5	±18	JG	3-135
uA709AM	General Purpose	1	0.3	2	200	45 Typ	±5	±18	J,JG,U,W	3-215
uA709M	General Purpose	1	0.3	5	500	45 Typ		±18	J,JG,U,W	3-215
LM108	High Performance	1	0.3	2	2	50	±5	±18	JG	3-25
LM101A	High Performance	1	0.5	2	75	50	±5	±22	FH,FK,JG,U,W	3-17
uA748M	General Purpose	1	0.5	5	500	±10	±2	±22	JG,U	3-229
TL060M	BIFET, Low Power	1	3.5	6	0.2	4	±1.5	±18	JG	3-103

industrial temperature range (values specified for T_A = 25°C)

DEVICE NUMBER	DESCRIPTION	B_1 (MHz) TYP	SR (V/μs) TYP	V_{IO} (mV) MAX	I_{IB} (nA) MAX	A_{VD} (V/mV) MIN	SUPPLY VOLTAGE (V) MIN	MAX	PACKAGES	PAGE
TL070I	BIFET, Low Noise	13	13	6	200	50	±3.5	±18	D,JG,P	3-125
TL080I	BIFET, Low Power	3	13	6	200	50	±3.5	±18	JG,P	3-135
LM208	High Performance	1	0.3	2	2	50	±5	±18	D,JG,P	3-25
LM201A	High Performance	1	0.5	2	75	50	±5	±22	D,JG,P,W	3-17
TL060I	BIFET, Low Power	1	3.5	6	0.2	4	±1.5	±18	D,JG,P	3-103

commercial temperature range (values specified for T_A = 25°C)

DEVICE NUMBER	DESCRIPTION	B_1 (MHz) TYP	SR V/μs TYP	V_{IO} (mV) MAX	I_{IB} (nA) MAX	A_{VD} (V/mV) MIN	SUPPLY VOLTAGE (V) MIN	MAX	PACKAGES	PAGE
TL070AC	BIFET, Low Noise	3	13	6	0.2	50	±3.5	±18	D,JG,P	3-125
TL070C	BIFET, Low Noise	3	13	10	0.2	25	±3.5	±18	D,JG,P	3-125
TL080AC	BIFET, Low Power	3	13	6	0.2	50	±3.5	±18	JG,P	3-135
TL080C	BIFET, Low Power	3	13	15	0.4	25	±3.5	±18	JG,P	3-135
uA709C	General Purpose	1	0.3	7.5	1500	15		±18	JG,P	3-215
LM308	High Performance	1	0.3	7.5	7	25	±5	±18	D,JG,P	3-25
uA748C	General Purpose	1	0.5	6	500	20	±2	±18	D,JG,P	3-229
TL060AC	BIFET, Low Power	1	3.5	6	0.2	4	±1.5	±18	D,JG,P	3-103
TL060BC	BIFET, Low Power	1	3.5	3	0.2	4	±1.5	±18	D,JG,P	3-103
TL060C	BIFET, Low Power	1	3.5	15	0.2	3	±1.5	±18	D,JG,P	3-103
LM301A	High Performance	1	7.5	50	75	25	±5	±18	D,JG,P,W	3-17

3

Operational Amplifiers

SELECTION GUIDE

internally compensated, single

military temperature range

(values specified for $T_A = 25°C$)

DEVICE NUMBER	DESCRIPTION	B_1 (MHz) TYP	SR (V/μs) TYP	V_{IO} (mV) MAX	I_{IB} (nA) MAX	A_{VD} (V/mV) MIN	SUPPLY VOLTAGE (V) MIN	MAX	PACKAGES	PAGE
TL291M	High Speed	20	50				±4	±18	JG	3-155
SE5534	High Performance	10	13	2	800	50	±3	±20	FH,FK,JG,U	3-91
SE5534A	High Performance	10	13	2	800	50	±3	±20	FH,FK,JG,U	3-91
LM110	Unity-Gain Voltage Follower	10	30	4	3	1	±5	±18	JG	3-27
TL081M	BIFET, General Purpose	3	13	6	0.2	25	±4	±18	FH,FK,JG	3-135
TL088M	BIFET, Low V_{IO}	3	13	3	0.4	50	±3.5	±18	JG,U	3-145
TL071M	BIFET, Low Noise	3	13	6	0.2	35	±5	±18	FH,FK,JG	3-125
TLC271AM	LinCMOS, Programmable	2.3	4.5	5	0.001 Typ	10	4	16	FH,FK,JG	3-165
TLC271BM	LinCMOS, Programmable	2.3	4.5	2	0.001 Typ	10	4	16	FH,FK,JG	3-165
TLC271M	LinCMOS, Programmable	2.3	4.5	10	0.001 Typ	10	4	16	FH,FK,JG	3-165
TLC277M	LinCMOS, High Bias	2.3	4.5	0.5	0.001 Typ	10	4	16	FH,FK,JG	3-201
uA741M	General Purpose	1	0.5	5	500	50	±2	±22	FH,FK,J,JG,U	3-223
LM107	High Performance	1	0.5	2	75	50	±2	±22	J,JG,U,W	3-21
TL061M	BIFET, Low Power	1	3.5	6	0.2	4	±1.5	±18	FH,FK,JG,U	3-103
TL066M	BIFET, Adjustable Low-Power	1	3.5	6	0.2	4	±1.2	±18	FH,FK,JG	3-113
OP-012A	Precision Low-Input Current	0.8	0.12	0.15	2	80	±5	±20	JG	3-71
OP-012B	Precision Low-Input Current	0.8	0.12	0.3	2	80	±5	±20	JG	3-71
OP-012C	Precision Low-Input Current	0.8	0.12	1	5	40	±5	±20	JG	3-71
TLC27M7M	LinCMOS, Programmable	0.7	0.6	0.5	0.001 Typ	20	4	16	FH,FK,JG	3-201
TL321M	High Performance	0.6	0.3	5	−150	50	3	30	FH,FK,JG	3-157
uA702M	General Purpose	0.5	11	2	500	6000	6 / −3	14 / −7	JG,U	3-201
TLC27L7M	LinCMOS, Programmable	0.1	0.04	0.5	0.001 Typ	30	4	16	FH,FK,JG	3-201

automotive temperature range

(values specified for $T_A = 25°C$)

DEVICE NUMBER	DESCRIPTION	B_1 (MHz) TYP	SR (V/μs) TYP	V_{IO} (mV) MAX	I_{IB} (nA) MAX	A_{VD} (V/mV) MIN	SUPPLY VOLTAGE (V) MIN	MAX	PACKAGES	PAGE
LM218	High Performance	15	70	4	250	50	±5	±20	D,JG,P	3-43
TLC271AI	LinCMOS, Programmable	2.3	4.5	5	0.001 Typ	7	3	16	D,JG,P	3-165
TLC271BI	LinCMOS, Programmable	2.3	4.5	2	0.001 TYP	7	3	16	D,JG,P	3-165
TLC271I	LinCMOS, Programmable	2.3	4.5	10	0.001 Typ	7	3	16	D,JG,P	3-165
TLC277I	LinCMOS, Programmable	2.3	4.5	0.5	0.001 Typ	10	3	16	D,JG,P	3-201
LM207	High Performance	1	0.5	2	75	50	±2	±22	J,N,W	3-21
TLC27M7I	LinCMOS, Medium Bias	0.7	0.6	0.5	0.001 Typ	20	3	16	D,JG,P	3-201
TLC27L7I	LinCMOS, Low Bias	0.1	0.04	0.5	0.001 Typ	30	3	16	D,JG,P	3-201
TLC261AI	LinCMOS, Programmable		12	5	0.001 Typ		2	16	D,JG,P	3-199
TLC261BI	LinCMOS, Programmable		12	2	0.001 Typ		2	16	D,JG,P	3-199
TLC261I	LinCMOS, Programmable		12	10	0.001 Typ		2	16	D,JG,P	3-199

3

Operational Amplifiers

TEXAS
INSTRUMENTS
POST OFFICE BOX 225012 • DALLAS, TEXAS 75265

internally compensated, single

industrial temperature range

(values specified for T_A = 25°C)

DEVICE NUMBER	DESCRIPTION	B_I (MHz) TYP	SR (V/µs) TYP	V_{IO} (mV) MAX	I_{IB} (nA) MAX	A_{VD} (V/mV) MIN	SUPPLY VOLTAGE (V) MIN	MAX	PACKAGES	PAGE
TL087I	BIFET, Low Offset	25	13	0.5	0.2	50	±4	±18	D,JG,P	3-145
LM210	Unity-gain Voltage Follower	20	30	4	3		±5	±18	JG,P	3-27
TL071I	BIFET, Low Noise	3	13	6	0.2	50	±3.5	±18	D,JG,P	3-125
TL081I	BIFET, General Purpose	3	13	6	0.2	50	±3.5	±18	JG,P	3-135
TL088I	BIFET, Low V_{IO}	3	13	1	0.2	50	±4	±18	D,JG,P	3-145
TL066I	BIFET, Adjustable Low-Power	1	3.5	6	0.2	4	±1.2	±18	D,JG,P	3-113
TL061I	BIFET, Low Power	1	3.5	6	0.2	4	±1.5	±18	D,JG,P	3-103
TL321I	High Performance	0.6	0.3	5	−150	50	3	30	JG,P	3-157

commercial temperature range

(values specified for T_A = 25°C)

DEVICE NUMBER	DESCRIPTION	B_I (MHz) TYP	SR (V/µs) TYP	V_{IO} (mV) MAX	I_{IB} (nA) MAX	A_{VD} (V/mV) MIN	SUPPLY VOLTAGE (V) MIN	MAX	PACKAGES	PAGE
LM310	Unity-Gain Voltage Follower	20	30	7.5	7		±5	±18	JG,P	3-27
TL291C	High Speed	20	50				±4	±18	JG,P	3-155
LM318	High Performance	15	70	10	250	25	±5	±20	D,JG,P	3-43
NE5534	High Performance	10	6	4	1500	25	±3	±20	JG,P	3-91
NE5534A	High Performance	10	6	4	1500	25	±3	±20	JG,P	3-91
OP-227E	Low Noise	8	2.8	0.08	±40	800			J,N	3-77
OP-227F	Low Noise	8	2.8	0.12	±55	800			J,N	3-77
OP-227G	Low Noise	8	2.8	0.18	±80	700			J,N	3-77
TL087C	BIFET, General Purpose	3	13	0.5	0.2	50	±5	±18	D,JG,P	3-145
TL071AC	BIFET, Low Noise	3	13	6	0.2	50	±3.5	±18	D,JG,P	3-125
TL071BC	BIFET, Low Noise	3	13	3	0.2	50	+3.5	±18	D,JG,P	3-125
TL071C	BIFET, Low Noise	3	13	10	0.2	25	±3.5	±18	D,JG,P	3-125
TL081AC	BIFET, General Purpose	3	13	6	0.2	50	±3.5	±18	JG,P	3-135
TL081BC	BIFET, General Purpose	3	13	3	0.2	50	±3.5	±18	JG,P	3-135
TL081C	BIFET, General Purpose	3	13	15	0.4	25	±3.5	±18	JG,P	3-135
TL088C	BIFET, Low V_{IO}	3	13	1	0.2	50	±4	±18	D,JG,P	3-145
TLC277C	LinCMOS, High Bias	2.3	4.5	0.5	0.001 Typ	10 Typ	3	16	D,JG,P	3-201
uA741C	General Purpose	1	0.5	6	500	20	±2	±18	D,JG,P	3-223
LM307	High Performance	1	0.5	7.5	250	25	±2	±18	D,J,JG,N,P,W	3-21
TL061AC	BIFET, Low Power	1	3.5	6	0.2	4	±1.5	±18	D,JG,P	3-103
TL061BC	BIFET, Low Power	1	3.5	3	0.2	4	±1.5	±18	D,JG,P	3-103
TL061C	BIFET, Low Power	1	3.5	15	0.2	3	±1.5	±18	D,JG,P	3-103
TL066AC	BIFET, Adjustable Low-Power	1	3.5	6	0.2	4	±1.2	±18	D,JG,P	3-113
TL066BC	BIFET, Adjustable Low-Power	1	3.5	3	0.2	4	±1.2	±18	D,JG,P	3-113
TL066C	BIFET, Adjustable Low-Power	1	3.5	15	0.4	3	±1.2	±18	D,JG,P	3-113
TL068C	BIFET, Buffer	1	7	15	0.4		±1.5	18	LP	3-123

3

Operational Amplifiers

TEXAS INSTRUMENTS
POST OFFICE BOX 225012 ● DALLAS, TEXAS 75265

internally compensated, single

commercial temperature range (continued)   (values specified for $T_A = 25°C$)

DEVICE NUMBER	DESCRIPTION	B$_l$ (MHz) TYP	SR (V/µs) TYP	V$_{IO}$ (mV) MAX	I$_{IB}$ (nA) MAX	A$_{VD}$ (V/mV) MIN	SUPPLY VOLTAGE (V)		PACKAGES	PAGE
							MIN	MAX		
TLC251AC	LinCMOS, Programmable	0.7	0.6	5	0.001 Typ	10 Typ	1	16	D,JG,P	3-165
TLC251BC	LinCMOS, Programmable	0.7	0.04	2	0.001 Typ	10 Typ	1	16	D,JG,P	3-165
TLC251C	LinCMOS, Programmable	0.7	0.04	10	0.001 Typ	10 Typ	1	16	D,JG,P	3-165
TLC271AC	LinCMOS, Programmable	0.7	0.04	5	0.001 Typ	10 Typ	3	16	D,JG,P	3-165
TLC271BC	LinCMOS, Programmable	0.7	0.04	3	0.001 Typ	10 Typ	3	16	D,JG,P	3-165
TLC271C	LinCMOS, Programmable	0.7	0.04	10	0.001 Typ	10 Typ	3	16	D,JG,P	3-165
TLC27M7C	LinCMOS, Medium Bias	0.7	0.6	0.5	0.001 Typ	20	3	16	D,JG,P	3-201
uA714C	Ultra-Low Offset Voltage	0.6	0.17	0.15	± 7	100	± 3	± 18	JG,P	3-219
uA714E	Ultra-Low Offset Voltage	0.6	0.17	0.075	± 4	200	± 3	± 18	JG,P	3-219
uA714L	Ultra-Low Offset Voltage	0.6	0.17	0.25	± 30	50	± 3	± 18	JG,P	3-219
TL321C	High Performance	0.6	0.3	7	− 250	25	3	30	JG,P	3-157
OP-07C	Ultra-Low Offset	0.6	0.3	0.15	± 7	100	± 3	± 18	JG,P	3-67
OP-07D	Ultra-Low Offset	0.6	0.3	0.15	± 12	400 Typ	± 3	± 18	JG,P	3-67
OP-07E	Ultra-Low Offset	0.6	0.3	0.75	+ 4	150	± 3	± 18	JG,P	3-67
OP-12E	Precision Low-Input Current	0.2	0.12	0.15	2	80	± 5	± 20	D,JG,P	3-71
OP-12F	Precision Low-Input Current	0.2	0.12	0.3	2	80	± 5	± 20	D,JG,P	3-71
OP-12G	Precision Low-Input Current	0.2	0.12	1	5	40	± 5	± 20	D,JG,P	3-71
TLC27L7C	LinCMOS, Low Bias	0.1	0.04	0.5	0.001 Typ	30	3	16	D,JG,P	3-201

3

Operational Amplifiers

TEXAS
INSTRUMENTS
POST OFFICE BOX 225012 • DALLAS, TEXAS 75265

internally compensated, dual

military temperature range (values specified for $T_A = 25°C$)

DEVICE NUMBER	DESCRIPTION	B_I (MHz) TYP	SR (V/μs) TYP	V_{IO} (mV) MAX	I_{IB} (nA) MAX	A_{VD} (V/mV) MIN	SUPPLY VOLTAGE (V) MIN	MAX	PACKAGES	PAGE
TL292M	High Frequency	20	50						JG	3-155
RM4558	High Performance	3	1.7	5	500	50		±22	JG	3-87
TL072M	BIFET, Low Noise	3	13	6	0.2	35	±3.5	±18	FH,FK,JG	3-125
TL082M	BIFET, General Purpose	3	13	16	0.2	25	±3.5	±18	FH,FK,JG	3-135
TL083M	BIFET, General Purpose	3	13	6	0.2	25	±3.5	±18	FH,FK,J	3-135
TL088M	BIFET, General Purpose	3	13	3	0.4	50	±3.5	±18	JG,U	3-145
TL288M	BIFET, General Purpose	3	13	3	0.4	50	±3.5	±18	JG,U	3-145
TLC272AM	LinCMOS, High Bias	2.3	4.5	5	0.001 Typ	10	4	16	FH,FK,JG	3-175
TLC272BM	LinCMOS, High Bias	2.3	4.5	2	0.001 Typ	10	4	16	FH,FK,JG	3-175
TLC272M	LinCMOS, High Bias	2.3	4.5	10	0.001 Typ	10	4	16	FH,FK,JG	3-175
TLC27L2AM	LinCMOS, Low Bias	2.3	4.5	5	0.001 Typ	30	4	16	FH,FK,JG	3-175
TLC27L2BM	LinCMOS, Low Bias	2.3	4.5	2	0.001 Typ	30	4	16	FH,FK,JG	3-175
TLC27L2M	LinCMOS, Low Bias	2.3	4.5	10	0.001 Typ	30	4	16	FH,FK,JG	3-175
TLC27M2AM	LinCMOS, Medium Bias	2.3	4.5	5	0.001 Typ	20	4	16	FH,FK,JG	3-175
TLC27M2BM	LinCMOS, Medium Bias	2.3	4.5	2	0.001 Typ	20	4	16	FH,FK,JG	3-175
TLC27M2M	LinCMOS, Medium Bias	2.3	4.5	10	0.001 Typ	20	4	16	FH,FK,JG	3-175
MC1558	General Purpose	1	0.5	5	500	50	±2	±22	FH,FK,JG,U	3-53
TL322M	Low Power	1	0.6	8	−500	200 Typ	±1.5	±18	JG	3-161
TL062M	BIFET, Low Power	1	3.5	6	0.2	4	±1.5	±18	FH,FK,JG,U	3-103
LM158	High Gain	0.6		5	−150	50	3	30	FH,FK,JG,U	3-36
TL022M	Low Power	0.5	0.5	5	100	72	±2	±22	U	3-95

3

Operational Amplifiers

SELECTION GUIDE

internally compensated, dual

automotive temperature range

(values specified for T_A = 25°C)

DEVICE NUMBER	DESCRIPTION	B_I (MHz) TYP	SR (V/µs) TYP	V_{IO} (mV) MAX	I_{IB} (nA) MAX	A_{VD} (V/mV) MIN	SUPPLY VOLTAGE (V)		PACKAGES	PAGE
							MIN	MAX		
LM2904	High Gain	5	1	7	−250	100 Typ	±3	±26	D,JG,P,U	3-36
RV4558	High Performance	3	1.7	6	500	50	±5	±18	D,JG,P	3-87
TLC272AI	LinCMOS, High Bias	2.3	4.5	5	0.002 Typ	10	3	16	D,JG,P	3-175
TLC272BI	LinCMOS, High Bias	2.3	4.5	2	0.002 Typ	10	3	16	D,JG,P	3-175
TLC272I	LinCMOS, High Bias	2.3	4.5	10	0.002 Typ	10	3	16	D,JG,P	3-175
TLC27L2AI	LinCMOS, Low Bias	2.3	4.5	5	0.002 Typ	30	3	16	D,JG,P	3-175
TLC27L2BI	LinCMOS, Low Bias	2.3	4.5	2	0.002 Typ	30	3	16	D,JG,P	3-175
TLC27L2I	LinCMOS, Low Bias	2.3	4.5	10	0.002 Typ	30	3	16	D,JG,P	3-175
TLC27M2AI	LinCMOS, Medium Bias	2.3	4.5	5	0.002 Typ	20	3	16	D,JG,P	3-175
TLC27M2BI	LinCMOS, Medium Bias	2.3	4.5	2	0.002 Typ	20	3	16	D,JG,P	3-175
TLC27M2I	LinCMOS, Medium Bias	2.3	4.5	10	0.002 Typ	20	3	16	D,JG,P	3-175
TL322I	Low Power	1	0.6	8	0.5	20	±1.5	±18	D,JG,P	3-157

industrial temperature range

(values specified for T_A = 25°C)

DEVICE NUMBER	DESCRIPTION	B_1 (MHz) TYP	SR (V/µs) TYP	V_{IO} (mV) MAX	I_{IB} (nA) MAX	A_{VD} (V/mV) MIN	SUPPLY VOLTAGE (V)		PACKAGES	PAGE
							MIN	MAX		
TL072I	BIFET, Low Noise	3	13	6	0.2	50	±3.5	±18	D,JG,P	3-125
TL082I	BIFET, General Purpose	3	13	6	0.2	50	±3.5	±18	JG,P	3-135
TL083I	BIFET, General Purpose	3	13	6	0.2	50	±3.5	±18	J,N	3-135
TL287I	BIFET, General Purpose	3	13	0.5	0.2	50	±3.5	±18	D,JG,P	3-145
TL288I	BIFET, General Purpose	3	13	0.5	0.2	50	±3.5	±18	D,JG,P	3-145
TL062I	BIFET, Low Power	1	3.5	6	0.2	4	±1.5	±18	D,JG,P	3-103
LM258	High Gain	0.6		5	−150	50	3	30	D,JG,P,U	3-36
LM258A	High Gain	0.6		3	−80	50	3	30	D,JG,P,U	3-36
TLC262AI	LinCMOS, Programmable		12	5	0.001 Typ		2	16	D,JG,P	3-199
TLC262BI	LinCMOS, Programmable		12	2	0.001 Typ		2	16	D,JG,P	3-199
TLC262I	LinCMOS, Programmable		12	10	0.001 Typ		2	16	D,JG,P	3-199

3

Operational Amplifiers

TEXAS
INSTRUMENTS

POST OFFICE BOX 225012 • DALLAS, TEXAS 75265

internally compensated, dual

commercial temperature range | (values specified for $T_A = 25°C$)

DEVICE NUMBER	DESCRIPTION	B_1 (MHz) TYP	SR (V/µs) TYP	V_{IO} (mV) MAX	I_{IB} (nA) MAX	A_{VD} (V/mV) MIN	SUPPLY VOLTAGE (V) MIN	MAX	PACKAGES	PAGE
TL292C	High Frequency	20	50				±4	±18	JG,P	3-155
NE5532	Low Noise	10	9	4	800	15	±3	±20	JG,P	3-63
NE5532A	Low Noise	10	9	4	800	15	±3	±20	JG,P	3-63
RC4559	High Performance	4	2	6	250	20		±18	D,P	3-101
RC4558	High Performance	3.5	1.7	5	500	50		±18	D,JG,P	3-103
TL072AC	BIFET, Low Noise	3	13	6	0.2	50	±3.5	±18	D,JG,P	3-145
TL072BC	BIFET, Low Noise	3	13	3	0.2	50	±3.5	±18	D,JG,P	3-145
TL072C	BIFET, Low Noise	3	13	10	0.2	25	±3.5	±18	D,JG,P	3-145
TL082AC	BIFET, General Purpose	3	13	6	0.2	50	±3.5	±18	JG,P	3-155
TL082BC	BIFET, General Purpose	3	13	3	0.2	50	±3.5	±18	JG,P	3-155
TL082C	BIFET, General Purpose	3	13	15	0.4	25	±3.5	±18	JG,P	3-155
TL083AC	BIFET, General Purpose	3	13	6	0.2	50	±3.5	±18	J,N	3-155
TL083C	BIFET, General Purpose	3	13	15	0.4	25	±3.5	±18	J,N	3-155
TL287C	BIFET, General Purpose	3	13	0.5	0.2	50	±3.5	±18	D,JG,P	3-155
TL288C	BIFET, General Purpose	3	13	0.5	0.2	50	±3.5	±18	D,JG,P	3-155
TLC252AC	LinCMOS, High Bias	2.3	4.5	5	0.001 Typ	10	1	16	D,JG,P	3-175
TLC252BC	LinCMOS, High Bias	2.3	4.5	2	0.001 Typ	10	1	16	D,JG,P	3-175
TLC252C	LinCMOS, High Bias	2.3	4.5	10	0.001 Typ	10	1	16	D,JG,P	3-175
TLC25L2AC	LinCMOS, Low Bias	2.3	4.5	5	0.001 Typ	30	1	16	D,JG,P	3-175
TLC25L2BC	LinCMOS, Low Bias	2.3	4.5	2	0.001 Typ	30	1	16	D,JG,P	3-175
TLC25L2C	LinCMOS, Low Bias	2.3	4.5	10	0.001 Typ	30	1	16	D,JG,P	3-175
TLC25M2AC	LinCMOS, Medium Bias	2.3	4.5	5	0.001 Typ	20	1	16	D,JG,P	3-175
TLC25M2BC	LinCMOS, Medium Bias	2.3	4.5	2	0.001 Typ	20	1	16	D,JG,P	3-175
TLC25M2C	LinCMOS, Medium Bias	2.3	4.5	10	0.001 Typ	20	1	16	D,JG,P	3-175
TLC272AC	LinCMOS, High Bias	2.3	4.5	5	0.001 Typ	10	3	16	D,JG,P	3-175
TLC272BC	LinCMOS, High Bias	2.3	4.5	2	0.001 Typ	10	3	16	D,JG,P	3-175
TLC272C	LinCMOS, High Bias	2.3	4.5	10	0.001 Typ	10	3	16	D,JG,P	3-175
TLC27L2AC	LinCMOS, Low Bias	2.3	4.5	5	0.001 Typ	20	3	16	D,JG,P	3-175
TLC27L2BC	LinCMOS, Low Bias	2.3	4.5	2	0.001 Typ	20	3	16	D,JG,P	3-175
TLC27L2C	LinCMOS, Low Bias	2.3	4.5	10	0.001 Typ	20	3	16	D,JG,P	3-175
TLC27M2AC	LinCMOS, Medium Bias	2.3	4.5	5	0.001 Typ	20	3	16	D,JG,P	3-175
TLC27M2BC	LinCMOS, Medium Bias	2.3	4.5	2	0.001 Typ	20	3	16	D,JG,P	3-175
TLC27M2C	LinCMOS, Medium Bias	2.3	4.5	10	0.001 Typ	20	3	16	D,JG,P	3-175
MC1458	General Purpose	1	0.5	6	500	20	±1.5	±18	D,JG,P,U	3-53
TL322C	Low Power	1	0.6	10	−500	20	±1.5	±18	D,JG,P	3-161
TL062AC	BIFET, Low Power	1	3.5	6	0.2	4	±1.2	±18	D,JG,P	3-103
TL062BC	BIFET, Low Power	1	3.5	3	0.2	4	±1.2	±18	D,JG,P	3-103
TL062C	BIFET, Low Power	1	3.5	15	0.2	3	±1.2	±18	D,JG,P	3-103
LM358	High Gain	0.6		7	−250	25	3	30	D,JG,P,U	3-36
LM358A	High Gain	0.6		3	−100	25	3	30	D,JG,P,U	3-36
TL022C	Low Power	0.5	0.5	5	250	60	±2	±18	JG,P	3-95

3

Operational Amplifiers

internally compensated, quad

military temperature range (values specified for T_A = 25°C)

DEVICE NUMBER	DESCRIPTION	B_1 (MHz) TYP	SR (V/µs) TYP	V_{IO} (mV) MAX	I_{IB} (nA) MAX	A_{VD} (V/mV) MIN	SUPPLY VOLTAGE (V) MIN	SUPPLY VOLTAGE (V) MAX	PACKAGES	PAGE
TL294M	High Frequency	20	50				±4	±18	J	3-155
RM4136	High Performance	3.5	1.7	4	400	50	±4	±22	FH,FK,J,W	3-83
TL074M	BIFET, Low Noise	3	13	9	0.2	35	±3.5	±18	FH,FK,J,W	3-125
TL084M	BIFET, General Purpose	3	13	9	0.2	25	±3.5	±18	FH,FK,J,W	3-135
TLC274AM	LinCMOS, High Bias	2.3	4.5	5	0.001 Typ	10	1	16	FH,FK,J	3-187
TLC274BM	LinCMOS, High Bias	2.3	4.5	2	0.001 Typ	10	1	16	FH,FK,J	3-187
TLC274M	LinCMOS, High Bias	2.3	4.5	10	0.001 Typ	10	1	16	FH,FK,J	3-187
LM148	General Purpose	1	0.5	5	100	50		±22	FH,FK,J	3-33
MC3503	General Purpose	1	0.6	5	−500	50	±1.5	±18	J	3-57
TL064M	BIFET, Low Power	1	3.5	9	0.2	4	±1.5	±18	FH,FK,J,W	3-103
TLC27M4AM	LinCMOS, Medium Bias	0.7	0.6	5	0.001 Typ	20	1	16	FH,FK,J	3-187
TLC27M4BM	LinCMOS, Medium Bias	0.7	0.6	2	0.001 Typ	20	1	16	FH,FK,J	3-187
TLC27M4M	LinCMOS, Medium Bias	0.7	0.6	10	0.001 Typ	20	1	16	FH,FK,J	3-187
LM124	General Purpose	0.6	0.5	5	−150	50	3	30	FH,FK,J,W	3-29
TL044M	Low Power	0.5	0.5	5	100	72	±2	±22	FH,FK,J,W	3-99
TLC27L4AM	LinCMOS, Low Bias	0.1	0.04	5	0.001 Typ	30	1	16	FH,FK,J	3-187
TLC27L4BM	LinCMOS, Low Bias	0.1	0.04	2	0.001 Typ	30	1	16	FH,FK,J	3-187
TLC27L4M	LinCMOS, Low Bias	0.1	0.04	10	0.001 Typ	30	1	16	FH,FK,J	3-187

automotive temperature range (values specified for T_A = T_A 25°C)

DEVICE NUMBER	DESCRIPTION	B_1 (MHz) TYP	SR (V/µs) TYP	V_{IO} (mV) MAX	I_{IB} (nA) MAX	A_{VD} (V/mV) MIN	SUPPLY VOLTAGE (V) MIN	SUPPLY VOLTAGE (V) MAX	PACKAGES	PAGE
RV4136	High Performance	3	1.7	6	500	20	±4.5	±32	D,J,N,W	3-83
LM2900	General Purpose	2.5	0.5		200	1.2	±4.5	±32	J,N	3-47
TLC274AI	LinCMOS, High Bias	2.3	4.5	5	0.001 Typ	10	1	16	D,J,N	3-187
TLC274BI	LinCMOS, High Bias	2.3	4.5	2	0.001 Typ	10	1	16	D,J,N	3-187
TLC274I	LinCMOS, High Bias	2.3	4.5	10	0.001 Typ	10	1	16	D,J,N	3-187
MC3303	General Purpose	1	0.6	8	−500	20	3	36	D,J,N	3-187
TLC27M4AI	LinCMOS, Medium Bias	0.7	0.6	5	0.001 Typ	20	1	16	D,J,N	3-187
TLC27M4BI	LinCMOS, Medium Bias	0.7	0.6	2	0.001 Typ	20	1	16	D,J,N	3-187
TLC27M4I	LinCMOS, Medium Bias	0.7	0.6	10	0.001 Typ	20	1	16	D,J,N	3-187
LM2902	General Purpose	0.6		7	−250	100 Typ	3	26	D,J,N,W	3-29
TLC27L4AI	LinCMOS, Low Bias	0.1	0.04	5	0.001 Typ	30	1	16	D,J,N	3-187
TLC27L4BI	LinCMOS, Low Bias	0.1	0.04	2	0.001 Typ	30	1	16	D,J,N	3-187
TLC27L4I	LinCMOS, Low Bias	0.1	0.04	10	0.001 Typ	30	1	16	D,J,N	3-187
TLC264AI	LinCMOS, Programmable		12	5	0.001 Typ		2	16	D,J,N	3-199
TLC264BI	LinCMOS, Programmable		12	2	0.001 Typ		2	16	D,J,N	3-199
TLC264I	LinCMOS, Programmable		12	10	0.001 Typ		2	16	D,J,N	3-199

3

Operational Amplifiers

TEXAS
INSTRUMENTS
POST OFFICE BOX 225012 • DALLAS, TEXAS 75265

internally compensated, quad

industrial temperature range   (values specified for $T_A = T_A$ 25°C)

DEVICE NUMBER	DESCRIPTION	B_1 (MHz) TYP	SR (V/μs) TYP	V_{IO} (mV) MAX	I_{IB} (nA) MAX	A_{VD} (V/mV) MIN	SUPPLY VOLTAGE (V)		PACKAGES	PAGE
							MIN	MAX		
TL074I	BIFET, Low Noise	3	13	6	0.2	50	±3.5	±18	D,J,N	3-125
TL084I	BIFET, General Purpose	3	13	6	0.2	50	±3.5	±18	J,N	3-125
LM248	General Purpose	1	0.5	6	200	25		±18	D,J,N	3-33
TL064I	BIFET, Low Power	1	3.5	6	0.2	4	±1.5	±18	D,J,N	3-103
LM224	General Purpose	0.6		5	−150	50	3	30	D,J,N,W	3-135

3

Operational Amplifiers

internally compensated, quad

commercial temperature range   (values specified for T_A = 25°C)

DEVICE NUMBER	DESCRIPTION	B_1 (MHz) TYP	SR (V/μs) TYP	V_{IO} (mV) MAX	I_{IB} (nA) MAX	A_{VD} (V/mV) MIN	SUPPLY VOLTAGE (V) MIN	SUPPLY VOLTAGE (V) MAX	PACKAGES	PAGE
TL294C	High Frequency	20	50				±4	±18	J,N	3-155
RC4136	High Performance	3	1.7	6	500	20	±4	±18	D,J,N,W	3-83
TL074AC	BIFET, Low Noise	3	13	6	0.2	50	±3.5	±18	D,J,N	3-125
TL074BC	BIFET, Low Noise	3	13	3	0.2	50	±3.5	±18	D,J,N	3-125
TL074C	BIFET, Low Noise	3	13	10	0.2	25	±3.5	±18	D,J,N	3-125
TL075C	BIFET, Low Noise	3	13	13	0.2	25	±3.5	±18	N	3-125
TL084AC	BIFET, General Purpose	3	13	6	0.2	50	±3.5	±18	J,N	3-135
TL084BC	BIFET, General Purpose	3	13	3	0.2	50	±3.5	±18	J,N	3-135
TL084C	BIFET, General Purpose	3	13	15	0.4	25	±3.5	±18	J,N	3-135
TL085C	BIFET, General Purpose	3	13	15	0.4	25	±3.5	±18	N	3-135
TL136C	High Performance	3	2	6	500	3 Typ	±4	±18	D,J,N	3-151
LM3900	General Purpose	2.5	0.5			1.2	±4.5	±18	J,N	3-47
TLC254AC	LinCMOS, High Bias	2.3	4.5	5	0.001 Typ	10	1	16	D,J,N	3-187
TLC254BC	LinCMOS, High Bias	2.3	4.5	2	0.001 Typ	10	1	16	D,J,N	3-187
TLC254C	LinCMOS, High Bias	2.3	4.5	10	0.001 Typ	10	1	16	D,J,N	3-187
TLC274AC	LinCMOS, High Bias	2.3	4.5	5	0.001 Typ	10	3	16	D,J,N	3-187
TLC274BC	LinCMOS, High Bias	2.3	4.5	2	0.001 Typ	10	3	16	D,J,N	3-187
TLC274C	LinCMOS, High Bias	2.3	4.5	10	0.001 Typ	10	3	16	D,J,N	3-187
LM348	General Purpose	1	0.5	6	200	25		±18	D,J,N	3-187
MC3403	General Purpose	1	0.6	10	−500	20	±1.5	±18	D,J,N	3-57
TL064AC	BIFET, Low Power	1	3.5	6	0.2	4	±1.5	±18	D,J,N	3-103
TL064BC	BIFET, Low Power	1	3.5	3	0.2	4	±1.5	±18	D,J,N	3-103
TL064C	BIFET, Low Power	1	3.5	15	0.2	3	±1.5	±18	D,J,N	3-103
TLC25M4AC	LinCMOS, Medium Bias	0.7	0.6	5	0.001 Typ	20	1	16	D,J,N	3-187
TLC25M4BC	LinCMOS, Medium Bias	0.7	0.6	2	0.001 Typ	20	1	16	D,J,N	3-187
TLC25M4C	LinCMOS, Medium Bias	0.7	0.6	10	0.001 Typ	20	1	16	D,J,N	3-187
TLC27M4AC	LinCMOS, Medium Bias	0.7	0.6	5	0.001 Typ	20	3	16	D,J,N	3-187
TLC27M4BC	LinCMOS, Medium Bias	0.7	0.6	2	0.001 Typ	20	3	16	D,J,N	3-187
TLC27M4C	LinCMOS, Medium Bias	0.7	0.6	10	0.001 Typ	20	3	16	D,J,N	3-187
LM324	General Purpose	0.6		7	−250	25	3	30	D,J,N,W	3-29
LM324A	General Purpose	0.6		7	−100	25	3	30	D,J,N,W	3-29
TL044C	General Purpose	0.5	0.5	5	250	60	±2	±18	J,N,W	3-99
TLC25L4AC	LinCMOS, Low Bias	0.1	0.04	5	0.001 Typ	30	1	16	D,J,N	3-187
TLC25L4BC	LinCMOS, Low Bias	0.1	0.04	2	0.001 Typ	30	1	16	D,J,N	3-187
TLC25L4C	LinCMOS, Low Bias	0.1	0.04	10	0.001 Typ	30	1	16	D,J,N	3-187
TLC27L4AC	LinCMOS, Low Bias	0.1	0.04	5	0.001 Typ	30	3	16	D,J,N	3-187
TLC27L4BC	LinCMOS, Low Bias	0.1	0.04	2	0.001 Typ	30	3	16	D,J,N	3-187
TLC27L4C	LinCMOS, Low Bias	0.1	0.04	10	0.001 Typ	30	3	16	D,J,N	3-187

3

Operational Amplifiers

TEXAS
INSTRUMENTS
POST OFFICE BOX 225012 • DALLAS, TEXAS 75265

Input Offset Voltage (V_{IO})

The d-c voltage that must be applied between the input terminals to force the quiescent d-c output voltage to zero or other level, if specified.

Average Temperature Coefficient of Input Offset Voltage (α_{VIO})

The ratio of the change in input offset voltage to the change in free-air temperature. This is an average value for the specified temperature range.

$$\alpha_{VIO} = \left[\frac{(V_{IO} @ T_{A(1)}) - (V_{IO} @ T_{A(2)})}{T_{A(1)} - T_{A(2)}} \right] \text{ where } T_{A(1)} \text{ and } T_{A(2)} \text{ are the specified temperature extremes.}$$

Input Offset Current (I_{IO})

The difference between the currents into the two input terminals with the output at zero volts.

Average Temperature Coefficient of Input Offset Current (α_{IIO})

The ratio of the change in input offset current to the change in free-air temperature. This is an average value for the specified temperature range.

$$\alpha_{IIO} = \left[\frac{(I_{IO} @ T_{A(1)}) - (I_{IO} @ T_{A(2)})}{T_{A(1)} - T_{A(2)}} \right] \text{ where } T_{A(1)} \text{ and } T_{A(2)} \text{ are the specified temperature extremes.}$$

Input Bias Current (I_{IB})

The average of the currents into the two input terminals with the output at zero volts.

Common-Mode Input Voltage (V_{IC})

The average of the two input voltages.

Common-Mode Input Voltage Range (V_{ICR})

The range of common-mode input voltage that if exceeded will cause the amplifier to cease functioning properly.

Differential Input Voltage (V_{ID})

The voltage at the noninverting input with respect to the inverting input.

Maximum Peak Output Voltage Swing (V_{OM})

The maximum positive or negative peak output voltage that can be obtained without waveform clipping when the quiescent d-c output voltage is zero.

Maximum Peak-to-Peak Output Voltage Swing (V_{OPP})

The maximum peak-to-peak output voltage that can be obtained without waveform clipping when the quiescent d-c output voltage is zero.

Large-Signal Voltage Amplification (A_V)

The ratio of the peak-to-peak output voltage swing to the change in input voltage required to drive the output.

Differential Voltage Amplification (A_{VD})

The ratio of the change in output voltage to the change in differential input voltage producing it.

3

Operational Amplifiers

GLOSSARY
OPERATIONAL AMPLIFIER TERMS AND DEFINITIONS

Maximum-Output-Swing Bandwidth (B_{OM})

The range of frequencies within which the maximum output voltage swing is above a specified value.

Unity-Gain Bandwidth (B_1)

The range of frequencies within which the open-loop voltage amplification is greater than unity.

Phase Margin (ϕ_m)

The absolute value of the open-loop phase shift between the output and the inverting input at the frequency at which the modulus of the open-loop amplification is unity.

Gain Margin (A_m)

The reciprocal of the open-loop voltage amplification at the lowest frequency at which the open-loop phase shift is such that the output is in phase with the inverting input.

Input Resistance (r_I)

The resistance between the input terminals with either input grounded.

Differential Input Resistance (r_{id})

The small-signal resistance between the two ungrounded input terminals.

Output Resistance (r_o)

The resistance between the output terminal and ground.

Input Capacitance (C_i)

The capacitance between the input terminals with either input grounded.

Common-Mode Input Impedance (z_{ic})

The parallel sum of the small-signal impedance between each input terminal and ground.

Output Impedance (z_o)

The small-signal impedance between the output terminal and ground.

Common-Mode Rejection Ratio (k_{CMR}, CMRR)

The ratio of differential voltage amplification to common-mode voltage amplification.
NOTE: This is measured by determining the ratio of a change in input common-mode voltage to the resulting change in input offset voltage.

Supply Voltage Sensitivity (k_{SVS}, $\Delta V_{IO}/\Delta V_{CC}$)

The absolute value of the ratio of the change in input offset voltage to the change in supply voltages producing it.
NOTES: 1. Unless otherwise noted, both supply voltages are varied symmetrically.
          2. This is the reciprocal of supply voltage rejection ratio.

Supply Voltage Rejection Ratio (k_{SVR}, $\Delta V_{CC}/\Delta V_{IO}$)

The absolute value of the ratio of the change in supply voltages to the change in input offset voltage.
NOTES: 1. Unless otherwise noted, both supply voltages are varied symmetrically.
          2. This is the reciprocal of supply voltage sensitivity.

3

Operational Amplifiers

TEXAS
INSTRUMENTS
POST OFFICE BOX 225012 • DALLAS, TEXAS 75265

Equivalent Input Noise Voltage (V_n)

The voltage of an ideal voltage source (having an internal impedance equal to zero) in series with the input terminals of the device that represents the part of the internally generated noise that can properly be represented by a voltage source.

Equivalent Input Noise Current (I_n)

The current of an ideal current source (having an internal impedance equal to infinity) in parallel with the input terminals of the device that represents the part of the internally generated noise that can properly be represented by a current source.

Average Noise Figure ($\overline{F}$)

The ratio of (1) the total output noise power within a designated output frequency band when the noise temperature of the input termination(s) is at the reference noise temperature, T_O, at all frequencies to (2) that part of (1) caused by the noise temperature of the designated signal-input termination within a designated signal-input frequency band.

Short-Circuit Output Current (I_{OS})

The maximum output current available from the amplifier with the output shorted to ground, to either supply, or to a specified point.

Supply Current (I_{CC})

The current into the V_{CC} or V_{CC+} terminal of an integrated circuit.

Total Power Dissipation (P_D)

The total d-c power supplied to the device less any power delivered from the device to a load.
NOTE: At no load: $P_D = V_{CC+} \cdot I_{CC+} + V_{CC-} \cdot I_{CC-}$.

Crosstalk Attenuation (V_{o1}/V_{o2})

The ratio of the change in output voltage of a driven channel to the resulting change in output voltage of another channel.

Rise Time (t_r)

The time required for an output voltage step to change from 10% to 90% of its final value.

Total Response Time (Settling Time) (t_{tot})

The time between a step-function change of the input signal level and the instant at which the magnitude of the output signal reaches for the last time a specified level range ($\pm \epsilon$) containing the final output signal level.

Overshoot Factor

The ratio of (1) the largest deviation of the output signal value from its final steady-state value after a step-function change of the input signal, to (2) the absolute value of the difference between the steady-state output signal values before and after the step-function change of the input signal.

Slew Rate (SR)

The average time rate of change of the closed-loop amplifier output voltage for a step-signal input.

3

Operational Amplifiers

3

**LINEAR
INTEGRATED
CIRCUITS**

**TYPES LM101A, LM201A, LM301A
HIGH-PERFORMANCE OPERATIONAL AMPLIFIERS**

D961, OCTOBER 1979—REVISED AUGUST 1983

- Low Input Currents
- Low Input Offset Parameters
- Frequency and Transient Response Characteristics Adjustable
- Short-Circuit Protection
- Offset-Voltage Null Capability
- No Latch-Up
- Wide Common-Mode and Differential Voltage Ranges
- Same Pin Assignments as uA709
- Designed to be Interchangeable with National Semiconductor LM101A and LM301A

D, JG, OR P DUAL-IN-LINE PACKAGE
(TOP VIEW)

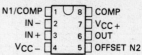

W FLAT PACKAGE
(TOP VIEW)

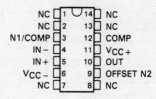

LM101A
U FLAT PACKAGE
(TOP VIEW)

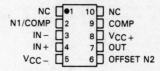

LM101A
FH OR FK CHIP-CARRIER PACKAGE
(TOP VIEW)

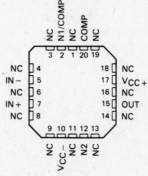

NC—No internal connection

description

The LM101A, LM201A, and LM301A are high-performance operational amplifiers featuring very low input bias current and input offset voltage and current to improve the accuracy of high-impedance circuits using these devices. The high common-mode input voltage range and the absence of latch-up make these amplifiers ideal for voltage-follower applications. The devices are protected to withstand short-circuits at the output. The external compensation of these amplifiers allows the changing of the frequency response (when the closed-loop gain is greater than unity) for applications requiring wider bandwidth or higher slew rate. A potentiometer may be connected between the offset-null inputs (N1 and N2), as shown in Figure 7, to null out the offset voltage.

The LM101A is characterized for operation over the full military temperature range of $-55\,°C$ to $125\,°C$, the LM201A is characterized for operation from $-25\,°C$ to $85\,°C$, and the LM301A is characterized for operation from $0\,°C$ to $70\,°C$.

symbol

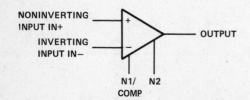

3

Operational Amplifiers

TYPES LM101A, LM201A, LM301A
HIGH-PERFORMANCE OPERATIONAL AMPLIFIERS

absolute maximum ratings over operating free-air temperature range (unless otherwise noted)

		LM101A	LM201A	LM301A	UNIT
Supply voltage V_{CC+} (see Note 1)		22	22	18	V
Supply voltage V_{CC-} (see Note 1)		−22	−22	−18	V
Differential input voltage (see Note 2)		±30	±30	±30	V
Input voltage (either input, see Notes 1 and 3)		±15	±15	±15	V
Voltage between either offset null terminal (N1/N2) and V_{CC-}		−0.5 to 2	−0.5 to 2	−0.5 to 2	V
Duration of output short-circuit (see Note 4)		unlimited	unlimited	unlimited	
Continuous total power dissipation at (or below) 25°C free-air temperature (see Note 5)		500	500	500	mW
Operating free-air temperature range		−55 to 125	−25 to 85	0 to 70	°C
Storage temperature range		−65 to 150	−65 to 150	−65 to 150	°C
Lead temperature 1,6 mm (1/16 inch) from case for 60 seconds	FH, FK, JG, U, or W package	300	300	300	°C
Lead temperature 1,6 mm (1/16 inch) from case for 10 seconds	D or P package		260	260	°C

NOTES: 1. All voltage values, unless otherwise noted, are with respect to the midpoint between V_{CC+} and V_{CC-}.
2. Differential voltages are at the noninverting input terminal with respect to the inverting input terminal.
3. The magnitude of the input voltage must never exceed the magnitude of the supply voltage or 15 volts, whichever is less.
4. The output may be shorted to ground or either power supply. For the LM101A only, the unlimited duration of the short-circuit applies at (or below) 125°C case temperature or 75°C free-air temperature. For the LM201A only, the unlimited duration of the short-circuit applies at (or below) 85°C case temperature or 75°C free air temperature.
5. For operation above 25°C free-air temperature, refer to Dissipation Derating Curves, Section 2. In the J and JG packages, LM101A chips are alloy-mounted; LM201A and LM301A chips are glass-mounted.

3

Operational Amplifiers

TEXAS
INSTRUMENTS
POST OFFICE BOX 225012 ● DALLAS, TEXAS 75265

electrical characteristics at specified free-air temperature, C_C = 30 pF (see Note 6)

PARAMETER		TEST CONDITIONS†		LM101A, LM201A MIN	TYP	MAX	LM301A MIN	TYP	MAX	UNIT
V_{IO}	Input offset voltage	V_O = 0 V	25°C		0.6	2		2	7.5	mV
			Full range			3			10	
α_{VIO}	Average temperature coefficient of input offset voltage	V_O = 0 V	Full range		3	15		6	30	µV/°C
I_{IO}	Input offset current		25°C		1.5	10		3	50	nA
			Full range			20			70	
α_{IIO}	Average temperature coefficient of input offset current	T_A = −55°C to 25°C			0.02	0.2				nA/°C
		T_A = 25°C to MAX			0.01	0.1				
		T_A = 0°C to 25°C						0.02	0.6	
		T_A = 25°C to 70°C						0.01	0.3	
I_{IB}	Input bias current		25°C		30	75		70	250	nA
			Full range			100			300	
V_{ICR}	Common-mode input voltage range	See Note 7	Full range	±15			±12			V
V_{OPP}	Maximum peak-to-peak output voltage swing	$V_{CC\pm}$ = ±15 V, R_L = 10 kΩ	25°C	24	28		24	28		V
			Full range	24			24			
		$V_{CC\pm}$ = ±15 V, R_L = 2 kΩ	25°C	20	26		20	26		
			Full range	20			20			
A_{VD}	Large-signal differential voltage amplification	$V_{CC\pm}$ = ±15 V, V_O = ±10 V, $R_L \geq$ 2 kΩ	25°C	50	200		25	200		V/mV
			Full range	25			15			
r_i	Input resistance		25°C	1.5	4		0.5	2		MΩ
CMRR	Common-mode rejection ratio	V_{IC} = V_{ICR} min	25°C	80	98		70	90		dB
			Full range	80			70			
k_{SVR}	Supply voltage rejection ratio ($\Delta V_{CC}/\Delta V_{IO}$)		25°C	80	98		70	96		dB
			Full range	80			70			
I_{CC}	Supply current	No Load, V_O = 0 V, See Note 7	25°C		1.8	3		1.8	3	mA
			MAX		1.2	2.5				

†All characteristics are measured under open-loop conditions with zero common-mode input voltage unless otherwise specified. Full range for LM101A is −55°C to 125°C, for LM201A is −25°C to 85°C, and for LM301A is 0°C to 70°C.

NOTES: 6. Unless otherwise noted, $V_{CC\pm}$ = ±5 V to ±20 V for LM101A and LM201A, and $V_{CC\pm}$ = ±5 V to ±15 V for LM301A. All typical values are at $V_{CC\pm}$ = ±15 V.

7. For LM101A and LM201A, $V_{CC\pm}$ = ±20 V. For LM301A, $V_{CC\pm}$ = ±15 V.

3

Operational Amplifiers

TYPICAL CHARACTERISTICS

INPUT OFFSET CURRENT
vs
FREE-AIR TEMPERATURE

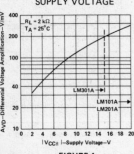

FIGURE 1

INPUT BIAS CURRENT
vs
FREE-AIR TEMPERATURE

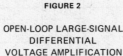

FIGURE 2

MAXIMUM PEAK-TO-PEAK
OUTPUT VOLTAGE (WITH
SINGLE-POLE COMPENSATION)
vs FREQUENCY

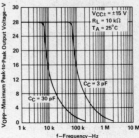

FIGURE 3

OPEN-LOOP LARGE-SIGNAL
DIFFERENTIAL
VOLTAGE AMPLIFICATION
vs
SUPPLY VOLTAGE

FIGURE 4

OPEN-LOOP LARGE-SIGNAL
DIFFERENTIAL
VOLTAGE AMPLIFICATION
vs
FREQUENCY

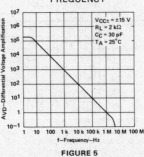

FIGURE 5

VOLTAGE-FOLLOWER
LARGE-SIGNAL PULSE RESPONSE

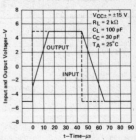

FIGURE 6

TYPICAL APPLICATION DATA

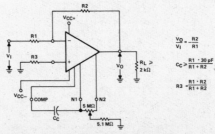

$$\frac{V_O}{V_I} = -\frac{R2}{R1}$$

$$C_C > \frac{R1 \cdot 30 \text{ pF}}{R1 + R2}$$

$$R3 = \frac{R1 \cdot R2}{R1 + R2}$$

FIGURE 7— INVERTING CIRCUIT WITH ADJUSTABLE GAIN,
SINGLE-POLE COMPENSATION, AND OFFSET ADJUSTMENT

3

Operational Amplifiers

Texas
INSTRUMENTS
POST OFFICE BOX 225012 • DALLAS, TEXAS 75265

**LINEAR
INTEGRATED
CIRCUITS**

**TYPES LM107, LM207, LM307
HIGH-PERFORMANCE OPERATIONAL AMPLIFIERS**

D962, DECEMBER 1970—REVISED AUGUST 1983

- Low Input Currents
- No Frequency Compensation Required
- Low Input Offset Parameters
- Short-Circuit Protection
- No Latch-Up
- Wide Common-Mode and Differential Voltage Ranges

description

The LM107, LM207, and LM307 are high-performance operational amplifiers featuring very low input bias current and input offset voltage and current to improve the accuracy of high-impedance circuits using these devices.

The high common-mode input voltage range and the absence of latch-up make these amplifiers ideal for voltage follower applications. The devices are short-circuit protected and the internal frequency compensation ensures stability without external components.

The LM107 is characterized for operation over the full military temperature range of −55°C to 125°C, the LM207 is characterized for operation from −25°C to 85°C, and the LM307 is characterized for operation from 0°C to 70°C.

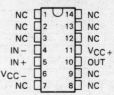

LM107 . . . J OR W PACKAGE
LM207, LM307 . . . W PACKAGE
(TOP VIEW)

NC	1	14	NC
NC	2	13	NC
NC	3	12	NC
IN−	4	11	$V_{CC}+$
IN+	5	10	OUT
$V_{CC}-$	6	9	NC
NC	7	8	NC

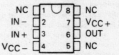

LM107 . . . JG PACKAGE
LM207, LM307 . . . D, JG, OR P PACKAGE
(TOP VIEW)

NC	1	8	NC
IN−	2	7	$V_{CC}+$
IN+	3	6	OUT
$V_{CC}-$	4	5	NC

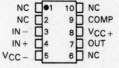

LM107 . . . U FLAT PACKAGE
(TOP VIEW)

NC	1	10	NC
NC	2	9	COMP
IN−	3	8	$V_{CC}+$
IN+	4	7	OUT
$V_{CC}-$	5	6	NC

NC—No internal connection

symbol

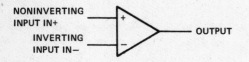

NONINVERTING INPUT IN+

INVERTING INPUT IN−

OUTPUT

3

Operational Amplifiers

**TEXAS
INSTRUMENTS**

POST OFFICE BOX 225012 ● DALLAS, TEXAS 75265

	LM107	LM207	LM307	UNIT
Supply voltage V_{CC+} (see Note 1)	22	22	18	V
Supply voltage V_{CC-} (see Note 1)	−22	−22	−18	V
Differential input voltage (see Note 2)	±30	±30	±30	V
Input voltage (either input, see Notes 1 and 3)	±15	±15	±15	V
Duration of output short-circuit (see Note 4)	unlimited	unlimited	unlimited	
Continuous total dissipation at (or below) 25°C free-air temperature (see Note 5)	500	500	500	mW
Operating free-air temperature range	−55 to 125	−25 to 85	0 to 70	°C
Storage temperature range	−65 to 150	−65 to 150	−65 to 150	°C
Lead temperature 1,6 mm (1/16 inch) from case for 60 seconds JG, U, or W package	300	300	300	°C
Lead temperature 1,6 mm (1/16 inch) from case for 10 seconds D or P package		260	260	°C

NOTES: 1. All voltage values, unless otherwise noted, are with respect to the midpoint between V_{CC+} and V_{CC-}.
2. Differential voltages are at the noninverting input terminal with respect to the inverting input terminal.
3. The magnitude of the input voltage must never exceed the magnitude of the supply voltage or 15 volts, whichever is less.
4. The output may be shorted to ground or either power supply. For the LM107 only, the unlimited duration of the short-circuit applies at (or below) 125°C case temperature or 75°C free-air temperature. For the LM207 only, the unlimited duration of the short-circuit applies at (or below) 85°C case temperature or 75°C free air temperature.
5. For operation above 25°C free-air temperature, refer to Dissipation Derating Curves, Section 2.

3

Operational Amplifiers

TEXAS
INSTRUMENTS
POST OFFICE BOX 225012 • DALLAS, TEXAS 75265

electrical characteristics at specified free-air temperature (see Note 6)

PARAMETER		TEST CONDITIONS†		LM107, LM207			LM307			UNIT
				MIN	TYP	MAX	MIN	TYP	MAX	
V_{IO}	Input offset voltage	$V_O = 0$	25°C		0.6	2		2	7.5	mV
			Full range			3			10	
α_{VIO}	Average temperature coefficient of input offset voltage	$V_O = 0$	Full range		3	15		6	30	µV/°C
I_{IO}	Input offset current	$V_O = 0$	25°C		1.5	10		3	50	nA
			Full range			20			70	
α_{IIO}	Average temperature coefficient of input offset current	$T_A = -55°C$ to 25°C			0.02	0.2				nA/°C
		$T_A = 25°C$ to MAX			0.01	0.1				
		$T_A = 0°C$ to 25°C						0.02	0.6	
		$T_A = 25°C$ to 70°C						0.01	0.3	
I_{IB}	Input bias current		25°C		30	75		70	250	nA
			Full range			100			300	
V_{ICR}	Common-mode input voltage range	See Note 7	Full range	±15			±12			V
V_{OPP}	Maximum peak-to-peak output voltage swing	$V_{CC\pm} = \pm15$ V, $R_L = 10$ kΩ	25°C	24	28		24	28		V
			Full range	24			24			
		$V_{CC\pm} = \pm15$ V, $R_L = 2$ kΩ	25°C	20	26		20	26		
			Full range	20			20			
A_{VD}	Large-signal differential voltage amplification	$V_{CC\pm} = \pm15$ V, $V_O = \pm10$ V, $R_L \geq 2$ kΩ	25°C	50	200		25	200		V/mV
			Full range	25			15			
r_i	Input resistance		25°C	1.5	4		0.5	2		MΩ
CMRR	Common-mode rejection ratio	$V_{IC} = V_{ICR}$ min	25°C	80	98		70	90		dB
			Full range	80			70			
k_{SVR}	Supply voltage rejection ratio ($\Delta V_{CC}/\Delta V_{IO}$)		25°C	80	98		70	96		dB
			Full range	80			70			
I_{CC}	Supply current	No Load, $V_O = 0$, See Note 7	25°C		1.8	3		1.8	3	mA
			MAX		1.2	2.5				

†All characteristics are measured under open-loop conditions with zero common-mode input voltage unless otherwise specified. Full range for LM107 is −55°C to 125°C, for LM207 is −25°C to 85°C, and for LM307 is 0°C to 70°C.

NOTES: 6. Unless otherwise noted $V_{CC\pm} = \pm5$ V to ±20 V for LM107 and LM207, and $V_{CC\pm} = \pm5$ V to ±15 V for LM307. All typical values are at $V_{CC\pm} = \pm15$ V.

7. For LM107 and LM207, $V_{CC\pm} = \pm20$ V. For LM307, $V_{CC\pm} = \pm15$ V.

3

Operational Amplifiers

D2808, OCTOBER 1983

- Input Bias Current
  LM108, LM208 . . . 3 nA Max
      LM308 . . . 10 nA Max

- Offset Current
  LM108, LM208 . . . 400 pA Max
      LM308 . . . 1500 pA Max

- Supply Current . . . 300 μA Typ

- Direct Replacement for
  National Semiconductor
  LM108, LM208, and LM308

LM108 . . . JG DUAL-IN-LINE PACKAGE
LM208, LM308 . . . D, JG, OR P DUAL-IN-LINE PACKAGE
(TOP VIEW)

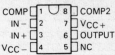

COMP	1	8	COMP2
IN –	2	7	VCC +
IN +	3	6	OUTPUT
VCC –	4	5	NC

NC – No internal connection

symbol

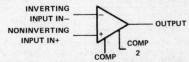

INVERTING
INPUT IN–

NONINVERTING
INPUT IN+

OUTPUT

COMP 2

COMP

description

The LM108, LM208, and LM308 are precision operational amplifiers featuring very low input bias current and input offset voltage and current to improve the accuracy of high-impedance circuits using these devices. In most cases these devices make it possible to eliminate offset adjustments for performances approaching that of chopper-stabilized amplifiers. As an example, the LM108 device is available with offset voltage less than 1-millivolt and temperature coefficient less than 5μV/°C over the entire military temperature range.

These devices are designed to withstand short-circuits at the output. The external compensation of these amplifiers allows changing the frequency response (when the closed-loop gain is greater than unity) for applications requiring wider bandwidth or higher slew rate.

The LM108 is characterized for operation over the full military temperature range of −55°C to 125°C. The LM208 is characterized for operation from −25°C to 85°C, and the LM308 is characterized for operation from 0°C to 70°C.

3

Operational Amplifiers

**TEXAS
INSTRUMENTS**

POST OFFICE BOX 225012 • DALLAS, TEXAS 75265

absolute maximum ratings over operating free-air temperature range (unless otherwise noted)

		LM108	LM208	LM308	UNIT
Supply voltage, V_{CC+} (see Note 1)		18	20	20	V
Supply voltage, V_{CC-} (see Note 1)		−18	−20	−20	V
Input voltage (see Note 2)		±15	±15	±15	V
Differential input current (see Note 3)		±10	±10	±10	mA
Duration of output short-circuit (see Notes 4 and 5)		Unlimited	Unlimited	Unlimited	
Continuous total power dissipation at (or below) 25 °C free-air temperature		500	500	500	mW
Operating temperature range		−55 to 125	−25 to 85	0 to 70	°C
Storage temperature range		−65 to 150	−65 to 150	−65 to 150	°C
Lead temperature 1,6 mm (1/16 inch) from case for 60 seconds	JG package	300	300	300	°C
Lead temperature 1,6 mm (1/16 inch) from case for 10 seconds	D or P package		260	260	°C

NOTES: 1. All voltage values, unless otherwise noted, are with respect to the midpoint between V_{CC+} and V_{CC-}.
2. The magnitude of the input voltage must never exceed the magnitude of the supply voltage or 15 volts, whichever is less.
3. The inputs are shunted with two opposite-facing base-emitter diodes for over-voltage protection. Therefore, excessive current will flow if a differential input voltage in excess of approximately 1 V is applied between the inputs unless some limiting resistance is used.
4. Differential voltages are at the noninverting input terminal with respect to the inverting input terminal.
5. The output may be shorted to ground or either power supply.

electrical characteristics at specified free-air temperature (see Note 6)

PARAMETER		TEST CONDITIONS†		LM108, LM208			LM308			UNIT
				MIN	TYP	MAX	MIN	TYP	MAX	
V_{IO}	Input offset voltage		25 °C		0.7	2		2	7.5	mV
			Full range			3			10	
α_{VIO}	Average temperature coefficient of input offset voltage		Full range		3	15		6	30	µV/°C
I_{IO}	Input offset current		25 °C		0.05	0.2		0.2	1	nA
			Full range			0.4			1.5	
α_{IIO}	Average temperature coefficient of input offset current		Full range		0.5	2.5		2	10	pA/°C
I_{IB}	Input bias current		25 °C		0.8	2		1.5	7	nA
			Full range			3			10	
V_{ICR}	Common-mode input voltage range	$V_{CC} = ±15$ V	Full range	±13.5			±14			V
V_{OM}	Maximum peak output voltage swing	$V_{CC} = 15$ V, $R_L = 10$ kΩ	Full range	±13	±14		±13	±14		V
A_{VD}	Large-signal differential voltage gain	$V_{CC} = ±15$ V, $V_O = ±10$ V, $R_L > 10$ kΩ	25 °C	50	300		25	300		V/mV
			Full range	25			15			
r_i	Input resistance		25 °C	30	70		10	40		MΩ
CMRR	Common-mode rejection ratio		Full range	85	100		80	100		dB
k_{SVR}	Supply voltage rejection ratio ($\Delta V_{CC}/\Delta V_{IO}$)		Full range	80	96		80	96		dB
I_{CC}	Supply current		25 °C		0.3	0.6		0.3	0.8	mA
			MAX		0.15	0.4				

† All characteristics are specified under open-loop conditions with zero common-mode input voltage unless otherwise noted. Full range is −55 °C to 125 °C for LM108, −25 °C to 85 °C for LM208, and 0 °C to 70 °C for LM308. For conditions shown as MAX, use the appropriate maximum value specified under absolute maximum ratings.
NOTE 6: Unless otherwise noted, $V_{CC±} = ±5$ V to ±18 V for LM108 and LM208, and $V_{CC} = ±5$ V to ±15 V for LM308.

TEXAS
INSTRUMENTS

POST OFFICE BOX 225012 • DALLAS, TEXAS 75265

1083

LINEAR
INTEGRATED
CIRCUITS

TYPES LM110, LM210, AND LM310
VOLTAGE FOLLOWERS

D2815, OCTOBER 1983

- Input Current . . . 10 nA Max
- Small-Signal Bandwidth . . . 20 MHz
- Slew Rate . . . 30 V/μs
- Supply Voltage Range . . . ±5 V to ±18 V
- Direct Replacements for National Semiconductor LM110, LM210, and LM310

LM110 . . . JG DUAL-IN-LINE PACKAGE
LM210, LM310 . . . JG OR P DUAL-IN-LINE PACKAGE
(TOP VIEW)

BAL1	1	8	BAL2
NC	2	7	$V_{CC}+$
IN +	3	6	OUT
$V_{CC}-$	4	5	BOOSTER

NC—No internal connection

description

The LM110 series are monolithic operational amplifiers internally connected as unity-gain non-inverting amplifiers. They use transistors in the input stage to get low bias current without sacrificing speed and they have internal frequency compensation and provision for offset balancing. Increased output swing under load can be obtained by connecting an external resistor between the Booster terminal and the $V_{CC}-$ terminal.

symbol

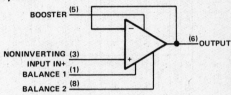

These devices are useful in fast sample-and-hold circuits, active filters, or as general purpose buffers. They are plug-in replacements for the LM102 series voltage followers, offering lower offset voltage, drift, bias current, and noise in addition to higher speed and wider operating voltage range.

The LM110 is characterized for operation over the full military temperature range of −55°C to 125°C. The LM210 is characterized for operation over the temperature range of −25°C to 85°C, and LM310 is characterized for operation over the temperature range of 0°C to 70°C.

Operational Amplifiers 3

absolute maximum ratings

	LM110	LM210	LM310	UNITS
Supply voltage, $V_{CC}+$ (see Note 1)	18	18	18	V
Supply voltage, $V_{CC}-$	−18	−18	−18	V
Input voltage (see Note 2)	±15	±15	±15	V
Duration of output short-circuit (see Notes 3 and 4)	Unlimited	Unlimited	Unlimited	
Continuous total dissipation at (or below) 25°C free-air temperature range (see Note 5)	500	500	500	mW
Operating free-air temperature range	−55 to 125	−25 to 85	0 to 70	°C
Storage temperature range	−65 to 150	−65 to 150	−65 to 150	°C
Lead temperature at 1,6 mm (1/16 inch) from case for 60 seconds — JG package	300	300	300	°C
Lead temperature at 1,6 mm (1/16 inch) from case for 10 seconds — P package		260	260	°C

NOTES: 1. All voltage values, unless otherwise noted, are with respect to the midpoint between $V_{CC}+$ and $V_{CC}-$.
2. The magnitude of the input voltage must never exceed the magnitude of the supply voltage or 15 volts, whichever is less.
3. The output may be shorted to ground or either power supply.
4. It is necessary to insert a resistor (R ≥ 2 kilohms) in series with the input when the amplifier is driven from low-impedance sources to prevent damage when the output is shorted.
5. For operation above 25°C free-air temperature, refer to Dissipation Derating Tables in Section 2. In the JG package, LM110 chips are alloy mounted, LM210 and LM310 chips are glass mounted.

TEXAS INSTRUMENTS
POST OFFICE BOX 225012 • DALLAS, TEXAS 75265

3 Operational Amplifiers

electrical characteristics at specified free-air temperatures, VCC± = ±5 V to ±18 V (unless otherwise noted)

PARAMETER		TEST CONDITIONS[†]	LM110 MIN	LM110 TYP	LM110 MAX	LM210 MIN	LM210 TYP	LM210 MAX	LM310 MIN	LM310 TYP	LM310 MAX	UNIT
V_{IO}	Input offset voltage	25°C		1.5	4		1.5	4		2.5	7.5	mV
		Full range			6			6			10	
α_{VIO}	Average temperature coefficient of input offset voltage	Full range		12			12			10		µV/°C
I_{IB}	Input bias current	25°C		1	3		1	3		2	7	nA
		Full range			10			10			10	
V_{OM}	Maximum peak output voltage swing	$V_{CC\pm} = \pm15$ V, $R_L = 10$ kΩ	Full range	±10			±10			±10			V
A_V	Large-signal voltage gain	$V_{CC\pm} = \pm15$ V, $V_O = \pm10$ V, $R_L = 8$ kΩ	25°C	0.999	0.9999		0.999	0.9999		0.999	0.9999		V/V
		$V_{CC\pm} = \pm15$ V, $V_O = \pm10$ V, $R_L = 10$ kΩ	Full range	0.999			0.999			0.999			
k_{SVR}	Supply voltage rejection ratio ($\Delta V_{CC}/\Delta V_{IO}$)	Full range	70	80		70	80		70	80		dB
r_i	Input resistance	25°C	10^{10}	10^{12}		10^{10}	10^{12}		10^{10}	10^{12}		Ω
r_o	Output resistance	25°C		0.75	2.5		0.75	2.5		0.75	2.5	Ω
C_i	Input capacitance	25°C		1.5			1.5			1.5		pF
I_{CC}	Supply current	$V_O = 0$, No load	25°C	3.9	5.5		3.9	5.5		3.9	5.5		mA
			MAX	2	4		2	4		2	4		

† All characteristics are specified under open-loop operation. Full range is −55°C to 125°C for LM110, −25°C to 85°C for LM210, and 0°C to 70°C for LM310. For conditions shown as MAX, use the appropriate value specified under absolute maximum ratings.

TEXAS INSTRUMENTS
POST OFFICE BOX 225012 • DALLAS, TEXAS 75265

1083

LINEAR
INTEGRATED
CIRCUITS

TYPES LM124, LM224, LM224A,
LM324, LM324A, LM2902
QUADRUPLE OPERATIONAL AMPLIFIERS
D1990, SEPTEMBER 1975—REVISED SEPTEMBER 1983

- **Wide Range of Supply Voltages:**
  Single Supply . . . 3 V to 30 V
  (LM2902 . . . 3 V to 26 V),
  or Dual Supplies

- **Low Supply Current Drain Independent of
  Supply Voltage . . . 0.7 mA Typ**

- **Common-Mode Input Voltage Range
  Includes Ground Allowing Direct Sensing
  near Ground**

- **Low Input Bias and Offset Parameters:**
  Input Offset Voltage . . . 3 mV Typ
  A Versions . . . 2 mV Typ
  Input Offset Current . . . 2 nA Typ
  Input Bias Current . . . 20 nA Typ
  A Versions . . . 15 nA Typ

- **Differential Input Voltage Range Equal to
  Maximum-Rated Supply Voltage . . . 32 V
  (26 V for LM2902)**

- **Open-Loop Differential Voltage
  Amplification . . . 100 V/mV Typ**

- **Internal Frequency Compensation**

description

These devices consist of four independent, high-gain frequency-compensated operational amplifiers that were designed specifically to operate from a single supply over a wide range of voltages. Operation from split supplies is also possible so long as the difference between the two supplies is 3 volts to 30 volts (for the LM2902, 3 volts to 26 volts), and Pin 4 is at least 1.5 volts more positive than the input common-mode voltage. The low supply current drain is independent of the magnitude of the supply voltage.

Applications include transducer amplifiers, d-c amplification blocks, and all the conventional operational amplifier circuits that now can be more easily implemented in single-supply-voltage systems. For example, the LM124 can be operated directly off of the standard five-volt supply that is used in digital systems and will easily provide the required interface electronics without requiring additional ±15-volt supplies.

The LM124 is characterized for operation over the full military temperature range of −55 °C to 125 °C. The LM2902 is characterized for operation from −40 °C to 85 °C, the LM224 and LM224A from −25 °C to 85 °C, and the LM324 and LM324A from 0 °C to 70 °C.

D, J, OR N DUAL-IN-LINE PACKAGE,
OR W FLAT PACKAGE
(TOP VIEW)

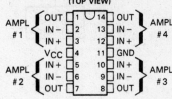

LM124
FH OR FK CHIP CARRIER PACKAGE
(TOP VIEW)

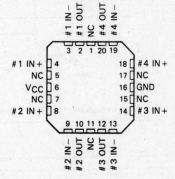

NC—No internal connection

symbol (each amplifier)

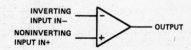

Operational Amplifiers

3

**TEXAS
INSTRUMENTS**
POST OFFICE BOX 225012 • DALLAS, TEXAS 75265

schematic (each amplifier)

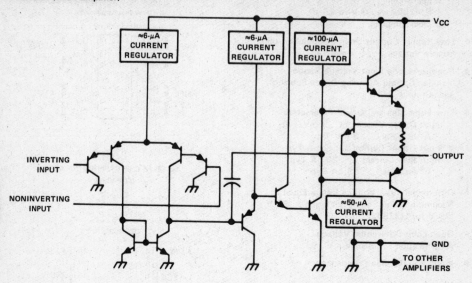

absolute maximum ratings over operating free-air temperature range (unless otherwise noted)

		LM124 LM224, LM224A, LM324, LM324A	LM2902	UNIT
Supply voltage, V_{CC} (see Note 1)		32	26	V
Differential voltage (see Note 2)		±32	±26	V
Input voltage range (either input)		−0.3 to 32	−0.3 to 26	V
Duration of output short-circuit (one amplifier) to ground at (or below) 25 °C free-air temperature ($V_{CC} \leq 15$ V) (see Note 3)		unlimited	unlimited	
Continuous total dissipation at (or below) 25 °C free-air temperature (see Note 4)	D or J package	900	900	mW
	N package	875	875	
Operating free-air temperature range	LM124	−55 to 125		°C
	LM224, LM224A	−25 to 85		
	LM324, LM324A	0 to 70		
	LM2902		−40 to 85	
Storage temperature range		−65 to 150	−65 to 150	°C
Lead temperature 1,6 mm (1/16 inch) from case for 60 seconds	FH, FK, J or W package	300	300	°C
Lead temperature 1,6 mm (1/16 inch) from case for 10 seconds	D or N package	260	260	°C

NOTES: 1. All voltage values, except differential voltages and V_{CC} specified for the measurement of I_{OS}, are with respect to the network ground terminal.
     2. Differential voltages are at the noninverting input terminal with respect to the inverting input terminal.
     3. Short circuits from outputs to V_{CC} can cause excessive heating and eventual destruction.
     4. For operation above 25 °C free-air temperature, refer to dissipation Derating Curves, Section 2. In the J package, LM124 chips are alloy-mounted; LM224, LM324, and LM2902 chips are glass-mounted.

TEXAS INSTRUMENTS
POST OFFICE BOX 225012 • DALLAS, TEXAS 75265

electrical characteristics at specified free-air temperature, V_{CC} = 5 V (unless otherwise noted)

PARAMETER	TEST CONDITIONS†		LM124, LM224			LM324			LM2902			UNIT
			MIN	TYP	MAX	MIN	TYP	MAX	MIN	TYP	MAX	
V_{IO} Input offset voltage	V_{CC} = 5 V to MAX, V_{IC} = V_{ICR} min, V_O = 1.4 V	25°C		3	5		3	7		3	7	mV
		Full range			7			9			10	
I_{IO} Input offset current	V_O = 1.4 V	25°C		2	30		2	50		2	50	nA
		Full range			100			150			200	
I_{IB} Input bias current	V_O = 1.4 V	25°C		-20	-150		-20	-250		-20	-250	nA
		Full range			-300			-500			-500	
V_{ICR} Common-mode input voltage range	V_{CC} = 5 V to MAX	25°C	0 to V_{CC}-1.5			0 to V_{CC}-1.5			0 to V_{CC}-1.5			V
		Full range	0 to V_{CC}-2			0 to V_{CC}-2			0 to V_{CC}-2			
V_{OH} High-level output voltage	R_L = 2 kΩ	25°C	V_{CC}-1.5			V_{CC}-1.5			V_{CC}-1.5			V
	R_L = 10 kΩ	25°C							V_{CC}-1.5			
	V_{CC} = MAX, R_L = 2 kΩ	Full range	26			26			22			
	V_{CC} = MAX, R_L = 10 kΩ	Full range	27	28		27	28		23	24		
V_{OL} Low-level output voltage	R_L ≤ 10 kΩ	Full range		5	20		5	20		5	100	mV
A_{VD} Large-signal differential voltage amplification	V_{CC} = 15 V, V_O = 1 V to 11 V, R_L ≥ 2 kΩ	25°C	50	100		25	100			100		V/mV
		Full range	25			15			15			
CMRR Common-mode rejection ratio	V_{IC} = V_{ICR} min	25°C	70	80		65	80		50	80		dB
kSVR Supply voltage rejection ratio ($\Delta V_{CC}/\Delta V_{IO}$)		25°C	65	100		65	100		50	100		dB
V_{o1}/V_{o2} Crosstalk attenuation	f = 1 kHz to 20 kHz	25°C		120			120			120		dB
I_O Output current	V_{CC} = 15 V, V_{ID} = 1 V, V_O = 0	25°C	-20	-30	-60	-20	-30	-60	-20	-30	-60	mA
		Full range	-10			-10			-10			
	V_{CC} = 15 V, V_{ID} = -1 V, V_O = 15 V	25°C	10	20		10	20		10	20		
		Full range	5			5			5			
	V_{ID} = -1 V, V_O = 200 mV	25°C	12	30		12	30			30		µA
I_{OS} Short-circuit output current	V_{CC} at 5 V, GND at -5 V, V_O = 0	25°C		±40	±60		±40	±60		±40	±60	mA
I_{CC} Supply current (four amplifiers)	V_{CC} = 2.5 V, No load	Full range		0.7	1.2		0.7	1.2		0.7	1.2	mA
	V_{CC} = MAX, V_O = 0.5 V_{CC}, No load	Full range		1.1	3		1.1	3		1.1	3	

† All characteristics are measured under open-loop conditions with zero common-mode input voltage unless otherwise specified. "MAX" V_{CC} for testing purposes is 26 V for LM2902, 30 V for the others. Full range is −55°C to 125°C for LM124, −25°C to 85°C for LM224, 0°C to 70°C for LM324, −40°C to 85°C for LM2902.

Operational Amplifiers

3

3 Operational Amplifiers

electrical characteristics at specified free-air temperature, V_{CC} = 5 V (unless otherwise noted)

PARAMETER	TEST CONDITIONS†		LM224A			LM324A			UNIT
			MIN	TYP	MAX	MIN	TYP	MAX	
V_{IO} Input offset voltage	V_{CC} = 5 V to 30 V, V_{IC} = V_{ICR} min, V_O = 1.4 V	25°C		2	3		2	3	mV
		Full range			4			5	
I_{IO} Input offset current	V_O = 1.4 V	25°C		2	15		2	30	nA
		Full range			30			75	
I_{IB} Input bias current	V_O = 1.4 V	25°C		−15	−80		−15	−100	nA
		Full range			−100			−200	
V_{ICR} Common-mode input voltage range	V_{CC} = 30 V	25°C	0 to V_{CC}−1.5			0 to V_{CC}−1.5			V
		Full range	0 to V_{CC}−2			0 to V_{CC}−2			V
V_{OH} High-level output voltage	R_L = 2 kΩ	Full range	26			26			V
	V_{CC} = 30 V, R_L = 2 kΩ	Full range	27	28		27	28		
	V_{CC} = 30 V, R_L = 10 kΩ	Full range							
V_{OL} Low-level output voltage	R_L ≤ 10 kΩ	Full range		5	20		5	20	mV
A_{VD} Large-signal differential voltage amplification	V_{CC} = 15 V, V_O = 1 V to 11 V, R_L ≥ 2 kΩ	25°C	50	100		25	100		V/mV
		Full range	25			15			
CMRR Common-mode rejection ratio	V_{IC} = V_{ICR} min	25°C	70	80		65	80		dB
k_{SVR} Supply voltage rejection ratio ($\Delta V_{CC}/\Delta V_{IO}$)		25°C	65	100		65	100		dB
V_{o1}/V_{o2} Crosstalk attenuation	f = 1 kHz to 20 kHz	25°C		120			120		dB
I_O Output current	V_{CC} = 15 V, V_{ID} = 1 V, V_O = 0	25°C	−20	−30	−60	−20	−30	−60	mA
		Full range	−10			−10			
	V_{CC} = 15 V, V_{ID} = −1 V, V_O = 5 V	25°C	10	20		10	20		
		Full range	5			5			
	V_{ID} = −1 V, V_O = 200 mV	25°C	12	30		12	30		µA
I_{OS} Short-circuit output current	V_{CC} at 5 V, GND at −5 V, V_O = 0	25°C		±40	±60		±40	±60	mA
I_{CC} Supply current (four amplifiers)	V_O = 2.5 V, No load			0.7	1.2		0.7	1.2	mA
	V_{CC} = 30 V, V_O = 15 V, No load	Full range		1.1	3		1.1	3	mA

†All characteristics are measured under open-loop conditions with zero common-mode input voltage unless otherwise specified. Full range is −25°C to 85°C for LM224A and 0°C to 70°C for LM324A.

TEXAS INSTRUMENTS
POST OFFICE BOX 225012 • DALLAS, TEXAS 75265

LINEAR
INTEGRATED
CIRCUITS

TYPES LM148, LM248, LM348
QUADRUPLE OPERATIONAL AMPLIFIERS

D2551, OCTOBER 1979—REVISED SEPTEMBER 1983

- uA741 Operating Characteristics
- Low Supply Current Drain . . . 0.6 mA Typ (per amplifier)
- Low Input Offset Voltage
- Low Input Offset Current
- Class AB Output Stage
- Input/Output Overload Protection
- Designed to be Interchangeable with National LM148, LM248, and LM348.

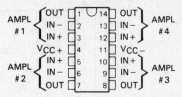

LM148 . . . J PACKAGE
LM248, LM348 . . . D, J, OR N PACKAGE
(TOP VIEW)

description

The LM148, LM248, and LM348 are quadruple, independent, high-gain, internally compensated operational amplifiers designed to have operating characteristics similar to the uA741. These amplifiers exhibit low supply current drain, and input bias and offset currents that are much less than those of the uA741.

The LM148 is characterized for operation over the full military temperature range of −55°C to 125°C, the LM248 is characterized for operation from −25°C to 85°C, and the LM348 is characterized for operation from 0°C to 70°C.

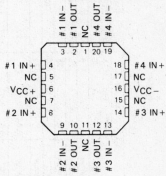

LM148 . . . FH OR FK PACKAGE
(TOP VIEW)

NC—No internal connection

symbol (each amplifier)

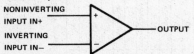

NONINVERTING INPUT IN+

INVERTING INPUT IN−

OUTPUT

3

Operational Amplifiers

absolute maximum ratings over operating free-air temperature range (unless otherwise noted)

		LM148	LM248	LM348	UNIT
Supply voltage V_{CC+} (see Note 1)		22	18	18	V
Supply voltage V_{CC-} (see Note 1)		−22	−18	−18	V
Differential input voltage (see Note 2)		44	36	36	V
Input voltage (either input, see Notes 1 and 3)		±22	±18	±18	V
Duration of output short-circuit (see Note 4)		unlimited	unlimited	unlimited	
Continuous total power dissipation at (or below)	D, FH, FK, or J package	900	900	900	mW
25°C free-air temperature (see Note 5)	N package		875	875	
Operating free-air temperature range		−55 to 125	−25 to 85	0 to 70	°C
Storage temperature range		−65 to 150	−65 to 150	−65 to 150	°C
Lead temperature 1,6 mm (1/16 inch) from case for 60 seconds	FH, FK, or J package	300	300	300	°C
Lead temperature 1,6 mm (1/16 inch) from case for 10 seconds	D or N package		260	260	°C

NOTES: 1. All voltage values, except differential voltages, are with respect to the midpoint between V_{CC+} and V_{CC-}.
2. Differential voltages are at the noninverting input terminal with respect to the inverting terminal.
3. The magnitude of the input voltage must never exceed the magnitude of the supply voltage or the value specified in the table, whichever is less.
4. The output may be shorted to ground or either power supply. Temperature and/or supply voltages must be limited to ensure that the dissipation rating is not exceeded.
5. For operation above 25°C free-air temperature, refer to Dissipation Derating Curves, Section 2. In the J package, LM148 chips are alloy mounted, LM248 and LM348 chips are glass mounted.

TEXAS
INSTRUMENTS

POST OFFICE BOX 225012 • DALLAS, TEXAS 75265

3

Operational Amplifiers

electrical characteristics, $V_{CC} \pm = \pm 15$ V

PARAMETER		TEST CONDITIONS†		LM148 MIN	LM148 TYP	LM148 MAX	LM248 MIN	LM248 TYP	LM248 MAX	LM348 MIN	LM348 TYP	LM348 MAX	UNIT
V_{IO}	Input offset voltage	$V_O = 0$	25°C		1	5		1	6		1	6	mV
			Full range			6			7.5			7.5	
I_{IO}	Input offset current	$V_O = 0$	25°C		4	25		4	50		4	50	nA
			Full range			75			125			100	
I_{IB}	Input bias current	$V_O = 0$	25°C		30	100		30	200		30	200	nA
			Full range			325			500			400	
V_{ICR}	Common-mode input voltage range		Full range	±12			±12			±12			V
V_{OM}	Maximum peak output voltage swing	$R_L = 10$ kΩ	25°C	±12	±13		±12	±13		±12	±13		V
		$R_L \geq 10$ kΩ	Full range	±12			±12			±12			
		$R_L = 2$ kΩ	25°C	±10	±12		±10	±12		±10	±12		
		$R_L \geq 2$ kΩ	Full range	±10			±10			±10			
A_{VD}	Large-signal differential voltage amplification	$V_O = \pm 10$ V, $R_L \geq 2$ kΩ	25°C	50	160		25	160		25	160		V/mV
			Full range	25			15			15			
r_i	Input resistance		25°C	0.8	2.5		0.8	2.5		0.8	2.5		MΩ
B_1	Unity-gain bandwidth	$A_{VD} = 1$	25°C		1			1			1		MHz
ϕ_M	Phase margin	$A_{VD} = 1$	25°C		60°			60°			60°		
CMRR	Common-mode rejection ratio	$V_{IC} = V_{ICR}$ min, $V_O = 0$	25°C	70	90		70	90		70	90		dB
			Full range	70			70			70			
k_{SVR}	Supply voltage rejection ratio ($\Delta V_{CC}\pm/\Delta V_{IO}$)	$V_{CC}\pm = \pm 9$ V to ± 15 V, $V_O = 0$	25°C	77	96		77	96		77	96		dB
			Full range	77			77			77			
I_{OS}	Short-circuit output current	$V_O = 0$	25°C		±25			±25			±25		mA
I_{CC}	Supply current (four amplifiers)	$V_O = 0$, No load	25°C		2.4	3.6		2.4	4.5		2.4	4.5	mA
V_{O1}/V_{O2}	Crosstalk attenuation	$f = 1$ Hz to 20 kHz	25°C		120			120			120		dB

†All characteristics are measured under open-loop conditions with zero common-mode input voltage unless otherwise specified. Full range for T_A is -55°C to 125°C for LM148, -25°C to 85°C for LM248, and 0°C to 70°C for LM348.

TEXAS INSTRUMENTS

POST OFFICE BOX 225012 • DALLAS, TEXAS 75265

operating characteristics, $V_{CC\pm} = \pm 15$ V, $T_A = 25\,^\circ$C

	PARAMETER	TEST CONDITIONS			MIN	TYP	MAX	UNIT
SR	Slew rate at unity gain	R_L = 2 kΩ,	C_L = 100 pF,	See Figure 1		0.5		V/μs

PARAMETER MEASUREMENT INFORMATION

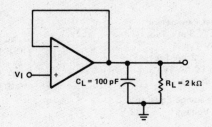

FIGURE 1—UNITY-GAIN AMPLIFIER

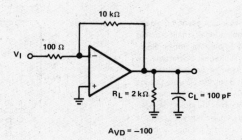

$A_{VD} = -100$

FIGURE 2—INVERTING AMPLIFIER

Operational Amplifiers

3

LINEAR
INTEGRATED
CIRCUITS

TYPES LM158, LM258, LM358
LM258A, LM358A, LM2904
DUAL OPERATIONAL AMPLIFIERS
D2231, JUNE 1976—REVISED AUGUST 1983

- **Wide Range of Supply Voltages:**
  Single Supply . . . 3 V to 30 V
  (LM2904 . . . 3 V to 26 V),
  or Dual Supplies

- **Low Supply Current Drain Independent of Supply Voltage . . . 0.7 mA Typ**

- **Common-Mode Input Voltage Range Includes Ground Allowing Direct Sensing near Ground**

- **Low Input Bias and Offset Parameters:**
  Input Offset Voltage . . . 3 mV Typ
  A Versions . . . 2 mV Typ
  Input Offset Current . . . 2 nA Typ
  Input Bias Current . . . 20 nA Typ
  A Versions . . . 15 nA Typ

- **Differential Input Voltage Range Equal to Maximum-Rated Supply Voltage . . . ±32 V (±26 V for LM2904)**

- **Open-Loop Differential Voltage Amplification . . . 100 V/mV Typ**

- **Internal Frequency Compensation**

description

These devices consist of two independent, high-gain, frequency-compensated operational amplifiers that were designed specifically to operate from a single supply over a wide range of voltages. Operation from split supplies is also possible so long as the difference between the two supplies is 3 volts to 30 volts (3 volts to 26 volts for the LM2904), and the V_{CC} pin is at least 1.5 volts more positive than the input common-mode voltage. The low supply current drain is independent of the magnitude of the supply voltage.

Applications include transducer amplifiers, d-c amplification blocks, and all the conventional operational amplifier circuits that now can be more easily implemented in single-supply-voltage systems. For example, these devices can be operated directly off of the standard five-volt supply that is used in digital systems and will easily provide the required interface electronics without requiring additional ±15-volt supplies.

The LM158 is characterized for operation over the full military temperature range of −55°C to 125°C. The LM258 and LM258A are characterized for operation from −25°C to 85°C, the LM358 and LM358A from 0° to 70°, and the LM2904 from −40°C to 85°C.

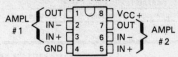

D, JG, OR P DUAL-IN-LINE PACKAGE
(TOP VIEW)

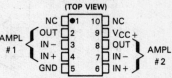

U FLAT PACKAGE
(TOP VIEW)

LM 158
FH OR FK CHIP CARRIER PACKAGE
(TOP VIEW)

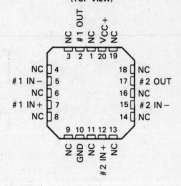

NC—No internal connection

symbol (each amplifier)

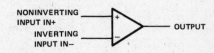

3

Operational Amplifiers

TEXAS
INSTRUMENTS
POST OFFICE BOX 225012 • DALLAS, TEXAS 75265

schematic (each amplifier)

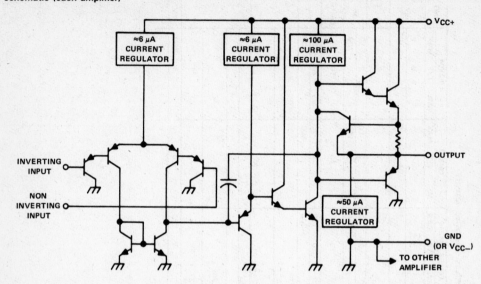

absolute maximum ratings over operating free-air temperature range (unless otherwise noted)

		LM158, LM258, LM258A LM358, LM358A	LM2904	UNIT
Supply voltage, V_{CC} (see Note 1)		32	26	V
Differential voltage (see Note 2)		±32	±26	V
Input voltage range (either input)		−0.3 to 32	−0.3 to 26	V
Duration of output short-circuit (one amplifier) to ground at (or below) 25°C free-air temperature ($V_{CC} \leq 15$ V) (see Note 3)		unlimited	unlimited	
Continuous total dissipation at (or below) 25°C free-air temperature (see Note 4)	D package	725	725	mW
	JG package (alloy-mounted chip)	1050		
	JG package (glass-mounted chip)	825	825	
	P package	725	725	
	U package	675	675	
Operating free-air temperature range	LM158	−55 to 125		°C
	LM258, LM258A	−25 to 85		
	LM358, LM358A	0 to 70		
	LM2904		−40 to 85	
Storage temperature range		−65 to 150	−65 to 150	°C
Lead temperature 1,6 mm (1/16 inch) from case for 60 seconds	FH, FK, JG, or U package	300	300	°C
Lead temperature 1,6 mm (1/16 inch) from case for 10 seconds	D or P package	260	260	°C

NOTES: 1. All voltage values, except differential voltages and V_{CC} specified for the measurement of i_{OS}, are with respect to the network ground terminal.
2. Differential voltages are at the noninverting input terminal with respect to the inverting input terminal.
3. Short circuits from outputs to V_{CC} can cause excessive heating and eventual destruction.
4. For operation above 25°C free-air temperature, refer to Dissipation Derating Curves, Section 2. In the JG package, LM158 chips are alloy-mounted; LM258, LM258A, LM358, LM358A, and LM2904 chips are glass-mounted.

3

Operational Amplifiers

TYPES LM158, LM258, LM358, LM2904
DUAL OPERATIONAL AMPLIFIERS

3

Operational Amplifiers

electrical characteristics at specified free-air temperature, V_{CC} = 5 V (unless otherwise noted)

PARAMETER		TEST CONDITIONS†		LM158, LM258			LM358			LM2904			UNIT
				MIN	TYP	MAX	MIN	TYP	MAX	MIN	TYP	MAX	
V_{IO}	Input offset voltage	V_{CC} = 5 V to MAX, V_{IC} = V_{ICR} min, V_O = 1.4 V	25°C		3	5		3	7		3	7	mV
			Full range			7			9			10	
α_{VIO}	Average temperature coefficient of input offset voltage		Full range		7			7			7		μV/°C
I_{IO}	Input offset current	V_O = 1.4 V	25°C		2	30		2	50		2	50	nA
			Full range			100			150			200	
α_{IIO}	Average temperature coefficient of input offset current		Full range		10			10			10		pA/°C
I_{IB}	Input bias current	V_O = 1.4 V	25°C		-20	-150		-20	-250		-20	-250	nA
			Full range			-300			-500			-500	
V_{ICR}	Common-mode input voltage range	V_{CC} = 5 V to MAX	25°C	0 to V_{CC}-1.5			0 to V_{CC}-1.5			0 to V_{CC}-1.5			V
			Full range	0 to V_{CC}-2			0 to V_{CC}-2			0 to V_{CC}-2			
V_{OH}	High-level output voltage	R_L ≥ 2 kΩ	25°C	V_{CC}-1.5			V_{CC}-1.5			V_{CC}-1.5			V
		R_L ≥ 10 kΩ	25°C										
		V_{CC} = MAX, R_L ≥ 2 kΩ	Full range	26			26			22			
		V_{CC} = MAX, R_L ≥ 10 kΩ	Full range	27	28		27	28		23	24		
V_{OL}	Low-level output voltage	R_L ≤ 10 kΩ	Full range		5	20		5	20		5	100	mV

TEXAS INSTRUMENTS

POST OFFICE BOX 225012 ● DALLAS, TEXAS 75265

electrical characteristics over operating free-air temperature range, $V_{CC} = 5$ V (unless otherwise noted)†

PARAMETER	TEST CONDITIONS	T_A	LM158, LM258 MIN	TYP	MAX	LM358 MIN	TYP	MAX	LM2904 MIN	TYP	MAX	UNIT
A_{VD} Large-signal differential voltage amplification	$V_{CC} = 15$ V, $V_O = 1$ V to 11 V, $R_L \geq 2$ kΩ	25°C	50	100		25	100			100		V/mV
		Full range	25			15			15			
CMRR Common-mode rejection ratio	$V_{CC} = 5$ V to MAX, $V_{IC} = V_{ICR}$ min	25°C	70	80		65	80		50	80		dB
k_{SVR} Supply voltage rejection ratio ($\Delta V_{CC}/\Delta V_{IO}$)	$V_{CC} = 5$ V to MAX	25°C	65	100		65	100		50	100		dB
V_{o1}/V_{o2} Crosstalk attenuation	$f = 1$ kHz to 20 kHz	25°C		120			120			120		dB
I_O Output current	$V_{CC} = 15$ V, $V_{ID} = 1$ V, $V_O = 0$	25°C	−20	−30		−20	−30		−20	−30		mA
		Full range	−10			−10			−10			
	$V_{CC} = 15$ V, $V_{ID} = -1$ V, $V_O = 15$ V	25°C	10	20		10	20		10	20		
		Full range	5			5			5			
	$V_{ID} = -1$ V, $V_O = 200$ mV	25°C	12	30		12	30			30		μA
I_{OS} Short-circuit output current	V_{CC} at 5 V, GND at −5 V, $V_O = 0$	25°C		±40	±60		±40	±60		±40	±60	mA
I_{CC} Supply current (two amplifiers)	$V_O = 2.5$ V, No load	Full range		0.7	1.2		0.7	1.2		0.7	1.2	mA
	$V_{CC} = $ MAX, $V_O = 0.5$ V_{CC-}, No load	Full range		1	2		1	2		1	2	

†All characteristics are measured under open-loop conditions with zero common-mode input voltage unless otherwise specified. "MAX" V_{CC} for testing purposes is 26 V for LM2904, 30 V for the others. Full range is −55°C to 125°C for LM158, −25°C to 85°C for LM258, 0°C to 70°C for LM358, and −40°C to 85°C for LM2904.

Operational Amplifiers

3

TYPES LM258A, LM358A
DUAL OPERATIONAL AMPLIFIERS

3 Operational Amplifiers

electrical characteristics at specified free-air temperature, VCC = 5 V (unless otherwise noted)

PARAMETER		TEST CONDITIONS†	LM258A MIN	LM258A TYP	LM258A MAX	LM358A MIN	LM358A TYP	LM358A MAX	UNIT
V_{IO}	Input offset voltage	V_{CC} = 5 v to 30 V, V_{IC} = V_{ICR} min, V_O = 1.4 V 25°C		2	3		2	3	mV
		Full range			4			5	
$α_{VIO}$	Average temperature coefficient of input offset voltage	Full range		7	15		7	20	µV/°C
I_{IO}	Input offset current	V_O = 1.4 V 25°C		2	15		2	30	nA
		Full range			30			75	
$α_{IIO}$	Average temperature coefficient of input offset current	Full range		10	200		10	300	pA/°C
I_{IB}	Input bias current	V_O = 1.4 V 25°C		-15	-80		-15	-100	nA
		Full range			-100			-200	
V_{ICR}	Common-mode input voltage range	V_{CC} 30 V 25°C	0 to V_{CC} – 1.5			0 to V_{CC} – 1.5			V
		Full range	0 to V_{CC} – 2			0 to V_{CC} – 2			
V_{OH}	High-level output voltage	R_L ≥ 2 kΩ 25°C	V_{CC} – 1.5			V_{CC} – 1.5			V
		V_{CC} = 30 V, R_L ≥ 2 kΩ Full range	26			26			
		V_{CC} = 30 V, R_L ≥ 10 kΩ Full range	27	28		27	28		
V_{OL}	Low-level output voltage	R_L ≤ 10 kΩ Full range		5	20		5	20	mV

TEXAS INSTRUMENTS
POST OFFICE BOX 225012 • DALLAS, TEXAS 75265

	PARAMETER	TEST CONDITIONS	Tₐ	LM258A MIN	TYP	MAX	LM358A MIN	TYP	MAX	UNIT
A_{VD}	Large-signal differential voltage amplification	$V_{CC} = 15$ V, $V_O = 1$ V to 11 V, $R_L = \geq 2$ kΩ	25°C	50	100		25	100		V/mV
			Full range	25			15			
CMRR	Common-mode rejection ratio		25°C	70	80		65	80		dB
k_{SVR}	Supply voltage rejection ratio ($\Delta V_{CC}/\Delta V_{IO}$)		25°C	65	100		65	100		dB
V_{O1}/V_{O2}	Crosstalk attenuation	$f = 1$ kHz to 20 kHz	25°C		120			120		dB
I_O	Output current	$V_{CC} = 15$ V, $V_{ID} = 1$ V, $V_O = 0$	25°C	20	30	60	20	30	60	mA
			Full range	10			10			
		$V_{CC} = 15$ V, $V_{ID} = -1$ V, $V_O = 15$ V	25°C	10	20		10	20		
			Full range	5			5			
		$V_{ID} = -1$ V, $V_O = 200$ mV	25°C	12	30		12	30		µA
I_{OS}	Short-circuit output current	V_{CC} at 5 V, GND at −5 V, $V_O = 0$	25°C		±40	±60		±40	±60	mA
I_{CC}	Supply current (two amplifiers)	$V_O = 2.5$ V, No load	Full range		0.7	1.2		0.7	1.2	mA
		$V_{CC} = 30$ V, $V_O = 15$ V, No load	Full range		1	2			2	

† All characteristics are measured under open-loop conditions with zero common-mode input voltage unless otherwise specified. Full range is −25°C to 85°C for LM258A and 0°C to 70°C for LM358A.

Operational Amplifiers

3

3

Operational Amplifiers

**LINEAR
INTEGRATED
CIRCUITS**

**TYPES LM218, LM318
HIGH-PERFORMANCE OPERATIONAL AMPLIFIERS**

D2219, JUNE 1976—REVISED AUGUST 1983

- Small-Signal Bandwidth . . . 15 MHz Typ
- Slew Rate . . . 50 V/μs Min
- Bias Current . . . 250 nA Max (LM218)
- Supply Voltage Range . . . ± 5 V to ± 20 V
- Internal Frequency Compensation
- Input and Output Overload Protection
- Same Pin Assignments as General-Purpose Operational Amplifiers

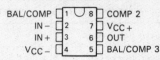

**D, JG, OR P DUAL-IN-LINE PACKAGE
(TOP VIEW)**

BAL/COMP	1		8	COMP 2
IN−	2		7	V_{CC+}
IN+	3		6	OUT
V_{CC-}	4		5	BAL/COMP 3

description

The LM218 and LM318 are precision, high-speed operational amplifiers designed for applications requiring wide bandwidth and high slew rate. They feature a factor-of-ten increase in speed over general purpose devices without sacrificing dc performance.

These operational amplifiers have internal unity-gain frequency compensation. This considerably simplifies their application since no external components are necessary for operation. However, unlike most internally compensated amplifiers, external frequency compensation may be added for optimum performance. For inverting applications, feed-forward compensation will boost the slew rate to over 150 V/μs and almost double the bandwidth. Overcompensation may be used with the amplifier for greater stability when maximum bandwidth is not needed. Further, a single capacitor may be added to reduce the settling time for 0.1% error band to under 1 μs.

symbol

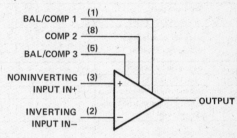

BAL/COMP 1 (1)
COMP 2 (8)
BAL/COMP 3 (5)
NONINVERTING INPUT IN+ (3)
INVERTING INPUT IN− (2)
OUTPUT

The high speed and fast settling time of these operational amplifiers make them useful in A/D converters, oscillators, active filters, sample and hold circuits, and general purpose amplifiers.

The LM218 is characterized for operation from −25°C to 85°C, and the LM318 is characterized for operation from 0°C to 70°C.

absolute maximum ratings over operating free-air temperature range (unless otherwise noted)

		LM218	LM318	UNIT
Supply voltage, V_{CC+} (see Note 1)		20	20	V
Supply voltage, V_{CC-} (see Note 1)		−20	−20	V
Input voltage (either input, see Notes 1 and 2)		±15	±15	V
Differential input current (see Note 3)		±10	±10	mA
Duration of output short-circuit (see Note 4)		unlimited	unlimited	
Continuous total power dissipation at (or below) 25°C free-air temperature (see Note 5)		500	500	mW
Operating free-air temperature range		−25 to 85	0 to 70	°C
Storage temperature range		−65 to 150	−65 to 150	°C
Lead temperature 1,6 mm (1/16 inch) from case for 60 seconds	JG package	300	300	°C
Lead temperature 1,6 mm (1/16 inch) from case for 10 seconds	D or P package	260	260	°C

NOTES: 1. All voltage values, unless otherwise noted, are with respect to the midpoint between V_{CC+} and V_{CC-}.
2. The magnitude of the input voltage must never exceed the magnitude of the supply voltage or 15 volts, whichever is less.
3. The inputs are shunted with two opposite-facing base-emitter diodes for over voltage protection. Therefore, excessive current will flow if a differential input voltage in excess of approximately 1 V is applied between the inputs unless some limiting resistance is used.
4. The output may be shorted to ground or either power supply. For the LM218 only, the unlimited duration of the short-circuit applies at (or below) 85°C case temperature or 75°C free-air temperature.
5. For operation above 25°C free-air temperature, refer to Dissipation Derating Curves, Section 2. In the JG package, LM218 and LM318 chips are glass-mounted.

**TEXAS
INSTRUMENTS**

POST OFFICE BOX 225012 • DALLAS, TEXAS 75265

3

Operational Amplifiers

3

Operational Amplifiers

electrical characteristics at specified free-air temperature (see Note 6)

PARAMETER		TEST CONDITIONS†		LM218			LM318			UNIT
				MIN	TYP	MAX	MIN	TYP	MAX	
V_{IO}	Input offset voltage	$V_O = 0$	25°C		2	4		4	10	mV
			Full range			6			15	
I_{IO}	Input offset current	$V_O = 0$	25°C		6	50		30	200	nA
			Full range			100			300	
I_{IB}	Input bias current	$V_O = 0$	25°C		120	250		150	250	nA
			Full range			500			750	
V_{ICR}	Common-mode input voltage range	$V_{CC\pm} = \pm15$ V	Full range	±11.5			±11.5			V
V_{OM}	Maximum peak output voltage swing	$V_{CC\pm} = \pm15$ V, $R_L = 2$ kΩ	Full range	±12	±13		±12	±13		V
A_{VD}	Large-signal differential voltage amplification	$V_{CC\pm} = \pm15$ V, $V_O = \pm10$ V, $R_L \geq 2$ kΩ	25°C	50	200		25	200		V/mV
			Full range	25			20			
B_1	Unity-gain bandwidth	$V_{CC\pm} = \pm15$ V	25°C		15			15		MHz
r_i	Input resistance		25°C	1	3		0.5	3		MΩ
CMRR	Common-mode rejection ratio	$V_{IC} = V_{ICR}$ min	Full range	80	100		70	100		dB
k_{SVR}	Supply voltage rejection ratio ($\Delta V_{CC}/\Delta V_{IO}$)		Full range	70	80		65	80		dB
I_{CC}	Supply current	No load, $V_O = 0$	25°C		5	8		5	10	mA
			Full range		4.5	7				

† All characteristics are measured under open-loop conditions with zero common-mode input voltage unless otherwise specified. Full range for LM218 is -25°C to 85°C and for LM318 is 0°C to 70°C.
NOTE 6: Unless otherwise noted, $V_{CC} = \pm5$ V to ±20 V. All typical values are at $V_{CC\pm} = \pm15$ V.

operating characteristics, $V_{CC+} = 15$ V, $V_{CC-} = -15$ V, $T_A = 25$°C

	PARAMETER	TEST CONDITIONS			MIN	TYP	MAX	UNIT
SR	Slew rate at unity gain	$\Delta V_I = 10$ V,	$C_L = 10$ pF,	See Figure 1	50	70		V/μs

parameter measurement information

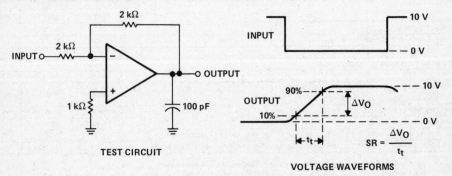

TEST CIRCUIT

VOLTAGE WAVEFORMS

$$SR = \frac{\Delta V_O}{t_t}$$

FIGURE 1—SLEW RATE

TEXAS INSTRUMENTS
POST OFFICE BOX 225012 • DALLAS, TEXAS 75265

schematic

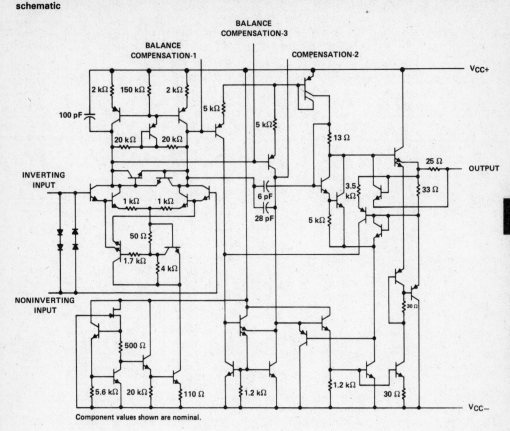

Component values shown are nominal.

Operational Amplifiers

3

LINEAR
INTEGRATED
CIRCUITS

TYPES LM2900, LM3900
QUADRUPLE OPERATIONAL AMPLIFIERS

D2531, JULY 1979—REVISED AUGUST 1983

- Wide Range of Supply Voltages, Single or Dual Supplies
- Wide Bandwidth
- Large Output Voltage Swing
- Output Short-Circuit Protection
- Internal Frequency Compensation
- Low Input Bias Current
- Designed to be Interchangeable with National Semiconductor LM2900 and LM3900, Respectively

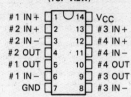

J OR N DUAL-IN-LINE PACKAGE
(TOP VIEW)

#1 IN+	1	14	V_{CC}
#2 IN+	2	13	#3 IN+
#2 IN−	3	12	#4 IN+
#2 OUT	4	11	#4 IN−
#1 OUT	5	10	#4 OUT
#1 IN−	6	9	#3 OUT
GND	7	8	#3 IN−

description

These devices consist of four independent, high-gain frequency-compensated Norton operational amplifiers that were designed specifically to operate from a single supply over a wide range of voltages. Operation from split supplies is also possible. The low supply current drain is essentially independent of the magnitude of the supply voltage. These devices provide wide bandwidth and large output voltage swing.

The LM2900 is characterized for operation from −40°C to 85°C, and the LM3900 is characterized for operation from 0°C to 70°C.

symbol (each amplifier)

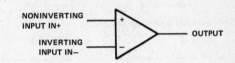

NONINVERTING
INPUT IN+

INVERTING
INPUT IN−

OUTPUT

schematic (each amplifier)

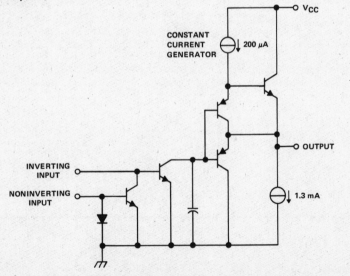

V_{CC}

CONSTANT
CURRENT
GENERATOR
↓ 200 μA

OUTPUT

INVERTING
INPUT

NONINVERTING
INPUT

↓ 1.3 mA

3

Operational Amplifiers

TEXAS
INSTRUMENTS
POST OFFICE BOX 225012 • DALLAS, TEXAS 75265

TYPES LM2900, LM3900
QUADRUPLE OPERATIONAL AMPLIFIERS

absolute maximum ratings over operating free-air temperature range (unless otherwise noted)

		LM2900	LM3900	UNIT
Supply voltage, V_{CC} (see Note 1)		32	32	V
Input current		20	20	mA
Duration of output short circuit (one amplifier) to ground at (or below) 25°C free-air temperature (see Note 2)		unlimited	unlimited	
Continuous total dissipation at (or below) 25°C free-air temperature (see Note 3)	J Package	1025	1025	mW
	N Package	875	875	
Operating free-air temperature range		−40 to 85	0 to 70	°C
Storage temperature range		−65 to 150	−65 to 150	°C
Lead temperature 1,6 mm (1/16 inch) from case for 60 seconds	J Package	300	300	°C
Lead temperature 1,6 mm (1/16 inch) from case for 10 seconds	N Package	260	260	°C

NOTES: 1. All voltage values, except differential voltages, are with respect to the network ground terminal.
2. Short circuits from outputs to V_{CC} can cause excessive heating and eventual destruction.
3. For operation above 25°C free-air temperature, refer to Dissipation Derating Curves, Section 2. In the J package, LM2900 and LM3900 chips are glass-mounted.

recommended operating comditions

	LM2900		LM3900		UNIT
	MIN	MAX	MIN	MAX	
Input current (see Note 4)		−1		−1	mA
Operating free-air temperature, T_A	−40	85	0	70	°C

NOTE 4: Clamp transistors are included that prevent the input voltages from swinging below ground more than approximately −0.3 volt. The negative input currents that may result from large signal overdrive with capacitive input coupling must be limited externally to values of approximately −1 mA. Negative input currents in excess of −4 mA will cause the output voltage to drop to a low voltage. These values apply for any one of the input terminals. If more than one of the input terminals are simultaneously driven negative, maximum currents are reduced. Common-mode current biasing can be used to prevent negative input voltages.

3

Operational Amplifiers

TEXAS INSTRUMENTS
POST OFFICE BOX 225012 • DALLAS, TEXAS 75265

electrical characteristics, V_{CC} = 15 V, T_A = 25°C (unless otherwise noted)

PARAMETER		TEST CONDITIONS[†]		LM2900 MIN	LM2900 TYP	LM2900 MAX	LM3900 MIN	LM3900 TYP	LM3900 MAX	UNIT
I_{IB}	Input bias current (inverting input)	I_{I+} = 0	T_A = 25°C		30	200		30	200	nA
			T_A = full range							
$\dfrac{I_{I-}}{I_{I+}}$	Mirror gain	I_{I+} = 20 μA to 200 μA, T_A = full range, See Note 5		0.9		1.1	0.9		1.1	μA/μA
	Change in mirror gain				2	5		2	15	%
	Mirror current	V_{I+} = V_{I-}, T_A = full range, See Note 5			10	500		10	500	μA
A_{VD}	Large-signal differential voltage amplification	V_O = 10 V, R_L = 10 kΩ, f = 100 Hz		1.2	2.8		1.2	2.8		V/mV
r_i	Input resistance (inverting input)				1			1		MΩ
r_o	Output resistance				8			8		kΩ
B_1	Unity-gain bandwidth (inverting input)				2.5			2.5		MHz
k_{SVR}	Supply voltage rejection ratio($\Delta V_{CC}/\Delta V_{IO}$)				70			70		dB
V_{OH}	High-level output voltage	I_{I+} = 0, I_{I-} = 0	R_L = 2 kΩ	13.5			13.5			V
			V_{CC} = 30 V, No load		29.5			29.5		
V_{OL}	Low-level output voltage	I_{I+} = 0, R_L = 2 kΩ	I_{I-} = 10 μA,		0.09	0.2		0.09	0.2	V
I_{OHS}	Short-circuit output current (output internally high)	I_{I+} = 0, V_O = 0	I_{I-} = 0,		−6	−18		−6	−10	mA
	Pull-down current			0.5	1.3		0.5	1.3		mA
I_{OL}	Low-level output current‡	I_{I-} = 5 μA, V_{OL} = 1 V			5			5		mA
I_{CC}	Supply current (four amplifiers)	No load			6.2	10		6.2	10	mA

† All characteristics are measured under open-loop conditions with zero common-mode voltage unless otherwise specified. Full range for T_A is −40°C to 85°C for LM2900, and 0°C to 70°C for LM3900.

‡ The output current-sink capability can be increased for large-signal conditions by overdriving the inverting input.

NOTE 5: These parameters are measured with the output balanced midway between V_{CC} and ground.

operating characteristics, $V_{CC\pm}$ = ±15 V, T_A = 25°C

PARAMETER		TEST CONDITIONS	MIN	TYP	MAX	UNIT
SR	Slew rate at unity gain — Low-to-high output	V_O = 10 V, C_L = 100 pF,		0.5		V/μs
	Slew rate at unity gain — High-to-low output	R_L = 2 kΩ		20		

3

Operational Amplifiers

TEXAS
INSTRUMENTS
POST OFFICE BOX 225012 • DALLAS, TEXAS 75265

TYPICAL CHARACTERISTICS†

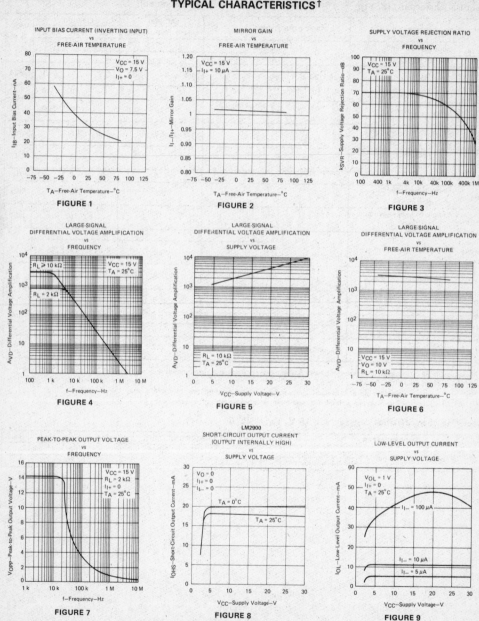

†Data at high and low temperatures are applicable only within the rated operating free-air temperature ranges of the various devices.

TEXAS
INSTRUMENTS
POST OFFICE BOX 225012 • DALLAS, TEXAS 75265

TYPICAL CHARACTERISTICS†

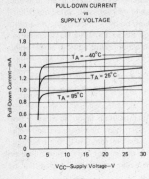

FIGURE 10

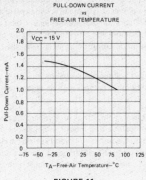

FIGURE 11

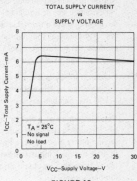

FIGURE 12

Operational Amplifiers

3

†Data at high and low temperatures are applicable only within the rated operating free-air temperature ranges of the various devices.

TYPICAL APPLICATION DATA

Norton (or current-differencing) amplifiers can be used in most standard general-purpose op-amp applications. Performance as a dc amplifier in a single-power-supply mode is not as precise as a standard integrated-circuit operational amplifier operating from dual supplies. Operation of the amplifier can best be understood by noting that input currents are differenced at the inverting input terminal and this current then flows through the external feedback resistor to produce the output voltage. Common-mode current biasing is generally useful to allow operating with signal levels near (or even below) ground.

Internal transistors clamp negative input voltages at pproximately −0.3 volt but the magnitude of current flow has to be limited by the external input network. For operation at high temperature, this limit should be approximately −100 microamperes.

Noise immunity of a Norton amplifier is less than that of standard bipolar amplifiers. Circuit layout is more critical since coupling from the output to the noninverting input can cause oscillations. Care must also be exercised when driving either input from a low-impedance source. A limiting resistor should be placed in series with the input lead to limit the peak input current. Current up to 20 milliamperes will not damage the device but the current mirror on the noninverting input will saturate and cause a loss of mirror gain at higher current levels, especially at high operating temperatures.

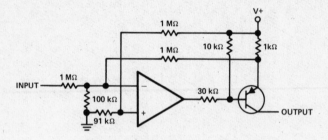

$I_O \approx 1$ mA per input volt

FIGURE 13—VOLTAGE-CONTROLLED CURRENT SOURCE

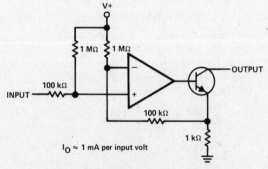

$I_O \approx 1$ mA per input volt

FIGURE 14—VOLTAGE-CONTROLLED CURRENT SINK

TEXAS
INSTRUMENTS
POST OFFICE BOX 225012 • DALLAS, TEXAS 75265

3

Operational Amplifiers

**LINEAR
INTEGRATED
CIRCUITS**

**TYPES MC1558, MC1458
DUAL GENERAL-PURPOSE OPERATIONAL AMPLIFIERS**

D972, FEBRUARY 1971—REVISED AUGUST 1983

- Short-Circuit Protection
- Wide Common-Mode and Differential Voltage Ranges
- No Frequency Compensation Required
- Low Power Consumption
- No Latch-up
- Designed to be Interchangeable with Motorola MC1558/MC1458 and Signetics S5558/N5558

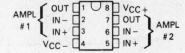

MC1558 . . . JG PACKAGE
MC1458 . . . D, JG, OR P PACKAGE
(TOP VIEW)

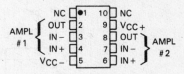

MC1558, MC1458 . . . U FLAT PACKAGE
(TOP VIEW)

description

The MC1558 and MC1458 are dual general-purpose operational amplifiers with each half electrically similar to uA741 except that offset null capability is not provided.

The high common-mode input voltage range and the absence of latch-up make these amplifiers ideal for voltage-follower applications. The devices are short-circuit protected and the internal frequency compensation ensures stability without external components.

The MC1558 is characterized for operation over the full military temperature range of −55 °C to 125 °C; the MC1458 is characterized for operation from 0 °C to 70 °C.

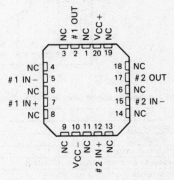

MC1558 . . . FH OR FK PACKAGE
(TOP VIEW)

NC—No internal connection

3

Operational Amplifiers

symbol (each amplifier)

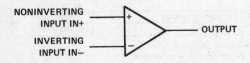

NONINVERTING
INPUT IN+

INVERTING
INPUT IN−

OUTPUT

TYPES MC1558, MC1458
DUAL GENERAL-PURPOSE OPERATIONAL AMPLIFIERS

schematic (each amplifier)

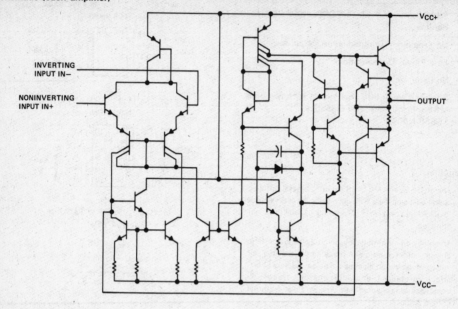

absolute maximum ratings over operating free-air temperature range (unless otherwise noted)

			MC1558	MC1458	UNIT
Supply voltage V_{CC+} (see Note 1)			22	18	V
Supply voltage V_{CC-} (see Note 1)			−22	−18	V
Differential input voltage (see Note 2)			±30	±30	V
Input voltage at either input (see Notes 1 and 3)			±15	±15	V
Duration of output short-circuit (see Note 4)			unlimited	unlimited	
Continuous total dissipation at (or below) 25°C free-air temperature (see Note 5)	Each amplifier		500	500	mW
	Total package	D, FH, FK, JG, or P package	680	680	
		U package	675	675	
Operating free-air temperature range			−55 to 125	0 to 70	°C
Storage temperature range			−65 to 150	−65 to 150	°C
Lead temperature 1,6 mm (1/16 inch) from case for 60 seconds	FH, FK, JG or U package		300	300	°C
Lead temperature 1,6 mm (1/16 inch) from case for 10 seconds	D or P package			260	°C

NOTES: 1. All voltage values, unless otherwise noted, are with respect to the midpoint between V_{CC+} and V_{CC-}.
2. Differential voltages are at the noninverting input terminal with respect to the inverting input terminal.
3. The magnitude of the input voltage must never exceed the magnitude of the supply voltage or 15 volts, whichever is less.
4. The output may be shorted to ground or either power supply. For the MC1558 only, the unlimited duration of the short-circuit applies at (or below) 125°C case temperature or 70°C free-air temperature.
5. For operation above 25°C free-air temperature, refer to Dissipation Derating Curves, Section 2. In the JG package, MC1558 chips are alloy mounted, MC1458 chips are glass mounted.

TEXAS
INSTRUMENTS
POST OFFICE BOX 225012 • DALLAS, TEXAS 75265

electrical characteristics at specified free-air temperature, V_{CC+} = 15 V, V_{CC-} = -15 V

PARAMETER		TEST CONDITIONS†		MC1558			MC1458			UNIT
				MIN	TYP	MAX	MIN	TYP	MAX	
V_{IO}	Input offset voltage	$V_O = 0$	25°C		1	5		1	6	mV
			Full range			6			7.5	
I_{IO}	Input offset current	$V_O = 0$	25°C		20	200		20	200	nA
			Full range			500			300	
I_{IB}	Input bias current	$V_O = 0$	25°C		80	500		80	500	nA
			Full range			1500			800	
V_{ICR}	Common-mode input voltage range		25°C	±12	±13		±12	±13		V
			Full range	±12			±12			
V_{OM}	Maximum peak output voltage swing	$R_L = 10\ k\Omega$	25°C	±12	±14		±12	±14		V
		$R_L \geq 10\ k\Omega$	Full range	±12			±12			
		$R_L = 2\ k\Omega$	25°C	±10	±13		±10	±13		
		$R_L \geq 2\ k\Omega$	Full range	±10			±10			
A_{VD}	Large-signal differential voltage amplification	$R_L \geq 2\ k\Omega$, $V_O = \pm 10\ V$	25°C	50	200		20	200		V/mV
			Full range	25			15			
B_{OM}	Maximum-output-swing bandwidth (closed-loop)	$R_L = 2\ k\Omega$, $V_O \geq \pm 10\ V$, $A_{VD} = 1$, THD $\leq$ 5%	25°C		14			14		kHz
B_1	Unity-gain bandwidth		25°C		1			1		MHz
ϕ_m	Phase margin	$A_{VD} = 1$	25°C		65°			65°		
A_m	Gain margin		25°C		11			11		dB
r_i	Input resistance		25°C	0.3	2		0.3	2		MΩ
r_o	Output resistance	$V_O = 0$, See Note 6	25°C		75			75		Ω
C_i	Input capacitance		25°C		1.4			1.4		pF
z_{ic}	Common-mode input impedance	$f = 20\ Hz$	25°C		200			200		MΩ
CMRR	Common-mode rejection ratio	$V_{IC} = V_{ICR}$ min, $V_O = 0$	25°C	70	90		70	90		dB
			Full Range	70			70			
k_{SVS}	Supply voltage sensitivity ($\Delta V_{IO}/\Delta V_{CC}$)	$V_{CC} = \pm 9\ V$ to $\pm 15\ V$, $V_O = 0$	25°C		30	150		30	150	µV/V
			Full range			150			150	
V_n	Equivalent input noise voltage (closed-loop)	$A_{VD} = 100$, $R_S = 0$, $f = 1\ kHz$, BW = 1 Hz	25°C		45			45		nV/$\sqrt{Hz}$
I_{OS}	Short-circuit output current		25°C		±25	±40		±25	±40	mA
I_{CC}	Supply current (both amplifiers)	No load, $V_O = 0$	25°C		3.4	5		3.4	5.6	mA
			Full range			6.6			6.6	
P_D	Total power dissipation (both amplifiers)	No load, $V_O = 0$	25°C		100	150		100	170	mW
			Full range			200			200	
V_{o1}/V_{o2}	Crosstalk attenuation		25°C		120			120		dB

†All characteristics are specified under open-loop operating conditions with zero common-mode input voltage unless otherwise specified. Full range for MC1558 is -55°C to 125°C and for MC1458 is 0°C to 70°C.

NOTE 6: This typical value applies only at frequencies above a few hundred hertz because of the effects of drift and thermal feedback.

Operational Amplifiers

3

TEXAS INSTRUMENTS
POST OFFICE BOX 225012 • DALLAS, TEXAS 75265

TYPES MC1558, MC1458
DUAL GENERAL-PURPOSE OPERATIONAL AMPLIFIERS

operating characteristics, V_{CC+} = 15 V, V_{CC-} = -15 V, T_A = 25°C

PARAMETER		TEST CONDITIONS		MC1558			MC1458			UNIT
				MIN	TYP	MAX	MIN	TYP	MAX	
t_r	Rise time	V_I = 20 mV,	R_L = 2 kΩ,		0.3			0.3		μs
	Overshoot factor	C_L = 100 pF,	See Figure 1		5%			5%		
SR	Slew rate at unity gain	V_I = 10 V,	R_L = 2 kΩ,							
		C_L = 100 pF,	See Figure 1		0.5			0.5		V/μs

PARAMETER MEASUREMENT INFORMATION

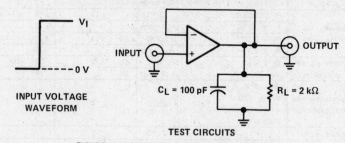

INPUT VOLTAGE
WAVEFORM

TEST CIRCUITS

FIGURE 1—RISE TIME, OVERSHOOT, AND SLEW RATE

3

Operational Amplifiers

TEXAS
INSTRUMENTS
POST OFFICE BOX 225012 • DALLAS, TEXAS 75265

883

- **Wide Range of Supply Voltages Single Supply . . . 3 V to 36 V or Dual Supplies**
- **Class AB Output Stage**
- **True Differential Input Stage**
- **Low Input Bias Current**
- **Internal Frequency Compensation**
- **Short-Circuit Protection**
- **Designed to be Interchangeable with Motorola MC3503, MC3303, MC3403**

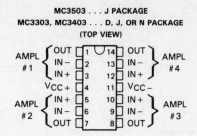

MC3503 . . . J PACKAGE
MC3303, MC3403 . . . D, J, OR N PACKAGE
(TOP VIEW)

description

The MC3503, MC3303, and the MC3403 are quadruple operational amplifiers similar in performance to the uA741 but with several distinct advantages. They are designed to operate from a single supply over a range of voltages from 3 volts to 36 volts. Operation from split supplies is also possible provided the difference between the two supplies is 3 volts to 36 volts. The common-mode input range includes the negative supply. Output range is from the negative supply to $V_{CC} - 1.5$ V. Quiescent supply currents are less than one-half those of the uA741.

The MC3503 is characterized for operation over the full military temperature range of $-55\,°C$ to $125\,°C$. The MC3303 is characterized for operation from $-40\,°C$ to $85\,°C$, and the MC3403 is characterized for operation from $0°$ to $70°$.

symbol (each amplifier)

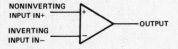

NONINVERTING INPUT IN+

INVERTING INPUT IN−

OUTPUT

3

Operational Amplifiers

TEXAS INSTRUMENTS

POST OFFICE BOX 225012 • DALLAS, TEXAS 75265

TYPES MC3503, MC3303, MC3403
QUADRUPLE LOW-POWER OPERATIONAL AMPLIFIERS

schematic (each amplifier)

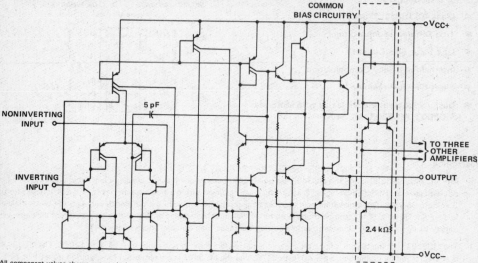

All component values shown are nominal

absolute maximum ratings over operating free-air temperature range (unless otherwise noted)

		MC3503	MC3303	MC3403	UNIT
Supply voltage V_{CC+} (see Note 1)		18	18	18	V
Supply voltage V_{CC-} (see Note 1)		−18	−18	−18	V
Supply voltage V_{CC+} with respect to V_{CC-}		36	36	36	V
Differential input voltage (see Note 2)		±36	±36	±36	V
Input voltage (see Notes 1 and 3)		±18	±18	±18	V
Continuous total power dissipation at (or below)	D package		960	960	
25 °C free-air temperature (see Note 4)	J package	1375	1025	1025	mW
	N package		875	875	
Operating free-air temperature range		−55 to 125	−40 to 85	0 to 70	°C
Storage temperature range		−65 to 150	−65 to 150	−65 to 150	°C
Lead temperature 1,6 mm (1/16 inch) from case for 60 seconds	J package	300	300	300	°C
Lead temperature 1,6 mm (1/16 inch) from case for 10 seconds	D or N package		260	260	°C

NOTES: 1. These voltage values are with respect to the midpoint between V_{CC+} and V_{CC-}.
2. Differential voltages are at the noninverting input terminal with respect to the inverting terminal.
3. Neither input must ever be more positive then V_{CC+} or more negative than V_{CC-}.
4. For operation above 25 °C free-air temperature, refer to Dissipation Derating Curves, Section 2. In the J package, MC3503 chips are alloy mounted, MC3303 and MC3403 chips are glass mounted.

electrical characteristics at specified free-air temperature; V_{CC+} = 14 V, V_{CC-} = 0 V for MC3303; $V_{CC\pm}$ = ±15 V for MC3403 and MC3503

PARAMETER		TEST CONDITIONS†		MC3503			MC3303			MC3403			UNIT
				MIN	TYP	MAX	MIN	TYP	MAX	MIN	TYP	MAX	
V_{IO}	Input offset voltage	See Note 5	25°C		2	5		2	8		2	10	mV
			Full range			6			10			12	
α_{VIO}	Temperature coefficient of input offset voltage	See Note 5	Full range		10			10			10		µV/°C
I_{IO}	Input offset current	See Note 5	25°C		30	50		30	75		30	50	nA
			Full range			200			250			200	
α_{IIO}	Temperature coefficient of input offset current	See Note 5	Full range		50			50			50		pA/°C
I_{IB}	Input bias current	See Note 5	25°C		−0.2	−0.5		−0.2	−0.5		−0.2	−0.5	µA
			Full range			−1.5			−1			−0.8	
V_{ICR}	Common-mode input voltage range‡		25°C	V_{CC-} to 13	V_{CC-} to 13.5		V_{CC-} to 12	V_{CC-} to 12.5		V_{CC-} to 13	V_{CC-} to 13.5		V
V_{OM}	Peak output voltage swing	R_L = 10 kΩ	25°C	±12	±13.5		12	12.5		±12	±13.5		V
		R_L = 2 kΩ	25°C	±10	±13		10	12		±10	±13		
		R_L = 2 kΩ	Full range	±10			10			±10			
A_{VD}	Large-signal differential voltage amplification	V_O = ±10 V,	25°C	50	200		20	200		20	200		V/mV
		R_L = 2 kΩ	Full range	25			15			15			
B_{OM}	Maximum-output-swing bandwidth	V_{OPP} = 20 V, A_{VD} = 1, THD ≤ 5%, R_L = 2 kΩ	25°C		9			9			9		kHz
B_1	Unity-gain bandwidth	V_O = 50 mV, R_L = 10 kΩ	25°C		1			1			1		MHz
ϕ_m	Phase margin	C_L = 200 pF, R_L = 2 kΩ	25°C		60°			60°			60°		
r_i	Input resistance	f = 20 Hz	25°C	0.3	1		0.3	1		0.3	1		MΩ
r_o	Output resistance	f = 20 Hz	25°C		75			75			75		Ω
CMRR	Common-mode rejection ratio	V_{IC} = V_{ICR} min	25°C	70	90		70	90		70	90		dB
k_{SVS}	Supply voltage sensitivity ($\Delta V_{IO}/\Delta V_{CC}$)	V_{CC} = ±2.5 to ±15 V	25°C		30	150		30	150		30	150	µV/V
I_{OS}	Short-circuit output current§		25°C	±10	±30	±45	±10	±30	±45	±10	±30	±45	mA
I_{CC}	Total supply current	No load, See Note 5	25°C		2.8	4		2.8	7		2.8	7	mA

† All characteristics are measured under open-loop conditions with zero common-mode voltage unless other specified. Full range for T_A −55°C to 125°C for MC3503, −40°C to 85°C for MC3303, and 0°C to 70°C for MC3403.
‡ The V_{ICR} limits are directly linked volt-for-volt to supply voltage, viz the positive limit is 2 volts less than V_{CC+}.
§ Temperature and/or supply voltages must be limited to ensure that the dissipation rating is not exceeded.
NOTE 5: V_{IO}, I_{IO}, I_{IB}, and I_{CC} are defined at V_O = 0 for MC3403 and MC3503, and V_O = 7 V for MC3303.

Operational Amplifiers

3

TEXAS
INSTRUMENTS
POST OFFICE BOX 225012 • DALLAS, TEXAS 75265

TYPES MC3503, MC3303, MC3403
QUADRUPLE LOW-POWER OPERATIONAL AMPLIFIERS

electrical characteristics, V_{CC+} = 5 V, V_{CC-} = 0 V, T_A = 25°C (unless otherwise noted)

PARAMETER		TEST CONDITIONS†	MC3503			MC3303			MC3403			UNIT
			MIN	TYP	MAX	MIN	TYP	MAX	MIN	TYP	MAX	
V_{IO}	Input offset voltage	V_O = 2.5 V		2	5			10		2	10	mV
I_{IO}	Input offset current	V_O = 2.5 V		30	50			75		30	50	nA
I_{IB}	Input bias current	V_O = 2.5 V		−0.2	−0.5			−0.5		−0.2	−0.5	pA
V_{OM}	Peak output voltage swing‡	R_L = 10 kΩ	3.3	3.5		3.3	3.5		3.3	3.5		V
		R_L = 10 kΩ, V_{CC+} = 5 V to 30 V	V_{CC+}−1.7			V_{CC+}−1.7			V_{CC+}−1.7			
A_{VD}	Large-signal differential voltage amplification	V_O = 1.7 V to 3.3 V, R_L = 2 kΩ	20	200		20	200		20	200		V/mV
k_{SVS}	Supply voltage sensitivity ($\Delta V_{IO}/\Delta V_{CC\pm}$)	V_{CC} = ±15 V to ±2.5 V			150			150			150	µV/V
I_{CC}	Supply current	No load, V_O = 2.5 V		2.5	4		2.5	7		2.5	7	mA
V_{o1}/V_{o2}	Crosstalk attenuation	f = 1 kHz to 20 kHz		120			120			120		dB

†All characteristics are measured under open-loop conditions with zero common-mode input voltage unless otherwise specified.
‡Output will swing essentially to ground.

operating characteristics, V_{CC+} = 14 V, V_{CC-} = 0 V for MC3303; $V_{CC\pm}$ = ±15 V for MC3403 and MC3503; T_A = 25°C, A_{VD} = 1 (unless otherwise noted)

PARAMETER		TEST CONDITIONS	MIN	TYP	MAX	UNIT
SR	Slew rate at unity gain	V_I = ±10 V, C_L = 100 pF, R_L = 2 kΩ, See Figure 1		0.6		V/µs
t_r	Rise time	ΔV_O = 50 mV, C_L = 100 pF, R_L = 10 kΩ, See Figure 1		0.35		µs
t_f	Fall time			0.35		µs
	Overshoot factor			20%		
	Crossover distortion	V_{IPP} = 30 mV, V_{OPP} = 2 V, f = 10 kHz		1%		

PARAMETER MEASUREMENT INFORMATION

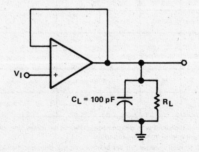

C_L = 100 pF R_L

V_I

FIGURE 1—UNITY-GAIN AMPLIFIER

TEXAS INSTRUMENTS
POST OFFICE BOX 225012 • DALLAS, TEXAS 75265

Operational Amplifiers

TYPICAL CHARACTERISTICS†

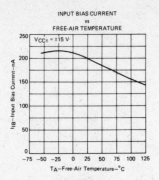

FIGURE 2

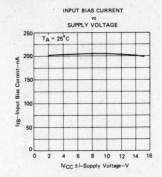

FIGURE 3

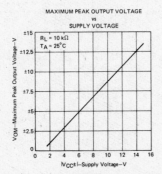

FIGURE 4

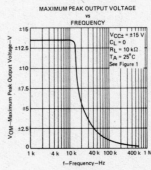

FIGURE 5

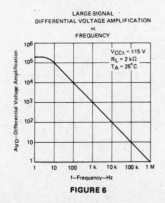

FIGURE 6

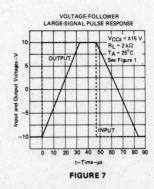

FIGURE 7

†Data at high and low temperatures are applicable only within the rated operating free-air temperature ranges of the various devices.

Operational Amplifiers

3

TEXAS INSTRUMENTS
POST OFFICE BOX 225012 • DALLAS, TEXAS 75265

**LINEAR
INTEGRATED
CIRCUITS**

**TYPES NE5532, NE5532A
DUAL LOW-NOISE OPERATIONAL AMPLIFIERS**

D2563, NOVEMBER 1979 – REVISED AUGUST 1983

- Equivalent Input Noise
  Voltage . . . . 5 nV/$\sqrt{Hz}$ Typ at 1 kHz

- Unity-Gain Bandwidth . . . . 10 MHz Typ

- Common-Mode Rejection Ratio . . . . 100 dB Typ

- High DC Voltage Gain . . . . 100 V/mV Typ

- Peak-to-Peak Output Voltage
  Swing . . . 32 V Typ with $V_{CC\pm}$ = ±18 V and
  R_L = 600 Ω

- High Slew Rate . . . . 9 V/µs Typ

- Wide Supply Voltage Range . . . ±3 V to ±20 V

- Designed to be Interchangeable with Signetics
  NE5532 and NE5532A

**NE5532, NE5532A . . . JG OR P
DUAL-IN-LINE PACKAGE
(TOP VIEW)**

```
        ┌───┬───┐
OUT   1 │   ● 8 │ VCC+
IN −  2 │       │ 7 OUT
IN +  3 │       │ 6 IN −
VCC − 4 │       │ 5 IN +
        └───────┘
```

description

The NE5532 and NE5532A are monolithic high-performance operational amplifiers combining excellent dc and ac characteristics. They feature very low noise, high output drive capability, high unity-gain and maximum-output-swing bandwidths, low distortion, high slew rate, input-protection diodes, and output short-circuit protection. These operational amplifiers are internally compensated for unity gain operation. The NE5532A has guaranteed maximum limits for equivalent input noise voltage.

The NE5532 and NE5532A are characterized for operation from 0°C to 70°C.

symbol (each amplifier)

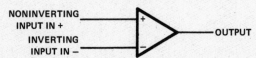

schematic (each amplifier)

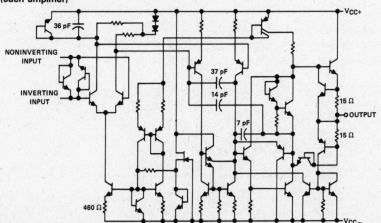

All component values shown are nominal.

**TEXAS
INSTRUMENTS**
POST OFFICE BOX 225012 • DALLAS, TEXAS 75265

Operational Amplifiers

3

TYPES NE5532, NE5532A
DUAL LOW-NOISE OPERATIONAL AMPLIFIERS

absolute maximum ratings over operating free-air temperature range (unless otherwise noted)

Supply voltage, V_{CC+} (see Note 1) . . . . 22 V

Supply voltage, V_{CC-} (see Note 1) . . . . −22 V

Input voltage, either input (see Notes 1 and 2) . . . . $V_{CC\pm}$

Input current (see Note 3) . . . . ±10 mA

Duration of output short-circuit (see Note 4) . . . . unlimited

Continuous total power dissipation at (or below) 25 °C free-air temperature (see Note 5):

    JG package . . . . 825 mW

    P package . . . . 725 mW

Operating free-air temperature range: NE5532, NE5532A . . . . 0 °C to 70 °C

Storage temperature range . . . . −65 °C to 150 °C

Lead temperature 1,6 mm (1/16 inch) from case for 60 seconds: JG package . . . . 300 °C

Lead temperature 1,6 mm (1/16 inch) from case for 10 seconds: P package . . . . 260 °C

NOTES: 1. All voltage values, except differential voltages, are with respect to the midpoint between V_{CC+} and V_{CC-}.
2. The magnitude of the input voltage must never exceed the magnitude of the supply voltage.
3. Excessive input current will flow if a differential input voltage in excess of approximately 0.6 V is applied between the inputs unless some limiting resistance is used.
4. The output may be shorted to ground or either power supply. Temperature and/or supply voltages must be limited to ensure the maximum dissipation rating is not exceeded.
5. For operation above 25 °C free-air temperature, refer to the Dissipation Derating Curves in Section 2. In the JG package, chips are glass-mounted.

3

Operational Amplifiers

TEXAS
INSTRUMENTS
POST OFFICE BOX 225012 • DALLAS, TEXAS 75265

electrical characteristics, $V_{CC\pm} = \pm 15$ V, $T_A = 25°C$ (unless otherwise noted)

PARAMETER		TEST CONDITIONS[†]		NE5532, NE5532A MIN	TYP	MAX	UNIT
V_{IO}	Input offset voltage	$V_O = 0$	$T_A = 25°C$		0.5	4	mV
			$T_A = 0°C$ to $70°C$			5	
I_{IO}	Input offset current	$T_A = 25°C$			10	150	nA
		$T_A = 0°C$ to $70°C$				200	
I_{IB}	Input bias current	$T_A = 25°C$			200	800	nA
		$T_A = 0°C$ to $70°C$				1000	
V_{ICR}	Common-mode input voltage range			± 12	± 13		V
V_{OPP}	Maximum peak-to-peak output voltage swing	$R_L \geq 600$ Ω	$V_{CC\pm} \pm 15$ V	24	26		V
			$V_{CC\pm} = \pm 18$ V	30	32		
A_{VD}	Large-signal differential voltage amplification	$R_L \geq 600$ Ω, $V_O = \pm 10$ V	$T_A = 25°C$	15	50		V/mV
			$T_A = 0°C$ to $70°C$	10			
		$R_L \geq 2$ kΩ, $V_O = \pm 10$ V	$T_A = 25°C$	25	100		
			$T_A = 0°C$ to $70°C$	15			
A_{vd}	Small-signal differential voltage amplification	$f = 10$ kHz			2.2		V/mV
B_{OM}	Maximum-output-swing bandwidth	$R_L = 600$ Ω,	$V_O = \pm 10$ V		140		kHz
		$R_L = 600$ Ω,	$V_{CC\pm} = \pm 18$ V, $V_O = \pm 14$ V		100		
B_1	Unity-gain bandwidth	$R_L = 600$ Ω,	$C_L = 100$ pF		10		MHz
r_i	Input resistance			30	300		kΩ
z_o	Output impedance	$A_{VD} = 30$ dB, $R_L = 600$ Ω, $f = 10$ kHz			0.3		Ω
CMRR	Common-mode rejection ratio	$V_{IC} = V_{ICR}$ min		70	100		dB
k_{SVR}	Supply voltage rejection ratio ($\Delta V_{CC\pm}/\Delta V_{IO}$)	$V_{CC\pm} = \pm 9$ V to ± 15 V, $V_O = 0$		80	100		dB
I_{OS}	Output short-circuit current				38		mA
I_{CC}	Total supply current	No load, $V_O = 0$			8	16	mA
V_{o1}/V_{o2}	Crosstalk attenuation	$V_{o1} = 10$ V peak, $f = 1$ kHz			110		dB

[†]All characteristics are measured under open-loop conditions with zero common-mode input voltage unless otherwise specified.

operating characteristics, $V_{CC\pm} = \pm 15$ V, $T_A = 25°C$

PARAMETER		TEST CONDITIONS	NE5532 MIN	TYP	MAX	NE5532A MIN	TYP	MAX	UNIT
SR	Slew rate at unity gain			9			9		V/μs
	Overshoot factor	$V_I = 100$ mV, $A_{VD} = 1$, $R_L = 600$ Ω, $C_L = 100$ pF		10%			10%		
V_n	Equivalent input noise voltage	$f = 30$ Hz		8			8	10	nV/$\sqrt{Hz}$
		$f = 1$ kHz		5			5	6	
I_n	Equivalent input noise current	$f = 30$ Hz		2.7			2.7		pA/$\sqrt{Hz}$
		$f = 1$ kHz		0.7			0.7		

3

Operational Amplifiers

LINEAR INTEGRATED CIRCUITS

- Ultra-Low Offset Voltage . . . 30 μV Typ (OP-07E)

- Ultra-Low Offset Voltage Temperature Coefficient . . . . 0.3 μV/°C Typ (OP-07E)

- Ultra-Low Noise

- No External Components Required

- Replaces Chopper Amplifiers at a Lower Cost

- Single-Chip Monolithic Fabrication

- Wide Input Voltage Range 0 to ±14 V Typ

- Wide Supply Voltage Range ±3 V to ±18 V

- Essentially Equivalent to Fairchild μA714 Operational Amplifiers

- Direct Replacement for PMI OP-07C, OP-07D, OP-07E

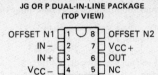

JG OR P DUAL-IN-LINE PACKAGE
(TOP VIEW)

OFFSET N1	1	8	OFFSET N2
IN –	2	7	V$_{CC}$+
IN +	3	6	OUT
V$_{CC}$ –	4	5	NC

NC—No internal connection

symbol

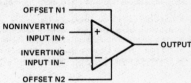

OFFSET N1
NONINVERTING INPUT IN+
INVERTING INPUT IN—
OFFSET N2
OUTPUT

description

These devices represent a breakthrough in operational amplifier performance. Low offset and long-term stability are achieved by means of a low-noise, chopperless, bipolar-input-transistor amplifier circuit. For most applications, no external components are required for offset nulling and frequency compensation. The true differential input, with a wide input voltage range and outstanding common-mode rejection, provides maximum flexibility and performance in high-noise environments and in noninverting applications. Low bias currents and extremely high input impedances are maintained over the entire temperature range. The OP-07 is unsurpassed for low-noise, high-accuracy amplification of very-low-level signals.

These devices are characterized for operation from 0°C to 70°C.

schematic

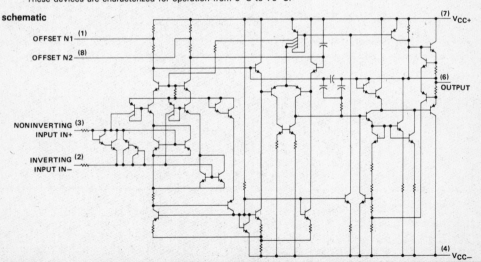

3

Operational Amplifiers

TEXAS INSTRUMENTS
POST OFFICE BOX 225012 • DALLAS, TEXAS 75265

absolute maximum ratings over operating free-air temperature range (unless otherwise noted)

Supply voltage V_{CC+} (see Note 1) .... 22 V

Supply voltage V_{CC-} .... −22 V

Differential input voltage (see Note 2) .... ±30 V

Input voltage (either input, see Note 3) .... ±22 V

Duration of output short circuit (see Note 4) .... unlimited

Continuous total dissipation at (or below) 25 °C free-air temperature (see Note 5) .... 500 mW

Operating free-air temperature range .... 0 °C to 70 °C

Storage temperature range .... −65 °C to 150 °C

Lead temperature 1,6 mm (1/16 inch) from case for 60 seconds: JG package .... 300 °C

Lead temperature 1,6 mm (1/16 inch) from case for 10 seconds: P package .... 260 °C

NOTES: 1. All voltage values, unless otherwise noted, are with respect to the midpoint between V_{CC+} and V_{CC-}.
2. Differential voltages are at the noninverting input terminal with respect to the inverting input terminal.
3. The magnitude of the input voltage must never exceed the magnitude of the supply voltage or 15 volts, whichever is less.
4. The output may be shorted to ground or either power supply.
5. For operation above 25 °C free-air temperature, refer to Dissipation Derating Curves in Section 2. In the JG package, these chips are glass-mounted.

3

Operational Amplifiers

TEXAS
INSTRUMENTS

POST OFFICE BOX 225012 • DALLAS, TEXAS 75265

electrical characteristics at specified free-air temperature, $V_{CC}\pm = \pm 15$ V (unless otherwise noted)

PARAMETER	TEST CONDITIONS†		OP-07C			OP-07D			OP-07E			UNIT
			MIN	TYP	MAX	MIN	TYP	MAX	MIN	TYP	MAX	
V_{IO} Input offset voltage	$V_O = 0$, $R_S = 50\,\Omega$	25°C		60	150		60	150		30	75	μV
	$V_O = 0$, $R_S = 50\,\Omega$	0°C to 70°C		85	250		85	250		45	130	μV
α_{VIO} Temperature coefficient of input offset voltage	$V_O = 0$, $R_S = 50\,\Omega$	0°C to 70°C		0.5	1.8		0.7	2.5		0.3	1.3	μV/°C
Long-term drift of input offset voltage	See Note 6			0.4			0.5			0.3		μV/mo
Offset adjustment range	$R_S = 20\,k\Omega$, See Figure 1	25°C		±4			±4			±4		mV
I_{IO} Input offset current		25°C		0.8	6		0.8	6		0.5	3.8	nA
		0°C to 70°C		1.6	8		1.6	8		0.9	5.3	nA
α_{IIO} Temperature coefficient of input offset current		0°C to 70°C		12	50		12	50		8	35	pA/°C
I_{IB} Input bias current		25°C		±1.8	±7		±2	±12		±1.2	±4	nA
		0°C to 70°C		±2.2	±9		±3	±14		±1.5	±5.5	nA
α_{IIB} Temperature coefficient of input bias current		0°C to 70°C		18	50		18	50		13	35	pA/°C
V_{ICR} Common-mode input voltage range		25°C	±13	±14		±13	±14		±13	±14		V
		0°C to 70°C	±13	±13.5		±13	±13.5		±13	±13.5		V
V_{OM} Peak output voltage	$R_L \geq 10\,k\Omega$	25°C	±12	±13		±12	±13		±12.5	±13		V
	$R_L \geq 2\,k\Omega$	25°C	±11.5	±12.8		±11.5	±12.8		±12	±12.8		V
	$R_L \geq 1\,k\Omega$	25°C		±12			±12			±12		V
	$R_L \geq 2\,k\Omega$	0°C to 70°C	±11	±12.6		±11	±12.6		±12	±12.6		V
A_{VD} Large-signal differential voltage amplification	$V_{CC}\pm = \pm 3\,V$, $V_O = \pm 0.5\,V$, $R_L \geq 500\,\Omega$	25°C	100	400		120	400		150	400		V/mV
	$V_O = \pm 10\,V$, $R_L \geq 2\,k\Omega$	25°C	120	400		100	400		200	500		
		0°C to 70°C	100	400		100	400		180	450		
B_1 Unity gain bandwidth		25°C		0.6			0.6			0.6		MHz
r_i Input resistance		25°C	8	33		7	31		15	50		MΩ
CMRR Common-mode rejection ratio	$V_{IC} = \pm 13\,V$, $R_S = 50\,\Omega$	25°C	100	120		94	110		106	123		dB
		0°C to 70°C	97	120		94	106		103	123		
k_{SVS} Supply voltage sensitivity ($\Delta V_{IO}/\Delta V_{CC}$)	$V_{CC}\pm = \pm 3\,V$ to $\pm 18\,V$, $R_S = 50\,\Omega$	25°C		7	32		7	32		5	20	μV/V
		0°C to 70°C		10	51		10	51		7	32	
P_D Power dissipation	$V_O = 0$, No load	25°C		80	150		80	150		75	120	mW
	$V_{CC}\pm = \pm 3\,V$, $V_O = 0$, No load	25°C		4	8		4	8		4	6	mW

†All characteristics are measured under open-loop conditions with zero common-mode input voltage unless otherwise noted.

NOTE 6: Since long-term drift cannot be measured on the individual devices prior to shipment, this specification is not intended to be a guarantee or warranty. It is an engineering estimate of the averaged trend line of drift versus time over extended periods after the first thirty days of operation.

Operational Amplifiers

3

TYPES OP-07C, OP-07D, OP-07E
ULTRA-LOW-OFFSET VOLTAGE OPERATIONAL AMPLIFIERS

operating characteristics at specified free-air temperature, $V_{CC\pm} = \pm 15$ V (unless otherwise noted)

PARAMETER		TEST CONDITIONS[†]		OP-7C			OP-7D			OP-7E			UNIT
				MIN	TYP	MAX	MIN	TYP	MAX	MIN	TYP	MAX	
V_n	Equivalent input noise voltage	$T_A = 25°C$	f = 10 Hz		10.5	20		10.5	20		10.3	18	nV/√Hz
			f = 100 Hz		10.2	13.5		10.3	13.5		10.0	13	
			f = 1 kHz		9.8	11.5		9.8	11.5		9.6	11	
V_{NPP}	Peak-to-peak equivalent input noise voltage	f = 0.1 Hz to 10 Hz, $T_A = 25°C$			0.38	0.65		0.38	0.65		0.35	0.6	μV
I_n	Equivalent input noise current	$T_A = 25°C$	f = 10 Hz		0.35	0.9		0.35	0.9		0.32	0.8	pA/√Hz
			f = 100 Hz		0.15	0.27		0.15	0.27		0.14	0.23	
			f = 1 kHz		0.13	0.18		0.13	0.18		0.12	0.17	
I_{NPP}	Peak-to-peak equivalent input noise current	f = 0.1 Hz to 10 Hz, $T_A = 25°C$			15	35		15	35		14	30	pA
SR	Slew rate	$R_L \geq 2$ kΩ, $T_A = 25°C$		0.1	0.3		0.1	0.3		0.1	0.3		V/μs

[†]All characteristics are measured under open-loop conditions with zero common-mode input voltage unless otherwise specified.

TYPICAL APPLICATION DATA

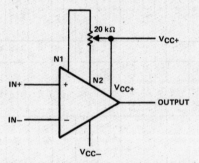

FIGURE 1—INPUT OFFSET VOLTAGE NULL CIRCUIT

3

Operational Amplifiers

TEXAS
INSTRUMENTS
POST OFFICE BOX 225012 • DALLAS, TEXAS 75265

LINEAR
INTEGRATED
CIRCUITS

TYPES OP-12A, OP-12B, OP-12C, OP-12E, OP-12F, OP-12G
PRECISION LOW-INPUT-CURRENT OPERATIONAL AMPLIFIERS

D2817, OCTOBER 1983

- Internally Frequency Compensated

- Improved Version of LM108

- Direct Replacement for PMI OP-12A, OP-12B, OP-12C, OP-12E, OP-12F, and OP-12G.

OP-12A, OP-12B, OP-12C . . . JG PACKAGE
OP-12E, OP-12F, OP-12G . . . D, JG, OR P PACKAGE
(TOP VIEW)

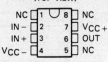

NC — No internal connection

description

The OP-12 devices are precision low-input-current internally compensated operational amplifiers. The devices are improved versions of the LM108 series. The OP-12 amplifiers exhibit low input bias current and input offset voltage and current to improve the accuracy of high-impedance circuits using these devices. The devices feature short-circuit protection and internal frequency compensation.

The OP-12A, OP-12B, and OP-12C are characterized for operation over the full military temperature range of −55°C to 125°C. The OP-12E, OP-12F, and OP-12G are characterized for operation from 0°C to 70°C.

symbol

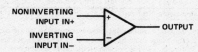

3

Operational Amplifiers

DEVICE FEATURES

PARAMETER	OP-12A OP-12E	OP-12B OP-12F	OP-12C OP-12G
Input offset voltage (Max)	150 μV	300 μV	1000 μV
Temperature coefficient of input offset voltage (Max)	2.5 μV/°C	3.5 μV/°C	10 μV/°C
Input offset current (Max)	200 pA	200 pA	500 pA
Input bias current (Max)	2 nA	2 nA	5 nA
Common-mode input voltage range	± 13 V	± 13 V	± 13 V
Power dissipation (Max)	6 mW	6 mW	8 mW

Copyright © 1983 by Texas Instruments Incorporated

TEXAS
INSTRUMENTS

POST OFFICE BOX 225012 • DALLAS, TEXAS 75265

absolute maximum ratings over free-air temperature range (unless otherwise noted)

	OP-12A, OP-12B OP-12C	OP-12E, OP-12F OP-12G	UNIT
Supply voltage, V_{CC+} (see Note 1)	20	18	V
Supply voltage, V_{CC-} (see Note 1)	−20	−18	V
Input voltage (either input, see Note 2)	±15	±15	V
Differential input current (see Note 3)	±10	±10	mA
Duration of output short circuit (see Note 4)	unlimited	unlimited	
Continuous total dissipation at (or below) 25°C free-air temperature (see Note 5)	500	500	mW
Operating free-air temperature range	−55 to 125	0 to 70	°C
Storage temperature range	−65 to 150	−65 to 150	°C
Lead temperature 1,6 mm (1/16 inch) from case for 60 seconds JG package	300	300	°C
Lead temperature 1,6 mm (1/16 inch) from case for 10 seconds D or P package		260	°C

NOTES: 1. All voltage values, except otherwise noted, are with respect to the midpoint between V_{CC+} and V_{CC-}.
2. The magnitude of the input voltage must never exceed the magnitude of the supply voltage or 15 volts, whichever is less.
3. The inputs are shunted with back-to-back diodes for input overvoltage protection. Therefore, excessive current will flow if a differential voltage in excess of 1 volt is applied between the inputs unless some limiting resistance is provided.
4. The output may be shorted to ground or to either supply. Temperature and/or supply voltages must be limited to ensure that the dissipation rating is not exceeded.
5. For operation above 25°C free-air temperature, refer to Dissipation Derating Curves, Section 2. In the JG packages, OP-12A, OP-12B, and OP-12C chips are alloy-mounted; OP-12E, OP-12F, and OP-12G chips are glass-mounted.

3

Operational Amplifiers

TEXAS
INSTRUMENTS
POST OFFICE BOX 225012 • DALLAS, TEXAS 75265

electrical characteristics at specified free-air temperature, $V_{CC}\pm = \pm20$ V for OP-12A and OP-12B, ±15 V for OP-12C (unless otherwise noted)

PARAMETER		TEST CONDITIONS[†]	OP-12A			OP-12B			OP-12C			UNIT
			MIN	TYP	MAX	MIN	TYP	MAX	MIN	TYP	MAX	
V_{IO}	Input offset voltage	25°C		0.07	0.15		0.18	0.3		0.25	1	mV
		−55°C to 125°C		0.12	0.35		0.28	0.6		0.4	2	
α_{VIO}	Average temperature coefficient of input offset voltage	$V_O = 0$, −55°C to 125°C		0.5	2.5		1	3.5		1.5	10	µV/°C
I_{IO}	Input offset current	25°C		0.05	0.2		0.05	0.2		0.08	0.5	nA
		−55°C to 125°C		0.12	0.4		0.12	0.4		0.18	1	
α_{IIO}	Average temperature coefficient of input offset current	$V_O = 0$, −55°C to 125°C		0.5	2.5		0.5	2.5		1	5	pA/°C
I_{IB}	Input bias current	25°C		0.8	2		0.8	2		1	5	nA
		−55°C to 125°C		1.2	3		1.2	3		1.8	10	
V_{ICR}	Common-mode input voltage range	$V_{CC} = \pm15$ V, 25°C	±13	±14		±13	±14		±13	±14		V
		−55°C to 125°C	±13	±14		±13	±14		±13	±14		
V_{OM}	Maximum peak output voltage swing	$V_{CC}\pm = \pm15$ V, $R_L \geq 10$ kΩ	±13	±14		±13	±14		±13	±14		V
		$V_{CC}\pm = \pm15$ V, $R_L \geq 2$ kΩ	±10	±12		±10	±12		±10	±12		
		$V_{CC}\pm = \pm15$ V, $R_L \geq 10$ kΩ	±13	±14		±13	±14		±13	±14		
		$V_{CC}\pm = \pm15$ V, $R_L \geq 5$ kΩ	±10	±13		±10	±13		±10	±12		
A_{VD}	Large-signal differential voltage amplification	$V_O = \pm10$ V, $R_L \geq 10$ kΩ	80	300		80	300		40	250		V/mV
		$V_O = \pm10$ V, $R_L \geq 2$ kΩ	50	150		50	150			100		
		$V_O = \pm10$ V, $R_L \geq 5$ kΩ, −55°C to 125°C	40	120		40	120		15	80		
B_1	Unity-gain bandwidth	$A_{VD} = 1$, 25°C		0.8			0.8			0.8		MHz
r_i	Input resistance	25°C	26	70		26	70		10	50		MΩ
r_o	Output resistance	25°C		200			200			200		Ω
CMRR	Common-mode rejection ratio	$V_{IC} = \pm13$ V, 25°C	104	120		104	120		84	116		dB
		−55°C to 125°C	100	116		100	116		80	112		
k_{SVR}	Supply voltage rejection ratio ($\Delta V_{CC}\pm / V_{IO}$)	$V_{CC} = \pm5$ V to ±15 V, 25°C	104	120		104	120		84	116		dB
		−55°C to 125°C	100	116		100	116		80	112		
P_D	Power dissipation	$V_{CC}\pm = \pm15$ V, $V_O = 0$, No load		9	18		9	18		15	24	mW
		$V_{CC}\pm = \pm5$ V, $V_O = 0$, No load		3	6		3	6		4	8	
I_{CC}	Supply current	$V_{CC}\pm = \pm15$ V, $V_O = 0$, No load, 25°C		0.3	0.6		0.3	0.6		0.4	0.8	mA

[†] All characteristics are specified under open-loop conditions with zero common-mode input voltage, unless otherwise noted.

Operational Amplifiers

3

TYPES OP-12A, OP-12B, OP-12C
PRECISION LOW-INPUT-CURRENT OPERATIONAL AMPLIFIERS

operating characteristics at 25 °C free-air temperature, $V_{CC\pm}$ = ±20 V for OP-12A and OP-12B, ±15 V for OP-12C (unless otherwise noted)

PARAMETER		TEST CONDITIONS[†]	OP-12A			OP-12B			OP-12C			UNIT
			MIN	TYP	MAX	MIN	TYP	MAX	MIN	TYP	MAX	
SR	Slew rate at unity gain	$R_L \geq 2$ kΩ		0.12			0.12			0.12		V/µs
V_n	Equivalent input noise voltage	f = 10 Hz		22			22			22		nV$\sqrt{Hz}$
		f = 100 Hz		21			21			21		
		f = 1000 Hz		20			20			20		
I_n	Equivalent input noise current	f = 10 Hz		0.15			0.15			0.15		pA$\sqrt{Hz}$
		f = 100 Hz		0.14			0.14			0.14		
		f = 1000 Hz		0.13			0.13			0.13		
V_{NPP}	Peak-to-peak input noise voltage	f = 0.1 Hz to 10 Hz		0.9			0.9			0.9		µV
I_{NPP}	Peak-to-peak input noise current	f = 0.1 Hz to 10 Hz		3			3			3		pA

[†]All characteristics are specified under open-loop conditions with zero common-mode input voltage, unless otherwise noted.

3

Operational Amplifiers

TEXAS
INSTRUMENTS
POST OFFICE BOX 225012 • DALLAS, TEXAS 75265

electrical characteristics at specified free-air temperature, $VCC\pm$ = ±20 V for OP-12E and OP-12F, ±15 V for OP-12G (unless otherwise noted)

PARAMETER		TEST CONDITIONS†		OP-12E			OP-12F			OP-12G			UNIT
				MIN	TYP	MAX	MIN	TYP	MAX	MIN	TYP	MAX	
V_{IO}	Input offset voltage	V_O = 0, R_S = 50 Ω	25°C		0.07	0.15		0.18	0.3		0.25	1	mV
			0°C to 70°C		0.1	0.26		0.23	0.45		0.32	1.4	
α_{VIO}	Average temperature coefficient of input offset voltage	V_O = 0	0°C to 70°C		0.5	2.5		1	3.5		1.5	10	µV/°C
I_{IO}	Input offset current	V_O = 0	25°C		0.05	0.2		0.05	0.2		0.08	0.5	nA
			0°C to 70°C		0.08	0.3		0.11	0.6		0.12	0.7	
α_{IIO}	Average temperature coefficient of input offset current	V_O = 0	0°C to 70°C		0.5	2.5		1	5		1	5	pA/°C
I_{IB}	Input bias current	V_O = 0	25°C		0.8	2		0.8	2		1	5	nA
			0°C to 70°C		1	2.6		1.2	5.2		1.4	6.5	
V_{ICR}	Common-mode input voltage range	VCC = ±15 V	25°C	±13	±14		±13	±14		±13	±14		V
V_{OM}	Maximum peak output voltage swing	$VCC\pm$ = ±15 V, R_L ≥ 10 kΩ	25°C	±13	±14		±13	±14		±13	±14		V
		$VCC\pm$ = ±15 V, R_L ≥ 2 kΩ		±10	±12		±10	±12		±10	±12		
		$VCC\pm$ = ±15 V, R_L ≥ 10 kΩ		±13	±14		±13	±14		±13	±14		
		$VCC\pm$ = ±15 V, R_L ≥ 5 kΩ	0°C to 70°C	±10	±12		±10	±12		±10	±12		
A_{VD}	Large-signal differential voltage amplification	V_O = ±10 V, R_L ≥ 10 kΩ	25°C	80	300		80	300		40	250		V/mV
		V_O = ±10 V, R_L ≥ 2 kΩ		50	150		50	150			100		
		V_O = ±10 V, R_L ≥ 10 kΩ		25	100		15	100			80		
		V_O = ±10 V, R_L ≥ 2 kΩ	0°C to 70°C	60	200		60	200		25	150		
B_{OM}	Maximum-output swing bandwidth	A_{VD} = 1	25°C		0.8			0.8			0.8		MHz
r_i	Input resistance		25°C	26	70		26	70		10	50		MΩ
r_o	Output resistance	V_O = 0	25°C		200			200			200		Ω
CMRR	Common-mode rejection ratio	V_{IC} = ±13 V, R_S = 50 Ω, V_O = 0	25°C	104	120		102	120		84	116		dB
			0°C to 70°C	100	116		100	116		80	112		
k_{SVR}	Supply voltage rejection ratio ($\Delta VCC\pm/V_{IO}$)	VCC = ±5 V to ±15 V, V_O = 0, R_S = 50 Ω	25°C	104	120		102	120		84	116		dB
			0°C to 70°C	100	116		100	116		80	112		
P_D	Power dissipation	$VCC\pm$ = ±15 V, V_O = 0, No load	25°C		9	18		9	18		15	24	mW
		$VCC\pm$ = ±5 V, V_O = 0, No load	25°C		3	6		3	6		4	8	
I_{CC}	Supply current	$VCC\pm$ = ±15 V, V_O = 0, No load	25°C		0.3	0.6		0.3	0.6		0.4	0.8	mA

†All characteristics are specified under open-loop conditions with zero common-mode input voltage, unless otherwise noted.

3

Operational Amplifiers

TEXAS INSTRUMENTS
POST OFFICE BOX 225012 • DALLAS, TEXAS 75265

operating characteristics at 25°C free-air temperature, $V_{CC\pm} = \pm20$ V for OP-12E and OP-12F, ±15 V for OP-12G (unless otherwise noted)

PARAMETER		TEST CONDITIONS[†]	OP-12E MIN	OP-12E TYP	OP-12E MAX	OP-12F MIN	OP-12F TYP	OP-12F MAX	OP-12G MIN	OP-12G TYP	OP-12G MAX	UNIT
SR	Slew rate at unity gain	$R_L \geq 2$ kΩ		0.12			0.12			0.12		V/μs
V_n	Equivalent input noise voltage	f = 10 Hz		22			22			22		nV$\sqrt{Hz}$
		f = 100 Hz		21			21			21		
		f = 1000 Hz		20			20			20		
I_n	Equivalent input noise current	f = 10 Hz		0.15			0.15			0.15		pA$\sqrt{Hz}$
		f = 100 Hz		0.14			0.14			0.14		
		f = 1000 Hz		0.13			0.13			0.13		
V_{NPP}	Peak-to-peak input noise voltage	f = 0.1 Hz to 10 Hz		0.9			0.9			0.9		μV
I_{NPP}	Peak-to-peak input noise current	f = 0.1 Hz to 10 Hz		3			3			3		pA

[†]All characteristics are specified under open-loop conditions with zero common-mode input voltage, unless otherwise noted.

3

Operational Amplifiers

TEXAS
INSTRUMENTS
POST OFFICE BOX 225012 • DALLAS, TEXAS 75265

**LINEAR
INTEGRATED
CIRCUITS**

**TYPES OP-227E, OP-227F, OP-227G
LOW-NOISE DUAL OPERATIONAL AMPLIFIERS**

D2805, OCTOBER 1983

- Low Offset Voltage . . . 20 μV Typ (OP-227E)

- Low Coefficient of Input Offset Voltage . . . 0.4 μV/°C Typ (OP-227F)

- Low Equivalent Input Noise Voltage:
  3 nV/$\sqrt{Hz}$ Typ at f = 1 kHz (OP-227E and OP-227F)
  0.2 μV p-p Max at f = 0.1 to 10 Hz (OP-227E and OP-227F)

- High Slew Rate . . . 2.8 V/μs Typ

- High Voltage Amplification . . . 1.8×10^6 Typ at $R_L \geq 2$ kΩ (OP-227E and OP-227F)

- Direct Replacements for PMI OP-227E, OP-227F, OP-227G

**J OR N DUAL-IN-LINE PACKAGE
(TOP VIEW)**

#1 OFFSET	1	14	#1 $V_{CC}+$
#1 OFFSET	2	13	#1 OUT
#1 IN−	3	12	#1 $V_{CC}-$
#1 IN+	4	11	#2 IN+
#2 $V_{CC}-$	5	10	#2 IN−
#2 OUT	6	9	#2 OFFSET
#2 $V_{CC}+$	7	8	#2 OFFSET

description

The OP-227 family of dual operational amplifiers offers low offset, low noise, and high slew rates. The temperature coefficient of input offset voltage is typically only 0.4 μv/°C for the OP-227F. Both the OP-227E and OP-227F achieve typical voltage gains of 1.8 million with a 2-kΩ load, and both have maximum peak-to-peak noise voltage of 0.2 microvolts and equivalent input noise voltage of typically 3 nV/$\sqrt{Hz}$. The 2.8-V/μs slew rate and 8-Megahertz bandwidth bring a high level of performance to instrumentation designs. These devices usually require no external components for offset nulling or frequency compensation.

The OP-227E, OP-227F, and OP-227G will be characterized for operation over the temperature range of −25°C to 85°C.

symbol (each amplifier)

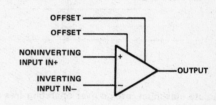

3

Operational Amplifiers

**TEXAS
INSTRUMENTS**

POST OFFICE BOX 225012 • DALLAS, TEXAS 75265

schematic (each half of amplifier)

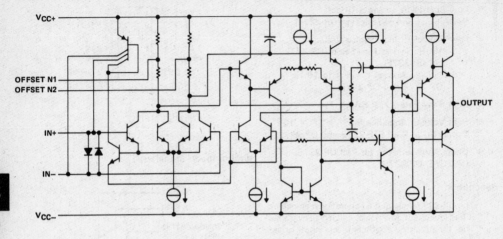

absolute maximum ratings over operating free-air temperature range (unless otherwise noted)

Supply voltages, V_{CC} (see Note 1) .... ± 22 V

Differential input current (see Note 2) .... ± 25 mA

Continuous total dissipation at (or below) 25°C free-air temperature (see Note 3):

    J package .... 1025 mW

    N package .... 875 mW

Operating free-air temperature range .... −25°C to 85°C

Storage temperature range .... −65°C to 150°C

Lead temperature 1,6 mm (1/16 inch) from case for 60 seconds: J package .... 300°C

Lead temperature 1,6 mm (1/16 inch) from case for 10 seconds: N package .... 260°C

NOTES: 1. All voltage values, except differential input voltage, are with respect to the midpoint between V_{CC+} and V_{CC-}.
       2. Because of the back-to-back diodes protecting the differential inputs, input current can become excessive if differential input voltage reaches approximately 0.7 volts.
       3. For operation above 25°C free-air temperature, refer to Dissipation Derating Curves, Section 2. In the J package, OP-227 chips are glass mounted.

TEXAS INSTRUMENTS

POST OFFICE BOX 225012 • DALLAS, TEXAS 75265

electrical characteristics at specified free-air temperature, VCC± = ±15 V (unless otherwise noted)

PARAMETER		TEST CONDITIONS†		OP-227E			OP-227F			OP-227G			UNIT
				MIN	TYP	MAX	MIN	TYP	MAX	MIN	TYP	MAX	
V_{IO}	Input offset voltage	$V_O = 0$, $R_S = 50\ \Omega$	25°C		20	80		40	120		60	180	μV
			-25°C to 85°C		40	140		60	200		85	280	μV
αV_{IO}	Temperature coefficient of input offset voltage	$R_p = 8\ k\Omega$ to $20\ k\Omega$ (nulled) Unnulled	-25°C to 85°C		0.5	1		0.4	1.5		0.5	1.8	μV/°C
	Long-term drift of input offset voltage	See Note 4			0.2	1		0.3	1.5		0.4	2	μV/mo
	Offset adjustment range	$R_p = 10\ k\Omega$	25°C		±4			±4			±4		mV
I_{IO}	Input offset current	$V_O = 0$	25°C		7	35		9	50		12	75	nA
			-25°C to 85°C		10	50		14	85		20	135	nA
I_{IB}	Input bias current	$V_O = 0$	25°C		±10	±40		±12	±55		±15	±80	nA
			-25°C to 85°C		±14	±60		±18	±95		±25	±150	nA
V_{ICR}	Common-mode input voltage range		25°C	±11	±12.3		±11	±12.3		±11	±12.3		V
			-25°C to 85°C	±10.5	±11.8		±10.5	±11.8		±10.5	±11.8		V
V_{OM}	Maximum output voltage swing	$R_L \geq 600\ \Omega$	25°C	±10	±11.5		±10	±11.5		±10	±11.5		V
		$R_L \geq 2\ k\Omega$	25°C	±12	±13.8		±12	±13.8		±11.5	±13.5		V
			-25°C to 85°C	±11.7	±13.6		±11.4	±13.5		±11	±13.3		V
A_{VD}	Large-signal differential voltage amplification	$V_{CC\pm} = \pm4\ V$, $V_O = \pm1\ V$, $R_L \geq 600\ \Omega$	25°C	250	700		250	700		200	500		V/mV
		$V_O = \pm10\ V$, $R_L \geq 1\ k\Omega$	25°C	800	1500		800	1500		700	1500		V/mV
		$V_O = \pm10\ V$, $R_L \geq 2\ k\Omega$		1000	1800		1000	1800		450	1000		V/mV
			-25°C to 85°C	750	1500		700	1500					V/mV
r_i	Input resistance	Differential-mode	25°C	1.5	6		1.2	5		0.8	4		MΩ
		Common-mode	25°C		3			2.5			2		GΩ
CMRR	Common-mode rejection ratio	$V_{IC} = \pm11\ V$	25°C	114	126		106	123		100	120		dB
		$V_{IC} = \pm10\ V$	-25°C to 85°C	110	124		102	121		96	118		dB
k_{SVS}	Supply voltage sensitivity ($\Delta V_{IO}/\Delta V_{CC\pm}$)	$V_{CC\pm} = \pm4\ V$ to $\pm18\ V$	25°C		1	10		1	10		2	20	μV/V
		$V_{CC\pm} = \pm4.5\ V$ to $\pm18\ V$	-25°C to 85°C		2	15		2	16		2	32	μV/V
P_D	Power dissipation (each amplifier)	$V_O = 0$	25°C		90	140		90	140		90	140	mW

†All characteristics are measured under open-loop conditions with zero common-mode input voltage unless otherwise noted.

NOTE 4: Since long-term drift cannot be measured on the individual devices prior to shipment, this specification is not intended to be a guarantee or warranty. It is an engineering estimate of the averaged trend line of drift versus time over extended periods after the first thirty days of operation.

3

Operational Amplifiers

TEXAS INSTRUMENTS
POST OFFICE BOX 225012 • DALLAS, TEXAS 75265

operating characteristics, $V_{CC} = \pm 15$ V, $T_A = 25\,°C$

PARAMETER		TEST CONDITIONS	OP-227E			OP-227F			OP-227G			UNIT
			MIN	TYP	MAX	MIN	TYP	MAX	MIN	TYP	MAX	
B_1	Unity-gain bandwidth		5	8		5	8		5	8		MHz
SR	Slew rate at unity gain		1.7	2.8		1.7	2.8		1.7	2.8		V/μs
V_{NPP}	Peak-to-peak equivalent input noise voltage	$f = 0.1$ Hz to 10 Hz, $R_S = 100\ \Omega$		0.08	0.2		0.08	0.2		0.09	0.28	μV
V_n	Equivalent input noise voltage	$f = 10$ Hz, $R_S = 100\ \Omega$		3.5	6		3.5	6		3.8	9	nV/√Hz
		$f = 30$ Hz, $R_S = 100\ \Omega$		3.1	4.7		3.1	4.7		3.3	5.9	
		$f = 1000$ Hz, $R_S = 100\ \Omega$		3	3.9		3	3.9		3.2	4.6	
I_n	Equivalent input noise current	$f = 10$ Hz		1.7	4.5		1.7	4.5		1.7		pA/√Hz
		$f = 30$ Hz		1	2.5		1	2.5		1		
		$f = 1$ kHz		0.4	0.7		0.4	0.7		0.4	0.7	
V_{o1}/V_{o2}	Crosstalk attenuation	$A_V = 100$	126	154		126	154		126	154		dB

3

Operational Amplifiers

TEXAS
INSTRUMENTS
POST OFFICE BOX 225012 • DALLAS, TEXAS 75265

**LINEAR
INTEGRATED
CIRCUITS**

**TYPE RC4559
DUAL HIGH-PERFORMANCE OPERATIONAL AMPLIFIER**

D2785, OCTOBER 1983

- Matched Gain and Offset Between Amplifiers
- Unity-Gain Bandwidth . . . 3 MHz Min
- Slew Rate . . . 1.5 V/ns Min
- Low Equivalent Input Noise Voltage
  . . . 2 $\mu V/\sqrt{Hz}$ Max (20 Hz to 20 kHz)
- No Frequency Compensation Required
- No Latch Up
- Wide Common-Mode Voltage Range
- Low Power Consumption
- Designed to be Interchangeable with Raytheon RC4559

**D OR P DUAL-IN-LINE PACKAGE
(TOP VIEW)**

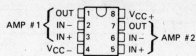

symbol (each amplifier)

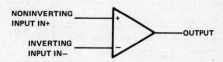

description

The RC4559 is a dual high-performance operational amplifier. The high common-mode input voltage and the absence of latch-up make this amplifier ideal for low-noise signal applications such as audio preamplifiers and signal conditioners. This amplifier features a guaranteed dynamic performance and output drive capability that far exceeds that of the general-purpose type amplifiers.

The RC4559 is characterized for operation from 0°C to 70°C.

absolute maximum ratings over operating free-air temperature range (unless otherwise noted)

Supply voltage V_{CC+} (see Note 1) . . . . 18 V
Supply voltage V_{CC-} (see Note 1) . . . . −18 V
Differential input voltage (see Note 2) . . . . ±30 V
Input voltage (any input, see Notes 1 and 3) . . . . ±15 V
Duration of output short-circuit to ground, one amplifier at a time (see Note 4) . . . . unlimited
Continuous total dissipation . . . . 500 mW
Operating free-air temperature range . . . . 0°C to 70°C
Storage temperature range . . . . −65°C to 125°C
Lead temperature 1,6 mm (1/16 inch) from case for 10 seconds . . . . 260°C

NOTES: 1. All voltage values, unless otherwise noted, are with respect to the zero reference level (ground) of the supply voltages where the zero reference level is the midpoint between V_{CC+} and V_{CC-}.
2. Differential voltages are at the noninverting input terminal with respect to the inverting input terminal.
3. The magnitude of the input voltage must never exceed the magnitude of the supply voltage or 15 volts, whichever is less.
4. Temperature and/or supply voltages must be limited to ensure that the dissipation rating is not exceeded.

Operational Amplifiers

3

**TEXAS
INSTRUMENTS**

POST OFFICE BOX 225012 • DALLAS, TEXAS 75265

TYPE RC4559
DUAL HIGH-PERFORMANCE OPERATIONAL AMPLIFIER

electrical characteristics at specified free-air temperature, $V_{CC+} = 15$ V, $V_{CC-} = -15$ V

	PARAMETER	TEST CONDITIONS[†]		MIN	TYP	MAX	UNIT
V_{IO}	Input offset voltage	$V_O = 0$	25 °C		2	6	mV
			0 °C to 70 °C			7.5	
I_{IO}	Input offset current	$V_O = 0$	25 °C		5	100	nA
			0 °C to 70 °C			200	
I_{IB}	Input bias current	$V_O = 0$	25 °C		40	250	nA
			0 °C to 70 °C			500	
V_I	Input voltage range		25 °C	±12	±13		V
V_{OM}	Maximum peak output voltage swing	$R_L \geq 3$ kΩ	25 °C	±12	±13		V
		$R_L = 600$ Ω	25 °C	±9.5	±10		
		$R_L \geq 2$ kΩ	0 °C to 70 °C	±10			
A_{VD}	Large-signal differential voltage amplification	$V_O = \pm10$ V,	25 °C	20	300		V/mV
		$R_L = 2$ kΩ	0 °C to 70 °C	15			
B_{OM}	Maximum output-swing bandwidth	$V_{OPP} = 20$ V, $R_L = 2$ kΩ	25 °C	24	32		kHz
B_1	Unity-gain bandwidth		25 °C	3	4		MHz
r_i	Input resistance		25 °C	0.3	1		MΩ
CMRR	Common-mode rejection ratio	$V_O = 0$	25 °C	80	100		dB
k_{SVS}	Supply voltage sensitivity ($\Delta V_{IO}/\Delta V_{CC}$)	$V_O = 0$	25 °C		10	75	μV/V
V_n	Equivalent input noise voltage (closed-loop)	$A_{VD} = 100$, $R_S = 1$ kΩ, $f = 20$ Hz to 20 kHz	25 °C		1.4	2	μV
I_n	Equivalent input noise current	$f = 20$ Hz to 20 kHz	25 °C		25		pA
I_{CC}	Supply current (both amplifiers)	No load, No signal	25 °C		3.3	5.6	mA
			0 °C		4	6.6	
			70 °C		3	5	
V_{o1}/V_{o2}	Crosstalk attenuation	$A_{VD} = 100$, $R_S = 1$ kΩ, $f = 10$ kHz	25 °C		90		dB
			25 °C		90		

[†]All characteristics are specified under open-loop operation, unless otherwise noted.

matching characteristics at $V_{CC+} = 15$ V, $V_{CC-} = -15$ V, $T_A = 25$ °C

	PARAMETER	TEST CONDITIONS	MIN	TYP	MAX	UNIT
V_{IO}	Input offset voltage	$V_O = 0$		±0.2		mV
I_{IO}	Input offset current	$V_O = 0$		±7.5		nA
I_{IB}	Input bias current	$V_O = 0$		±15		nA
A_{VD}	Large-signal differential voltage amplification	$V_O = \pm10$ V, $R_L = 2$ kΩ		±1		dB

operating characteristics, $V_{CC+} = 15$ V, $V_{CC-} = -15$ V, $T_A = 25$ °C

	PARAMETER	TEST CONDITIONS		MIN	TYP	MAX	UNIT
t_r	Rise time	$V_I = 20$ mV,	$R_L = 2$ kΩ,		80		μs
	Overshoot	$C_L = 100$ pF			18 %		
SR	Slew rate at unity gain	$V_I = 10$ V, $C_L = 100$ pF	$R_L = 2$ kΩ,	1.5	2		V/μs

TEXAS INSTRUMENTS
POST OFFICE BOX 225012 • DALLAS, TEXAS 75265

TYPES RM4136, RV4136, RC4136
QUAD HIGH-PERFORMANCE OPERATIONAL AMPLIFIERS

D2142, MARCH 1976—REVISED SEPTEMBER 1983

- Continuous-Short-Circuit Protection
- Wide Common-Mode and Differential Voltage Ranges
- No Frequency Compensation Required
- Low Power Consumption
- No Latch-Up
- Unity Gain Bandwidth 3 MHz Typical
- Gain and Phase Match Between Amplifiers
- Designed to be Interchangeable with Raytheon RM4136, RV4136, and RC4136
- Low Noise . . . 8 nV/$\sqrt{Hz}$ Typ at 1 kHz

D, J, OR N DUAL-IN-LINE
OR W FLAT PACKAGE
(TOP VIEW)

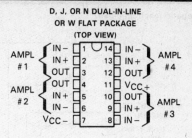

RM4136
FH OR FK CHIP CARRIER PACKAGE
(TOP VIEW)

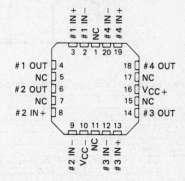

NC—No internal connection

description

The RM4136, RV4136, and RC4136 are quad high-performance operational amplifiers with each amplifier electrically similar to uA741 except that offset null capability is not provided.

The high common-mode input voltage range and the absence of latch-up make these amplifiers ideal for voltage-follower applications. The devices are short-circuit protected and the internal frequency compensation ensures stability without external components.

The RM4136 is characterized for operation over the full military temperature range of −55°C to 125°C, the RV4136 is characterized for operation from −40°C to 85°C, and the RC4136 is characterized for operation from 0°C to 70°C.

symbol (each amplifier)

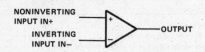

NONINVERTING INPUT IN+

INVERTING INPUT IN−

OUTPUT

Operational Amplifiers

TEXAS
INSTRUMENTS
POST OFFICE BOX 225012 • DALLAS, TEXAS 75265

TYPES RM4136, RV4136, RC4136
QUAD HIGH-PERFORMANCE OPERATIONAL AMPLIFIERS

schematic (each amplifier)

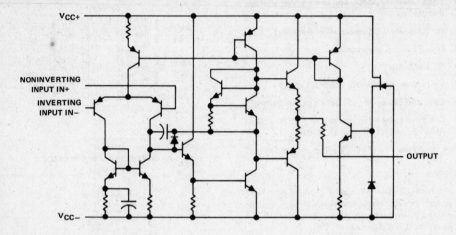

absolute maximum ratings over operating free-air temperature range (unless otherwise noted)

		RM4136	RV4136	RC4136	UNIT
Supply voltage V_{CC+} (see Note 1)		22	18	18	V
Supply voltage V_{CC-} (see Note 1)		−22	−18	−18	V
Differential input voltage (see Note 2)		±30	±30	±30	V
Input voltage (any input, see Notes 1 and 3)		±15	±15	±15	V
Duration of output short-circuit to ground, one amplifier at a time (see Note 4)		unlimited	unlimited	unlimited	
Continuous total dissipation at (or below) 25°C free-air temperature (see Note 5)		800	800	800	mW
Operating free-air temperature range		−55 to 125	−40 to 85	0 to 70	°C
Storage temperature range		−65 to 150	−65 to 150	−65 to 150	°C
Lead temperature 1,6 mm (1/16 inch) from case for 60 seconds	FH, FK, J, or W package	300	300	300	°C
Lead temperature 1,6 mm (1/16 inch) from case for 10 seconds	D or N package		260	260	

NOTES: 1. All voltage values, unless otherwise noted, are with respect to the midpoint between V_{CC+} and V_{CC-}.
     2. Differential voltages are at the noninverting input terminal with respect to the inverting input terminal.
     3. The magnitude of the input voltage must never exceed the magnitude of the supply voltage or 15 volts, whichever is less.
     4. Temperature and/or supply voltages must be limited to ensure that the dissipation rating is not exceeded.
     5. For operation above 25°C free-air temperature, refer to Dissipation Derating Curves, Section 2. In the J package, RM4136 chips are alloy-mounted; RV4136 and RC4136 chips are glass-mounted.

TEXAS INSTRUMENTS

POST OFFICE BOX 225012 ● DALLAS, TEXAS 75265

electrical characteristics at specified free-air temperature, $V_{CC+} = 15$ V, $V_{CC-} = -15$ V

PARAMETER		TEST CONDITIONS[†]		RM4136 MIN	TYP	MAX	RV4136 MIN	TYP	MAX	RC4136 MIN	TYP	MAX	UNIT
V_{IO}	Input offset voltage	$V_O = 0$	25°C		0.5	4		0.5	6		0.5	6	mV
			Full range			6			7.5			7.5	
I_{IO}	Input offset current	$V_O = 0$	25°C		5	150		5	200		5	200	nA
			Full range			500			500			300	
I_{IB}	Input bias current	$V_O = 0$	25°C		140	400		140	500		140	500	nA
			Full range			1500			1500			800	
V_I	Input voltage range		25°C	±12	±14		±12	±14		±12	±14		V
V_{OM}	Maximum peak output voltage swing	$R_L = 10$ kΩ	25°C	±12	±14		±12	±14		±12	±14		V
		$R_L = 2$ kΩ	25°C	±10	±13		±10	±13		±10	±13		
		$R_L \geq 2$ kΩ	Full range	±10			±10			±10			
A_{VD}	Large-signal differential voltage amplification	$V_O = ±10$ V, $R_L \geq 2$ kΩ	25°C	50	350		20	300		20	300		V/mV
			Full range	25			15			15			
B_1	Unity-gain bandwidth		25°C		3.5			3			3		MHz
r_i	Input resistance		25°C	0.3	5		0.3	5		0.3	5		MΩ
CMRR	Common-mode rejection ratio	$V_O = 0$, $R_S = 50$ Ω	25°C	70	90		70	90		70	90		dB
k_{SVS}	Supply voltage sensitivity ($\Delta V_{IO}/\Delta V_{CC}$)	$V_{CC} = ±9$ V to ±15 V, $V_O = 0$	25°C		30	150		30	150		30	150	μV/V
V_n	Equivalent input noise voltage (closed-loop)	$A_{VD} = 100$, BW = 1 Hz, f = 1 kHz, $R_S = 100$ Ω	25°C		8			8			8		nV√Hz
I_{CC}	Supply current (All four amplifiers)	$V_O = 0$, No load	25°C		5	11.3		5	11.3		5	11.3	mA
			MIN T_A		6	13.3		6	13.7		6	13.7	
			MAX T_A		4.5	10		4.5	10		4.5	10	
P_D	Total power dissipation (All four amplifiers)	$V_O = 0$, No load	25°C		150	340		150	340		150	340	mW
			MIN T_A		180	400		180	400		180	400	
			MAX T_A		135	300		135	300		135	300	
V_{o1}/V_{o2}	Crosstalk attenuation	$A_{VD} = 100$, f = 10 kHz, $R_S = 1$ kΩ	25°C		105			105			105		dB

[†]All characteristics are measured under open-loop conditions with zero common-mode input voltage unless otherwise specified. Full range is −55°C to 125°C for RM4136, −40°C to 85°C for RV4136, and 0°C to 70°C for RC4136.

operating characteristics, $V_{CC+} = 15$ V, $V_{CC-} = -15$ V, $T_A = 25$°C

PARAMETER		TEST CONDITIONS		RM4136 MIN	TYP	MAX	RV4136, RC4136 MIN	TYP	MAX	UNIT
t_r	Rise Time	$V_I = 20$ mV, $R_L = 2$ kΩ, $C_L = 100$ pF			0.13			0.13		μs
	Overshoot factor				5%			5%		
SR	Slew rate at unity gain	$V_I = 10$ V, $R_L = 2$ kΩ, $C_L = 100$ pF			1.7			1.7		V/μs

3

Operational Amplifiers

Operational Amplifiers

**LINEAR
INTEGRATED
CIRCUITS**

**TYPES RM4558, RV4558, RC4558
DUAL HIGH-PERFORMANCE OPERATIONAL AMPLIFIERS**

D2141, MARCH 1976—REVISED FEBRUARY 1984

- Continuous-Short-Circuit Protection
- Wide Common-Mode and Differential Voltage Ranges
- No Frequency Compensation Required
- Low Power Consumption
- No Latch-up
- Unity Gain Bandwidth 3 MHz Typical
- Gain and Phase Match Between Amplifiers
- Low Noise . . . 8 nV/$\sqrt{Hz}$ Typ at 1 kHz
- Designed to be Interchangeable with Raytheon RM4558, RV4558, and RC4558

D, JG, OR P DUAL-IN-LINE PACKAGE
(TOP VIEW)

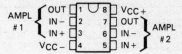

AMPL #1 { OUT [1 8] VCC+
IN− [2 7] OUT
IN+ [3 6] IN− } AMPL #2
VCC− [4 5] IN+

description

The RM4558, RV4558, and RC4558 are dual general-purpose operational amplifiers with each half electrically similar to uA741 except that offset null capability is not provided.

The high common-mode input voltage range and the absence of latch-up make these amplifiers ideal for voltage-follower applications. The devices are short-circuit protected and the internal frequency compensation ensures stability without external components.

The RM4558 is characterized for operation over the full military temperature range of −55 °C to 125 °C; the RV4558 is characterized for operation from −40 °C to 85 °C; and the RC4558 is characterized for operation from 0 °C to 70 °C.

schematic (each amplifier)

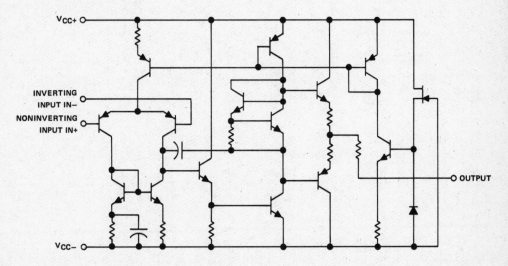

284

**TEXAS
INSTRUMENTS**
POST OFFICE BOX 225012 • DALLAS, TEXAS 75265

Operational Amplifiers

3

absolute maximum ratings over operating free-air temperature range (unless otherwise noted)

		RM4558	RV4558	RC4558	UNIT
Supply voltage V_{CC+} (see Note 1)		22	18	18	V
Supply voltage V_{CC-} (see Note 1)		-22	-18	-18	V
Differential input voltage (see Note 2)		±30	±30	±30	V
Input voltage (any input, see Notes 1 and 3)		±15	±15	±15	V
Duration of output short-circuit to ground, one amplifier at a time (see Note 4)		unlimited	unlimited	unlimited	
Continuous total dissipation at (or below) 25°C free-air temperature (see Note 5)		680	680	680	mW
Operating free-air temperature range		-55 to 125	-40 to 85	0 to 70	°C
Storage temperature range		-65 to 150	-65 to 150	-65 to 150	°C
Lead temperature 1,6 mm (1/16 inch) from case for 60 seconds	JG package	300	300	300	°C
Lead temperature 1,6 mm (1/16 inch) from case for 10 seconds	D or P package		260	260	°C

NOTES: 1. All voltage values, unless otherwise noted, are with respect to the midpoint between V_{CC+} and V_{CC-}.
2. Differential voltages are at the noninverting input terminal with respect to the inverting input terminal.
3. The magnitude of the input voltage must never exceed the magnitude of the supply voltage or 15 volts, whichever is less.
4. Temperature and/or supply voltages must be limited to ensure that the dissipation rating is not exceeded.
5. For operation above 25°C free-air temperature, refer to Dissipation Derating Curves, Section 2. In the JG packages, RM4558 chips are alloy mounted; RV4558 and RC4558 chips are glass mounted.

3

Operational Amplifiers

TEXAS
INSTRUMENTS
POST OFFICE BOX 225012 • DALLAS, TEXAS 75265

284

electrical characteristics at specified free-air temperature, V_{CC+} = 15 V, CCC_- = −15 V

PARAMETER		TEST CONDITIONS†	RM4558 MIN	RM4558 TYP	RM4558 MAX	RV4558 MIN	RV4558 TYP	RV4558 MAX	RC4558 MIN	RC4558 TYP	RC4558 MAX	UNIT
V_{IO} Input offset voltage		V_O = 0, 25°C		0.5	5		0.5	6		0.5	6	mV
		Full range			6			7.5			7.5	
I_{IO} Input offset current		V_O = 0, 25°C		5	200		5	200		5	200	nA
		Full range			500			500			300	
I_{IB} Input bias current		V_O = 0, 25°C		140	500		140	500		150	500	nA
		Full range			1500			1500			800	
V_{ICR} Common-mode input voltage range		25°C	±12	±14		±12	±14		±12	±14		V
V_{OM} Maximum output voltage swing		R_L = 10 kΩ, 25°C	±12	±14		±12	±14		±12	±14		V
		R_L = 2 kΩ, 25°C	±10	±13		±10	±13		±10	±13		
		R_L ≥ 2 kΩ, Full range	±10			±10			±10			
A_{VD} Large-signal differential voltage amplification		R_L ≥ 2 kΩ, V_O = ±10 V, 25°C	50	350		20	300		20	300		V/mV
		Full range	25			15			15			
B_1 Unity-gain bandwidth		25°C	2	3.5			3			3		MHz
r_i Input resistance		25°C	0.3	5		0.3	5		0.3	5		MΩ
CMRR Common-mode rejection ratio		25°C	70	90		70	90		70	90		dB
k_{SVS} Supply voltage sensitivity ($\Delta V_{IO}/\Delta V_{CC}$)		V_{CC} = ±15 V to ±9 V, 25°C		30	150		30	150		30	150	μV/V
V_n Equivalent input noise voltage (closed-loop)		A_{VD} = 100, R_S = 100 Ω, f = 1 kHz, BW = 1 Hz, 25°C		8			8			8		nV/√Hz
I_{CC} Supply current (Both amplifiers)		No load, V_O = 0, 25°C		2.5	5.6		2.5	5.6		2.5	5.6	mA
		MIN T_A		3	6.6		3	6.6		3	6.6	
		MAX T_A		2	5		2.3	5		2.3	5	
P_D Total power dissipation (Both amplifiers)		No load, V_O = 0, 25°C		75	170		75	170		75	170	mW
		MIN T_A		90	200		90	200		90	200	
		MAX T_A		60	150		70	150		70	150	
V_{o1}/V_{o2} Crosstalk attenuation	Open loop	R_S = 1 kΩ, 25°C		85			85			85		dB
	A_{VD} = 100	f = 10 kHz		105			105			105		

† All characteristics are measured under open-loop conditions with zero common-mode input voltage unless otherwise specified. Full range is −55°C to 125°C for RM4558, −40°C to 85°C for RV4558, and 0°C to 70°C for RC4558.

3

Operational Amplifiers

TEXAS INSTRUMENTS
POST OFFICE BOX 225012 ● DALLAS, TEXAS 75265

operating characteristics, V_{CC+} = 15 V, V_{CC-} = −15 V, T_A = 25°C

PARAMETER		TEST CONDITIONS	RM4558			RV4558			RC4558			UNIT
			MIN	TYP	MAX	MIN	TYP	MAX	MIN	TYP	MAX	
t_r	Rise time	V_I = 20 mV, R_L = 2 kΩ,		0.13			0.13			0.13		ns
	Overshoot	C_L = 100 pF		5%			5%			5%		
SR	Slew rate at unity gain	V_I = 10 V, R_L = 2 kΩ, C_L = 100 pF	1.3	1.7		1.3	1.7		1.3	1.7		V/µs

Texas
Instruments
POST OFFICE BOX 225012 • DALLAS, TEXAS 75265

TYPES SE5534, SE5534A, NE5534, NE5534A LOW-NOISE OPERATIONAL AMPLIFIERS

D2532, JULY 1979—REVISED AUGUST 1983

- Equivalent Input Noise Voltage
  3.5 nV/$\sqrt{\text{Hz}}$ Typ
- Unity-Gain Bandwidth ...... 10 MHz Typ
- Common-Mode Rejection Ratio
  100 dB Typ
- High DC Voltage Gain .... 100 V/mV Typ
- Peak-to-Peak Output Voltage Swing
  32 V Typ with $V_{CC\pm}$ = ±18 V and
  R_L = 600 Ω
- High Slew Rate .... 13 V/μs Typ
- Wide Supply Voltage Range
  ±3 V to ±20 V
- Low Harmonic Distortion
- Designed to be Interchangeable with Signetics
  SE5534, SE5534A, NE5534, and NE5534A

SE5534, SE5534A . . . JG
NE5534, NE5534A . . . JG or P
DUAL-IN-LINE PACKAGE
(TOP VIEW)

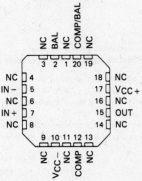

SE5534, SE5534A
U FLAT PACKAGE
(TOP VIEW)

SE5534, SE5534A
FH OR FK CHIP CARRIER PACKAGE
(TOP VIEW)

NC—No internal connection

3

Operational Amplifiers

symbol

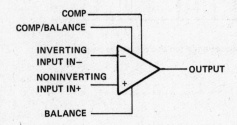

description

The SE5534, SE5534A, NE5534, and NE5534A are monolithic high-performance operational amplifiers combining excellent dc and ac characteristics. Some of the features include very low noise, high output drive capability, high unity-gain and maximum-output-swing bandwidths, low distortion, and high slew rate.

These operational amplifiers are internally compensated for a gain equal to or greater than three. Optimization of the frequency response for various applications can be obtained by use of an external compensation capacitor between COMP and COMP/BAL. The devices feature input-protection diodes, output short-circuit protection, and offset-voltage nulling capability.

The SE5534A and NE5534A have guaranteed maximums on equivalent input noise voltage.

The SE5534 and SE5534A are characterized for operation over the full military temperature range of −55°C to 125°C; the NE5534 and NE5534A are characterized for operation from 0°C to 70°C.

TYPES SE5534, SE5534A, NE5534, NE5534A
LOW-NOISE OPERATIONAL AMPLIFIERS

schematic

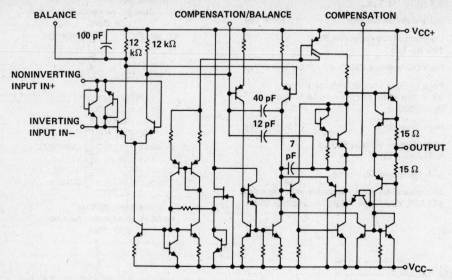

All component values shown are nominal.

absolute maximum ratings over operating free-air temperature range (unless otherwise noted)

Supply voltage, V_{CC+} (see Note 1) .... 22 V
Supply voltage, V_{CC-} (see Note 1) .... −22 V
Input voltage either input (see Notes 1 and 2) .... V_{CC+}
Input current (see Note 3) .... ±10 mA
Duration of output short-circuit (see Note 4) .... unlimited
Continuous total power dissipation at (or below) 25°C free-air temperature (see Note 5)
    FH package (see Note 6) .... 1200 mW
    FK package (see Note 6) .... 1375 mW
    SE5534, SE5534A in JG package .... 1050 mW
    NE5534, NE5534A in JG package .... 825 mW
    P package .... 725 mW
    U package .... 675 mW
Operating free-air temperature range: SE5534, SE5534A .... −55°C to 125°C
    NE5534, NE5534A .... 0°C to 70°C
Storage temperature range .... −65°C to 150°C
Lead temperature 1,6 mm (1/16 inch) from case for 60 seconds: FH, FK, JG, or U package .... 300°C
Lead temperature 1,6 mm (1/16 inch) from case for 10 seconds: P package .... 260°C

NOTES:
1. All voltage values, except differential voltages, are with respect to the midpoint between V_{CC+} and V_{CC-}.
2. The magnitude of the input voltage must never exceed the magnitude of the supply voltage.
3. Excessive current will flow if a differential input voltage in excess of approximately 0.6 V is applied between the inputs unless some limiting resistance is used.
4. The output may be shorted to ground or either power supply. Temperature and/or supply voltages must be limited to ensure the maximum dissipation rating is not exceeded.
5. For operation above 25°C free-air temperature, refer to the Dissipation Derating Curves, Section 2. In the JG package, SE5534 and SE5534A chips are alloy-mounted; NE5534 and NE5534A chips are glass-mounted.
6. For FH and FK packages, power rating and derating factor will vary with actual mounting technique used. The values stated here are believed to be conservative.

TEXAS
INSTRUMENTS
POST OFFICE BOX 225012 • DALLAS, TEXAS 75265

electrical characteristics, $V_{CC\pm} = \pm 15$ V, $T_A = 25\,°C$ (unless otherwise noted)

PARAMETER		TEST CONDITIONS[†]		SE5534, SE5534A MIN	TYP	MAX	NE5534, NE5534A MIN	TYP	MAX	UNIT
V_{IO}	Input offset voltage	$V_O = 0$, $R_S = 50\ \Omega$	$T_A = 25\,°C$		0.5	2		0.5	4	mV
			T_A = full range			3			5	
I_{IO}	Input offset current	$V_O = 0$	$T_A = 25\,°C$		10	200		20	300	nA
			T_A = full range			500			400	
I_{IB}	Input bias current	$V_O = 0$	$T_A = 25\,°C$		400	800		500	1500	nA
			T_A = full range			1500			2000	
V_{ICR}	Common-mode input voltage range			± 12	± 13		± 12	± 13		V
V_{OPP}	Maximum peak-to-peak output voltage swing	$R_L \geq 600\ \Omega$	$V_{CC\pm} = \pm 15$ V	24	26		24	26		V
			$V_{CC\pm} = \pm 18$ V	30	32		30	32		
A_{VD}	Large-signal differential voltage amplification	$V_O = \pm 10$ V, $R_L \geq 600\ \Omega$	$T_A = 25\,°C$	50	100		25	100		V/mV
			T_A = full range	25			15			
A_{vd}	Small-signal differential voltage amplification	$f = 10$ kHz	$C_C = 0$		6			6		V/mV
			$C_C = 22$ pF		2.2			2.2		
B_{OM}	Maximum-output-swing bandwidth	$V_O = \pm 10$ V,	$C_C = 0$		200			200		kHz
		$V_O = \pm 10$ V,	$C_C = 22$ pF		95			95		
		$V_{CC\pm} = \pm 18$ V, $V_O = \pm 14$ V, $R_L = 600\ \Omega$,	$C_C = 22$ pF		70			70		
B_1	Unity-gain bandwidth	$C_C = 22$ pF,	$C_L = 100$ pF		10			10		MHz
r_i	Input resistance			50	100		30	100		kΩ
z_o	Output impedance	$A_{VD} = 30$ dB, $C_C = 22$ pF,	$R_L = 600\ \Omega$, $f = 10$ kHz		0.3			0.3		Ω
CMRR	Common-mode rejection ratio	$V_O = 0$, $R_S = 50\ \Omega$	$V_{IC} = V_{ICR}$ min,	80	100		70	100		dB
k_{SVR}	Supply voltage rejection ratio ($\Delta V_{CC}/\Delta V_{IO}$)	$V_{CC\pm} = \pm 9$ V to ± 15 V, $V_O = 0$,	$R_S = 50\ \Omega$	86	100		80	100		dB
I_{OS}	Output short-circuit current				38			38		mA
I_{CC}	Supply current	No load, $V_O = 0$	$T_A = 25\,°C$		4	6.5		4	8	mA
			T_A = full range			9				

[†]All characteristics are measured under open-loop conditions with zero common-mode input voltage unless otherwise specified. Full range for $T_A = -55\,°C$ to $125\,°C$ for SE5534 and SE5534A and $0\,°C$ to $70\,°C$ for NE5534 and NE5534A.

operating characteristics, $V_{CC\pm} = \pm 15$ V, $T_A = 25\,°C$

PARAMETER		TEST CONDITIONS		SE5534, NE5534 MIN	TYP	MAX	SE5534A, NE5534A MIN	TYP	MAX	UNIT
SR	Slew rate at unity gain	$C_C = 0$			13			13		V/µs
		$C_C = 22$ pF			6			6		
t_r	Rise time	$V_I = 50$ mV, $R_L = 600\ \Omega$, $C_L = 100$ pF	$A_{VD} = 1$, $C_C = 22$ pF,		20			20		ns
	Overshoot factor				20%			20%		
t_r	Rise time	$V_I = 50$ mV, $R_L = 600\ \Omega$, $C_L = 500$ pF	$A_{VD} = 1$, $C_C = 47$ pF,		50			50		ns
	overshoot factor				35%			35%		
V_n	Equivalent input noise voltage	$f = 30$ Hz			7			5.5	7	nV/√Hz
		$f = 1$ kHz			4			3.5	4.5	
I_n	Equivalent input noise current	$f = 30$ Hz			2.5			1.5		pA/√Hz
		$f = 1$ kHz			0.6			0.4		
F	Average noise figure	$R_S = 5$ kΩ, $f = 10$ Hz to 20 kHz						0.9		dB

3

Operational Amplifiers

TYPICAL CHARACTERISTICS†

NORMALIZED INPUT BIAS CURRENT
and INPUT OFFSET CURRENT
vs
FREE-AIR TEMPERATURE

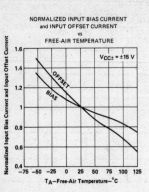

FIGURE 1

MAXIMUM PEAK-TO-PEAK OUTPUT VOLTAGE
vs
FREQUENCY

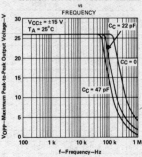

FIGURE 2

LARGE-SIGNAL
DIFFERENTIAL VOLTAGE AMPLIFICATION
vs
FREQUENCY

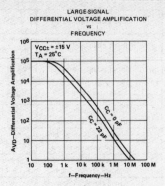

FIGURE 3

NORMALIZED SLEW RATE and
UNITY-GAIN BANDWIDTH
vs
SUPPLY VOLTAGE

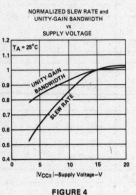

FIGURE 4

NORMALIZED SLEW RATE and
UNITY-GAIN BANDWIDTH
vs
FREE-AIR TEMPERATURE

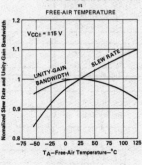

FIGURE 5

TOTAL HARMONIC DISTORTION
vs
FREQUENCY

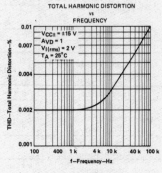

FIGURE 6

EQUIVALENT INPUT NOISE VOLTAGE
vs
FREQUENCY

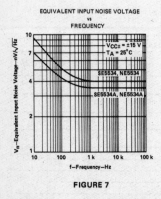

FIGURE 7

EQUIVALENT INPUT NOISE CURRENT
vs
FREQUENCY

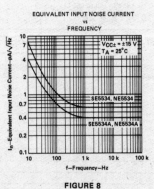

FIGURE 8

TOTAL EQUIVALENT INPUT NOISE VOLTAGE
vs
SOURCE RESISTANCE

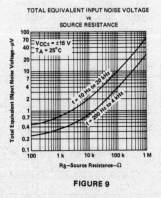

FIGURE 9

†Data at high and low temperatures are applicable only within the rated operating free-air temperature ranges of the various devices.

3

Operational Amplifiers

**TEXAS
INSTRUMENTS**

POST OFFICE BOX 225012 • DALLAS, TEXAS 75265

TYPES TL022M, TL022C
DUAL LOW-POWER OPERATIONAL AMPLIFIERS

D1661, SEPTEMBER 1973—REVISED AUGUST 1983

- **Very Low Power Consumption**
- **Power Dissipation with ±2-V Supplies . . . 170 μW Typ**
- **Low Input Bias and Offset Currents**
- **Output Short-Circuit Protection**
- **Low Input Offset Voltage**
- **Internal Frequency Compensation**
- **Latch-Up-Free Operation**
- **Popular Dual Op-Amp Pin-Out**

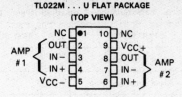

TL022M . . . U FLAT PACKAGE
(TOP VIEW)

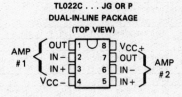

TL022C . . . JG OR P
DUAL-IN-LINE PACKAGE
(TOP VIEW)

NC—No internal connection

description

The TL022 is a dual low-power operational amplifier designed to replace higher power devices in many applications without sacrificing system performance. High input impedance, low supply currents, and low equivalent input noise voltage over a wide range of operating supply voltages result in an extremely versatile operational amplifier for use in a variety of analog applications including battery-operated circuits. Internal frequency compensation, absence of latch-up, high slew rate, and output short-circuit protection assure ease of use.

The TL022M is characterized for operation over the full military temperature range of −55°C to 125°C; the TL022C is characterized for operation from 0°C to 70°C.

symbol (each amplifier)

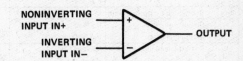

absolute maximum ratings over operating free-air temperature range (unless otherwise noted)

			TL022M	TL022C	UNIT
Supply voltage V_{CC+} (see Note 1)			22	18	V
Supply voltage V_{CC-} (see Note 1)			−22	−18	V
Differential input voltage (see Note 2)			±30	±30	V
Input voltage (any input, see Notes 1 and 3)			±15	±15	V
Duration of output short-circuit (see Note 4)			unlimited	unlimited	
Continuous total dissipation at (or below) 25°C free-air temperature range (see Note 5)	Each amplifier		500	500	mW
	Total package	JG or P package	680	680	
		U package	675	675	
Operating free-air temperature range			−55 to 125	0 to 70	°C
Storage temperature range			−65 to 150	−65 to 150	°C
Lead temperature 1,6 mm (1/16 inch) from case for 60 seconds		JG or U package	300	300	°C
Lead temperature 1,6 mm (1/16 inch) from case for 10 seconds		P package		260	°C

NOTES: 1. All voltage values, unless otherwise noted, are with respect to the midpoint between V_{CC+} and V_{CC-}.
2. Differential voltages are at the noninverting input terminal with respect to the inverting input terminal.
3. The magnitude of the input voltage must never exceed the magnitude of the supply voltage or 15 volts, whichever is less.
4. The output may be shorted to ground or either power supply. For the TL022M only, the unlimited duration of the short-circuit applies at (or below) 125°C case temperature or 75°C free-air temperature.
5. For operation above 25°C free-air temperature, refer to Dissipation Derating Curves in Section 2. In the JG package, TL022M chips are alloy-mounted; TI022C chips are glass-mounted.

TEXAS
INSTRUMENTS
POST OFFICE BOX 225012 • DALLAS, TEXAS 75265

3

Operational Amplifiers

electrical characteristics at specified free-air temperature, V_{CC+} = 15 V, V_{CC-} = −15 V

PARAMETER		TEST CONDITIONS[†]		TL022M			TL022C			UNIT
				MIN	TYP	MAX	MIN	TYP	MAX	
V_{IO}	Input offset voltage	V_O = 0,	25°C		1	5		1	5	mV
		R_S = 50 Ω	Full range			6			7.5	
I_{IO}	Input offset current	V_O = 0	25°C		5	40		15	80	nA
			Full range			100			200	
I_{IB}	Input bias current	V_O = 0	25°C		50	100		100	250	nA
			Full range			250			400	
V_{ICR}	Common-mode input voltage range		25°C	±12	±13		±12	±13		V
			Full range	±12			±12			
V_{OPP}	Maximum peak-to-peak output voltage swing	R_L = 10 kΩ	25°C	20	26		20	26		V
		R_L ≥ 10 kΩ	Full range	20			20			
A_{VD}	Large-signal differential voltage amplication	R_L ≥ 10 kΩ,	25°C	72	86		60	80		dB
		V_O = ±10 V	Full range	72			60			
B_1	Unity-gain bandwidth		25°C		0.5			0.5		MHz
CMRR	Common-mode rejection ratio	V_{IC} = V_{ICR} min,	25°C	60	72		60	72		dB
		R_S = 50 Ω	Full range	60			60			
k_{SVS}	Supply voltage sensitivity ($\Delta V_{IO}/\Delta V_{CC}$)	V_{CC} = ±9 V to ±15 V,	25°C		30	150		30	200	µV/V
		R_S = 50 Ω	Full range			150			200	
V_n	Equivalent input noise voltage	A_{VD} = 20 dB, B = 1 Hz, f = 1 kHz	25°C		50			50		nV/√Hz
I_{OS}	Short-circuit output current		25°C		±6			±6		mA
I_{CC}	Supply current (both amplifiers)	No load,	25°C		130	200		130	250	µA
		V_O = 0	Full range			200			250	
P_D	Total dissipation (both amplifiers)	No load,	25°C		3.9	6		3.9	7.5	mW
		V_O = 0	Full range			6			7.5	

[†]All characteristics are measured under open-loop conditions with zero common-mode input voltage unless otherwise specified. Full range for TL022M is −55°C to 125°C and for TL022C is 0°C to 70°C.

operating characteristics, V_{CC+} = 15 V, V_{CC-} = −15 V, T_A = 25°C

PARAMETER		TEST CONDITIONS		TL022M			TL022C			UNIT
				MIN	TYP	MAX	MIN	TYP	MAX	
t_r	Rise time	V_I = 20 mV,	R_L = 10 kΩ,		0.3			0.3		µs
	Overshoot factor	C_L = 100 pF,	See Figure 1		5%			5%		
SR	Slew rate at unity gain	V_I = 10 V,	R_L = 10 kΩ,		0.5			0.5		V/µs
		C_L = 100 pF,	See Figure 1							

TEXAS
INSTRUMENTS
POST OFFICE BOX 225012 • DALLAS, TEXAS 75265

PARAMETER MEASUREMENT INFORMATION

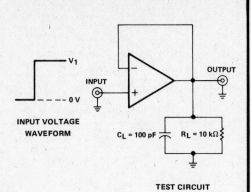

INPUT VOLTAGE
WAVEFORM

C_L = 100 pF R_L = 10 kΩ

TEST CIRCUIT

FIGURE 1–RISE TIME, OVERSHOOT FACTOR,
AND SLEW RATE

TYPICAL CHARACTERISTICS

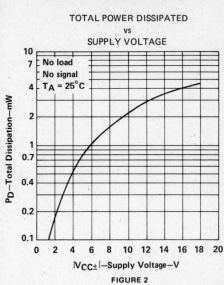

TOTAL POWER DISSIPATED
vs
SUPPLY VOLTAGE

No load
No signal
T_A = 25°C

P_D–Total Dissipation—mW

$|V_{CC\pm}|$–Supply Voltage–V

FIGURE 2

schematic

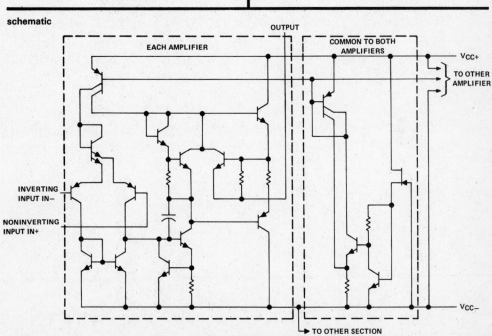

OUTPUT

EACH AMPLIFIER

COMMON TO BOTH
AMPLIFIERS

V_{CC+}

TO OTHER
AMPLIFIER

INVERTING
INPUT IN–

NONINVERTING
INPUT IN+

V_{CC-}

TO OTHER SECTION

3

Operational Amplifiers

TYPES TL044M, TL044C
QUAD LOW-POWER OPERATIONAL AMPLIFIERS

D1662, SEPTEMBER 1973—REVISED AUGUST 1983

- Very Low Power Consumption
- Typical Power Dissipation with ±2-V Supplies . . . 340 μW
- Low Input Bias and Offset Currents
- Output Short-Circuit Protection
- Low Input Offset Voltage
- Internal Frequency Compensation
- Latch-Up-Free Operation
- Power Applied in Pairs

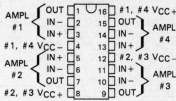

J OR N DUAL-IN-LINE
OR W FLAT PACKAGE
(TOP VIEW)

Pins 4 and 12 are internally connected together in the N package only.

description

The TL044 is a quad low-power operational amplifier designed to replace higher-power devices in many applications without sacrificing system performance. High input impedance, low supply currents, and low equivalent input noise voltage over a wide range of operating supply voltages result in an extremely versatile operational amplifier for use in a variety of analog applications including battery-operated circuits. Internal frequency compensation, absence of latch-up, high slew rate, and output short-circuit protection assure ease of use. Power may be applied separately to Section A (amplifiers 1 and 4) or Section B (amplifiers 2 and 3) while the other pair remains unpowered.

The TL044M is characterized for operation over the full military temperature range of −55°C to 125°C; the TL044C is characterized for operation from 0°C to 70°C.

TL044M . . . FH OR FK PACKAGE
(TOP VIEW)

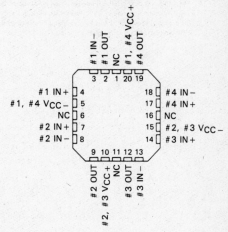

NC—No internal connection

symbol (each amplifier)

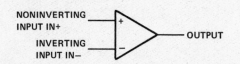

NONINVERTING
INPUT IN+

INVERTING
INPUT IN−

OUTPUT

Operational Amplifiers

3

TEXAS
INSTRUMENTS
POST OFFICE BOX 225012 • DALLAS, TEXAS 75265

absolute maximum ratings over operating free-air temperature range (unless otherwise noted)

		TL044M	TL044C	UNIT
Supply voltage V_{CC+} (see Note 1)		22	18	V
Supply voltage V_{CC-} (see Note 1)		−22	−18	V
Differential input voltage (see Note 2)		±30	±30	V
Input voltage (any input, see Notes 1 and 3)		±15	±15	V
Duration of output short-circuit (see Note 4)		unlimited	unlimited	
Continuous total dissipation at (or below) 25°C	Each amplifier	500	500	mW
free-air temperature range (see Note 5)	Total package	680	680	
Operating free-air temperature range		−55 to 125	0 to 70	°C
Storage temperature range		−65 to 150	−65 to 150	°C
Lead temperature 1,6 mm (1/16 inch) from case for 60 seconds	FH, FK, J, or W package	300	300	°C
Lead temperature 1,6 mm (1/16 inch) from case for 10 seconds	N package		260	°C

NOTES: 1. All voltage values, unless otherwise noted, are with respect to the midpoint between V_{CC+} and V_{CC-}.
2. Differential voltages are at the noninverting input terminal with respect to the inverting input terminal.
3. The magnitude of the input voltage must never exceed the magnitude of the supply voltage or 15 volts, whichever is less.
4. The output may be shorted to ground or either power supply. For the TL044M only, the unlimited duration of the short-circuit applies at (or below) 125°C case temperature or 85°C free-air temperature.
5. For operation above 25°C free-air temperature, refer to Dissipation Derating Curves in Section 2. In the J package, TL044M chips are alloy-mounted; TL044C chips are glass-mounted.

3

Operational Amplifiers

TEXAS
INSTRUMENTS
POST OFFICE BOX 225012 ● DALLAS, TEXAS 75265

electrical characteristics at specified free-air temperature, V_{CC+} = 15 V, V_{CC-} = −15 V

PARAMETER		TEST CONDITIONS[†]		TL044M			TL044C			UNIT
				MIN	TYP	MAX	MIN	TYP	MAX	
V_{IO}	Input offset voltage	V_O = 0,	25°C		1	5		1	5	mV
		R_S = 50 Ω	Full range			6			7.5	
I_{IO}	Input offset current	V_O = 0	25°C		5	40		15	80	nA
			Full range			100			200	
I_{IB}	Input bias current	V_O = 0	25°C		50	100		100	250	nA
			Full range			250			400	
V_{ICR}	Common-mode input voltage range		25°C	±12	±13		±12	±13		V
			Full range	±12			±12			
V_{OPP}	Maximum peak-to-peak output voltage swing	R_L = 10 kΩ	25°C	20	26		20	26		V
		R_L ≥ 10 kΩ	Full range	20			20			
A_{VD}	Large-signal differential voltage amplification	R_L ≥ 10 kΩ,	25°C	72	86		60	80		dB
		V_O = ±10 V	Full range	72			60			
B_1	Unity-gain bandwidth		25°C		0.5			0.5		MHz
CMRR	Common-mode rejection ratio	V_{IC} = V_{ICR} min,	25°C	60	72		60	72		dB
		V_O = 0, R_S = 50 Ω	Full range	60			60			
k_{SVS}	Supply voltage sensitivity ($\Delta V_{IO}/\Delta V_{CC}$)	V_{CC} = ±9 V to ±15 V,	25°C		30	150		30	200	μV/V
		V_O = 0, R_S = 50 Ω	Full range			150			200	
V_n	Equivalent input noise voltage	A_{VD} = 20 dB, B = 1 Hz, f = 1 kHz	25°C		50			50		nV/$\sqrt{Hz}$
I_{OS}	Short-circuit output current		25°C		±6			±6		mA
I_{CC}	Supply current (four amplifiers)	No load,	25°C		250	400		250	500	μA
		V_O = 0	Full range			400			500	
P_D	Total dissipation (four amplifiers)	No load,	25°C		7.5	12		7.5	15	mW
		V_O = 0	Full range			12			15	

[†]All characteristics are measured under open-loop conditions with zero common-mode input voltage, unless otherwise specified. Full range for TL044M is −55°C to 125°C and for TL044C is 0°C to 70°C.

operating characteristics, V_{CC+} = 15 V, V_{CC-} = −15 V, T_A = 25°C

PARAMETER		TEST CONDITIONS		TL044M			TL044C			UNIT
				MIN	TYP	MAX	MIN	TYP	MAX	
t_r	Rise time	V_I = 20 mV,	R_L = 10 kΩ,		0.3			0.3		μs
	Overshoot factor	C_L = 100 pF,	See Figure 1		5%			5%		
SR	Slew rate at unity gain	V_I = 10 V, C_L = 100 pF,	R_L = 10 kΩ, See Figure 1		0.5			0.5		V/μs

3

Operational Amplifiers

3

Operational Amplifiers

TYPES TL060, TL060A, TL060B, TL061, TL061A, TL061B, TL062, TL062A, TL062B, TL064, TL064A, TL064B
LOW-POWER JFET-INPUT OPERATIONAL AMPLIFIERS

D2392, NOVEMBER 1978—REVISED AUGUST 1983

20 DEVICES COVER MILITARY, INDUSTRIAL, AND COMMERCIAL TEMPERATURE RANGES

- Very Low Power Consumption
- Typical Supply Current . . . 200 μA (per Amplifier)
- Wide Common-Mode and Differential Voltage Ranges
- Low Input Bias and Offset Currents

- Output Short-Circuit Protection
- High Input Impedance . . . JFET-Input Stage
- Internal Frequency Compensation (Except TL060)
- Latch-Up Free Operation
- High Slew Rate . . . . 3.5 V/μs Typ

description

The JFET-input operational amplifiers of the TL061 series are designed as low-power versions of the TL081 series amplifiers. They feature high input impedance, wide bandwidth, high slew rate, and low input offset and bias currents. The TL061 series features the same terminal assignments as the TL071 and TL081 series. Each of these JFET-input operational amplifiers incorporates well-matched, high-voltage JFET and bipolar transistors in a monolithic integrated circuit.

Device types with an "M" suffix are characterized for operation over the full military temperature range of −55°C to 125°C, those with an "I" suffix are characterized for operation from −25°C to 85°C, and those with a "C" suffix are characterized for operation from 0°C to 70°C.

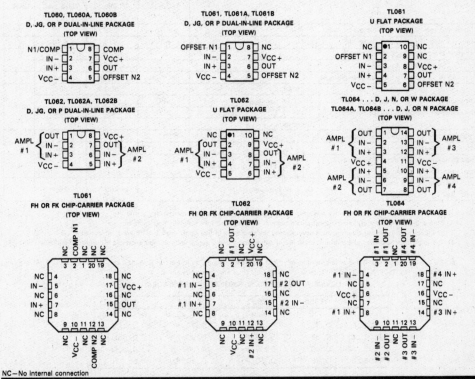

NC—No internal connection

3

Operational Amplifiers

**TEXAS
INSTRUMENTS**
POST OFFICE BOX 225012 • DALLAS, TEXAS 75265

symbol (each amplifier)

schematic (each amplifier)

NONINVERTING INPUT
IN+

OUTPUT

INVERTING INPUT
IN−

N1 N2
OFFSET NULL/COMPENSATION
TL060 AND TL061 ONLY

C1 = 10 pF ON TL061, TL062, AND TL064 ONLY
COMPONENT VALUES SHOWN ARE NOMINAL

absolute maximum ratings over operating free-air temperature range (unless otherwise noted)

		TL06_M	TL06_I	TL06_C TL06_AC TL06_BC	UNIT
Supply voltage, V_{CC+} (see Note 1)		18	18	18	V
Supply voltage, V_{CC-} (see Note 1)		−18	−18	−18	V
Differential input voltage (see Note 2)		±30	±30	±30	V
Input voltage (see Notes 1 and 3)		±15	±15	±15	V
Duration of output short circuit (see Note 4)		unlimited	unlimited	unlimited	
Continuous total dissipation at (or below) 25°C free-air temperature (see Note 5)	D package		680	680	mW
	FH or FK package	680			
	J, JG, N, P, or W package	680	680	680	
	U package	675			
Operating free-air temperature range		−55 to 125	−25 to 85	0 to 70	°C
Storage temperature range		−65 to 150	−65 to 150	−65 to 150	°C
Lead temperature 1,6 mm (1/16 inch) from case for 60 seconds	J, JG, U, FH, FK, or W package	300	300	300	°C
Lead temperature 1,6 mm (1/16 inch) from case for 10 seconds	D, N, or P package		260	260	°C

NOTES: 1. All voltage values, except differential voltages, are with respect to the midpoint between V_{CC+} and V_{CC-}.
2. Differential voltages are at the noninverting input terminal with respect to the inverting input terminal.
3. The magnitude of the input voltage must never exceed the magnitude of the supply voltage or 15 volts, whichever is less.
4. The output may be shorted to ground or to either supply. Temperature and/or supply voltages must be limited to ensure that the dissipation rating is not exceeded.
5. For operation above 25°C free-air temperature, refer to Dissipation Derating Curves, Section 2. In the J and JG packages, TL06_M chips are alloy-mounted; TL06_I, TL06_C, TL06_AC, and TL06_BC chips are glass-mounted.

DEVICE TYPES, SUFFIX VERSIONS, AND PACKAGES

	TL060	TL061	TL062	TL064
TL06_M	JG	FH, FK, JG, U	FH, FK, JG, U	FH, FK, J, W
TL06_I	D, JG, P	D, JG, P	D, JG, P	D, J, N
TL06_C	D, JG, P	D, JG, P	D, JG, P	D, J, N
TL06_AC	D, JG, P	D, JG, P	D, JG, P	D, J, N
TL06_BC	D, JG, P	D, JG, P	D, JG, P	D, J, N

TEXAS INSTRUMENTS
POST OFFICE BOX 225012 • DALLAS, TEXAS 75265

3
Operational Amplifiers

electrical characteristics, $V_{CC\pm} = \pm 15$ V (unless otherwise noted)

PARAMETER		TEST CONDITIONS[†]		TL060M TL061M TL062M			TL064M			UNIT
				MIN	TYP	MAX	MIN	TYP	MAX	
V_{IO}	Input offset voltage	$V_O = 0$, $R_S = 50\ \Omega$	$T_A = 25\,°C$		3	6		3	9	mV
			$T_A = -55\,°C$ to $125\,°C$			9			15	
α_{VIO}	Temperature coefficient of input offset voltage	$V_O = 0$, $R_S = 50\ \Omega$, $T_A = -55\,°C$ to $125\,°C$			10			10		$\mu V/°C$
I_{IO}	Input offset current[‡]	$V_O = 0$	$T_A = 25\,°C$		5	100		5	100	pA
			$T_A = -55\,°C$ to $125\,°C$			20			20	nA
I_{IB}	Input bias current	$V_O = 0$	$T_A = 25\,°C$		30	200		30	200	pA
			$T_A = -55\,°C$ to $125\,°C$			50			50	nA
V_{ICR}	Common-mode input voltage range	$T_A = 25\,°C$		± 11.5	± 12		± 11.5	± 12		V
V_{OM}	Maximum peak output voltage swing	$R_L = 10\ k\Omega$, $T_A = 25\,°C$		± 10	± 13.5		± 10	± 13.5		V
		$R_L \geq 10\ k\Omega$, $T_A = -55\,°C$ to $125\,°C$		± 10			± 10			
A_{VD}	Large-signal differential voltage amplification	$V_O = \pm 10$ V, $T_A = 25\,°C$		4	6		4	6		V/mV
		$R_L \geq 10\ k\Omega$, $T_A = -55\,°C$ to $125\,°C$		4			4			
B_1	Unity-gain bandwidth	$R_L = 10\ k\Omega$, $T_A = 25\,°C$			1			1		MHz
r_i	Input resistance	$T_A = 25\,°C$			10^{12}			10^{12}		Ω
CMRR	Common-mode rejection ratio	$V_{IC} = V_{ICR}$ min, $V_O = 0$, $R_S = 50\ \Omega$, $T_A = 25\,°C$		80	86		80	86		dB
k_{SVR}	Supply voltage rejection ratio ($\Delta V_{CC\pm}/\Delta V_{IO}$)	$V_{CC} = \pm 15$ V to ± 9 V, $V_O = 0$, $R_S = 50\ \Omega$, $T_A = 25\,°C$		80	95		80	95		dB
P_D	Total power dissipation (each amplifier)	No load, $V_O = 0$, $T_A = 25\,°C$			6	7.5		6	7.5	mW
I_{CC}	Supply current (each amplifier)	No load, $V_O = 0$, $T_A = 25\,°C$			200	250		200	250	μA
V_{o1}/V_{o2}	Crosstalk attenuation	$A_{VD} = 100$, $T_A = 25\,°C$			120			120		dB

[†]All characteristics are measured under open-loop conditions with zero common-mode voltage unless otherwise specified.

[‡]Input bias currents of a FET-input operational amplifier are normal junction reverse currents, which are temperature sensitive as shown in Figure 17. Pulse techniques must be used that will maintain the junction temperature as close to the ambient temperature as possible.

3

Operational Amplifiers

TYPES TL060, TL060A, TL060B, TL061, TL061A, TL061B, TL062, TL062A, TL062B, TL064, TL064A, TL064B, LOW-POWER JFET-INPUT OPERATIONAL AMPLIFIERS

electrical characteristics, $V_{CC}\pm$ = ±15 V (unless otherwise noted)

PARAMETER	TEST CONDITIONS[†]	TL060I TL061I TL062I TL064I MIN	TYP	MAX	TL060C TL061C TL062C TL064C MIN	TYP	MAX	TL060AC TL061AC TL062AC TL064AC MIN	TYP	MAX	TL060BC TL061BC TL062BC TL064BC MIN	TYP	MAX	UNIT
V_{IO} Input offset voltage	$V_O=0$, $R_S=50\,\Omega$, $T_A=25°C$		3	6		3	15		3	6		2	3	mV
	$T_A=$ full range			9			20			7.5			5	
α_{VIO} Temperature coefficient of input offset voltage	$V_O=0$, $R_S=50\,\Omega$, $T_A=$ full range		10			10			10			10		μV/°C
I_{IO} Input offset current‡	$V_O=0$, $T_A=25°C$		5	100		5	200		5	100		5	100	pA
	$T_A=$ full range			10			2			2			2	nA
I_{IB} Input bias current‡	$V_O=0$, $T_A=25°C$		30	200		30	200		30	200		30	200	pA
	$T_A=$ full range			20			10			7			7	nA
V_{ICR} Common-mode input voltage range	$T_A=25°C$	±11.5	±12		±11	±12		±11.5	±12		±11.5	±12		V
V_{OM} Maximum peak output voltage swing	$R_L=10\,k\Omega$, $T_A=25°C$	±10	±13.5		±10	±13.5		±10	±13.5		±10	±13.5		V
	$R_L\geq10\,k\Omega$, $T_A=$ full range	±10			±10			±10			±10			
A_{VD} Large-signal differential voltage amplification	$V_O=\pm10\,V$, $R_L=10\,k\Omega$, $T_A=25°C$	4	6		3	6		4	6		4	6		V/mV
	$T_A=$ full range	4			3			4			4			
B_1 Unity-gain bandwidth	$T_A=25°C$		1			1			1			1		MHz
r_i Input resistance	$T_A=25°C$		10^{12}			10^{12}			10^{12}			10^{12}		Ω
CMRR Common-mode rejection ratio	$V_{IC}=V_{ICR}$ min, $V_O=0$, $R_S=50\,\Omega$, $T_A=25°C$	80	86		70	86		80	86		80	86		dB
k_{SVR} Supply voltage rejection ratio ($\Delta V_{CC}\pm/\Delta V_{IO}$)	$V_{CC}=\pm15\,V$ to $\pm9\,V$, $V_O=0$, $R_S=50\,\Omega$, $T_A=25°C$	80	95		70	95		80	95		80	95		dB
P_D Total power dissipation (each amplifier)	No load, $V_O=0$, $T_A=25°C$		6	7.5		6	7.5		6	7.5		6	7.5	mW
I_{CC} Supply current (each amplifier)	No load, $V_O=0$, $T_A=25°C$		200	250		200	250		200	250		200	250	μA
V_{O1}/V_{O2} Crosstalk attenuation	$A_{VD}=100$, $T_A=25°C$		120			120			120			120		dB

†All characteristics are measured under open-loop conditions with zero common-mode voltage unless otherwise specified. Full range for T_A is −25°C to 85°C for TL06_I and 0°C to 70°C for TL06_C, TL06_AC, and TL06_BC.

‡Input bias currents of a FET-input operational amplifier are normal junction reverse currents, which are temperature sensitive as shown in Figure 17. Pulse techniques must be used that will maintain the junction temperature as close to the ambient temperature as possible.

3 Operational Amplifiers

TEXAS
INSTRUMENTS

POST OFFICE BOX 225012 • DALLAS, TEXAS 75265

TYPES TL060, TL060A, TL060B, TL061, TL061A, TL061B, TL062, TL062A, TL062B, TL064, TL064A, TL064B
LOW-POWER JFET-INPUT OPERATIONAL AMPLIFIERS

operating characteristics, $V_{CC\pm} = \pm15$ V, $T_A = 25\,°C$

	PARAMETER	TEST CONDITIONS		MIN	TYP	MAX	UNIT
SR	Slew rate at unity gain	$V_I = 10$ V,	$R_L = 10$ kΩ,	2	3.5		$V/\mu s$
		$C_L = 100$ pF,	See Figure 1				
t_r	Rise time	$V_I = 20$ mV,	$R_L = 10$ kΩ,		0.2		μs
	Overshoot factor	$C_L = 100$ pF,	See Figure 1		10%		
V_n	Equivalent input noise voltage	$R_S = 100$ Ω,	$f = 1$ kHz		42		$nV/\sqrt{Hz}$

PARAMETER MEASUREMENT INFORMATION

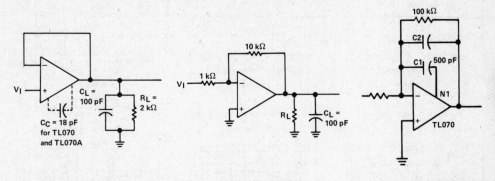

FIGURE 1—UNITY-GAIN AMPLIFIER

FIGURE 2—GAIN-OF-10
INVERTING AMPLIFIER

FIGURE 3—FEED-FORWARD
COMPENSATION

INPUT OFFSET VOLTAGE NULL CIRCUITS

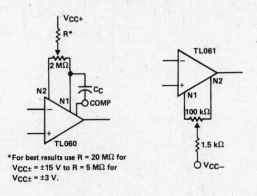

*For best results use R = 20 MΩ for
$V_{CC\pm} = \pm15$ V to R = 5 MΩ for
$V_{CC\pm} = \pm3$ V.

FIGURE 4

FIGURE 5

Operational Amplifiers

3

TYPICAL CHARACTERISTICS[†]

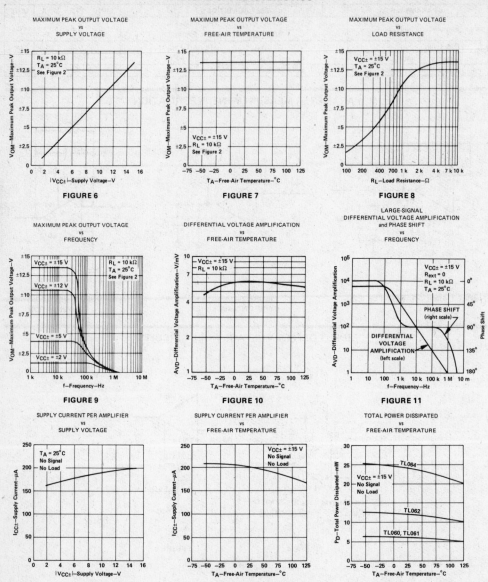

† Data at high and low temperatures are applicable only within the rated operating free-air temperature ranges of the various devices. A 10-pF compensation capacitor is used with TL060 and TL060A.

3

Operational Amplifiers

TYPICAL CHARACTERISTICS†

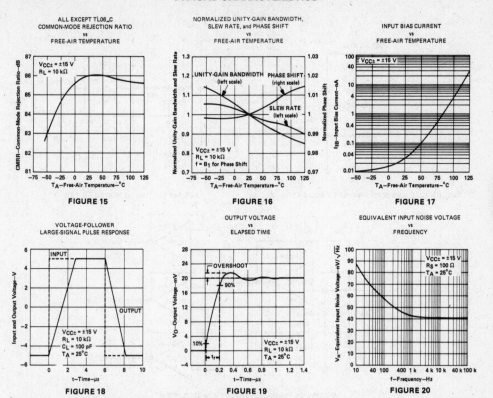

FIGURE 15

FIGURE 16

FIGURE 17

FIGURE 18

FIGURE 19

FIGURE 20

† Data at high and low temperatures are applicable only within the rated operating free-air temperature ranges of the various devices. A 10-pF compensation capacitor is used with TL060 and TL060A.

TYPICAL APPLICATION DATA

FIGURE 21—INSTRUMENTATION AMPLIFIER

TEXAS
INSTRUMENTS
POST OFFICE BOX 225012 • DALLAS, TEXAS 75265

Operational Amplifiers

3

TYPICAL APPLICATION DATA

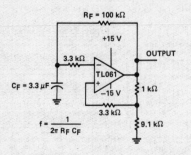

$$f = \frac{1}{2\pi\, R_F\, C_F}$$

FIGURE 22—0.5-Hz SQUARE-WAVE OSCILLATOR

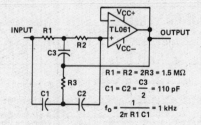

$$R1 = R2 = 2R3 = 1.5\ M\Omega$$
$$C1 = C2 = \frac{C3}{2} = 110\ pF$$
$$f_o = \frac{1}{2\pi\, R1\, C1} = 1\ kHz$$

FIGURE 23—HIGH-Q NOTCH FILTER

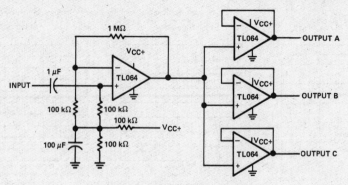

FIGURE 24—AUDIO DISTRIBUTION AMPLIFIER

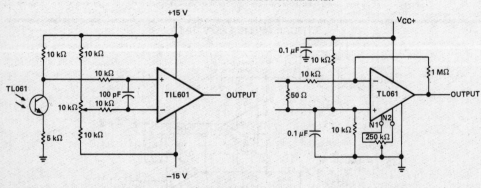

FIGURE 25—LOW-LEVEL LIGHT DETECTOR PREAMPLIFIER

FIGURE 26—AC AMPLIFIER

3

Operational Amplifiers

TEXAS
INSTRUMENTS

POST OFFICE BOX 225012 • DALLAS, TEXAS 75265

TYPICAL APPLICATION DATA

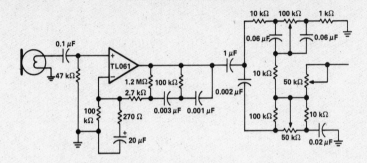

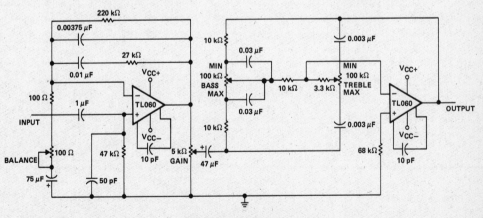

FIGURE 27—MICROPHONE PREAMPLIFIER WITH TONE CONTROL

3

Operational Amplifiers

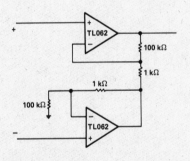

FIGURE 28—INSTRUMENTATION AMPLIFIER

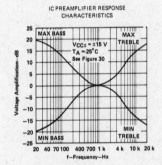

IC PREAMPLIFIER RESPONSE CHARACTERISTICS

FIGURE 29

FIGURE 30—IC PREAMPLIFIER

TEXAS INSTRUMENTS
POST OFFICE BOX 225012 • DALLAS, TEXAS 75265

TYPES TL066M, TL066I, TL066C, TL066AC, TL066BC
ADJUSTABLE LOW-POWER JFET-INPUT OPERATIONAL AMPLIFIERS

D2494, FEBRUARY 1979—REVISED AUGUST 1983

5 DEVICES COVER COMMERCIAL, INDUSTRIAL, AND MILITARY TEMPERATURE RANGES

- Very Low, Adjustable ("Programmable") Power Consumption
- Adjustable Supply Current . . . 5 μA to 200 μA
- Very Low Input Bias and Offset Currents
- Wide Supply Range . . . ±1.2 V to ±18 V
- Wide Common-Mode and Differential Voltage Range
- Output Short-Circuit Protection
- High Input Impedance . . . JFET-Input Stage
- Unity-Gain Bandwidth . . . 1 MHz Typ (100 kHz at 25 μW)
- High Slew Rate . . . 3.5 V/μs Typ
- Internal Frequency Compensation
- Latch-Up-Free Operation

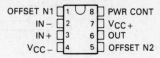

TL066M . . . JG PACKAGE
TL066I, TL066C, TL066AC, TL066BC . . . D, JG, OR P PACKAGE
(TOP VIEW)

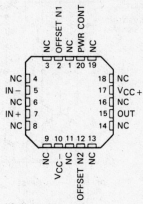

TL066M . . . FH OR FK PACKAGE
(TOP VIEW)

NC—No internal connection

description

The TL066, TL066A, and TL066B are JFET-input operational amplifiers similar to the TL061 with the additional feature of being power-adjustable. They feature very low input offset and bias currents, high input impedance, wide bandwidth, and high slew rate. The power-control feature permits the amplifiers to be adjusted to require as little as 25 microwatts of power. This type of amplifier, which provides for changing several characteristics by varying one external element, is sometimes referred to as being "programmable." The JFET-input stage combined with the adjustable-low-power feature results in superior bandwidth and slew rate performance compared to low-power bipolar-input devices.

symbol

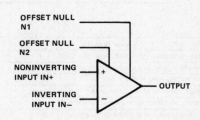

The TL066M is characterized for operation over the full military temperature range of −55°C to 125°C; the TL066I is characterized for operation from −25°C to 85°C; the TL066C, TL066AC, and TL066BC are characterized for operation from 0°C to 70°C.

Operational Amplifiers

3

**TEXAS
INSTRUMENTS**
POST OFFICE BOX 225012 • DALLAS, TEXAS 75265

TYPES TL066M, TL066I, TL066C, TL066AC, TL066BC
ADJUSTABLE LOW-POWER JFET-INPUT OPERATIONAL AMPLIFIERS

schematic

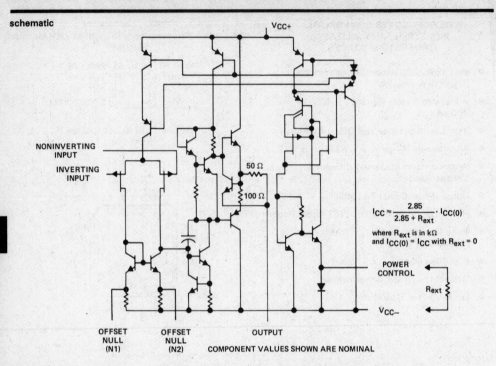

$$I_{CC} \approx \frac{2.85}{2.85 + R_{ext}} \cdot I_{CC(0)}$$

where R_{ext} is in kΩ
and $I_{CC(0)} = I_{CC}$ with $R_{ext} = 0$

COMPONENT VALUES SHOWN ARE NOMINAL

absolute maximum ratings over operating free-air temperature (unless otherwise noted)

		TL066M	TL066I	TL066C TL066AC TL066BC	UNIT
Supply voltage, V_{CC+} (see Note 1)		18	18	18	V
Supply voltage, V_{CC-} (see Note 1)		−18	−18	−18	V
Differential input voltage (see Note 2)		±30	±30	±30	V
Input voltage (see Notes 1 and 3)		±15	±15	±15	V
Voltage between power-control terminal and V_{CC-}		±0.5	±0.5	±0.5	V
Duration of output short circuit (see Note 4)		unlimited	unlimited	unlimited	
Continuous total dissipation at (or below) 25°C free-air temperature (see Note 5)		680	680	680	mW
Operating free-air temperature range		−55 to 125	−25 to 85	0 to 70	°C
Storage temperature range		−65 to 150	−65 to 150	−65 to 150	°C
Lead temperature 1,6 mm (1/16 inch) from case for 60 seconds	FH, FK, or JG package	300	300	300	°C
Lead temperature 1,6 mm (1/16 inch) from case for 10 seconds	D or P package		260	260	°C

NOTES: 1. All voltage values, except differential voltages, are with respect to the midpoint between V_{CC+} and V_{CC-}.
2. Differential voltages are at the noninverting input terminal with respect to the inverting input terminal.
3. The magnitude of the input voltage must never exceed the magnitude of the supply voltage or 15 volts, whichever is less.
4. The output may be shorted to ground or to either supply. Temperature and/or supply voltages must be limited to ensure that the dissipation rating is not exceeded.
5. For operation above 25°C free-air temperature, refer to Dissipation Derating Curves in Section 2. In the JG package, the TL066I, TL066C, TL066AC, and TL066BC chips are glass-mounted; the TL066M chips are alloy-mounted.

**TEXAS
INSTRUMENTS**

POST OFFICE BOX 225012 • DALLAS, TEXAS 75265

electrical characteristics, VCC = ± 15 V

PARAMETER	TEST CONDITIONS[†]	TL066M MIN	TYP	MAX	TL066I MIN	TYP	MAX	TL066C MIN	TYP	MAX	UNIT
V_{IO} Input offset voltage	$V_O = 0$, $R_S = 50\ \Omega$, $T_A = 25°C$		3	6		3	6		3	15	mV
	$V_O = 0$, $R_S = 50\ \Omega$, $T_A = $ full range			9			9			20	
α_{VIO} Temperature coefficient of input offset voltage	$V_O = 0$, $R_S = 50\ \Omega$, $T_A = $ full range		10			10			10		$\mu V/°C$
I_{IO} Input offset current‡	$V_O = 0$, $T_A = 25°C$		5	100		5	100		5	200	pA
	$V_O = 0$, $T_A = $ full range			20			10			5	nA
I_{IB} Input bias current‡	$V_O = 0$, $T_A = 25°C$		30	200		30	200		30	400	pA
	$V_O = 0$, $T_A = $ full range			50			20			10	nA
V_{ICR} Common-mode input voltage range	$T_A = 25°C$	±11.5			±11.5			±11			V
V_{OM} Maximum peak output voltage swing	$T_A = 25°C$, $R_L \geq 10\ k\Omega$	±10	±13.5		±10	±13.5		±10	±13.5		V
	$T_A = $ full range, $R_L \geq 10\ k\Omega$	±10	±13.5		±10	±13.5		±10	±13.5		
A_{VD} Large-signal differential voltage amplification	$R_L \geq 10\ k\Omega$, $V_O = \pm 10\ V$, $T_A = 25°C$	4	6		4	6		3	6		V/mV
	$R_L \geq 10\ k\Omega$, $V_O = \pm 10\ V$, $T_A = $ full range	4			3			3			
B_1 Unity-gain bandwidth	$T_A = 25°C$		1			1			1		MHz
r_i Input resistance	$T_A = 25°C$		10^{12}			10^{12}			10^{12}		Ω
r_o Output resistance	$T_A = 25°C$, $f = 1\ kHz$		220			220			220		Ω
CMRR Common-mode rejection ratio	$V_{IC} = V_{ICR}$ min, $V_O = 0$, $R_S = 50\ \Omega$, $T_A = 25°C$	80	86		80	86		70	76		dB
k_{SVR} Supply voltage rejection ratio ($\Delta V_{CC} \pm/\Delta V_{IO}$)	$V_{CC} = \pm 9\ V$ to $\pm 15\ V$, $V_O = 0$, $R_S = 50\ \Omega$, $T_A = 25°C$	80	95		80	95		70	95		dB
P_D Total power dissipation	$V_O = 0$, No load, $T_A = 25°C$		6	7.5		6	7.5		6	7.5	mW
I_{CC} Supply current	$V_O = 0$, No load, $T_A = 25°C$		200	250		200	250		200	250	μA

† All characteristics are measured under open-loop conditions with zero common-mode input voltage unless otherwise specified. Full range of T_A is −55°C to 125°C for TL066M; −25°C to 85°C for TL066I; and 0°C to 70°C for TL066C. The electrical parameters are measured with the power-control terminal (pin 8) connected to V_{CC}−.

‡ Input bias currents of a FET-input operational amplifier are normal junction reverse currents, which are temperature-sensitive. Pulse techniques must be used that will maintain the junction temperature as close to the ambient temperature as is possible.

3

Operational Amplifiers

TEXAS
INSTRUMENTS
POST OFFICE BOX 225012 • DALLAS, TEXAS 75265

electrical characteristics, $V_{CC} = \pm 15$ V

PARAMETER		TEST CONDITIONS†	TL066AC MIN	TL066AC TYP	TL066AC MAX	TL066BC MIN	TL066BC TYP	TL066BC MAX	UNIT
V_{IO}	Input offset voltage	$V_O = 0$, $R_S = 50\ \Omega$, $T_A = 25\,°C$		3	6		2	3	mV
		$V_O = 0$, $R_S = 50\ \Omega$, $T_A =$ full range			7.5			5	
α_{VIO}	Temperature coefficient of input offset voltage	$V_O = 0$, $R_S = 50\ \Omega$, $T_A =$ full range		10			10		$\mu V/°C$
I_{IO}	Input offset current‡	$V_O = 0$, $T_A = 25\,°C$		5	100		5	100	pA
		$V_O = 0$, $T_A =$ full range			3			3	nA
I_{IB}	Input bias current‡	$V_O = 0$, $T_A = 25\,°C$		30	200		30	200	pA
		$V_O = 0$, $T_A =$ full range			7			7	nA
V_{ICR}	Common-mode input voltage range	$T_A = 25\,°C$	± 11.5			± 11.5			V
V_{OM}	Maximum peak output voltage swing	$T_A = 25\,°C$, $R_L \geq 10\ k\Omega$	± 10	± 13.5		± 10	± 13.5		V
		$T_A =$ full range, $R_L \geq 10\ k\Omega$	± 10	± 13.5		± 10	± 13.5		
A_{VD}	Large-signal differential voltage amplification	$R_L \geq 10\ k\Omega$, $V_O = \pm 10$ V, $T_A = 25\,°C$	4	6		4	6		V/mV
		$R_L \geq 10\ k\Omega$, $V_O = \pm 10$ V, $T_A =$ full range	4			4			
B_1	Unity-gain bandwidth	$T_A = 25\,°C$, $R_L = 10\ k\Omega$		1			1		MHz
r_i	Input resistance	$T_A = 25\,°C$		10^{12}			10^{12}		Ω
r_o	Output resistance	$T_A = 25\,°C$, $f = 1$ kHz		220			220		Ω
CMRR	Common-mode rejection ratio	$V_{IC} = V_{ICR}$ min, $V_O = 0$, $R_S = 50\ \Omega$, $T_A = 25\,°C$	80	86		80	86		dB
k_{SVR}	Supply voltage rejection ratio ($\Delta V_{CC\pm}/\Delta V_{IO}$)	$V_{CC} = \pm 9$ V to ± 15 V, $V_O = 0$, $R_S = 50\ \Omega$, $T_A = 25\,°C$	80	95		80	95		dB
P_D	Total power dissipation	No load, $V_O = 0$, $T_A = 25\,°C$		6	7.5		6	7.5	mW
I_{CC}	Supply current	No load, $V_O = 0$, $T_A = 25\,°C$		200	250		200	250	μA

†All characteristics are measured under open-loop conditions with zero common-mode input voltage unless otherwise specified. Full range of $T_A = -55\,°C$ to $125\,°C$ for TL066M; $-25\,°C$ to $85\,°C$ for TL066I; and $0\,°C$ to $70\,°C$ for TL066C, TL066AC, and TL066BC. The electrical parameters are measured with the power-control terminal (pin 8) connected to V_{CC-}.

‡Input bias currents of a FET-input operational amplifier are normal junction reverse currents, which are temperature-sensitive. Pulse techniques must be used that will maintain the junction temperature as close to the ambient temperature as is possible.

3

Operational Amplifiers

Texas Instruments

POST OFFICE BOX 225012 • DALLAS, TEXAS 75265

TYPES TL066M, TL066I, TL066C, TL066AC, TL066BC
ADJUSTABLE LOW-POWER JFET-INPUT OPERATIONAL AMPLIFIERS

operating characteristics, $V_{CC\pm} = \pm15$ V, $T_A = 25°C$, $R_{ext} = 0$

	PARAMETER	TEST CONDITIONS		MIN	TYP	MAX	UNIT
SR	Slew rate at unity gain	$V_I = 10$ V,	$R_L = 10$ kΩ,	2	3.5		V/µs
		$C_L = 100$ pF,	See Figure 1				
t_r	Rise time	$V_I = 20$ mV,	$R_L = 10$ kΩ		0.2		µs
	Overshoot factor	$C_L = 100$ pF,	See Figure 1		10%		
V_n	Equivalent input noise voltage	$R_S = 100$ Ω,	$f = 1$ kHz		42		nV/$\sqrt{Hz}$

PARAMETER MEASUREMENT INFORMATION

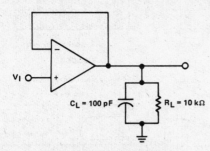

FIGURE 1—UNITY-GAIN AMPLIFIER

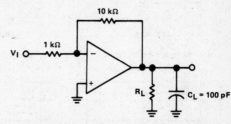

FIGURE 2—GAIN-OF-10 INVERTING AMPLIFIER

INPUT OFFSET VOLTAGE NULL CIRCUIT

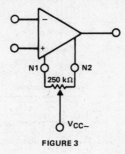

FIGURE 3

3

Operational Amplifiers

TEXAS INSTRUMENTS
POST OFFICE BOX 225012 • DALLAS, TEXAS 75265

TYPES TL066M, TL066I, TL066C, TL066AC, TL066BC
ADJUSTABLE LOW-POWER JFET-INPUT OPERATIONAL AMPLIFIERS

TYPICAL CHARACTERISTICS†

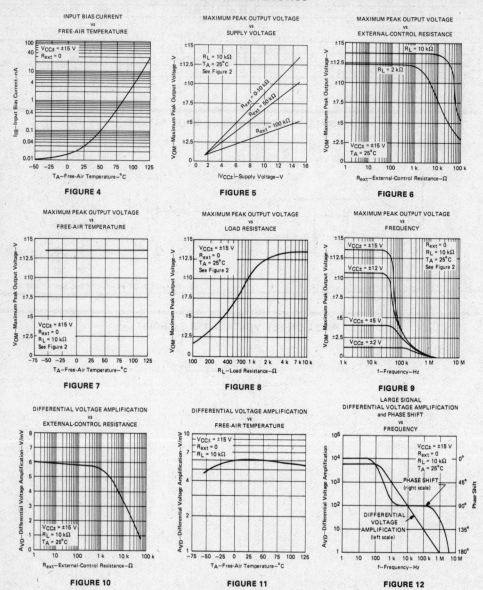

FIGURE 4 — INPUT BIAS CURRENT vs FREE-AIR TEMPERATURE

FIGURE 5 — MAXIMUM PEAK OUTPUT VOLTAGE vs SUPPLY VOLTAGE

FIGURE 6 — MAXIMUM PEAK OUTPUT VOLTAGE vs EXTERNAL-CONTROL RESISTANCE

FIGURE 7 — MAXIMUM PEAK OUTPUT VOLTAGE vs FREE-AIR TEMPERATURE

FIGURE 8 — MAXIMUM PEAK OUTPUT VOLTAGE vs LOAD RESISTANCE

FIGURE 9 — MAXIMUM PEAK OUTPUT VOLTAGE vs FREQUENCY

FIGURE 10 — DIFFERENTIAL VOLTAGE AMPLIFICATION vs EXTERNAL-CONTROL RESISTANCE

FIGURE 11 — DIFFERENTIAL VOLTAGE AMPLIFICATION vs FREE-AIR TEMPERATURE

FIGURE 12 — LARGE SIGNAL DIFFERENTIAL VOLTAGE AMPLIFICATION and PHASE SHIFT vs FREQUENCY

†Data at high and low temperatures are applicable only within the rated free-air temperature ranges of the various devices.

3

Operational Amplifiers

3-118

TEXAS
INSTRUMENTS
POST OFFICE BOX 225012 • DALLAS, TEXAS 75265

883

TYPICAL CHARACTERISTICS†

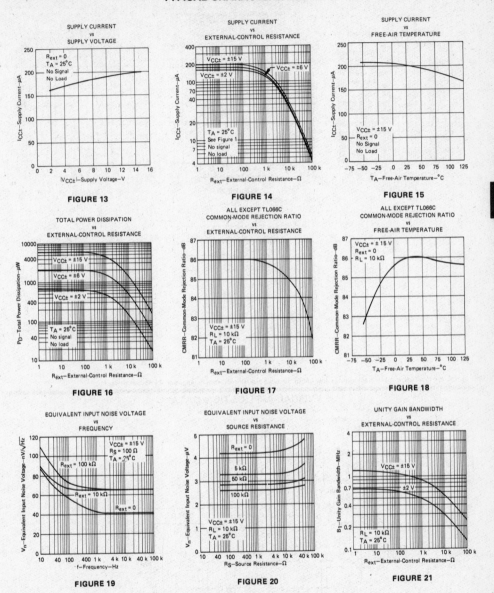

3

Operational Amplifiers

†Data at high and low temperatures are applicable only within the rated free-air temperature ranges of the various devices.

TEXAS INSTRUMENTS

POST OFFICE BOX 225012 • DALLAS, TEXAS 75265

TYPES TL066M, TL066I, TL066C, TL066AC, TL066BC
ADJUSTABLE LOW-POWER JFET-INPUT OPERATIONAL AMPLIFIERS

TYPICAL CHARACTERISTICS†

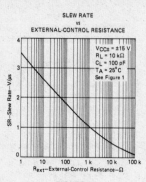

FIGURE 22

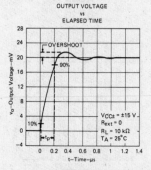

FIGURE 23

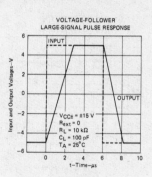

FIGURE 24

FIGURE 25

†Data at high and low temperatures are applicable only within the rated free-air temperature ranges of the various devices.

TYPICAL APPLICATION DATA

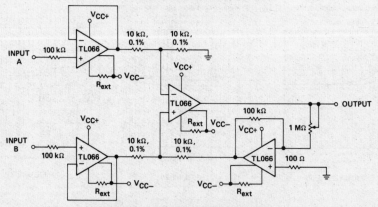

FIGURE 26—INSTRUMENTATION AMPLIFIER

3

Operational Amplifiers

TEXAS INSTRUMENTS
POST OFFICE BOX 225012 • DALLAS, TEXAS 75265

883

TYPICAL APPLICATION DATA

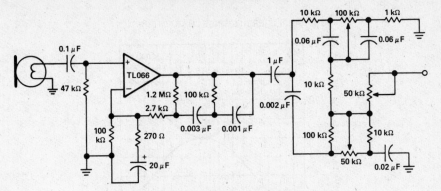

FIGURE 27—MICROPHONE PREAMPLIFIER WITH TONE CONTROL

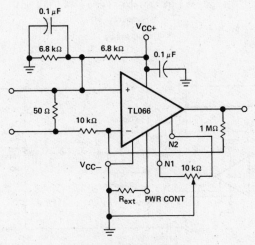

FIGURE 28—AC AMPLIFIER

3

Operational Amplifiers

TYPICAL APPLICATION DATA

IC PREAMPLIFIER RESPONSE CHARACTERISTICS

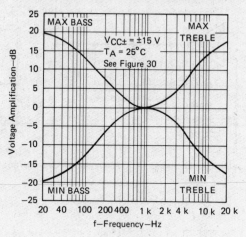

FIGURE 29

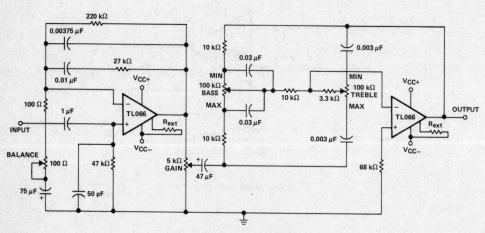

FIGURE 30—IC PREAMPLIFIER

TEXAS
INSTRUMENTS
POST OFFICE BOX 225012 • DALLAS, TEXAS 75265

**LINEAR
INTEGRATED
CIRCUITS**

**TYPE TL068C
3-PIN VOLTAGE FOLLOWER WITH JFET INPUT**

D2660, AUGUST 1983

- Standard TO-92 Package
- Supply Current .... 300 μA Max
- Wide Input/Output Voltage Range
- Low Input Bias Current
- Output Short-Circuit Protection
- High-Impedance Input . . . JFET Input Stage
- Internal Frequency Compensation
- Latch-Up-Free Operation

LP PACKAGE

(TOP VIEW)

OUTPUT
V_{EE}
INPUT

description

The TL068C is a JFET-input unity-gain amplifier featuring high input impedance, wide bandwidth, and low input bias current. A current-sourcing load such as a pull-up resistor is required for circuit operation.

The TL068C is characterized for operation over the commercial temperature range of 0°C to 70°C.

schematic

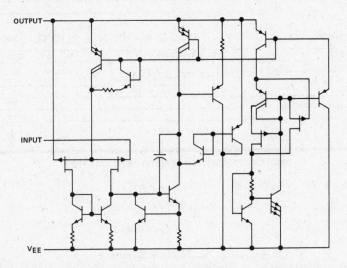

absolute maximum ratings over operating free-air temperature range (unless otherwise noted)

Voltage from output to V$_{EE}$ . . . . 36 V
Voltage from input to V$_{EE}$ . . . . 36 V
Voltage from input to output . . . . 30 V
Duration of short circuit (see Note 1) . . . . Unlimited
Continuous total dissipation at (or below) 25°C free-air temperature (see Note 2) . . . . 775 mW
Operating free-air temperature range . . . . 0°C to 70°C
Storage temperature range . . . . −65°C to 150°C
Lead temperature 1,6 mm (1/16 inch) from case for 10 seconds . . . . 260°C

NOTES: 1. The output may be shorted to any point as long as the voltage from output to V$_{EE}$ does not exceed 36 V. Temperature and/or V$_{EE}$ must be limited to ensure that the dissipation rating is not exceeded.
2. For operation above 25°C free-air temperature, refer to Dissipation Derating Curves in Section 2.

Copyright © 1983 by Texas Instruments Incorporated

POST OFFICE BOX 225012 • DALLAS, TEXAS 75265

Operational Amplifiers

3

TYPE TL068C
3-PIN VOLTAGE FOLLOWER WITH JFET INPUT

electrical characteristics, $V_{EE} = -15$ V, $V_+ = +15$ V, $T_A = 25°C$ (unless otherwise noted)

	PARAMETER	TEST CONDITIONS		MIN	TYP	MAX	UNITS
V_{IO}	Input offset voltage	$I_O = 2$ mA			3	15	mV
I_{IB}	Input bias current		$T_A = 25°C$		30	400	pA
			$T_A = 0°C$ to 70°C			10	nA
V_{IR}	Input voltage range	$V_{EE} = -15$ V, $V_+ = 15$ V, $R_L = 10$ kΩ		12 to −11.5	13.5 to −12		V
A_V	Large-signal voltage amplification	$V_{EE} = -15$V, $V_+ = 15$ V, $R_L = 10$ kΩ		0.999	0.9997		V/V
k_{SVR}	Supply voltage rejection ratio	$V_{EE} = -15$ V to +10 V		70	78		dB
k_{LCS}	Load-circuit sensitivity $(\Delta V_{IO}/\Delta I_O)$	$I_O = 0.5$ mA to 5 mA			2	4	mV/mA
I_{OS}	Short-circuit output current				25		mA
I_{EE}	Supply current				−125	−300	µA

operating characteristics, $V_{EE} = -15$ V, $V_+ = 15$ V, $T_A = 25°C$, $R_L = 10$ kΩ, $C_L = 100$ pF

PARAMETER	TEST CONDITIONS		MIN	TYP	MAX	UNITS
Bandwidth				1		MHz
Slew rate	$V_O = ±10$ V	Positive-going edge		7		V/µs
		Negative-going edge			100	
Rise time	$V_O = 100$ mV			130		ns
Overshoot				20%		

PARAMETER MEASUREMENT INFORMATION

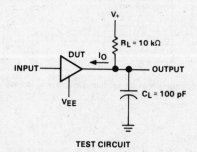

TEST CIRCUIT

3

Operational Amplifiers

TEXAS INSTRUMENTS
POST OFFICE BOX 225012 • DALLAS, TEXAS 75265

883

LINEAR
INTEGRATED
CIRCUITS

**TYPES TL070, TL070A, TL071, TL071A, TL071B,
TL072, TL072A, TL072B, TL074, TL074A, TL074B, TL075
LOW-NOISE JFET-INPUT OPERATIONAL AMPLIFIERS**

D2393, SEPTEMBER 1978—REVISED SEPTEMBER 1983

19 DEVICES COVER COMMERCIAL, INDUSTRIAL, AND MILITARY TEMPERATURE RANGES

- Low Power Consumption
- Wide Common-Mode and Differential Voltage Ranges
- Low Input Bias and Offset Currents
- Output Short-Circuit Protection
- Low Total Harmonic Distortion 0.003% Typ

- Low Noise . . . Vn = 18 nV$\sqrt{\text{Hz}}$ Typ
- High Input Impedance . . . JFET-Input Stage
- Internal Frequency Compensation (Except TL070, TL070A)
- Latch-Up-Free Operation
- High Slew Rate . . . 13 V/μs Typ

description

The JFET-input operational amplifiers on the TL07_ series are designed as low-noise versions of the TL08_ series amplifiers with low input bias and offset currents and fast slew rate. The low harmonic distortion and low noise make the TL07_ series ideally suited as amplifiers for high-fidelity and audio preamplifier applications. Each amplifier features JFET-inputs (for high input impedance) coupled with bipolar output stages all integrated on a single monolithic chip.

Device types with an "M" suffix are characterized for operation over the full military temperature range of −55 °C to 125 °C, those with an "I" suffix are characterized for operation from −25 °C to 85 °C, and those with a "C" suffix are characterized for operation from 0 °C to 70 °C.

TL070, TL070A
D, JG, OR P DUAL-IN-LINE PACKAGE
(TOP VIEW)

N1/COMP	1	8	COMP
IN−	2	7	VCC+
IN+	3	6	OUT
VCC−	4	5	OFFSET N2

TL071, TL071A, TL071B
D, JG, OR P DUAL-IN-LINE PACKAGE
(TOP VIEW)

N1/OFFSET	1	8	NC
IN−	2	7	VCC+
IN+	3	6	OUT
VCC−	4	5	OFFSET N2

TL072, TL072A, TL072B
D, JG, OR P DUAL-IN-LINE PACKAGE
(TOP VIEW)

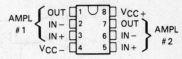

TL074, TL074A, TL074B
D, J, OR N DUAL-IN-LINE
OR W FLAT PACKAGE
(TOP VIEW)

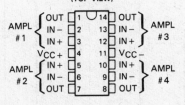

TL075
N DUAL-IN-LINE PACKAGE
(TOP VIEW)

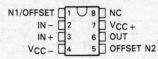

NC—No internal connection

Operational Amplifiers

3

TYPES TL070, TL070A, TL071, TL071A, TL071B,
TL072, TL072A, TL072B, TL074, TL074A, TL074B, TL075
LOW-NOISE JFET-INPUT OPERATIONAL AMPLIFIERS

TL071
FH OR FK CHIP-CARRIER PACKAGE
(TOP VIEW)

TL072
FH OR FK CHIP-CARRIER PACKAGE
(TOP VIEW)

TL074
FH OR FK CHIP-CARRIER PACKAGE
(TOP VIEW)

NC—No internal connection

schematic (each amplifier)

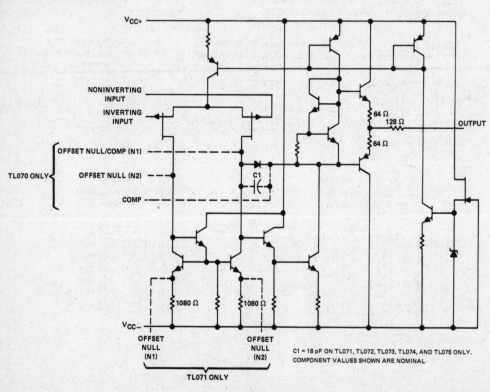

C1 = 18 pF ON TL071, TL072, TL073, TL074, AND TL075 ONLY.
COMPONENT VALUES SHOWN ARE NOMINAL

Texas
Instruments
POST OFFICE BOX 225012 • DALLAS, TEXAS 75265

symbols

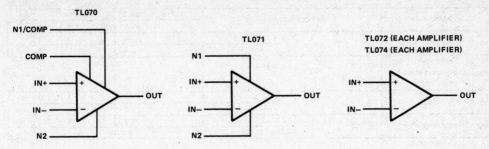

DEVICE TYPES, SUFFIX VERSIONS, AND PACKAGES					
	TL070	TL071	TL072	TL074	TL075
TL07_M	*	FH, FK, JG	FH, FK, JG	FH, FK, J, W	*
TL07_I	D, JG, P	D, JG, P	D, JG, P	D, J, N	*
TL07_C	D, JG, P	D, JG, P	D, JG, P	D, J, N	N
TL07_AC	D, JG, P	D, JG, P	D, JG, P	D, J, N	*
TL07_BC	*	D, JG, P	D, JG, P	D, J, N	*

*These combinations are not defined by this data sheet.

absolute maximum ratings over operating free-air temperature range (unless otherwise noted)

		TL07_M	TL07_I	TL07_C TL07_AC TL07_BC	UNIT
Supply voltage, V_{CC+} (see Note 1)		18	18	18	V
Supply voltage, V_{CC-} (see Note 1)		-18	-18	-18	V
Differential input voltage (see Note 2)		±30	±30	±30	V
Input voltage (see Notes 1 and 3)		±15	±15	±15	V
Duration of output short circuit (see Note 4)		unlimited	unlimited	unlimited	
Continuous total dissipation at (or below) 25°C free-air temperature (see Note 5)		680	680	680	mW
Operating free-air temperature range		-55 to 125	-25 to 85	0 to 70	°C
Storage temperature range		-65 to 150	-65 to 150	-65 to 150	°C
Lead temperature 1,6 mm (1/16 inch) from case for 60 seconds	J, JG, JH, FK, or W package	300	300	300	°C
Lead temperature 1,6 mm (1/16 inch) from case for 10 seconds	D, N, or P package		260	260	°C

NOTES: 1. All voltage values, except differential voltages, are with respect to the midpoint between V_{CC+} and V_{CC-}.
      2. Differential voltages are at the noninverting input terminal with respect to the inverting input terminal.
      3. The magnitude of the input voltage must never exceed the magnitude of the supply voltage or 15 volts, whichever is less.
      4. The output may be shorted to ground or to either supply. Temperature and/or supply voltages must be limited to ensure that the dissipation rating is not exceeded.
      5. For operation above 25°C free-air temperature, refer to Dissipation Derating Curves, Section 2. In the J and JG packages, TL07_M chips are alloy-mounted; TL07_I, TL07_C, TL07_AC, and TL07_BC chips are glass mounted.

3

Operational Amplifiers

electrical characteristics, $V_{CC\pm} = \pm 15$ V (unless otherwise noted)

PARAMETER		TEST CONDITIONS[†]		TL071M, TL072M			TL074M			UNIT
				MIN	TYP	MAX	MIN	TYP	MAX	
V_{IO}	Input offset voltage	$V_O = 0$, $R_S = 50\ \Omega$	$T_A = 25°C$		3	6		3	9	mV
			$T_A = -55°C$ to $125°C$			9			15	
α_{VIO}	Temperature coefficient of input offset voltage	$V_O = 0$, $R_S = 50\ \Omega$, $T_A = -55°$ to $125°C$			10			10		$\mu V/°C$
I_{IO}	Input offset current[‡]	$V_O = 0$	$T_A = 25°C$		5	100		5	100	pA
			$T_A = -55°C$ to $125°C$			20			20	nA
I_{IB}	Input bias current[‡]	$V_O = 0$	$T_A = 25°C$		30	200		30	200	pA
			$T_A = -55°C$ to $125°C$			50			20	nA
V_{ICR}	Common-mode input voltage range	$T_A = 25°C$		± 11	± 12		± 11	± 12		V
V_{OM}	Maximum peak output voltage swing	$T_A = 25°C$, $R_L = 10\ k\Omega$		± 12	± 13.5		± 12	± 13.5		V
		$T_A = -55°C$ to $125°C$	$R_L \geq 10\ k\Omega$	± 12			± 12			
			$R_L \geq 2\ k\Omega$	± 10	± 12		± 10	± 12		
A_{VD}	Large-signal differential voltage amplification	$V_O = \pm 10$ V, $R_L \geq 2\ k\Omega$, $T_A = 25°C$		35	200		35	200		V/mV
		$V_O = \pm 10$ V, $R_L \geq 2\ k\Omega$, $T_A = -55°C$ to $125°C$		15			15			
B_1	Unity-gain bandwidth	$T_A = 25°C$			3			3		MHz
r_i	Input resistance	$T_A = 25°C$			10^{12}			10^{12}		Ω
CMRR	Common-mode rejection ratio	$V_{IC} = V_{ICR}$ min, $V_O = 0$, $R_S = 50\ \Omega$, $T_A = 25°C$		80	86		80	86		dB
k_{SVR}	Supply voltage rejection ratio ($\Delta V_{CC\pm}/\Delta V_{IO}$)	$V_{CC} = \pm 15$ V to ± 9 V, $V_O = 0$, $R_S = 50\ \Omega$, $T_A = 25°C$		80	86		80	86		dB
I_{CC}	Supply current (per amplifier)	No load, $V_O = 0$, $T_A = 25°C$			1.4	2.5		1.4	2.5	mA
V_{o1}/V_{o2}	Crosstalk attenuation	$A_{VD} = 100$, $T_A = 25°C$			120			120		dB

[†]All characteristics are measured under open-loop conditions with zero common-mode input voltage unless otherwise specified.

[‡]Input bias currents of a FET-input operational amplifier are normal junction reverse currents, which are temperature sensitive as shown in Figure 18. Pulse techniques must be used that will maintain the junction temperatures as close to the ambient temperature as is possible.

3

Operational Amplifiers

TEXAS
INSTRUMENTS
POST OFFICE BOX 225012 • DALLAS, TEXAS 75265

3

Operational Amplifiers

electrical characteristics, VCC± = ±15 V (unless otherwise noted)

PARAMETER		TEST CONDITIONS†	TL070I TL071I TL072I TL074I			TL070C TL071C TL072C TL075C			TL070AC TL071AC TL072AC TL074AC			TL071BC TL072BC TL074BC			UNIT
			MIN	TYP	MAX	MIN	TYP	MAX	MIN	TYP	MAX	MIN	TYP	MAX	
V_{IO}	Input offset voltage	$V_O = 0$, $R_S = 50\,\Omega$, $T_A = 25°C$		3	6		3	10		3	6		2	3	mV
		$V_O = 0$, $R_S = 50\,\Omega$, $T_A =$ full range			8			13			7.5			5	
α_{VIO}	Temperature coefficient of input offset voltage	$R_S = 50\,\Omega$, $T_A =$ full range		10			10			10			10		$\mu V/°C$
I_{IO}	Input offset current‡	$V_O = 0$, $T_A = 25°C$		5	100		5	100		5	100		5	100	pA
		$V_O = 0$, $T_A =$ full range			10			2			2			2	nA
I_{IB}	Input bias current‡	$V_O = 0$, $T_A = 25°C$		30	200		30	200		30	200		30	200	pA
		$V_O = 0$, $T_A =$ full range			20			7			7			7	nA
V_{ICR}	Common-mode input voltage range	$T_A = 25°C$	±11	±12		±11	±12		±11	±12		±11	±12		V
V_{OM}	Maximum peak output voltage swing	$T_A = 25°C$, $R_L = 10\,k\Omega$	±12	±13.5		±12	±13.5		±12	±13.5		±12	±13.5		V
		$R_L \geq 10\,k\Omega$, $T_A =$ full range	±12			±12			±12			±12			
		$R_L \geq 2\,k\Omega$, $T_A = 25°C$	±10	±12		±10	±12		±10	±12		±10	±12		
A_{VD}	Large-signal differential voltage amplification	$V_O = \pm10\,V$, $R_L \geq 2\,k\Omega$, $T_A = 25°C$	50	200		25	200		50	200		50	200		V/mV
		$V_O = \pm10\,V$, $R_L \geq 2\,k\Omega$, $T_A =$ full range	25			15			25			25			
B_1	Unity-gain bandwidth	$T_A = 25°C$		3			3			3			3		MHz
r_i	Input resistance	$T_A = 25°C$		10^{12}			10^{12}			10^{12}			10^{12}		Ω
CMRR	Common-mode rejection ratio	$V_{IC} = V_{ICR}$ min, $V_O = 0$, $R_S = 50\,\Omega$, $T_A = 25°C$	80	86		70	86		80	86		80	86		dB
k_{SVR}	Supply voltage rejection ratio ($\Delta V_{CC}\pm/\Delta V_{IO}$)	$V_{CC} = \pm15\,V$ to $\pm9\,V$, $V_O = 0$, $R_S = 50\,\Omega$, $T_A = 25°C$	80	86		70	86		80	86		80	86		dB
I_{CC}	Supply current (per amplifier)	No load, $V_O = 0$, $T_A = 25°C$		1.4	2.5		1.4	2.5		1.4	2.5		1.4	2.5	mA
V_{o1}/V_{o2}	Crosstalk attenuation	$A_{VD} = 100$, $T_A = 25°C$		120			120			120			120		dB

†All characteristics are measured under open-loop conditions with zero common-mode input voltage unless otherwise specified. Full range for T_A is 25°C to 85°C for TL07_I and 0°C to 70°C for TL07_C, TL07_AC, and TL07_BC.
‡Input bias currents of a FET-input operational amplifier are normal junction reverse currents, which are temperature sensitive as shown in Figure 18. Pulse techniques must be used that will maintain the junction temperatures as close to the ambient temperature as is possible.

TYPES TL070, TL070A, TL071, TL071A, TL071B, TL072, TL072A, TL072B, TL074, TL074A, TL074B, TL075
LOW-NOISE JFET-INPUT OPERATIONAL AMPLIFIERS

operating characteristics, $V_{CC\pm} = \pm15$ V, $T_A = 25\,^\circ$C

	PARAMETER	TEST CONDITIONS		TL07_M MIN	TL07_M TYP	TL07_M MAX	ALL OTHERS MIN	ALL OTHERS TYP	ALL OTHERS MAX	UNIT
SR	Slew rate at unity gain	$V_I = 10$ V, $C_L = 100$ pF,	$R_L = 2$ kΩ, See Figure 1	10	13		8	13		V/μs
t_r	Rise time	$V_I = 20$ mV, $C_L = 100$ pF,	$R_L = 2$ kΩ, See Figure 1		0.1			0.1		μs
	Overshoot factor				10			10		%
V_n	Equivalent input noise voltage	$R_S = 100$ Ω	f = 1 kHz		18			18		nV/$\sqrt{Hz}$
			f = 10 Hz to 10 kHz		4			4		μV
I_n	Equivalent input noise current	$R_S = 100$ Ω,	f = 1 kHz		0.01			0.01		pA/$\sqrt{Hz}$
THD	Total harmonic distortion	$V_{O(rms)} = 10$ V, $R_S \leq 1$ kΩ, $R_L \geq 2$ kΩ,	f = 1 kHz		0.003			0.003		%

PARAMETER MEASUREMENT INFORMATION

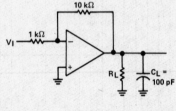

FIGURE 1—UNITY-GAIN AMPLIFIER   FIGURE 2—GAIN-OF-10 INVERTING AMPLIFIER

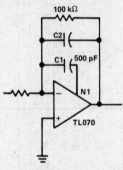

FIGURE 3—FEED-FORWARD COMPENSATION

INPUT OFFSET VOLTAGE NULL CIRCUITS

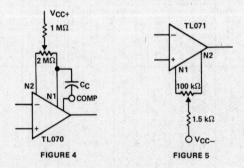

FIGURE 4   FIGURE 5

TYPICAL CHARACTERISTICS[†]

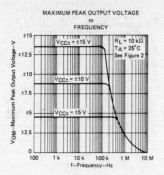

FIGURE 6

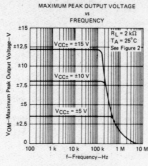

FIGURE 7

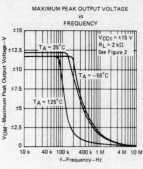

FIGURE 8

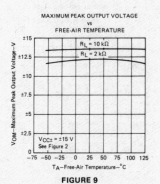

FIGURE 9

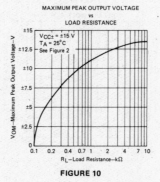

FIGURE 10

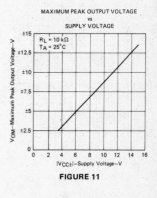

FIGURE 11

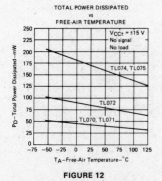

FIGURE 12

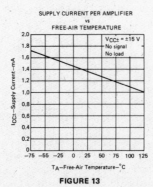

FIGURE 13

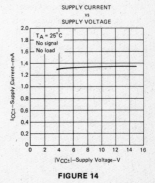

FIGURE 14

[†] Data at high and low temperatures are applicable only within the rated operating free-air temperature ranges of the various devices. A 18-pF compensation capacitor is used with TL070 and TL070A.

Operational Amplifiers — 3

TYPICAL CHARACTERISTICS†

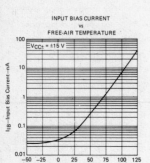

LARGE-SIGNAL
DIFFERENTIAL VOLTAGE AMPLIFICATION
vs
FREE-AIR TEMPERATURE

FIGURE 15

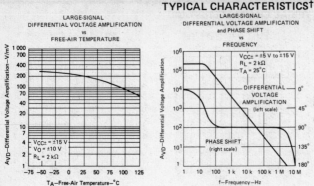

LARGE-SIGNAL
DIFFERENTIAL VOLTAGE AMPLIFICATION
and PHASE SHIFT
vs
FREQUENCY

FIGURE 16

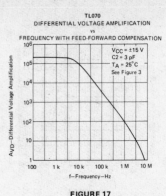

TL070
DIFFERENTIAL VOLTAGE AMPLIFICATION
vs
FREQUENCY WITH FEED-FORWARD COMPENSATION

FIGURE 17

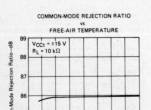

INPUT BIAS CURRENT
vs
FREE-AIR TEMPERATURE

FIGURE 18

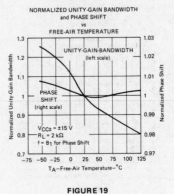

NORMALIZED UNITY-GAIN BANDWIDTH
and PHASE SHIFT
vs
FREE-AIR TEMPERATURE

FIGURE 19

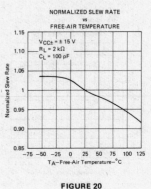

NORMALIZED SLEW RATE
vs
FREE-AIR TEMPERATURE

FIGURE 20

COMMON-MODE REJECTION RATIO
vs
FREE-AIR TEMPERATURE

FIGURE 21

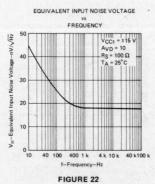

EQUIVALENT INPUT NOISE VOLTAGE
vs
FREQUENCY

FIGURE 22

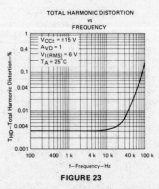

TOTAL HARMONIC DISTORTION
vs
FREQUENCY

FIGURE 23

† Data at high and low temperatures are applicable only with the rated operating free-air temperature ranges of the vaious devices. A 18-pF compensation capacitor is used with TL070 and TL070A.

TEXAS
INSTRUMENTS
POST OFFICE BOX 225012 • DALLAS, TEXAS 75265

Operational Amplifiers

3

TYPICAL CHARACTERISTICS

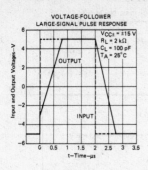

VOLTAGE-FOLLOWER
LARGE-SIGNAL PULSE RESPONSE

$V_{CC\pm} = \pm15\ V$
$R_L = 2\ k\Omega$
$C_L = 100\ pF$
$T_A = 25°C$

FIGURE 24

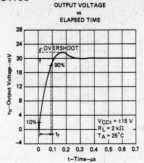

OUTPUT VOLTAGE
vs
ELAPSED TIME

$V_{CC\pm} = \pm15\ V$
$R_L = 2\ k\Omega$
$T_A = 25°C$

FIGURE 25

3

Operational Amplifiers

TYPICAL APPLICATION DATA

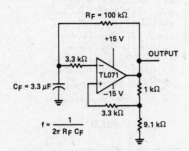

$R_F = 100\ k\Omega$

$3.3\ k\Omega$
$C_F = 3.3\ \mu F$
TL071
+15 V
−15 V
$3.3\ k\Omega$
OUTPUT
$1\ k\Omega$
$9.1\ k\Omega$

$$f = \frac{1}{2\pi\ R_F\ C_F}$$

FIGURE 26—0.5-Hz SQUARE-WAVE OSCILLATOR

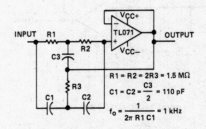

INPUT R1
V_{CC+}
TL071
V_{CC-}
R2
C3
R3
C1 C2
OUTPUT

$R1 = R2 = 2R3 = 1.5\ M\Omega$

$$C1 = C2 = \frac{C3}{2} = 110\ pF$$

$$f_o = \frac{1}{2\pi\ R1\ C1} = 1\ kHz$$

FIGURE 27—HIGH-Q NOTCH FILTER

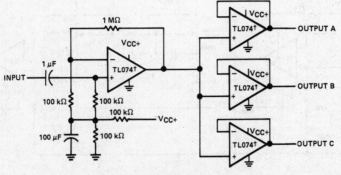

$1\ M\Omega$
V_{CC+}
$1\ \mu F$
INPUT
TL074†
$100\ k\Omega$
$100\ k\Omega$
$100\ k\Omega$
$100\ \mu F$
$100\ k\Omega$
$-V_{CC+}$

V_{CC+}
TL074†
OUTPUT A

V_{CC+}
TL074†
OUTPUT B

V_{CC+}
TL074†
OUTPUT C

† or TL075

FIGURE 28—AUDIO DISTRIBUTION AMPLIFIER

TYPICAL APPLICATION DATA

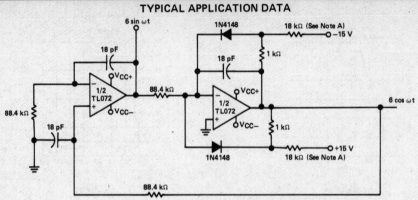

Note A: These resistor values may be adjusted for a symmetrical output.

FIGURE 29—100-KHz QUADRATURE OSCILLATOR

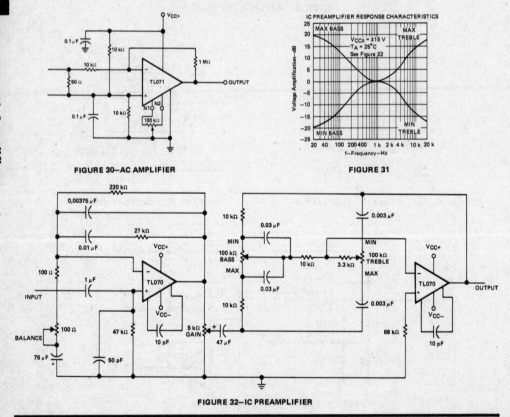

FIGURE 30—AC AMPLIFIER

IC PREAMPLIFIER RESPONSE CHARACTERISTICS

$V_{CC\pm} = \pm15$ V
$T_A = 25°C$
See Figure 32

FIGURE 31

FIGURE 32—IC PREAMPLIFIER

TEXAS
INSTRUMENTS
POST OFFICE BOX 225012 • DALLAS, TEXAS 75265

3

Operational Amplifiers

**LINEAR
INTEGRATED
CIRCUITS**

**TYPES TL080 THRU TL085, TL080A THRU TL084A
TL081B, TL082B, TL084B
JFET-INPUT OPERATIONAL AMPLIFIERS**
D2297, FEBRUARY 1977–REVISED SEPTEMBER 1983

24 DEVICES COVER MILITARY, INDUSTRIAL AND COMMERCIAL TEMPERATURE RANGES

- Low-Power Consumption
- Wide Common-Mode and Differential Voltage Ranges
- Low Input Bias and Offset Currents
- Output Short-Circuit Protection
- Low Total Harmonic Distortion . . . 0.003% TYP

- High Input Impedance . . . JFET-Input Stage
- Internal Frequency Compensation (Except TL080, TL080A)
- Latch-Up-Free Operation
- High Slew Rate . . . 13 V/μs Typ

**TL080, TL080A
JG OR P DUAL-IN-LINE PACKAGE
(TOP VIEW)**

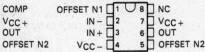

N1/COMP	1	8	COMP
IN−	2	7	VCC+
IN+	3	6	OUT
VCC−	4	5	OFFSET N2

**TL081, TL081A, TL081B
JG OR P DUAL-IN-LINE PACKAGE
(TOP VIEW)**

OFFSET N1	1	8	NC
IN−	2	7	VCC+
IN+	3	6	OUT
VCC−	4	5	OFFSET N2

**TL082, TL082A, TL082B
JG OR P DUAL-IN-LINE PACKAGE
(TOP VIEW)**

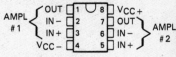

AMPL #1 { OUT | 1 | 8 | VCC+
IN− | 2 | 7 | OUT
IN+ | 3 | 6 | IN− } AMPL #2
VCC− | 4 | 5 | IN+

**TL081M . . . FH OR FK
CHIP CARRIER PACKAGE
(TOP VIEW)**

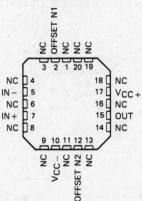

**TL082M . . . FH OR FK
CHIP CARRIER PACKAGE
(TOP VIEW)**

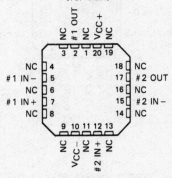

NC—No internal connection

3

Operational Amplifiers

DEVICE TYPES, SUFFIX VERSIONS, AND PACKAGES

	TL080	TL081	TL082	TL083	TL084	TL085
TL08_M	JG	FH, FK, JG	FH, FK, JG	FH, FK, J	FH, FK, J, W	*
TL08_I	JG, P	JG, P	JG, P	J, N	J, N	*
TL08_C	JG, P	JG, P	JG, P	J, N	J, N	N
TL08_AC	JG, P	JG, P	JG, P	J, N	J, N	*
TL08_BC	*	JG, P	JG, P	*	J, N	*

*These combinations are not defined by this data sheet.

**TEXAS
INSTRUMENTS**
POST OFFICE BOX 225012 • DALLAS, TEXAS 75265

TYPES TL080 THRU TL085, TL080A THRU TL084A
TL081B, TL082B, TL084B
JFET-INPUT OPERATIONAL AMPLIFIERS

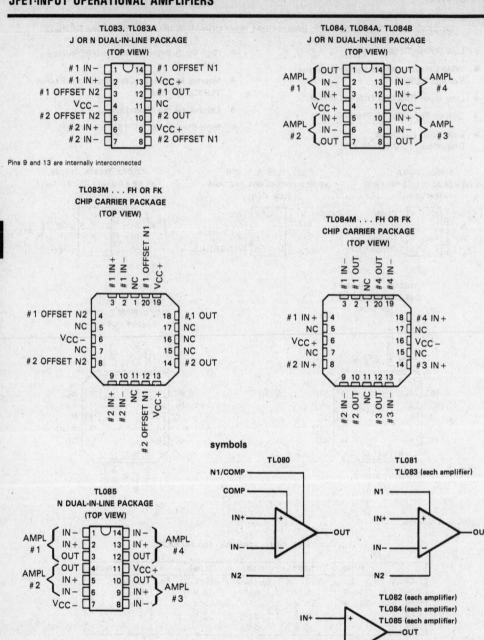

TL083, TL083A
J OR N DUAL-IN-LINE PACKAGE
(TOP VIEW)

#1 IN−	1	14	#1 OFFSET N1
#1 IN+	2	13	V_{CC}+
#1 OFFSET N2	3	12	#1 OUT
V_{CC}−	4	11	NC
#2 OFFSET N2	5	10	#2 OUT
#2 IN+	6	9	V_{CC}+
#2 IN−	7	8	#2 OFFSET N1

Pins 9 and 13 are internally interconnected

TL084, TL084A, TL084B
J OR N DUAL-IN-LINE PACKAGE
(TOP VIEW)

AMPL #1: OUT (1), IN− (2), IN+ (3)
V_{CC}+ (4)
AMPL #2: IN+ (5), IN− (6), OUT (7)
AMPL #4: OUT (14), IN− (13), IN+ (12)
V_{CC}− (11)
AMPL #3: IN+ (10), IN− (9), OUT (8)

TL083M . . . FH OR FK
CHIP CARRIER PACKAGE
(TOP VIEW)

TL084M . . . FH OR FK
CHIP CARRIER PACKAGE
(TOP VIEW)

TL085
N DUAL-IN-LINE PACKAGE
(TOP VIEW)

AMPL #1: IN− (1), IN+ (2), OUT (3)
AMPL #2: OUT (4), IN+ (5), IN− (6)
V_{CC}− (7)
AMPL #4: IN− (14), IN+ (13), OUT (12)
V_{CC}+ (11)
AMPL #3: OUT (10), IN+ (9), IN− (8)

symbols

TL080
N1/COMP
COMP
IN+
IN−
N2
OUT

TL081
TL083 (each amplifier)
N1
IN+
IN−
N2
OUT

TL082 (each amplifier)
TL084 (each amplifier)
TL085 (each amplifier)
IN+
IN−
OUT

NC—No internal connection

TEXAS
INSTRUMENTS
POST OFFICE BOX 225012 • DALLAS, TEXAS 75265

3

Operational Amplifiers

description

The TL08_ JFET-input operational amplifier family is designed to offer a wider selection than any previously developed operational amplifier family. Each of these JFET-input operational amplifiers incorporates well-matched, high-voltage JFET and bipolar transistors in a monolithic integrated circuit. The devices feature high slew rates, low input bias and offset currents, and low offset voltage temperature coefficient. Offset adjustment and external compensation options are available within the TL08_ family.

Device types with an "M" suffix are characterized for operation over the full military temperature range of −55 °C to 125 °C, those with an "I" suffix are characterized for operation from −25 °C to 85 °C, and those with a "C" suffix are characterized for operation from 0 °C to 70 °C.

schematic (each amplifier)

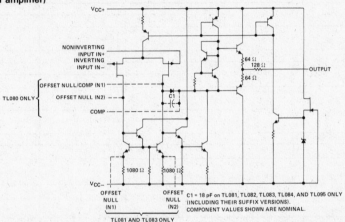

$C1 = 18$ pF on TL081, TL082, TL083, TL084, AND TL095 ONLY (INCLUDING THEIR SUFFIX VERSIONS). COMPONENT VALUES SHOWN ARE NOMINAL.

absolute maximum ratings over operating free-air temperature range (unless otherwise noted)

		TL08_M	TL08_I	TL08_C TL08_AC TL08_BC	UNIT
Supply voltage, V_{CC+} (see Note 1)		18	18	18	V
Supply voltage, V_{CC-} (see Note 1)		−18	−18	−18	V
Differential input voltage (see Note 2)		±30	±30	±30	V
Input voltage (see Notes 1 and 3)		±15	±15	±15	V
Duration of output short circuit (see Note 4)		unlimited	unlimited	unlimited	
Continuous total dissipation at (or below) 25 °C free-air temperature (see Note 5)		680	680	680	mW
Operating free-air temperature range		−55 to 125	−25 to 85	0 to 70	°C
Storage temperature range		−65 to 150	−65 to 150	−65 to 150	°C
Lead temperature 1,6 mm (1/16 inch) from case for 60 seconds	FH, FK, J, JG, or W package	300	300	300	°C
Lead temperature 1,6 mm (1/16 inch) from case for 10 seconds	N or P package		260	260	°C

NOTES: 1. All voltage values, except differential voltages, are with respect to the midpoint between V_{CC+} and V_{CC-}.
2. Differential voltages are at the noninverting input terminal with respect to the inverting input terminal.
3. The magnitude of the input voltage must never exceed the magnitude of the supply voltage or 15 volts, whichever is less.
4. The output may be shorted to ground or to either supply. Temperature and/or supply voltages must be limited to ensure that the dissipation rating is not exceeded.
5. For operation above 25 °C free-air temperature, refer to Dissipation Derating Curves in Section 2. In the J and JG packages, TL08_M chips are alloy-mounted; TL08_I, TL08_C, TL08_AC, and TL08_BC chips are glass-mounted.

3

Operational Amplifiers

electrical characteristics, $V_{CC\pm} = \pm15$ V (unless otherwise noted)

PARAMETER		TEST CONDITIONS[†]		TL080M, TL081M TL082M, TL083M			TL084M			UNIT
				MIN	TYP	MAX	MIN	TYP	MAX	
V_{IO}	Input offset voltage	$V_O = 0$, $R_S = 50\ \Omega$	$T_A = 25°C$		3	6		3	9	mV
			$T_A = -55°C$ to $125°C$			9			15	
α_{VIO}	Temperature coefficient of input offset voltage	$V_O = 0$, $T_A = -55°C$ to $125°C$	$R_S = 50\ \Omega$,		10			10		$\mu V/°C$
I_{IO}	Input offset current[‡]	$V_O = 0$	$T_A = 25°C$		5	100		5	100	pA
			$T_A = -55°C$ to $125°C$			20			20	nA
I_{IB}	Input bias current[‡]	$V_O = 0$	$T_A = 25°C$		30	200		30	200	pA
			$T_A = -55°C$ to $125°C$			50			20	nA
V_{ICR}	Common-mode input voltage range	$T_A = 25°C$		±11	±12		±11	±12		V
V_{OM}	Maximum peak output voltage swing	$T_A = 25°C$,	$R_L = 10\ k\Omega$	±12	±13.5		±12	±13.5		V
		$T_A = -55°C$ to $125°C$	$R_L \geq 10\ k\Omega$	±12			±12			
			$R_L \geq 2\ k\Omega$	±10	±12		±10	±12		
A_{VD}	Large-signal differential voltage amplification	$V_O = \pm10$ V, $T_A = 25°C$	$R_L \geq 2\ k\Omega$,	25	200		25	200		V/mV
		$V_O = \pm10$ V, $T_A = -55°C$ to $125°C$	$R_L \geq 2\ k\Omega$,	15			15			
B_1	Unity-gain bandwidth	$T_A = 25°C$			3			3		MHz
r_i	Input resistance	$T_A = 25°C$			10^{12}			10^{12}		Ω
CMRR	Common-mode rejection ratio	$V_{IC} = V_{ICR}$ min, $R_S = 50\ \Omega$,	$V_O = 0$, $T_A = 25°C$	80	86		80	86		dB
k_{SVR}	Supply voltage rejection ratio ($\Delta V_{CC\pm}/\Delta V_{IO}$)	$V_{CC} = \pm15$ V to ±9 V, $R_S = 50\ \Omega$,	$V_O = 0$, $T_A = 25°C$	80	86		80	86		dB
I_{CC}	Supply current (per amplifier)	No load, $T_A = 25°C$	$V_O = 0$,		1.4	2.8		1.4	2.8	mA
V_{o1}/V_{o2}	Crosstalk attenuation	$A_{VD} = 100$,	$T_A = 25°C$		120			120		dB

[†]All characteristics are measured under open-loop conditions with zero common-mode input voltage unless otherwise specified.

[‡]Input bias currents of a FET-input operational amplifier are normal junction reverse currents, which are temperature sensitive as shown in Figure 18. Pulse techniques must be used that will maintain the junction temperatures as close to the ambient temperature as is possible.

3

Operational Amplifiers

TEXAS
INSTRUMENTS
- POST OFFICE BOX 225012 • DALLAS, TEXAS 75265

electrical characteristics, VCC± = ±15 V (unless otherwise noted)

PARAMETER	TEST CONDITIONS[†]	TL080I TL081I TL082I TL083I TL084I MIN	TYP	MAX	TL080C TL081C TL082C TL083C TL084C TL085C MIN	TYP	MAX	TL080AC TL081AC TL082AC TL083AC TL084AC MIN	TYP	MAX	TL081BC TL082BC TL084BC MIN	TYP	MAX	UNIT
V_{IO} Input offset voltage	$V_O = 0$, $R_S = 50\ \Omega$ $T_A = 25°C$		3	6		3	15		3	6		2	3	mV
	$V_O = 0$, $R_S = 50\ \Omega$, T_A = full range			9			20			7.5			5	
α_{VIO} Temperature coefficient of input offset voltage	$V_O = 0$, $R_S = 50\ \Omega$, T_A = full range		10			10			10			10		$\mu V/°C$
I_{IO} Input offset current[‡]	$V_O = 0$ $T_A = 25°C$		5	100		5	200		5	100		5	100	pA
	$V_O = 0$ T_A = full range			10			10			2			2	nA
I_{IB} Input bias current[‡]	$V_O = 0$ $T_A = 25°C$		30	200		30	400		30	200		30	200	pA
	$V_O = 0$ T_A = full range			20			10			7			7	nA
V_{ICR} Common-mode input voltage range	$T_A = 25°C$	±11	±12		±11	±12		±11	±12		±11	±12		V
V_{OM} Maximum peak output voltage swing	$T_A = 25°C$, $R_L = 10\ k\Omega$	±12	±13.5		±12	±13.5		±12	±13.5		±12	±13.5		V
	$R_L \geq 10\ k\Omega$ T_A = full range	±12			±12			±12			±12			
	$R_L \geq 2\ k\Omega$, $T_A = 25°C$	±10	±12		±10	±12		±10	±12		±10	±12		
A_{VD} Large-signal differential voltage amplification	$V_O = \pm10\ V$, $R_L \geq 2\ k\Omega$, $T_A = 25°C$	50	200		25	200		50	200		50	200		V/mV
	$V_O = \pm10\ V$, T_A = full range	25			15			25			25			
B_1 Unity-gain bandwidth	$T_A = 25°C$		3			3			3			3		MHz
r_i Input resistance	$T_A = 25°C$		10^{12}			10^{12}			10^{12}			10^{12}		Ω
CMRR Common-mode rejection ratio	$V_{IC} = V_{ICR}$ min, $V_O = 0$, $R_S = 50\ \Omega$, $T_A = 25°C$	80	86		70	86		80	86		80	86		dB
k_{SVR} Supply voltage rejection ratio ($\Delta V_{CC} \pm/\Delta V_{IO}$)	$V_{CC} = \pm15\ V$ to $\pm9\ V$, $V_O = 0$, $R_S = 50\ \Omega$, $T_A = 25°C$	80	86		70	86		80	86		80	86		dB
I_{CC} Supply current (per amplifier)	No load, $V_O = 0$, $T_A = 25°C$		1.4	2.8		1.4	2.8		1.4	2.8		1.4	2.8	mA
V_{o1}/V_{o2} Crosstalk attenuation	$A_{VD} = 100$, $T_A = 25°C$		120			120			120			120		dB

[†]All characteristics are measured under open-loop conditions with zero common-mode input voltage unless otherwise specified. Full range for T_A is 25°C to 85°C for TL08_I and 0°C to 70°C for TL08_C, TL08_AC, and TL08_BC. Pulse techniques must be used that will maintain the junction temperatures as close to the ambient temperature as is possible.

[‡]Input bias currents of a FET-input operational amplifier are normal junction reverse currents, which are temperature sensitive as shown in Figure 18.

3

Operational Amplifiers

operating characteristics, $V_{CC\pm} = \pm 15$ V, $T_A = 25\,°C$

	PARAMETER	TEST CONDITIONS		MIN	TYP	MAX	UNIT
SR	Slew rate at unity gain	$V_I = 10$ V, $C_L = 100$ pF,	$R_L = 2$ kΩ, See Figure 1	8	13		V/μs
t_r	Rise time	$V_I = 20$ mV,	$R_L = 2$ kΩ,		0.1		μs
	Overshoot factor	$C_L = 100$ pF,	See Figure 1		10%		
V_n	Equivalent input noise voltage	$R_S = 100$ Ω	$f = 1$ kHz		18		nV/$\sqrt{\text{Hz}}$
			$f = 10$ Hz to 10 kHz		4		μV
I_n	Equivalent input noise current	$R_S = 100$ Ω,	$f = 1$ kHz		0.01		pA/$\sqrt{\text{Hz}}$
THD	Total harmonic distortion	$V_{O(rms)} = 10$ V, $R_L \geq 2$ kΩ,	$R_S \leq 1$ kΩ, $f = 1$ kHz		0.003%		

PARAMETER MEASUREMENT INFORMATION

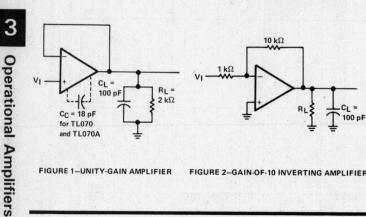

FIGURE 1–UNITY-GAIN AMPLIFIER   FIGURE 2–GAIN-OF-10 INVERTING AMPLIFIER   FIGURE 3–FEED-FORWARD
                                                                                        COMPENSATION

INPUT OFFSET VOLTAGE NULL CIRCUITS

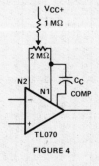

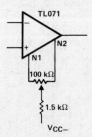

FIGURE 4   FIGURE 5

3 Operational Amplifiers

Texas
Instruments
POST OFFICE BOX 225012 • DALLAS, TEXAS 75265

TYPICAL CHARACTERISTICS†

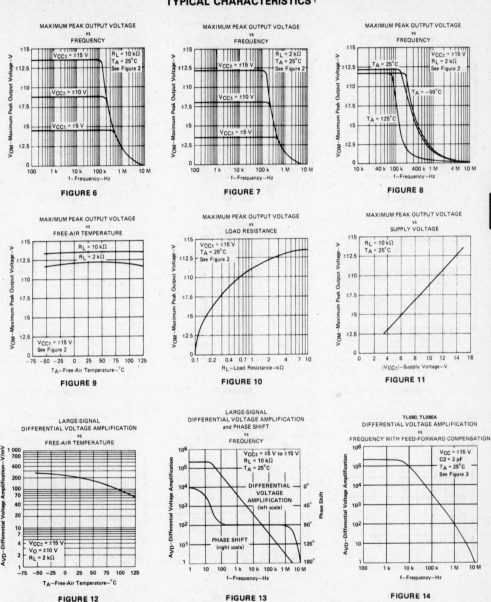

FIGURE 6

FIGURE 7

FIGURE 8

FIGURE 9

FIGURE 10

FIGURE 11

FIGURE 12

FIGURE 13

FIGURE 14

†Data at high and low temperatures are applicable only within the rated operating free-air temperature ranges of the various devices. A 12-pF compensation capacitor is used with TL080 and TL080A.

Operational Amplifiers

3

TEXAS
INSTRUMENTS
POST OFFICE BOX 225012 • DALLAS, TEXAS 75265

TYPICAL CHARACTERISTICS †

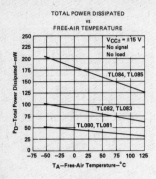

TOTAL POWER DISSIPATED
vs
FREE-AIR TEMPERATURE

FIGURE 15

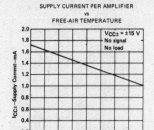

SUPPLY CURRENT PER AMPLIFIER
vs
FREE-AIR TEMPERATURE

FIGURE 16

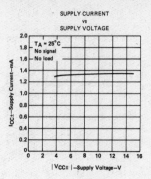

SUPPLY CURRENT
vs
SUPPLY VOLTAGE

FIGURE 17

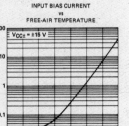

INPUT BIAS CURRENT
vs
FREE-AIR TEMPERATURE

FIGURE 18

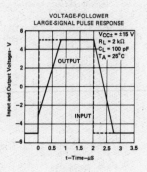

VOLTAGE-FOLLOWER
LARGE-SIGNAL PULSE RESPONSE

FIGURE 19

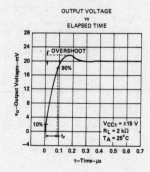

OUTPUT VOLTAGE
vs
ELAPSED TIME

FIGURE 20

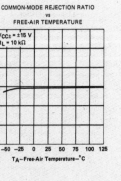

COMMON-MODE REJECTION RATIO
vs
FREE-AIR TEMPERATURE

FIGURE 21

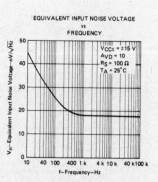

EQUIVALENT INPUT NOISE VOLTAGE
vs
FREQUENCY

FIGURE 22

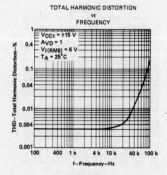

TOTAL HARMONIC DISTORTION
vs
FREQUENCY

FIGURE 23

†Data at high and low temperatures are applicable only within the rated operating free-air temperature ranges of the various devices. A 12-pF compensation capacitor is used with TL080 and TL080A.

98

Operational Amplifiers

3

TYPICAL APPLICATION DATA

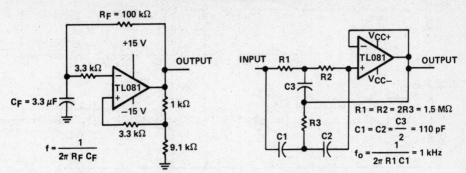

FIGURE 24—0.5-Hz SQUARE-WAVE OSCILLATOR

FIGURE 25—HIGH-Q NOTCH FILTER

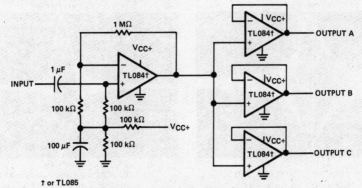

† or TL085

FIGURE 26—AUDIO DISTRIBUTION AMPLIFIER

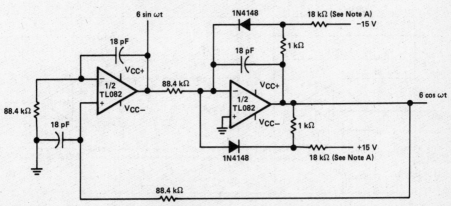

NOTE A: These resistor values may be adjusted for a symmetrical output.

FIGURE 27—100-kHz QUADRATURE OSCILLATOR

Operational Amplifiers

3

Texas Instruments

POST OFFICE BOX 225012 • DALLAS, TEXAS 75265

TYPICAL APPLICATION DATA

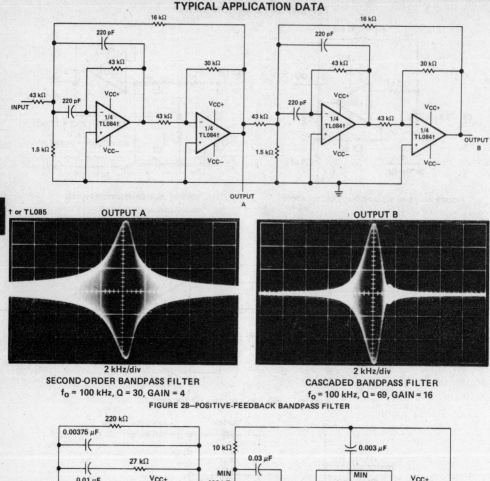

† or TL085

OUTPUT A

OUTPUT B

2 kHz/div

SECOND-ORDER BANDPASS FILTER
$f_o = 100$ kHz, Q = 30, GAIN = 4

2 kHz/div

CASCADED BANDPASS FILTER
$f_o = 100$ kHz, Q = 69, GAIN = 16

FIGURE 28–POSITIVE-FEEDBACK BANDPASS FILTER

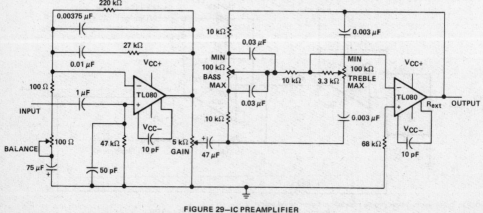

FIGURE 29–IC PREAMPLIFIER

TEXAS
INSTRUMENTS

POST OFFICE BOX 225012 • DALLAS, TEXAS 75265

983

TYPES TL087, TL088, TL287, TL288
JFET-INPUT OPERATIONAL AMPLIFIERS

D2484, MARCH 1979—REVISED AUGUST 1983

- Low Input Offset Voltage . . . 0.5 mV Max
- Low Power Consumption
- Wide Common-Mode and Differential Voltage Ranges
- Low Input Bias and Offset Currents
- Output Short-Circuit Protection

- High Input Impedance . . . JFET-Input Stage
- Internal Frequency Compensation
- Latch-Up-Free Operation
- High Slew Rate . . . 13 V/µs Typ
- Low Total Harmonic Distortion . . . 0.003% Typ

description

These JFET-input operational amplifiers incorporate well-matched high-voltage JFET and bipolar transistors in a monolithic integrated circuit. They feature low input offset voltage, high slew rate, low input bias and offset current, and low temperature coefficient of input offset voltage. Offset-voltage adjustment is provided for the TL087 and TL088.

Device types with an ''M'' suffix are characterized for operation over the full military temperature range of −55°C to 125°C, those with an ''I'' suffix are characterized for operation from −25°C to 85°C, and those with a ''C'' suffix are characterized for operation from 0°C to 70°C.

TL087, TL088
D, JG, OR P DUAL-IN-LINE PACKAGE
(TOP VIEW)

TL088M
U FLAT PACKAGE
(TOP VIEW)

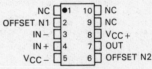

TL287, TL288
D, JG, OR P DUAL-IN-LINE PACKAGE
(TOP VIEW)

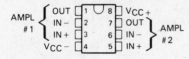

TL288M
U FLAT PACKAGE
(TOP VIEW)

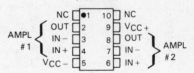

NC—No internal connection

symbol (each amplifier)

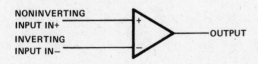

NONINVERTING INPUT IN+
INVERTING INPUT IN−
OUTPUT

3

Operational Amplifiers

**TEXAS
INSTRUMENTS**
POST OFFICE BOX 225012 • DALLAS, TEXAS 75265

TYPES TL087, TL088, TL287, TL288
JFET-INPUT OPERATIONAL AMPLIFIERS

absolute maximum ratings over operating free-air temperature range (unless otherwise noted)

		TL088M TL288M	TL087I TL088I TL287I TL288I	TL087C TL088C TL287C TL288C	UNIT
Supply voltage, $V_{CC}+$ (see Note 1)		18	18	18	V
Supply voltage, $V_{CC}-$ (see Note 1)		−18	−18	−18	V
Differential input voltage (see Note 2)		±30	±30	±30	V
Input voltage (see Notes 1 and 3)		±15	±15	±15	V
Duration of output short circuit (see Note 4)		unlimited	unlimited	unlimited	
Continuous total dissipation at (or below) 25°C free-air temperature (see Note 5)	JG or P package	680	680	680	mW
	U package	675			
	D package		725	725	
Operating free-air temperature range		−55 to 125	−25 to 85	0 to 70	°C
Storage temperature range		−65 to 150	−65 to 150	−65 to 150	°C
Lead temperature 1,6 mm (1/16 inch) from case for 60 seconds	JG or U package	300	300	300	°C
Lead temperature 1,6 mm (1/16 inch) from case for 10 seconds	D or P package		260	260	°C

NOTES: 1. All voltage values, except differential voltages, are with respect to the midpoint between $V_{CC}+$ and $V_{CC}-$.
2. Differential voltages are at the noninverting input terminal with respect to the inverting input terminal.
3. The magnitude of the input voltage must never exceed the magnitude of the supply voltage or 15 volts, whichever is less.
4. The output may be shorted to ground or to either supply. Temperature and/or supply voltages must be limited to ensure that the dissipation rating is not exceeded.
5. For operation above 25°C free-air temperature, refer to Dissipation Derating Curves, Section 2. In the JG package, TL088M and TL288M chips are alloy-mounted; TL087I, TL088I, TL287I, TL288I, TL087C, TL088C, TL287C, and TL288C chips are glass-mounted.

3

Operational Amplifiers

3-146

TEXAS
INSTRUMENTS
POST OFFICE BOX 225012 • DALLAS, TEXAS 75265

88

electrical characteristics, VCC± = ±15 V

PARAMETER	TEST CONDITIONS†		TL088M TL288M MIN	TYP	MAX	TL087I TL088I TL287I TL288I MIN	TYP	MAX	TL087C TL088C TL287C TL288C MIN	TYP	MAX	UNIT
V_{IO} Input offset voltage	RS = 50 Ω, VO = 0, TA = 25°C	TL087, TL287					0.1	0.5		0.1	0.5	mV
	RS = 50 Ω, VO = 0, TA = 25°C	TL088, TL288		0.1	3		0.1	1		0.1	1	
	RS = 50 Ω, VO = 0, TA = full range	TL087, TL287						2			1.5	
	TA = full range	TL088, TL288			6			3			2.5	
α_{VIO} Temperature coefficient of input offset voltage	RS = 50 Ω, TA = full range			10			10			10		μV/°C
I_{IO} Input offset current	TA = 25°C			5	100		5	100		5	100	pA
	TA = full range				25			3			2	nA
I_{IB} Input bias current‡	TA = 25°C			30	400		30	200		30	200	pA
	TA = full range				100			20			7	nA
V_{ICR} Common-mode input voltage range	TA = 25°C		VCC− +4 to VCC+ −4			VCC− +4 to VCC+ −4			VCC− +4 to VCC+ −4			V
V_{OPP} Maximum-peak-to-peak output voltage swing	RL = 10 kΩ, TA = 25°C		24	27		24	27		24	27		V
	RL ≥ 10 kΩ, TA = full range		24			24			24			
	RL ≥ 2 kΩ		20			20			20			
A_{VD} Large-signal differential voltage amplification	RL ≥ 2 kΩ, VO = ±10 V, TA = 25°C		50	200		50	200		50	200		V/mV
	RL ≥ 2 kΩ, VO = ±10 V, TA = full range		25			25			25			
B_1 Unity-gain bandwidth	TA = 25°C			3			3			3		MHz
r_i Input resistance	TA = 25°C			10^{12}			10^{12}			10^{12}		Ω
CMRR Common-mode rejection ratio	RS = 50 Ω, VIC = VICR min, TA = 25°C		80	95		80	95		80	95		dB
k_{SVR} Supply voltage rejection ratio (ΔVCC±/ΔVIO)	RS = 50 Ω, VCC± = ±9 V to ±15 V, TA = 25°C		80	95		80	95		80	95		dB
I_{CC} Supply current (per amplifier)	No load, VO = 0, TA = 25°C			1.4	2.8		1.4	2.8		1.4	2.8	mA

† All characteristics are measured under open-loop conditions with zero common-mode input voltage unless otherwise specified. Full range for T$_A$ is −55°C to 125°C for TL_88M; −25°C to 85°C for TL_8_I; and 0°C to 70°C for TL_8_C.

‡ Input bias currents of a FET-input operational amplifier are normal junction reverse currents, which are temperature sensitive. Pulse techniques must be used that will maintain the junction temperature as close to the ambient temperature as possible.

3

Operational Amplifiers

operating characteristics $V_{CC} = \pm 15$ V, $T_A = 25^\circ$C

	PARAMETER	TEST CONDITIONS		MIN	TYP	MAX	UNIT
SR	Slew rate at unity gain	$V_I = 10$ V,	$R_L = 2$ kΩ,	8	13		V/μs
		$C_L = 100$ pF,	$A_{VD} = 1$				
t_r	Rise time	$V_I = 20$ mV,	$R_L = 2$ kΩ,		0.1		μs
	Overshoot factor	$C_L = 100$ pF,	$A_{VD} = 1$		10%		
V_n	Equivalent input noise voltage	$R_S = 100$ Ω,	$f = 1$ kHz		18		nV/$\sqrt{Hz}$

TYPICAL CHARACTERISTICS[†]

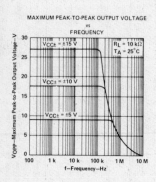

FIGURE 1

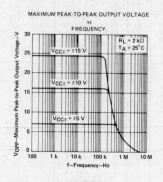

FIGURE 2

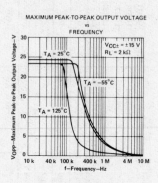

FIGURE 3

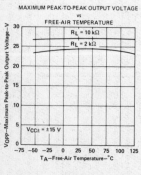

FIGURE 4

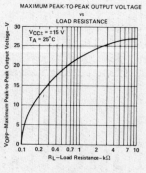

FIGURE 5

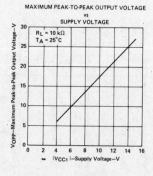

FIGURE 6

[†]Data at high and low temperatures are applicable only within the rated operating free-air temperature ranges of the various devices.

TEXAS
INSTRUMENTS
POST OFFICE BOX 225012 • DALLAS, TEXAS 75265

3

Operational Amplifiers

88

TYPICAL CHARACTERISTICS†

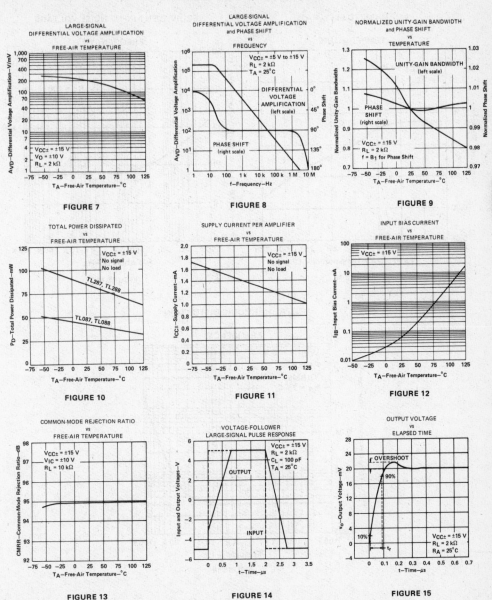

†Data at high and low temperatures are applicable only within the rated operating free-air temperature ranges of the various devices.

TEXAS INSTRUMENTS
POST OFFICE BOX 225012 • DALLAS, TEXAS 75265

Operational Amplifiers

3

TYPES TL087, TL088, TL287, TL288
JFET-INPUT OPERATIONAL AMPLIFIERS

TYPICAL CHARACTERISTICS†

NORMALIZED SLEW RATE
vs
TEMPERATURE

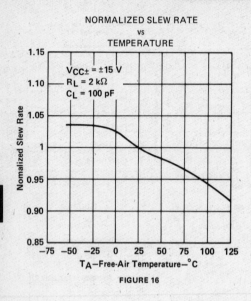

FIGURE 16

EQUIVALENT INPUT NOISE VOLTAGE
vs
FREQUENCY

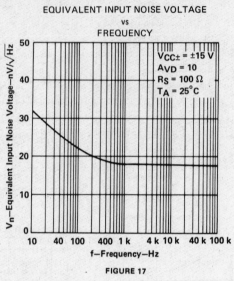

FIGURE 17

TOTAL HARMONIC DISTORTION
vs
FREQUENCY

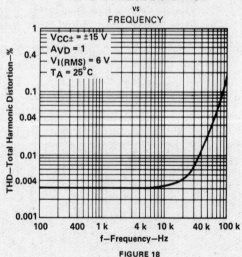

FIGURE 18

†Data at high and low temperatures are applicable only within the rated operating free-air temperature ranges of the various devices.

TEXAS
INSTRUMENTS
POST OFFICE BOX 225012 • DALLAS, TEXAS 75265

883

TYPE TL136C
QUAD HIGH-PERFORMANCE OPERATIONAL AMPLIFIER

D2604, NOVEMBER 1981—REVISED AUGUST 1983

- **Continuous-Short Circuit Protection**
- **Wide Common-Mode and Differential Voltage Ranges**
- **No Frequency Compensation Required**
- **Low Power Consumption**
- **No Latch-up**
- **Unity-Gain Bandwidth 3 MHz Typical**
- **Gain and Phase Match Between Amplifiers**

D, J OR N DUAL-IN-LINE PACKAGE
(TOP VIEW)

AMPL #1	OUT	1 · 14	OUT	AMPL #4
	IN−	2 · 13	IN−	
	IN+	3 · 12	IN+	
	V$_{CC}$+	4 · 11	V$_{CC}$−	
AMPL #2	IN+	5 · 10	IN+	AMPL #3
	IN−	6 · 9	IN−	
	OUT	7 · 8	OUT	

description

The TL136C is a quad high-performance operational amplifier with each amplifier electrically similar to uA741 except that offset null capability is not provided.

The high common-mode input voltage range and the absence of latch-up make these amplifiers ideal for voltage-follower applications. The devices are short-circuit protected and the internal frequency compensation ensures stability without external components.

The TL136C is characterized for operation from 0 °C to 70 °C.

symbol (each amplifier)

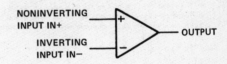

NONINVERTING INPUT IN+

INVERTING INPUT IN−

OUTPUT

3

Operational Amplifiers

schematic (each amplifier)

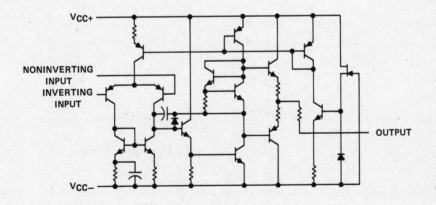

883

TEXAS INSTRUMENTS

POST OFFICE BOX 225012 • DALLAS, TEXAS 75265

TYPE TL136C
QUAD HIGH-PERFORMANCE OPERATIONAL AMPLIFIER

absolute maximum ratings over operating free-air temperature range (unless otherwise noted)

Supply voltage V_{CC+} (see Note 1)	18 V
Supply voltage V_{CC-} (see Note 1)	−18 V
Differential input voltage (see Note 2)	±30 V
Input voltage (any input, see Notes 1 and 3)	±15 V
Duration of output short-circuit to ground, one amplifier at a time (see Note 4)	unlimited
Continuous total dissipation at (or below) 25 °C free-air temperature (see Note 5)	800 mW
Operating free-air temperature range	0 °C to 70 °C
Storage temperature range	−65 °C to 150 °C
Lead temperature 1,6 mm (1/16 inch) from case for 60 seconds: J package	300 °C
Lead temperature 1,6 mm (1/16 inch) from case for 10 seconds: D, N package	260 °C

NOTES: 1. All voltage values, unless otherwise noted, are with respect to the midpoint between V_{CC+} and V_{CC-}.
2. Differential voltages are at the noninverting input terminal with respect to the inverting input terminal.
3. The magnitude of the input voltage must never exceed the magnitude of the supply voltage or 15 volts, whichever is less.
4. Temperature and/or supply voltages must be limited to ensure that the dissipation rating is not exceeded.
5. For operation above 25 °C free-air temperature, refer to Dissipation Derating Curves in Section 2. In the J package, the chips are glass-mounted.

3

Operational Amplifiers

TEXAS
INSTRUMENTS
POST OFFICE BOX 225012 • DALLAS, TEXAS 75265

electrical characteristics at specified free-air temperature, $V_{CC+} = 15$ V, $V_{CC-} = -15$ V

	PARAMETER	TEST CONDITIONS[†]		MIN	TYP	MAX	UNIT
V_{IO}	Input offset voltage	$V_O = 0$, $R_S = 50$ Ω	25°C		0.5	6	mV
			0°C to 70°C			7.5	
I_{IO}	Input offset current	$V_O = 0$	25°C		5	200	nA
			0°C to 70°C			300	
I_{IB}	Input bias current	$V_O = 0$	25°C		40	500	nA
			0°C to 70°C			800	
V_{ICR}	Common-mode input voltage range		25°C	±12	±14		V
V_{OPP}	Maximum peak-to-peak output voltage swing	$R_L = 10$ kΩ	25°C	24	28		V
		$R_L = 2$ kΩ	25°C	20	26		
		$R_L \geq 2$ kΩ	0°C to 70°C	20			
A_{VD}	Large-signal differential voltage amplification	$R_L \geq 2$ kΩ, $V_O = ±10$ V	25°C	20	300		V/mV
			0°C to 70°C	15			
B_1	Unity-gain bandwidth		25°C		3		MHz
r_i	Input resistance		25°C	0.3	5		MΩ
CMRR	Common-mode rejection ratio	$V_{IC} = V_{ICR}$ min, $R_S = 50$ Ω	25°C	70	90		dB
k_{SVS}	Supply voltage sensitivity ($\Delta V_{IO}/\Delta V_{CC}$)	$V_{CC\pm} = ±9$ V to ± 15 V, $R_S = 50$ Ω	25°C		30	150	µV/V
V_n	Equivalent input noise voltage (closed-loop)	$A_{VD} = 100$, $R_S = 100$ Ω, $f = 1$ kHz, BW = 1 Hz	25°C		7.5		nV/√Hz
I_{CC}	Supply current (All four amplifiers)	No load, $V_O = 0$ V	25°C		5	11.3	mA
			0°C		6	13.7	
			70°C		4.5	11.3	
P_D	Total power dissipation (All four amplifiers)	No load, $V_O = 0$ V	25°C		150	340	mW
			0°C		180	400	
			70°C		135	300	
$V_{o1}V_{o2}$	Crosstalk attenuation Open loop A_{VD}	$R_S = 1$ kΩ, $f = 10$ kHz	25°C		105		dB
			25°C		105		

[†]All characteristics are measured under open-loop conditions with zero common-mode input voltage unless otherwise specified.

operating characteristics, $V_{CC+} = 15$ V, $V_{CC-} = -15$ V, $T_A = 25$°C

	PARAMETER	TEST CONDITIONS		MIN	TYP	MAX	UNIT
t_r	Rise time	$V_I = 20$ mV, $C_L = 100$ pF	$R_L = 2$ kΩ,		0.13		µs
SR	Slew rate at unity gain	$V_I = 10$ V, $C_L = 100$ pF	$R_L = 2$ kΩ,		2.0		V/µs

3

Operational Amplifiers

TEXAS
INSTRUMENTS
POST OFFICE BOX 225012 • DALLAS, TEXAS 75265

**LINEAR
INTEGRATED
CIRCUITS**

**TYPES TL291, TL292, TL294
HIGH-FREQUENCY OPERATIONAL AMPLIFIERS**

D2782, SEPTEMBER 1983

- Small-Signal Unity-Gain Bandwidth
  . . . 20 MHz Typ

- Noninverting Slew Rate . . . 50 V/μs Typ
  (Unity-Gain Follower)

- Internal Frequency Compensation

- Full-Power Bandwidth at V_{OPP} = 20 V
  . . . 400 kHz Typ

- Open-Loop Gain at Full-Power Bandwidth,
  V_{OPP} = 20 V . . . 34 dB Typ

- Output Short-Circuit Protection

- TL291 Has Offset Null Capability

- Pinout is Same as Standard General Purpose
  Operational Amplifiers

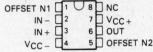

TL291
JG OR P DUAL-IN-LINE PACKAGE
(TOP VIEW)

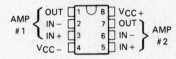

TL292
JG OR P DUAL-IN-LINE PACKAGE
(TOP VIEW)

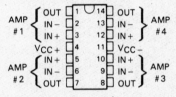

TL294
J OR N DUAL-IN-LINE PACKAGE
(TOP VIEW)

NC—No internal connection

symbol (each amplifier)

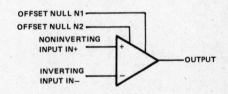

description

These devices are high-speed operational amplifiers designed for applications requiring wide bandwidth and a fast slew rate. These monolithic circuits incorporate new high-frequency P-N-P transistors that eliminate the need for large feed-forward capacitors required in previous moderately high-frequency designs to pass the signal around slow lateral P-N-P stages.

These operational amplifiers have a typical full-power bandwidth of 400 kilohertz for a 20-volt peak-to-peak output swing. because of the higher 20-megahertz unity-gain bandwidth, the typical open-loop gain at the 400-kilohertz full-power bandwidth is a very respectable 34 decibels.

The TL291 single-channel operational amplifier pinout includes offset nulling, which is easily accomplished by connecting a potentiometer across the offset null pins with the wiper connected to the V_{CC}– pin.

The TL291M, TL292M, and TL294M will be characterized for operation over the full military temperature range of −55 °C to 125 °C. The TL291C, TL292C, and TL294C will be characterized for operation from 0 °C to 70 °C.

Operational Amplifiers

3

**TEXAS
INSTRUMENTS**

POST OFFICE BOX 225012 • DALLAS, TEXAS 75265

**LINEAR
INTEGRATED
CIRCUITS**

**TYPES TL321M, TL321I, TL321C
OPERATIONAL AMPLIFIERS**

D2343, APRIL 1977—REVISED AUGUST 1983

- Wide Range of Supply Voltages Single Supply . . . 3 V to 30 V or Dual Supplies
- Low Supply Current Drain Independent of Supply Voltage . . . 0.8 mA Typ
- Common-Mode Input Voltage Range Includes Ground Allowing Direct Sensing near Ground
- Low Input Bias and Offset Parameters
  Input Offset Voltage . . . 2 mV TYP
  Input Offset Current . . . 3 nA Typ (TL321M)
  Input bias Current . . . 45 nA Typ
- Differential Input Voltage Range Equal to Maximum-Rated Supply Voltage . . . ±32 V
- Open-Loop Differential Voltage Amplification . . . 100 V/mV Typ
- Internal Frequency Compensation

**TL321M . . . JG
TL321I, TL321C . . . JG OR P
DUAL-IN-LINE PACKAGE
(TOP VIEW)**

NC	1	8	NC
IN−	2	7	VCC
IN+	3	6	OUT
GND	4	5	NC

**TL321M . . . FH OR FK
CHIP CARRIER PACKAGE
(TOP VIEW)**

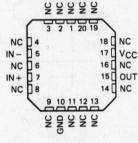

NC—No internal connection

symbol

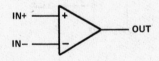

IN+ + OUT
IN− −

description

The TL321 is a high-gain, frequency-compensated operational amplifier that was designed specifically to operate from a single supply over a wide range of voltages. Operation from split supplies is also possible so long as the difference between the two supplies is 3 volts to 30 volts and Pin 7 is at least 1.5 volts more positive than the input common-mode voltage. The low supply current drain is independent of the magnitude of the supply voltage.

Applications include transducer amplifiers, d-c amplification blocks, and all the conventional operational amplifier circuits that now can be more easily implemented in single-supply-voltage systems. For example, the TL321 can be operated directly off of the standard five-volt supply that is used in digital systems and will easily provide the required interface electronics without requiring additional ±15-volt supplies.

The TL321M is characterized for operation over the full military temperature range of −55°C to 125°C. The TL321I is characterized for operation from −25°C to 85°C. The TL321C is characterized for operation from 0°C to 70°C.

3

Operational Amplifiers

**TEXAS
INSTRUMENTS**

POST OFFICE BOX 225012 • DALLAS, TEXAS 75265

schematic

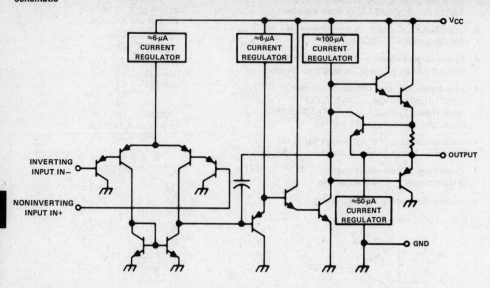

absolute maximum ratings over operating free-air temperature range (unless otherwise noted)

Supply voltage, V_{CC} (see Note 1) .... 32 V

Differential input voltage (see Note 2) .... ±32 V

Input voltage range (either input) .... −0.3 V to 32 V

Duration of output short-circuit to ground at (or below) 25 °C free-air temperature
(V_{CC} ≤ 15 V) (see Note 3) .... unlimited

Continuous total dissipation at (or below) 25 °C free-air temperature (see Note 4) .... 680 mW

Operating free-air temperature range: TL321M .... −55 °C to 125 °C

TL321I .... −25 °C to 85 °C

TL321C .... 0 °C to 70 °C

Storage temperature range .... −65 °C to 150 °C

Lead temperature 1,6 mm (1/16 inch) from case for 60 seconds: FH, FK, or JG package .... 300 °C

Lead temperature 1,6 mm (1/16 inch) from case for 10 seconds: P package .... 260 °C

NOTES: 1. All voltage values, except differential voltages, are with respect to the network ground terminal.
2. Differential voltages are at the noninverting input terminal with respect to the inverting input terminal.
3. Short circuits from the output to V_{CC} can cause excessive heating and eventual destruction.
4. For operation above 25 °C free-air temperature, refer to Dissipation Derating Curves, Section 2. In the JG package, TL321M chips are alloy-mounted; TL321I and TL321C chips are glass-mounted.

TEXAS
INSTRUMENTS
POST OFFICE BOX 225012 • DALLAS, TEXAS 75265

3

Operational Amplifiers

electrical characteristics at specified free-air temperature, V_{CC} = 5 V (unless otherwise noted)

PARAMETER		TEST CONDITIONS†		TL321M, TL321I			TL321C			UNIT
				MIN	TYP	MAX	MIN	TYP	MAX	
V_{IO}	Input offset voltage	$V_{IC} = V_{ICR}$ min, V_{CC} = 5 V to 30 V, V_O = 1.4 V, R_S = 50 kΩ	25°C		2	5		2	7	mV
			Full range			7			9	
I_{IO}	Input offset current	V_O = 1.4 V	25°C		3	30		5	50	nA
			Full range			100			150	
I_{IB}	Input bias current	V_O = 1.4 V,	25°C		−45	−150		−45	−250	nA
			Full range			−300			−500	
V_{ICR}	Common-mode input voltage range	V_{CC} = 5 V to 30 V	25°C	0 to $V_{CC}-1.5$			0 to $V_{CC}-1.5$			V
			Full range	0 to $V_{CC}-2$			0 to $V_{CC}-2$			
V_{OH}	High-level output voltage	V_{CC} = 30 V, R_L = 2 kΩ	Full range	26			26			V
		V_{CC} = 30 V, $R_L \geq 10$ kΩ	Full range	27	28		27	28		
V_{OL}	Low-level output voltage	$R_L \geq 2$ kΩ	25°C	3.5			3.5			mV
		$R_L \leq 10$ kΩ	Full range		5	20		5	20	
A_{VD}	Large-signal differential voltage amplification	V_{CC} = 15 V, V_O = 1 V to 11 V, $R_L \geq 2$ kΩ	25°C	50	100		25	100		V/mV
			Full range	25			15			
CMRR	Common-mode rejection ratio	$V_{IC} = V_{ICR}$ min, R_S = 50 Ω	25°C	70	85		65	85		dB
k_{SVR}	Supply voltage rejection ratio($\Delta V_{CC}/\Delta V_{IO}$)	V_{CC} = 5 V to 30 V, R_S = 50 Ω	25°C	65	100		65	100		dB
I_O	Output current	Source V_{CC} = 15 V, V_{ID} = 1 V, V_O = 0	25°C	−25	−40		−20	−40		mA
			Full range	−10	−20		−10	−20		
		Sink V_{CC} = 15 V, V_{ID} = −1 V, V_O = 15 V	25°C	10	20		10	20		
			Full range	5	8		5	8		
		V_{ID} = −1 V, V_O = 200 mV	25°C	12	50		12	50		μA
I_{CC}	Supply current	No load V_O = 15 V, V_{CC} = 30 V	Full range			2			2	mA
		No load V_O = 2.5 V, V_{CC} = 5 V	Full range		0.4	1			1	

†All characteristics are measured under open-loop conditions with zero common-mode input voltage unless otherwise specified. Full range is −55°C to 125°C for TL321M, −25°C to 85°C for TL321I, and 0°C to 70°C for TL321C.

3

Operational Amplifiers

TEXAS
INSTRUMENTS
POST OFFICE BOX 225012 • DALLAS, TEXAS 75265

TYPES TL322M, TL322I, TL322C
DUAL LOW-POWER OPERATIONAL AMPLIFIERS

D2567, OCTOBER 1979—REVISED SEPTEMBER 1983

- **Wide Range of Supply Voltages Single Supply . . . 3 V to 36 V or Dual Supplies**
- **Class AB Output Stage**
- **True Differential Input Stage**
- **Low Input Bias Current**
- **Internal Frequency Compensation**
- **Short-Circuit Protection**

TL322M . . . JG PACKAGE
TL322I, TL322C . . . D, JG, OR P PACKAGE
(TOP VIEW)

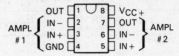

description

The TL322M, TL322I, and the TL322C are dual operational amplifiers similar in performance to the MC3403. They are designed to operate from a single supply over a range of voltages from 3 volts to 36 volts. Operation from split supplies is also possible provided the difference between the two supplies is 3 volts to 36 volts. The common-mode input range includes the negative supply. Output range is from the negative supply to V_{CC-} 1.5 V. Quiescent supply currents per amplifier are typically less than one-half those of the uA741.

The TL322M is characterized for operation over the full military temperature range of $-55\,°C$ to $125\,°C$. The TL322I is characterized for operation from $-40\,°C$ to $85\,°C$. The TL322C is characterized for operation from $0\,°C$ to $70\,°C$.

symbol (each amplifier)

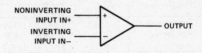

absolute maximum ratings over operating free-air temperature range (unless otherwise noted)

		TL322M	TL322I	TL322C	UNIT
Supply voltage V_{CC+} (see Note 1)		18	18	18	V
Supply voltage V_{CC-} (see Note 1)		-18	-18	-18	V
Supply voltage V_{CC+} with respect to V_{CC-}		36	36	36	V
Differential input voltage (see Note 2)		±36	±36	±36	V
Input voltage (see Notes 1 and 3)		±18	±18	±18	V
Continuous total power dissipation at (or below) 25 °C free-air temperature (see Note 4)		680	680	680	mW
Operating free-air temperature range		-55 to 125	-40 to 85	0 to 70	°C
Storage temperature range		-65 to 150	-65 to 150	-65 to 150	°C
Lead temperature 1,6 mm (1/16 inch) from case for 60 seconds	JG package	300	300	300	°C
Lead temperature 1,6 mm (1/16 inch) from case for 10 seconds	D or P package		260	260	°C

NOTES: 1. These voltage values are with respect to the midpoint between V_{CC+} and V_{CC-}.
2. Differential voltages are at the noninverting input terminal with respect to the inverting input terminal.
3. Neither input must ever be more positive than V_{CC+} or more negative than V_{CC-}.
4. For operation above 25 °C free-air temperature, refer to Dissipation Derating Curves in Section 2. In the JG package, TL322M chips are alloy mounted and TL322I and TL322C chips are glass mounted.

Operational Amplifiers

3

TEXAS INSTRUMENTS
POST OFFICE BOX 225012 • DALLAS, TEXAS 75265

electrical characteristics at specified free-air temperature, $V_{CC\pm} = \pm 15$ V (unless otherwise noted)

PARAMETER		TEST CONDITIONS†		TL322M MIN	TYP	MAX	TL322I MIN	TYP	MAX	TL322C MIN	TYP	MAX	UNIT
V_{IO}	Input offset voltage	$V_O = 0$, $R_S = 50\ \Omega$	25°C		2	8		2	8		2	10	mV
			Full range			10			10			12	
α_{VIO}	Temperature coefficient of input offset voltage	$V_O = 0$, $R_S = 50\ \Omega$	25°C		10			10			10		$\mu V/°C$
I_{IO}	Input offset current	$V_O = 0$	25°C		30	75		30	75		30	50	nA
			Full range			250			250			200	
α_{IIO}	Temperature coefficient of input offset current	$V_O = 0$	25°C		50			50			50		$pA/°C$
I_{IB}	Input bias current	$V_O = 0$	25°C		−0.2	−0.5		−0.2	−0.5		−0.2	−0.5	μA
			Full range			−1.15			−1			−0.8	
V_{ICR}	Common-mode input voltage range‡		25°C	$V_{CC}-$ to 13	$V_{CC}-$ to 13.5		$V_{CC}-$ to 13	$V_{CC}-$ to 13.5		$V_{CC}-$ to 13	$V_{CC}-$ to 13.5		V
V_{OM}	Peak output voltage swing	$R_L = 10\ k\Omega$	25°C	±12	±13.5		±12	±12.5		±12	±13.5		V
		$R_L = 2\ k\Omega$	25°C	±10	±13		±10	±12		±10	±13		
			Full range	±10			±10			±10			
A_{VD}	Large-signal differential voltage amplification	$V_O = \pm 10$ V, $R_L = 2\ k\Omega$	25°C		200		20	200		20	200		V/mV
			Full range	25			15			15			
B_{OM}	Maximum-output swing bandwidth	$V_{OPP} = 20$ V, $A_{VD} = 1$, THD ≤ 5%, $R_L = 2\ k\Omega$	25°C		9			9			9		kHz
B_1	Unity-gain bandwidth	$V_O = 50$ mV, $R_L = 10\ k\Omega$	25°C		1			1			1		MHz
ϕ_m	Phase margin	$R_L = 2\ k\Omega$, $C_L = 200$ pF	25°C		60°			60°			60°		
r_i	Input resistance	$f = 20$ Hz	25°C	0.3	1		0.3	1		0.3	1		$M\Omega$
r_o	Output resistance	$f = 20$ Hz	25°C		75			75			75		Ω
CMRR	Common-mode rejection ratio	$V_{IC} = V_{ICR}$ min, $R_S = 50\ \Omega$	25°C	70	90		70	90		70	90		dB
k_{SVS}	Supply voltage sensitivity ($\Delta V_{IO}/\Delta V_{CC}$)	$V_{CC} = \pm 2.5$ V to ± 15 V, $R_S = 50\ \Omega$	25°C		30	150		30	150		30	150	$\mu V/V$
I_{OS}	Short-circuit output current§	$V_O = 0$	25°C	±10	±30	±45	±10	±30	±45	±10	±30	±45	mA
I_{CC}	Total supply current	$V_O = 0$, No load	25°C		1.4	2.5		1.4	4		1.4	4	mA

†All characteristics are specified under open-loop conditions unless otherwise noted. Full range for T_A is −55°C to 125°C for TL322M; −40°C to 85°C for TL322I, and 0°C to 70°C for TL322C.

‡The V_{ICR} limits are directly linked volt-for-volt to supply voltage, viz the positive limit is 2 volts less than V_{CC+}.

§Temperature and/or supply voltages must be limited to ensure that the dissipation rating is not exceeded.

3

Operational Amplifiers

TEXAS
INSTRUMENTS
POST OFFICE BOX 225012 • DALLAS, TEXAS 75265

electrical characteristics, V_{CC+} = 5 V, V_{CC-} = 0 V, T_A = 25°C (unless otherwise noted)

	PARAMETER	TEST CONDITIONS[†]	TL322M MIN	TL322M TYP	TL322M MAX	TL322I MIN	TL322I TYP	TL322I MAX	TL322C MIN	TL322C TYP	TL322C MAX	UNIT
V_{IO}	Input offset voltage	V_O = 2.5 V, R_S = 50 Ω		2	8			8		2	10	mV
I_{IO}	Input offset current	V_O = 2.5 V		30	75			75		30	50	nA
I_{IB}	Input bias current			−0.2	−0.5			−0.5		−0.2	−0.5	pA
V_{OM}	Peak output voltage swing§	R_L = 10 kΩ	3.3	3.5		3.3	3.5		3.3	3.5		V
		R_L = 10 kΩ, V_{CC+} = 5 V to 30 V	V_{CC+}−1.7			V_{CC+}−1.7			V_{CC+}−1.7			
A_{VD}	Large-signal differential voltage amplification	V_O = 1.7 V to 3.3 V, R_L = 2 kΩ	20	200		20	200		20	200		V/mV
k_{SVS}	Supply voltage sensitivity ($\Delta V_{IO}/\Delta V_{CC\pm}$)	V_{CC} = ±2.5 V to ±15 V			150			150			150	μV/V
I_{CC}	Supply current	V_O = 2.5 V, No load		1.2	2.5		1.2	4		1.2	4	mA
V_{o1}/V_{o2}	Crosstalk attenuation	A_{VD} = 100, f = 1 kHz to 20 kHz		120			120			120		dB

[†]All characteristics are specified under open-loop conditions.
§Output will swing essentially to ground.

switching characteristics: $V_{CC\pm}$ = ±15 V, A_{VD} = 1, T_A = 25°C (unless otherwise noted)

	PARAMETER	TEST CONDITIONS			MIN	TYP	MAX	UNIT
SR	Slew rate at unity gain	V_I = ±10 V,	C_L = 100 pF,	See Figure 1		0.6		V/μs
t_r	Rise time	ΔV_O = 50 mV, See Figure 1	C_L = 100 pF,	R_L = 10 kΩ,		0.35		μs
t_f	Fall time					0.35		μs
	Overshoot factor					20%		
	Crossover distortion	V_{IPP} = 30 mV,	V_{OPP} = 2 V,	f = 10 kHz		1%		

3

Operational Amplifiers

PARAMETER MEASUREMENT INFORMATION

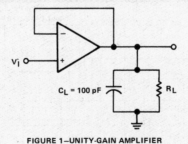

FIGURE 1—UNITY-GAIN AMPLIFIER

Operational Amplifiers

3

TYPICAL CHARACTERISTICS†

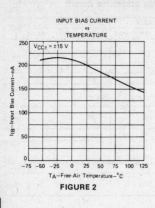

INPUT BIAS CURRENT
vs
TEMPERATURE

$V_{CC\pm} = \pm15\ V$

FIGURE 2

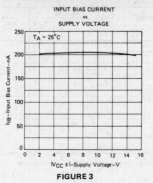

INPUT BIAS CURRENT
vs
SUPPLY VOLTAGE

$T_A = 25°C$

FIGURE 3

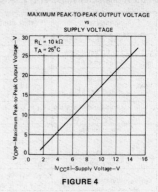

MAXIMUM PEAK-TO-PEAK OUTPUT VOLTAGE
vs
SUPPLY VOLTAGE

$R_L = 10\ k\Omega$
$T_A = 25°C$

FIGURE 4

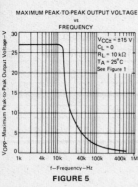

MAXIMUM PEAK-TO-PEAK OUTPUT VOLTAGE
vs
FREQUENCY

$V_{CC\pm} = \pm15\ V$
$C_L = 0$
$R_L = 10\ k\Omega$
$T_A = 25°C$
See Figure 1

FIGURE 5

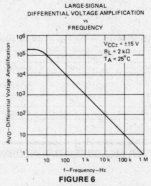

LARGE-SIGNAL
DIFFERENTIAL VOLTAGE AMPLIFICATION
vs
FREQUENCY

$V_{CC\pm} = \pm15\ V$
$R_L = 2\ k\Omega$
$T_A = 25°C$

FIGURE 6

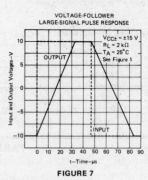

VOLTAGE-FOLLOWER
LARGE-SIGNAL PULSE RESPONSE

$V_{CC\pm} = \pm15\ V$
$R_L = 2\ k\Omega$
$T_A = 25°C$
See Figure 1

FIGURE 7

†Data at high and low temperatures are applicable only within the rated operating free-air temperature ranges of the various devices.

schematic (each amplifier)

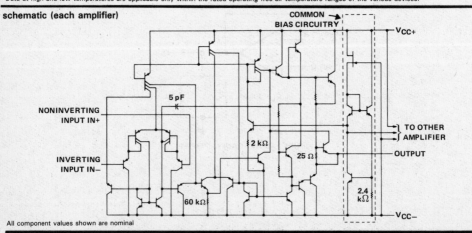

All component values shown are nominal

TEXAS INSTRUMENTS

POST OFFICE BOX 225012 • DALLAS, TEXAS 75265

LINEAR
INTEGRATED
CIRCUITS

TYPES TLC251, TLC251A, TLC251B, TLC271, TLC271A, TLC271B
PROGRAMMABLE LOW-POWER LinCMOS™ OPERATIONAL AMPLIFIERS

D2751, JULY 1983—REVISED NOVEMBER 1983

- **Wide Range of Supply Voltages:**
  1 V to 16 V (TLC251C)
  3 V to 16 V (TLC271C, TLC271I)
  4 V to 16 V (TLC271M)

- **True Single Supply Operation**

- **Common-Mode Input Voltage Range Includes the Negative Rail**

- **Low Noise . . . 30 nV/$\sqrt{\text{Hz}}$ Typ at 1 kHz (High Bias)**

D, JG, OR P DUAL-IN-LINE PACKAGE
(TOP VIEW)

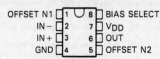

FH OR FK PACKAGE
(TOP VIEW)

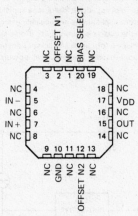

NC—No internal connection

description

The TLC251 and TLC271 series are low-cost, low-power programmable operational amplifiers designed to operate with single or dual supplies. Unlike traditional metal-gate CMOS op amps, these devices utilize Texas Instruments silicon-gate LinCMOS™ process, giving them stable input offset voltages without sacrificing the advantages of metal-gate CMOS. This series of parts is available in selected grades of input offset voltage and can be nulled with one external potentiometer. Because the input common-mode range extends to the negative rail and the power consumption is extremely low, this family is ideally suited for battery-powered or energy-conserving applications. A bias-select pin can be used to program one of three ac performance and power-dissipation levels to suit the application. The TLC251 offers the same operation sd the TLC271, but also features guaranteed operation down to a 1 V supply. Both devices are stable at unity gain.

symbol

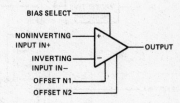

TEMPERATURE RANGES AND PACKAGES

SERIES	TEMPERATURE RANGE	PACKAGES
TLC251 _ C Types	0 °C to 70°	JG, P, D
TLC271 _ C Types	0 °C to 70 °C	JG, P, D
TLC271 _ I Types	−40 °C to 85 °C	JG, P, D
TLC271 _ M Types	−55 °C to 125 °C	JG, FH, FK

DEVICE FEATURES

PARAMETER	LOW BIAS	MEDIUM BIAS	HIGH BIAS
Supply current (Typ)	10 μA	150 μA	1000 μA
Slew rate (Typ)	0.04 V/μs	0.6 V/μs	4.5 V/μs
Input offset voltage (Max)			
. . . Standard types	10 mV	10 mV	10 mV
. . . A-suffix types	5 mV	5 mV	5 mV
. . . B-suffix types	2 mV	2 mV	2 mV
Offset voltage drift (Typ)	0.1 μV/month[†]	0.1 μV/month[†]	0.1 μV/month[†]
Offset voltage temperature coefficient (Typ)	0.7 μV/°C	2 μV/°C	5 μV/°C
Input bias current (Typ)	1 pA	1 pA	1 pA
Input offset current (Typ)	1 pA	1 pA	1 pA

[†]The long-term drift value applies after the first month.

Copyright © 1983 by Texas Instruments Incorporated

TEXAS INSTRUMENTS

POST OFFICE BOX 225012 • DALLAS, TEXAS 75265

3

Operational Amplifiers

TYPES TLC251, TLC251A, TLC251B, TLC271, TLC271A, TLC271B
PROGRAMMABLE LOW-POWER LinCMOS™ AMPLIFIERS

description (continued)

These devices have internal electrostatic discharge (ESD) protection circuits that will prevent catastrophic failures at voltages up to 2000 volts as tested under MIL-STD-883B, Method 3015.1. However, care should be exercised in handling these devices as exposure to ESD may result in a degradation of the device parametric performance.

Because of the extremely high input impedance and low input bias and offset currents, applications for the TLC251 and TLC271 series include many areas that have previously been limited to BIFET and NFET product types. Any circuit using high-impedance elements and requiring small offset errors is a good candidate for cost-effective use of these devices. Many features associated with bipolar technology are available with LinCMOS operational amplifiers without the power penalties of traditional bipolar devices. General applications such as transducer interfacing, analog calculations, amplifier blocks, active filters, and signal buffering are all easily designed with the TLC271. Remote and inaccessible equipment applications are possible using the low-voltage and low-power capabilities of the TLC251. In addition, by driving the bias-select input with a logic signal from a microprocessor, these operational amplifiers can have software-controlled performance and power consumption. The TLC251 is well suited to solve the difficult problems associated with single-battery and solar-cell-powered applications.

schematic

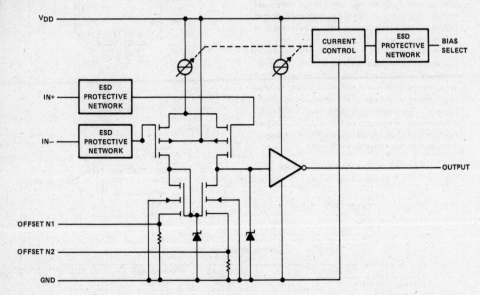

TEXAS
INSTRUMENTS

POST OFFICE BOX 225012 • DALLAS, TEXAS 75265

TYPES TLC251, TLC251A, TLC251B, TLC271, TLC271A, TLC271B
PROGRAMMABLE LOW-POWER LinCMOS™ AMPLIFIERS

absolute maximum ratings over operating free-air temperature (unless otherwise noted)

Supply voltage, V_{DD} (see Note 1) . . . . 18 V
Differential input voltage (see Note 2) . . . . ±18 V
Input voltage range (any input) . . . . −0.3 V to 18 V
Duration of short-circuit at (or below) 25 °C free-air temperature (see Note 3) . . . . unlimited
Continuous total dissipation at (or below) 25 °C free-air temperature (see Note 4):
    D package . . . . 725 mW
    FH package (see Note 5) . . . . 1200 mW
    FK package (see Note 5) . . . . 1375 mW
    JG package . . . . 1050 mW
    P package . . . . 725 mW
Operating free-air temperature range: TLC271M, TLC271AM, TLC271BM . . . . −55 °C to 125 °C
                     TLC271I, TLC271AI, TLC271BI . . . . −40 °C to 85 °C
                     TLC251C, TLC251AC, TLC251BC,
                     TLC271C, TLC271AC, TLC271BC . . . . 0 °C to 70 °C
Storage temperature range . . . . −65 °C to 150 °C
Lead temperature 1,6 mm (1/16) inch from the case for 60 seconds: JG package . . . . 300 °C
Lead Temperature 1,6 mm (1/16 inch) from case for 10 seconds: D or P package . . . . 260 °C

NOTES: 1. All voltage values, except differential voltages, are with respect to network ground terminal.
      2. Differential voltages are at the noninverting input terminal, with respect to the inverting input terminal.
      3. The output may be shorted to either supply. Temperature and/or supply voltages must be limited to ensure the maximum dissipation rating is not exceeded.
      4. For operation above 25 °C free-air temperature, refer to Dissipation Derating Table below.
      5. For FH and FK packages, power rating and rerating factor will vary with actual mounting technique used. The values stated here are believed to be conservative.

DISSIPATION DERATING TABLE

PACKAGE	POWER RATING	DERATING FACTOR	ABOVE T_A
D	725 mW	5.8 mW/°C	25 °C
FH	1200 mW	9.6 mW/°C	25 °C
FK	1375 mW	11.0 mW/°C	25 °C
JG	1050 mW	8.4 mW/°C	25 °C
P	725 mW	5.8 mW/°C	25 °C

recommended operating conditions

		TLC251 TLC251A TLC251B			TLC271 TLC271A TLC271B			UNIT
		MIN	NOM	MAX	MIN	NOM	MAX	
Supply voltage, V_{DD}	M-suffix types				4		16	V
	I-suffix types				3		16	
	C-suffix types	1		16	3		16	
Common-mode input voltage, V_{IC}	V_{DD} = 1 V	0		0.2				V
	V_{DD} = 4 V	0		3	0		3	
	V_{DD} = 10 V	−0.05		9	−0.05		9	
	V_{DD} = 16 V	−0.05		14	−0.05		14	
Operating free-air temperature, T_A	M-suffix types				−55		125	°C
	I-suffix types				−40		85	
	C-suffix types	0		70	0		70	
Bias Select pin voltage		See application notes						

3   Operational Amplifiers

TEXAS
INSTRUMENTS
POST OFFICE BOX 225012 • DALLAS, TEXAS 75265

electrical characteristics at specified free-air temperature, V_{DD} = 10 V (unless otherwise noted)

PARAMETER		TEST CONDITIONS†	BIAS	TLC271_M MIN	TYP	MAX	TLC271_I MIN	TYP	MAX	TLC251_C, TLC271_C MIN	TYP	MAX	UNIT
V_{IO} Input offset voltage	TLC251_, TLC271_	V_O = 1.4 V, R_S = 50 Ω	25°C Any			10			10			10	mV
			Full range			12			13			12	
	TLC251A_, TLC271A_		25°C Any			5			5			5	
			Full range			6.5			7			6.5	
	TLC251B_, TLC271B_		25°C Any			2			2			2	
			Full range			3			3.5			3	
α_{VIO} Average temperature coefficient of input offset voltage			Full range Low		0.7			0.7			0.7		μV/°C
			Medium		2			2			2		
			High		5			5			5		
I_{IO} Input offset current		V_{IC} = 5 V, V_O = 5 V	25°C Any		1			1			1		pA
			Full range		15000			200			100		
I_{IB} Input bias current		V_{IC} = 5 V, V_O = 5 V	25°C Any		1			1			1		pA
			Full range		35000			300			150		
V_{ICR} Common-mode input voltage range			25°C Any	−0.2 to 9			−0.2 to 9			−0.2 to 9			V
V_{OM} Peak output voltage range‡		V_{ID} = 100 mV	25°C Any	8	8.6		8	8.6		8	8.6		V
			Full range	7.8			7.8			7.8			
A_{VD} Large-signal differential voltage amplification		V_O = 1 to 6 V, R_S = 50 Ω	25°C Low	30	500		30	500		30	500		V/mV
			Medium	20	280		20	280		20	280		
			High	10	40		10	40		10	40		
			Full range Low	20			20			25			
			Medium	10			10			15			
			High	7			7			7.5			
CMRR Common-mode rejection ratio		V_O = 1.4 V, V_{IC} = V_{ICR} min	25°C Any	70	88		70	88		70	88		dB
k_{SVR} Supply voltage rejection ratio ($\Delta V_{CC}/\Delta V_{IO}$)		V_{DD} = 5 to 10 V, V_O = 1.4 V	25°C Low	70	88		70	88		70	88		dB
			Medium	70	88		70	88		70	88		
			High	65	82		65	82		65	82		
I_{OS} Short-circuit output current		V_O = 0, V_{ID} = 100 mV, V_O = V_{DD}, V_{ID} = −100 mV	25°C Any		−55			−55			−55		mA
					15			15			15		
$I_{IH(SEL)}$ High-level input current to bias select		$V_{I(SEL)}$ = 0 V	25°C High		10.5			10.5			10.5		μA
$I_{IL(SEL)}$ Low-level input current to bias select		$V_{I(SEL)}$ = 10 V	25°C Low		1.3			1.3			1.3		μA
I_{DD} Supply current		No load, V_O = 5 V, V_{IC} = 5 V	25°C Low		10	20		10	20		10	20	μA
			Medium		150	300		150	300		150	300	
			High		1000	2000		1000	2000		1000	2000	
			Full range Low			40			40			30	
			Medium			500			500			400	
			High			3000			2500			2200	

†All characteristics are measured under open-loop conditions with zero common-mode input voltage unless otherwise specified. Full range for T_A is −55°C to 125°C for TLC2__M, −40°C to 85°C for TLC2__J, and 0°C to 70°C for TLC2__C. Unless otherwise noted, an output load resistor is connected from the output to ground and has the following values: for low bias R_L = 1 MΩ, for medium bias R_L = 100 kΩ, and for high bias R_L = 10 kΩ.
‡The output will swing to the potential of the ground pin.

Texas Instruments
POST OFFICE BOX 225012 • DALLAS, TEXAS 75265

3 Operational Amplifiers

electrical characteristics at specified free-air temperature, V_{DD} = 1 V

PARAMETER			TEST CONDITIONS[†]		BIAS	TLC251_C			UNIT
						MIN	TYP	MAX	
V_{IO}	Input offset voltage	TLC251C	V_O = 0.2 V, R_S = 50 Ω	25 °C	Any			10	mV
				0 °C to 70 °C				12	
		TLC251AC		25 °C	Any			5	
				0 °C to 70 °C				6.5	
		TLC251BC		25 °C	Any			2	
				0 °C to 70 °C				3	
α_{VIO}	Average Temperature Coefficient of Input Offset Voltage			0 °C to 70 °C	Any		1		$\mu V/°C$
I_{IO}	Input offset current		V_O = 0.2 V	25 °C	Any		1		pA
				0 °C to 70 °C				100	
I_{IB}	Input bias current		V_O = 0.2 V	25 °C	Any		1		pA
				0 °C to 70 °C				150	
V_{ICR}	Common-mode input voltage range			25 °C	Any	0 to 0.2			V
V_{OM}	Peak output voltage swing[‡]		V_{ID} = 100 mV	25 °C	Any		450		mV
A_{VD}	Large-signal differential voltage amplification		V_O = 100 to 300 mV, R_S = 50 Ω	25 °C	Low		20		V/mV
					High		10		
CMRR	Common-mode rejection ratio		R_S = 50 Ω, V_O = 0.2 V, V_{IC} = V_{IC} min	25 °C	Any		77		dB
I_{DD}	Supply current		V_O = 0.2 V, No load	25 °C	Low		2		μA
					High		12		

[†]All characteristics are measured under open-loop conditions with zero common-mode input voltage unless otherwise specified. Unless otherwise noted, an output load resistor is connected from the output to ground and has the following values: for low bias R_L = 1 MΩ, for medium bias R_L = 100 kΩ, and for high bias R_L = 10 kΩ.

[‡]The output will swing to the potential of the ground pin.

operating characteristics, V_{DD} = 1 V, T_A = 25 °C

PARAMETER		TEST CONDITIONS	BIAS	TLC251_C			UNIT
				MIN	TYP	MAX	
B_1	Unity-gain bandwidth	C_L = 10 pF	Low		12		kHz
			High		75		
SR	Slew rate at unity gain	See Figure 1	Low		0.001		V/μs
			High		0.01		
	Overshoot factor	See Figure 1	Low		35%		
			High		30%		

3

Operational Amplifiers

TYPES TLC251, TLC251A, TLC251B, TLC271, TLC271A, TLC271B
PROGRAMMABLE LOW-POWER LinCMOS™ AMPLIFIERS

operating characteristics, V_{DD} = 10 V, T_A = 25°C

PARAMETER		TEST CONDITIONS	BIAS	TLC2_M			TLC2_I			TLC2_C			UNIT
				MIN	TYP	MAX	MIN	TYP	MAX	MIN	TYP	MAX	
B_1	Unity-gain bandwidth	A_V = 40 dB, C_L = 10 pF, R_S = 50Ω	Low		0.1			0.1			0.1		MHz
			Medium		0.7			0.7			0.7		
			High		2.3			2.3			2.3		
SR	Slew rate at unity gain	See Figure 1	Low		0.04			0.04			0.04		V/µs
			Medium		0.6			0.6			0.6		
			High		4.5			4.5			4.5		
	Overshoot factor	See Figure 1	Low		30%			30%			30%		
			Medium		35%			35%			35%		
			High		35%			35%			35%		
ϕ_m	Phase margin at unity gain	A_V = 40 dB, R_S = 100 Ω, C_L = pF	Low		43°			43°			43°		
			Medium		43°			43°			43°		
			High		50°			50°			50°		
V_n	Equivalent input noise voltage	f = 1 kHz, R_S = 100 Ω	Low		70			70			70		nV/√Hz
			Medium		38			38			38		
			High		30			30			30		

PARAMETER MEASUREMENT INFORMATION

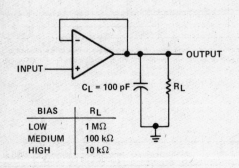

BIAS	R_L
LOW	1 MΩ
MEDIUM	100 kΩ
HIGH	10 kΩ

FIGURE 1—UNITY-GAIN AMPLIFIER

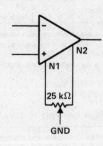

FIGURE 2—INPUT OFFSET VOLTAGE NULL CIRCUIT

Texas
INSTRUMENTS
POST OFFICE BOX 225012 • DALLAS, TEXAS 75265

TYPICAL CHARACTERISTICS

SUPPLY CURRENT
vs
BIAS SELECT PIN VOLTAGE

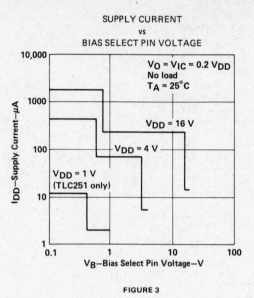

FIGURE 3

SUPPLY CURRENT
vs
SUPPLY VOLTAGE

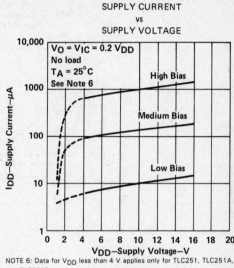

NOTE 6: Data for V_{DD} less than 4 V applies only for TLC251, TLC251A, and TLC251B.

FIGURE 4

SUPPLY CURRENT
vs
FREE-AIR TEMPERATURE

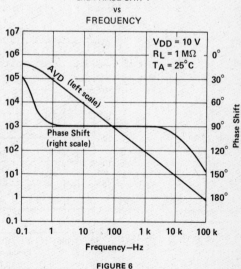

FIGURE 5

LOW BIAS
LARGE-SIGNAL
DIFFERENTIAL VOLTAGE AMPLIFICATION
and PHASE SHIFT
vs
FREQUENCY

FIGURE 6

Operational Amplifiers

3

TYPICAL CHARACTERISTICS
MEDIUM BIAS
LARGE-SIGNAL
DIFFERENTIAL VOLTAGE AMPLIFICATION
and PHASE SHIFT
vs
FREQUENCY

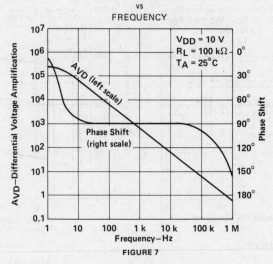

FIGURE 7

HIGH BIAS
LARGE-SIGNAL
DIFFERENTIAL VOLTAGE AMPLIFICATION
and PHASE SHIFT
vs
FREQUENCY

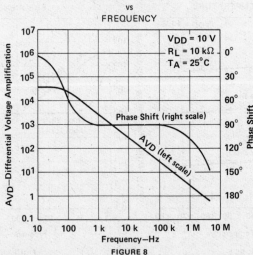

FIGURE 8

3

Operational Amplifiers

TEXAS
INSTRUMENTS
POST OFFICE BOX 225012 • DALLAS, TEXAS 75265

118

TYPICAL APPLICATION INFORMATION

latchup avoidance

Junction-isolated CMOS circuits have an inherent parasitic PNPN structure that can function as an SCR. Under certain conditions, this SCR may be triggered into a low-impedance state, resulting in excessive supply current. To avoid such conditions, no voltage greater than 0.3 V beyond the supply rails should be applied to any pin. In general, the op amp supplies should be applied simultaneously with, or before, application of any input signals.

using the bias select pin

The TLC251 and TLC271 have a bias select pin that allows the selection of one of three I_{DD} conditions (10, 150, and 1000 μA typical). This allows the user to trade-off power and ac performance. As shown in the typical supply current (I_{DD}) versus supply voltage (V_{DD}) curves (Figure 4), the I_{DD} varies only slightly from 4 to 16 V. Below 4 V, the I_{DD} varies more significantly. Note that the I_{DD} values in the medium and low-bias modes at V_{DD} = 1 V are typically 2 μA, and in the high mode are typically 12μA. The following table shows the recommended bias select pin connections at V_{DD} = 10 V:

RECOMMENDED BIAS SELECT PIN USE AT V_{DD} = 10 V

BIAS MODE	AC PERFORMANCE	BIAS SELECT CONNECTION[†]	TYPICAL I_{DD}[§]
Low	Low	V_{DD}	10 μA
Medium	Medium	0.8 V to 9.2 V	150 μA
High	High	Ground pin	1000 μA

[†]The Bias Select pin may also be controlled by external circuitry to conserve power, etc. For information regarding the bias select pin, see Figure 3 in the typical characteristics curves.

[§]For I_{DD} characteristics at voltages other than 10 V, see Figure 4 in the typical characteristics curves.

output stage considerations

The amplifier's output stage consists of a source-follower-connected pullup transistor and an open-drain pulldown transistor. The high-level output voltage (V_{OH}) is virtually independent of the I_{DD} selection, and increases with higher values of V_{DD} and reduced output loading. The low-level output voltage (V_{OL}) decreases with reduced output current and higher input common-mode voltage. With no load, V_{OL} is essentially equal to the GND pin potential.

input offset nulling

Both the TLC251 and TLC271 offer external offset null control. Nulling may be achieved by adjusting a 25-kΩ potentiometer connected between the offset null terminals with the wiper connected to the device GND pin as shown in Figure 2. The amount of nulling range varies with the bias selection. At I_{DD} settings of 150 and 1000 μA (medium and high bias), the nulling range will allow the maximum offset specified to be trimmed to zero. In low bias or when the TLC251 is used below 4 V, total nulling may not be possible on all units.

supply configurations

Even though the TLC251 and TLC271 are characterized for single-supply operation, they can be used effectively in a split-supply configuration when the input common-mode voltage (V_{ICR}), output swing (V_{OL} and V_{OH}), and supply voltage limits are not exceeded.

circuit layout precautions

The user is cautioned that whenever extremely high circuit impedances are used, care must be exercised in layout, construction, board cleanliness, and supply filtering to avoid hum and noise pickup, as well as excessive dc leakages.

3

Operational Amplifiers

TEXAS
INSTRUMENTS
POST OFFICE BOX 225012 • DALLAS, TEXAS 75265

Operational Amplifiers

- **Wide Range of Supply Voltages:**
  1 V to 16 V (TLC252C)
  3 V to 16 V (TLC272C, TLC272I)
  4 V to 16 V (TLC272M)

- **True Single-Supply Operation**

- **Common-Mode Input Voltage Includes the Negative Rail**

- **Low Noise . . . 30 nV/√Hz Typ at f = 1 kHz (High-Bias Versions)**

D, JG, OR P DUAL-IN-LINE PACKAGE
(TOP VIEW)

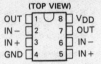

```
OUT  [1  U  8] VDD
IN−  [2     7] OUT
IN+  [3     6] IN−
GND  [4     5] IN+
```

FH OR FK PACKAGE
(TOP VIEW)

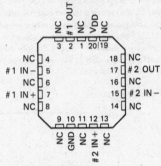

NC—No internal connection

symbol (each amplifier)

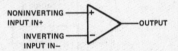

NONINVERTING INPUT IN+ → +
INVERTING INPUT IN− → −
→ OUTPUT

description

The TLC252 and TLC272 series are low-cost, low-power dual operational amplifiers designed to operate with single or dual supplies. These devices utilize the Texas Instruments silicon gate LinCMOS™ process, giving them stable input offset voltages that are available in selected grades of 2, 5 or 10 mV maximum, very high input impedances, and extremely low input offset and bias currents. Because the input common-mode range extends to the negative rail and the power consumption is extremely low, this series is ideally suited for battery-powered or energy-conserving applications. The TLC252 types offer guaranteed operation down to a 1-V supply. All devices are unity-gain stable and have excellent noise characteristics.

3

Operational Amplifiers

DEVICE FEATURES

PARAMETER	TLC25L2 TLC27L2 (LOW BIAS)	TLC25M2 TLC27M2 (MEDIUM BIAS)	TLC252 TLC272 (HIGH BIAS)
Supply current (Typ)	20 μA	300 μA	2000 μA
Slew rate (Typ)	0.04 V/μs	0.6 V/μs	4.5 V/μs
Input offset voltage (Max)			
. . . Standard types	10 mV	10 mV	10 mV
. . . A-suffix types	5 mV	5 mV	5 mV
. . . B-suffix types	2 mV	2 mV	2 mV
Offset voltage drift (Typ)	0.1 μV/month[†]	0.1 μV/month[†]	0.1 μV/month[†]
Offset voltage temperature coefficient (Typ)	0.7 μV/°C	2 μV/°C	5 μV/°C
Input bias current (Typ)	1 pA	1 pA	1 pA
Input offset current (Typ)	1 pA	1 pA	1 pA

[†]The offset voltage drift applies after the first month only.

TEMPERATURE RANGES AND PACKAGES

TYPES	TEMPERATURE RANGE	PACKAGES
TLC25_2_C	0°C to 70°	JG, P, D
TLC27_2_C	0°C to 70°C	JG, P, D
TLC27_2_I	−40°C to 85°C	JG, P, D
TLC27_2_M	−55°C to 125°C	JG, FH, FK

TEXAS INSTRUMENTS
POST OFFICE BOX 225012 • DALLAS, TEXAS 75265

description (continued)

These devices have internal electrostatic discharge (ESD) protection circuits that will prevent catastrophic failures at voltages up to 2000 volts as tested under MIL-STD-883B, Method 3015.1. However, care should be exercised in handling these devices as exposure to ESD may result in a degradation of the device parametric performance.

Because of the extremely high input impedance and low input bias and offset currents, applications for the TLC252 and TLC272 series include many areas that have previously been limited to BIFET and NFET product types. Any circuit using high-impedance elements and requiring small offset errors is a good candidate for cost-effective use of these devices. Many features associated with bipolar technology are available with LinCMOS™ operational amplifiers without the power penalties of traditional bipolar devices. General applications such as transducer interfacing, analog calculations, amplifier blocks, active filters, and signal buffering are all easily designed with the TLC252 and TLC272 series. Remote and inaccessible equipment applications are possible using the low-voltage and low-power capabilities of the TLC252. The TLC252 types are well suited to solve the difficult problems associated with single-battery and solar-cell-powered applications. This series includes devices that are characterized for commercial, industrial, and military temperature ranges and are available in 8-pin plastic and ceramic dual-in-line (DIP) packages, small outline (D) package, and chip carrier (FH, FK) packages.

schematic (each amplifier)

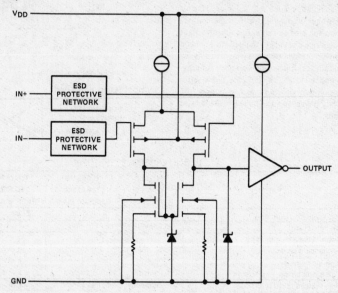

1083

TEXAS
INSTRUMENTS
POST OFFICE BOX 225012 • DALLAS, TEXAS 75265

3

Operational Amplifiers

absolute maximum ratings over operating free-air temperature range (unless otherwise noted)

Supply voltage, V_{DD} (see Note 1)	18 V
Differential input voltage (see Note 2)	±18 V
Input voltage range (any input)	−0.3 V to 18 V
Duration of short-circuit at (or below) 25°C free-air temperature (see Note 3)	unlimited
Continuous total dissipation at (or below) 25°C free-air temperature (see Note 4):	
D package	725 mW
FH package (see Note 5)	1200 mW
FK package (see Note 5)	1375 mW
JG package	1050 mW
P package	725 mW
Operating free-air temperature range: TLC27_2_M	−55°C to 125°C
TLC27_2_I	−40°C to 85°C
TLC25_2_C, TLC27_2_C	0°C to 70°C
Storage temperature range	−65°C to 150°C
Lead temperature 1,6 mm (1/16 inch) from the case for 60 seconds: JG package	300°C
Lead temperature 1,6 mm (1/16 inch) from the case for 10 seconds: D or P package	260°C

NOTES: 1. All voltage values, except differential voltages, are with respect to network ground terminal.
2. Differential voltages are at the noninverting input terminal with respect to the inverting input terminal.
3. The output may be shorted to either supply. Temperature and/or supply voltages must be limited to ensure the maximum dissipation rating is not exceeded.
4. For operation above 25°C free-air temperature, refer to the Dissipation Derating Table.
5. For FH and FK packages, power rating and derating factor will vary with the actual mounting technique used. The values stated here are believed to be conservative.

DISSIPATION DERATING TABLE

PACKAGE	POWER RATING	DERATING FACTOR	ABOVE T_A
D	725 mW	5.8 mW/°C	25°C
FH	1200 mW	9.6 mW/°C	25°C
FK	1375 mW	11 mW/°C	25°C
JG	1050 mW	8.4 mW/°C	25°C
P	725 mW	5.8 mW/°C	25°C

recommended operating conditions

		TLC25_2 TLC25_2A TLC25_2B			TLC27_2 TLC27_2A TLC27_2B			UNIT
		MIN	NOM	MAX	MIN	NOM	MAX	
Supply voltage, V_{DD}	M-suffix types				4		16	
	I-suffix types				3		16	V
	C-suffix types	1		16	3		16	
Common-mode input voltage, V_{IC}	V_{DD} = 1 V	0		0.2				
	V_{DD} = 4 V	0		3	0		3	
	V_{DD} = 10 V	−0.05		9	−0.05		9	V
	V_{DD} = 16 V	−0.05		14	−0.05		14	
Operating free-air temperature, T_A	M-suffix types				−55		125	
	I-suffix types				−40		85	°C
	C-suffix types	0		70	0		70	

O83

M-SUFFIX TYPES

electrical characteristics at specified free-air temperature, V_{DD} = 10 V (unless otherwise noted)

PARAMETER			TEST CONDITIONS[†]	TLC272_M MIN	TYP	MAX	TLC27L2_M MIN	TYP	MAX	TLC27M2_M MIN	TYP	MAX	UNIT
V_{IO}	Input offset voltage	TLC27_2M	V_O = 1.4 V, R_S = 50 Ω	25°C		10			10			10	mV
				−55°C to 125°C		12			12			12	
		TLC27_2AM		25°C		5			5			5	
				−55°C to 125°C		6.5			6.5			6.5	
		TLC27_2BM		25°C		2			2			2	
				−55°C to 125°C		3.5			3.5			3.5	
α_{VIO}	Average temperature coefficient of input offset voltage			−55°C to 125°C	5			0.7			2		μV/°C
I_{IO}	Input offset current		V_{IC} = 5 V, V_O = 5 V	25°C		1			1			1	pA
				−55°C to 125°C		15			15			15	nA
I_{IB}	Input bias current		V_{IC} = 5 V, V_O = 5 V	25°C		1			1			1	pA
				−55°C to 125°C		35			35			35	nA
V_{ICR}	Common-mode input voltage range			25°C	−0.2 to 9			−0.2 to 9			−0.2 to .9			V
V_{OM}	Peak output voltage swing[‡]		V_{ID} = 100 mV	25°C	8	8.6		8	8.6		8	8.6		V
				−55°C to 125°C	7.8			7.8			7.8			
A_{VD}	Large-signal differential voltage amplification		V_O = 1 to 6 V, R_S = 50 Ω	25°C	10	40		30	500		20	280		V/mV
				−55° to 125°C	7			20			10			
CMRR	Common-mode rejection ratio		V_O = 1.4 V, V_{IC} = V_{ICR} min	25°C	70	88		70	88		70	88		dB
k_{SVR}	Supply voltage rejection ratio (Δ V_{CC}/Δ V_{IO})		V_{DD} = 5 to 10 V, V_O = 1.4 V	25°C	65	82		70	88		70	88		dB
I_{OS}	Short-circuit output current		V_O = 0, V_{ID} = 100 mV	25°C		−55			−55			−55		mA
			V_O = V_{DD}, V_{ID} = −100 mV			15			15			15		
I_{DD}	Supply current (each amplifier)		No load, V_O = 5 V, V_{IC} = 5 V	25°C		1000	2000		10	20		150	300	μA
				−55°C to 125°C			3000			40			500	

† All characteristics are measured under open-loop conditions with zero common-mode input voltage unless otherwise specified. Unless otherwise noted, an output load resistor is connected form the output to the ground pin.

‡ The output will swing to the potential of the ground pin.

1083

TEXAS INSTRUMENTS

POST OFFICE BOX 225012 • DALLAS, TEXAS 75265

I-SUFFIX TYPES

electrical characteristics at specified free-air temperature, V_{DD} = 10 V (unless otherwise noted)

PARAMETER			TEST CONDITIONS[†]	TLC272_I			TLC27L2_I			TLC27M2_I			UNIT
				MIN	TYP	MAX	MIN	TYP	MAX	MIN	TYP	MAX	
V_{IO}	Input offset voltage	TLC27_2I	V_O = 1.4 V, R_S = 50 Ω										mV
			25°C			10			10			10	
			−40°C to 85°C			13			13			13	
		TLC27_2AI	25°C			5			5			5	
			−40°C to 85°C			7			7			7	
		TLC27_2BI	25°C			2			2			2	
			−40°C to 85°C			3.5			3.5			3.5	
α_{VIO}	Average temperature coefficient of input offset voltage		−40°C to 85°C		5			0.7			2		µV/°C
I_{IO}	Input offset current		V_{IC} = 5 V, V_O = 5 V										pA
			25°C		1			1			1		
			−40°C to 85°C			200			200			200	
I_{IB}	Input bias current		V_{IC} = 5 V, V_O = 5 V										pA
			25°C		1			1			1		
			−40°C to 85°C			300			300			300	
V_{ICR}	Common-mode input voltage range		25°C	−0.2 to 9			−0.2 to 9			−0.2 to 9			V
V_{OM}	Peak output voltage swing[‡]	V_{ID} = 100 mV	25°C	8	8.6		8	8.6		8	8.6		V
			−40°C to 85°C	7.8			7.8			7.8			
A_{VD}	Large-signal differential voltage amplification	V_O = 1 to 6 V, R_S = 50 Ω	25°C	10	40		30	500		20	280		V/mV
			−40°C to 85°C	7			20			10			
CMRR	Common-mode rejection ratio	V_O = 1.4 V, V_{IC} = V_{ICR} min	25°C	70	88		70	88		70	88		dB
k_{SVR}	Supply voltage rejection ratio ($\Delta V_{CC}/\Delta V_{IO}$)	V_{DD} = 5 to 10 V, V_O = 1.4 V	25°C	65	82		70	88		70	88		dB
I_{OS}	Short-circuit output current	V_O = 0, V_{ID} = 100 mV	25°C		−55			−55			−55		mA
		V_O = V_{DD}, V_{ID} = −100 mV			15			15			15		
I_{DD}	Supply current (each amplifier)	No load, V_O = 5 V, V_{IC} = 5 V	25°C		1000	2000		10	20		150	300	µA
			−40°C to 85°C			2500			40			500	

[†] All characteristics are measured under open-loop conditions with zero common-mode input voltage unless otherwise specified. Unless otherwise noted, an output load resistor is connected from the output to the ground pin.
[‡] The output will swing to the potential of the ground pin.

3

Operational Amplifiers

TYPES TLC252, TLC25L2, TLC25M2, TLC272, TLC27L2, TLC27M2
LinCMOS™ DUAL OPERATIONAL AMPLIFIERS

C-SUFFIX TYPES

electrical characteristics at specified free-air temperature, V_{DD} = 10 V (unless otherwise noted)

PARAMETER			TEST CONDITIONS†	TLC252_C, TLC272_C			TLC25L2_C, TLC27L2_C			TLC25M2_C, TLC27M2_C			UNIT
				MIN	TYP	MAX	MIN	TYP	MAX	MIN	TYP	MAX	
V_{IO}	Input offset voltage	TLC2_2C	V_O = 1.4 V, R_S = 50 Ω	25°C		10			10			10	mV
				0°C to 70°C		12			12			12	
		TLC2_2AC		25°C		5			5			5	
				0°C to 70°C		6.5			6.5			6.5	
		TLC2_2BC		25°C		2			2			2	
				0°C to 70°C		3			3			3	
α_{VIO}	Average temperature coefficient of input offset voltage		0°C to 70°C		5			0.7			2		µV/°C
I_{IO}	Input offset current		V_{IC} = 5 V, V_O = 5 V	25°C	1			1			1		pA
				0°C to 70°C		100			100			100	
I_{IB}	Input bias current		V_{IC} = 5 V, V_O = 5 V	25°C	1			1			1		pA
				0°C to 70°C		150			150			150	
V_{ICR}	Common-mode input voltage range		25°C	−0.2 to 9			−0.2 to 9			−0.2 to 9			V
V_{OM}	Peak output voltage swing‡		V_{ID} = 100 mV	25°C	8	8.6		8	8.6		8	8.6	V
				0°C to 70°C	7.8			7.8			7.8		
A_{VD}	Large-signal differential voltage amplification	V_O = 1 to 6 V, R_S = 50 Ω	25°C	10	40		30	500		20	280		V/mV
			0° to 70°C	7.5			25			15			
CMRR	Common-mode rejection ratio	V_O = 1.4 V, V_{IC} = V_{ICR} min	25°C	70	88		70	88		70	88	dB
k_{SVR}	Supply voltage rejection ratio (Δ V_{CC}/Δ V_{IO})	V_{DD} = 5 to 10 V, V_O = 1.4 V	25°C	65	82		70	88		70	88	dB
I_{OS}	Short-circuit output current	V_O = 0, V_{ID} = 100 mV / V_O = V_{DD}, V_{ID} = −100 mV	25°C		−55			−55			−55	mA
					15			15			15	
I_{DD}	Supply current (each amplifier)	No load, V_O = 5 V, V_{IC} = 5 V	25°C		1000	2000		10	20		150	300	µA
			0°C to 70°C		2200			30			400		

† All characteristics are measured under open-loop conditions with zero common-mode input voltage unless otherwise specified. Unless otherwise noted, an output load resistor is connected from the output to the ground pin.

‡ The output will swing to the potential of the ground pin.

TEXAS
INSTRUMENTS
POST OFFICE BOX 225012 • DALLAS, TEXAS 75265

1083

C-SUFFIX TYPES

electrical characteristics at specified free-air temperature, V_{DD} = 1 V (unless otherwise noted)

PARAMETER			TEST CONDITIONS†		TLC252_C			TLC25L2_C			TLC25M2_C			UNIT
					MIN	TYP	MAX	MIN	TYP	MAX	MIN	TYP	MAX	
V_{IO}	Input offset voltage	TLC25_2C	V_O = 0.2 V, R_S = 50 Ω	25°C			10			10			10	mV
				0°C to 70°C			12			12			12	
		TLC25_2AC		25°C			5			5			5	
				0°C to 70°C			6.5			6.5			6.5	
		TLC25_2BC		25°C			2			2			2	
				0°C to 70°C			3			3			3	
α_{VIO}	Average temperature coefficient of input offset voltage			0°C to 70°C		1			1			1		μV/°C
I_{IO}	Input offset current		V_O = 0.2 V	25°C		1			1			1		pA
				0°C to 70°C			100			100			100	
I_{IB}	Input bias current		V_O = 0.2 V	25°C		1			1			1		pA
				0°C to 70°C			150			150			150	
V_{ICR}	Common-mode input voltage range			25°C	0 to 0.2			0 to 0.2			0 to 0.2			V
V_{OM}	Peak output voltage swing‡		V_{ID} = 100 mV	25°C		450			450			450		mV
A_{VD}	Large-signal differential voltage amplification		V_O = 100 to 300 mV, R_S = 50 Ω	25°C		10			20			20		V/mV
CMRR	Common-mode rejection ratio		V_O = 0.2 V, V_{IC} = V_{ICR} min	25°C		77			77			77		dB
I_{DD}	Supply current (each amplifier)		No load, V_O = 0.2 V	25°C		12			2			2		μA

† All characteristics are measured under open-loop conditions with zero common-mode input voltage unless otherwise specified. Unless otherwise noted, an output load resistor is connected form the output to the ground pin.
‡ The output will swing to the potential of the ground pin.

operating characteristics, V_{DD} = 1 V, T_A = 25°C

PARAMETER		TEST CONDITIONS	TLC252_C			TLC25L2_C			TLC25M2_C			UNIT
			MIN	TYP	MAX	MIN	TYP	MAX	MIN	TYP	MAX	
B_1	Unity-gain bandwidth	A_V = 40 dB, C_L = 10 pF, R_S = 50 Ω		75			12			12		kHz
SR	Slew rate at unity gain	See Figure 1		0.01			0.001			0.001		V/μs
	Overshoot factor	See Figure 1		30%			35%			35%		

3

Operational Amplifiers

TEXAS
INSTRUMENTS
POST OFFICE BOX 225012 • DALLAS, TEXAS 75265

TYPES TLC252, TLC25L2, TLC25M2, TLC272, TLC27L2, TLC27M2
LinCMOS™ DUAL OPERATIONAL AMPLIFIERS

operating characteristics, $V_{DD} = 10$ V, $T_A = 25°C$

PARAMETER		TEST CONDITIONS	TLC252__C TLC272__M TLC272__I TLC272__C			TLC25L2__C TLC27L2__M TLC27L2__I TLC27L2__C			TLC25M2__C TLC27M2__M TLC27M2__I TLC27M2__C			UNIT
			MIN	TYP	MAX	MIN	TYP	MAX	MIN	TYP	MAX	
B_1	Unity-gain bandwidth	$A_V = 40$ dB, $C_L = 10$ pF, $R_S = 50 \, \Omega$		2.3			0.1			0.7		MHz
SR	Slew rate at unity gain	See Figure 1		4.5			0.04			0.6		V/µs
	Overshoot factor	See Figure 1		35%			30%			35%		
ϕ_m	Phase margin at unity gain	$A_V = 40$ dB, $R_S = 100 \, \Omega$, $C_L = 10$ pF		50°			43°			43°		
V_n	Equivalent input noise voltage	$f = 1$ kHz, $R_S = 100 \, \Omega$		30			70			38		nV/√Hz
V_{o1}/V_{o2}	Cross talk attenuation	$A_V = 100$		120			120			120		dB

PARAMETER MEASUREMENT INFORMATION

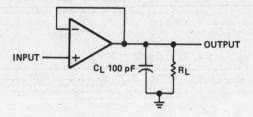

FIGURE 1—UNITY GAIN AMPLIFIER

3

Operational Amplifiers

TEXAS
INSTRUMENTS
POST OFFICE BOX 225012 • DALLAS, TEXAS 75265

1083

TYPICAL CHARACTERISTICS

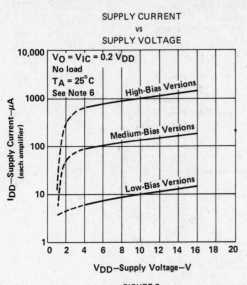

SUPPLY CURRENT
vs
SUPPLY VOLTAGE

FIGURE 2

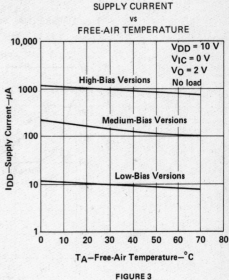

SUPPLY CURRENT
vs
FREE-AIR TEMPERATURE

FIGURE 3

NOTE 6: Data for V_{DD} less than 4 V does not apply for the TLC272 series.

LOW-BIAS VERSIONS
LARGE-SIGNAL
DIFFERENTIAL VOLTAGE AMPLIFICATION
and PHASE SHIFT
vs
FREQUENCY

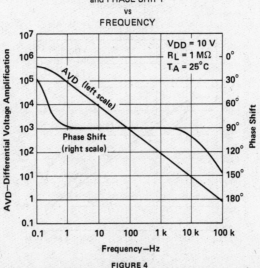

FIGURE 4

Operational Amplifiers

3

TEXAS
INSTRUMENTS
POST OFFICE BOX 225012 ● DALLAS, TEXAS 75265

TYPICAL CHARACTERISTICS

MEDIUM-BIAS VERSIONS
LARGE-SIGNAL
DIFFERENTIAL VOLTAGE AMPLIFICATION
and PHASE SHIFT
vs
FREQUENCY

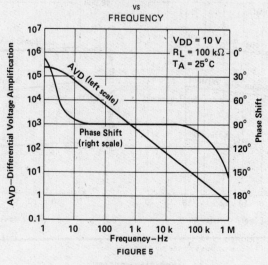

FIGURE 5

HIGH-BIAS VERSIONS
LARGE-SIGNAL
DIFFERENTIAL VOLTAGE AMPLIFICATION
and PHASE SHIFT
vs
FREQUENCY

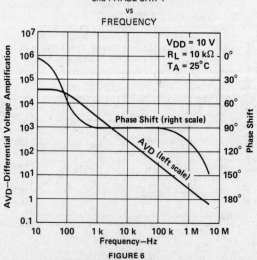

FIGURE 6

Texas Instruments
POST OFFICE BOX 225012 • DALLAS, TEXAS 75265

1083

TYPICAL APPLICATION INFORMATION

latchup avoidance

Junction-isolated CMOS circuits have an inherent parasitic PNPN structure that can function as an SCR. Under certain conditions, this SCR may be triggered into a low-impedance state, resulting in excessive supply current. To avoid such conditions, no voltage greater than 0.3 V beyond the supply rails should be applied to any pin. In general, the op amp supplies should be established simultaneously with, or before, any input signals are applied.

output stage considerations

The amplifier's output stage consists of a source follower connected pullup transistor and an open drain pulldown transistor. The high-level output voltage (V_{OH}) is virtually independent of the I_{DD} selection, and increases with higher values of V_{DD} and reduced output loading. The low-level output voltage (V_{OL}) decreases with reduced output current and higher input common-mode voltage. With no load, V_{OL} is essentially equal to the GND pin potential.

supply configurations

Even though the TLC252 and TLC272 are characterized for single-supply operation, they can be used effectively in a split supply configuration if the input common-mode voltage (V_{ICR}), output swing (V_{OL} and V_{OH}), and supply voltage limits are not exceeded.

circuit layout precautions

The user is cautioned that whenever extremely high circuit impedances are used, care must be exercised in layout, construction, board cleanliness, and supply filtering to avoid hum and noise pickup, as well as excessive DC leakages.

3

Operational Amplifiers

Operational Amplifiers

TYPES TLC254, TLC25L4, TLC25M4, TLC274, TLC27L4, TLC27M4
LinCMOS™ QUAD OPERATIONAL AMPLIFIERS

D2753, JUNE 1983–REVISED NOVEMBER 1983

- **Wide Range of Supply Voltages:**
  1 V to 16 V (TLC254C)
  3 V to 16 V (TLC274C, TLC274I)
  4 V to 16 V (TLC274M)

- **True Single-Supply Operation**

- **Common-Mode Input Voltage Includes the Negative Rail**

- **Low Noise . . . 30 nV/$\sqrt{Hz}$ Typ at f = 1 kHz (High-Bias Versions)**

description

The TLC254 and TLC274 series are low-cost, low-power quad operational amplifiers designed to operate with single or dual supplies. These devices utilize the Texas Instruments silicon-gate LinCMOS™ process, giving them stable input offset voltages that are available in selected grades of 2, 5 or 10 mV maximum, very high input impedances, and extremely low input offset and bias currents. Because the input common-mode range extends to the negative rail and the power consumption is extremely low, this series is ideally suited for battery-powered or energy-conserving applications. The TLC254 types offer guaranteed operation down to a 1-V supply. All devices are unity-gain stable and have excellent noise characteristics.

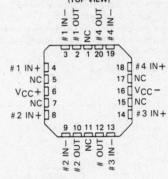

D, J, OR N DUAL IN-LINE PACKAGE
(TOP VIEW)

FH OR FK PACKAGE
(TOP VIEW)

NC—No internal connection

symbol (each amplifier)

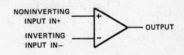

NONINVERTING INPUT IN+

INVERTING INPUT IN−

OUTPUT

TEMPERATURE RANGES AND PACKAGES

TYPES	TEMPERATURE RANGE	PACKAGES
TLC25_4_C	0°C to 70°	J, N, D
TLC27_4_C	0°C to 70°C	J, N, D
TLC27_4_I	−40°C to 85°C	J, N, D
TLC27_4_M	−55°C to 125°C	J, FH, FK

DEVICE FEATURES

PARAMETER	TLC25L4 TLC27L4 (LOW BIAS)	TLC25M4 TLC27M4 (MEDIUM BIAS)	TLC254 TLC274 (HIGH BIAS)
Supply current (Typ)	40 μA	600 μA	4000 μA
Slew rate (Typ)	0.04 V/μs	0.6 V/μs	4.5 V/μs
Input offset voltage (Max)			
. . . Standard types	10 mV	10 mV	10 mV
. . . A-suffix types	5 mV	5 mV	5 mV
. . . B-suffix types	2 mV	2 mV	2 mV
Offset voltage drift (Typ)	0.1 μV/month[†]	0.1 μV/month[†]	0.1 μV/month[†]
Offset voltage temperature coefficient (Typ)	0.7 μV/°C	2 μV/°C	5 μV/°C
Input bias current (Typ)	1 pA	1 pA	1 pA
Input offset current (Typ)	1 pA	1 pA	1 pA

[†]The long-term drift value applies after the first month.

Operational Amplifiers

3

TEXAS INSTRUMENTS
POST OFFICE BOX 225012 • DALLAS, TEXAS 75265

description (continued)

These devices have internal electrostatic discharge (ESD) protection circuits that will prevent catastrophic failures at voltages up to 2000 volts as tested under MIL-STD-883B, Method 3015.1. However, care should be exercised in handling these devices as exposure to ESD may result in a degradation of the device parametric performance.

Because of the extremely high input impedance and low input bias and offset currents, applications for the TLC254 and TLC274 series include many areas that have previously been limited to BIFET and NFET product types. Any circuit using high-impedance elements and requiring small offset errors is a good candidate for cost-effective use of these devices. Many features associated with bipolar technology are available with LinCMOS™ operational amplifiers without the power penalties of traditional bipolar devices. General applications such as transducer interfacing, analog calculations, amplifier blocks, active filters, and signal buffering are all easily designed with the TLC254 and TLC274 series. Remote and inaccessible equipment applications are possible using the low-voltage and low-power capabilities of the TLC254. The TLC254 types are well suited to solve the difficult problems associated with single-battery and solar-cell-powered applications. This series includes devices that are characterized for commercial, industrial, and military temperature ranges and are available in 14-pin plastic and ceramic dual-in-line (DIP) packages, small outline (D) package, and chip carrier (FH, FK) packages.

schematic (each amplifier)

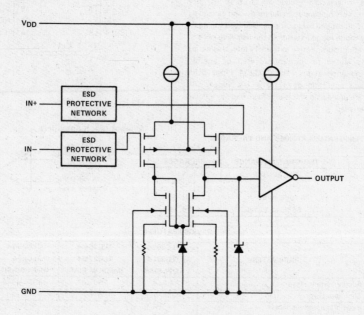

TEXAS
INSTRUMENTS

POST OFFICE BOX 225012 • DALLAS, TEXAS 75265

absolute maximum ratings over operating free-air temperature range (unless otherwise noted)

Supply voltage, V_{DD} (see Note 1)	18 V
Differential input voltage (see Note 2)	±18 V
Input voltage range (any input)	−0.3 V to 18 V
Duration of short-circuit at (or below) 25°C free-air temperature (see Note 3)	unlimited

Continuous total dissipation at (or below) 25°C free-air temperature (see Note 4):

D package	950 mW
FH package (see Note 5)	1200 mW
FK package (see Note 5)	1375 mW
J package	1375 mW
N package	875 mW

Operating free-air temperature range: TLC27_4_M	−55°C to 125°C
TLC27_4_I	−25°C to 85°C
TLC25_4_C, TLC27_4_C	0°C to 70°C
Storage temperature range	−65°C to 150°C
Lead temperature 1,6 mm (1/16 inch) from the case for 60 seconds: J package	300°C
Lead temperature 1,6 mm (1/16 inch) from the case for 10 seconds: D or N package	260°C

NOTES: 1. All voltage values, except differential voltages, are with respect to network ground terminal.
2. Differential voltages are at the noninverting input terminal with respect to the inverting input terminal.
3. The output may be shorted to either supply. Temperature and/or supply voltages must be limited to ensure the maximum dissipation rating is not exceeded.
4. For operation above 25°C free-air temperature, refer to the Dissipation Derating Table.
5. For FH and FK packages, power rating and derating factor will vary with the actual mounting technique used. The values stated hee are belived to be conservative.

DISSIPATION DERATING TABLE

PACKAGE	POWER RATING	DERATING FACTOR	ABOVE T_A
D	950 mW	7.6 mW/°C	25°C
FH	1200 mW	9.6 mW/°C	25°C
FK	1375 mW	11 mW/°C	25°C
J	1375 mW	11 mW/°C	25°C
N	875 mW	7 mW/°C	25°C

recommended operating conditions

		TLC25_4 TLC25_4A TLC25_4B			TLC27_4 TLC27_4A TLC27_4B			UNIT
		MIN	NOM	MAX	MIN	NOM	MAX	
Supply voltage, V_{DD}	M-suffix types				4		16	
	I-suffix types				3		16	V
	C-suffix types	1		16	3		16	
Common-mode input voltage, V_{IC}	$V_{DD} = 1$ V	0		0.2				
	$V_{DD} = 4$ V	0		3	0		3	V
	$V_{DD} = 10$ V	−0.05		9	−0.05		9	
	$V_{DD} = 16$ V	−0.05		14	−0.05		14	
Operating free-air temperature, T_A	M-suffix types				−55		125	
	I-suffix types				−40		85	°C
	C-suffix types	0		70	0		70	

TEXAS INSTRUMENTS
POST OFFICE BOX 225012 • DALLAS, TEXAS 75265

TYPES TLC274, TLC27L4, TLC27M4
LinCMOS™ QUAD OPERATIONAL AMPLIFIERS

M-SUFFIX TYPES

electrical characteristics at specified free-air temperature, V_{DD} = 10 V (unless otherwise noted)

PARAMETER		TEST CONDITIONS[†]		TLC274_M			TLC27L4_M			TLC27M4_M			UNIT
				MIN	TYP	MAX	MIN	TYP	MAX	MIN	TYP	MAX	
V_{IO} Input offset voltage	TLC27_4M	V_O = 1.4 V, R_S = 50 Ω	25°C			10			10			10	mV
			−55°C to 125°C			12			12			12	
	TLC27_4AM		25°C			5			5			5	
			−55°C to 125°C			6.5			6.5			6.5	
	TLC27_4BM		25°C			2			2			2	
			−55°C to 125°C			3.5			3.5			3.5	
α_{VIO}	Average temperature coefficient of input offset voltage		−55°C to 125°C		5			0.7			2		μV/°C
I_{IO}	Input offset current	V_{IC} = 5 V, V_O = 5 V	25°C		1			1			1		pA
			−55°C to 125°C			15			15			15	nA
I_{IB}	Input bias current	V_{IC} = 5 V, V_O = 5 V	25°C		1			1			1		pA
			−55°C to 125°C			35			35			35	nA
V_{ICR}	Common-mode input voltage range		25°C	−0.2 to 9			−0.2 to 9			−0.2 to 9			V
V_{OM}	Peak output voltage swing[‡]	V_{ID} = 100 mV	25°C	8	8.6		8	8.6		8	8.6		V
			−55°C to 125°C	7.8			7.8			7.8			
A_{VD}	Large-signal differential voltage amplification	V_O = 1 to 6 V, R_S = 50 Ω	25°C	10	40		30	500		20	280		V/mV
			−55° to 125°C	7			20			10			
CMRR	Common-mode rejection ratio	V_O = 1.4 V, V_{IC} = V_{ICR} min	25°C	70	88		70	88		70	88		dB
k_{SVR}	Supply voltage rejection ratio (ΔV_{CC}/ΔV_{IO})	V_{DD} = 5 to 10 V, V_O = 1.4 V	25°C	65	82		70	88		70	88		dB
I_{OS}	Short-circuit output current	V_O = 0, V_{ID} = 100 mV	25°C		−55			−55			−55		mA
		V_O = V_{DD}, V_{ID} = −100 mV			15			15			15		
I_{DD}	Supply current (each amplifier)	No load, V_O = 5 V,	25°C		1000	2000		10	20		150	300	μA
		V_{IC} = 5 V	−55°C to 125°C			3000			40			500	

[†] All characteristics are measured under open-loop conditions with zero common-mode input voltage unless otherwise specified. Unless otherwise noted, an output load resistor is connected form the output to the ground pin.
[‡] The output will swing to the potential of the ground pin.

TEXAS
INSTRUMENTS
POST OFFICE BOX 225012 • DALLAS, TEXAS 75265

118

I-SUFFIX TYPES

electrical characteristics at specified free-air temperature, V_{DD} = 10 V (unless otherwise noted)

PARAMETER			TEST CONDITIONS†	TLC274_I			TLC27L4_I			TLC27M4_I			UNIT
				MIN	TYP	MAX	MIN	TYP	MAX	MIN	TYP	MAX	
V_{IO}	Input offset voltage	TLC27_4I	V_O = 1.4 V, R_S = 50 Ω			10			10			10	mV
						13			13			13	
		TLC27_4AI				5			5			5	
			−40°C to 85°C			7			7			7	
		TLC27_4BI				2			2			2	
						3.5			3.5			3.5	
α_{VIO}	Average temperature coefficient of input offset voltage		−40°C to 85°C		5			0.7			2		µV/°C
I_{IO}	Input offset current	V_{IC} = 5 V, V_O = 5 V	25°C		1			1			1		pA
			−40°C to 85°C			200			200			200	
I_{IB}	Input bias current	V_{IC} = 5 V, V_O = 5 V	25°C		1			1			1		pA
			−40°C to 85°C			300			300			300	
V_{ICR}	Common-mode input voltage range		25°C	−0.2 to 9			−0.2 to 9			−0.2 to 9			V
V_{OM}	Peak output voltage swing‡	V_{ID} = 100 mV	25°C	8	8.6		8	8.6		8	8.6		V
			−40°C to 85°C	7.8			7.8			7.8			
A_{VD}	Large-signal differential voltage amplification	V_O = 1 to 6 V, R_S = 50 Ω	25°C	10	40		30	500		20	280		V/mV
			−40°C to 85°C	7			20			10			
CMRR	Common-mode rejection ratio	V_O = 1.4 V, V_{IC} = V_{ICR} min	25°C	70	88		70	88		70	88		dB
k_{SVR}	Supply voltage rejection ratio ($\Delta V_{CC}/\Delta V_{IO}$)	V_{DD} = 5 to 10 V, V_O = 1.4 V	25°C	65	82		70	88		70	88		dB
I_{OS}	Short-circuit output current	V_O = 0, V_{ID} = 100 mV	25°C		−55			−55			−55		mA
		V_O = V_{DD}, V_{ID} = −100 mV			15			15			15		
I_{DD}	Supply current (each amplifier)	No load, V_O = 5 V, V_{IC} = 5 V	25°C		1000	2000		10	20		150	300	µA
			−40°C to 85°C			2500			40			500	

† All characteristics are measured under open-loop conditions with zero common-mode input voltage unless otherwise specified. Unless otherwise noted, an output load resistor is connected from the output to the ground pin.

‡ The output will swing to the potential of the ground pin.

3

Operational Amplifiers

TEXAS
INSTRUMENTS
POST OFFICE BOX 225012 • DALLAS, TEXAS 75265

C-SUFFIX TYPES

electrical characteristics at specified free-air temperature, V_{DD} = 10 V (unless otherwise noted)

PARAMETER			TEST CONDITIONS[†]	TLC254_C, TLC274_C MIN	TYP	MAX	TLC25L4_C, TLC27L4_C MIN	TYP	MAX	TLC25M4_C, TLC27M4_C MIN	TYP	MAX	UNIT
V_{IO}	Input offset voltage	TLC2_4C	V_O = 1.4 V, R_S = 50 Ω	25°C		10			10			10	mV
				0°C to 70°C		12			12			12	
		TLC2_4AC		25°C		5			5			5	
				0°C to 70°C		6.5			6.5			6.5	
		TLC2_4BC		25°C		2			2			2	
				0°C to 70°C		3			3			3	
α_{VIO}	Average temperature coefficient of input offset voltage		0°C to 70°C		5			0.7			2		μV/°C
I_{IO}	Input offset current		V_{IC} = 5 V, V_O = 5 V	25°C	1			1			1		pA
				0°C to 70°C		100			100			100	
I_{IB}	Input bias current		V_{IC} = 5 V, V_O = 5 V	25°C	1			1			1		pA
				0°C to 70°C		150			150			150	
V_{ICR}	Common-mode input voltage range		25°C	−0.2 to 9			−0.2 to 9			−0.2 to 9			V
V_{OM}	Peak output voltage swing[‡]		V_{ID} = 100 mV	25°C	8	8.6		8	8.6		8	8.6	V
				0°C to 70°C	7.8			7.8			7.8		
A_{VD}	Large-signal differential voltage amplification		V_O = 1 to 6 V, R_S = 50 Ω	25°C	10	40		30	500		20	280	V/mV
				0° to 70°C	7.5			25			15		
CMRR	Common-mode rejection ratio		V_O = 1.4 V, V_{IC} = V_{ICR} min	25°C	70	88		70	88		70	88	dB
k_{SVR}	Supply voltage rejection ratio ($\Delta V_{CC}/\Delta V_{IO}$)		V_{DD} = 5 to 10 V, V_O = 1.4 V	25°C	65	82		70	88		70	88	dB
I_{OS}	Short-circuit output current		V_O = 0, V_{ID} = 100 mV	25°C		−55			−55			−55	mA
			V_O = V_{DD}, V_{ID} = −100 mV			15			15			15	
I_{DD}	Supply current (each amplifier)		No load, V_O = 5 V, V_{IC} = 5 V	25°C		1000	2000		10	20		150	300	μA
				0°C to 70°C		2200			30			400	

† All characteristics are measured under open-loop conditions with zero common-mode input voltage unless otherwise specified. Unless otherwise noted, an output load resistor is connected from the output to the ground pin.
‡ The output will swing to the potential of the ground pin.

3

Operational Amplifiers

TEXAS
INSTRUMENTS
POST OFFICE BOX 225012 • DALLAS, TEXAS 75265

11

C-SUFFIX TYPES

electrical characteristics at specified free-air temperature, V_{DD} = 1 V (unless otherwise noted)

PARAMETER			TEST CONDITIONS†	TLC254_C MIN	TLC254_C TYP	TLC254_C MAX	TLC25L4_C MIN	TLC25L4_C TYP	TLC25L4_C MAX	TLC25M4_C MIN	TLC25M4_C TYP	TLC25M4_C MAX	UNIT
V_{IO}	Input offset voltage	TLC25_4C	V_O = 0.2 V, R_S = 50 Ω	25°C		10			10			10	mV
				0°C to 70°C		12			12			12	
		TLC25_4AC		25°C		5			5			5	
				0°C to 70°C		6.5			6.5			6.5	
		TLC25_4BC		25°C		2			2			2	
				0°C to 70°C		3			3			3	
α_{VIO}	Average temperature coefficient of input offset voltage		0°C to 70°C		1			1			1		μV/°C
I_{IO}	Input offset current	V_O = 0.2	25°C		1			1			1		pA
			0°C to 70°C			100			100			100	
I_{IB}	Input bias current	V_O = 0.2	25°C		1			1			1		pA
			0°C to 70°C			150			150			150	
V_{ICR}	Common-mode input voltage range	25°C	0 to 0.2			0 to 0.2			0 to 0.2			V
V_{OM}	Peak output voltage swing‡	V_{ID} = 100 mV	25°C		450			450			450		mV
A_{VD}	Large-signal differential voltage amplification	V_O = 100 to 300 mV, R_S = 50 Ω	25°C	10			20			20			V/mV
CMRR	Common-mode rejection ratio	V_O = 0.2 V, V_{IC} = V_{ICR} min	25°C	77			77			77			dB
I_{DD}	Supply current (each amplifier)	No load, V_O = 0.2 V	25°C		12			2			2		μA

† All characteristics are measured under open-loop conditions with zero common-mode input voltage unless otherwise specified. Unless otherwise noted, an output load resistor is connected form the output to the ground pin.

‡ The output will swing to the potential of the ground pin.

operating characteristics, V_{DD} = 1 V, T_A = 25°C

PARAMETER		TEST CONDITIONS	TLC254_C MIN	TLC254_C TYP	TLC254_C MAX	TLC25L4_C MIN	TLC25L4_C TYP	TLC25L4_C MAX	TLC25M4_C MIN	TLC25M4_C TYP	TLC25M4_C MAX	UNIT
B_1	Unity-gain bandwidth	A_V = 40 dB, C_L = 10 pF, R_S = 50 Ω		75			12			12		kHz
SR	Slew rate at unity gain	See Figure 1		0.01			0.001			0.001		V/μs
	Overshoot factor	See Figure 1		30%			35%			35%		

3

Operational Amplifiers

operating characteristics, V_{DD} = 10 V, T_A = 25°C

PARAMETER		TEST CONDITIONS	TLC254__C TLC274__M TLC274__I TLC274__C			TLC25L4__C TLC27L4__M TLC27L4__I TLC27L4__C			TLC25M4__C TLC27M4__M TLC27M4__I TLC27M4__C			UNIT
			MIN	TYP	MAX	MIN	TYP	MAX	MIN	TYP	MAX	
B_1	Unity-gain bandwidth	A_V = 40 dB, C_L = 10 pF, R_S = 50 Ω		2.3			0.1			0.7		MHz
SR	Slew rate at unity gain	See Figure 1		4.5			0.04			0.6		V/µs
	Overshoot factor	See Figure 1		35%			30%			35%		
ϕ_m	Phase margin at unity gain	A_V = 40 dB, R_S = 100 Ω, C_L = 10 pF		50°			43°			43°		
V_n	Equivalent input noise voltage	f = 1 kHz, R_S = 100 Ω		30			70			38		nV/√Hz
V_{O1}/V_{O2}	Cross talk attenuation	A_V = 100		120			120			120		dB

PARAMETER MEASUREMENT INFORMATION

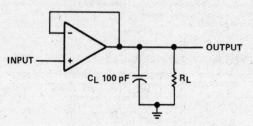

FIGURE 1—UNITY-GAIN AMPLIFIER

TEXAS INSTRUMENTS
POST OFFICE BOX 225012 • DALLAS, TEXAS 75265

TYPICAL CHARACTERISTICS

SUPPLY CURRENT
vs
SUPPLY VOLTAGE

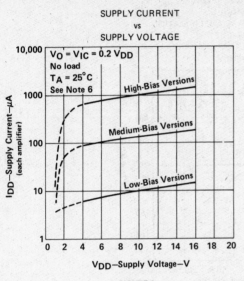

FIGURE 2

NOTE 6: Data for V_{DD} less than 4 V does not apply for the TLC274 series.

SUPPLY CURRENT
vs
FREE-AIR TEMPERATURE

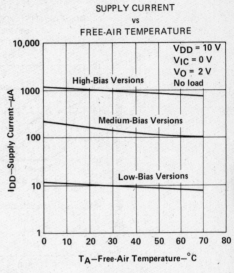

FIGURE 3

LOW-BIAS VERSIONS
LARGE-SIGNAL
DIFFERENTIAL VOLTAGE AMPLIFICATION
and PHASE SHIFT
vs
FREQUENCY

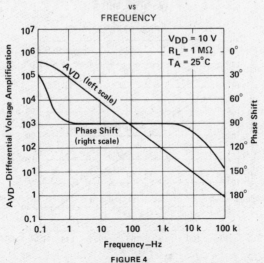

FIGURE 4

Operational Amplifiers

3

TYPICAL CHARACTERISTICS

MEDIUM-BIAS VERSIONS
LARGE-SIGNAL
DIFFERENTIAL VOLTAGE AMPLIFICATION
and PHASE SHIFT
vs
FREQUENCY

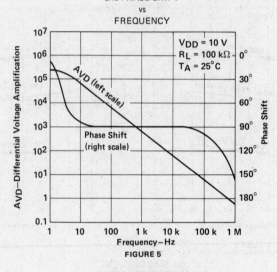

FIGURE 5

HIGH-BIAS VERSIONS
LARGE-SIGNAL
DIFFERENTIAL VOLTAGE AMPLIFICATION
and PHASE SHIFT
vs
FREQUENCY

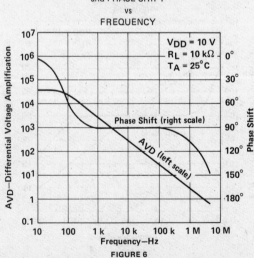

FIGURE 6

3

Operational Amplifiers

TEXAS
INSTRUMENTS
POST OFFICE BOX 225012 • DALLAS, TEXAS 75265

118

TYPICAL APPLICATION INFORMATION

latchup avoidance

Junction-isolated CMOS circuits have an inherent parasitic PNPN structure that can function as an SCR. Under certain conditions, this SCR may be triggered into a low-impedance state, resulting in excessive supply current. To avoid such conditions, no voltage greater than 0.3 V beyond the supply rails should be applied to any pin. In general, the op amp supplies should be established simultaneously with, or before, any input signals are applied.

output stage considerations

The amplifier's output stage consists of a source-follower-connected pullup transistor and an open-drain pulldown transistor. The high-level output voltage (V_{OH}) is virtually independent of the I_{DD} selection, and increases with higher values of V_{DD} and reduced output loading. The low-level output voltage (V_{OL}) decreases with reduced output current and higher input common-mode voltage. With no load, V_{OL} is essentially equal to the GND pin potential.

supply configurations

Even though the TLC254 and TLC274 are characterized for single-supply operation, they can be used effectively in a split-supply configuration if the input common-mode voltage (V_{ICR}), output swing (V_{OL} and V_{OH}), and supply voltage limits are not exceeded.

circuit layout precautions

The user is cautioned that whenever extremely high circuit impedances are used, care must be exercised in layout, construction, board cleanliness, and supply filtering to avoid hum and noise pickup, as well as excessive DC leakages.

3

Operational Amplifiers

3

LINEAR
INTEGRATED
CIRCUITS

TYPES TLC261, TLC262, TLC264
PROGRAMMABLE LOW-POWER LinCMOS™ OPERATIONAL AMPLIFIERS

D2827, JANUARY 1984

- Wide Range of Supply Voltage: 2 V to 16 V
- True Single-Supply Operation
- Designed for Performance Similar to Popular BIFET Op Amps
- Common-Mode Input Voltage Includes the Negative Rail
- Slew Rate . . . 12 V/μs Typ
- High Input Impedance . . . 10^{12} Ω Typ

DEVICE FEATURES

PARAMETER	FEATURE
Supply current per channel (Typ)	2.5 mA
Slew rate (Typ)	12 V/μs
Input offset voltage (Max)	
. . . Standard types	10 mV
. . . A-suffix types	5 mV
. . . B-suffix types	2 mV
Offset voltage drift (Typ)	0.1 μV/month[†]
Offset voltage temperature coefficient (Typ)	5 μV/°C
Input bias current (Typ)	1 pA
Input offset current (Typ)	1 pA
Operating temperature range	−40°C to 85°C

[†]The long-term drift value applies after the first month.

description

The TLC261, TLC262, and TLC264 LinCMOS™ operational amplifiers are designed to provide a true single-supply alternative to the popular BIFET op amps. The negative supply rail is included in both input and output common-mode voltage ranges. In addition, these devices feature input offset voltage selection.

Unlike traditional metal-gate CMOS op amps, these devices utilize the Texas Instruments silicon-gate LinCMOS™ process giving them stable input offset voltages without sacrificing the advantage of metal-gate CMOS. Because the input common-mode range extends to the negative rail and the power consumption is extremely low, this family is ideally suited for battery-powered or energy-conserving applications. All devices are stable at unity gain.

TLC261
D, JG, OR P DUAL-IN-LINE PACKAGE
(TOP VIEW)

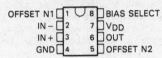

TLC262
D, JG, OR P DUAL-IN-LINE PACKAGE
(TOP VIEW)

TLC264
D, J, OR N DUAL-IN-LINE PACKAGE
(TOP VIEW)

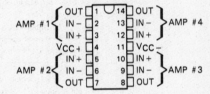

symbol

TLC261

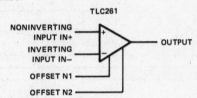

symbol (each amplifier)

TLC262, TLC264

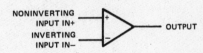

Operational Amplifiers

3

Copyright © 1984 by Texas Instruments Incorporated

TEXAS
INSTRUMENTS

POST OFFICE BOX 225012 • DALLAS, TEXAS 75265

absolute maximum ratings over operating free-air temperature (unless otherwise noted)

Supply voltage, V_{DD} (see Note 1) . . . . 18 V
Differential input voltage (see Note 2) . . . . ±18 V
Input voltage range (any input) . . . . −0.3 V to 18 V
Duration of short-circuit at (or below) 25 °C free-air temperature (see Note 3) . . . . unlimited
Continuous total dissipation at (or below) 25 °C free-air temperature (see Note 4):
    TLC261ID, TLC262ID . . . . 725 mW
    TLC261IJG, TLC262IJG . . . . 825 mW
    TLC261IP, TLC262IP . . . . 725 mW
    TLC264ID . . . . 950 mW
    TLC264IJ . . . . 1025 mW
    TLC264IN . . . . 875 mW
Operating free-air temperature range . . . . −40 °C to 85 °C
Storage temperature range . . . . −65 °C to 150 °C
Lead temperature 1,6 mm (1/16 inch) from case for 60 seconds: J or JG package . . . . 300 °C
Lead temperature 1,6 mm (1/16 inch) from case for 10 seconds: D, N, or P package . . . . . . 260 °C

NOTES: 1. All voltage values, except differential voltages, are with respect to network ground terminal.
2. Differential voltages are at the noninverting input terminal with respect to the inverting input terminal.
3. The output may be shorted to either supply. Temperature and/or supply voltages must be limited to ensure the maximum dissipation rating is not exceeded.
4. For operation above 25 °C free-air temperature, refer to Dissipation Derating Table below. In the J and JG packages, these chips are glass mounted.

DISSIPATION DERATING TABLE

PACKAGE	POWER RATING	DERATING FACTOR	ABOVE T_A
D (8-Pin)	725 mW	5.8 mW/°C	25 °C
D (14-Pin)	950 mW	7.6 mW/°C	25 °C
J (glass mounted)	1025 mW	8.2 mW/°C	25 °C
JG (glass mounted)	825 mW	6.6 mW/°C	25 °C
N	875 mW	7 mW/°C	25 °C
P	725 mW	5.8 mW/°C	25 °C

recommended operating conditions

		MIN	NOM	MAX	UNIT
Supply voltage, V_{DD}		2		16	V
Common-mode input voltage, V_{IC}	$V_{DD} = 2$ V	0		1.2	V
	$V_{DD} = 4$ V	0		3	
	$V_{DD} = 10$ V	−0.05		9	
	$V_{DD} = 16$ V	−0.05		14	
Operating free-air temperature, T_A		−40		85	°C

TEXAS
INSTRUMENTS
POST OFFICE BOX 225012 • DALLAS, TEXAS 75265

TYPES TLC277, TLC27L7, TLC27M7
LinCMOS™ DUAL OPERATIONAL AMPLIFIERS

D2798, OCTOBER 1983

- **Wide Range of Supply Voltages**
  3 V to 16 V (TLC277C, TLC277I)
  4 V to 16 V (TLC277M)

- **True Single-Supply Operation**

- **Common-Mode Input Voltage Includes the Negative Rail**

- **Low Noise . . . 30 nV/√Hz Typ at f = 1 kHz (TLC277)**

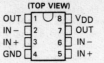

D, JG, OR P DUAL-IN-LINE PACKAGE
(TOP VIEW)

```
OUT  [1    8] VDD
IN-  [2    7] OUT
IN+  [3    6] IN-
GND  [4    5] IN+
```

description

The members of the TLC277 family are low-offset-voltage, low-power, dual operational amplifiers designed to operate with single or dual supplies. These devices utilize the Texas Instruments silicon-gate LinCMOS™ process, providing stable input offset voltages, very high input impedances, and extremely low input offset and bias currents. This series is ideally suited for battery-powered or energy-saving applications because the input common-mode range includes the negative rail and the power comsumption is extremely low.

These devices have internal electrostatic discharge (ESD) protection circuits that will prevent catastrophic failures at voltages up to 2000 volts as tested under MIL-STD-883B, Method 3015.1. However, care should be exercised in handling these devices as exposure to ESD may result in a degradation of the device parametric performance.

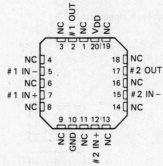

FH OR FK PACKAGE
(TOP VIEW)

NC—No internal connection

symbol (each amplifier)

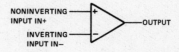

NONINVERTING INPUT IN+
INVERTING INPUT IN−
OUTPUT

DEVICE FEATURES

PARAMETER	TLC27L7	TLC27M7	TLC277
Supply current (Typ)	20 μA	300 μA	2000 μA
Slew rate (Typ)	0.04 V/μs	0.6 V/μs	4.5 V/μs
Input offset voltage (Max)	0.5 mV	0.5 mV	0.5 mV
Offset voltage drift (Typ)	0.1 μV/month†	0.1 μV/month†	0.1 μV/month†
Input bias current (Typ)	1 pA	1 pA	1 pA
Input offset current (Typ)	1 pA	1 pA	1 pA

†The long-term drift value applies after the first month.

DEVICE TYPES AND PACKAGES

DEVICE TYPE	PACKAGE
TLC277M	FH, FK, JG
TLC27L7M	FH, FK, JG
TLC27M7M	FH, FK, JG
TLC277I	D, JG, P
TLC27L7I	D, JG, P
TLC27M7I	D, JG, P
TLC277C	D, JG, P
TLC27L7C	D, JG, P
TLC27M7C	D, JG, P

Operational Amplifiers

3

Copyright © 1983 by Texas Instruments Incorporated

TEXAS INSTRUMENTS
POST OFFICE BOX 225012 • DALLAS, TEXAS 75265

TYPES TLC277, TLC27L7, TLC27M7
LinCMOS™ DUAL OPERATIONAL AMPLIFIERS

description (continued)

Because of the extremely high input impedance and low input bias and offset currents, applications for the TLC277 series include many areas that have previously been limited to BIFET and NFET product types. Any circuit using high-impedance elements and requiring small offset errors is a good candidate for cost-effective use of these devices. Many features associated with bipolar technology are available with LinCMOS™ operational amplifiers without the power penalties of traditional bipolar devices. General applications such as transducer interfacing, analog calculations, amplifier blocks, active filters, and signal buffering are all easily designed with the TLC277 series. This series includes devices that will be characterized for commercial, industrial, and military temperature ranges and will be available in 8-pin plastic and ceramic dual-in-line (P, JG) packages, small-outline (D) package, and chip carrier (FH, FK) packages.

schematic (each amplifier)

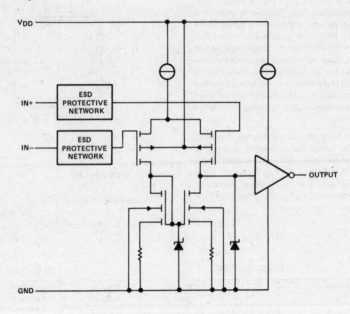

TEXAS INSTRUMENTS
POST OFFICE BOX 225012 ● DALLAS, TEXAS 75265

absolute maximum ratings over operating free-air temperature range (unless otherwise noted)

Supply voltage, V_{DD} (see Note 1) .... 18 V
Differential input voltage (see Note 2) .... ±18 V
Input voltage range (any input) .... −0.3 V to 18 V
Duration of short-circuit at (or below) 25 °C free-air temperature (see Note 3) .... unlimited
Continuous total dissipation at (or below) 25 °C free-air temperature (see Note 4)
   D package .... 725 mW
   FH package (see Note 5) .... 1200 mW
   FK package (see Note 5) .... 1375 mW
   JG package .... 1050 mW
   P package .... 725 mW
Operating free-air temperature range: TLC277M, TLC27L7M, TLC27M7M .... −55 °C to 125 °C
         TLC277I, TLC27L7I, TLC27M7I .... −40 °C to 85 °C
         TLC277C, TLC27L7C, TLC27M7C .... 0 °C to 70 °C
Storage temperature range .... −65 °C to 150 °C
Lead temperature 1,6 mm (1/16 inch) from case for 60 seconds: FH, FK, or JG package .... 300 °C
Lead temperature 1,6 mm (1/16 inch) from case for 10 seconds: D or P package .... 260 °C

NOTES: 1. All voltage values, except differential voltages, are with respect to network ground terminal.
     2. Differential voltages are at the noninverting input terminal with respect to the inverting input terminal.
     3. The output may be shorted to either supply. Temperature and/or supply voltages must be limited to ensure the maximum dissipation rating is not exceeded.
     4. For operation above 25 °C free-air temperature, refer to the Dissipation Derating Table.
     5. For FH and FK packages, power rating and derating factor will vary with the actual mounting technique used. The values stated here are believed to be conservative.

DISSIPATION DERATING TABLE

PACKAGE	POWER RATING	DERATING FACTOR	ABOVE T_A
D	725 mW	5.8 mW/°C	25 °C
FH	1200 mW	9.6 mW/°C	25 °C
FK	1375 mW	11 mW/°C	25 °C
JG	1050 mW	8.4 mW/°C	25 °C
P	725 mW	5.8 mW/°C	25 °C

recommended operating conditions

		MIN	NOM	MAX	UNIT
Supply voltage, V_{DD}	M-suffix types	4		16	V
	I-suffix and C-suffix types	3		16	
Common-mode input voltage, V_{IC}	$V_{DD} = 4$ V	0		3	
	$V_{DD} = 10$ V	−0.2		9	V
	$V_{DD} = 16$ V	−0.2		14	
Operating free-air temperature, T_A	M-suffix types	−55		125	
	I-suffix types	−40		85	°C
	C-suffix types	0		70	

Operational Amplifiers   3

3

Operational Amplifiers

M-SUFFIX TYPES

electrical characteristics at specified free-air temperature, V_{DD} = 10 V (unless otherwise noted)

	PARAMETER	TEST CONDITIONS		TLC277M MIN	TYP	MAX	TLC27L7M MIN	TYP	MAX	TLC27M7M MIN	TYP	MAX	UNIT
V_{IO}	Input offset voltage	V_O = 1.4 V, R_S = 50 Ω	25°C			0.5			0.5			0.5	mV
			−55°C to 125°C			2			2			2	
α_{VIO}	Average temperature coefficient of input offset voltage		−55°C to 125°C		5			0.7			2		µV/°C
I_{IO}	Input offset current	V_{IC} = 5 V, V_O = 5 V	25°C		1			1			1		pA
			−55°C to 125°C			15			15			15	nA
I_{IB}	Input bias current	V_{IC} = 5 V, V_O = 5 V	25°C		1			1			1		pA
			−55°C to 125°C			35			35			35	nA
V_{ICR}	Common-mode input voltage range		25°C	−0.2 to		9	−0.2 to		9	−0.2 to		9	V
V_{OM}	Peak output voltage swing‡	V_{ID} = 100 mV	25°C	8	8.6		8	8.6		8	8.6		V
			−55°C to 125°C	7.8			7.8			7.8			
A_{VD}	Large-signal differential voltage amplification	V_O = 1 to 6 V, R_S = 50 Ω	25°C	10	40		30	500		20	280		V/mV
			−55°C to 125°C	7			20			10			
CMRR	Common-mode rejection ratio	V_O = 1.4 V, V_{IC} = V_{ICR} min	25°C	70	88		70	88		70	88		dB
k_{SVR}	Supply voltage rejection ratio ($\Delta V_{CC}/\Delta V_{IO}$)	V_{DD} = 5 to 10 V, V_O = 1.4 V	25°C	65	82		70	88		70	88		dB
I_{OS}	Short-circuit output current	V_O = 0, V_{ID} = 100 mV, V_{DD}, V_{ID} = −100 mV	25°C		−55			−55			−55		mA
					15			15			15		
I_{DD}	Supply current (each amplifier)	No load, V_O = 5 V, V_{IC} = 5 V	25°C		1000	2000		10	20		150	300	µA
			−55°C to 125°C			3000			40			500	

† All characteristics are measured under open-loop conditions with zero common-mode input voltage unless otherwise specified. Unless otherwise noted, an output load resistor is connected from the output to the ground pin.

‡ The output will swing to the potential of the ground pin.

TEXAS INSTRUMENTS

POST OFFICE BOX 225012 • DALLAS, TEXAS 75265

I-SUFFIX TYPES

electrical characteristics at specified free-air temperature, V_{DD} = 10 V (unless otherwise noted)

PARAMETER		TEST CONDITIONS†	TLC277I MIN	TYP	MAX	TLC27L7I MIN	TYP	MAX	TLC27M7I MIN	TYP	MAX	UNIT
V_{IO}	Input offset voltage	V_O = 1.4 V, R_S = 50 Ω	25°C		0.5			0.5			0.5	mV
			−40°C to 85°C		2			2			2	
α_{VIO}	Average temperature coefficient of input offset voltage		−40°C to 85°C	5			0.7			2		µV/°C
I_{IO}	Input offset current	V_{IC} = 5 V, V_O = 5 V	25°C	1			1			1		pA
			−40°C to 85°C		200			200			200	
I_{IB}	Input bias current	V_{IC} = 5 V, V_O = 5 V	25°C	1			1			1		pA
			−40°C to 85°C		300			300			300	
V_{ICR}	Common-mode input voltage range		25°C	−0.2 to 9			−0.2 to 9			−0.2 to 9		V
V_{OM}	Peak output voltage swing‡	V_{ID} = 100 mV	25°C	8	8.6		8	8.6		8	8.6	V
			−40°C to 85°C	7.8			7.8			7.8		
A_{VD}	Large-signal differential voltage amplification	V_O = 1 to 6 V, R_S = 50 Ω	25°C	10	40		30	500		20	280	V/mV
			−40°C to 85°C	7			20			10		
CMRR	Common-mode rejection ratio	V_O = 1.4 V, V_{IC} = V_{ICR} min	25°C	70	88		70	88		70	88	dB
k_{SVR}	Supply voltage rejection ratio ($\Delta V_{CC}/\Delta V_{IO}$)	V_{DD} = 5 to 10 V, V_O = 1.4 V	25°C	65	82		70	88		70	88	dB
I_{OS}	Short-circuit output current	V_O = 0, V_{ID} = 100 mV, V_O = V_{DD}, V_{ID} = −100 mV	25°C		−55			−55			−55	mA
I_{DD}	Supply current (each amplifier)	No load, V_O = 5 V, V_{IC} = 5 V	25°C	1000	2000		10	20		150	300	µA
			−40°C to 85°C		3000			40			500	

† All characteristics are measured under open-loop conditions with zero common-mode input voltage unless otherwise specified. Unless otherwise noted, an output load resistor is connected from the output to the ground pin.

‡ The output will swing to the potential of the ground pin.

3

Operational Amplifiers

TEXAS
INSTRUMENTS
POST OFFICE BOX 225012 • DALLAS, TEXAS 75265

3

Operational Amplifiers

C-SUFFIX TYPES

electrical characteristics at specified free-air temperature, V_{DD} = 10 V (unless otherwise noted)

PARAMETER		TEST CONDITIONS†		TLC277C			TLC27L7C			TLC27M7C			UNIT
				MIN	TYP	MAX	MIN	TYP	MAX	MIN	TYP	MAX	
V_{IO}	Input offset voltage	V_O = 1.4 V, R_S = 50 Ω	25°C			0.5			0.5			0.5	mV
			0°C to 70°C			1.5			1.5			1.5	
α_{VIO}	Average temperature coefficient of input offset voltage		0°C to 70°C		5			0.7			2		µV/°C
I_{IO}	Input offset current	V_{IC} = 5 V, V_O = 5 V	25°C		1			1			1		pA
			0°C to 70°C			100			100			100	pA
I_{IB}	Input bias current	V_{IC} = 5 V, V_O = 5 V	25°C		1			1			1		pA
			0°C to 70°C			150			150			150	pA
V_{ICR}	Common-mode input voltage range		25°C	-0.2 to 9			-0.2 to 9			-0.2 to 9			V
V_{OM}	Peak output voltage swing‡	V_{ID} = 100 mV	25°C	8	8.6		8	8.6		8	8.6		V
			0°C to 70°C	7.8			7.8			7.8			V
A_{VD}	Large-signal differential voltage amplification	V_O = 1 to 6 V, R_S = 50 Ω	25°C	10	40		30	500		20	280		V/mV
			0°C to 70°C	7.5			25			15			
CMRR	Common-mode rejection ratio	V_O = 1.4 V, V_{IC} = V_{ICR} min	25°C	70	88		70	88		70	88		dB
k_{SVR}	Supply voltage rejection ratio ($\Delta V_{CC}/\Delta V_{IO}$)	V_{DD} = 5 to 10 V, V_O = 1.4 V	25°C	65	82		70	88		70	88		dB
I_{OS}	Short-circuit output current	V_O = 0, V_{ID} = 100 mV V_O = V_{DD}, V_{ID} = -100 mV	25°C		-55			-55			-55		mA
I_{DD}	Supply current (each amplifier)	No load, V_O = 5 V, V_{IC} = 5 V	25°C		1000	2000		10	20		150	300	µA
			0°C to 70°C			2200			30			400	

†All characteristics are measured under open-loop conditions with zero common-mode input voltage unless otherwise specified. Unless otherwise noted, an output load resistor is connected from the output to the ground pin.

‡The output will swing to the potential of the ground pin.

1083

TEXAS INSTRUMENTS
POST OFFICE BOX 225012 • DALLAS, TEXAS 75265

operating characteristics, V_{DD} = 10 V, T_A = 25°C

	PARAMETER	TEST CONDITIONS	TLC277M TLC277I TLC277C			TLC27L7M TLC27L7I TLC27L7C			TLC27M7M TLC27M7I TLC27M7C			UNIT
			MIN	TYP	MAX	MIN	TYP	MAX	MIN	TYP	MAX	
B_1	Unity-gain bandwidth	A_V = 40 dB, C_L = 10 pF, R_S = 50 Ω		2.3			0.1			0,7		MHz
SR	Slew rate at unity gain	See Figure 1		4.5			0.04			0.6		V/μs
	Overshoot factor	See Figure 1		35%			30%			35%		
ϕ_m	Phase margin at unity gain	A_V = 40 dB, R_S = 100 Ω, C_L = 10 pF		50°			43°			43°		
V_n	Equivalent input noise voltage	f = 1 kHz, R_S = 100 Ω		30			70			38		nV/√Hz
V_{o1}/V_{o2}	Crosstalk attenuation	A_V = 100		120			120			120		dB

3

Operational Amplifiers

PARAMETER MEASUREMENT INFORMATION

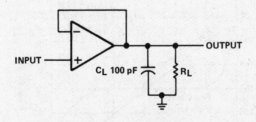

FIGURE 1—UNITY GAIN AMPLIFIER

TEXAS INSTRUMENTS
POST OFFICE BOX 225012 • DALLAS, TEXAS 75265

TYPES TLC277, TLC27L7, TLC27M7
LinCMOS™ DUAL OPERATIONAL AMPLIFIERS

TYPICAL CHARACTERISTICS

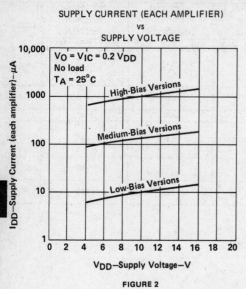

SUPPLY CURRENT (EACH AMPLIFIER)
vs
SUPPLY VOLTAGE

$V_O = V_{IC} = 0.2 V_{DD}$
No load
$T_A = 25°C$

High-Bias Versions

Medium-Bias Versions

Low-Bias Versions

I_{DD}—Supply Current (each amplifier)—μA

V_{DD}—Supply Voltage—V

FIGURE 2

SUPPLY CURRENT (EACH AMPLIFIER)
vs
FREE-AIR TEMPERATURE

$V_{DD} = 10 V$
$V_{IC} = 0 V$
$V_O = 2 V$
No load

High-Bias Versions

Medium-Bias Versions

Low-Bias Versions

I_{DD}—Supply Current (each amplifier)—μA

T_A—Free-Air Temperature—°C

FIGURE 3

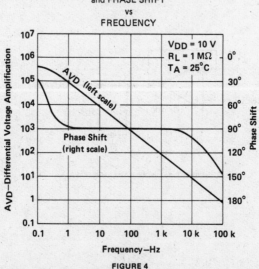

LOW-BIAS VERSIONS
LARGE-SIGNAL
DIFFERENTIAL VOLTAGE AMPLIFICATION
and PHASE SHIFT
vs
FREQUENCY

$V_{DD} = 10 V$
$R_L = 1 M\Omega$
$T_A = 25°C$

AVD (left scale)

Phase Shift
(right scale)

A_{VD}—Differential Voltage Amplification

Phase Shift

Frequency—Hz

FIGURE 4

TEXAS
INSTRUMENTS
POST OFFICE BOX 225012 • DALLAS, TEXAS 75265

TYPICAL CHARACTERISTICS

MEDIUM-BIAS VERSIONS
LARGE-SIGNAL
DIFFERENTIAL VOLTAGE AMPLIFICATION
and PHASE SHIFT
vs
FREQUENCY

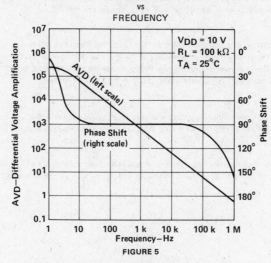

FIGURE 5

HIGH-BIAS VERSIONS
LARGE-SIGNAL
DIFFERENTIAL VOLTAGE AMPLIFICATION
and PHASE SHIFT
vs
FREQUENCY

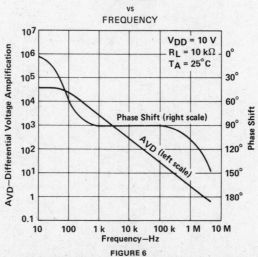

FIGURE 6

3

Operational Amplifiers

TYPES TLC277, TLC27L7, TLC27M7
LinCMOS™ DUAL OPERATIONAL AMPLIFIERS

TYPICAL APPLICATION INFORMATION

latchup avoidance

Junction-isolated CMOS circuits have an inherent parasitic PNPN structure that can function as an SCR. Under certain conditions, this SCR may be triggered into a low-impedance state, resulting in excessive supply current. To avoid such conditions, no voltage greater than 0.3 V beyond the supply rails should be applied to any pin. In general, the op amp supplies should be established simultaneously with, or before, any input signals are applied.

output stage considerations

The amplifier's output stage consists of a source follower connected pullup transistor and an open drain pulldown transistor. The high-level output voltage (V_{OH}) is virtually independent of the I_{DD} selection, and increases with higher values of V_{DD} and reduced output loading. The low-level output voltage (V_{OL}) decreases with reduced output current and higher input common-mode voltage. With no load, V_{OL} is essentially equal to the GND pin potential.

supply configurations

Even though the TLC277 is characterized for single-supply operation, it can be used effectively in a split-supply configuration if the input common-mode voltage (V_{ICR}), output swing (V_{OL} and V_{OH}), and supply voltage limits are not exceeded.

circuit layout precautions

The user is cautioned that whenever extremely high circuit impedances are used, care must be exercised in layout, construction, board cleanliness, and supply filtering to avoid hum and noise pickup, as well as excessive DC leakages.

3 Operational Amplifiers

TEXAS INSTRUMENTS
POST OFFICE BOX 225012 • DALLAS, TEXAS 75265

D1004, JUNE 1976—REVISED OCTOBER 1983

- Open-Loop Voltage Amplification
  . . . 3600 Typ
- CMRR . . . 100 dB Typ
- Designed to be Interchangeable with
  Fairchild μA702

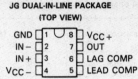

JG DUAL-IN-LINE PACKAGE
(TOP VIEW)

GND	1	8	VCC+
IN−	2	7	OUT
IN+	3	6	LAG COMP
VCC−	4	5	LEAD COMP

description

The uA702 is a high-gain, wideband operational amplifier having differential inputs and single-ended emitter-follower outputs. Provisions are incorporated within the circuit whereby external components may be used to compensate the amplifier for stable operation under various feedback or load conditions. Component matching, inherent in silicon monolithic circuit-fabrication techniques, produces an amplifier with low-drift and low-offset characteristics. The uA702 is particularly useful for applications requiring transfer or generation of linear and nonlinear functions up to a frequency of 30 MHz.

The uA702 is characterized for operation over the full military temperature range of −55°C to 125°C.

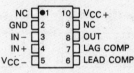

U FLAT PACKAGE
(TOP VIEW)

NC	1	10	VCC+
GND	2	9	NC
IN−	3	8	OUT
IN+	4	7	LAG COMP
VCC−	5	6	LEAD COMP

NC—No internal connection

symbol

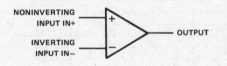

NONINVERTING INPUT IN+

INVERTING INPUT IN−

OUTPUT

schematic

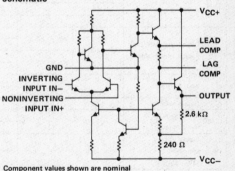

VCC+

LEAD COMP

LAG COMP

GND
INVERTING INPUT IN−
NONINVERTING INPUT IN+

OUTPUT

2.6 kΩ

240 Ω

VCC−

Component values shown are nominal

Operational Amplifiers

3

absolute maximum ratings over operating free-air temperature range (unless otherwise noted)

Supply voltage V_{CC+} (see Note 1) . . . . 14 V
Supply voltage V_{CC-} (see Note 1) . . . . −7 V
Differential input voltage (see Note 2) . . . . ±5 V
Input voltage (either input, see Notes 1 and 3) . . . . −6 V to 1.5 V
Peak output current ($t_W \leq 1$ s) . . . . 50 mA
Continuous total dissipation at (or below) 70°C free-air temperature (see Note 4) . . . . 300 mW
Operating free-air temperature range . . . . −55°C to 125°C
Storage temperature range . . . . −65°C to 150°C
Lead temperature 1,6 mm (1/16 inch) from case for 60 seconds . . . . 300°C

NOTES: 1. All voltage values, unless otherwise noted, are with respect to the network ground terminal.
2. Differential voltages are at the noninverting input terminal with respect to the inverting input terminal.
3. The magnitude of the input voltage must never exceed the magnitude of the lesser of the two supply voltages.
4. For operation above 70°C free-air temperature, refer to Dissipation Derating Curves, Section 2. In the JG packages, uA702M chips are alloy-mounted.

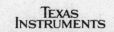

TEXAS
INSTRUMENTS

POST OFFICE BOX 225012 • DALLAS, TEXAS 75265

electrical characteristics at specified free-air temperature

PARAMETER		TEST CONDITIONS[†]		$V_{CC+} = 12$ V $V_{CC-} = -6$ V			$V_{CC+} = 6$ V $V_{CC-} = -3$ V			UNIT
				MIN	TYP	MAX	MIN	TYP	MAX	
V_{IO}	Input offset voltage	$R_S \leq 2$ kΩ	25°C		0.5	2		0.7	3	mV
			Full range			3			4	
α_{VIO}	Average temperature coefficient of input offset voltage	$R_S = 50$ Ω	−55°C to 25°C		2	10		3	15	µV/°C
			25°C to 125°C		2.5	10		3.5	15	
I_{IO}	Input offset current		25°C		0.2	0.5		0.12	0.5	µA
			−55°C		0.4	1.5		0.3	1.5	
			125°C		0.08	0.5		0.05	0.5	
α_{IIO}	Average temperature coefficient of input offset current		−55°C to 25°C		3	16		2	13	nA/°C
			25°C to 125°C		1	5		0.7	4	
I_{IB}	Input bias current		25°C		2	5		1.2	3.5	µA
			−55°C		4.3	10		2.6	7.5	
V_{ICR}	Common-mode input voltage range	Positive swing	25°C	0.5	1		0.5	1		V
		Negative swing		−4	−5		−1.5	−2		
V_{OM}	Maximum peak output voltage swing	$R_L \geq 100$ kΩ	25°C	±5	±5.3		±2.5	±2.7		V
			Full range	±5			±2.5			
		$R_L = 10$ kΩ	25°C	±3.5	±4		±1.5	±2		
		$R_L \geq 10$ kΩ	Full range	±3.5			±1.5			
A_{VD}	Large-signal differential voltage amplification	$R_L \geq 100$ kΩ, $V_O = ±5$ V	25°C	2500	3600	6000				
			Full range	2000		7000				
		$V_O = ±2.5$ V	25°C				600	900	1500	
			Full range				500		1750	
r_i	Input resistance		25°C	16	40		22	67		kΩ
			Full range	6			8			
r_o	Output resistance	$V_O = 0$, See Note 5	25°C		200	500		300	700	Ω
CMRR	Common-mode rejection ratio	$R_S \leq 2$ kΩ	25°C	80	100		80	100		dB
			Full range	70			70			
k_{SVS}	Supply voltage sensitivity ($\Delta V_{IO}/\Delta V_{CC}$)	$R_S \leq 2$ kΩ	25°C		75			75		µV/V
			Full range			200			200	
I_{CC}	Supply current	No load, No signal	25°C		5	6.7		2.1	3.3	mA
			−55°C		5	7.5		2.1	3.9	
			125°C		4.4	6.7		1.7	3.3	
P_D	Total power dissipation	No load, No signal	25°C		90	120		19	30	mW
			−55°C		90	135		19	35	
			125°C		80	120		15	30	

[†]All characteristics are specified under open-loop operation. Full range is −55°C to 125°C.
NOTE 5: This typical value applies only at frequencies above a few hundred hertz because of the effects of drift and thermal feedback.

3

Operational Amplifiers

TEXAS
INSTRUMENTS
POST OFFICE BOX 225012 • DALLAS, TEXAS 75265

108

operating characteristics V_{CC+} = 12 V, V_{CC-} = −6 V, T_A = 25°C

	PARAMETER	TEST FIGURE	TEST CONDITIONS		MIN	TYP	MAX	UNIT
t_r	Rise time	1	V_I = 10 mV,	C_L = 0		25	120	ns
		2	V_I = 1 mV			10	30	ns
	Overshoot factor	1	V_I = 10 mV,	C_L = 100 pF		10%	50%	
		2	V_I = 1 mV			20%	40%	
SR	Slew rate	1	V_I = 6 V,	C_L = 100 pF		1.7		V/μs
		2	V_I = 100 mV			11		

PARAMETER MEASUREMENT INFORMATION

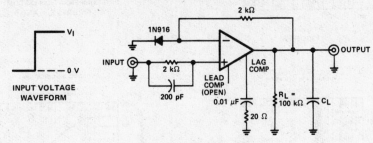

FIGURE 1—UNITY-GAIN AMPLIFIER

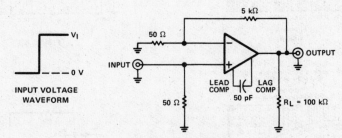

FIGURE 2—GAIN-OF-100 AMPLIFIER

3

Operational Amplifiers

TYPICAL CHARACTERISTICS

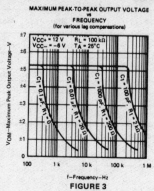

FIGURE 3

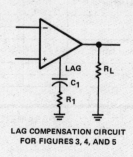

LAG COMPENSATION CIRCUIT
FOR FIGURES 3, 4, AND 5

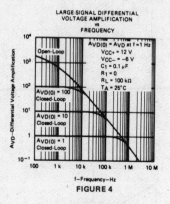

FIGURE 4

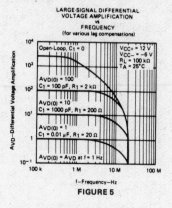

FIGURE 5

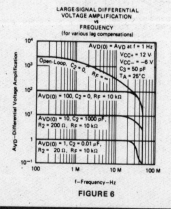

FIGURE 6

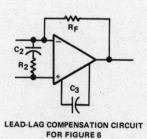

LEAD-LAG COMPENSATION CIRCUIT
FOR FIGURE 6

3

Operational Amplifiers

TEXAS
INSTRUMENTS
POST OFFICE BOX 225012 • DALLAS, TEXAS 75265

1083

LINEAR INTEGRATED CIRCUITS

TYPES uA709AM, uA709M, uA709C
GENERAL-PURPOSE OPERATIONAL AMPLIFIERS

D942, FEBRUARY 1971—REVISED AUGUST 1983

- Common-Mode Input Range . . . ±10 V Typical
- Designed to be Interchangeable with Fairchild μA709A, μA709, and μA709C
- Maximum Peak-to-Peak Output Voltage Swing . . . 28-V Typical with 15-V Supplies

uA709AM, uA709M . . . J OR W PACKAGE
(TOP VIEW)

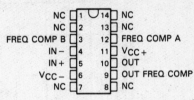

uA709AM, uA709M . . . JG PACKAGE
uA709C . . . JG OR P PACKAGE
(TOP VIEW)

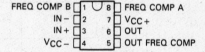

description

These circuits are general-purpose operational amplifiers, each having high-impedance differential inputs and a low-impedance output. Component matching, inherent with silicon monolithic circuit-fabrication techniques, produces an amplifier with low-drift and low-offset characteristics. Provisions are incorporated within the circuit whereby external components may be used to compensate the amplifier for stable operation under various feedback or load conditions. These amplifiers are particularly useful for applications requiring transfer or generation of linear or nonlinear functions.

The uA709A circuit features improved offset characteristics, reduced input-current requirements, and lower power dissipation when compared to the uA709 circuit. In addition, maximum values of the average temperature coefficients of offset voltage and current are guaranteed.

The uA709AM and uA709M are characterized for operation over the full military temperature range of −55°C to 125°C. The uA709C is characterized for operation from 0°C to 70°C.

uA709AM, uA709M . . . U FLAT PACKAGE
(TOP VIEW)

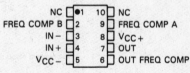

NC—No internal connection

symbol

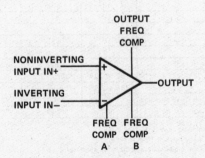

Operational Amplifiers

3

TEXAS INSTRUMENTS
POST OFFICE BOX 225012 • DALLAS, TEXAS 75265

3-215

TYPES uA709AM, uA709M, uA709C
GENERAL-PURPOSE OPERATIONAL AMPLIFIERS

schematic

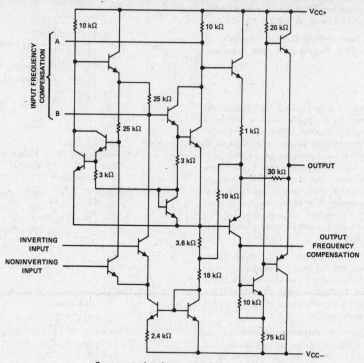

Component values shown are nominal.

absolute maximum ratings over operating free-air temperature range (unless otherwise noted)

		uA709AM uA709M	uA709C	UNIT
Supply voltage V_{CC+} (see Note 1)		18	18	V
Supply voltage V_{CC-} (see Note 1)		−18	−18	V
Differential input voltage (see Note 2)		±5	±5	V
Input voltage (either input, see Notes 1 and 3)		±10	±10	V
Duration of output short-circuit (see Note 4)		5	5	s
Continuous total dissipation at (or below) 70°C free-air temperature (see Note 5)		300	300	mW
Operating free-air temperature range		−55 to 125	0 to 70	°C
Storage temperature range		−65 to 150	−65 to 150	°C
Lead temperature 1,6 mm (1/16 inch) from case for 60 seconds	J, JG, U, or W package	300	300	°C
Lead temperature 1,6 mm (1/16 inch) from case for 10 seconds	P package		260	°C

NOTES: 1. All voltage values, unless otherwise noted, are with respect to the midpoint between V_{CC+} and V_{CC-}.
      2. Differential voltages are at the noninverting input terminal with respect to the inverting input terminal.
      3. The magnitude of the input voltage must never exceed the magnitude of the supply voltage or 10 volts, whichever is less.
      4. The output may be shorted to ground or either power supply.
      5. For operation of uA709AM and uA709M above 70°C free-air temperature, refer to the Dissipation Derating Curves, Section 2. In the J and JG packages, uA709AM and uA709M chips are alloy-mounted; uA709C chips are glass-mounted.

TEXAS INSTRUMENTS
POST OFFICE BOX 225012 • DALLAS, TEXAS 75265

electrical characteristics at specified free-air temperature, $V_{CC\pm} = \pm 9$ V to ± 15 V (unless otherwise noted)

PARAMETER		TEST CONDITIONS[†]		uA709AM			uA709M			UNIT
				MIN	TYP[‡]	MAX	MIN	TYP[‡]	MAX	
V_{IO}	Input offset voltage	$V_O = 0$, $R_S \leq 10$ kΩ	25°C		0.6	2		1	5	mV
			Full range			3			6	
α_{VIO}	Average temperature coefficient of input offset voltage	$V_O = 0$, $R_S = 50$ Ω	Full range		1.8	10		3		µV/°C
		$V_O = 0$, $R_S = 10$ kΩ	Full range		4.8	25		6		
I_{IO}	Input offset current	$V_O = 0$	25°C		10	50		50	200	nA
			−55°C		40	250		100	500	
			125°C		3.5	50		20	200	
α_{IIO}	Average temperature coefficient of input offset current	$V_O = 0$	−55°C to 25°C		0.45	2.8				nA/°C
			25°C to 125°C		0.08	0.5				
I_{IB}	Input bias current	$V_O = 0$	25°C		0.1	0.2		0.2	0.5	µA
			−55°C		0.3	0.6		0.5	1.5	
V_{ICR}	Common-mode input voltage range	$V_{CC\pm} = \pm 15$ V	25°C	±8	±10		±8	±10		V
			Full range	±8			±8			
V_{OPP}	Maximum peak-to-peak output voltage swing	$V_{CC\pm} = \pm 15$ V, $R_L \geq 10$ kΩ	25°C	24	28		24	28		V
			Full range	24			24			
		$V_{CC\pm} = \pm 15$ V, $R_L = 2$ kΩ	25°C	20	26		20	26		
		$V_{CC\pm} = \pm 15$ V, $R_L \geq 2$ kΩ	Full range	20			20			
A_{VD}	Large-signal differential voltage amplification	$V_{CC\pm} = \pm 15$ V, $R_L \geq 2$ kΩ, $V_O = \pm 10$ V	25°C		45			45		V/mV
			Full range	25		70	25		70	
r_i	Input resistance		25°C	350	750		150	400		kΩ
			−55°C	85	185		40	100		
r_o	Output resistance	$V_O = 0$ See Note 6	25°C		150			150		Ω
CMRR	Common-mode rejection ratio	$V_{IC} = V_{ICR}$ min	25°C	80	110		70	90		dB
			Full range	80			70			
k_{SVS}	Power supply sensitivity ($\Delta V_{IO}/\Delta V_{CC}$)	$V_{CC} = \pm 9$ V to ± 15 V	25°C		40	100		25	150	µV/V
			Full range			100			150	
I_{CC}	Supply current	$V_{CC\pm} = \pm 15$ V, No load, $V_O = 0$	25°C		2.5	3.6		2.6	5.5	mA
			−55°C		2.7	4.5				
			125°C		2.1	3				
P_D	Total power dissipation	$V_{CC\pm} = \pm 15$ V, No load, $V_O = 0$	25°C		75	108		78	165	mW
			−55°C		81	135				
			125°C		63	90				

[†] All characteristics are specified under open-loop with zero common-mode input voltage unless otherwise specified. Full range for uA709AM and uA709M is −55°C to 125°C.

[‡] All typical values are at $V_{CC\pm} = \pm 15$ V.

NOTE 6: This typical value applies only at frequencies above a few hundred hertz because of the effects of drift and thermal feedback.

3

Operational Amplifiers

883

electrical characteristics at specified free-air temperature (unless otherwise noted $V_{CC\pm} = \pm15$ V)

PARAMETER		TEST CONDITIONS†		uA709C			UNIT
				MIN	TYP	MAX	
V_{IO}	Input offset voltage	$V_{CC\pm} = \pm9$ V to ±15 V, $V_O = 0$	25°C		2	7.5	mV
			Full range			10	
I_{IO}	Input offset current	$V_{CC\pm} = \pm9$ V to ±15 V, $V_O = 0$	25°C		100	500	nA
			Full range			750	
I_{IB}	Input bias current	$V_{CC\pm} = \pm9$ V to ±15 V, $V_O = 0$	25°C		0.3	1.5	μA
			Full range			2	
V_{ICR}	Common-mode input voltage range		25°C	±8	±10		V
V_{OPP}	Maximum peak-to-peak output voltage swing	$R_L \geq 10$ kΩ	25°C	24	28		V
			Full range	24			
		$R_L = 2$ kΩ	25°C	20	26		
		$R_L \geq 2$ kΩ	Full range	20			
A_{VD}	Large-signal differential voltage amplification	$R_L \leq 2$ kΩ, $V_O = \pm10$ V	25°C	15	45		V/mV
			Full range	12			
r_i	Input resistance		25°C	50	250		kΩ
			Full range	35			
r_o	Output resistance	$V_O = 0$, See Note 6	25°C		150		Ω
CMRR	Common-mode rejection ratio	$V_{IC} = V_{ICR}$ min	25°C	65	90		dB
k_{SVS}	Supply voltage sensitivity	$V_{CC} = \pm9$ V to ±15 V	25°C		25	200	μV/V
P_D	Total power dissipation	$V_O = 0$ No load	25°C		80	200	mW

†All characteristics are specified under open-loop operation with zero volts common-mode voltage unless otherwise specified. Full range for uA709C is 0°C to 70°C.

NOTE 6: This typical value applies only at frequencies above a few hundred hertz because of the effects of drift and thermal feedback.

operating characteristics $V_{CC\pm} = \pm9$ V to ±15 V, $T_A = 25$°C

PARAMETER		TEST CONDITIONS		uA709AM uA709M uA709C			UNIT
				MIN	TYP	MAX	
t_r	Rise time	$V_I = 20$ mV, $R_L = 2$ kΩ, See Figure 1	$C_L = 0$		0.3	1	μs
	Overshoot factor		$C_L = 100$ pF		6%	30%	

PARAMETER MEASUREMENT INFORMATION

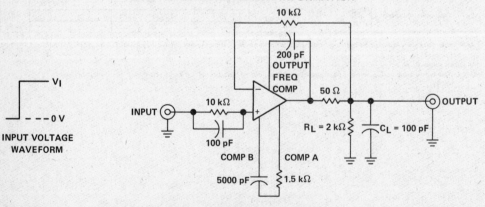

FIGURE 1—RISE TIME AND SLEW RATE

TEXAS INSTRUMENTS

POST OFFICE BOX 225012 • DALLAS, TEXAS 75265

88

- Ultra-Low Offset Voltage . . . 30 μV Typ (uA714E)

- Ultra-Low Offset Voltage Temperature Coefficient . . . 0.3 μV/°C Typ (uA714E)

- Ultra-Low Noise

- No External Components Required

- Replaces Chopper Amplifiers at a Lower Cost

- Single-Chip Monolithic Fabrication

- Wide Input Voltage Range 0 to ±14 V Typ

- Wide Supply Voltage Range ±3 V to ±18 V

- Essentially Equivalent to PMI OP-07 Series Operational Amplifiers

- Direct Replacements for Fairchild μA714C, μA714E, μA714L

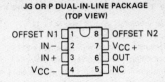

JG OR P DUAL-IN-LINE PACKAGE
(TOP VIEW)

OFFSET N1	1	8	OFFSET N2
IN−	2	7	V$_{CC}$+
IN+	3	6	OUT
V$_{CC}$−	4	5	NC

NC—No internal connection

symbol

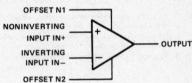

OFFSET N1
NONINVERTING INPUT IN+
INVERTING INPUT IN−
OFFSET N2
OUTPUT

description

These devices represent a breakthrough in operational amplifier performance. Low offset and long-term stability are achieved by means of a low-noise, chopperless, bipolar-input-transistor amplifier circuit. For most applications, no external components are required for offset nulling and frequency compensation. The true differential input, with a wide input voltage range and outstanding common-mode rejection, provides maximum flexibility and performance in high-noise environments and in noninverting applications. Low bias currents and extremely high input impedances are maintained over the entire temperature range. The uA714 is unsurpassed for low-noise, high-accuracy amplification of very-low-level signals.

These devices are characterized for operation from 0°C to 70°C.

schematic

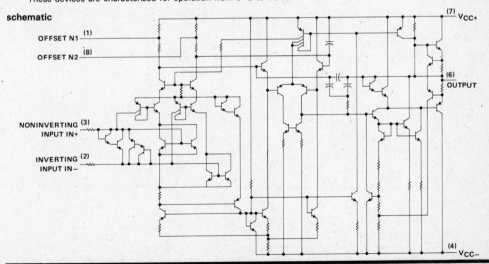

OFFSET N1 (1)
OFFSET N2 (8)
NONINVERTING INPUT IN+ (3)
INVERTING INPUT IN− (2)
(7) V$_{CC}$+
(6) OUTPUT
(4) V$_{CC}$−

Operational Amplifiers

3

absolute maximum ratings over operating free-air temperature range (unless otherwise noted)

Supply voltage V_{CC+} (see Note 1)	22 V
Supply voltage V_{CC-}	−22 V
Differential input voltage (see Note 2)	±30 V
Input voltage (either input, see Note 3)	±22 V
Duration of output short circuit (see Note 4)	unlimited
Continuous total dissipation at (or below) 25 °C free-air temperature (see Note 5)	500 mW
Operating free-air temperature range	0 °C to 70 °C
Storage temperature range	−65 °C to 150 °C
Lead temperature 1,6 mm (1/16 inch) from case for 60 seconds: JG package	300 °C
Lead temperature 1,6 mm (1/16 inch) from case for 10 seconds: P package	260 °C

NOTES: 1. All voltage values, unless otherwise noted, are with respect to the midpoint between V_{CC+} and V_{CC-}.
2. Differential voltages are at the noninverting input terminal with respect to the inverting input terminal.
3. The magnitude of the input voltage must never exceed the magnitude of the supply voltage or 15 volts, whichever is less.
4. The output may be shorted to ground or either power supply.
5. For operation above 25 °C free-air temperature, refer to Dissipation Derating Curves in Section 2. In the JG package, these chips are glass-mounted.

3

Operational Amplifiers

TEXAS
INSTRUMENTS
POST OFFICE BOX 225012 • DALLAS, TEXAS 75265

electrical characteristics at specified free-air temperature, $V_{CC} \pm = \pm 15$ V (unless otherwise noted)

PARAMETER	TEST CONDITIONS†		uA714C MIN	TYP	MAX	uA714E MIN	TYP	MAX	uA714L MIN	TYP	MAX	UNIT
V_{IO} Input offset voltage	$V_O = 0$, $R_S = 50\ \Omega$	25°C		60	150		30	75		100	250	µV
	$V_O = 0$, $R_S = 50\ \Omega$	0°C to 70°C		85	250		45	130			400	µV
α_{VIO} Temperature coefficient of input offset voltage	$V_O = 0$, $R_S = 50\ \Omega$	0°C to 70°C		0.5	1.8		0.3	1.3		1	3	µV/°C
Long-term drift of input offset voltage	See Note 6	25°C		0.4	2		0.3	1.5		0.5	3	µV/mo
Offset adjustment range	$R_S = 20\ k\Omega$, See Figure 1	25°C		±4			±4			±4		mV
I_{IO} Input offset current		25°C		0.8	6		0.5	3.8		5	20	nA
		0°C to 70°C		1.6	8		0.9	5.3		8	40	nA
α_{IIO} Temperature coefficient of input offset current		0°C to 70°C		12	50		8	35		20	100	pA/°C
I_{IB} Input bias current		25°C		±1.8	±7		±1.2	±4		±6	±30	nA
		0°C to 70°C		±2.2	±9		±1.5	±5.5		±15	±60	nA
α_{IIB} Temperature coefficient of input bias current		0°C to 70°C		18	50		13	35		35	150	pA/°C
V_{ICR} Common-mode input voltage range		25°C	±13	±14		±13	±14		±13	±14		V
		0°C to 70°C	±13	±13.5		±13	±13.5		±13	±13.5		V
V_{OM} Peak output voltage	$R_L \geq 10\ k\Omega$	25°C	±12	±13		±12.5	±13		±12	±13		V
	$R_L \geq 2\ k\Omega$	25°C	±11.5	±12.8		±12	±12.8		±11	±12.8		V
	$R_L \geq 1\ k\Omega$	25°C		±12		±10.5	±12			±12		V
	$R_L \geq 2\ k\Omega$	0°C to 70°C	±11	±12.6		±12	±12.6		±10	±12.6		V
A_{VD} Large-signal differential voltage amplification	$V_{CC\pm} = \pm 3$ V, $V_O = \pm 0.5$ V, $R_L \geq 500\ k\Omega$	25°C	100	400		150	500		50	150		V/mV
		25°C	120	400		200	500		100	300		V/mV
	$V_O = \pm 10$ V, $R_L = 2\ k\Omega$	0°C to 70°C	100	400		180	450		80	400		V/mV
B_1 Unity gain bandwidth		25°C		0.6			0.6			0.6		MHz
r_i Input resistance		25°C	8	33		15	50		8	33		MΩ
CMRR Common-mode rejection ratio	$V_{IC} = \pm 13$ V, $R_S = 50\ \Omega$	25°C	100	120		106	123		100	120		dB
		0°C to 70°C	97	120		103	123		94	120		dB
k_{SVR} Supply voltage rejection ratio $(\Delta V_{CC} / \Delta V_{IO})$	$V_{CC\pm} = \pm 3$ V to ± 18 V, $R_S = 50\ \Omega$	25°C	90	104		94	107		90	104	32	dB
		0°C to 70°C	86	100		90	104		83	100		dB
P_D Power dissipation	$V_O = 0$, No load	25°C		80	150		75	120		100	180	mW
	$V_{CC\pm} = \pm 3$ V, $V_O = 0$, No load	25°C		4	8		4	6		5	12	mW

† All characteristics are measured under open-loop conditions with zero common-mode input voltage unless otherwise noted.

NOTE 6: Since long-term drift cannot be measured on the individual devices prior to shipment, this specification is not intended to be a guarantee or warranty. It is an engineering estimate of the averaged trend line of drift versus time over extended periods after the first thirty days of operation.

3

Operational Amplifiers

TYPES uA714C, uA714E, uA714L
ULTRA-LOW-OFFSET VOLTAGE OPERATIONAL AMPLIFIERS

operating characteristics at specified free-air temperature, $V_{CC\pm} = \pm 15$ V (unless otherwise noted)

	PARAMETER	TEST CONDITIONS†		uA714C			uA714E			uA714L			UNIT
				MIN	TYP	MAX	MIN	TYP	MAX	MIN	TYP	MAX	
V_n	Equivalent input noise voltage	$T_A = 25°C$	f = 10 Hz		10.5	20		10.3	18		10.5		nV/√Hz
			f = 100 Hz		10.2	13.5		10	13		10.2		
			f = 1 kHz		9.8	11.5		9.6	11		9.8		
V_{NPP}	Peak-to-peak equivalent input noise voltage	f = 0.1 Hz to 10 Hz, $T_A = 25°C$			0.38	0.65		0.35	0.6				μV
I_n	Equivalent input noise current	$T_A = 25°C$	f = 10 Hz		0.35	0.9		0.32	0.8		0.35	0.8	pA/√Hz
			f = 100 Hz		0.15	0.27		0.14	0.23		0.15	0.23	
			f = 1 kHz		0.13	0.18		0.12	0.17		0.13	0.17	
I_{NPP}	Peak-to-peak equivalent input noise current	f = 0.1 Hz to 10 Hz, $T_A = 25°C$			15	35		14	30		15		pA
SR	Slew rate	$R_L \geq 2$ kΩ, $T_A = 25°C$			0.17			0.17			0.17		V/μs

†All characteristics are measured under open-loop conditions with zero common-mode input voltage unless otherwise specified.

TYPICAL APPLICATION DATA

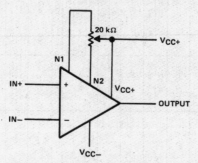

FIGURE 1—INPUT OFFSET VOLTAGE NULL CIRCUIT

TEXAS
INSTRUMENTS
POST OFFICE BOX 225012 • DALLAS, TEXAS 75265

3

Operational Amplifiers

LINEAR
INTEGRATED
CIRCUITS

TYPES uA741M, uA741C
GENERAL-PURPOSE OPERATIONAL AMPLIFIER

D920, NOVEMBER 1970–REVISED AUGUST 1983

- Short-Circuit Protection
- Offset-Voltage Null Capability
- Large Common-Mode and
  Differential Voltage Ranges
- No Frequency Compensation Required
- Low Power Consumption
- No Latch-up
- Designed to be Interchangeable with Fairchild
  μA741M, μA741C

description

The uA741 is a general-purpose operational amplifier featuring offset-voltage null capability.

The high common-mode input voltage range and the absence of latch-up make the amplifier ideal for voltage-follower applications. The device is short-circuit protected and the internal frequency compensation ensures stability without external components. A low potentiometer may be connected between the offset null inputs to null out the offset voltage as shown in Figure 2.

The uA741M is characterized for operation over the full military temperature range of $-55\,°C$ to $125\,°C$; the uA741C is characterized for operation from $0\,°C$ to $70\,°C$.

symbol

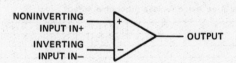

NONINVERTING
INPUT IN+

INVERTING
INPUT IN−

OUTPUT

uA741M . . . J PACKAGE
(TOP VIEW)

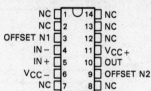

NC	1	14 NC
NC	2	13 NC
OFFSET N1	3	12 NC
IN−	4	11 $V_{CC}+$
IN+	5	10 OUT
$V_{CC}-$	6	9 OFFSET N2
NC	7	8 NC

uA741M . . . JG PACKAGE
uA741C . . . D, P, OR JG PACKAGE
(TOP VIEW)

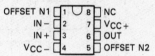

OFFSET N1	1	8 NC
IN−	2	7 $V_{CC}+$
IN+	3	6 OUT
$V_{CC}-$	4	5 OFFSET N2

uA741M . . . U FLAT PACKAGE
(TOP VIEW)

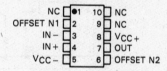

NC	1	10 NC
OFFSET N1	2	9 NC
IN−	3	8 $V_{CC}+$
IN+	4	7 OUT
$V_{CC}-$	5	6 OFFSET N2

uA741M . . . FH, FK PACKAGE
(TOP VIEW)

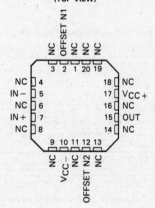

NC—No internal connection

Operational Amplifiers

3

schematic

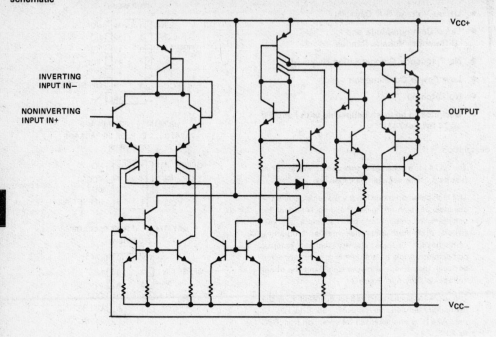

absolute maximum ratings over operating free-air temperature range (unless otherwise noted)

		uA741M	uA741C	UNIT
Supply voltage V_{CC+} (see Note 1)		22	18	V
Supply voltage V_{CC-} (see Note 1)		−22	−18	V
Differential input voltage (see Note 2)		±30	±30	V
Input voltage any input (see Notes 1 and 3)		±15	±15	V
Voltage between either offset null terminal (N1/N2) and V_{CC-}		±0.5	±0.5	V
Duration of output short-circuit (see Note 4)		unlimited	unlimited	
Continuous total power dissipation at (or below) 25°C free-air temperature (see Note 5)		500	500	mW
Operating free-air temperature range		−55 to 125	0 to 70	°C
Storage temperature range		−65 to 150	−65 to 150	°C
Lead temperature 1,6 mm (1/16 inch) from case for 60 seconds	FH, FK, J, JG, or U package	300	300	°C
Lead temperature 1,6 mm (1/16 inch) from case for 10 seconds	D, N or P package		260	°C

NOTES: 1. All voltage values, unless otherwise noted, are with respect to the midpoint between V_{CC+} and V_{CC-}.
2. Differential voltages are at the noninverting input terminal with respect to the inverting input terminal.
3. The magnitude of the input voltage must never exceed the magnitude of the supply voltage or 15 volts, whichever is less.
4. The output may be shorted to ground or either power supply. For the uA741M only, the unlimited duration of the short-circuit applies at (or below) 125°C case temperature or 75°C free-air temperature.
5. For operation above 25°C free-air temperature, refer to Dissipation Derating Curves, Section 2. In the J and JG packages, uA741M chips are alloy mounted; uA741C chips are glass mounted.

TEXAS
INSTRUMENTS
POST OFFICE BOX 225012 ● DALLAS, TEXAS 75265

electrical characteristics at specified free-air temperature, V_{CC+} = 15 V, V_{CC-} = −15 V

PARAMETER		TEST CONDITIONS[†]		uA741M			uA741C			UNIT
				MIN	TYP	MAX	MIN	TYP	MAX	
V_{IO}	Input offset voltage	$V_O = 0$	25°C		1	5		1	6	mV
			Full range			6			7.5	
$\Delta V_{IO(adj)}$	Offset voltage adjust range	$V_O = 0$	25°C		±15			±15		mV
I_{IO}	Input offset current	$V_O = 0$	25°C		20	200		20	200	nA
			Full range			500			300	
I_{IB}	Input bias current	$V_O = 0$	25°C		80	500		80	500	nA
			Full range			1500			800	
V_{ICR}	Common-mode input voltage range		25°C	±12	±13		±12	±13		V
			Full range	±12			±12			
V_{OM}	Maximum peak output voltage swing	$R_L = 10$ kΩ	25°C	±12	±14		±12	±14		V
		$R_L \geq 10$ kΩ	Full range	±12			±12			
		$R_L = 2$ kΩ	25°C	±10	±13		±10	±13		
		$R_L \geq 2$ kΩ	Full range	±10			±10			
A_{VD}	Large-signal differential voltage amplification	$R_L \geq 2$ kΩ	25°C	50	200		20	200		V/mV
		$V_O = \pm10$ V	Full range	25			15			
r_i	Input resistance		25°C	0.3	2		0.3	2		MΩ
r_o	Output resistance	$V_O = 0$, See Note 6	25°C		75			75		Ω
C_i	Input capacitance		25°C		1.4			1.4		pF
CMRR	Common-mode rejection ratio	$V_{IC} = V_{ICR}$ min	25°C	70	90		70	90		dB
			Full range	70			70			
k_{SVS}	Supply voltage sensitivity ($\Delta V_{IO}/\Delta V_{CC}$)	$V_{CC} = \pm9$ V to ±15 V	25°C		30	150		30	150	μV/V
			Full range			150			150	
I_{OS}	Short-circuit output current		25°C		±25	±40		±25	±40	mA
I_{CC}	Supply current	No load, $V_O = 0$	25°C		1.7	2.8		1.7	2.8	mA
			Full range			3.3			3.3	
P_D	Total power dissipation	No load, $V_O = 0$	25°C		50	85		50	85	mW
			Full range			100			100	

[†]All characteristics are measured under open-loop conditions with zero common-mode input voltage unless otherwise specified. Full range for uA741M is −55°C to 125°C and for uA741C is 0°C to 70°C.

NOTE 6: This typical value applies only at frequencies above a few hundred hertz because of the effects of drift and thermal feedback.

3

Operational Amplifiers

TYPES uA741M, uA741C
GENERAL-PURPOSE OPERATIONAL AMPLIFIERS

operating characteristics, V_{CC+} = 15 V, V_{CC-} = −15 V, T_A = 25°C

PARAMETER		TEST CONDITIONS	uA741M			uA741C			UNIT
			MIN	TYP	MAX	MIN	TYP	MAX	
t_r	Rise time	V_I = 20 mV, R_L = 2 kΩ,		0.3			0.3		μs
	Overshoot factor	C_L = 100 pF, See Figure 1		5%			5%		
SR	Slew rate at unity gain	V_I = 10 V, R_L = 2 kΩ, C_L = 100 pF, See Figure 1		0.5			0.5		V/μs

PARAMETER MEASUREMENT INFORMATION

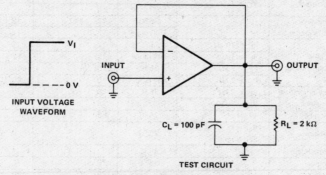

INPUT VOLTAGE WAVEFORM

TEST CIRCUIT

FIGURE 1—RISE TIME, OVERSHOOT, AND SLEW RATE

TYPICAL APPLICATION DATA

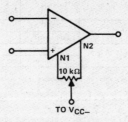

FIGURE 2 — INPUT OFFSET VOLTAGE NULL CIRCUIT

3 Operational Amplifiers

TEXAS
INSTRUMENTS
POST OFFICE BOX 225012 • DALLAS, TEXAS 75265

TYPICAL CHARACTERISTICS

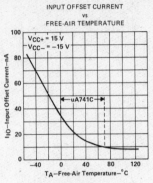

INPUT OFFSET CURRENT
vs
FREE-AIR TEMPERATURE

FIGURE 3

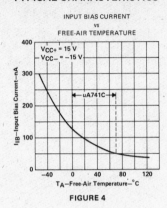

INPUT BIAS CURRENT
vs
FREE-AIR TEMPERATURE

FIGURE 4

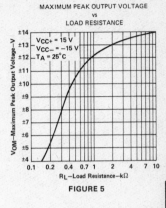

MAXIMUM PEAK OUTPUT VOLTAGE
vs
LOAD RESISTANCE

FIGURE 5

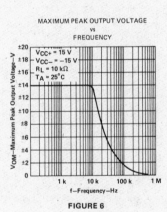

MAXIMUM PEAK OUTPUT VOLTAGE
vs
FREQUENCY

FIGURE 6

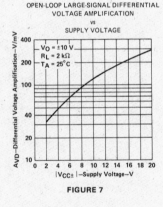

OPEN-LOOP LARGE-SIGNAL DIFFERENTIAL
VOLTAGE AMPLIFICATION
vs
SUPPLY VOLTAGE

FIGURE 7

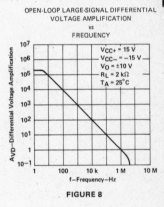

OPEN-LOOP LARGE-SIGNAL DIFFERENTIAL
VOLTAGE AMPLIFICATION
vs
FREQUENCY

FIGURE 8

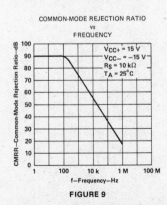

COMMON-MODE REJECTION RATIO
vs
FREQUENCY

FIGURE 9

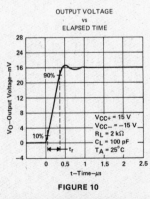

OUTPUT VOLTAGE
vs
ELAPSED TIME

FIGURE 10

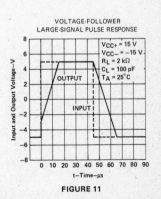

VOLTAGE-FOLLOWER
LARGE-SIGNAL PULSE RESPONSE

FIGURE 11

Operational Amplifiers 3

TYPES uA748M, uA748C
GENERAL-PURPOSE OPERATIONAL AMPLIFIERS

D921, DECEMBER 1970—REVISED AUGUST 1983

- Frequency and Transient Response Characteristics Adjustable
- Short-Circuit Protection
- Offset-Voltage Null Capability
- Wide Common-Mode and Differential Voltage Ranges
- Low Power Consumption
- No Latch-Up
- Same Pin Assignments as uA709

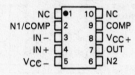

uA748M . . . JG
uA748C . . . D, JG, OR P
DUAL-IN-LINE PACKAGE
(TOP VIEW)

N1/COMP	1		8	COMP
IN –	2		7	$V_{CC}+$
IN +	3		6	OUT
V_{CC} –	4		5	N2

uA748M . . . U FLAT PACKAGE
(TOP VIEW)

NC	1		10	NC
N1/COMP	2		9	COMP
IN –	3		8	$V_{CC}+$
IN +	4		7	OUT
V_{CC} –	5		6	N2

NC—No internal connection

description

The uA748 is a general-purpose operational amplifier that offers the same advantages and attractive features as the uA741 except for internal compensation. External compensation can be as simple as a 30-pF capacitor for unity-gain conditions and, when the closed-loop gain is greater than one, can be changed to obtain wider bandwidth or higher slew rate. This circuit features high gain, large differential and common-mode input voltage range, and output short-circuit protection. Input offset voltage adjustment can be provided by connecting a variable resistor between the offset null pins as shown in Figure 12.

The uA748M is characterized for operation over the full military temperature range of $-55\,°C$ to $125\,°C$; the uA748C is characterized for operation from $0\,°C$ to $70\,°C$.

symbol

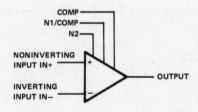

COMP
N1/COMP
N2

NONINVERTING INPUT IN+ +

INVERTING INPUT IN– –

OUTPUT

Operational Amplifiers

3

TEXAS
INSTRUMENTS
POST OFFICE BOX 225012 • DALLAS, TEXAS 75265

TYPES uA748M, uA748C
GENERAL-PURPOSE OPERATIONAL AMPLIFIERS

schematic

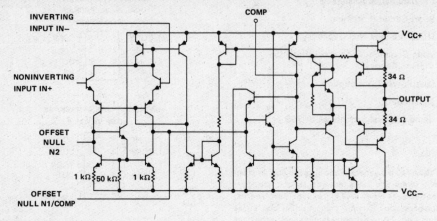

Resistor values shown are nominal.

absolute maximum ratings over operating free-air temperature (unless otherwise noted)

		uA748M	uA748C	UNIT
Supply voltage V_{CC+} (see Note 1)		22	18	V
Supply voltage V_{CC-} (see Note 1)		−22	−18	V
Differential input voltage (see Note 2)		±30	±30	V
Input voltage (either input, see Notes 1 and 3)		±15	±15	V
Voltage between either offset null terminal (N1/N2) and V_{CC-}		−0.5 to 2	−0.5 to 2	V
Duration of output short-circuit (see Note 4)		unlimited	unlimited	
Continuous total power dissipation at (or below) 25°C free-air temperature (see Note 5)		500	500	mW
Operating free-air temperature range		−55 to 125	0 to 70	°C
Storage temperature range		−65 to 150	−65 to 150	°C
Lead temperature 1,6 mm (1/16 inch) from case for 60 seconds	JG or U package	300	300	°C
Lead temperature 1,6 mm (1/16 inch) from case for 10 seconds	D or P package		260	°C

NOTES: 1. All voltage values, unless otherwise noted, are with respect to the midpoint between V_{CC+} and V_{CC-}.
2. Differential voltages are at the noninverting input terminal with respect to the inverting input terminal.
3. The magnitude of the input voltage must never exceed the magnitude of the supply voltage or 15 volts, whichever is less.
4. The output may be shorted to ground or either power supply. For the uA748M only, the unlimited duration of the short-circuit applies at (or below) 125°C case temperature or 75°C free-air temperature.
5. For operation above 25°C free-air temperature, refer to Dissipation Derating Curves, Section 2. In the J and JG package, uA748M chips are alloy-mounted; uA748C chips are glass-mounted.

TEXAS INSTRUMENTS
POST OFFICE BOX 225012 • DALLAS, TEXAS 75265

electrical characteristics at specified free-air temperature, V_{CC+} = 15 V, V_{CC-} = −15 V, C_C = 30 pF

PARAMETER		TEST CONDITIONS[†]		uA748M			uA748C			UNIT
				MIN	TYP	MAX	MIN	TYP	MAX	
V_{IO}	Input offset voltage	$V_O = 0$	25°C		1	5		1	6	mV
			Full range			6			7.5	
I_{IO}	Input offset current	$V_O = 0$	25°C		20	200		20	200	nA
			Full range			500			300	
I_{IB}	Input bias current	$V_O = 0$	25°C		80	500		80	500	nA
			Full range			1500			800	
V_{ICR}	Common-mode input voltage range		25°C	±12	±13		±12	±13		V
			Full range	±12			±12			
V_{OM}	Maximum peak output voltage swing	$R_L = 10$ kΩ	25°C	±12	±14		±12	±14		V
		$R_L ≥ 10$ kΩ	Full range	±12			±12			
		$R_L = 2$ kΩ	25°C	±10	±13		±10	±13		
		$R_L ≥ 2$ kΩ	Full range	±10			±10			
A_{VD}	Large-signal differential voltage amplification	$R_L ≥ 2$ kΩ, $V_O = ±10$ V	25°C	50	200		20	200		V/mV
			Full range	25			15			
r_i	Input resistance		25°C	0.3	2		0.3	2		MΩ
r_o	Output resistance	$V_O = 0$, See Note 6	25°C		75			75		Ω
C_i	Input capacitance		25°C		1.4			1.4		pF
CMRR	Common-mode rejection ratio	$V_{IC} = V_{ICR}$ min, $V_O = 0$	25°C	70	90		70	90		dB
			Full range	70			70			
k_{SVS}	Supply voltage sensitivity ($\Delta V_{IO}/\Delta V_{CC}$)	$V_{CC} = ±9$ V to ±15 V, $V_O = 0$	25°C		30	150		30	150	μV/V
			Full range			150			150	
I_{OS}	Short-circuit output current		25°C		±25	±40		±25	±40	mA
I_{CC}	Supply current	No load, $V_O = 0$	25°C		1.7	2.8		1.7	2.8	mA
			Full range			3.3			3.3	
P_D	Total power dissipation	No load, $V_O = 0$	25°C		50	85		50	85	mW
			Full range			100			100	

[†]All characteristics are measured under open-loop conditions with zero common-mode input voltage unless otherwise specified. Full range for uA748M is −55°C to 125°C and for uA748C is 0°C to 70°C.
NOTE 6: This typical value applies only at frequencies above a few hundred hertz because of the effects of drift and thermal feedback.

operating characteristics, V_{CC+} = 15 V, V_{CC-} = −15 V, T_A = 25°C

PARAMETER		TEST CONDITIONS		uA748M			uA748C			UNIT
				MIN	TYP	MAX	MIN	TYP	MAX	
t_r	Rise time	$V_I = 20$ mV, $R_L = 2$ kΩ, $C_L = 100$ pF, $C_C = 30$ pF, See Figure 1			0.3			0.3		μs
	Overshoot factor				5%			5%		
SR	Slew rate at unity gain	$V_I = 10$ V, $R_L = 2$ kΩ, $C_L = 100$ pF, $C_C = 30$ pF, See Figure 1			0.5			0.5		V/μs

3

Operational Amplifiers

PARAMETER MEASUREMENT INFORMATION

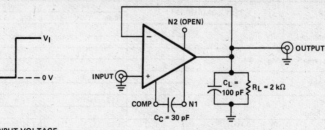

INPUT VOLTAGE
WAVEFORM

TEST CIRCUIT

FIGURE 1—RISE TIME, OVERSHOOT, AND SLEW RATE

TYPICAL CHARACTERISTICS

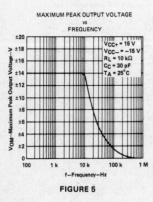

INPUT OFFSET CURRENT
vs
FREE-AIR TEMPERATURE

FIGURE 2

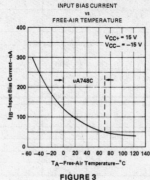

INPUT BIAS CURRENT
vs
FREE-AIR TEMPERATURE

FIGURE 3

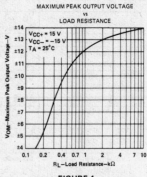

MAXIMUM PEAK OUTPUT VOLTAGE
vs
LOAD RESISTANCE

FIGURE 4

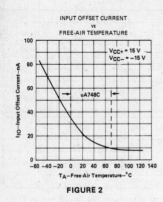

MAXIMUM PEAK OUTPUT VOLTAGE
vs
FREQUENCY

FIGURE 5

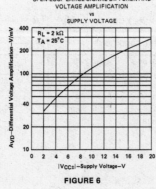

OPEN-LOOP LARGE-SIGNAL DIFFERENTIAL
VOLTAGE AMPLIFICATION
vs
SUPPLY VOLTAGE

FIGURE 6

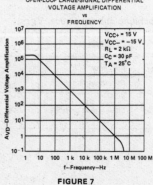

OPEN-LOOP LARGE-SIGNAL DIFFERENTIAL
VOLTAGE AMPLIFICATION
vs
FREQUENCY

FIGURE 7

3

Operational Amplifiers

Texas
INSTRUMENTS

POST OFFICE BOX 225012 • DALLAS, TEXAS 75265

TYPICAL CHARACTERISTICS

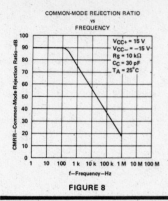

COMMON-MODE REJECTION RATIO
vs
FREQUENCY

V_{CC+} = 15 V
V_{CC-} = −15 V
R_S = 10 kΩ
C_C = 30 pF
T_A = 25°C

FIGURE 8

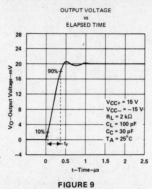

OUTPUT VOLTAGE
vs
ELAPSED TIME

V_{CC+} = 15 V
V_{CC-} = −15 V
R_L = 2 kΩ
C_L = 100 pF
C_C = 30 pF
T_A = 25°C

FIGURE 9

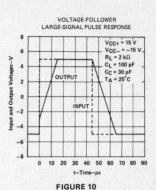

VOLTAGE-FOLLOWER
LARGE-SIGNAL PULSE RESPONSE

V_{CC+} = 15 V
V_{CC-} = −15 V
R_L = 2 kΩ
C_L = 100 pF
C_C = 30 pF
T_A = 25°C

FIGURE 10

3

Operational Amplifiers

TYPICAL APPLICATION DATA

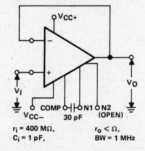

r_i = 400 MΩ,
C_i = 1 pF,

$r_o < \Omega$,
BW = 1 MHz

FIGURE 11—UNITY-GAIN VOLTAGE FOLLOWER

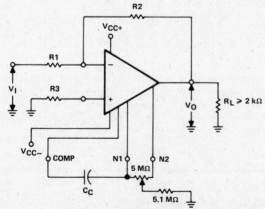

$$\frac{V_O}{V_I} = -\frac{R2}{R1}$$

$$C_C \geq \frac{R1 \cdot 30 \text{ pF}}{R1 + R2}$$

$$R3 = \frac{R1 \cdot R2}{R1 + R2}$$

**FIGURE 12—INVERTING CIRCUIT WITH ADJUSTABLE GAIN,
COMPENSATION, AND OFFSET ADJUSTMENT**

General Information   1

Thermal Information   2

Operational Amplifiers   3

Voltage Comparators   4

Special Functions   5

Voltage Regulators   6

Data Acquisition   7

Appendix   A

4

Voltage Comparators

VOLTAGE COMPARATORS

military temperature range (values specified at $T_A = 25\,°C$)

DEVICE NUMBER	TYPE	REMARKS	V_{IO} MAX (mV)	I_{IB} MAX (μA)	I_{OL} MIN (mA)	RESPONSE TIME TYP (ns)	POWER SUPPLIES		PACKAGES	PAGE
							$V_{CC}+$ NOM (V)	$V_{CC}-$ NOM (V)		
uA710M			2	20	2	40	12	−6	J,JG,U	4-87
LM106		Strobe	2	45	100	28	12	−6	J,JG,W	4-9
LM111		Strobe	3	0.1	8	115	15	−15	FH,FK,J,JG,U	4-15
TL510M	Single	Strobe	2	15	2	30			FH,FK,JG,U	4-45
TL810M		Improved TL710M	2	15	2	30			JG,U	4-67
TL710M			5	75	1.6	40	15	−6	FH,FK,JG,U	4-59
TL331M		V_{CC}: 12 V to 36 V	5	−0.1	6	300	5	0	JG	4-37
TL506M		Strobes	2	20	100	28	12	−6	J,W	4-39
TL820M		Dual TL810M	2	15	2	30	12	−6	J	4-79
TL514M	Dual	Dual TL510M	2	15	2	30	12	−6	FH,FK,J,W	4-51
uA711M		Strobes	3.5	75	0.5	40	12	−6	J,U	4-91
TL811M		Strobes	3.5	20	0.5	33	12	−6	J,U	4-73
LM193		V_{CC}: 2 V to 36 V	5	0.1	6	300	5	0	FH,FK,JG	4-29
TLC372M		LinCMOS	10			200	5	0	JG	4-83
LM139A		V_{CC}: 2 V to 36 V	2	−0.1	6	300	5	0	D,J,N	4-25
LM139	Quad	V_{CC}: 2 V to 36 V	5	−0.1	6	300	5	0	FH,FK,J	4-25
TLC374M		LinCMOS	10			200	5	0	J	4-85

automotive temperature range (values specified at $T_A = 25\,°C$)

DEVICE NUMBER	TYPE	REMARKS	V_{IO} MAX (mV)	I_{IB} MAX (μA)	I_{OL} MIN (mA)	RESPONSE TIME TYP (ns)	POWER SUPPLIES		PACKAGES	PAGE
							$V_{CC}+$ NOM (V)	$V_{CC}-$ NOM (V)		
LM2903	Dual	V_{CC}: 2 V to 36 V	7	0.25	6	300	5	0	D,JG,P	4-29
LM3302	Quad	V_{CC}: 2 V to 36 V	20	−0.5	6	300	5	0	D,J,N	4-35
LM2901		V_{CC}: 2 V to 36 V	15	−0.4	6	300	5	0	D,J,N	4-25

industrial temperature range (values specified at $T_A = 25\,°C$)

DEVICE NUMBER	TYPE	REMARKS	V_{IO} MAX (mV)	I_{IB} MAX (μA)	I_{OL} MIN (mA)	RESPONSE TIME TYP (ns)	POWER SUPPLIES		PACKAGES	PAGE
							$V_{CC}+$ NOM (V)	$V_{CC}-$ NOM (V)		
LM206		Strobe	2	45	100	28	12	−6	D,J,JG,N,P	4-9
LM211	Single	Strobe	3	0.1	8	115	15	−15	D,JG,P	4-15
TL331I			5	−0.1	6	300	5	0	D,JG,P	4-37
LM293A			2	0.25	6	300	5	0	D,JG,P	4-29
LM219	Dual		4	0.5	3.2	80	5	0	J,N	4-33
LM293			5	0.25	6	300	5	0	D,JG,P	4-29
LM239A	Quad		4	−0.4	6	300	5	0	D,J,N	4-25
LM239			9	−0.4	6	300	5	0	D,J,N	4-25

4

Voltage Comparators

TEXAS
INSTRUMENTS
POST OFFICE BOX 225012 • DALLAS, TEXAS 75265

commercial temperature range (values specified at $T_A = 25\,^\circ C$)

DEVICE NUMBER	TYPE	REMARKS	V_{IO} MAX (mV)	I_{IB} MAX (μA)	I_{OL} MIN (mA)	RESPONSE TIME TYP (ns)	POWER SUPPLIES $V_{CC}+$ NOM (V)	$V_{CC}-$ NOM (V)	PACKAGES	PAGE
TL510C	Single	Strobe	3.5	20	1.6	30			JG,P	4-45
TL810C		Improved TL710C	3.5	20	1.6	30			JG,P	4-67
LM306		Strobe	5	40	100	28	12	−6	D,J,JG,N,P	4-9
TL331C		V_{CC}: 2 V to 36 V	5	−0.25	6	300	5	0	D,JG,P	4-37
LM311		Strobe	7.5	0.25	8	115	15	−15	D,JG,P	4-15
TL710C			7.5	100		40	12	−6	J,JG,N,P,U	4-59
TL721			±100			12 Max	0	−5.2	JG,P	4-65
TL712		Output enable				25			JG,P	4-63
LM393A	Dual	V_{CC}: 2 V to 36 V	2	0.25	6	300	5	0	D,JG,P	4-29
TL820C		Dual TL810C	3.5	20	1.6	30	12	−6	J,N	4-79
TL514C		Dual TL510C	3.5	20	1.6	30	12	−6	J,N	4-51
uA711C		Strobes	5	100	0.5	40	12	−6	J,N	4-91
TL811C		Improved uA711C	5	30	0.5	33	12	−6	J,N	4-73
TL506C		Strobes	5	25	100	28	12	−6	J,N	4-39
LM393		V_{CC}: 2 V to 36 V	5	0.25	6	300	5	0	D,JG,P	4-29
LM319			8	1	3.2	80	5	0	J,N	4-33
TLC372C		LinCMOS	10			200	5	0	JG,P	4-83
LM339	Quad		5	−0.15	6	300	5	0	D,J,N	4-25
LM339A			2	−0.15	6	300	5	0	D,J,N	4-25
TLC374C		LinCMOS	10			200	5	0	D,J,N	4-85

4

Voltage Comparators

Texas
INSTRUMENTS
POST OFFICE BOX 225012 ● DALLAS, TEXAS 75265

Input Offset Voltage (V_{IO})

The d-c voltage that must be applied between the input terminals to force the quiescent d-c output voltage to the specified level.

Average Temperature Coefficient of Input Offset Voltage (αV_{IO})

The ratio of the change in input offset voltage to the change in free-air temperature. This is an average value for the specified temperature range.

$$\alpha V_{IO} = \left[\frac{(V_{IO} @ T_{A(1)}) - (V_{IO} @ T_{A(2)})}{T_{A(1)} - T_{A(2)}} \right] \text{ where } T_{A(1)} \text{ and } T_{A(2)} \text{ are the specified temperature extremes.}$$

Input Offset Current (I_{IO})

The difference between the currents into the two input terminals with the output at the specified level.

Average Temperature Coefficient of Input Offset Current (αI_{IO})

The ratio of the change in input offset current to the change in free-air temperature. This is an average value for the specified temperature range.

$$\alpha I_{IO} = \left[\frac{(I_{IO} @ T_{A(1)}) - (I_{IO} @ T_{A(2)})}{T_{A(1)} - T_{A(2)}} \right] \text{ where } T_{A(1)} \text{ and } T_{A(2)} \text{ are the specified temperature extremes.}$$

Input Bias Current (I_{IB})

The average of the currents into the two input terminals with the output at the specified level.

High-Level Strobe Current ($I_{IH(S)}$)

The current flowing into or out of* the strobe at a high-level voltage.

Low-Level Strobe Current ($I_{IL(S)}$)

The current flowing out of* the strobe at a low-level voltage.

High-Level Strobe Voltage ($V_{IH(S)}$)

For a device having an active-low strobe, a voltage within the range that is guaranteed not to interfere with the operation of the comparator.

Low-Level Strobe Voltage ($V_{IL(S)}$)

For a device having an active-low strobe, a voltage within the range that is guaranteed to force the output high or low, as specified, independently of the differential inputs.

Input Voltage Range (V_I)

The range of voltage that if exceeded at either input terminal will cause the comparator to cease functioning properly.

Common-Mode Input Voltage (V_{IC})

The average of the two input voltages.

*Current out of a terminal is given as a negative value.

Voltage Comparators

4

GLOSSARY

Common-Mode Input Voltage Range (V_{ICR})

The range of common-mode input voltage that if exceeded will cause the comparator to cease functioning properly.

Differential Input Voltage (V_{ID})

The voltage at the noninverting input with respect to the inverting input.

Differential Input Voltage Range (V_{ID})

The range of voltage between the two input terminals that if exceeded will cause the comparator to cease functioning properly.

Differential Voltage Amplification (A_{VD})

The ratio of the change in output to the change in differential input voltage producing it with the common-mode input voltage held constant.

High-Level Output Voltage (V_{OH})

The voltage at an output with input conditions applied that according to the product specification will establish a high level at the output.

Low-Level Output Voltage (V_{OL})

The voltage at an output with input conditions applied that according to the product specification will establish a low level at the output.

High-Level Output Current, (I_{OH})

The current into* an output with input conditions applied that according to the product specification will establish a high level at the output.

Low-Level Output Current, (I_{OL})

The current into* an output with input conditions applied that according to the product specification will establish a low level at the output.

Output Resistance (r_O)

The resistance between an output terminal and ground.

Common-Mode Rejection Ration (k_{CMR}, CMRR)

The ratio of differential voltage amplification to common-mode voltage amplification.
NOTE: This is measured by determining the ratio of a change in input common-mode voltage to the resulting change in input offset voltage.

Supply Current (I_{CC+}, I_{CC-})

The current into* the V_{CC+} or V_{CC-} terminal of an integrated circuit.

Total Power Dissipation (P_D)

The total d-c power supplied to the device less any power delivered from the device to a load.
NOTE: At no load: $P_D = V_{CC+} \cdot I_{CC+} + V_{CC-} \cdot I_{CC-}$.

---

*Current out of a terminal is given as a negative value.

TEXAS
INSTRUMENTS
POST OFFICE BOX 225012 • DALLAS, TEXAS 75265

Response Time

The interval between the application of an input step function and the instant at when the output crosses the logic threshold voltage.

NOTE: The input step drives the comparator from some initial condition sufficient to saturate the output (or in the case of high-to-low-level response time, to turn the output off) to an input level just barely in excess of that required to bring the output back to the logic threshold voltage. This excess is referred to as the voltage overdrive.

Strobe Release Time

The time required for the output to rise to the logic threshold voltage after the strobe terminal has been driven from its active logic level to its inactive logic level.

Voltage Comparators

4

4

Voltage Comparators

LINEAR
INTEGRATED
CIRCUITS

TYPES LM106, LM206, LM306
DIFFERENTIAL COMPARATORS WITH STROBES

D1108, OCTOBER 1979—REVISED JULY 1983

- Fast Response Times
- Improved Gain and Accuracy
- Fan-Out to 10 Series 54/74 TTL Loads
- Strobe Capability
- Short-Circuit and Surge Protection
- Designed to be Interchangeable with National Semiconductor LM106, LM206, and LM306

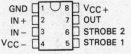

J OR N DUAL-IN-LINE OR W FLAT PACKAGE (TOP VIEW)

NC	1	14	NC
GND	2	13	NC
IN+	3	12	NC
IN−	4	11	$V_{CC}+$
NC	5	10	NC
$V_{CC}-$	6	9	OUT
STROBE 1	7	8	STROBE 2

NC—No internal connection

D, JG OR P DUAL-IN-LINE PACKAGE (TOP VIEW)

GND	1	8	$V_{CC}+$
IN+	2	7	OUT
IN−	3	6	STROBE 2
$V_{CC}-$	4	5	STROBE 1

description

The LM106, LM206, and LM306 are high-speed voltage comparators with differential inputs, a low-impedance high-sink-current (100 mA) output, and two strobe inputs. These devices detect low-level analog or digital signals and can drive digital logic or lamps and relays directly. Short-circuit protection and surge-current limiting is provided.

The circuit is similar to a TL810 with gated output. A low-level input at either strobe causes the output to remain high regardless of the differential input. When both strobe inputs are either open or at a high logic level, the output voltage is controlled by the differential input voltage. The circuit will operate with any negative supply voltage between −3 volts and −12 volts with little difference in performance.

The LM106 is characterized for operation over the full military temperature range of −55°C to 125°C, the LM206 is characterized for operation from −25°C to 85°C, and the LM306 from 0°C to 70°C.

functional block diagram

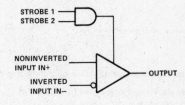

STROBE 1
STROBE 2

NONINVERTED INPUT IN+

INVERTED INPUT IN−

OUTPUT

Voltage Comparators

4

TEXAS
INSTRUMENTS
POST OFFICE BOX 225012 • DALLAS, TEXAS 75265

schematic

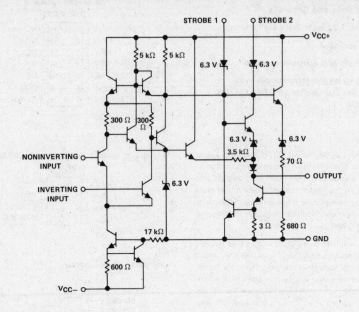

Resistor values are nominal in ohms.

absolute maximum ratings over operating free-air temperature range (unless otherwise noted)

Supply voltage V_{CC+} (see Note 1)	15 V
Supply voltage V_{CC-} (see Note 1)	−15 V
Differential input voltage (see Note 2)	±5 V
Input voltage (either input, see Notes 1 and 3)	±7 V
Strobe voltage range (see Note 1)	0 V to V_{CC+}
Output voltage (see Note 1)	24 V
Voltage from output to V_{CC-}	30 V
Duration of output short-circuit (see Note 4)	10 s
Continuous total power dissipation at (or below) 25°C free-air temperature (see Note 5)	600 mW
Operating free-air temperature range: LM106 Circuits	−55°C to 125°C
LM206 Circuits	−25°C to 85°C
LM306 Circuits	0°C to 70°C
Storage temperature range	−65°C to 150°C
Lead temperature 1,6 mm (1/16 inch) from case for 60 seconds: J, JG or W package	300°C
Lead temperature 1,6 mm (1/16 inch) from case for 10 seconds: D, N, or P package	260°C

NOTES: 1. All voltage values, except differential voltages and the voltage from the output to V_{CC-}, are with respect to the network ground terminal.
2. Differential voltages are at the noninverting input terminal with respect to the inverting input terminal.
3. The magnitude of the input voltage must never exceed the magnitude of the supply voltage or 7 volts, whichever is less.
4. The output may be shorted to ground or either power supply.
5. For operation above 25°C free-air temperature, refer to Dissipation Derating Curves, Section 2. In the J and JG packages, LM106 chips are alloy-mounted; LM206 and LM306 chips are glass-mounted.

TEXAS
INSTRUMENTS
POST OFFICE BOX 225012 • DALLAS, TEXAS 75265

electrical characteristics at specified free-air temperature, V_{CC+} = 12 V, V_{CC-} = -3 V to 12 V (unless otherwise noted)

PARAMETER		TEST CONDITIONS†		LM106, LM206 MIN	TYP	MAX	LM306 MIN	TYP	MAX	UNIT
V_{IO}	Input offset voltage	$R_S \le 200\ \Omega$, See Note 6	25°C		0.5‡	2		1.6‡	5	mV
			Full range			3			6.5	
α_{VIO}	Average temperature coefficient of input offset voltage	$R_S = 50\ \Omega$, See Note 6	Full range		3	10		5	20	µV/°C
I_{IO}	Input offset current	See Note 6	25°C		0.7‡	3		1.8‡	5	µA
			MIN		2	7		1	7.5	
			MAX		0.4	3		0.5	5	
α_{IIO}	Average temperature coefficient of input offset current	See Note 6	MIN to 25°C		15	75		24	100	nA/°C
			25°C to MAX		5	25		15	50	
I_{IB}	Input bias current	V_O = 0.5 V to 5 V	MIN to 25°C			45			40	µA
			25°C to MAX		7‡	20		16‡	25	
$I_{IL(S)}$	Low-level strobe current	$V_{(strobe)}$ = 0.4 V	Full range		-1.7‡	-3.2		-1.7‡	-3.2	mA
$V_{IH(S)}$	High-level strobe voltage		Full range	2.2			2.2			V
$V_{IL(S)}$	Low-level strobe voltage		Full range			0.9			0.9	V
V_{ICR}	Common-mode input voltage range	V_{CC-} = -7 V to -12 V	Full range	±5			±5			V
V_{ID}	Differential input voltage range		Full range	±5			±5			V
A_{VD}	Large-signal differential voltage amplification	No load, V_O = 0.5 V to 5 V	25°C		40‡			40‡		V/mV
V_{OH}	High-level output voltage	I_{OH} = -400 µA, V_{ID} = 5 mV	Full range	2.5		5.5				V
		I_{OH} = -400 µA, V_{ID} = 8 mV	Full range				2.5		5.5	
V_{OL}	Low-level output voltage	I_{OL} = 100 mA, V_{ID} = -5 mV	25°C		0.8‡	1.5				V
		I_{OL} = 100 mA, V_{ID} = -7 mV	25°C					0.8‡	2	
		I_{OL} = 50 mA, V_{ID} = -5 mV	Full range			1				
		I_{OL} = 50 mA, V_{ID} = -8 mV	Full range						1	
		I_{OL} = 16 mA, V_{ID} = -5 mV	Full range			0.4				
		I_{OL} = 16 mA, V_{ID} = -8 mV	Full range						0.4	
I_{OH}	High-level output current	V_{OH} = 8 V to 24 V, V_{ID} = 5 mV	MIN to 25°C		0.02‡	1				µA
			25°C to MAX			100				
		V_{ID} = 7 mV	MIN to 25°C					0.02‡	2	
		V_{ID} = 8 mV	25°C to MAX						100	
I_{CC+}	Supply current from V_{CC+}	V_{ID} = -5 mV, No load	Full range		6.6‡	10		6.6‡	10	mA
I_{CC-}	Supply current from V_{CC-}	No load	Full range		-1.9‡	-3.6		-1.9‡	-3.6	mA

†Unless otherwise noted, all characteristics are measured with both strobes open.

‡ These typical values are at V_{CC+} = 12 V, V_{CC-} = -6 V, T_A = 25°C. Full range (MIN to MAX) for LM106 is -55°C to 125°C; for LM206 is -25°C to 85°C; and for LM306 is 0°C to 70°C.

NOTE 6: The offset voltages and offset currents given are the maximum values required to drive the output down to the low range (V_{OL}) or up to the high range (V_{OH}). Thus these parameters actually define an error band and take into account the worst-case effects of voltage gain and input impedance.

switching characteristics, V_{CC+} = 12 V, V_{CC-} = -6 V, T_A = 25°C

PARAMETER	TEST CONDITIONS†	LM106, LM206 MIN	TYP	MAX	LM306 MIN	TYP	MAX	UNIT
Response time, low-to-high-level output	R_L = 390 Ω to 5 V, C_L = 15 pF, See Note 7		28	40		28	40	ns

NOTE 7: The response time specified is for a 100-mV input step with 5-mV overdrive and is the interval between the input step function and the instant when the output crosses 1.4 V.

4

Voltage Comparators

TEXAS INSTRUMENTS
POST OFFICE BOX 225012 • DALLAS, TEXAS 75265

Voltage Comparators

4

TYPICAL CHARACTERISTICS‡

INPUT OFFSET CURRENT
vs
FREE-AIR TEMPERATURE

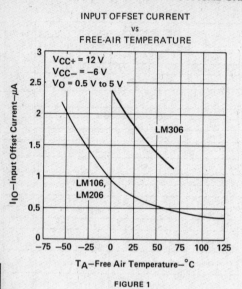

FIGURE 1

INPUT BIAS CURRENT
vs
FREE-AIR TEMPERATURE

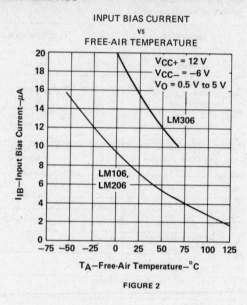

FIGURE 2

HIGH-LEVEL OUTPUT VOLTAGE
vs
FREE-AIR TEMPERATURE

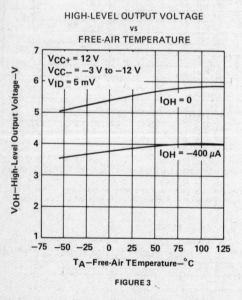

FIGURE 3

LOW-LEVEL OUTPUT VOLTAGE
vs
FREE-AIR TEMPERATURE

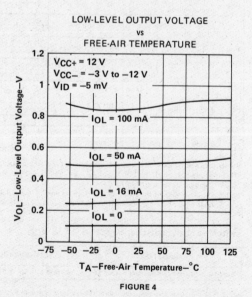

FIGURE 4

‡Data for free-air temperature outside the range specified in the absolute maximum ratings for LM206 or LM306 is not applicable for those types.

TEXAS INSTRUMENTS
POST OFFICE BOX 225012 • DALLAS, TEXAS 75265

TYPICAL CHARACTERISTICS‡

VOLTAGE TRANSFER CHARACTERISTICS

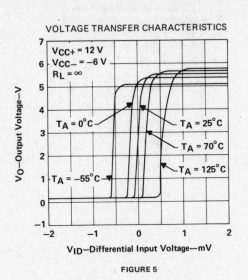

FIGURE 5

OUTPUT CURRENT vs DIFFERENTIAL INPUT VOLTAGE

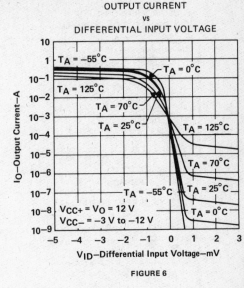

FIGURE 6

LARGE-SIGNAL DIFFERENTIAL VOLTAGE AMPLIFICATION vs FREE-AIR TEMPERATURE

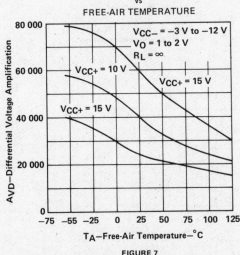

FIGURE 7

SHORT-CIRCUIT OUTPUT CURRENT vs FREE-AIR TEMPERATURE

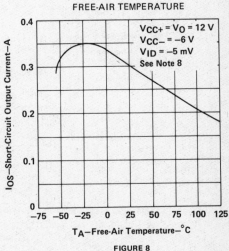

FIGURE 8

‡Data for free-air temperature outside the range specified in the absolute maximum ratings for LM206 or LM306 is not applicable for those types.
NOTE 8: This parameter was measured using a single 5-ms pulse.

Voltage Comparators

4

TEXAS INSTRUMENTS
POST OFFICE BOX 225012 • DALLAS, TEXAS 75265

TYPICAL CHARACTERISTICS‡

OUTPUT RESPONSE FOR
VARIOUS INPUT OVERDRIVES

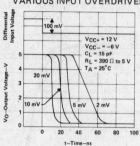

FIGURE 9

OUTPUT RESPONSE FOR
VARIOUS INPUT OVERDRIVES

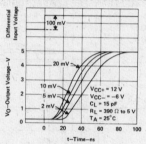

FIGURE 10

SUPPLY CURRENT FROM VCC+
vs
SUPPLY VOLTAGE VCC+

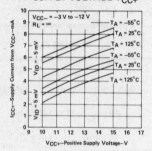

FIGURE 11

SUPPLY CURRENT FROM VCC−
vs
SUPPLY VOLTAGE VCC−

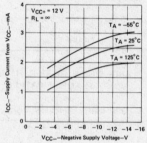

FIGURE 12

TOTAL POWER DISSIPATION
vs
FREE-AIR TEMPERATURE

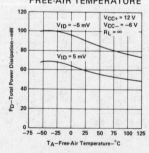

FIGURE 13

‡Data for free-air temperature outside the range specified in the absolute maximum ratings for LM206 or LM306 is not applicable for those types.

TEXAS
INSTRUMENTS
POST OFFICE BOX 225012 ● DALLAS, TEXAS 75265

Voltage Comparators

4

LINEAR INTEGRATED CIRCUITS

TYPES LM111, LM211, LM311
DIFFERENTIAL COMPARATORS WITH STROBES

D1312, SEPTEMBER 1973–REVISED AUGUST 1983

- ● **Fast Response Times**
- ● **Strobe Capability**
- ● **Designed to be Interchangeable with National Semiconductor LM111, LM211, and LM311**
- ● **Maximum Input Bias Current . . . 300 nA**
- ● **Maximum Input Offset Current . . . 70 nA**
- ● **Can Operate From Single 5-V Supply**

description

The LM111, LM211, and LM311 are single high-speed voltage comparators. These devices are designed to operate from a wide range of power supply voltage, including ± 15-volt supplies for operational amplifiers and 5-volt supplies for logic systems. The output levels are compatible with most TTL and MOS circuits. These comparators are capable of driving lamps or relays and switching voltages up to 50 volts at 50 milliamperes. All inputs and outputs can be isolated from system ground. The outputs can drive loads referenced to ground, $V_{CC}+$ or $V_{CC}-$. Offset balancing and strobe capability are available and the outputs can be wire-OR connected. If the strobe is low, the output will be in the off state regardless of the differential input. Although slower than the TL506 and TL514, these devices are not as sensitive to spurious oscillations.

The LM111 is characterized for operation over the full military range of −55°C to 125°C. The LM211 is characterized for operation from −25°C to 85°C, and the LM311 is characterized for operation from 0°C to 70°C.

functional block diagram

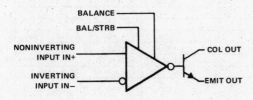

LM111, LM211, LM311
J DUAL-IN-LLINE PACKAGE
(TOP VIEW)

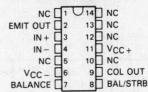

LM111 . . . JG DUAL-IN-LINE PACKAGE
LM211, LM311 . . . D, JG, OR P DUAL-IN-LINE PACKAGE
(TOP VIEW)

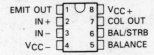

LM111 . . . U FLAT PACKAGE
(TOP VIEW)

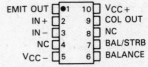

LM111 . . . FH OR FK CHIP CARRIER PACKAGE
(TOP VIEW)

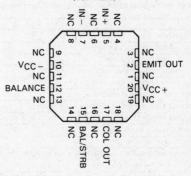

NC—No internal connection

4

Voltage Comparators

883

TEXAS
INSTRUMENTS
POST OFFICE BOX 225012 ● DALLAS, TEXAS 75265

4-15

schematic

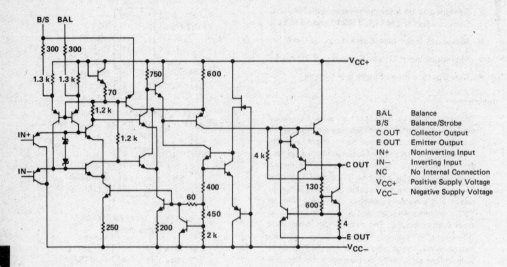

Resistor values shown are nominal and in ohms.

BAL	Balance
B/S	Balance/Strobe
C OUT	Collector Output
E OUT	Emitter Output
IN+	Noninverting Input
IN−	Inverting Input
NC	No Internal Connection
V_{CC+}	Positive Supply Voltage
$V_{CC−}$	Negative Supply Voltage

absolute maximum ratings over operating free-air temperature range (unless otherwise noted)

	LM111	LM211	LM311	UNIT
Supply voltage, V_{CC+} (see Note 1)	18	18	18	V
Supply voltage, $V_{CC−}$ (see Note 1)	−18	−18	−18	V
Differential input voltage (see Note 2)	±30	±30	±30	V
Input voltage (either input, see Notes 1 and 3)	±15	±15	±15	V
Voltage from emitter output to $V_{CC−}$	30	30	30	V
Voltage from collector output to $V_{CC−}$	50	50	40	V
Duration of output short-circuit (see Note 4)	10	10	10	s
Continuous total dissipation at (or below) 25°C free-air temperature (see Note 5)	500	500	500	mW
Operating free-air temperature range	−55 to 125	−25 to 85	0 to 70	°C
Storage temperature range	−65 to 150	−65 to 150	−65 to 150	°C
Lead temperature 1,6 mm (1/16 inch) from case for 10 seconds J, JG, FH, FK, or U package	300	300	300	°C
Lead temperature 1,6 mm (1/16 inch) from case for 60 seconds D or P package		260	260	°C

NOTES: 1. All voltage values, unless otherwise noted, are with respect to the midpoint between V_{CC+} and $V_{CC−}$.
　　　　2. Differential voltages are at the noninverting input terminal with respect to the inverting input terminal.
　　　　3. The magnitude of the input voltage must never exceed the magnitude of the supply voltage or ±15 volts, whichever is less.
　　　　4. The output may be shorted to ground or either power supply.
　　　　5. For operation above 25°C free-air temperature, refer to Dissipation Derating Curves, Section 2. In the J and JG packages, LM111 chips are alloy mounted, LM211 and LM311 chips are glass mounted.

Texas Instruments

POST OFFICE BOX 225012 • DALLAS, TEXAS 75265

883

electrical characteristics at specified free-air temperature, $V_{CC\pm} = \pm15$ V (unless otherwise noted)

PARAMETER		TEST CONDITIONS[†]		LM111, LM211			LM311			UNIT
				MIN	TYP[‡]	MAX	MIN	TYP[‡]	MAX	
V_{IO}	Input offset voltage	See Note 6	25°C		0.7	3		2	7.5	mV
			Full range			4			10	
I_{IO}	Input offset current	See Note 6	25°C		4	10		6	50	nA
			Full range			20			70	
I_{IB}	Input bias current	$V_O = 1$ V to 14 V	25°C		75	100		100	250	nA
			Full range			150			300	
$I_{IL(S)}$	Low-level strobe current (see Note 7)	$V_{(strobe)} = 0.3$ V, $V_{ID} \leq -10$ mV	25°C		−3			−3		mA
V_{ICR}	Common-mode input voltage range		Full range	13 to −14.5	13.8 to −14.7		13 to −14.5	13.8 to −14.7		V
A_{VD}	Large-signal differential voltage amplification	$V_O = 5$ V to 35 V, $R_L = 1$ kΩ	25°C	40	200		40	200		V/mV
I_{OH}	High-level (collector) output current	$I_{strobe} = -3$ mA, $V_{ID} = 5$ mV, $V_{OH} = 35$ V	25°C		0.2	10				nA
		$V_{OH} = 35$ V	Full range			0.5				μA
		$V_{ID} = 10$ mV	25°C					0.2	50	nA
V_{OL}	Low-level (collector-to-emitter) output voltage	$I_{OL} = 50$ mA	$V_{ID} = -5$ mV	25°C		0.75	1.5				V
			$V_{ID} = -10$ mV	25°C				0.75	1.5		
		$V_{CC+} = 4.5$ V, $V_{CC-} = 0$, $I_{OL} = 8$ mA	$V_{ID} = -6$ mV	Full range		0.23	0.4				
			$V_{ID} = -10$ mV	Full range				0.23	0.4		
I_{CC+}	Supply current from V_{CC+}, output low	$V_{ID} = -10$ mV, No load	25°C		5.1	6		5.1	7.5	mA
I_{CC-}	Supply current from V_{CC-}, output high	$V_{ID} = 10$ mV, No load	25°C		−4.1	−5		−4.1	−5	mA

[†]Unless otherwise noted, all characteristics are measured with the balance and balance/strobe terminals open and the emitter output grounded. Full range for LM111 is −55°C to 125°C, for LM211 is −25°C to 85°C, and for LM311 is 0°C to 70°C.

[‡]All typical values are at $T_A = 25$°C.

NOTES: 6. The offset voltages and offset currents given are the maximum values required to drive the collector output up to 14 V or down to 1 V with a pull-up resistor of 7.5 kΩ to V_{CC+}. Thus these parameters actually define an error band and take into account the worst-case effects of voltage gain and input impedance.

7. The strobe should not be shorted to ground; it should be current driven at −3 to −5 mA, e.g., see Figures 13 and 27.

switching characteristics, $V_{CC+} = 15$ V, $V_{CC-} = -15$ V, $T_A = 25$°C

PARAMETER	TEST CONDITIONS	MIN	TYP	MAX	UNIT
Response time, low-to-high-level output	$R_C = 500$ Ω to 5 V, $C_L = 5$ pF, See Note 8		115		ns
Response time, high-to-low-level output			165		ns

NOTE 8: The response time specified is for a 100-mV input step with 5-mV overdrive and is the interval between the input step function and the instant when the output crosses 1.4 V.

4

Voltage Comparators

TEXAS
INSTRUMENTS
POST OFFICE BOX 225012 • DALLAS, TEXAS 75265

Voltage Comparators

4

TYPICAL CHARACTERISTICS

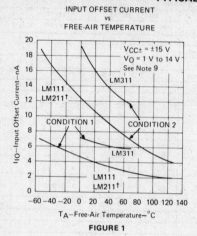

INPUT OFFSET CURRENT
vs
FREE-AIR TEMPERATURE

FIGURE 1

INPUT BIAS CURRENT
vs
FREE-AIR TEMPERATURE

FIGURE 2

VOLTAGE TRANSFER CHARACTERISTICS

FIGURE 3

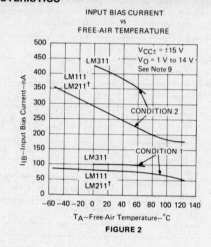

COLLECTOR OUTPUT TRANSFER CHARACTERISTIC
TEST CIRCUIT FOR FIGURE 3

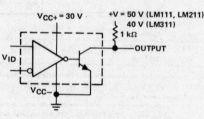

EMITTER OUTPUT TRANSFER CHARACTERISTIC
TEST CIRCUIT FOR FIGURE 3

†Data at high and low temperatures are applicable only within the rated operating free-air temperature ranges of the various devices.
NOTE 9:   Condition 1 is with the balance and balance/strobe terminals open. Condition 2 is with the balance and balance/strobe terminals connected to V_{CC+}.

4-18

TYPICAL CHARACTERISTICS

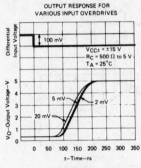

FIGURE 4

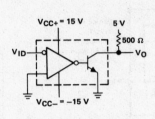

TEST CIRCUIT FOR FIGURES 4 AND 5

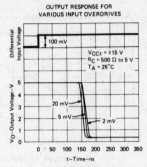

FIGURE 5

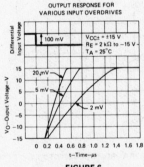

FIGURE 6

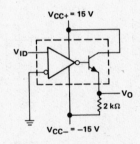

TEST CIRCUIT FOR FIGURES 6 AND 7

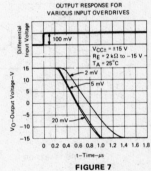

FIGURE 7

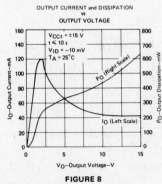

FIGURE 8

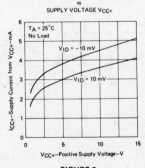

FIGURE 9

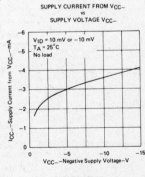

FIGURE 10

Voltage Comparators

4

TYPICAL APPLICATION DATA

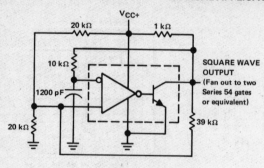

FIGURE 11—100-kHz
FREE-RUNNING MULTIVIBRATOR

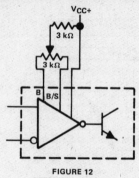

FIGURE 12
OFFSET BALANCING

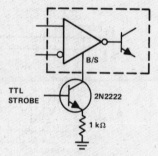

FIGURE 13—STROBING

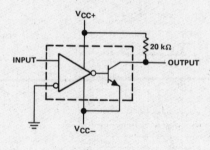

FIGURE 14—ZERO-CROSSING DETECTOR

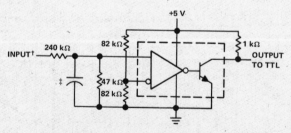

†Resistor values shown are for a 0-to-30-V logic swing and a
15-V threshold.
‡May be added to control speed and reduce susceptibility to
noise spikes.

FIGURE 15—TTL INTERFACE WITH HIGH-LEVEL LOGIC

TEXAS
INSTRUMENTS
POST OFFICE BOX 225012 • DALLAS, TEXAS 75265

TYPICAL APPLICATION DATA

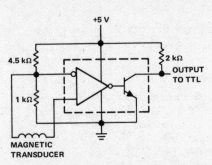

FIGURE 16—DETECTOR FOR MAGNETIC TRANSDUCER

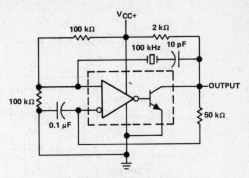

FIGURE 17—100-kHz CRYSTAL OSCILLATOR

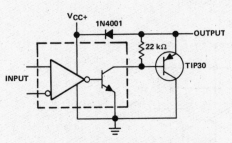

FIGURE 18—COMPARATOR AND SOLENOID DRIVER

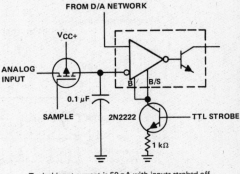

Typical input current is 50 pA with inputs strobed off.

FIGURE 19—STROBING BOTH INPUT AND
OUTPUT STAGES SIMULTANEOUSLY

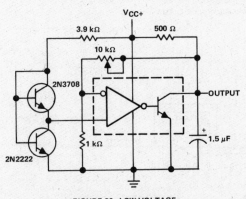

FIGURE 20—LOW-VOLTAGE
ADJUSTABLE REFERENCE SUPPLY

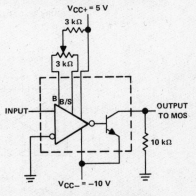

FIGURE 21—ZERO-CROSSING
DETECTOR DRIVING MOS LOGIC

Voltage Comparators

4

Texas
Instruments
POST OFFICE BOX 225012 • DALLAS, TEXAS 75265

TYPICAL APPLICATION DATA

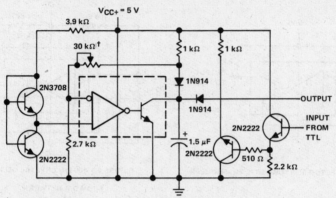

†Adjust to set clamp level.

FIGURE 22—PRECISION SQUARER

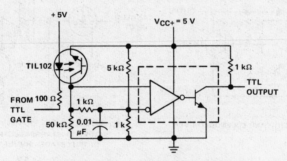

FIGURE 23—DIGITAL TRANSMISSION ISOLATOR

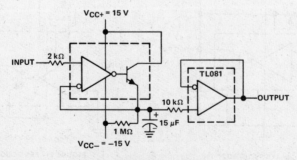

FIGURE 24—POSITIVE-PEAK DETECTOR

TEXAS INSTRUMENTS

POST OFFICE BOX 225012 • DALLAS, TEXAS 75265

TYPICAL APPLICATION DATA

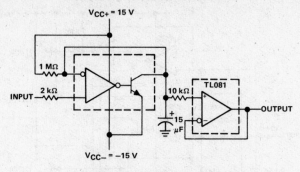

FIGURE 25—NEGATIVE-PEAK DETECTOR

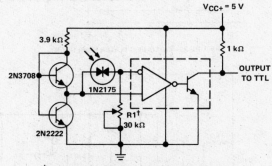

†R1 sets the comparison level. At comparison, the photodiode has less than 5 mV across it decreasing dark current
by an order of magnitude.

FIGURE 26—PRECISION PHOTODIODE COMPARATOR

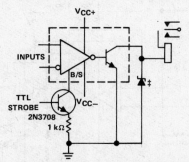

‡Transient voltage and inductive kickback protection

FIGURE 27—RELAY DRIVER WITH STROBE

4

Voltage Comparators

TYPICAL APPLICATION DATA

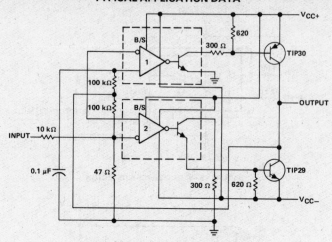

FIGURE 28—SWITCHING POWER AMPLIFIER

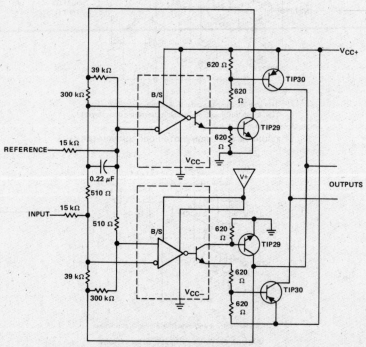

FIGURE 29—SWITCHING POWER AMPLIFIERS

TEXAS
INSTRUMENTS

POST OFFICE BOX 225012 • DALLAS, TEXAS 75265

4

Voltage Comparators

LINEAR INTEGRATED CIRCUITS

TYPES LM139, LM239, LM339, LM139A LM239A, LM339A, LM2901 QUADRUPLE DIFFERENTIAL COMPARATORS

D1979, OCTOBER 1979—REVISED AUGUST 1983

- Single Supply or Dual Supplies
- Wide Range of Supply Voltage . . 2 to 36 V
- Low Supply Current Drain Independent of Supply Voltage . . . . 0.8 mA Typ
- Low Input Bias Current . . . . . . 25 nA Typ
- Low Input Offset Current . . . . . . 3 nA Typ (LM139)
- Low Input Offset Voltage . . . . . 2 mV Typ
- Common-Mode Input Voltage Range Includes Ground
- Differential Input Voltage Range Equal to Maximum-Rated Supply Voltage . . . ±36 V
- Low Output Saturation Voltage
- Output Compatible with TTL, MOS, and CMOS

D, J, OR N DUAL-IN-LINE PACKAGE
(TOP VIEW)

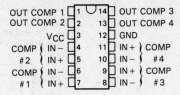

LM139, LM139A
FH, OR FK CHIP CARRIER PACKAGE
(TOP VIEW)

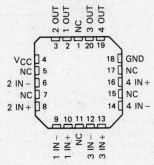

NC—No internal connection

description

These devices consist of four independent voltage comparators that are designed to operate from a single power supply over a wide range of voltages. Operation from dual supplies is also possible so long as the difference between the two supplies is 2 volts to 36 volts and pin 3 is at least 1.5 volts more positive than the input common-mode voltage. Current drain is independent of the supply voltage. The outputs can be connected to other open-collector outputs to achieve wired-AND relationships.

symbol (each comparator)

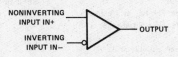

Voltage Comparators

4

TEXAS
INSTRUMENTS
POST OFFICE BOX 225012 • DALLAS, TEXAS 75265

schematic (each comparator)

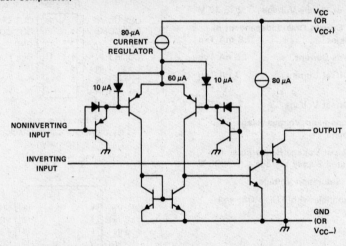

absolute maximum ratings over operating free-air temperature range (unless otherwise noted)

Supply voltage, V_{CC} (see Note 1)	36 V
Differential input voltage (see Note 2)	±36 V
Input voltage range (either input)	−0.3 V to 36 V
Output voltage	36 V
Output current	20 mA
Duration of output short-circuit to ground (see Note 3)	unlimited

Continuous total dissipation at (or below) 25 °C free-air temperature (see Note 4):

D, FH, FK, or J package	900 mW
J package	875 mW

Operating free-air temperature range:

LM139, LM139A	−55 °C to 125 °C
LM239, LM239A	−25 °C to 85 °C
LM339, LM339A	0 °C to 70 °C
LM2901	−40 °C to 85 °C

Storage temperature range	−65 °C to 150 °C
Lead temperature 1,6 mm (1/16 inch) from case for 60 seconds: FH, FK, or J package	300 °C
Lead temperature 1,6 mm (1/16 inch) from case for 10 seconds: D or N package	260 °C

NOTES: 1. All voltage values, except differential voltages, are with respect to the network ground terminal.
2. Differential voltages are at the noninverting input terminal with respect to the inverting input terminal.
3. Short circuits from outputs to V_{CC} can cause excessive heating and eventual destruction.
4. For operation above 25 °C free-air temperature, refer to Dissipation Derating Curves, Section 2. In the J package, LM139 and LM139A chips are alloy-mounted; LM239, LM239A, LM339, LM339A, and LM2901 chips are glass-mounted.

4

Voltage Comparators

TEXAS
INSTRUMENTS
POST OFFICE BOX 225012 • DALLAS, TEXAS 75265

electrical characteristics at specified free-air temperature, V_{CC} = 5 V (unless otherwise noted)

PARAMETER		TEST CONDITIONS†		LM139			LM139A			UNIT
				MIN	TYP	MAX	MIN	TYP	MAX	
V_{IO}	Input offset voltage	V_{CC} = 5 V to 30 V,	25°C		2	5		1	2	mV
		V_{IC} = V_{ICR} min, V_O = 1.4 V	−55°C to 125°C			9			4	
I_{IO}	Input offset current	V_O = 1.4 V	25°C		3	25		3	25	nA
			−55°C to 125°C			100			100	
I_{IB}	Input bias current	V_O = 1.4 V	25°C		−25	−100		−25	−100	nA
			−55°C to 125°C			−300			−300	
V_{ICR}	Common-mode input voltage range		25°C	0 to		V_{CC}−1.5	0 to		V_{CC}−1.5	V
			−55°C to 125°C	0 to		V_{CC}−2	0 to		V_{CC}−2	
A_{VD}	Large-signal differential voltage amplification	V_{CC} = 15 V, V_O = 1.4 V to 11.4 V, R_L ≥ 15 kΩ to V_{CC}	25°C		200		50	200		V/mV
I_{OH}	High-level output current	V_{ID} = 1 V, V_{OH} = 5 V	25°C		0.1			0.1		nA
		V_{ID} = 1 V, V_{OH} = 30 V	−55°C to 125°C			1			1	μA
V_{OL}	Low-level output voltage	V_{ID} = −1 V, I_{OL} = 4 mA	25°C		150	400		150	400	mV
I_{OL}	Low-level output current	V_{ID} = −1 V, V_{OL} = 1.5 V	25°C	6	16	700	6	16	700	mA
I_{CC}	Supply current (four comparators)	V_O = 2.5 V, No load	25°C		0.8	2		0.8	2	mA

†All characteristics are measured with zero common-mode input voltage unless otherwise specified.

switching characteristics, V_{CC} = 5 V, T_A = 25°C

PARAMETER	TEST CONDITIONS		MIN	TYP	MAX	UNIT
Response time	R_L connected to 5 V through 5.1 kΩ, C_L = 15 pF,§ See Note 5	100-mV input step with 5-mV overdrive		1.3		μs
		TTL-level input step		0.3		

§C_L includes probe and jig capacitance.
NOTE 5: The response time specified is the interval between the input step function and the instant when the output crosses 1.4 V.

Voltage Comparators

4

TYPES LM239, LM339, LM239A, LM339A, LM2901
QUADRUPLE DIFFERENTIAL COMPARATORS

electrical characteristics at specified free-air temperature, V_{CC} = 5 V (unless otherwise noted)

PARAMETER	TEST CONDITIONS†		LM239, LM339 MIN	TYP	MAX	LM239A, LM339A MIN	TYP	MAX	LM2901 MIN	TYP	MAX	UNIT
V_{IO} Input offset voltage	V_{CC} = 5 V to 30 V, V_{IC} = V_{ICR} min, V_O = 1.4 V	25°C		2	5		1	2		2	7	mV
		Full range			9			4			15	
I_{IO} Input offset current	V_O = 1.4 V	25°C		5	50		5	50		5	50	nA
		Full range			150			150			200	
I_{IB} Input bias current	V_O = 1.4 V	25°C		−25	−250		−25	−250		−25	−250	nA
		Full range			−400			−400			−500	
V_{ICR} Common-mode input voltage range		25°C	0 to V_{CC}−1.5			0 to V_{CC}−1.5			0 to V_{CC}−1.5			V
		Full range	0 to V_{CC}−2			0 to V_{CC}−2			0 to V_{CC}−2			
A_{VD} Large-signal differential voltage amplification	V_{CC} = 15 V, V_O = 1.4 V to 11.4 V, R_L ≥ 15 kΩ to V_{CC}	25°C		200		50	200		25	100		V/mV
I_{OH} High-level output current	V_{ID} = 1 V, V_{OH} = 5 V	25°C		0.1	1		0.1	1		0.1	1	nA
	V_{OH} = 30 V	Full range										μA
V_{OL} Low-level output voltage	V_{ID} = −1 V, I_{OL} = 4 mA	25°C		150	400		150	400		150	500	mV
		Full range			700			700			700	
I_{OL} Low-level output current	V_{ID} = −1 V, V_{OL} = 1.5 V	25°C	6	16		6	16		6	16		mA
I_{CC} Supply current (four comparators)	V_O = 2.5 V, No load	25°C		0.8	2		0.8	2		0.8	2	mA
	V_{CC} = 30 V, V_O = 15 V, No load	25°C								1	2.5	

† Full range (MIN to MAX) for LM239 and LM239A is −25°C to 85°C, for LM339 and LM339A is 0°C to 70°C, and for LM2901 is −40°C to 85°C. All characteristics are measured with zero common-mode input voltage unless otherwise specified.

switching characteristics, V_{CC} = 5 V, T_A = 25°C

PARAMETER	TEST CONDITIONS		MIN	TYP	MAX	UNIT
Response time	R_L connected to 5 V through 5.1 kΩ, C_L = 15 pF,§ See Note 5	100-mV input step with 5-mV overdrive		1.3		μs
		TTL-level input step		0.3		μs

§C_L includes probe and jig capacitance.
NOTE 5: The response time specified is the interval between the input step function and the instant when the output crosses 1.4 V.

TEXAS
INSTRUMENTS
POST OFFICE BOX 225012 • DALLAS, TEXAS 75265

LINEAR
INTEGRATED
CIRCUITS

TYPES LM193, LM293, LM393, LM293A, LM393A, LM2903
DUAL DIFFERENTIAL COMPARATORS

D2232, JUNE 1976—REVISED AUGUST 1983

- Single Supply or Dual Supplies
- Wide Range of Supply
  Voltage . . . . 2 to 36 Volts
- Low Supply Current Drain Independent of
  Supply Voltage . . . . 0.5 mA Typ
- Low Input Bias Current . . . . . . 25 nA Typ
- Low Input Offset Current . . . . . . 3 nA Typ
  (LM193)
- Low Input Offset Voltage . . . . . 2 mV Typ
- Common-Mode Input Voltage Range
  Includes Ground
- Differential Input Voltage Range Equal to
  Maximum-Rated Supply Voltage . . . ±36 V
- Low Output Saturation Voltage
- Output Compatible with TTL, MOS, and
  CMOS

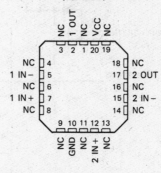

D, JG OR P DUAL-IN-LINE PACKAGE
(TOP VIEW)

NC—No internal connection

description

These devices consist of two independent voltage comparators that are designed to operate from a single power supply over a wide range of voltages. Operation from dual supplies is also possible so long as the difference between the two supplies is 2 volts to 36 volts and pin 8 is at least 1.5 volts more positive than the input common-mode voltage. Current drain is independent of the supply voltage. The outputs can be connected to other open-collector outputs to achieve wired-AND relationships.

symbol (each comparator)

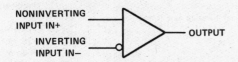

Voltage Comparators

4

TYPES LM193, LM293, LM393, LM293A, LM393A, LM2903
DUAL DIFFERENTIAL COMPARATORS

schematic (each comparator)

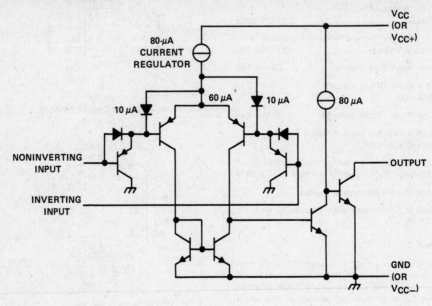

Current values shown are nominal.

absolute maximum ratings over operating free-air temperature range (unless otherwise noted)

Supply voltage, V_{CC} (see Note 1) . . . . 36 V
Differential input voltage (see Note 2) . . . . ±36 V
Input voltage range (either input) . . . . −0.3 V to 36 V
Output voltage . . . . 36 V
Output current . . . . 20 mA
Duration of output short-circuit to ground (see Note 3 . . . . unlimited
Continuous total dissipation at (or below) 25°C free-air temperature (see Note 4):
    D or P package . . . . 725 mW
    JG package (glass-mounted chip) . . . . 825 mW
    FH, FK, or JG (alloy mounted chip) package . . . . 900 mW
Operating free-air temperature range: LM193 . . . . −55°C to 125°C
                            LM293, LM293A . . . . −25°C to 85°C
                            LM393, LM393A . . . . 0°C to 70°C
                            LM2903 . . . . −40°C to 85°C
Storage temperature range . . . . −65°C to 150°C
Lead temperature 1,6 mm (1/16 inch) from case for 60 seconds: JG package . . . . 300°C
Lead temperature 1,6 mm (1/16 inch) from case for 10 seconds: D or P package . . . . 260°C

NOTES: 1. All voltage values, except differential voltages, are with respect to the network ground terminal.
          2. Differential voltages are at the noninverting input terminal with respect to the inverting input terminal.
          3. Short circuits from outputs to V_{CC} can cause excessive heating and eventual destruction.
          4. For operation above 25°C free-air temperature, refer to Dissipation Derating Curves, Section 2. In the JG package, LM193 chips are alloy-mounted;
             LM293, LM293A, LM393, LM393A, and LM2903 chips are glass-mounted.

**TEXAS
INSTRUMENTS**

POST OFFICE BOX 225012 ● DALLAS, TEXAS 75265

electrical characteristics at specified free-air temperature, $V_{CC} = 5$ V (unless otherwise noted)

PARAMETER	TEST CONDITIONS		LM193 MIN	LM193 TYP	LM193 MAX	LM293, LM393 MIN	LM293, LM393 TYP	LM293, LM393 MAX	LM293A, LM393A MIN	LM293A, LM393A TYP	LM293A, LM393A MAX	LM2903 MIN	LM2903 TYP	LM2903 MAX	UNIT
V_{IO} Input offset voltage	$V_{CC} = 5$ V to 30 V, $V_{IC} = V_{ICR}$, $V_O = 1.4$ V	25 °C		2	5		2	5		1	2		2	7	mV
		Full range			9			9			4			15	
I_{IO} Input offset current	$V_O = 1.4$ V	25 °C		3	25		5	50		5	50		5	50	nA
		Full range			100			150			150			200	
I_{IB} Input bias current	$V_O = 1.4$ V	25 °C		25	100		25	250		25	250		25	250	nA
		Full range			300			400			400			500	
V_{ICR} Common-mode input voltage range‡		25 °C	0 to		$V_{CC}-1.5$	0 to		$V_{CC}-1.5$	0 to		$V_{CC}-1.5$	0 to		$V_{CC}-1.5$	V
		Full range	0 to		$V_{CC}-2$	0 to		$V_{CC}-2$	0 to		$V_{CC}-2$	0 to		$V_{CC}-2$	
A_{VD} Large-signal differential voltage amplification	$V_{CC} = 15$ V, $V_O = 1.4$ V to 11.4 V, $R_L \geq 15$ kΩ to V_{CC}	25 °C	50	200			200		50	200			25	100	V/mV
I_{OH} High-level output current	$V_{OH} = 5$ V, $V_{ID} = 1$ V	25 °C		0.1			0.1			0.1			0.1		nA
	$V_{OH} = 30$ V, $V_{ID} = 1$ V	Full range			1			1			1			1	µA
V_{OL} Low-level output voltage	$I_{OL} = 4$ mA, $V_{ID} = 1$ V	25 °C		150	400		150	400		150	400		150	400	mV
		Full range			700			700			700			700	
I_{OL} Low-level output current	$V_{OL} = 1.5$ V, $V_{ID} = 1$ V	25 °C	6			6			6			6			mA
I_{CC} Supply current	$R_L = \infty$ $\begin{cases} V_{CC} = 5\text{ V}, \\ V_O = 2.5\text{ V} \end{cases}$	25 °C		0.8	1		0.8	1		0.8	1		0.8	1	mA
	$\begin{cases} V_{CC} = 30\text{ V}, \\ V_O = 15\text{ V} \end{cases}$	Full range			2.5			2.5			2.5			2.5	

† Full range (MIN to MAX) for LM193 is −55 °C to 125 °C, for the LM293 and LM293A is −25 °C to 85 °C, for the LM393 and LM393A is 0 °C to 70 °C, and for LM2903 is −40 °C to 85 °C. All characteristics are measured with zero common-mode input voltage unless otherwise specified.
‡ The voltage at either input or common-mode should not be allowed to go negative by more than 0.3 V. The upper end of the common-mode voltage range is V_{CC+} −1.5 V, but either or both inputs can go to 30 V without damage.

switching characteristics, $V_{CC} = 5$ V, $T_A = 25$ °C

PARAMETER	TEST CONDITIONS		MIN	TYP	MAX	UNIT
Response time	R_L connected to 5 V through 5.1 kΩ, $C_L = 15$ pF§, See Note 5	100-mV input step with 5-mV overdrive		1.3		µs
		TTL-level input step		0.3		µs

§ C_L includes probe and jig capacitance.
NOTE 5: The response time specified is the interval between the input step function and the instant when the output crosses 1.4 V.

4

Voltage Comparators

Voltage Comparators

- Can Operate from Single 5-V Supply
- Fast Response Time . . . 80 ns Typ with $V_{CC} = \pm 15$ V
- Low Input Bias Current Over Temperature Range
- Inputs and Outputs Can Be Isolated from System Ground
- High Common-Mode Slew Rate
- Outputs Compatible with TTL Circuits

J OR N DUAL-IN-LINE PACKKAGE
(TOP VIEW)

NC	1	14 NC
NC	2	13 NC
#1 GND	3	12 #1 OUT
#1 IN+	4	11 $V_{CC}+$
#1 IN−	5	10 #2 IN−
$V_{CC}−$	6	9 #2 IN+
#2 OUT	7	8 #2 GND

NC—No internal connection

description

The LM219 and LM319 each consists of two high-speed precision comparators that operate over a wide range of supply voltages. These comparators are fully specified for power supplies up to ±15 volts, but are specifically designed to operate from a single 5-volt digital logic supply. Due to the uncommitted collector at the outputs, the LM219 and LM319 are compatible with TTL circuits. These comparators are also well-suited for driving lamps and relays at currents up to 25 milliamperes. The LM219 series features faster response times but greater power dissipation than the LM111 series.

The LM219 is characterized for operation over the temperature range of −25 °C to 85 °C; the LM319 is characterized for operation over the temperature range of 0 °C to 70 °C.

symbol (each comparator)

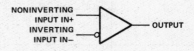

NONINVERTING INPUT IN+
INVERTING INPUT IN−
OUTPUT

absolute maximum ratings over free-air temperature range (unless otherwise noted)

Supply voltage, $V_{CC}+$ to $V_{CC}−$	36 V
Supply voltage, $V_{CC}+$ (see Note 1)	18 V
Supply voltage, $V_{CC}−$ (see Note 1).	−25 V
Differential input voltage (see Note 2)	±5 V
Input voltage (either input, see Note 3)	±15 V
Voltage from output to $V_{CC}−$	36 V
Duration of output short-circuit (see Note 4)	10 s
Continuous total power dissipation at (or below) 25 °C free-air temperature (see Note 5) . . .	500 mW
Operating free-air temperature range: LM219	−25 °C to 85 °C
LM319	0 °C to 70 °C
Storage temperature range	−65 °C to 150 °C
Lead temperature 1,6 mm (1/16 inch) from case for 60 seconds: J package	300 °C
Lead temperature 1,6 mm (1/16 inch) from case for 10 seconds: N package	260 °C

NOTES: 1. All voltage values, except differential voltages, are with respect to the appropriate comparator ground terminal unless otherwise specified.
   2. Differential voltages are at the noninverting input terminal with respect to the inverting input terminal.
   3. The magnitude of the input voltage must never exceed the magnitude of the supply voltage or 15 volts, whichever is less.
   4. The output may be shorted to ground or to either power supply.
   5. For operation above 25 °C free-air temperature, refer to Dissipation Derating Curves, Section 2.

4

Voltage Comparators

Copyright © 1983 by Texas Instruments Incorporated

TEXAS INSTRUMENTS

POST OFFICE BOX 225012 • DALLAS, TEXAS 75265

electrical characteristics at specified free-air temperature, $V_{CC\pm} = \pm 15$ V (unless otherwise noted)

PARAMETER		TEST CONDITIONS[†]		LM219			LM319			UNIT
				MIN	TYP	MAX	MIN	TYP	MAX	
V_{IO}	Input offset voltage	See Note 6	25°C		0.7	4		2	8	mV
			Full range			7			10	
I_{IO}	Input offset current	See Note 6	25°C		30	75		80	200	nA
			Full range			100			300	
I_{IB}	Input bias current		25°C		150	500		250	1000	nA
			Full range			1000			1200	
V_{ICR}	Common-mode input voltage range	$V_{CC+} = 5$ V, $V_{CC-} = 0$	Full range	±12	±13			±13		V
			Full range	1 to 3			1 to 3			
A_{VD}	Large-signal differential voltage amplification	$V_O = 1$ V to 4 V, $V_{CC+} = 5$ V, $V_{CC-} = 0$, $R_L = 2$ kΩ	25°C	10	40		8	40		V/mV
V_{OL}	Low-level output voltage	$I_{OL} = 25$ mA, $V_{ID} = -5$ mV	25°C		0.75	1.5				V
		$V_{ID} = -10$ mV	25°C					0.75	1.5	
		$V_{CC+} = 4.5$ V, $V_{ID} = -6$ mV	0°C to 85°C		0.23[‡]	0.4				
		$V_{CC-} = 0$, $V_{ID} = -10$ mV	0°C to 70°C					0.3[‡]	0.4	
		$I_{OL} = 3.2$ mA, $V_{ID} = -6$ mV	-25°C to 0°C			0.6				
I_{OH}	High-level output current	$V_{CC+} = 15$ V, $V_{ID} = 5$ mV	25°C		0.2	2				μA
		$V_{CC-} = 0$, $V_{ID} = 10$ mV	25°C					0.2	10	
		$V_{OH} = 35$ V, $V_{ID} = 7$ mV	-25°C to 85°C		1[‡]	10				
I_{CC+}	Positive supply current	$V_{CC+} = 5$ V, $V_{CC-} = 0$	25°C		4.3			4.3		mA
			25°C		8	11.5		8	12.5	
I_{CC-}	Negative supply current		25°C		-3	-4.5		-3	-5	mA

[†]Full range is -25°C to 85°C for the LM219 and 0°C to 70°C for the LM319.
[‡]These typical values are at worst-case temperature.
NOTE 6: Both the offset voltages and the offset currents are the maximum values needed to drive the output to within 1 volt of either supply with a 1-mA load. These parameters define an error band that includes the worst-case effects of voltage amplification and input impedance.

switching characteristics, $V_{CC} = -15$ V, $T_A = 25$°C

PARAMETER	TEST CONDITIONS	LM219			LM319			UNIT
		MIN	TYP	MAX	MIN	TYP	MAX	
Response time	See Note 7		80			80		ns

NOTE 7: The response time specified is for a 100-mV input step with 5-mV overdrive and is the interval between the input step function and the instant when the output crosses 1.4 V.

4

Voltage Comparators

TEXAS
INSTRUMENTS
POST OFFICE BOX 225012 • DALLAS, TEXAS 75265

**LINEAR
INTEGRATED
CIRCUITS**

**TYPE LM3302
QUADRUPLE DIFFERENTIAL COMPARATOR**

D2402, OCTOBER 1977—REVISED AUGUST 1983

- Single Supply or Dual Supplies
- Wide Range of Supply Voltage . . . 2 to 28 Volts
- Low Supply Current Drain Independent of Supply Voltage . . . 0.8 mA Typ
- Low Input Bias Current . . . 25 nA Typ
- Low Input Offset Current . . . 3 nA Typ
- Low Input Offset Voltage . . . 3 mV Typ
- Common-Mode Input Voltage Range Includes Ground
- Differential Input Voltage Range Equal to Maximum-Rated Supply Voltage . . . ±28 V
- Low Output Saturation Voltage
- Output Compatible with TTL, MOS, and CMOS

D, J, OR N DUAL-IN-LINE PACKAGE
(TOP VIEW)

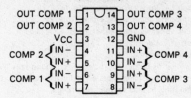

symbol (each comparator)

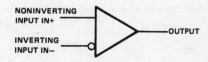

description

This device consists of four independent voltage comparators that are designed to operate from a single power supply over a wide range of voltages. Operation from dual supplies is also possible so long as the difference between the two supplies is 2 volts to 28 volts and pin 3 is at least 1.5 volts more positive than the input common-mode voltage. Current drain is independent of the supply voltage. The outputs can be connected to other open-collector outputs to achieve wired-AND relationships.

schematic (each comparator)

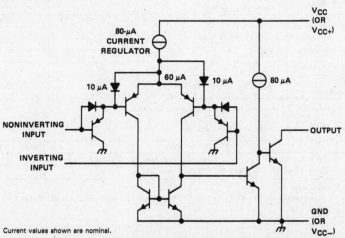

Current values shown are nominal.

Voltage Comparators

4

absolute maximum ratings over operating free-air temperature range (unless otherwise noted)

Supply voltage, V_{CC} (see Note 1)	28 V
Differential input voltage (see Note 2)	± 28 V
Input voltage range (either input)	−0.3 V to 28 V
Output voltage	28 V
Output current	20 mA
Duration of output short-circuit to ground (see Note 3)	unlimited
Continuous total dissipation at (or below) 25 °C free-air temperature (see Note 4)	500 mW
Operating free-air temperature range	−40 °C to 85 °C
Storage temperature range	−65 °C to 150 °C
Lead temperature 1,6 mm (1/16 inch) from case for 60 seconds: J package	300 °C
Lead temperature 1,6 mm (1/16 inch) from case for 10 seconds: D or N package	260 °C

NOTES: 1. All voltage values, except differential voltages, are with respect to the network ground terminal.
2. Differential voltages are at the noninverting input terminal with respect to the inverting input terminal.
3. Short circuits from the output to V_{CC} can cause excessive heating and eventual destruction.
4. For operation above 25 °C free-air temperature, refer to Dissipation Derating Curves, Section 2. In the J package, LM3302 chips are glass-mounted.

electrical characteristics at specified free-air temperature, V_{CC} = 5 V (unless otherwise noted)

PARAMETER		TEST CONDITIONS[†]		MIN	TYP	MAX	UNIT
V_{IO}	Input offset voltage	V_{CC} = 5 V to 28 V, V_{IC} = V_{ICR} min, V_O = 1.4 V	25 °C		3	20	mV
			−40 °C to 85 °C			40	
I_{IO}	Input offset current	V_O = 1.4 V	25 °C		3	100	nA
			−40 °C to 85 °C			300	
I_{IB}	Input bias current		25 °C		−25	−500	nA
			−40 °C to 85 °C			−1000	
V_{ICR}	Common-mode input voltage range		25 °C	0 to V_{CC}−1.5			V
			−40 °C to 85 °C	0 to V_{CC}−2			
A_{VD}	Large-signal differential voltage amplification	V_{CC} = 15 V, V_O = 1.4 V to 11.4 V, R_L = 15 kΩ to V_{CC}	25 °C	2	30		V/mV
I_{OH}	High-level output current	V_{ID} = 1 V, V_{OH} = 5 V	25 °C		0.1		nA
			−40 °C to 85 °C			1	µA
V_{OL}	Low-level output voltage	V_{ID} = −1 V, I_{OL} = 4 mA	25 °C		150	500	mV
			−40 °C to 85 °C			700	
I_{OL}	Low-level output current	V_{ID} = 1 V, V_{OL} = 1.5 V	25 °C	6	16		mA
I_{CC}	Supply current (four comparators)	V_O = 2.5 V, No load	25 °C		0.8	2	mA

[†]All characteristics are measured with zero common-mode input voltage unless otherwise specified.

switching characteristics, V_{CC} = 5 V, T_A = 25 °C

PARAMETER	TEST CONDITIONS		MIN	TYP	MAX	UNIT
Response time	R_L = 5.1 kΩ to 5 V, C_L = 15 pF[‡], See Note 5	100-mV input step with 5 mV overdrive		1.3		µs
		TTL-level input step		0.3		

[‡]C_L includes probe and jig capacitance.
NOTE 5: The response time specified is the interval between the input step function and the instant when the output crosses 1.4 V.

TEXAS
INSTRUMENTS

POST OFFICE BOX 225012 • DALLAS, TEXAS 75265

4

Voltage Comparators

LINEAR INTEGRATED CIRCUITS

- Single Supply or Dual Supplies
- Wide Range of Supply Voltage . . . 2 to 36 Volts
- Low Supply Current Drain Independent of Supply Voltage . . . 0.8 mA Typ
- Low Input Bias Current . . . 25 nA Typ
- Low Input Offset Current . . . 3 nA Typ (TL331M)
- Low Input Offset Voltage . . . 2 mV Typ
- Common-Mode Input Voltage Range Includes Ground
- Differential Input Voltage Range Equal to Maximum-Rated Supply Voltage . . . ±36 V
- Low Output Saturation Voltage
- Output Compatible with TTL, MOS, and CMOS

D, JG OR P
DUAL-IN-LINE PACKAGE
(TOP VIEW)

NC	1	8	NC
IN –	2	7	V_{CC}
IN +	3	6	OUT
GND	4	5	NC

NC—No internal connection

description

The TL331 is a voltage comparator that is designed to operate from a single power supply over a wide range of voltages. Operation from dual supplies is also possible so long as the difference between the two supplies is 2 volts to 36 volts and pin 7 is at least 1.5 volts more positive than the input common-mode voltage. Current drain is independent of the supply voltage.

The TL331M is characterized for operation over the full military temperature range of −55°C to 125°C. The TL331I is characterized for operation from −25°C to 85°C. The TL331C is characterized for operation from 0°C to 70°C.

schematic

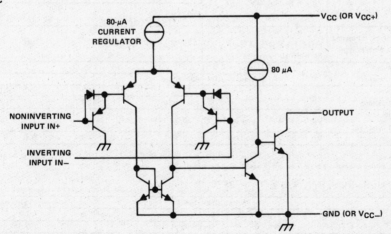

Current values shown are nominal.

Voltage Comparators

4

TEXAS INSTRUMENTS
POST OFFICE BOX 225012 • DALLAS, TEXAS 75265

absolute maximum ratings over operating free-air temperature range (unless otherwise noted)

Supply voltage, V_{CC} (see Note 1) . . . . 36 V
Differential input voltage (see Note 2) . . . . ±36 V
Input voltage range (either input) . . . . −0.3 V to 36 V
Output voltage . . . . 36 V
Output current . . . . 20 mA
Duration of output short-circuit to ground (see Note 3) . . . . unlimited
Continuous total dissipation at (or below) 25 °C free-air temperature (see Note 4) . . . . 680 mW
Operating free-air temperature range: TL331M . . . . −55 °C to 125 °C
TL331I . . . . −25 °C to 85 °C
TL331C . . . . 0 °C to 70 °C
Storage temperature range . . . . −65 °C to 150 °C
Lead temperature 1,6 mm (1/16 inch) from case for 60 seconds: JG package . . . . 300 °C
Lead temperature 1,6 mm (1/16 inch) from case for 10 seconds: D or P package . . . . 260 °C

NOTES: 1. All voltage values, except differential voltages, are with respect to the network ground terminal.
2. Differential voltages are at the noninverting input terminal with respect to the inverting input terminal.
3. Short circuits from the output to V_{CC} can cause excessive heating and eventual destruction.
4. For operation above 25 °C free-air temperature, refer to Dissipation Derating Curves, Section 2. In the JG package, TL331M chips are alloy-mounted; TL331I and TL331C chips are glass-mounted.

electrical characteristics at specified free-air temperature, V_{CC} = 5 V (unless otherwise noted)

PARAMETER		TEST CONDITIONS[†]		TL331M, TL331I			TL331C			UNIT
				MIN	TYP	MAX	MIN	TYP	MAX	
V_{IO}	Input offset voltage	V_{CC} = 5 V to 30 V, V_{IC} = V_{ICR} min, V_O = 1.4 V	25 °C		2	5		2	5	mV
			Full range			9			9	
I_{IO}	Input offset current	V_O = 1.4 V	25 °C		3	25		5	50	nA
			Full range			100			150	
I_{IB}	Input bias current		25 °C		−25	−100		−25	−250	nA
			Full range			−300			−400	
V_{ICR}	Common-mode input voltage range	V_{CC} = 5 V to 30 V	25 °C	0 to $V_{CC}-1.5$			0 to $V_{CC}-1.5$			V
			Full range	0 to $V_{CC}-2$			0 to $V_{CC}-2$			
A_{VD}	Large-signal differential voltage amplification	V_{CC} = 15 V, V_O = 1.4 V to 11.4 V, R_L = 15 kΩ to V_{CC}	25 °C		200			200		V/mV
I_{OH}	High-level output current	V_{ID} = 1 V	V_{OH} = 5 V, 25 °C		0.1			0.1		nA
			V_{OH} = 30 V, Full range			1			1	µA
V_{OL}	Low-level output voltage	V_{ID} = −1 V, I_{OL} = 4 mA	25 °C		150	400		150	400	mV
			Full range			700			700	
I_{OL}	Low-level output current	V_{ID} = −1 V, V_{OL} = 1.5 V	25 °C	6			6			mA
I_{CC}	Supply current	V_O = 2.5 V, No load	25 °C		0.5	0.8		0.5	0.8	mA

† Full range (MIN to MAX) for TL331M is −55 °C to 125 °C, for the TL331I is −25 °C to 85 °C, and for the TL331C is 0 °C to 70 °C. All characteristics are measured with zero common-mode input voltage unless otherwise specified.

switching characteristics, V_{CC} = 5 V, T_A = 25 °C

PARAMETER		TEST CONDITIONS		MIN	TYP	MAX	UNIT
Response time	R_L connected to 5 V through 5.1 kΩ, C_L = 15 pF,[‡] See Note 5	100-mV input step with 5-mV overdrive			1.3		µs
		TTL-level input step			0.3		

‡ C_L includes probe and jig capacitance.
NOTE 5: The response time specified is the interval between the input step function and the instant when the output crosses 1.4 V.

TEXAS
INSTRUMENTS
POST OFFICE BOX 225012 • DALLAS, TEXAS 75265

4

Voltage Comparators

LINEAR
INTEGRATED
CIRCUITS

TYPES TL506M, TL506C
DUAL DIFFERENTIAL COMPARATORS WITH STROBES

D1208, MARCH 1971—REVISED AUGUST 1983

- Each Comparator Identical to LM106 or LM306 with Common V_{CC+}, V_{CC-}, and Ground Connections
- Improved Gain and Accuracy
- Fan-Out to 10 Series 54/74 TTL Loads
- Strobe Capability
- Short-Circuit and Surge Protection
- Fast Response Times

TL506M . . . J OR W PACKAGE
TL506C . . . J OR N PACKAGE
(TOP VIEW)

#1 STRB A	1	14	#1 STRB B
#1 IN −	2	13	GND
#1 IN +	3	12	#1 OUT
V_{CC-}	4	11	V_{CC+}
#2 IN +	5	10	#2 OUT
#2 IN −	6	9	NC
#2 STRB A	7	8	#2 STRB B

NC—No internal connection

description

The TL506 is a dual high-speed comparator, with each half having differential inputs, a low-impedance output with high-sink-current capability (100 mA), and two strobe inputs. This device detects low-level analog or digital signals and can drive digital logic or lamps and relays directly. Short-circuit protection and surge-current limiting is provided.

The circuit is similar to a TL810 with gated output. A low-level input at either strobe causes the output to remain high regardless of the differential input. When both strobe inputs are either open or at a high logic level, the output voltage is controlled by the differential input voltage. The circuit will operate with any negative supply voltage between −3 V and −12 V with little difference in performance.

The TL506M is characterized for operation over the full military temperature range of −55 °C to 125 °C; the TL506C is characterized for operation from 0 °C to 70 °C.

functional block diagram (each comparator)

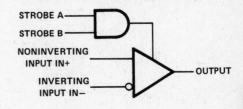

4

Voltage Comparators

TEXAS
INSTRUMENTS
POST OFFICE BOX 225012 • DALLAS, TEXAS 75265

schematic (each comparator)

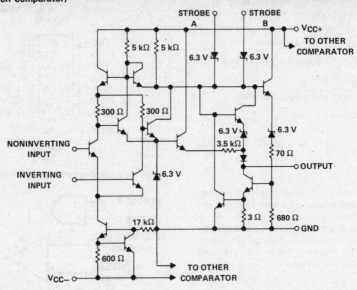

absolute maximum ratings over operating free-air temperature range (unless othewise noted)

Supply voltage V_{CC+} (see Note 1) . . . . 15 V
Supply voltage V_{CC-} (see Note 1) . . . . −15 V
Differential input voltage (see Note 2) . . . . ±5 V
Input voltage (any input, see Notes 1 and 3) . . . . ±7 V
Strobe voltage range (see Note 1) . . . . 0 V to V_{CC+}
Output voltage (see Note 1) . . . . 24 V
Voltage from output to V_{CC-} . . . . 30 V
Duration of output short-circuit (see Note 4) . . . . 10 s
Continuous total dissipation at (or below) 25 °C free-air temperature (see Note 5):
    J package (TL506MJ) . . . . 1375 mW
    J package (TL506CJ) . . . . 1025 mW
    N package . . . . 875 mW
    W package . . . . 1000 mW
Operating free-air temperature range: TL506M . . . . −55 °C to 125 °C
                                    TL506C . . . . 0 °C to 70 °C
Storage temperature range . . . . −65 °C to 150 °C
Lead temperature 1,6 mm (1/16 inch) from case for 60 seconds: J or W package . . . . 300 °C
Lead temperature 1,6 mm (1/16 inch) from case for 10 seconds: N package . . . . 260 °C

NOTES: 1. All voltage values, except differential voltages and the voltage from the output to V_{CC-}, are with respect to the network ground terminal.
       2. Differential voltages are at the noninverting input terminal with respect to the inverting input terminal.
       3. The magnitude of the input voltage must never exceed the magnitude of the supply voltage or 7 V, whichever is less.
       4. One output at a time may be shorted to ground or either power supply.
       5. For operation above 25 °C free-air temperature, refer to Dissipation Derating Curves, Section 2. In the J package, TL506M chips are alloy mounted; TL506C chips are glass mounted.

TEXAS INSTRUMENTS
POST OFFICE BOX 225012 • DALLAS, TEXAS 75265

electrical characteristics at specified free-air temperature, V_{CC+} = 12 V, V_{CC-} = −3 V to −12 V (unless otherwise noted)

PARAMETER		TEST CONDITIONS†		TL506M			TL506C			UNIT
				MIN	TYP	MAX	MIN	TYP	MAX	
V_{IO}	Input offset voltage	See Note 6	25°C		0.5‡	2		1.6‡	5	mV
			Full range			3			6.5	
α_{VIO}	Average temperature coefficient of input offset voltage	See Note 6	Full range		3	10		5	20	µV/°C
I_{IO}	Input offset current	See Note 6	25°C		0.7‡	3		1.8‡	5	µA
			MIN		2	7		1	7.5	
			MAX		0.4	3		0.5		
α_{IIO}	Average temperature coefficient of input offset current	See Note 6	MIN to 25°C		15	75		24	100	nA/°C
			25°C to MAX		5	25		15	50	
I_{IB}	Input bias current	V_O = 0.5 V to 5 V	25°C		7‡	20		16‡	25	µA
			Full range			45			40	
$I_{IL(S)}$	Low-level strobe current	$V_{(strobe)}$ = 0.4 V	Full range		−1.7‡	−3.3		−1.7‡	−3.3	mA
$V_{IH(S)}$	High-level strobe voltage		Full range	2.5			2.5			V
$V_{IL(S)}$	Low-level strobe voltage		Full range			0.9			0.9	V
V_{ICR}	Common-mode input voltage range	V_{CC-} = −7 V to −12 V	Full range	±5			±5			V
V_{ID}	Differential input voltage range		Full range	±5			±5			V
A_{VD}	Large-signal differential voltage amplification	No load, V_O = 0.5 V to 5 V	25°C		40 000‡			40 000‡		
V_{OH}	High-level output voltage	V_{ID} = 5 mV, I_{OH} = −400 µA	Full range	2.5		5.5	2.5		5.5	V
V_{OL}	Low-level output voltage	V_{ID} = −5 mV, I_{OL} = 100 mA	25°C		0.8‡	1.5		0.8‡	2	V
		V_{ID} = −5 mV, I_{OL} = 50 mA	Full range			1			1	
		V_{ID} = −5 mV, I_{OL} = 16 mA	Full range			0.4			0.4	
I_{OH}	High-level output current	V_{ID} = 5 mV, V_{OH} = 8 V to 24 V	25°C		0.02‡	1		0.02‡	2	µA
			Full range			100			100	
I_{CC+}	Supply current from V_{CC+}	V_{ID} = −5 mV, See Note 7	Full range		13.9‡	20		13.9‡	20	mA
I_{CC-}	Supply current from V_{CC-}	See Note 7	Full range		3.2‡	7.2		3.2‡	7.2	mA

†Unless otherwise noted, all characteristics are measured with the strobe open. Full range (MIN to MAX) for TL506M is −55°C to 125°C and for the TL506C is 0°C to 70°C.

‡These typical values are at V_{CC+} = 12 V, V_{CC-} = −6 V, T_A = 25°C.

NOTES: 6. The offset voltages and offset currents given are the maximum values required to drive the output down to the low range (V_{OL}) or up to the high range (V_{OH}). Thus these parameters actually define an error band and take into account the worst-case effects of voltage gain and input impedance.

7. Power supply currents are measured with the respective noninverting inputs and inverting inputs of both comparators connected in parallel. The outputs are open.

switching characteristics, V_{CC+} = 12 V, V_{CC-} = −6 V, T_A = 25°C

PARAMETER	TEST CONDITIONS†		TL506M			TL506C		UNIT
		MIN	TYP	MAX	MIN	TYP	MAX	
Response time, low-to-high-level output	R_L = 390 Ω to 5 V, C_L = 15 pF, See Note 8		28	40		28		ns

NOTE 8: The response time specified is for a 100-mV input step with 5-mV overdrive and is the interval between the input step function and the instant when the output crosses 1.4 V.

4

Voltage Comparators

TEXAS
INSTRUMENTS
POST OFFICE BOX 225012 • DALLAS, TEXAS 75265

TYPICAL CHARACTERISTICS§

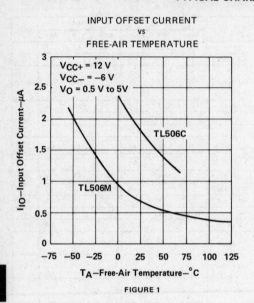

INPUT OFFSET CURRENT
vs
FREE-AIR TEMPERATURE

FIGURE 1

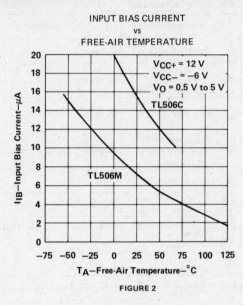

INPUT BIAS CURRENT
vs
FREE-AIR TEMPERATURE

FIGURE 2

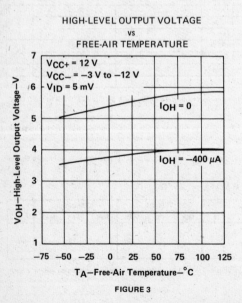

HIGH-LEVEL OUTPUT VOLTAGE
vs
FREE-AIR TEMPERATURE

FIGURE 3

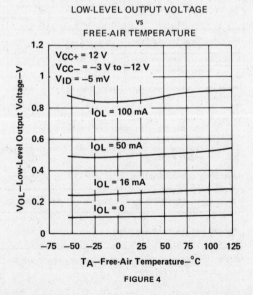

LOW-LEVEL OUTPUT VOLTAGE
vs
FREE-AIR TEMPERATURE

FIGURE 4

§Data for temperatures below 0°C and above 70°C is applicable to TL506M circuits only.

TEXAS
INSTRUMENTS
POST OFFICE BOX 225012 • DALLAS, TEXAS 75265

Voltage Comparators

4

TYPICAL CHARACTERISTICS§

VOLTAGE TRANSFER CHARACTERISTICS

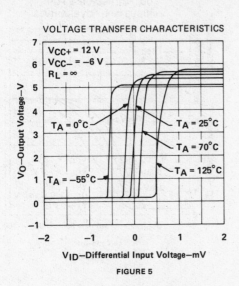

FIGURE 5

OUTPUT CURRENT vs DIFFERENTIAL INPUT VOLTAGE

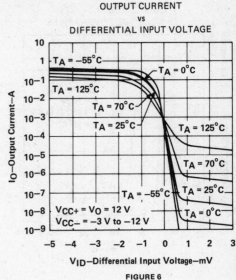

FIGURE 6

LARGE-SIGNAL DIFFERENTIAL VOLTAGE AMPLIFICATION vs FREE-AIR TEMPERATURE

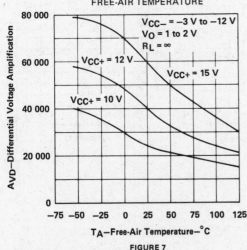

FIGURE 7

SHORT-CIRCUIT OUTPUT CURRENT vs FREE-AIR TEMPERATURE

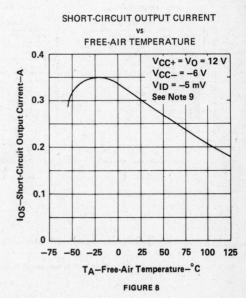

FIGURE 8

§Data for temperatures below 0°C and above 70°C is applicable to TL506M circuits only.
NOTE 9: This parameter was measured using a single 5-ms pulse.

Voltage Comparators

4

TYPES TL506M, TL506C
DUAL DIFFERENTIAL COMPARATORS WITH STROBES

TYPICAL CHARACTERISTICS§

OUTPUT RESPONSE FOR
VARIOUS INPUT OVERDRIVES

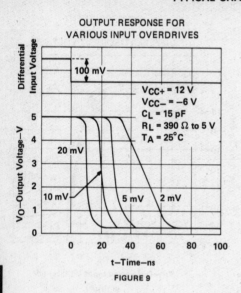

FIGURE 9

OUTPUT RESPONSE FOR
VARIOUS INPUT OVERDRIVES

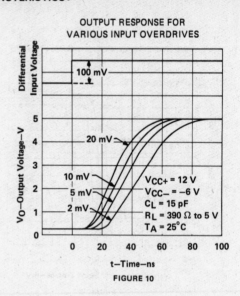

FIGURE 10

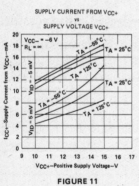

FIGURE 11

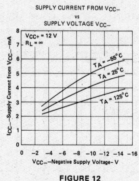

FIGURE 12

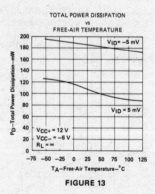

FIGURE 13

§Data for temperatures below 0°C and above 70°C is applicable to TL506M circuits only.

TEXAS
INSTRUMENTS
POST OFFICE BOX 225012 • DALLAS, TEXAS 75265

**LINEAR
INTEGRATED
CIRCUITS**

**TYPES TL510M, TL510C
DIFFERENTIAL COMPARATORS WITH STROBE**

D991, MARCH 1971–REVISED NOVEMBER 1983

- Low Offset Characteristics
- High Differential Voltage Amplification
- Fast Response Times
- Output Compatible with Most TTL Circuits

JG OR P
DUAL-IN-LINE PACKAGE
(TOP VIEW)

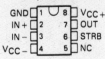

description

The TL510 monolithic high-speed voltage comparator is an improved version of the TL710 with an extra stage added to increase voltage amplification and accuracy, and a strobe input for greater flexibility. Typical voltage amplification is 33,000. Since the ouput cannot be more positive than the strobe, a low-level input at the strobe will cause the output to go low regardless of the differential input. Component matching, inherent in integrated circuit fabrication techniques, produces a comparator with low-drift and low-offset characteristics. These circuits are particularly useful for applications requiring an amplitude discriminator, memory sense amplifier, or a high-speed limit detector.

The TL510M is characterized for operation over the full military temperature range of −55°C to 125°C; the TL510C is characterized for operation from 0°C to 70°C.

TL510M . . . U FLAT PACKAGE
(TOP VIEW)

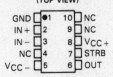

TL510M . . . FH OR FK
CHIP CARRIER PACKAGE
(TOP VIEW)

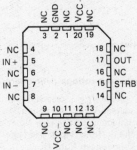

NC – No internal connection

functional block diagram (positive logic)

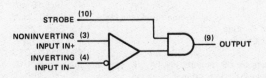

**TEXAS
INSTRUMENTS**
POST OFFICE BOX 225012 • DALLAS, TEXAS 75265

schematic

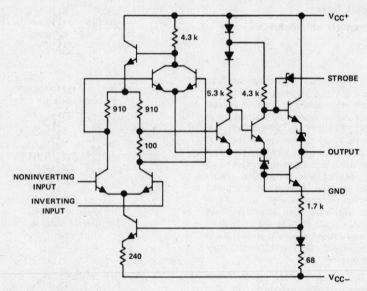

Resistor values shown are nominal in ohms.
Component values shown are nominal.

absolute maximum ratings over operating free-air temperature range (unless otherwise noted)

Supply voltage V_{CC+} (see Note 1) . . . . 14 V
Supply voltage V_{CC-} (see Note 1) . . . . −7 V
Differential input voltage (see Note 2) . . . . ±5 V
Input voltage (either input, see Note 1) . . . . ±7 V
Strobe voltage (see Note 1) . . . . 6 V
Peak output current ($t_W \leq 1$ s) . . . . 10 mA
Continuous total power dissipation at (or below) 70°C free-air temperature (see Note 3) . . . . 300 mW
Operating free-air temperature range: TL510M Circuits . . . . −55°C to 125°C
                                                              TL510C Circuits . . . . 0°C to 70°C
Storage temperature range . . . . −65°C to 150°C
Lead temperature 1,6 mm (1/16 inch) from case for 60 seconds: FH, FK, JG, or U package . . . 300°C
Lead temperature 1,6 mm (1/16 inch) from case for 10 seconds: P package . . . . 260°C

NOTES: 1. All voltage values, except differential voltages, are with respect to the network ground terminal.
2. Differential voltages are at the noninverting input terminal with respect to the inverting input terminal.
3. For operation of the TL510M above 70°C free-air temperature, refer to Dissipation Derating Curves, Section 2. In the JG package, TL510M chips are alloy mounted and TL510C chips are glass mounted.

Texas Instruments
POST OFFICE BOX 225012 ● DALLAS, TEXAS 75265

electrical characteristics at specified free-air temperature, V_{CC+} = 12 V, V_{CC-} = −6 V (unless otherwise noted)

PARAMETER		TEST CONDITIONS†		TL510M			TL510C			UNIT
				MIN	TYP	MAX	MIN	TYP	MAX	
V_{IO}	Input offset voltage	$R_S \leq 200\ \Omega$, See Note 4	25°C		0.6	2		1.6	3.5	mV
			Full range			3			4.5	
α_{VIO}	Average temperature coefficient of input offset voltage	$R_S = 50\ \Omega$, See Note 4	MIN to 25°C		3	10		3	20	µV/°C
			25°C to MAX		3	10		3	20	
I_{IO}	Input offset current	See Note 4	25°C		0.75	3		1.8	5	µA
			MIN		1.8	7			7.5	
			MAX		0.25	3				
α_{IIO}	Average temperature coefficient of input offset current	See Note 4	MIN to 25°C		15	75		24	100	nA/°C
			25°C to MAX		5	25		15	50	
I_{IB}	Input bias current	See Note 4	25°C		7	15		7	20	µA
			MIN		12	25		9	30	
$I_{IH(S)}$	High-level strobe current	$V_{(strobe)} = 5$ V, $V_{ID} = -5$ mV	25°C			±100			±100	µA
$I_{IL(S)}$	Low-level strobe current	$V_{(strobe)} = -100$ mV, $V_{ID} = 5$ mV	25°C		−1	−2.5		−1	−2.5	mA
V_{ICR}	Common-mode input voltage range	$V_{CC-} = -7$ V	Full range	±5			±5			V
V_{ID}	Differential input voltage range		Full range	±5			±5			V
A_{VD}	Large-signal differential voltage amplification	No load, $V_O = 0$ to 2.5 V	25°C	12.5	33		10	33		V/mV
			Full range	10			8			
V_{OH}	High-level output voltage	$V_{ID} = 5$ mV, $I_{OH} = 0$	Full range		4‡	5		4‡	5	V
		$V_{ID} = 5$ mV, $I_{OH} = -5$ mA	Full range	2.5	3.6‡		2.5	3.6‡		
V_{OL}	Low-level output voltage	$V_{ID} = -5$ mV, $I_{OL} = 0$	Full range	−1	−0.5‡	0§	−1	−0.5‡	0§	V
		$V_{(strobe)} = 0.3$ V, $V_{ID} = 5$ mV, $I_{OL} = 0$	Full range	−1		0§	−1		0§	V
I_{OL}	Low-level output current	$V_{ID} = -5$ mV, $V_O = 0$	25°C	2	2.4		1.6	2.4		mA
			MIN	1	2.3		0.5	2.4		
			MAX	0.5	2.3		0.5	2.4		
r_o	Output resistance	$V_O = 1.4$ V	25°C		200			200		Ω
CMRR	Common-mode rejection ratio	$R_S \leq 200\ \Omega$	Full range	80	100‡		70	100‡		dB
I_{CC+}	Supply current from V_{CC+}	$V_{ID} = -5$ mV, No load	Full range		5.5‡	9		5.5‡	9	mA
I_{CC-}	Supply current from V_{CC-}		Full range		−3.5‡	−7		−3.5‡	−7	mA
P_D	Total power dissipation		Full range		90‡	150		90‡	150	mW

†Unless otherwise noted, all characteristics are measured with the strobe open. Full range (MIN to MAX) for TL510M is −55°C to 125°C and for the TL510C is 0°C to 70°C.

‡These typical values are at T_A = 25°C.

§The algebraic convention, where the most-positive (least negative) limit is designated as maximum, is used in this data sheet for logic levels only, e.g., when 0 V is the maximum, the minimum limit is a more-negative voltage.

NOTE 4: These characteristics are verified by measurements at the following temperatures and output voltage levels: for TL510M, V_O = 1.8 V at T_A = −55°C, V_O = 1.4 V at T_A = 25°C, and V_O = 1 V at T_A = 125°C; for TL510C, V_O = 1.5 V at T_A = 0°C, V_O = 1.4 V at 25°C, and V_O = 1.2 V at T_A = 70°C. These output voltage levels where selected to approximate the logic threshold voltages of the types of digital logic circuits these comparators are intended to drive.

4

Voltage Comparators

TEXAS INSTRUMENTS
POST OFFICE BOX 225012 • DALLAS, TEXAS 75265

TYPES TL510M, TL510C
DIFFERENTIAL COMPARATORS WITH STROBE

switching characteristics, V_{CC+} = 12 V, V_{CC-} = -6 V, T_A = 25°C

PARAMETER	TEST CONDITIONS			MIN	TYP	MAX	UNIT
Response time	$R_L = \infty$	C_L = 5 pF,	See Note 5		30	80	ns
Strobe release time	$R_L = \infty$	C_L = 5 pF,	See Note 6		5	25	ns

NOTES: 5. The response time specified is for a 100-mV input step with 5-mV overdrive.
   6. For testing purposes, the input bias conditions are selected to produce an output voltage of 1.4 V. A 5-mV overdrive is then added to the input bias voltage to produce an output voltage that rises above 1.4 V. The time interval is measured from the 50% point of the strobe voltage curve to the point where the overdriven output voltage crosses the 1.4 V level.

TYPICAL CHARACTERISTICS

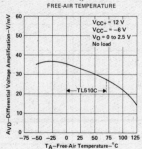

LARGE-SIGNAL DIFFERENTIAL
VOLTAGE AMPLIFICATION
vs
FREE-AIR TEMPERATURE

FIGURE 1

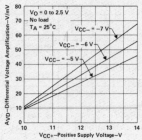

LAREG-SIGNAL DIFFERENTIAL
VOLTAGE AMPLIFICATION
vs
SUPPLY VOLTAGE

FIGURE 2

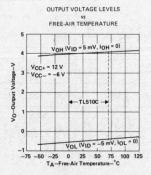

OUTPUT VOLTAGE LEVELS
vs
FREE-AIR TEMPERATURE

FIGURE 3

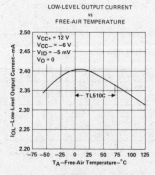

LOW-LEVEL OUTPUT CURRENT
vs
FREE-AIR TEMPERATURE

FIGURE 4

TEXAS
INSTRUMENTS
POST OFFICE BOX 225012 • DALLAS, TEXAS 75265

TYPICAL CHARACTERISTICS

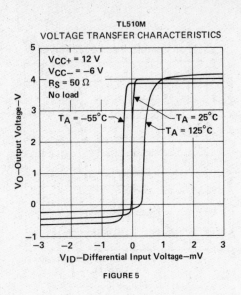

TL510M
VOLTAGE TRANSFER CHARACTERISTICS

FIGURE 5

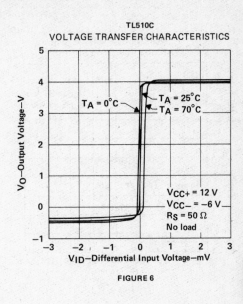

TL510C
VOLTAGE TRANSFER CHARACTERISTICS

FIGURE 6

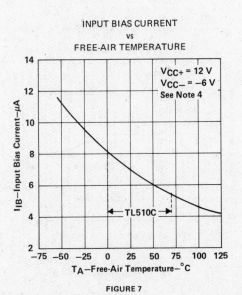

INPUT BIAS CURRENT
vs
FREE-AIR TEMPERATURE

FIGURE 7

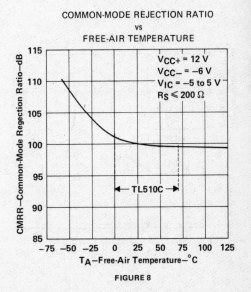

COMMON-MODE REJECTION RATIO
vs
FREE-AIR TEMPERATURE

FIGURE 8

4

Voltage Comparators

TEXAS
INSTRUMENTS
POST OFFICE BOX 225012 • DALLAS, TEXAS 75265

TYPICAL CHARACTERISTICS

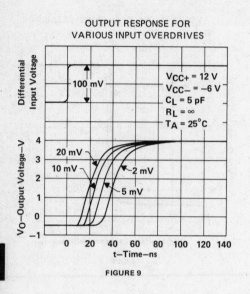

OUTPUT RESPONSE FOR
VARIOUS INPUT OVERDRIVES

FIGURE 9

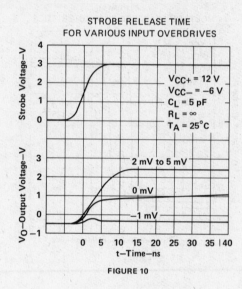

STROBE RELEASE TIME
FOR VARIOUS INPUT OVERDRIVES

FIGURE 10

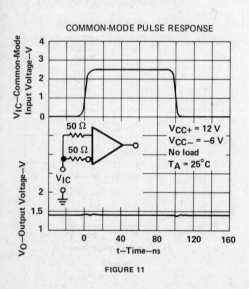

COMMON-MODE PULSE RESPONSE

FIGURE 11

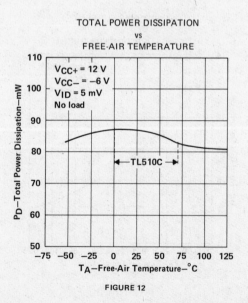

TOTAL POWER DISSIPATION
vs
FREE-AIR TEMPERATURE

FIGURE 12

**TEXAS
INSTRUMENTS**

POST OFFICE BOX 225012 • DALLAS, TEXAS 75265

Voltage Comparators

4

118-

LINEAR
INTEGRATED
CIRCUITS

TYPES TL514M, TL514C
DUAL DIFFERENTIAL COMPARATORS WITH STROBES

D999, OCTOBER 1977–REVISED OCTOBER 1983

- **Fast Response Times**
- **High Differential Voltage Amplification**
- **Low Offset Characteristics**
- **Outputs Compatible with Most TTL Circuits**

description

The TL514 is an improved version of the TL720 dual high-speed voltage comparator. When compared with the TL720, these circuits feature higher amplification (typically 33,000) due to an extra amplification stage, increased accuracy because of lower offset characteristics, and greater flexibility with the addition of a strobe to each comparator. Since the output cannot be more positive than the strobe, a low-level input at the strobe will cause the output to go low regardless of the differential input.

These circuits are especially useful in applications requiring an amplitude discriminator, memory sense amplifier, or a high-speed limit detector. The TL514M is characterized for operation over the full military temperature range of $-55\,°C$ to $125\,°C$, the TL514C is characterized for operation from $0\,°C$ to $70\,°C$.

symbol (each comparator)

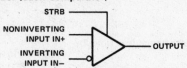

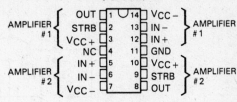

TL514M . . . J OR W PACKAGE
TL514C . . . J OR N PACKAGE
(TOP VIEW)

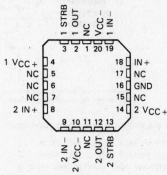

TL514M
FH OR FK CHIP CARRIER
(TOP VIEW)

NC—No internal connection

4

Voltage Comparators

TYPES TL514M, TL514C
DUAL DIFFERENTIAL COMPARATORS WITH STROBES

schematic (each comparator)

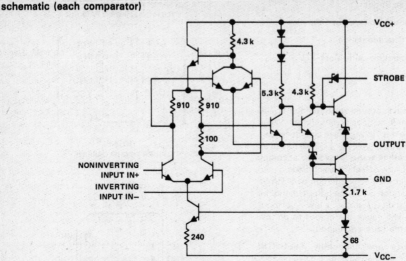

Resistor values shown are nominal in ohms.
Component values shown are nominal.

absolute maximum ratings over operating free-air temperature range (unless otherwise noted)

Supply voltage V_{CC+} (see Note 1) . . . . 14 V
Supply voltage V_{CC-} (see Note 1) . . . . −7 V
Differential input voltage (see Note 2) . . . . ±5 V
Input voltage (any input, see Note 1) . . . . ±7 V
Strobe voltage (see Note 1) . . . . 6 V
Peak output current ($t_W \leq 1$ s) . . . . 10 mA
Continuous total dissipation at (or below) 25 °C free-air temperature (see Note 3):
                    each comparator . . . . 300 mW
                    total package . . . . 600 mW
Operating free-air temperature range: TL514M Circuits . . . . −55 °C to 125 °C
                                  TL514C Circuits . . . . 0 °C to 70 °C
Storage temperature range . . . . −65 °C to 150 °C
Lead temperature 1,6 mm (1/16 inch) from case for 60 seconds: FH, FK, J, or W package . . . 300 °C
Lead temperature 1,6 mm (1/16 inch) from case for 10 seconds: N package . . . . 260 °C

NOTES: 1. All voltage values, except differential voltages, are with respect to the network ground terminal.
       2. Differential voltages are at the noninverting input terminal with respect to the inverting input terminal.
       3. For operation above 25 °C free-air temperature, refer to Dissipation Derating Curves, Section 2. In the J package, TL514M chips are alloy mounted and TL514C chips are glass mounted.

4

Voltage Comparators

11

electrical characteristics at specified free-air temperature, $V_{CC+} = 12$ V, $V_{CC-} = -6$ V

PARAMETER		TEST CONDITIONS†		TL514M			TL514C			UNIT
				MIN	TYP	MAX	MIN	TYP	MAX	
V_{IO}	Input offset voltage	$R_S \leq 200\ \Omega$, See Note 4	25°C		0.6	2		1.6	3.5	mV
			Full range			3			4.5	
α_{VIO}	Average temperature coefficient of input offset voltage	$R_S = 50\ \Omega$, See Note 4	MIN to 25°C		3	10		3	20	μV/°C
			25°C to MAX		3	10		3	20	
I_{IO}	Input offset current	See Note 4	25°		0.75	3		1.8	5	μA
			MIN		1.8	7			7.5	
			MAX		0.25	3			7.5	
α_{IIO}	Average temperature coefficient of input offset current	See Note 4	MIN to 25°C		15	75		24	100	nA/°C
			25°C to MAX		5	25		15	50	
I_{IB}	Input bias current	See Note 4	25°C		7	15		7	20	μA
			MIN		12	25		9	30	
$I_{IL(S)}$	High-level strobe current	$V_{(strobe)} = 5$ V, $V_{ID} = -5$ mV	25°C			±100			±100	μA
$I_{IH(S)}$	Low-level strobe current	$V_{(strobe)} = -100$ mV, $V_{ID} = 5$ mV	25°C		-1	-2.5		-1	-2.5	mA
V_{ICR}	Common-mode input voltage range	$V_{CC-} = -7$ V,	Full range	±5			±5			V
V_{ID}	Differential input voltage range		Full range	±5			±5			V
A_{VD}	Large-signal differential voltage amplification	No load, $V_O = 0$ to 2.5 V	25°C	12.5	33		10	33		V/mV
			Full range	10			8			
V_{OH}	High-level output voltage	$V_{ID} = 5$ mV, $I_{OH} = 0$	Full range		4§	5		4§	5	V
		$V_{ID} = 5$ mV, $I_{OH} = -5$ mA	Full range	2.5	3.6§		2.5	3.6§		
V_{OL}	Low-level output voltage	$V_{ID} = -5$ mV, $I_{OL} = 0$	Full range	-1	-0.5§	0‡	-1	-0.5§	0‡	V
		$V_{(strobe)} = 0.3$ V, $V_{ID} = 5$ mV, $I_{OL} = 0$	Full range	-1		0‡	-1		0‡	V
I_{OL}	Low-level output current	$V_{ID} = -5$ mV, $V_O = 0$	25°C	2	2.4		1.6	2.4		mA
			MIN	1	2.3		0.5	2.4		
			MAX	0.5	2.3		0.5	2.4		
r_o	Output resistance	$V_O = 1.4$ V	25°C		200			200		Ω
CMRR	Common-mode rejection ratio	$R_S \leq 200\ \Omega$	Full range	80	100§		70	100§		dB
I_{CC+}	Supply current from V_{CC+}¶	$V_{ID} = -5$ mV, No load	Full range		5.5§	9		5.5§	9	mA
I_{CC-}	Supply current from V_{CC-}¶		Full range		-3.5§	-7		-3.5§	-7	mA
P_D	Total power dissipation¶		Full range		90§	150		90§	150	mW

† Unless otherwise noted, all characteristics are measured with the strobe open. Full range (MIN to MAX) for TL514M is -55°C to 125°C and for the TL514C is 0°C to 70°C.

‡ The algebraic convention where the most-positive (least-negative) limit is designated as maximum, is used in this data sheet for logic levels only, e.g., when 0 V is the maximum, the minimum limit is a more-negative voltage.

§ These typical values are at $T_A = 25$°C.

¶ Supply current and power dissipation limits apply for each comparator.

NOTE 4: These characteristics are verified by measurements at the following temperatures and output voltage levels: for TL514M, $V_O = 1.8$ V at $T_A = -55$°C, $V_O = 1.4$ V at $T_A = 25$°C, and $V_O = 1$ V at $T_A = 125$°C; for TL514C, $V_O = 1.5$ V at $T_A = 0$°C, $V_O = 1.4$ V at 25°C, and $V_O = 1.2$ V at $T_A = 70$°C. These output voltage levels were selected to approximate the logic threshold voltages of the types of digital logic circuits these comparators are intended to drive.

4

Voltage Comparators

TYPES TL514M, TL514C
DUAL DIFFERENTIAL COMPARATORS WITH STROBES

switching characteristics, V_{CC+} = 12 V, V_{CC-} = −6 V, T_A = 25°C

PARAMETER	TEST CONDITIONS			MIN	TYP	MAX	UNIT
Response time	$R_L = \infty$	C_L = 5 pF,	See Note 5		30	80	ns
Strobe release time	$R_L = \infty$	C_L = 5 pF,	See Note 6		5	25	ns

NOTES: 5. The response time specified is for a 100-mV input step with 5 mV overdrive.
6. For testing purposes, the input bias conditions are selected to produce an output voltage of 1.4 V. A 5-mV overdrive is then added to the input bias voltage to produce an output voltage that rises above 1.4 V. The time interval is measured from the 50% point of the strobe voltage curve to the point where the overdriven output voltage crosses the 1.4 V level.

TYPICAL CHARACTERISTICS

LARGE-SIGNAL DIFFERENTIAL
VOLTAGE AMPLIFICATION
vs
FREE-AIR TEMPERATURE

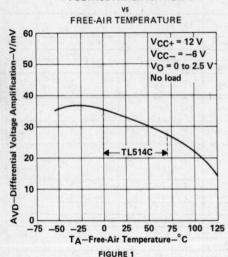

FIGURE 1

LARGE-SIGNAL DIFFERENTIAL
VOLTAGE AMPLIFICATION
vs
SUPPLY VOLTAGE

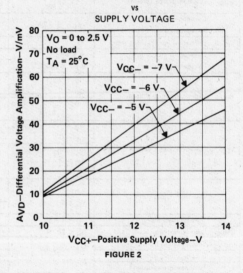

FIGURE 2

TEXAS
INSTRUMENTS
POST OFFICE BOX 225012 • DALLAS, TEXAS 75265

TYPICAL CHARACTERISTICS

OUTPUT VOLTAGE LEVELS
vs
FREE-AIR TEMPERATURE

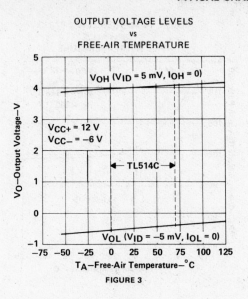

FIGURE 3

LOW-LEVEL OUTPUT CURRENT
vs
FREE-AIR TEMPERATURE

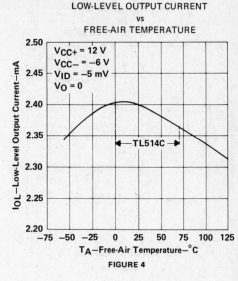

FIGURE 4

TL514M
VOLTAGE TRANSFER CHARACTERISTICS

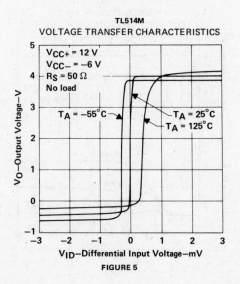

FIGURE 5

TL514C
VOLTAGE TRANSFER CHARACTERISTICS

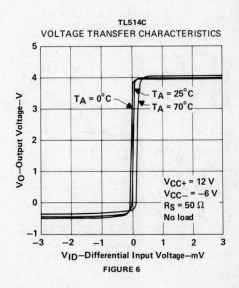

FIGURE 6

4

Voltage Comparators

TYPICAL CHARACTERISTICS

INPUT BIAS CURRENT
vs
FREE-AIR TEMPERATURE

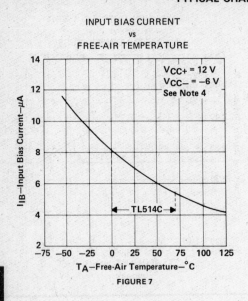

FIGURE 7

COMMON-MODE REJECTION RATIO
vs
FREE-AIR TEMPERATURE

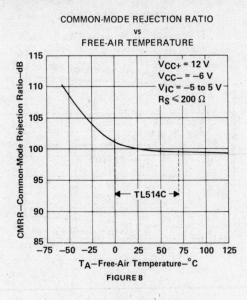

FIGURE 8

OUTPUT RESPONSE FOR
VARIOUS INPUT OVERDRIVES

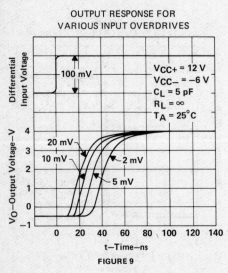

FIGURE 9

STROBE RELEASE TIME
FOR VARIOUS INPUT OVERDRIVES

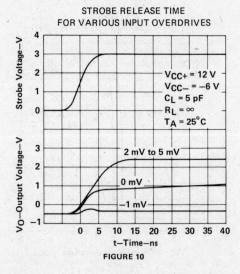

FIGURE 10

Voltage Comparators

4

NOTE 4: These characteristics are verified by measurements at the following temperatures and output voltage levels: for TL514M, $V_O = 1.8$ V at $T_A = -55°C$, $V_O = 1.4$ V at $T_A = 25°C$, and $V_O = 1$ V at $T_A = 125°C$; for TL514C, $V_O = 1.5$ V at $T_A = 0°C$, $V_O = 1.4$ V at $T_A = 25°C$, and $V_O = 1.2$ V at $T_A = 70°C$. These output voltage levels were selected to approximate the logic threshold voltages of the types of digital logic ciructs these comparators are intended to drive.

TEXAS
INSTRUMENTS
POST OFFICE BOX 225012 ● DALLAS, TEXAS 75265

TYPICAL CHARACTERISTICS

COMMON-MODE PULSE RESPONSE

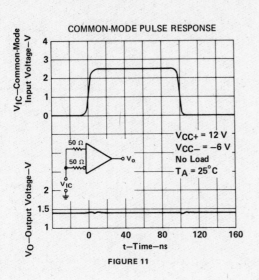

FIGURE 11

TOTAL POWER DISSIPATION
vs
FREE-AIR TEMPERATURE

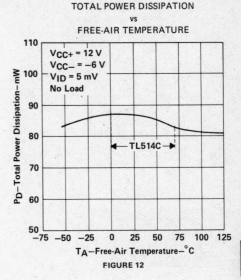

FIGURE 12

4

Voltage Comparators

4

Voltage Comparators

LINEAR
INTEGRATED
CIRCUITS

TYPES TL710M, TL710C
DIFFERENTIAL COMPARATORS
D2229, FEBRUARY 1971—REVISED AUGUST 1983

- Fast Response Times
- Low Offset Characteristics
- Output Compatible with Most TTL Circuits

description

The TL710 is a monolithic high-speed comparator having differential inputs and a low-impedance output. Component matching, inherent in silicon integrated circuit fabrication techniques, produces a comparator with a low-drift and low-offset characteristics. These circuits are especially useful for applications requiring an amplitude discriminator, memory sense amplifier, or a high-speed voltage comparator. The TL710M is characterized for operation over the full military temperature range of −55°C to 125°C; the TL710C is characterized for operation from 0°C to 70°C.

schematic

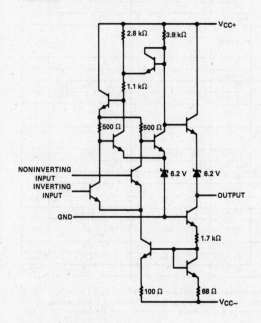

Component values shown are nominal.

J OR N DUAL-IN-LINE PACKAGE
(TOP VIEW)

NC	1	14	NC
GND	2	13	NC
IN+	3	12	NC
IN−	4	11	V$_{CC}$+
NC	5	10	NC
V$_{CC}$−	6	9	OUT
NC	7	8	NC

JG OR P DUAL-IN-LINE PACKAGE
(TOP VIEW)

GND	1	8	V$_{CC}$+
IN+	2	7	OUT
IN−	3	6	NC
V$_{CC}$−	4	5	NC

U FLAT PACKAGE
(TOP VIEW)

GND	1	10	NC
IN+	2	9	NC
IN−	3	8	V$_{CC}$+
NC	4	7	NC
V$_{CC}$−	5	6	OUT

NC—No internal connection

symbol

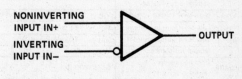

4

Voltage Comparators

TEXAS
INSTRUMENTS
POST OFFICE BOX 225012 • DALLAS, TEXAS 75265

absolute maximum ratings over operating free-air temperature range (unless otherwise noted)

		TL710M	TL710C	UNIT
Supply voltage V_{CC+} (see Note 1)		14	14	V
Supply voltage V_{CC-} (see Note 1)		−7	−7	V
Differential input voltage (see Note 2)		±5	±5	V
Input voltage (either input, see Note 1)		±7	±7	V
Peak output current ($t_w \leq 1$ s)		10	10	mA
Continuous total power dissipation at (or below) 70°C free-air temperature (see Note 3)		300	300	mW
Operating free-air temperature range		−55 to 125	0 to 70	°C
Storage temperature range		−65 to 150	−65 to 150	°C
Lead temperature 1,6 mm (1/16 inch) from case for 60 seconds	J, JG or U package	300	300	°C
Lead temperature 1,6 mm (1/16 inch) from case for 10 seconds	N or P package		260	°C

NOTES: 1. All voltage values, except differential voltages, are with respect to the network ground terminal.
2. Differential voltages are at the noninverting input terminal with respect to the inverting input terminal.
3. For operation of the TL710M above 70°C free-air temperature, refer to Dissipation Derating Curves, Section 2. In the J and JG packages, TL710M chips are alloy-mounted; TL710C chips are glass-mounted.

electrical characteristics at specified free-air temperature, V_{CC+} = 12 V, V_{CC-} = −6 V

PARAMETER		TEST CONDITIONS†		TL710M MIN	TYP	MAX	TL710C MIN	TYP	MAX	UNIT
V_{IO}	Input offset voltage	$R_S \leq 200\ \Omega$, See Note 4	25°C		2	5		2	7.5	mV
			Full range			6			10	
α_{VIO}	Average temperature coefficient of input offset voltage	$R_S \leq 200\ \Omega$, See Note 4	Full range		5			7.5		µV/°C
I_{IO}	Input offset current	See Note 4	25°C		1	10		1	15	µA
			Full range			20			25	
I_{IB}	Input bias current	See Note 4	25°C		25	75		25	100	µA
			Full range			150			150	
V_{ICR}	Common-mode input voltage range	$V_{CC} = -7$ V	25°C	±5			±5			V
V_{ID}	Differential input voltage range		25°C	±5			±5			V
A_{VD}	Large-signal differential voltage amplification	No load, See Note 4	25°C	750	1500		700	1500		V/V
			Full range	500			500			
V_{OH}	High-level output voltage	$V_{ID} = 15$ mV, $I_{OH} = -0.5$ mA	25°C	2.5	3.2	4	2.5	3.2	4	V
V_{OL}	Low-level output voltage	$V_{ID} = -15$ mV, $I_{OL} = 0$	25°C	−1	−0.5	0‡	−1	−0.5	0‡	V
I_{OL}	Low-level output current	$V_{ID} = -15$ mV, $V_O = 0$	25°C	1.6	2.5					mA
r_o	Output resistance	$V_O = 1.4$ V	25°C		200			200		Ω
CMRR	Common-mode rejection ratio	$R_S \leq 200\ \Omega$	25°C	70	90		65	90		dB
I_{CC+}	Supply current from V_{CC+}	$V_{ID} = -5$ V to 5 V	25°C		5.4	10.1		5.4		mA
I_{CC-}	Supply current from V_{CC-}	(−10 mV for typ)	25°C		−3.8	−8.9		−3.8		mA
P_D	Total power dissipation	No load	25°C		88	175		88		mW

NOTE 4: These characteristics are verified by measurements at the following temperatures and output voltage levels: for TL710M, V_O = 1.8 V at T_A = −55°C, V_O = 1.4 V at T_A = 25°C, and V_O = 1 V at T_A = 125°C; for TL710C, V_O = 1.5 V at T_A = 0°C, V_O = 1.4 V at T_A = 25°C, and V_O = 1.2 V at T_A = 70°C. These output voltage levels were selected to approximate the logic threshold voltages of the types of digital logic circuits these comparators are intended to drive.

†Full range for TL710M is −55°C to 125°C and for TL710C is 0°C to 70°C.
‡The algebraic convention where the most-positive (least-negative) limit is designated as maximum is used in this data sheet for logic levels only, e.g., when 0 V is the maximum, the minimum limit is a more-negative voltage.

TEXAS INSTRUMENTS
POST OFFICE BOX 225012 • DALLAS, TEXAS 75265

switching characteristics, V_{CC+} = 12 V, V_{CC-} = −6 V, T_A = 25°C

PARAMETER	TEST CONDITIONS	TL710M TYP	TL710C TYP	UNIT
Response time	No load, See Note 5	40	40	ns

NOTE 5: The response time specified is for a 100-mV input step with 5-mV overdrive and is the interval between the input step function and the instant
when the output crosses 1.4 V.

TYPICAL CHARACTERISTICS

TL710M
VOLTAGE TRANSFER CHARACTERISTICS

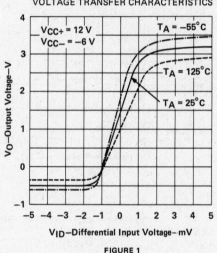

FIGURE 1

TL710C
VOLTAGE TRANSFER CHARACTERISTICS

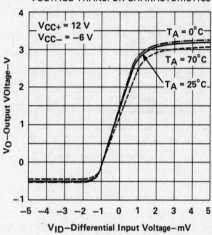

FIGURE 2

4

Voltage Comparators

4

Voltage Comparators

TYPICAL CHARACTERISTICS

OUTPUT VOLTAGE
vs
FREE-AIR TEMPERATURE

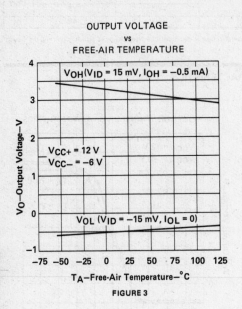

FIGURE 3

TOTAL POWER DISSIPATION
vs
FREE-AIR TEMPERATURE

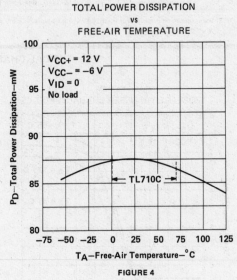

FIGURE 4

OUTPUT RESPONSE FOR VARIOUS
INPUT OVERDRIVES

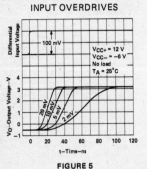

FIGURE 5

OUTPUT RESPONSE FOR VARIOUS
INPUT OVERDRIVES

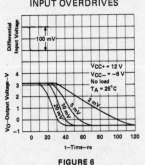

FIGURE 6

COMMON-MODE PULSE RESPONSE
vs
ELAPSED TIME

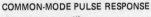
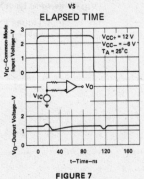

FIGURE 7

TEXAS INSTRUMENTS
POST OFFICE BOX 225012 • DALLAS, TEXAS 75265

- Operates from a Single 5-V Supply
- 0 to 5 V Common-Mode Input Voltage Range
- Self-Biased Inputs
- Complementary 3-State Outputs
- Enable Capability
- 5-mV Typical Hysteresis
- 25-ns Typical Response Times

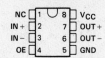

JG OR P DUAL-IN-LINE PACKAGE
(TOP VIEW)

NC	1	8	V_{CC}
IN +	2	7	OUT +
IN −	3	6	OUT −
OE	4	5	GND

NC—No Internal connection

description

The TL712 is a single high-speed voltage comparator fabricated with bipolar Schottky process technology. The circuit has differential analog inputs and complementary 3-state TTL-compatible logic outputs with symmetrical switching characteristics. When the output enable, OE, is low, both outputs are in the high-impedance state. This device operates from a singe 5-V supply and is useful as a disk memory read-chain data comparator.

The TL712 is characterized for operation from 0 °C to 70 °C.

functional block diagram

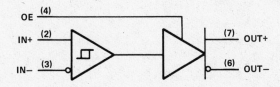

4

Voltage Comparators

TEXAS INSTRUMENTS

POST OFFICE BOX 225012 ● DALLAS, TEXAS 75265

4

Voltage Comparators

LINEAR
INTEGRATED
CIRCUITS

TYPE TL721
DIFFERENTIAL COMPARATOR

D2781, FEBRUARY 1984

- Operates From a Single −5.2-V Power Supply
- Self-Biased Inputs
- Common-Mode Input Voltage Range 0 to −5.2 V
- MECL III and MECL 10 000 Compatible
- Complementary ECL-Compatible Outputs
- Hysteresis . . . 5 mV Typ
- Response Times . . . 10 ns Typ

JG OR P DUAL-IN-LINE PACKAGE
(TOP VIEW)

NC	1	8	GND
IN−	2	7	OUT−
IN+	3	6	OUT+
NC	4	5	V_EE

NC—No internal connection

description

The TL721 is a single high-speed voltage comparator fabricated with bipolar Schottky[†] process technology. The circuit has differential analog inputs and complementary ECL-compatible logic outputs with symmetrical switching characteristics. The device operates from a single −5.2-volt supply and is useful as a disk memory read-chain data comparator.

The TL721 is characterized for operation from 0 °C to 70 °C.

symbol

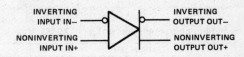

INVERTING INPUT IN−

NONINVERTING INPUT IN+

INVERTING OUTPUT OUT−

NONINVERTING OUTPUT OUT+

Voltage Comparators

4

†Integrated Schottky-Barrier diode-clamped transistor is patented by Texas Instruments. U.S. Patent Number 3,463,975.

TEXAS
INSTRUMENTS

POST OFFICE BOX 225012 • DALLAS, TEXAS 75265

electrical characteristics at $T_A = 25\,^\circ C$, $V_{EE} = -5.2$ V

PARAMETER		TEST CONDITIONS	MIN	TYP	MAX	UNIT
V_T	Threshold voltage (V_T+ and V_T-)	$V_{IC} = V_{ICR}$ min	-100[†]		100	mV
$V_T+ - V_T-$	Hysteresis			5	10	mV
V_{OH}	High-level output voltage	$V_{ID} = 100$ mV, $R_L = 50\ \Omega$ to -2 V	-0.96[†]		-0.81	V
V_{OL}	Low-level output voltage	$V_{ID} = -100$ mV, $R_L = 50\ \Omega$ to -2 V	-1.85[†]		-1.65	V
V_{ICR}	Common-mode input voltage range		0 to -5.2			V
r_{in}	Input resistance		4			kΩ
I_{EE}	Supply current	$V_{ID} = 0$, No load		-13	-17	mA

[†]The algebraic convention, where the more-negative limit is designated as minimum, is used in this data sheet for input threshold and output voltage levels only.

switching characteristics at $T_A = 25\,^\circ C$, $V_{EE} = -5.2$ V

PARAMETER		TEST CONDITIONS	MIN	TYP	MAX	UNIT
t_{PLH}	Propagation delay time, low-to-high-level output	$\Delta V_{ID} = +200$ mV to -200 mV or -200 mV to $+200$ mV, $R_L = 50\ \Omega$ to -2 V			12	ns
t_{PHL}	Propagation delay time, high-to-low-level output				12	ns

4

Voltage Comparators

TEXAS
INSTRUMENTS
POST OFFICE BOX 225012 • DALLAS, TEXAS 75265

- Low Offset Characteristics
- High Differential Voltage Amplification
- Fast Response Times
- Output Compatible with Most TTL Circuits

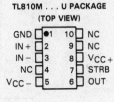

TL810M . . . JG PACKAGE
TL810C . . . JG OR P PACKAGE
(TOP VIEW)

GND	1	8	$V_{CC}+$
IN+	2	7	OUT
IN−	3	6	NC
$V_{CC}−$	4	5	NC

TL810M . . . U PACKAGE
(TOP VIEW)

GND	1	10	NC
IN+	2	9	NC
IN−	3	8	$V_{CC}+$
NC	4	7	STRB
$V_{CC}−$	5	6	OUT

NC—No internal connection

description

The TL810 is an improved version of the TL710 high-speed voltage comparator with an extra stage added to increase voltage amplification and accuracy. Typical amplification is 33,000. Component matching, inherent in monolithic integrated circuit fabrication techniques, produces a comparator with low-drift and low-offset characteristics. These circuits are particularly useful for applications requiring an amplitude discriminator, memory sense amplifier, or a high-speed limit detector.

The TL810M is characterized for operation over the full military temperature range of −55°C to 125°C; the TL810C is characterized for operation from 0°C to 70°C.

symbol

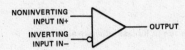

NONINVERTING INPUT IN+
INVERTING INPUT IN−
OUTPUT

4

Voltage Comparators

schematic

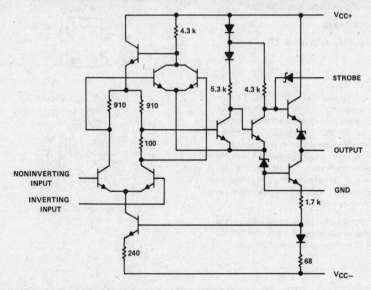

Resistor values shown are nominal in ohms.

absolute maximum ratings over operating free-air temperature range (unless otherwise noted)

Supply voltage $V_{CC}+$ (see Note 1) . . . . 14 V

Supply voltage $V_{CC}-$ (see Note 1) . . . . −7 V

Differential input voltage (see Note 2) . . . . ±5 V

Input voltage (either input, see Note 1) . . . . ±7 V

Peak output current ($t_w \leq 1$ s) . . . . 10 mA

Continuous total power dissipation at (or below) 70 °C free-air temperature (see Note 3) . . . . 300 mW

Operating free-air temperature range: TL810M Circuits . . . . −55 °C to 125 °C

TL810C Circuits . . . . 0 °C to 70 °C

Storage temperature range . . . . −65 °C to 150 °C

Lead temperature 1,6 mm (1/16 inch) from case for 60 seconds: JG or U package . . . . 300 °C

Lead temperature 1,6 mm (1/16 inch) from case for 10 seconds: P package . . . . 260 °C

NOTES: 1. All voltage values, except differential voltages, are with respect to the network ground terminal.
2. Differential voltages are at the noninverting input terminal with respect to the inverting input terminal.
3. For operation of the TL810M above 70 °C free-air temperature, refer to dissipation Derating Curves, Section 2. In the JG package, TL810M chips are alloy-mounted; TL810C chips are glass-mounted.

4

Voltage Comparators

TEXAS
INSTRUMENTS
POST OFFICE BOX 225012 • DALLAS, TEXAS 75265

electrical characteristics at specified free-air temperature, V_{CC+} = 12 V, V_{CC-} = -6 V (unless otherwise noted)

PARAMETER		TEST CONDITIONS[†]		TL810M			TL810C			UNIT
				MIN	TYP	MAX	MIN	TYP	MAX	
V_{IO}	Input offset voltage	R_S ≤ 200 Ω, See Note 4	25°C		0.6	2		1.6	3.5	mV
			Full range			3			4.5	
αVIO	Average temperature coefficient of input offset voltage	R_S = 50 Ω, See Note 4	MIN to 25°C		3	10		3	20	µV/°C
			25°C to MAX		3	10		3	20	
I_{IO}	Input offset current	See Note 4	25°C		0.75	3		1.8	5	µA
			MIN		1.8	7			7.5	
			MAX		0.25	3			7.5	
αIIO	Average temperature coefficient of input offset current	See Note 4	MIN to 25°C		15	75		24	100	nA/°C
			25°C to MAX		5	25		15	50	
I_{IB}	Input bias current	See Note 4	25°C		7	15		7	20	µA
			MIN		12	25		9	30	
V_{ICR}	Common-mode input voltage range	V_{CC-} = -7 V	Full range	±5			±5			V
A_{VD}	Large-signal differential voltage amplification	No load, V_O = 0 to 2.5 V	25°C	12.5	33		10	33		V/mV
			Full range	10			8			
V_{OH}	High-level output voltage	V_{ID} = 5 mV, I_{OH} = 0	Full range		4§	5		4§	5	V
		V_{ID} = 5 mV, I_{OH} = -5 mA	Full range	2.5	3.6§		2.5	3.6§		
V_{OL}	Low-level output voltage	V_{ID} = -5 mV, I_{OL} = 0	Full range	-1	-0.5§	0‡	-1	-0.5§	0‡	V
I_{OL}	Low-level output current	V_{ID} = -5 mV, V_O = 0	25°C	2	2.4		1.6	2.4		mA
			MIN	1	2.3		0.5	2.4		
			MAX	0.5	2.3		0.5	2.4		
r_o	Output resistance	V_O = 1.4 V	25°C		200			200		Ω
CMRR	Common-mode rejection ratio	R_S ≤ 200Ω	Full range	80	100§		70	100§		dB
I_{CC+}	Supply current from V_{CC+}	V_{ID} = -5 mV, No load	Full range		5.5§	9		5.5§	9	mA
I_{CC-}	Supply current from V_{CC-}		Full range		-3.5§	-7		-3.5§	-7	mA
P_D	Total power dissipation		Full range		90§	150		90§	150	mW

[†]Full range (MIN to MAX) for TL810M is -55°C to 125°C and for the TL810C is 0°C to 70°C.

‡The algebraic convention, where the most-positive (least-negative) limit is designated as maximum, is used in this data sheet for logic levels only, e.g., when 0 V is the maximum, the minimum limit is a more-negative voltage.

§These typical values are at T_A = 25°C.

NOTE 4: These characteristics are verified by measurements at the following temperatures and output voltage levels: for TL810M, V_O = 1.8 V at T_A = -55°C, V_O = 1.4 V at T_A = 25°C, and V_O = 1 V at T_A = 125°C; for TL810C, V_O = 1.5 V at T_A = 0°C, V_O = 1.4 V at 25°C, and V_O = 1.2 V at T_A = 70°C. These output voltage levels were selected to approximate the logic threshold voltages of the types of digital logic circuits these comparators are intended to drive.

switching characteristics, V_{CC+} = 12 V, V_{CC-} = -6 V, T_A = 25°C

PARAMETER	TEST CONDITIONS			MIN	TYP	MAX	UNIT
Response time	R_L = ∞,	C_L = 5 pF,	See Note 5		30	80	ns

NOTE 5: The response time specified is for a 100-mV input step with 5-mV overdrive and is the interval between the input step function and the instant when the output crosses 1.4 V.

Voltage Comparators

4

TYPICAL CHARACTERISTICS

LARGE-SIGNAL DIFFERENTIAL
VOLTAGE AMPLIFICATION
vs
FREE-AIR TEMPERATURE

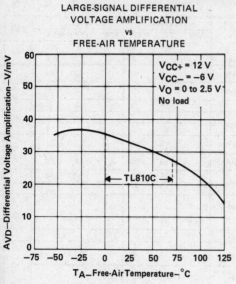

FIGURE 1

LARGE-SIGNAL DIFFERENTIAL
VOLTAGE AMPLIFICATION
vs
SUPPLY VOLTAGE

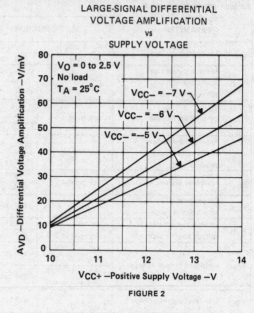

FIGURE 2

OUTPUT VOLTAGE LEVELS
vs
FREE-AIR TEMPERATURE

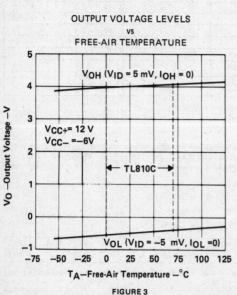

FIGURE 3

LOW-LEVEL OUTPUT CURRENT
vs
FREE-AIR TEMPERATURE

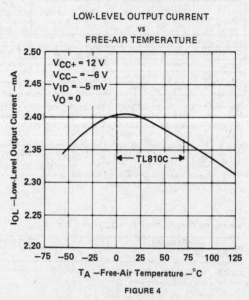

FIGURE 4

TEXAS
INSTRUMENTS
POST OFFICE BOX 225012 • DALLAS, TEXAS 75265

88

4

Voltage Comparators

TYPICAL CHARACTERISTICS

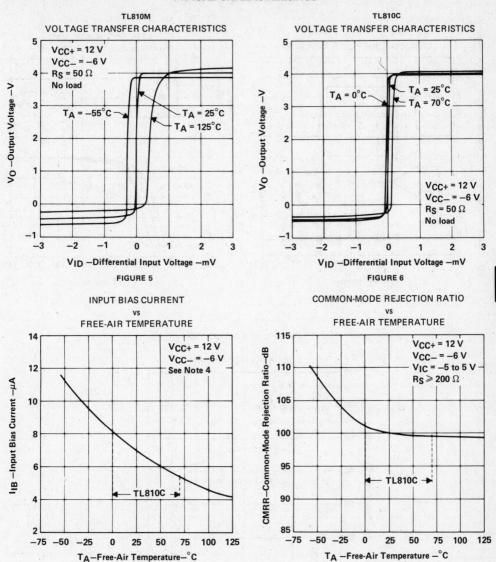

NOTE 4: These characteristics are verified by measurements at the following temperatures and output voltage levels: for TL810M, V_O = 1.8 V at T_A = −55°C, V_O = 1.4 V at T_A = 25°C, and V_O = 1 V at T_A = 125°C; for TL810C, V_O = 1.5 V at T_A = 0°C, V_O = 1.4 V at 25°C, and V_O = 1.2 V at T_A = 70°C. These output voltage levels were selected to approximate the logic threshold voltages of the types of digital logic circuits these comparators are intended to drive.

Voltage Comparators

TYPICAL CHARACTERISTICS

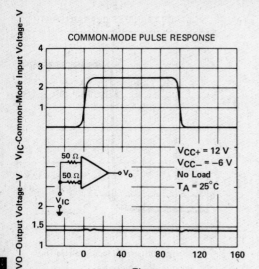

FIGURE 9

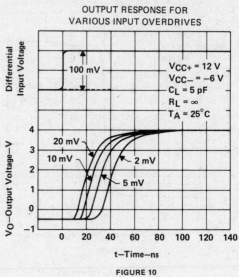

FIGURE 10

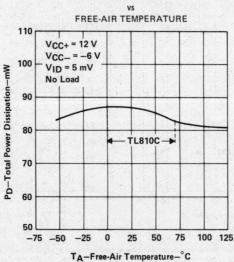

FIGURE 11

TEXAS
INSTRUMENTS

POST OFFICE BOX 225012 • DALLAS, TEXAS 75265

4

Voltage Comparators

- **Fast Response Times**
- **Improved Voltage Amplification and Offset Characteristics**
- **Output Compatible with Most TTL Circuits**

description

The TL811 is an improved version of the TL711 high-speed dual-channel voltage comparator. Voltage amplification is higher (typically 17,500) due to an extra stage, increasing the temperature accuracy. The output pulse width may be "stretched" by varying the capacitive loading.

Each channel has differential inputs, a strobe input, and an output in common with the other channel. When either strobe is taken low, it inhibits the associated channel. If both strobes are simultaneously low, the output will be low regardless of the conditions applied to the differential inputs.

These dual-channel voltage comparators are particularly attractive for applications requiring an amplitude-discriminating sense amplifier with an adjustable threshold voltage.

The TL811M is characterized for operation over the full military range of −55°C to 125°C; the TL811C is characterized for operation from 0°C to 70°C.

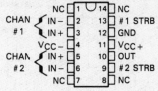

TL811M . . . J DUAL-IN-LINE PACKAGE
TL811C . . . J OR N DUAL-IN-LINE PACKAGE
(TOP VIEW)

TL811M . . . U FLAT PACKAGE
(TOP VIEW)

NC—No internal connection

functional block diagram

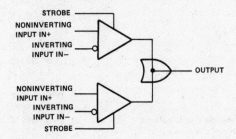

TEXAS INSTRUMENTS
POST OFFICE BOX 225012 • DALLAS, TEXAS 75265

4-73

Voltage Comparators

4

schematic

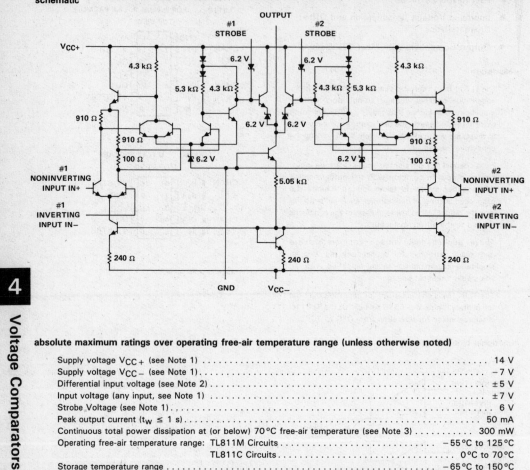

absolute maximum ratings over operating free-air temperature range (unless otherwise noted)

Supply voltage V_{CC+} (see Note 1) . . . . 14 V

Supply voltage V_{CC-} (see Note 1) . . . . −7 V

Differential input voltage (see Note 2) . . . . ±5 V

Input voltage (any input, see Note 1) . . . . ±7 V

Strobe Voltage (see Note 1) . . . . 6 V

Peak output current ($t_W \leq 1$ s) . . . . 50 mA

Continuous total power dissipation at (or below) 70°C free-air temperature (see Note 3) . . . . 300 mW

Operating free-air temperature range: TL811M Circuits . . . . −55°C to 125°C

TL811C Circuits . . . . 0°C to 70°C

Storage temperature range . . . . −65°C to 150°C

Lead temperature 1,6 mm (1/16 inch) from case for 60 seconds: J or U package . . . . 300°C

Lead temperature 1,6 mm (1/16 inch) from case for 10 seconds: N package . . . . 260°C

NOTES: 1. All voltage values, except differential voltages, are with respect to the network ground terminal.
2. Differential voltages are at the noninverting input terminal with respect to the inverting input terminal.
3. For operation of the TL811M above 70°C free-air temperature, refer to Dissipation Derating Curves, Section 2. In the J package, the TL811M chips are alloy-mounted; TL810C chips are glass-mounted.

TEXAS INSTRUMENTS

POST OFFICE BOX 225012 • DALLAS, TEXAS 75265

electrical characteristics at specified free-air temperature, V_{CC+} = 12 V, V_{CC-} = −6 V (unless otherwise noted)

PARAMETER		TEST CONDITIONS†		TL811M MIN	TYP	MAX	TL811C MIN	TYP	MAX	UNIT
V_{IO}	Input offset voltage	V_{IC} = 0, See Note 4	25°C		1	3.5		1	5	mV
			Full range			4.5			6	
		See Note 4	25°C		1	5		1	7.5	
			Full range			6			10	
α_{VIO}	Average temperature coefficient of input offset voltage	V_{IC} = 0, See Note 4	Full range		5			5		μV/°C
I_{IO}	Input offset current	See Note 4	25°C		0.5	3		0.5	5	μA
			Full range			5			10	
I_{IB}	Input bias current	See Note 4	25°C		7	20		7	30	μA
			Full range			30			50	
$I_{IL(S)}$	Low-level strobe current	$V_{(strobe)}$ = −100 mV	25°C		−1.2	−2.5		−1.2	−2.5	mA
V_{ICR}	Common-mode input voltage range	V_{CC-} = −7 V	25°C	±5			±5			V
V_{ID}	Differential input voltage range		25°C	±5			±5			V
A_{VD}	Large-signal differential voltage amplification	V_O = 0 to 2.5 V, No load	25°C	12.5	17.5		10	17.5		V/mV
			Full range	8			5			
V_{OH}	High-level output voltage	V_{ID} = 10 mV, I_{OH} = 0	25°C		4	5		4	5	V
		V_{ID} = 10 mV, I_{OH} = −5 mA	25°C	2.5	3.6		2.5	3.6		
V_{OL}	Low-level output voltage	V_{ID} = −10 mV, I_{OL} = 0	25°C	−1	−0.4	0‡	−1	−0.4	0‡	V
		V_{ID} = 10 mV, $V_{(strobe)}$ = 0.3 V, I_{OL} = 0	25°C	−1		0‡	1		0‡	
I_{OL}	Low-level output current	V_{ID} = −10 mV, V_O = 0	25°C	0.5	0.8		0.5	0.8		mA
r_o	Output resistance	V_O = 1.4 V	25°C		200			200		Ω
CMRR	Common-mode rejection ratio		25°C	70	90		65	90		dB
I_{CC+}	Supply current from V_{CC+}	V_{ID} = −5 to 5 V	25°C		6.5			6.5		mA
I_{CC-}	Supply current from V_{CC-}	(10 mV for typ),	25°C		−2.7			−2.7		mA
P_D	Total power dissipation	No load, See Note 5	25°C		94	150		94	200	mW

† Unless otherwise noted, all characteristics are measured with the strobe of the channel under test open, the strobe of the other channel is grounded. Full range for TL811M is −55°C to 125°C and for the TL811C is 0°C to 70°C.

‡The algebraic convention, where the most-positive (lease-negative) limit is designated as maximum, is used in this data sheet for logic levels only, e.g., when 0 V is the maximum, the minimum limit is a more-negative voltage.

NOTES: 4. These characteristics are verified by measurements at the following temperatures and output voltage levels: for TL811M, V_O = 1.8 V at T_A = −55°C, V_O = 1.4 V at T_A = 25°C, and V_O = 1 V at T_A = 125°C; for TL811C, V_O = 1.5 V at T_A = 0°C, V_O = 1.4 V at T_A = 25°C, and V_O = 1.2 V at 70°C. These output voltage levels were selected to approximate the logic threshold voltages of the types of digital logic circuits these comparators are intended to drive.

5. The strobes are alternately grounded.

4

Voltage Comparators

TEXAS
INSTRUMENTS
POST OFFICE BOX 225012 ● DALLAS, TEXAS 75265

switching characteristics, $V_{CC+} = 12$ V, $V_{CC-} = -6$ V, $T_A = 25\,^{\circ}$C

PARAMETER	TEST CONDITIONS	TL811M			TL811C			UNIT
		MIN	TYP	MAX	MIN	TYP	MAX	
Response time	$R_L = \infty$, $C_L = 5$ pF, See Note 6		33	80		33		ns
Strobe release time	$R_L = \infty$, $C_L = 5$ pF, See Note 7		5	25		5		ns

NOTES: 6. The response time specified is for a 100-mV input step with 5-mV overdrive and is the interval between the input step function and the instant when the output crosses 1.4 V.

7. For testing purposes, the input bias conditions are selected to produce an output voltage of 1.4 V. A 5-mV overdrive is then added to the input bias voltage to produce an output voltage that rises above 1.4 V. The time interval is measured from the 50% point on the strobe voltage waveform to the instant when the overdriven output voltage crosses the 1.4-V level.

TYPICAL CHARACTERISTICS

LARGE-SIGNAL DIFFERENTIAL
VOLTAGE AMPLIFICATION
vs
FREE-AIR TEMPERATURE

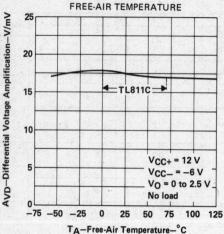

FIGURE 1

LARGE-SIGNAL DIFFERENTIAL
VOLTAGE AMPLIFICATION
vs
SUPPLY VOLTAGE

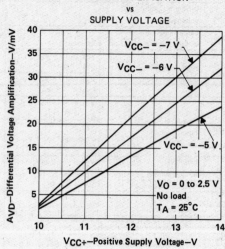

FIGURE 2

TEXAS
INSTRUMENTS
POST OFFICE BOX 225012 • DALLAS, TEXAS 75265

TYPICAL CHARACTERISTICS

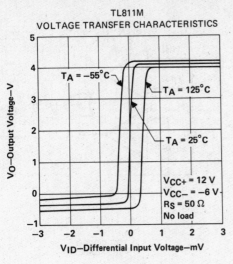

FIGURE 3

FIGURE 4

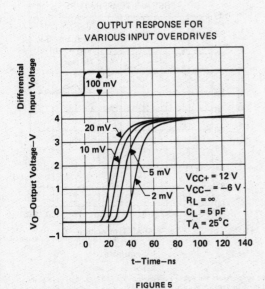

FIGURE 5

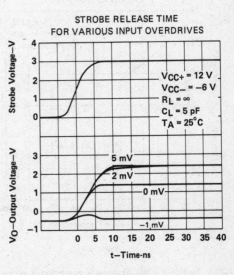

FIGURE 6

TEXAS
INSTRUMENTS
POST OFFICE BOX 225012 • DALLAS, TEXAS 75265

Voltage Comparators

4

TYPES TL811M, TL811C
DUAL-CHANNEL DIFFERENTIAL COMPARATORS WITH STROBES

TYPICAL CHARACTERISTICS

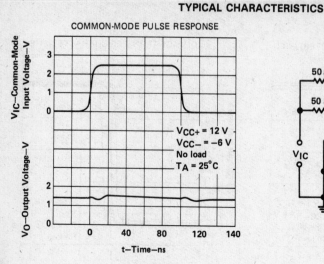

COMMON-MODE PULSE RESPONSE

V_{CC+} = 12 V
V_{CC-} = −6 V
No load
T_A = 25°C

FIGURE 7

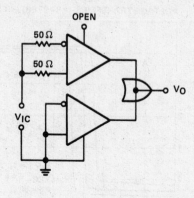

TEST CIRCUIT
FOR FIGURE 7

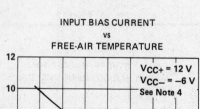

INPUT BIAS CURRENT
vs
FREE-AIR TEMPERATURE

V_{CC+} = 12 V
V_{CC-} = −6 V
See Note 4

TL811C

FIGURE 8

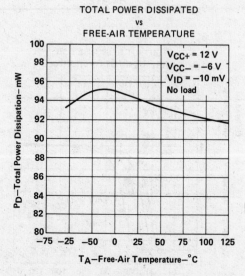

TOTAL POWER DISSIPATED
vs
FREE-AIR TEMPERATURE

V_{CC+} = 12 V
V_{CC-} = −6 V
V_{ID} = −10 mV
No load

FIGURE 9

NOTE 4: These characteristics are verified by measurements at the following temperatures and output voltage levels: for TL811M, V_O = 1.8 V at T_A = −55°C, V_O = 1.4 V at T_A = 25°C, and V_O = 1 V at T_A = 125°C; for TL811C, V_O = 1.5 V at T_A = 0°C, V_O = 1.4 V at T_A = 25°C, and V_O = 1.2 V at 70°C. These output voltage levels were selected to approximate the logic threshold voltages of the types of digital logic circuits these comparators are intended to drive.

TEXAS INSTRUMENTS
POST OFFICE BOX 225012 • DALLAS, TEXAS 75265

4 Voltage Comparators

- Fast Response Times
- High Differential Voltage Amplification
- Low Offset Characteristics
- Outputs Compatible with Most TTL Circuits

description

The TL820 is an improved version of the TL720 dual high-speed voltage comparator. Each comparator has differential inputs and a low-impedance output. When compared with the TL720, these circuits feature high amplification (typically 33,000) due to an extra amplification stage and increased accuracy because of lower offset characteristics. They are particularly useful in applications requiring an amplitude discriminator, memory sense amplifier, or a high-speed limit detector.

The TL820M is characterized for operation over the full military temperature range of −55°C to 125°C; the TL820C is characterized for operation from 0°C to 70°C.

symbol (each comparator)

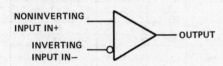

J OR N DUAL-IN-LINE PACKAGE (TOP VIEW)

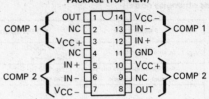

NC—No internal connection

schematic (each comparator)

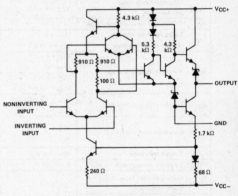

Component values shown are nominal.

absolute maximum ratings over operating free-air temperature range (unless otherwise noted)

Supply voltage V_{CC+} (see Note 1)	14 V
Supply voltage V_{CC-} (see Note 1)	−7 V
Differential input voltage (see Note 2)	±5 V
Input voltage (any input, see Note 1)	±7 V
Peak output current ($t_w \le 1$ s)	10 mA
Continuous total power dissipation at (or below) 70°C free-air temperature: each comparator	300 mW
total package (see Note 3)	600 mW
Operating free-air temperature range: TL820M Circuits	−55°C to 125°C
TL820C Circuits	0°C to 70°C
Storage temperature range	−65°C to 150°C
Lead temperature 1,6 mm (1/16 inch) from case for 60 seconds: J package	300°C
Lead temperature 1,6 mm (1/16 inch) from case for 10 seconds: N package	260°C

NOTES: 1. All voltage values, except differential voltages, are with respect to the network ground terminal.
2. Differential voltages are at the noninverting input terminal with respect to the inverting input terminal.
3. For operation of the TL820M above 70°C free-air temperature, refer to Dissipation Derating Curves, Section 2. In the J package, TL820M chips are alloy-mounted, TL820C chips are glass-mounted.

Voltage Comparators

4

TYPES TL820M, TL820C
DUAL DIFFERENTIAL COMPARATORS

electrical characteristics at specified free-air temperature, V_{CC+} = 12 V, V_{CC-} = −6 V (unless otherwise noted)

PARAMETER		TEST CONDITIONS[†]		TL820M			TL820C			UNIT
				MIN	TYP	MAX	MIN	TYP	MAX	
V_{IO}	Input offset voltage	R_S < 200 Ω, See Note 4	25°C		0.6	2		1.6	3.5	mV
			Full range			3			4.5	
α_{VIO}	Average temperature coefficient of input offset voltage	R_S = 50 Ω, See Note 4	MIN to 25°C		3	10		3	20	µV/°C
			25°C to MAX		3	10		3	20	
I_{IO}	Input offset current	See Note 4	25°C		0.75	3		1.8	5	µA
			MIN		1.8	7			7.5	
			MAX		0.25	3			7.5	
α_{IIO}	Average temperature coefficient of input offset current	See Note 4	MIN to 25°C		15	75		24	100	nA/°C
			25°C to MAX		5	25		15	50	
I_{IB}	Input bias current	See Note 4	25°C		7	15		7	20	µA
			MIN		12	25		9	30	
V_{ICR}	Common-mode input voltage range	V_{CC-} = −7 V	Full range	±5			±5			V
V_{ID}	Differential input voltage range		Full range	±5			±5			V
A_{VD}	Large-signal differential voltage amplification	No load, V_O = 0 to 2.5 V	25°C	12.5	33		10	33		V/mV
			Full range	10			8			
V_{OH}	High-level output voltage	V_{ID} = 5 mV, I_{OH} = 0	Full range		4[§]	5		4[§]	5	V
		V_{ID} = 5 mV, I_{OH} = −5 mA	Full range	2.5	3.6[§]		2.5	3.6[§]		
V_{OL}	Low-level output voltage	V_{ID} = −5 mV, I_{OL} = 0	Full range	−1	−0.5[§]	0[‡]	−1	−0.5[§]	0[‡]	V
I_{OL}	Low-level output current	V_{ID} = −5 mV, V_O = 0	25°C	2	2.4		1.6	2.4		mA
			MIN	1	2.3		0.5	2.4		
			MAX	0.5	2.3		0.5	2.4		
r_o	Output resistance	V_O = 1.4 V	25°C		200			200		Ω
CMRR	Common-mode rejection ratio	R_S < 200 Ω	Full range	80	100[§]		70	100[§]		dB
I_{CC+}	Supply current from V_{CC+} (each comparator)		Full range		5.5[§]	9		5.5[§]	9	mA
I_{CC-}	Supply current from V_{CC-} (each comparator)	V_{ID} = −5 mV, No load	Full range		−3.5[§]	−7		−3.5[§]	−7	mA
P_D	Total power dissipation (each comparator)		Full range		90[§]	150		90[§]	150	mW

[†]Full range (MIN to MAX) for TL820M is −55°C to 125°C and for the TL820C is 0°C to 70°C.
[‡]The algebraic convention where the most-positive (least-negative) limit is designated as maximum is used in this data sheet for logic levels only, e.g., when 0 V is the maximum, the minimum limit is a more-negative voltage.
[§]These typical values are at T_A = 25°C.
NOTE 4: These characteristics are verified by measurements at the following temperatures and output voltage levels: for TL820M, V_O = 1.8 V at T_A = −55°C, V_O = 1.4 V at T_A = 25°C, and V_O = 1 V at T_A = 125°C; for TL820C, V_O = 1.5 V at T_A = 0°C, V_O = 1.4 V at 25°C, and V_O = 1.2 V at T_A = 70°C. These output voltage levels were selected to approximate the logic threshold voltages of the types of digital logic circuits these comparators are intended to drive.

switching characteristics, V_{CC+} = 12 V, V_{CC-} = −6 V, T_A = 25°C

PARAMETER	TEST CONDITIONS	MIN	TYP	MAX	UNIT
Response time	R_L = ∞, C_L = 5 pF, See Note 5		30	80	ns

NOTE 5: The response time specified is for a 100-mV input step with 5-mV overdrive and is the interval between the input step function and the instant when the output crosses 1.4 V.

4

Voltage Comparators

TEXAS
INSTRUMENTS
POST OFFICE BOX 225012 • DALLAS, TEXAS 75265

TYPICAL CHARACTERISTICS

LARGE-SIGNAL DIFFERENTIAL
VOLTAGE AMPLIFICATION
VS
FREE-AIR TEMPERATURE

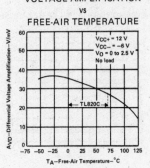

FIGURE 1

LARGE-SIGNAL DIFFERENTIAL
VOLTAGE AMPLIFICATION
VS
SUPPLY VOLTAGE

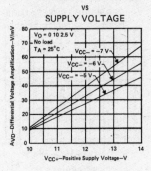

FIGURE 2

OUTPUT VOLTAGE LEVELS
VS
FREE-AIR TEMPERATURE

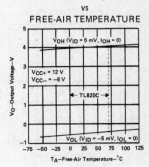

FIGURE 3

LOW-LEVEL OUTPUT CURRENT
VS
FREE-AIR TEMPERATURE

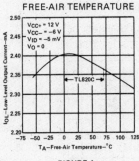

FIGURE 4

TL820M
VOLTAGE TRANSFER
CHARACTERISTICS

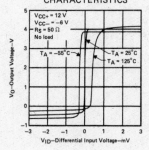

FIGURE 5

TL820C
VOLTAGE TRANSFER
CHARACTERISTICS

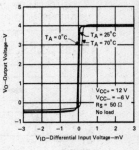

FIGURE 6

4

Voltage Comparators

**TEXAS
INSTRUMENTS**
POST OFFICE BOX 225012 • DALLAS, TEXAS 75265

TYPES TL820M, TL820C
DUAL DIFFERENTIAL COMPARATORS

TYPICAL CHARACTERISTICS

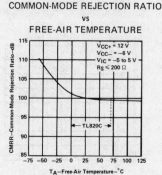

INPUT BIAS CURRENT
vs
FREE-AIR TEMPERATURE

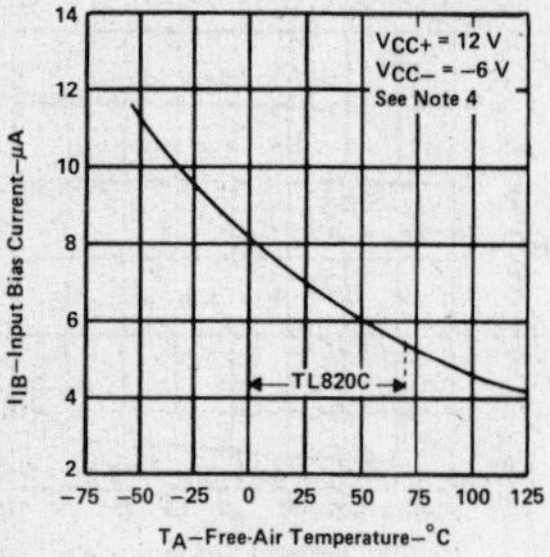

FIGURE 7

COMMON-MODE REJECTION RATIO
vs
FREE-AIR TEMPERATURE

FIGURE 8

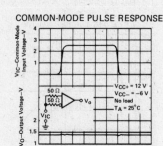

COMMON-MODE PULSE RESPONSE

FIGURE 9

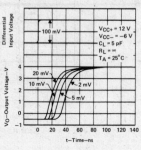

OUTPUT RESPONSE FOR
VARIOUS INPUT OVERDRIVES

FIGURE 10

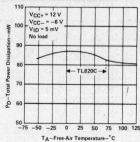

TOTAL POWER DISSIPATION
vs
FREE-AIR TEMPERATURE

FIGURE 11

NOTE 4: These characteristics are verified by measurements at the following temperatures and output voltage levels: for TL820M, V_O = 1.8 V at T_A = − 55°C, V_O = 1.4 V at T_A = 25°C, and V_O = 1 V at T_A = 125°C; for TL820C, V_O = 1.5 V at T_A = 0°C, V_O = 1.4 V at 25°C, and V_O = 1.2 V at T_A = 70°C. These output voltage levels were selected to approximate the logic threshold voltages of the types of digital logic circuits these comparators are intended to drive.

4

Voltage Comparators

TEXAS
INSTRUMENTS
POST OFFICE BOX 225012 • DALLAS, TEXAS 75265

**LINEAR
INTEGRATED
CIRCUITS**

**TYPES TLC372M, TLC372C
DUAL LinCMOS™ DIFFERENTIAL COMPARATORS**

D2821, NOVEMBER 1983

- Single- or Dual-Supply Operation
- Wide Range of Supply Voltages
  2 to 18 Volts
- Very Low Supply Current Drain
  0.2 mA Typ
- Fast Response Time . . . 200 ns Typ for
  TTL-Level Input Step
- Built-In ESD Protection
- High Input Impedance . . . 10^{12} Ω Typ
- Extremely Low Input Bias Current
  1 pA Typ
- Ultra-Stable Low Input Offset Voltage
- Common-Mode Input Voltage Range
  Includes Ground
- Output Compatible with TTL, MOS, and
  CMOS

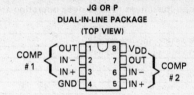

JG OR P
DUAL-IN-LINE PACKAGE
(TOP VIEW)

symbol (each comparator)

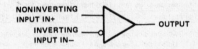

description

This device is fabricated using LinCMOS™ technology and consists of two independent voltage comparators designed to operate from a single power supply. Operation from dual supplies is also possible so long as the difference between the two supplies is 2 to 18 volts. Each of these devices features extremely high input impedance (typically greater than 10^{12} ohms) allowing direct interfacing with high-impedance sources. The outputs are n-channel open-drain configurations, and can be connected to achieve positive-logic wired-AND relationships.

These devices have internal electrostatic discharge (ESD) protection circuits that will prevent catastrophic failures at voltages up to 2000 volts as tested under MIL-STD-883B, Method 3015.1. However, care should be exercised in handling these devices as exposure to ESD may result in a degradation of the device parametric performance.

The TLC372M is characterized for operation over the full military temperature range of −55°C to 125°C. The TLC372C is characterized for operation from 0°C to 70°C.

Voltage Comparators

4

**TEXAS
INSTRUMENTS**

POST OFFICE BOX 225012 • DALLAS, TEXAS 75265

absolute maximum ratings over operating free-air temperature range (unless otherwise noted)

		TLC374M	TLC374C	UNIT
Supply voltage, V_{DD} (see Note 1)		18	18	V
Differential input voltage (see Note 2)		±18	±18	V
Input voltage, V_I		18	18	V
Output voltage, V_O		18	18	V
Output current, I_O		20	20	mA
Duration of output short-circuit to ground (see Note 3)		unlimited	unlimited	
Continuous total dissipation at (or below) 25 °C free-air temperature (see Note 4)		500	500	mW
Operating free-air temperature range		−55 to 125	0 to 70	°C
Storage temperature range		−65 to 150	−65 to 150	°C
Lead temperature 1,6 mm (1/16 inch) from case for 60 seconds	JG package	300	300	°C
Lead temperature 1,6 mm (1/16 inch) from case for 10 seconds	D or P package		260	°C

NOTES: 1. All voltage values, except differential voltages, are with respect to network ground terminal.
2. Differential voltages are at the noninverting input terminal with respect to the inverting input terminal.
3. Short circuits from outputs to V_{DD} can cause excessive heating and eventual destruction.
4. For operation above 25 °C free-air temperature, refer to Dissipation Derating Curves, Section 2. In the J package, TLC374C chips are glass mounted and TLC374M chips are alloy mounted.

electrical characteristics at specified free-air temperature, V_{DD} = 5 V (unless otherwise noted)

PARAMETER		TEST CONDITIONS†		TLC372M			TLC372C			UNIT
				MIN	TYP	MAX	MIN	TYP	MAX	
V_{IO}	Input offset voltage	$V_{IC} = V_{ICR}$ min, See Note 5	25 °C		2	10		2	10	mV
			Full range			12			12	
I_{IO}	Input offset current	See Note 5	25 °C		1			1		pA
			Full range			10			0.3	nA
I_{IB}	Input bias current		25 °C		1					pA
			Full range			20			0.6	nA
V_{ICR}	Common-mode input voltage range		25 °C		0 to $V_{CC}-1.5$			0 to $V_{CC}-1.5$		V
			Full range		0 to $V_{CC}-2$			0 to $V_{CC}-2$		
A_{VD}	Large-signal differential voltage amplification	V_{DD} = 15 V, $R_L \geq$ 15 kΩ to V_{DD}	25 °C		200			200		v/mV
I_{OH}	High-level output current	V_{ID} = 1 V	V_{OH} = 5 V	25 °C		0.1			0.1		nA
			V_{OH} = 15 V	Full range			1			1	µA
V_{OL}	Low-level output voltage	V_{ID} = −1 V, I_{OL} 4 mA	25 °C	150	400		150	400		mV
			Full range			700			700	
I_{OL}	Low-level output current	V_{ID} = −1 V, V_{OL} = 1.5 V	25 °C	6	16		6	16		mA
I_{DD}	Supply current (two comparators)	V_{ID} = 1 V, No load	25 °C		0.2			0.2		mA

†All characteristics are measured with zero common-mode input voltage unless otherwise specified.
NOTE 5: The offset voltages and offset currents given are the maximum values required to drive the output up to 4 V or down to 400 mV with a pull-up resistor of 2.5 kΩ to V_{DD}. Thus, these parameters actually define an error band and take into account the worst-case effects of voltage gain and input impedance. Full range for T_A is −55 °C to 125 °C for TLC372M, 0 °C to 70 °C for TLC372C.

switching characteristics, V_{DD} = 5 V, T_A = 25 °C

PARAMETER	TEST CONDITIONS		MIN	TYP	MAX	UNIT
Response time	R_L connected to 5 V through 5.1 kΩ, C_L = 15 pF‡, See Note 6	100-mV input step with 5-mV overdrive		650		ns
		TTL-level input step		200		

‡C_L includes probe and jig capacitance.
NOTE 6: The response time specified is the interval between the input step function and the instant when the output crosses 1.4 V.

TEXAS
INSTRUMENTS
POST OFFICE BOX 225012 • DALLAS, TEXAS 75265

TYPES TLC374M, TLC374C
QUADRUPLE LinCMOS™ DIFFERENTIAL COMPARATORS

D2783, NOVEMBER 1983

- Single- or Dual-Supply Operation
- Wide Range of Supply Voltages
  2 to 18 volts
- Very Low Supply Current Drain
  0.4 mA Typ
- Fast Response Time . . . 200 ns Typ for
  TTL-Level Input Step
- Built-In ESD Protection
- High Input Impedance . . . 10^{12} Ω Typ
- Extremely Low Input Bias Current
  1 pA Typ
- Ultra-Stable Low Input Offset Voltage
- Common-Mode Input Voltage Range
  Includes Ground
- Output Compatible with TTL, MOS, and
  CMOS

TLC374M . . . J DUAL-IN-LINE PACKAGE
TLC374C . . . D, J, OR N DUAL-IN-LINE PACKAGE
(TOP VIEW)

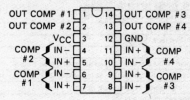

symbol (each comparator)

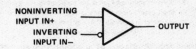

description

This device is fabricated using LinCMOS™ technology and consists of four independent voltage comparators designed to operate from a single power supply. Operation from dual supplies is also possible so long as the difference between the two supplies is 2 to 18 volts. Each of these devices features extremely high input impedance (typically greater than 10^{12} ohms) allowing direct interfacing with high-impedance sources. The outputs are n-channel open-drain configurations, and can be connected to achieve positive-logic wired-AND relationships. The TLC374C is designed as a pin-compatible, functional replacement for the LM339, offering twice the speed while consuming typically one-half of the power.

These devices have internal electrostatic discharge (ESD) protection circuits that will prevent catastrophic failures at voltages up to 2000 volts as tested under MIL-STD-883B, Method 3015.1. However, care should be exercised in handling these devices as exposure to ESD may result in a degradation of the device parametric performance.

The TLC374M is characterized for operation over the full military temperature range of −55°C to 125°C. The TLC374C is characterized for operation from 0°C to 70°C.

4

Voltage Comparators

Copyright © 1983 by Texas Instruments Incorporated

**TEXAS
INSTRUMENTS**

POST OFFICE BOX 225012 • DALLAS, TEXAS 75265

TYPES TLC374M, TLC374C
QUADRUPLE LinCMOS™ DIFFERENTIAL COMPARATORS

absolute maximum ratings over operating free-air temperature range (unless otherwise noted)

		TLC374M	TLC374C	UNIT
Supply voltage (see Note 1)		18	18	V
Differential input voltage (see Note 2)		±18	±18	V
Input voltage, V_I		18	18	V
Output voltage, V_O		18	18	V
Output current, I_O		20	20	mA
Duration of output short-circuit to ground (see Note 3)		unlimited	unlimited	
Continuous total dissipation at (or below) 25 °C free-air temperature (see Note 4)		500	500	mW
Operating free-air temperature range		−55 to 125	0 to 70	°C
Storage temperature range		−65 to 150	−65 to 150	°C
Lead temperature 1,6 mm (1/16 inch) from case for 60 seconds	JG package	300	300	°C
Lead temperature 1,6 mm (1/16 inch) from case for 10 seconds	D or P package		260	°C

NOTES: 1. All voltage values, except differential voltages, are with respect to network ground terminal.
2. Differential voltages are at the noninverting input terminal with respect to the inverting input terminal.
3. Short circuits from outputs to V_{CC} can cause excessive heating and eventual destruction.
4. For operation above 25 °C free-air temperature, refer to Dissipation Derating Curves, Section 2. In the J package, TLC374C chips are glass mounted and TLC374M chips are alloy mounted.

electrical characteristics at specified free-air temperature, V_{CC} = 5 V (unless otherwise noted)

PARAMETER		TEST CONDITIONS[†]		TLC374M			TLC374C			UNIT
				MIN	TYP	MAX	MIN	TYP	MAX	
V_{IO}	Input offset voltage	$V_{IC} = V_{ICR}$ min, See Note 5	25 °C		2	10		2	10	mV
			Full range			12			12	
I_{IO}	Input offset current	See Note 5	25 °C		1			1		pA
			Full range			10			0.3	nA
I_{IB}	Input bias current		25 °C		1			1		pA
			Full range			20			0.6	nA
V_{ICR}	Common-mode input voltage range		25 °C	0 to $V_{CC}-1.5$			0 to $V_{CC}-1.5$			V
			Full range	0 to $V_{CC}-2$			0 to $V_{CC}-2$			
A_{VD}	Large-signal differential voltage amplification	V_{CC} = 15 V, $R_L \geq$ 15 kΩ to V_{CC}	25 °C		200			200		v/mV
I_{OH}	High-level output current	V_{ID} = 1 V	V_{OH} = 5 V	25 °C		0.1			0.1		nA
			V_{OH} = 15 V	Full range			1			1	µA
V_{OL}	Low-level output voltage	V_{ID} = −1 V, I_{OL} = 4 mA	25 °C	150	400		150	400		mV
			Full range			700			700	
I_{OL}	Low-level output current	V_{ID} = −1 V, V_{OL} = 1.5 V	25 °C	6	16		6	16		mA
I_{CC}	Supply current (four comparators)	V_{ID} = −1 V, No load	25 °C		0.4	1		0.4	1	mA

[†]All characteristics are measured with zero common-mode input voltage unless otherwise specified.

NOTE 5: The offset voltages and offset currents given are the maximum values required to drive the output up to 4 V or down to 400 mV with a pull-up resistor of 2.5 kΩ to V_{CC}. Thus, these parameters actually define an error band and take into account the worst-case effects of voltage gain and input impedance. Full range for T_A is −55 °C to 125 °C for TLC374M, 0 °C to 70 °C for TLC374C.

switching characteristics, V_{CC} = 5 V, T_A = 25 °C

PARAMETER	TEST CONDITIONS		MIN	TYP	MAX	UNIT
Response time	R_L connected to 5 V through 5.1 kΩ, C_L = 15 pF[‡], See Note 6	100-mV input step with 5-mV overdrive		0.9		µs
		TTL-level input step		0.2		

[‡]C_L includes probe and jig capacitance.
NOTE 6: The response time specified is the interval between the input step function and the instant when the output crosses 1.4 V.

TEXAS
INSTRUMENTS
POST OFFICE BOX 225012 ● DALLAS, TEXAS 75265

- Fast Response Times
- Low Offset Characteristics
- Output Compatible with Most TTL Circuits
- Designed to be Interchangeable with Fairchild μA710

description

The uA710 is a monolithic high-speed comparator having differential inputs and a low-impedance output. Component matching, inherent in silicon integrated circuit fabrication techniques, produces a comparator with low-drift and low-offset characteristics. This circuit is especially useful for applications requiring an amplitude discriminator, memory sense amplifier, or a high-speed voltage comparator. The uA710M is characterized for operation over the full military temperature range of −55 °C to 125 °C.

schematic

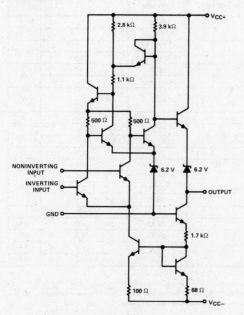

Component values shown are nominal.

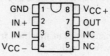

J DUAL-IN-LINE PACKAGE
(TOP VIEW)

NC	1	14	NC
GND	2	13	NC
IN+	3	12	NC
IN−	4	11	$V_{CC}+$
NC	5	10	NC
$V_{CC}−$	6	9	OUT
NC	7	8	NC

JG DUAL-IN-LINE PACKAGE
(TOP VIEW)

GND	1	8	$V_{CC}+$
IN+	2	7	OUT
IN−	3	6	NC
$V_{CC}−$	4	5	NC

U FLAT PACKAGE
(TOP VIEW)

GND	1	10	NC
IN+	2	9	NC
IN−	3	8	$V_{CC}+$
NC	4	7	NC
$V_{CC}−$	5	6	OUT

NC—No internal connection

symbol

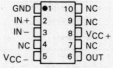

4

Voltage Comparators

TEXAS INSTRUMENTS
POST OFFICE BOX 225012 • DALLAS, TEXAS 75265

TYPE uA710M
DIFFERENTIAL COMPARATOR

absolute maximum ratings over operating free-air temperature range (unless otherwise noted)

Supply voltage V_{CC+} (see Note 1) .... 14 V
Supply voltage V_{CC-} (see Note 1) .... −7 V
Differential input voltage (see Note 2) .... ±5 V
Input voltage at either input (see Note 1) .... ±7 V
Peak output current ($t_W \leq 1$ s) .... 10 mA
Continuous total power dissipation at (or below) 25 °C free-air temperature (see Note 3) .. 300 mW
Operating free-air temperature range .... −55 °C to 125 °C
Storage temperature range .... −65 °C to 150 °C
Lead temperature 1,6 mm (1/16 inch) from case for 60 seconds .... 300 °C

NOTES: 1. All voltage values, except differential voltages, are with respect to the network ground terminal.
2. Differential voltages are at the noninverting input terminal with respect to the inverting input terminal.
3. For operation above 25 °C free-air temperature, refer to the Dissipation Derating Curves in Section 2. In the J and JG packages, uA710M chips are alloy mounted.

electrical characteristics at specified free-air temperature, V_{CC+} = 12 V, V_{CC-} = −6 V (unless otherwise noted)

	PARAMETER	TEST CONDITIONS†			MIN	TYP	MAX	UNIT
V_{IO}	Input offset voltage	$R_S \leqslant 200\ \Omega$,	See Note 4	25°C		0.6	2	mV
				Full range			3	
α_{VIO}	Average temperature coefficient of input offset voltage	$R_S \leqslant 50\ \Omega$,	See Note 4	Full range		3	10	µV/°C
I_{IO}	Input offset current	See Note 4		25°C		0.75	3	µA
				Full range			7	
α_{IIO}	Average temperature coefficient of input offset current	See Note 4		−55°C to 25°C		5	25	nA/°C
				25°C to 125°C		15	75	
I_{IB}	Input bias current	See Note 4		25°C		13	20	µA
				Full range			45	
V_{ICR}	Common-mode input voltage range	V_{CC-} = −7 V		25°C	±5			V
V_{ID}	Differential input voltage range			25°C	±5			V
A_{VD}	Large-signal differential voltage amplification	No load,	See Note 4	25°C	1250	1700		
				Full range	1000			
V_{OH}	High-level output voltage	V_{ID} = 5 mV,	I_{OH} = −5 mA	25°C	2.5	3.2	4	V
V_{OL}	Low-level output voltage	V_{ID} = −5 mV,	I_{OL} = 0	25°C	−1	−0.5	6‡	V
I_{OL}	Low-level output current	V_{ID} = −5 mV,	V_O = 0	25°C	2	2.5		mA
				−55°C	1	2.3		
				125°C	0.5	1.7		
r_o	Output resistance	V_O = 1.4 V		25°C		200		Ω
CMRR	Common-mode rejection ratio	$R_S \leqslant 200\ \Omega$		25°C	80	100		dB
I_{CC+}	Supply current from V_{CC+}	V_{ID} = −5 V to 5 V		25°C		5.2	9	mA
I_{CC-}	Supply current from V_{CC-}	(−10 mV for typ),		25°C		−4.6	−7	mA
P_D	Total power dissipation	No load		25°C		90	150	mW

NOTE 4: These characteristics are verified by measurements at the following temperatures and output voltage levels: V_O = 1.8 V at T_A = −55°C, V_O = 1.4 V at T_A = 25°C, and V_O = 1 V at T_A = 125°C. These output voltage levels were selected to approximate the logic threshold voltages of the types of digital logic circuits these comparators are intended to drive.

† Full range for uA710M is −55°C to 125°C.

‡ The algebraic convention where the more-positive (less-negative) limit is designated as maximum is used in this data sheet for logic levels only, e.g., when 0 V is the maximum, the minimum limit is a more-negative voltage.

4

Voltage Comparators

TEXAS INSTRUMENTS
POST OFFICE BOX 225012 • DALLAS, TEXAS 75265

switching characteristics, V_{CC+} = 12 V, V_{CC-} = −6 V, T_A = 25°C

PARAMETER	TEST CONDITIONS		TYP	UNIT
Response time	No load,	See Note 5	40	ns

NOTE 5: The response time specified is for a 100-mV input step with 5-mV overdrive and is the interval between the input step function and the instant when the output crosses 1.4 V.

TYPICAL CHARACTERISTICS

OUTPUT RESPONSE FOR VARIOUS
INPUT OVERDRIVES

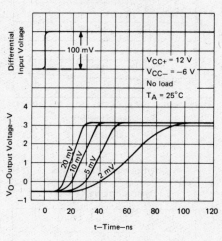

FIGURE 1

OUTPUT RESPONSE FOR VARIOUS
INPUT OVERDRIVES

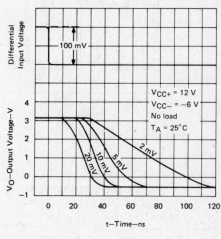

FIGURE 2

Voltage Comparators

4

TYPICAL CHARACTERISTICS

COMMON-MODE PULSE RESPONSE
vs
ELAPSED TIME

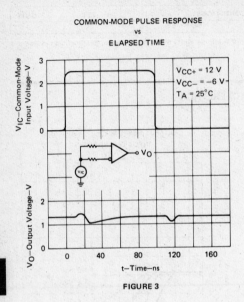

FIGURE 3

OUTPUT VOLTAGE
vs
FREE-AIR TEMPERATURE

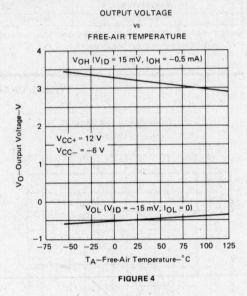

FIGURE 4

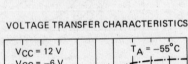

VOLTAGE TRANSFER CHARACTERISTICS

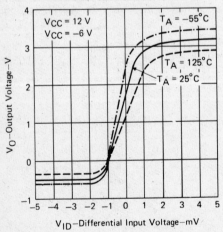

FIGURE 5

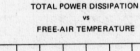

TOTAL POWER DISSIPATION
vs
FREE-AIR TEMPERATURE

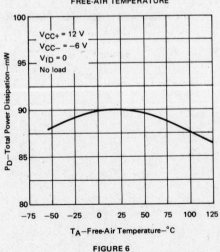

FIGURE 6

TEXAS
INSTRUMENTS
POST OFFICE BOX 225012 • DALLAS, TEXAS 75265

LINEAR
INTEGRATED
CIRCUITS

TYPES uA711M, uA711C
DUAL-CHANNEL DIFFERENTIAL COMPARATORS WITH STROBES

D977, FEBRUARY 1971—REVISED OCTOBER 1979

- Fast Response Times
- Low Offset Characteristics
- Output Compatible with Most TTL Circuits
- Designed to be Interchangeable with Fairchild μA711 and μA711C

description

The uA711 is a high-speed dual-channel comparator with differential inputs and a low-impedance output. Component matching, inherent with silicon monolithic circuit fabrication techniques, produces a comparator circuit with low-drift and low-offset characteristics. An independent strobe input is provided for each of the two channels, which when taken low, inhibits the associated channel. If both strobes are simultaneously low, the output will be low regardless of the conditions applied to the differential inputs. The comparator output pulse duration can be "stretched" by varying the capacitive loading. These dual comparators are particularly useful for applications requiring an amplitude-discriminating sense amplifier with an adjustable threshold voltage.

The uA711M is characterized for operation over the full military temperature range of −55°C to 125°C; the uA711C is characterized for operation from 0°C to 70°C.

J OR N DUAL-IN-LINE PACKAGE
(TOP VIEW)

NC	1	14	NC
#1 IN−	2	13	#1 STRB
#1 IN+	3	12	GND
$V_{CC}-$	4	11	$V_{CC}+$
#2 IN+	5	10	OUT
#2 IN−	6	9	#2 STB
NC	7	8	NC

U FLAT PACKAGE
(TOP VIEW)

#1 IN−	1	10	#1 STRB
#1 IN+	2	9	GND
$V_{CC}-$	3	8	$V_{CC}+$
#2 IN+	4	7	OUT
#2 IN−	5	6	#2 STB

NC — No internal connection

functional block diagram

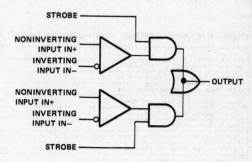

STROBE

NONINVERTING INPUT IN+
INVERTING INPUT IN−

NONINVERTING INPUT IN+
INVERTING INPUT IN−

STROBE

OUTPUT

Voltage Comparators

4

TEXAS
INSTRUMENTS

schematic

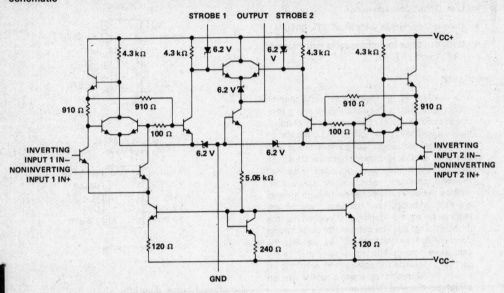

Component values shown are nominal.

absolute maximum ratings over operating free-air temperature range (unless otherwise noted)

	uA711M	uA711C	UNIT
Supply voltage V_{CC+} (see Note 1)	14	14	V
Supply voltage V_{CC-} (see Note 1)	−7	−7	V
Differential input voltage (see Note 2)	±5	±5	V
Input voltage (any input, see Note 1)	±7	±7	V
Strobe voltage (see Note 1)	6	6	V
Peak output current ($t_w \le 1$ s)	50	50	mA
Continuous total power dissipation at (or below) 70°C free-air temperature (see Note 3)	300	300	mW
Operating free-air temperature range	−55 to 125	0 to 70	°C
Storage temperature range	−65 to 150	−65 to 150	°C
Lead temperature 1,6 mm (1/16 inch) from case for 60 seconds J or U package	300	300	°C
Lead temperature 1,6 mm (1/16 inch) from case for 10 seconds N package		260	°C

NOTES: 1. All voltage values, except differential voltages, are with respect to the network ground terminal.
2. Differential voltages are at the noninverting input terminal with respect to the inverting input terminal.
3. For operation of uA711M above 70°C free-air temperature, refer to Dissipation Derating Curves, Section 2. In the J package, uA711M chips are alloy mounted, uA711C chips are glass mounted.

4

Voltage Comparators

Texas Instruments

POST OFFICE BOX 225012 • DALLAS, TEXAS 75265

electrical characteristics at specified free-air temperature, $V_{CC+} = 12$ V, $V_{CC-} = -6$ V (unless otherwise noted)

PARAMETER		TEST CONDITIONS†		uA711M MIN	uA711M TYP	uA711M MAX	uA711C MIN	uA711C TYP	uA711C MAX	UNIT
V_{IO}	Input offset voltage	$R_S \leqslant 200\ \Omega$, $V_{IC} = 0$, See Note 4	25°C		1	3.5		1	5	mV
			Full range			4.5			6	
		$R_S \leqslant 200\ \Omega$, See Note 4	25°C		1	5		1	7.5	
			Full range			6			10	
α_{VIO}	Average temperature coefficient of input offset voltage	$R_S \leqslant 200\ \Omega$, $V_{IC} = 0$, See Note 4	Full range		5			5		$\mu V/°C$
I_{IO}	Input offset current	See Note 4	25°C		0.5	10		0.5	15	μA
			Full range			20			25	
I_{IB}	Input bias current	See Note 4	25°C		25	75		25	100	μA
			Full range			150			150	
$I_{IL(S)}$	Low-level strobe current	$V_{(strobe)} = 0$, $V_{ID} = 10$ mV	25°C		-1.2	-2.5		-1.2	-2.5	mA
V_{ICR}	Common-mode input voltage range	$V_{CC-} = -7$ V	25°C	±5			±5			V
V_{ID}	Differential input voltage range		25°C	±5			±5			V
A_{VD}	Large-signal differential voltage amplification	No load, $V_O = 0$ to 2.5 V	25°C	750	1500		700	1500		
			Full range	500			500			
V_{OH}	High-level output voltage	$V_{ID} = 10$ mV, $I_{OH} = 0$	25°C		4.5	5		4.5	5	V
		$V_{ID} = 10$ mV, $I_{OH} = -5$ mA	25°C	2.5	3.5		2.5	3.5		
V_{OL}	Low-level output voltage	$V_{ID} = -10$ mV, $I_{OL} = 0$	25°C	-1	-0.5	0‡	-1	-0.5	0‡	V
		$V_{ID} = 10$ mV, $V_{(strobe)} = 0.3$ V, $I_{OL} = 0$	25°C	-1		0‡	-1		0‡	
I_{OL}	Low-level output current	$V_{ID} = -10$ mV, $V_O = 0$	25°C	0.5	0.8		0.5	0.8		mA
r_o	Output resistance	$V_O = 1.4$ V	25°C		200			200		Ω
CMRR	Common-mode rejection ratio	$R_S \leqslant 200\ \Omega$	25°C	70	90		65	90		dB
I_{CC+}	Supply current from V_{CC+}	$V_{ID} = -5$ V to 5 V (-10 mV for typ),	25°C		9			9		mA
I_{CC-}	Supply current from V_{CC-}	Strobes alternately grounded,	25°C		-4			-4		mA
P_D	Total power dissipation	No load	25°C		130	200		130	230	mW

† Unless otherwise noted, all characteristics are measured with the strobe of the channel under test open. The strobe of the other channel is grounded. Full range for uA711M is $-55°C$ to 125°C and for the uA711C is 0°C to 70°C.

‡ The algebraic convention, where the most-positive (least-negative) limit is designated as maximum, is used in this data sheet for logic levels only, e.g., when 0 V is the maximum, the minimum limit is a more-negative voltage.

NOTE 4: These characteristics are verified by measurements at the following temperatures and output voltage levels: for uA711M, $V_O = 1.8$ V at $T_A = -55°C$, $V_O = 1.4$ V at $T_A = 25°C$, and $V_O = 1$ V at $T_A = 125°C$; for uA711C, $V_O = 1.5$ V at $T_A = 0°C$, $V_O = 1.4$ V at $T_A = 25°C$, and $V_O = 1.2$ V at 70°C. These output voltage levels were selected to approximate the logic threshold voltages of the types of digital logic circuits these comparators are intended to drive.

switching characteristics, $V_{CC+} = 12$ V, $V_{CC-} = -6$ V, $T_A = 25°C$

PARAMETER	TEST CONDITIONS		uA711M MIN	uA711M TYP	uA711M MAX	uA711C MIN	uA711C TYP	uA711C MAX	UNIT
Response time	No load,	See Note 5		40	80		40		ns
Strobe release time	No load,	See Note 6		7	25		7		ns

NOTES: 5. The response time specified is for a 100-mV input step with 5-mV overdrive and is the interval between the input step function and the instant when the output crosses 1.4 V.

6. For testing purposes, the input bias conditions are selected to produce an output voltage of 1.4 V. A 5-mV overdrive is then added to the input bias voltage to produce an output voltage that rises above 1.4 V. The time interval is measured from the 50% point on the strobe voltage waveform to the instant when the overdriven output voltage crosses the 1.4-V level.

4

Voltage Comparators

TYPICAL CHARACTERISTICS

LARGE-SIGNAL DIFFERENTIAL VOLTAGE AMPLIFICATION vs FREE-AIR TEMPERATURE

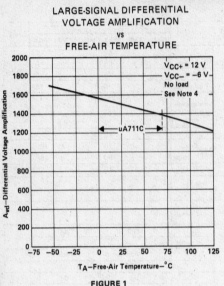

FIGURE 1

LARGE-SIGNAL DIFFERENTIAL VOLTAGE AMPLIFICATION vs SUPPLY VOLTAGE

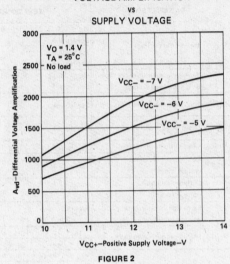

FIGURE 2

INPUT BIAS CURRENT vs FREE-AIR TEMPERATURE

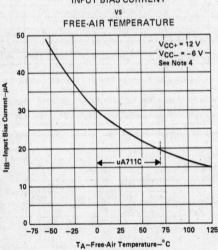

FIGURE 3

TOTAL POWER DISSIPATION vs FREE-AIR TEMPERATURE

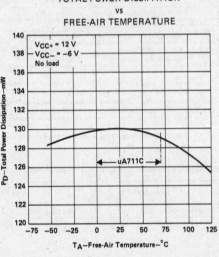

FIGURE 4

NOTE 4: These characteristics are verified by measurements at the following temperatures and output voltage levels: for uA711M, V_O = 1.8 V at T_A = −55°C, V_O = 1.4 V at T_A = 25°C, and V_O = 1 V at T_A = 125°C; for uA711C, V_O = 1.5 V at T_A = 0°C, V_O = 1.4 V at T_A = 25°C, and V_O = 1.2 V at 70°C. These output voltage levels were selected to approximate the logic threshold voltages of the types of digital logic circuits these comparators are intended to drive.

4

Voltage Comparators

TEXAS
INSTRUMENTS
POST OFFICE BOX 225012 • DALLAS, TEXAS 75265

TYPICAL CHARACTERISTICS

uA711M
VOLTAGE TRANSFER
CHARACTERISTIC

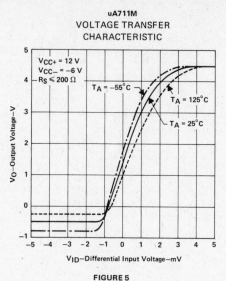

FIGURE 5

uA711C
VOLTAGE TRANSFER
CHARACTERISTICS

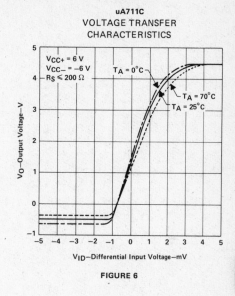

FIGURE 6

OUTPUT RESPONSE FOR
VARIOUS INPUT OVERDRIVES

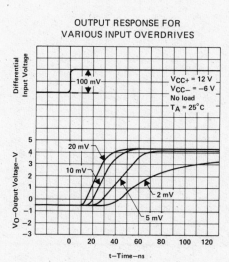

FIGURE 7

STROBE RELEASE TIME
FOR VARIOUS INPUT OVERDRIVES

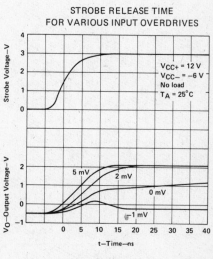

FIGURE 8

Voltage Comparators

4

TEXAS
INSTRUMENTS
POST OFFICE BOX 225012 • DALLAS, TEXAS 75265

General Information 1

Thermal Information 2

Operational Amplifiers 3

Voltage Comparators 4

Special Functions 5

Voltage Regulators 6

Data Acquisition 7

Appendix A

Special Functions

5

SPECIAL FUNCTIONS

precision timers

commercial temperature range

(values specified for T_A = 25 °C)

DEVICE NUMBER	DESCRIPTION	TIMING FROM	TIMING TO	OUTPUT CURRENT	PACKAGES	PAGE
NE555	Single Timer	1 μs	1 s	± 200 mA	D,JG,P	5-21
NE556	Dual Timer	1 μs	1 s	± 200 mA	D,J,N	5-31
TLC551C	LinCMOS, Single High-Speed Timer	1 μs	1 s	100 mA −10 mA	D,N	5-89
TLC552C	LinCMOS, Dual High-Speed Timer	1 μs	1 s	100 mA −10 mA	D,N	5-93
TLC555C	LinCMOS, Single High-Speed Timer	1 μs	1 s	100 mA −10 mA	D,JG,P	5-97
TLC556C	LinCMOS, Dual High-Speed Timer	1 μs	1 s	100 mA −10 mA	D,N	5-97
uA2240C	Programmable Timer/Counter	10 μs	Days	4 mA	N	5-109

automotive temperature range

(values specified for T_A = 25 °C)

DEVICE NUMBER	DESCRIPTION	TIMING FROM	TIMING TO	OUTPUT CURRENT	PACKAGES	PAGE
SA555	Single Timer	1 μs	1 s	± 200 mA	D,JG,P	5-21
SA556	Single Timer	1 μs	1 s	± 200 mA	D,J,N	5-31

military temperature range

(values specified for T_A = 25 °C)

DEVICE NUMBER	DESCRIPTION	TIMING FROM	TIMING TO	OUTPUT CURRENT	PACKAGES	PAGE
SE555	Single Timer	1 μs	1 s	± 200 mA	FH,FK,JG	5-21
SE555C	Single Timer	1 μs	1 s	± 200 mA	FH,FK,JG	5-21
SE556	Single Timer	1 μs	1 s	± 200 mA	FH,FK,J	5-31
SE556C	Single Timer	1 μs	1 s	± 200 mA	D,J,N	5-31
TLC555M	LinCMOS, Single High-Speed Timer	1 μs	1 s	100 mA −10 mA	JG	5-97
TLC556M	LinCMOS, Dual High-Speed Timer	1 μs	1 s	100 mA −10 mA	J	5-97

current mirrors

(values specified for T_A = 25 °C)

DEVICE NUMBER	TYPE	TEMP RANGE	CURRENT RATIO INPUT TO OUTPUT	INPUT CURRENT RANGE	PACKAGES	PAGE
TL010C	Programmable	0 °C to 70 °C	3:1 to 1:15	Variable	P	5-49
TL010I	Programmable	−40 °C to 85 °C	3:1 to 1:15	Variable	P	5-49
TL011C	Fixed	0 °C to 70 °C	1:1	1 μA to 1 mA	LP	5-53
TL011I	Fixed	−40 °C to 85 °C	1:1	1 μA to 1 mA	LP	5-53
TL012C	Fixed	0 °C to 70 °C	1:2	1 μA to 1 mA	LP	5-53
TL012I	Fixed	−40 °C to 85 °C	1:2	1 μA to 1 mA	LP	5-53
TL014C	Fixed	0 °C to 70 °C	1:4	1 μA to 1 mA	LP	5-53
TL014I	Fixed	−40 °C to 85 °C	1:4	1 μA to 1 mA	LP	5-53
TL021C	Fixed	0 °C to 70 °C	1:2	2 μA to 2 mA	LP	5-53
TL021I	Fixed	−40 °C to 85 °C	1:2	2 μA to 2 mA	LP	5-53

Special Functions

5

SELECTION GUIDE

Hall-effect sensor

(values specified for T_A = 25 °C)

DEVICE NUMBER	DESCRIPTION	TEMP RANGE	SENSITIVITY	LINEAR RANGE (GAUSS)	PACKAGES	PAGE
TL173C	Linear Hall-Effect Sensor	0 °C to 70 °C	1.5 mV/Gauss	±500	LP,LU	5-69
TL173I	Linear Hall-Effect Sensor	−20 °C to 85 °C	1.5 mV/Gauss	±500	LP,LU	5-69

Hall-effect switches

(values specified for T_A = 25 °C)

DEVICE NUMBER	DESCRIPTION	TEMP RANGE	SWITCHING RANGE (GAUSS)	MAXIMUM HYSTERESIS (GAUSS)	PACKAGES	PAGE
TL170C	General-Purpose	0 °C to 70 °C	+350 to −350	200	LP,LU	5-65
TL172C	Normally-Off	0 °C to 70 °C	+600 to +100	230	LP,LU	5-67
TL160	Special-Purpose	0 °C to 70 °C	Programmable	Programmable	LP,LU	5-63

sonar ranging functions

(values specified for T_A = 25 °C)

DEVICE NUMBER	DESCRIPTION	APPLICATION	PACKAGES	PAGE
TL851	Controller Circuit	Control integrated circuit for use in a sonar ranging module. Capable of driving 50-kHz transducers with a simple interface.	N	5-79
TL852	Receiver Circuit	Receiver integrated circuit for use in a sonar ranging module.	N	5-83
SN28827	Sonar Ranging Module	Sonar ranging module for measuring distances from a range of 6 inches to 35 feet. Uses the TL851 and TL852.		5-43

floppy-disk control circuits

(values specified for T_A = 25 °C)

DEVICE NUMBER	DESCRIPTION	PACKAGES	PAGE
MC3469	Write Controller	N	5-9
MC3470	Read-Amplifier System	N	5-11
MC3471	Write Controller and Head Driver	N	5-19
TL030	Four-Head Disk-Memory Read Amplifier	N	5-61
TL712	Disk-Memory Read-Chain Data	JG,P	4-63
TL721	Disk-Memory Read-Chain Data Comparator Compatible with MECL III and MECL 1000	JG,P	4-65

TEXAS
INSTRUMENTS
POST OFFICE BOX 225012 • DALLAS, TEXAS 75265

differential video amplifiers

commercial temperature range

(values specified for T_A = 25 °C)

DEVICE NUMBER	DESCRIPTION	BANDWIDTH	GAIN	PACKAGES	PAGE
MC1445	Amplifier with 2 multiplexed inputs, wide AGC range	60 MHz	100 Max	J,N	5-7
NE592	Amplifier with internal frequency compensation and adjustable/selectable gain options.	90 MHz	0 to 600	N	5-35
NE592A	Similar to NE592 but with tighter gain distribution.	90 MHz	0 to 600	N	5-35
TL026	Amplifier with a wide AGC range	60 MHz	100	JG,P	5-59
TL592	Similar to NE592 but in an 8-pin package.	90 MHz	0 to 600	P	5-73
TL592A	Similar to NE592A but in an 8-pin package.	90 MHz	0 to 600	P	5-73
TL592B	Low-noise version of NE592 and TL592.	90 MHz	0 to 600	N,P	5-77
uA733C	Amplifier with internal frequency compensation.	200 MHz	10, 100, 400	J,U	5-101

military temperature range

(values specified for T_A = 25 °C)

DEVICE NUMBER	DESCRIPTION	BANDWIDTH	GAIN	PACKAGES	PAGE
MC1545	Amplifier with 2 multiplexed inputs, wide AGC range	60 MHz	100 Max	J,N	5-7
SE592	Amplifier with internal frequency compensation and adjustable/selectable gain options.	90 MHz	0 to 600	N	5-35
uA733M	Amplifier with internal frequency compensation.	200 MHz	10, 100, 400	J,U	5-101

Special Functions

5

Special Functions

5

LINEAR
INTEGRATED
CIRCUITS

TYPES MC1545, MC1445
GATE-CONTROLLED 2-CHANNEL-INPUT VIDEO AMPLIFIER

D2572, JANUARY 1980—REVISED NOVEMBER 1983

- Differential Inputs and Outputs
- Channel Select Time . . . 20 ns Typ
- Bandwidth Typically 50 MHz
- 16-dB Minimum Gain
- Common-Mode Rejection Typically 85 dB
- Broadband Noise Typically 25 μV

MC1545 . . . J DUAL-IN-LINE OR
W FLAT PACKAGE
MC1445 . . . J OR N DUAL-IN-LINE PACKAGE
(TOP VIEW)

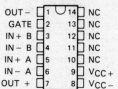

```
OUT −  [ 1  U  14 ]  NC
GATE   [ 2     13 ]  NC
IN + B [ 3     12 ]  NC
IN − B [ 4     11 ]  NC
IN + A [ 5     10 ]  NC
IN − A [ 6      9 ]  VCC +
OUT +  [ 7      8 ]  VCC −
```

NC—No internal connection

description

The MC1545 and MC1445 are general-purpose, gated, dual-channel wideband amplifiers designed for use in video-signal mixing and switching. Channel selection is accomplished by control of the voltage level at the gate. A high logic level selects channel A; a low logic level selects channel B. The unselected channel will have a gain of one or less.

The MC1545 is characterized for operation over the full military operating temperature range of −55°C to 125°C. The MC1445 is characterized for operation from 0°C to 70°C.

FUNCTION TABLE

GATE INPUT	SELECT
H	Channel A
L	Channel B

block diagram

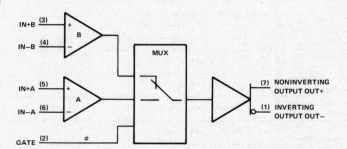

IN+B (3)
IN−B (4)
B

IN+A (5)
IN−A (6)
A

MUX

GATE (2) #

(7) NONINVERTING OUTPUT OUT+
(1) INVERTING OUTPUT OUT−

absolute maximum ratings over operating free-air temperature range (unless otherwise noted)

		MC1545	MC1445	UNIT
Supply voltage V_{CC+} (see Note 1)		+12	+12	V
Supply voltage V_{CC-} (see Note 1)		−12	−12	V
Differential input voltage (see Note 2)		±5	±5	V
Output current		±25	±25	mA
Continuous total dissipation at (or below) 25°C free-air temperature (see Note 3)		675	675	mW
Operating free-air temperature range		−55 to 125	0 to 75	°C
Storage temperature range		−65 to 150	−65 to 150	°C
Lead temperature 1,6 mm (1/16 inch) from case for 60 seconds	J or W package	300	300	°C
Lead temperature 1,6 mm (1/16 inch) from case for 10 seconds	N package	260	260	°C

NOTES: 1. Voltage values, except differential input voltage, are with respect to the midpoint of V_{CC+} and V_{CC-}.
2. Differential input voltages are measured at a noninverting input terminal with respect to the appropriate inverting input terminal.
3. For operation above 25°C free-air temperature, refer to the Dissipation Derating Curves, Section 2. In the J package, MC1545 chips are alloy mounted; MC1445 chips are glass mounted.

TEXAS
INSTRUMENTS
POST OFFICE BOX 225012 • DALLAS, TEXAS 75265

Special Functions

5

electrical characteristics at V_{CC+} = 5 V, V_{CC-} = -5 V, T_A = 25°C

PARAMETER		TEST CONDITIONS	MC1545			MC1445			UNIT
			MIN	TYP	MAX	MIN	TYP	MAX	
A_{VS}	Large-signal single-ended voltage amplification	f = 125 kHz, V_i = 20 mV	16	19	21	16	19.5	23	dB
BW	Bandwidth	V_i = 20 mV	40	50			50		MHz
V_{IO}	Input offset voltage			1	5			7.5	mV
I_{IO}	Input offset current			2			2		µA
I_{IB}	Input bias current			15	25		15	30	µA
V_{ICR}	Common-mode voltage range			±2.5			±2.5		V
V_{OQ}	Quiescent output voltage			0.1			0.1		V
ΔV_{OQ}	Change in quiescent output voltage	Gate input change from 5 V to 0 V		±15			±15		mV
V_{OPP}	Maximum peak-to-peak output voltage swing	f = 50 kHz, R_L = 1 kΩ	1.5	2.5		1.5	2.5		V
z_i	Input impedance	f = 50 kHz	4	10		3	10		kΩ
z_o	Output impedance	f = 50 kHz		25			25		Ω
CMRR	Common-mode rejection ratio	f = 50 kHz		85			85		dB
V_n	Broadband equivalent input noise voltage	BW = 5 Hz to 10 MHz, R_S = 50 Ω		25			25		µV
V_{TH}	High-level gate threshold voltage	$A_{VS(A)}$ ≥ 16 dB, $A_{VS(B)}$ ≤ 0 dB		1.5	2.2		1.3	3	V
V_{TL}	Low-level gate threshold voltage	$A_{VS(B)}$ ≥ 16 dB, $A_{VS(A)}$ ≤ 0 dB	0.4	0.7		0.2	0.4		V
I_{IH}	High-level gate current	V_I = 5 V			2			4	µA
I_{IL}	Low-level gate current	V_I = 0			2.5			4	mA
t_{PLH}	Propagation delay time, low-to-high-level output	ΔV_I = 20 mV, 50% to 50%		6.5	10		6.5		ns
t_{PHL}	Propagation delay time, high-to-low-level output	ΔV_I = 20 mV, 50% to 50%		6.3	10		6.3		ns
t_{TLH}	Transition time, low-to-high-level output	ΔV_I = 20 mV, 10% to 90%		6.5	15		6.5		ns
t_{THL}	Transition time, high-to-low-level output	ΔV_I = 20 mV, 10% to 90%		7	15		7		ns
I_{CC+}	Supply current from V_{CC+}	No load, No signal		7	11		7	15	mA
I_{CC-}	Supply current from V_{CC-}	No load, No signal		-7	-11		-7	-15	mA
P_D	Power dissipation	No load, No signal		70	110		70	150	mW

Special Functions

5

TEXAS
INSTRUMENTS
POST OFFICE BOX 225012 ● DALLAS, TEXAS 75265

- Designed for Straddle-Erase Heads
- Head Selection with Current Steering Through Write Head and Erase Coil in Write Mode
- Provides High-Impedance (Read Data Enable) During Read Mode
- Write Current (with Trimmed Internal Resistor and R_{ext} = 10 kΩ) . . . 3 A
- Write-Current Select Input Provides for Inner/Outer Track Compensation
- Degauss Period Externally Adjustable
- Specified with ± 10% Logic Supply and Head Supply (V_{BB}) from 10.8 V to 26.4 V
- Minimizes External Component Requirements
- Designed to be Interchangeable with Motorola MC3469P

N DUAL-IN-LINE PACKAGE
(TOP VIEW)

V_{ref}	1	16	CT1
I_{ref}	2	15	V_{BB}
GND	3	14	CT0
$\overline{WG}$	4	13	$\overline{ERASE0}$
$\overline{WD}$	5	12	COIL GND
R/W2	6	11	ERASE1
R/W1	7	10	V_{CC}
WR CUR SEL	8	9	HEADSEL

Special Functions

5

description

The MC3469 is a write-current controller designed to provide the entire interface for straddle-erase floppy-disk heads. The write current can be varied over a wide range by varying the value of an external resistor in series with a laser-trimmed internal resistor, and inner-track compensation is provided through the WR CUR SEL pin. A constant write current of 3 mA is provided by a 10-kΩ resistor between V_{ref} and I_{ref} when WR CUR SEL is low and 4 mA (an increase of 33%) when WR CUR SEL is high. Provisions are also made for adjusting the duration of the degaussing cycle that occurs at the end of each write operation.

The MC3469 will be characterized for operation from 0°C to 70°C.

functional block diagram

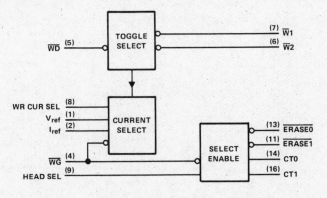

Copyright © 1983 by Texas Instruments Incorporated

TEXAS INSTRUMENTS

POST OFFICE BOX 225012 • DALLAS, TEXAS 75265

FUNCTION TABLE

FUNCTION	INPUTS		OUTPUTS			
	WG	HEADSEL	ERASE0	ERASE1	CT0	CT1
Write Head 0	L	H	L	H	H	L
Write Head 1	L	L	H	L	L	L
Write disable	H	X	H	H	NC	NC

L ≡ Low logic level, H ≡ High logic level
X ≡ Irrelevant, NC ≡ Maintains previous state

TABLE OF PIN FUNCTIONS

SIGNATURE	PIN NUMBER	DESCRIPTION
HEAD SEL	9	Selects Head 0 or Head 1
$\overline{WG}$	4	Active-low write enable
$\overline{WD}$	5	Serial data input
WR CUR SEL	8	Selects inner/outer track write current value
V_{ref}	1	Write-current external program resistor terminals
I_{ref}	2	
CT0	14	Totem-pole output to Head 0 center tap
$\overline{ERASE0}$	13	Open-collector erase output to Head 0
CT1	16	Totem-pole output to Head 1 center tap
$\overline{ERASE1}$	11	Open-collector erase output to Head 1
$\overline{W2}$	6	Constant-current push-pull write outputs
$\overline{W1}$	7	
V_{CC}	10	Positive power supply
GND	12, 3	Power supply return

TEXAS
INSTRUMENTS
POST OFFICE BOX 225012 • DALLAS, TEXAS 75265

LINEAR
INTEGRATED
CIRCUITS

TYPE MC3470
FLOPPY DISK READ-AMPLIFIER SYSTEM

D2759, NOVEMBER 1983

- Combines All Read-Amplifier Active Circuitry into One Monolithic Circuit

- Guaranteed Maximum Peak Shift of 5%

- Designed to be Interchangeable with Motorola MC3470

N DUAL-IN-LINE PACKAGE
(TOP VIEW)

AMPLIFIER {	1		18	V_CC2
INPUTS	2		17	AMPLIFIER
OFFSET {	3		16	} OUTPUTS
DECOUPLING {	4		15	DIFFERENTIATOR
GND	5		14	} INPUTS
CX1	6		13	DIFFERENTIATOR
CX/RX1	7		12	} COMPONENTS
CX2	8		11	V_CC1
CX/RX2	9		10	DATA OUTPUT

description

The MC3470 is a monolithic read-amplifier system containing all the active circuitry necessary for obtaining digital information from floppy disk storage. It is designed to accept the ac differential signal from the magnetic head and produce a digital output pulse corresponding to each peak of the input signal. The gain stage amplifies the input waveform and applies it to an external filter network, enabling the active differentiator and time domain filter to produce the desired output.

The MC3470 is characterized for operation from 0°C to 70°C.

Special Functions

5

functional block diagram

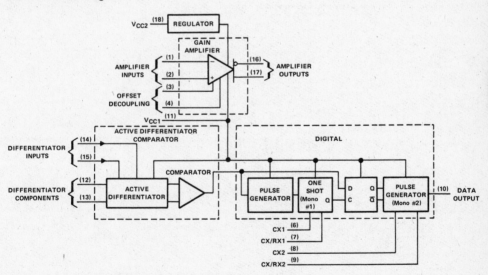

Copyright © 1983 by Texas Instruments Incorporated

TEXAS
INSTRUMENTS
POST OFFICE BOX 225012 ● DALLAS, TEXAS 75265

TYPE MC3470
FLOPPY DISK READ-AMPLIFIER SYSTEM

absolute maximum ratings over operating temperature range (unless otherwise noted)

Supply voltage, V_{CC1} (see Note 1) .... 7 V
Supply voltage, V_{CC2} .... 16 V
Input voltage range (amplifier inputs) .... −0.2 V to 7 V
Output voltage, V_O (data output) .... −0.2 V to 7 V
Operating free-air temperature range .... 0°C to 70°C
Storage temperature range .... −65°C to 150°C

NOTE 1: All voltage values are with respect to network ground terminal.

recommended operating conditions

		MIN	NOM	MAX	UNIT
Supply voltage V_{CC1}		4.75	5	5.25	V
Supply voltage V_{CC2}		10	12	14	V
Timing capacitor CX1 (see Note 2)		150		680	pF
Timing capacitor CX2		100		800	pF
Timing resistors RX1 and RX2		1.5		10	kΩ
Timing of digital section	Monostable no. 1	500		4000	ns
	Monostable no. 2	150		1000	
Operating free-air temperature, T_A		0		70	°C

NOTE 2: To minimize current transients, CX1 should be kept as small as convenient.

TEXAS
INSTRUMENTS
POST OFFICE BOX 225012 • DALLAS, TEXAS 75265

electrical characteristics over recommended ranges of supply voltages and operating free-air temperature (unless otherwise noted)

gain amplifier section

	PARAMETER	TEST CONDITIONS		MIN	TYP†	MAX	UNIT
A_{VD}	Differential voltage amplification	V_{id} = 5 mV rms,	f = 200 kHz	80	100	120	V/V
I_{IB}	Input bias current				−10	−25	μA
V_{ICR}	Common-mode input voltage range	THD ≤ 5%		−0.1 to 1.5			V
V_{IDR}	Differential input voltage range	THD ≤ 5%		±25			mV
V_{OPP}	Peak-to-peak differential output voltage				3	4	V
V_{OC}	Common-mode output voltage	V_I = 0 V,	V_{ID} = 0 V		3		V
V_{OD}	Differential output offset voltage	V_I = 0 V, T_A = 25°C	V_{ID} = 0 V,			0.4	V
I_O	Output current (each amplifier output)	To ground			−8		mA
		From V_{CC1}		2.8	4		
r_i	Input resistance	T_A = 25°C		100	250		kΩ
r_o	Output resistance (single-ended)	V_{CC1} = 5 V, T_A = 25°C	V_{CC2} = 12 V,		15		Ω
BW	Bandwidth (3 dB)	V_{id} = 2 mV rms, V_{CC2} = 12 V,	V_{CC1} = 5 V, T_A = 25°C	5			MHz
CMRR	Common-mode rejection ratio	V_{CC1} = 5 V, f = 100 kHz, T_A = 25°C	V_{IPP} = 200 mV, A_{VD} = 40 dB,	50			dB
k_{SVR}	Supply voltage rejection ratio	A_{VD} = 40 dB, T_A = 25°C	V_{CC1} = 5 ± 0.25 V, V_{CC2} = 12 V	50			dB
			V_{CC1} = 5 V, V_{CC2} = 12 ± 2 V	60			
V_n	Equivalent input noise voltage	BW = 10 Hz to 1 MHz, T_A = 25°C			15		μV

active-differentiator section

	PARAMETER	TEST CONDITIONS		MIN	TYP†	MAX	UNIT
I_{sink}	Sink current at pins 12 and 13	V_{OD} = V_{CC1}		1	1.4		mA
	Peak shift	V_{CC1} = 5 V, V_{IDPP} = 1 V, I_{cap} = 500 μA,	V_{CC2} = 12 V, f = 250 kHz, See Figure 1			5%	
r_{id}	Differential input resistance				30		kΩ
r_{od}	Differential output resistance				40		Ω

†All typical values are at V_{CC1} = 5 V, V_{CC2} = 12 V, T_A = 25°C.

Special Functions

5

TEXAS
INSTRUMENTS
POST OFFICE BOX 225012 ● DALLAS, TEXAS 75265

digital section

	PARAMETER	TEST CONDITIONS	MIN	TYP†	MAX	UNIT
V_{OH}	High-level output voltage (pin 10)	V_{CC1} = 4.75 V, $\quad$ V_{CC2} = 12 V, I_{OH} = −0.4 mA	2.7			V
V_{OL}	Low-level output voltage (pin 10)	V_{CC1} = 4.75 V, $\quad$ V_{CC2} = 12 V, I_{OL} = 8 mA			0.5	V
I_{CC1}	Supply current from V_{CC1}	V_{CC1} = 5.25 V		35	50	mA
I_{CC2}	Supply current from V_{CC2}	V_{CC2} = 14 V		4.5	10	mA

timing characteristics over recommended ranges of supply voltages and operating free-air temperature (unless otherwise noted) (see Figure 2)

	PARAMETER	TEST CONDITIONS	MIN	TYP†	MAX	UNIT
t_r	Rise time (pin 10)				20	ns
t_f	Fall time (pin 10)				25	ns
	Timing accuracy of monostable no. 1 compared to 0.625 RX1 · CX1 + 200 ns	RX1 = 1.5 kΩ to 10 kΩ, CX1 = 150 pF to 680 pF	85%		115%	
	Timing accuracy of monostable no. 2 compared to 0.625 RX2 · CX2	RX2 = 1.5 kΩ to 10 kΩ, CX2 = 100 pF to 800 pF	85%		115%	

†All typical values are at V_{CC1} = 5 V, V_{CC2} = 12 V, T_A = 25°C.

Special Functions

5

TEXAS
INSTRUMENTS
POST OFFICE BOX 225012 • DALLAS, TEXAS 75265

PARAMETER MEASUREMENT INFORMATION

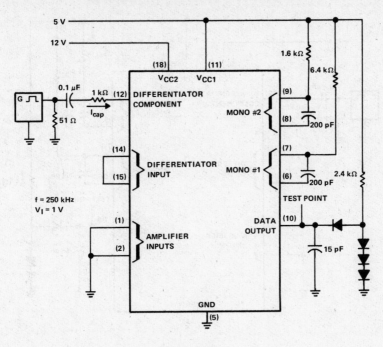

TEST CIRCUIT

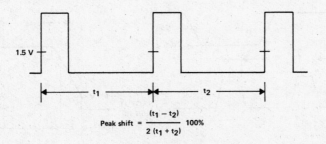

$$\text{Peak shift} = \frac{(t_1 - t_2)}{2 \, (t_1 + t_2)} \, 100\%$$

VOLTAGE WAVEFORMS

FIGURE 1—PEAK SHIFT

Special Functions

5

TEXAS INSTRUMENTS

POST OFFICE BOX 225012 • DALLAS, TEXAS 75265

TYPE MC3470
FLOPPY DISK READ-AMPLIFIER SYSTEM

PARAMETER MEASUREMENT INFORMATION

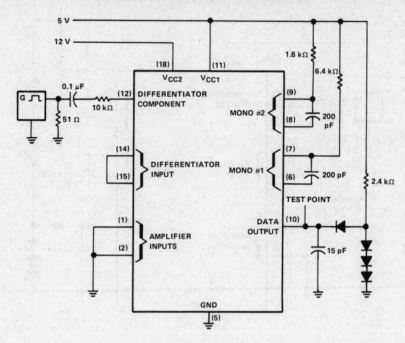

TEST CIRCUIT

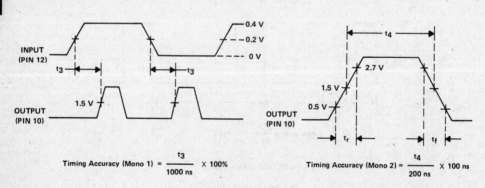

$$\text{Timing Accuracy (Mono 1)} = \frac{t_3}{1000 \text{ ns}} \times 100\%$$

$$\text{Timing Accuracy (Mono 2)} = \frac{t_4}{200 \text{ ns}} \times 100 \text{ ns}$$

VOLTAGE WAVEFORMS

FIGURE 2—TIMING ACCURACY

Special Functions

5

TEXAS
INSTRUMENTS
POST OFFICE BOX 225012 • DALLAS, TEXAS 75265

TYPICAL CHARACTERISTICS

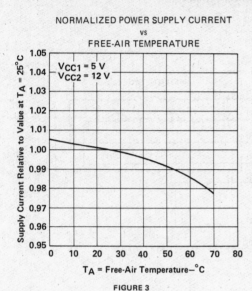

NORMALIZED POWER SUPPLY CURRENT
vs
FREE-AIR TEMPERATURE

V_{CC1} = 5 V
V_{CC2} = 12 V

Supply Current Relative to Value at T_A = 25°C

T_A = Free-Air Temperature—°C

FIGURE 3

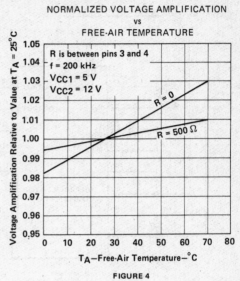

NORMALIZED VOLTAGE AMPLIFICATION
vs
FREE-AIR TEMPERATURE

R is between pins 3 and 4
f = 200 kHz
V_{CC1} = 5 V
V_{CC2} = 12 V

R = 0

R = 500 Ω

Voltage Amplification Relative to Value at T_A = 25°C

T_A—Free-Air Temperature—°C

FIGURE 4

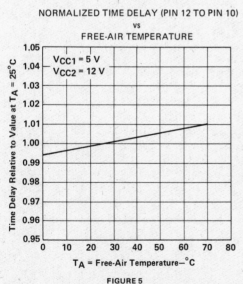

NORMALIZED TIME DELAY (PIN 12 TO PIN 10)
vs
FREE-AIR TEMPERATURE

V_{CC1} = 5 V
V_{CC2} = 12 V

Time Delay Relative to Value at T_A = 25°C

T_A = Free-Air Temperature—°C

FIGURE 5

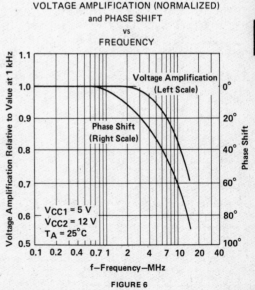

VOLTAGE AMPLIFICATION (NORMALIZED)
and PHASE SHIFT
vs
FREQUENCY

Voltage Amplification
(Left Scale)

Phase Shift
(Right Scale)

V_{CC1} = 5 V
V_{CC2} = 12 V
T_A = 25°C

Voltage Amplification Relative to Value at 1 kHz

Phase Shift

f—Frequency—MHz

FIGURE 6

Special Functions

5

TEXAS
INSTRUMENTS
POST OFFICE BOX 225012 • DALLAS, TEXAS 75265

TYPICAL APPLICATION INFORMATION

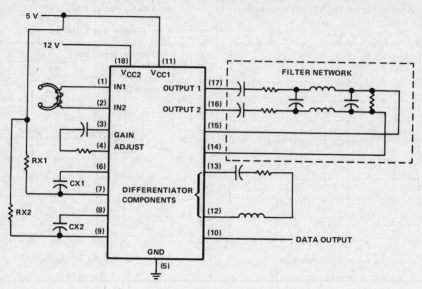

FIGURE 7

Special Functions

5

TEXAS
INSTRUMENTS
POST OFFICE BOX 225012 • DALLAS, TEXAS 75265

LINEAR
INTEGRATED
CIRCUITS

TYPE MC3471
FLOPPY DISK WRITE CONTROLLER/HEAD DRIVER

D2797, NOVEMBER 1983

- Provides Entire Interface Between Inputs and the Write and Erase Heads in Floppy Disk Systems

- Can be Used with Either Straddle-Erase or Tunnel-Erase Heads

- Head Selection, with Current Steering Through Write Head and Erase Coil in Write Mode

- Adjustable On-Chip Delay of Erase Timing

- Read-Write Current Select Input Provides Inner/Outer Track Compensation

- Minimizes Requirement for External Components

- Direct Replacement for Motorola MC3471

N DUAL-IN-LINE PACKAGE
(TOP VIEW)

V_{ref}	1	20	CENTER TAP 1
I_{ref}	2	19	V_{BB}
GND	3	18	CENTER TAP 0
WRITE GATE	4	17	ERASE 0
WRITE DATA	5	16	COIL GND
R/W2	6	15	ERASE 1
R/W1	7	14	V_{CC}
IRW SEL	8	13	HEAD SEL
NC	9	12	D1
INHIBIT	10	11	D2

NC—No internal connection

description

The MC3471 is a monolithic integrated write controller/head driver designed to provide the entire interface between the write data and head-control inputs and the heads (write and erase) for either tunnel-erase or straddle-erase floppy disk systems.

Provisions are made for selecting a range of accurately controlled write currents by varying the value of an external resistor connected between pins 1 and 2. Provisions for head selection during both read and write operations are also made. Degaussing the read/write head can be accomplished at the end of each write operation by a capacitor attached from pin 1 to ground; the degaussing period is controlled by the value of this capacitor. There are additional provisions for adjusting inner/outer track compensation, and the delay from write gate to erase turn-on and turn-off.

Erase delays are controlled by driving the delay inputs D1 and D2 with standard TTL open-collector logic (microprocessor compatible), or by using the external RC mode in which the delay is one time constant.

In addition, the INHIBIT output is provided to indicate when the heads are active during write, degauss, or erase.

The MC3471 will be characterized for operation from 0°C to 70°C.

Special Functions

5

TEXAS
INSTRUMENTS

POST OFFICE BOX 225012 • DALLAS, TEXAS 75265

LINEAR
INTEGRATED
CIRCUITS

TYPES SE555, SE555C, SA555, NE555
PRECISION TIMERS

D1669, SEPTEMBER 1973—REVISED OCTOBER 1983

- Timing from Microseconds to Hours
- Astable or Monostable Operation
- Adjustable Duty Cycle
- TTL-Compatible Output Can Sink or Source up to 200 mA
- Functionally Interchangeable with the Signetics SE555, SE555C, SA555, NE555; Have Same Pinout

NE555, SE555, SE555C . . . JG DUAL-IN-LINE PACKAGE
SA555, NE555 . . . D, JG, OR P DUAL-IN-LINE PACKAGE
(TOP VIEW)

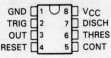

GND	1		8	V_{CC}
TRIG	2		7	DISCH
OUT	3		6	THRES
RESET	4		5	CONT

SE555, SE555C . . . FH OR FK CHIP CARRIER PACKAGE
(TOP VIEW)

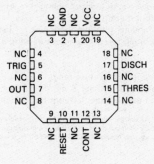

NC—No internal connection

description

These devices are monolithic timing circuits capable of producing accurate time delays or oscillation. In the time-delay or monostable mode of operation, the timed interval is controlled by a single external resistor and capacitor network. In the astable mode of operation, the frequency and duty cycle may be independently controlled with two external resistors and a single external capacitor.

The threshold and trigger levels are normally two-thirds and one-third, respectively, of V_{CC}. These levels can be altered by use of the control voltage terminal. When the trigger input falls below the trigger level, the flip-flop is set and the output goes high. If the trigger input is above the trigger level and the threshold input is above the threshold level, the flip-flop is reset and the output is low. The reset input can override all other inputs and can be used to initiate a new timing cycle. When the reset input goes low, the flip-flop is reset and the output goes low. Whenever the output is low, a low-impedance path is provided between the discharge terminal and ground.

The output circuit is capable of sinking or sourcing current up to 200 milliamperes. Operation is specified for supplies of 5 to 15 volts. With a 5-volt supply, output levels are compatible with TTL inputs.

The SE555 and SE555C are characterized for operation over the full military range of −55°C to 125°C. The SA555 is characterized for operation from −40°C to 85°C, and the NE555 is characterized for operation from 0°C to 70°C.

functional block diagram

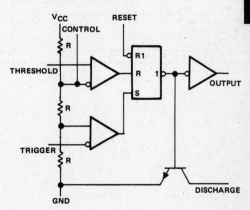

Reset can override Trigger, which can override Threshold.

Special Functions

5

TEXAS
INSTRUMENTS

POST OFFICE BOX 225012 • DALLAS, TEXAS 75265

FUNCTION TABLE

RESET	TRIGGER VOLTAGE†	THRESHOLD VOLTAGE†	OUTPUT	DISCHARGE SWITCH
Low	Irrelevant	Irrelevant	Low	On
High	< 1/3 V_{DD}	Irrelevant	High	Off
High	> 1/3 V_{DD}	> 2/3 V_{DD}	Low	On
High	> 1/3 V_{DD}	< 2/3 V_{DD}	As previously established	

†Voltage levels shown are nominal.

absolute maximum ratings over operating free-air temperature range (unless otherwise noted)

Supply voltage, V_{CC} (see Note 1) . . . . 18 V
Input voltage (control voltage, reset, threshold, trigger) . . . . V_{CC}
Output current . . . . ±225 mA
Continuous total dissipation at (or below) 25°C free-air temperature (see Note 2) . . . . 600 mW
Operating free-air temperature range: SE555, SE555C . . . . −55°C to 125°C
                                    SA555 . . . . −40°C to 85°C
                                    NE555 . . . . 0°C to 70°C
Storage temperature range . . . . −65°C to 150°C
Lead temperature 1,6 mm (1/16 inch) from case for 60 seconds: FH, FK, or JG package . . . 300°C
Lead temperature 1,6 mm (1/16 inch) from case for 10 seconds: D or P package . . . . 260°C

NOTES: 1. All voltage values are with respect to network ground terminal.
          2. For operation above 25°C free-air temperature, refer to Dissipation Derating Curves, Section 2. In the JG package, SE555 and SE555C chips are alloy mounted, SA555 and NE555 chips are glass mounted.

recommended operating conditions

	SE555		SE555C		SA555		NE555		UNIT
	MIN	MAX	MIN	MAX	MIN	MAX	MIN	MAX	
Supply voltage, V_{CC}	4.5	18	4.5	16	4.5	16	4.5	16	V
Input voltage (control voltage, reset, threshold, trigger)	V_{CC}		V_{CC}		V_{CC}		V_{CC}		V
Output current	±200		±200		±200		±200		mA
Operating free-air temperature, T_A	−55	125	−55	125	−40	85	0	70	°C

TEXAS
INSTRUMENTS
POST OFFICE BOX 225012 ● DALLAS, TEXAS 75265

Special Functions

5

electrical characteristics at 25 °C free-air temperature, V_{CC} = 5 V to 15 V (unless otherwise noted)

PARAMETER	TEST CONDITIONS		SE555			SE555C, SA555 NE555			UNIT
			MIN	TYP	MAX	MIN	TYP	MAX	
Threshold voltage level	V_{CC} = 15 V		9.4	10	10.6	8.8	10	11.2	V
	V_{CC} = 5 V		2.7	3.3	4	2.4	3.3	4.2	
Threshold current (see Note 3)				30	250		30	250	nA
Trigger voltage level	V_{CC} = 15 V		4.8	5	5.2	4.5	5	5.6	V
	V_{CC} = 5 V		1.45	1.67	1.9	1.1	1.67	2.2	
Trigger current	Trigger at 0 V			0.5	0.9		0.5	2	µA
Reset voltage level			0.4	0.7	1	0.4	0.7	1	V
Reset current	Reset at V_{CC}			0.1	0.4		0.1	0.4	mA
	Reset at 0 V			−0.4	−1		−0.4	−1	
Discharge switch off-state current				20	100		20	100	nA
Control voltage (open circuit)	V_{CC} = 15 V		9.6	10	10.4	9	10	11	V
	V_{CC} = 5 V		2.9	3.3	3.8	2.6	3.3	4	
Low-level output voltage	V_{CC} = 15 V	I_{OL} = 10 mA		0.1	0.15		0.1	0.25	V
		I_{OL} = 50 mA		0.4	0.5		0.4	0.75	
		I_{OL} = 100 mA		2	2.25		2	3.2	
		I_{OL} = 200 mA		2.5			2.5		
	V_{CC} = 5 V	I_{OL} = 5 mA		0.05	0.15		0.05	0.25	
		I_{OL} = 8 mA		0.1	0.2		0.25	0.3	
High-level output voltage	V_{CC} = 15 V	I_{OH} = −100 mA	13	13.3		12.75	13.3		V
		I_{OH} = −200 mA		12.5			12.5		
	V_{CC} = 5 V	I_{OH} = −100 mA	3	3.3		2.75	3.3		
Supply current	Output low, No load	V_{CC} = 15 V		10	12		10	15	mA
		V_{CC} = 5 V		3	5		3	6	
	Output high, No load	V_{CC} = 15 V		9	10		9	13	
		V_{CC} = 5 V		2	4		2	5	

NOTE 3: This parameter influences the maximum value of the timing resistors R_A and R_B in the circuit of Figure 13. For example, when V_{CC} = 5 V the maximum value is R = R_A + R_B ≈ 3.4 MΩ and for V_{CC} = 15 V the maximum value is 10 MΩ.

operating characteristics, V_{CC} = 5 V and 15 V

PARAMETER		TEST CONDITIONS[†]	SE555			SE555C, SA555 NE555			UNIT
			MIN	TYP	MAX	MIN	TYP	MAX	
Initial error of timing interval[‡]	Each timer, monostable[§]	T_A = 25 °C		0.5	1.5		1	3	%
	Each timer, astable[¶]			1.5			2.25		
Temperature coefficient of timing interval	Each timer, monostable[§]	T_A = MIN to MAX		30	100		50		ppm/°C
	Each timer, astable[¶]			90			150		
Supply voltage sensitivity of timing interval	Each timer, monostable[§]	T_A = 25 °C		0.05	0.2		0.1	0.5	%/V
	Each timer, astable[¶]			0.15			0.3		
Output pulse rise time		C_L = 15 pF, T_A = 25 °C		100	200		100	300	ns
Output pulse fall time				100	200		100	300	

[†]For conditions shown as MIN or MAX, use the appropriate value specified under recommended operating conditions.
[‡]Timing interval error is defined as the difference between the measured value and the nominal value computed by the formula: t_w = 1.1 R_AC.
[§]Values specified are for a device in a monostable circuit similar to Figure 10, with component values as follow: R_A = 2 kΩ to 100 kΩ, C = 0.1 µF.
[¶]Vallues specified are for a device in an astable circuit similar to Figure 1, with component values as follow: R_A = 1 kΩ to 100 kΩ, C = 0.1 µF.

Special Functions

5

TEXAS
INSTRUMENTS
POST OFFICE BOX 225012 • DALLAS, TEXAS 75265

Special Functions

5

TYPICAL CHARACTERISTICS†

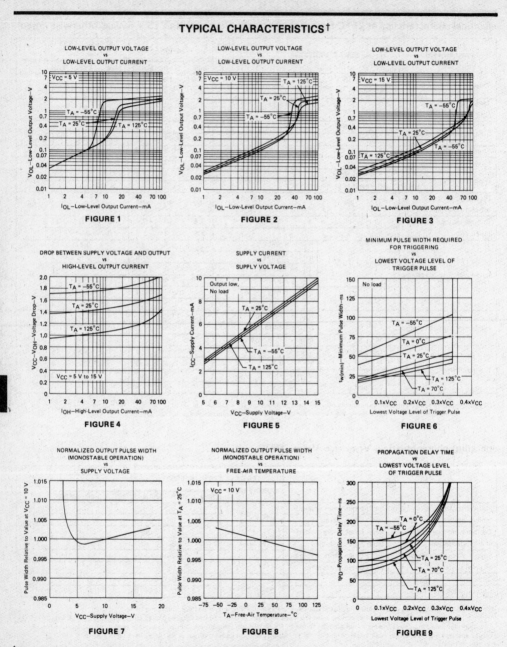

FIGURE 1

FIGURE 2

FIGURE 3

FIGURE 4

FIGURE 5

FIGURE 6

FIGURE 7

FIGURE 8

FIGURE 9

†Data for temperatures below 0°C and above 70°C are applicable for SE555 circuits only.

TEXAS
INSTRUMENTS
POST OFFICE BOX 225012 • DALLAS, TEXAS 75265

TYPICAL APPLICATION DATA

monostable operation

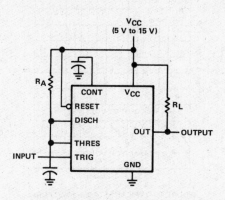

FIGURE 10—CIRCUIT FOR MONOSTABLE OPERATION

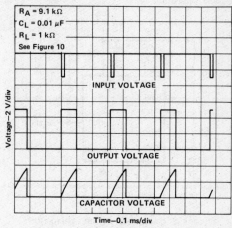

FIGURE 11—TYPICAL MONOSTABLE WAVEFORMS

For monostable operation, any of these timers may be connected as shown in Figure 10. If the output is low, application of a negative-going pulse to the trigger input sets the flip-flop ($\overline{Q}$ goes low), drives the output high, and turns off Q1. Capacitor C is then charged through R_A until the voltage across the capacitor reaches the threshold voltage of the threshold input. If the trigger input has returned to a high level, the output of the threshold comparator will reset the flip-flop ($\overline{Q}$ goes high), drive the output low, and discharge C through Q1.

Monostable operation is initiated when the trigger input voltage falls below the trigger threshold. Once initiated, the sequence will complete only if the trigger input is high at the end of the timing interval. Because of the threshold level and saturation voltage of Q1, the output pulse duration is approximately $t_W = 1.1\ R_A C$. Figure 12 is a plot of the time constant for various values of R_A and C. The threshold levels and charge rates are both directly proportional to the supply voltage, V_{CC}. The timing interval is therefore independent of the supply voltage, so long as the supply voltage is constant during the time interval.

Applying a negative-going trigger pulse simultaneously to the reset and trigger terminals during the timing interval will discharge C and re-initiate the cycle, commencing on the positive edge of the reset pulse. The output is held low as long as the reset pulse is low. When the reset input is not used, it should be connected to V_{CC} to prevent false triggering.

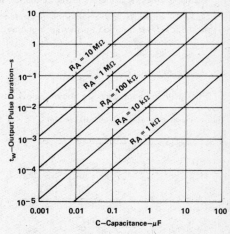

FIGURE 12—OUTPUT PULSE DURATION vs CAPACITANCE

Special Functions

5

TYPICAL APPLICATION DATA

astable operation

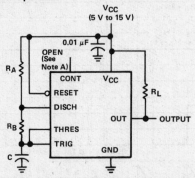

NOTE A: Decoupling the control voltage input to ground with a capacitor may improve operation. This should be evaluated for individual applications.

FIGURE 13—CIRCUIT FOR ASTABLE OPERATION

FIGURE 14—TYPICAL ASTABLE WAVEFORMS

Addition of a second resistor, R_B, to the circuit of Figure 10, as shown in Figure 13, and connection of the trigger input to the threshold input will cause the timer to self-trigger and run as a multivibrator. The capacitor C will charge through R_A and R_B then discharge through R_B only. The duty cycle may be controlled, therefore, by the values of R_A and R_B.

This astable connection results in capacitor C charging and discharging between the threshold-voltage level ($\approx 0.67 \cdot V_{CC}$) and the trigger-voltage level ($\approx 0.33 \cdot V_{CC}$). As in the monostable circuit, charge and discharge times (and therefore the frequency and duty cycle) are independent of the supply voltage.

Figure 14 shows typical waveforms generated during astable operation. The output high-level duration t_H and low-level duration t_L may be found by:

$$t_H = 0.693 \, (R_A + R_B) \, C$$

$$t_L = 0.693 \, (R_B) \, C$$

Other useful relationships are shown below.

$$\text{period} = t_H + t_L = 0.693 \, (R_A + 2R_B) \, C$$

$$\text{frequency} \approx \frac{1.44}{(R_A + 2R_B) \, C}$$

$$\text{Output driver duty cycle} = \frac{t_L}{t_H + t_L} = \frac{R_B}{R_A + 2R_B}$$

$$\text{Output waveform duty cycle} = \frac{t_H}{t_H + t_L} = 1 - \frac{R_B}{R_A + 2R_B}$$

$$\text{Low-to-high ratio} = \frac{t_L}{t_H} = \frac{R_B}{R_A + R_B}$$

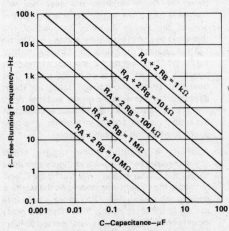

FIGURE 15—FREE-RUNNING FREQUENCY

TEXAS
INSTRUMENTS
POST OFFICE BOX 225012 • DALLAS, TEXAS 75265

TYPICAL APPLICATION DATA

missing-pulse detector

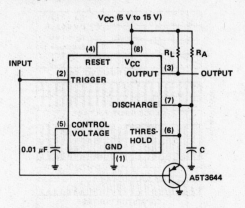

FIGURE 16—CIRCUIT FOR MISSING-PULSE DETECTOR

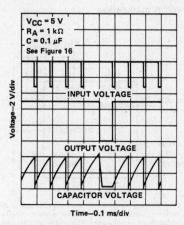

FIGURE 17—MISSING-PULSE DETECTOR WAVEFORMS

The circuit shown in Figure 16 may be utilized to detect a missing pulse or abnormally long spacing between consecutive pulses in a train of pulses. The timing interval of the monostable circuit is continuously retriggered by the input pulse train as long as the pulse spacing is less than the timing interval. A longer pulse spacing, missing pulse, or terminated pulse train will permit the timing interval to be completed, thereby generating an output pulse as illustrated in Figure 17.

frequency divider

By adjusting the length of the timing cycle, the basic circuit of Figure 10 can be made to operate as a frequency divider. Figure 18 illustrates a divide-by-3 circuit that makes use of the fact that retriggering cannot occur during the timing cycle.

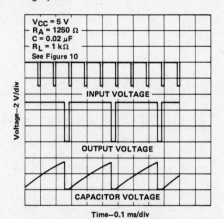

FIGURE 18—DIVIDE-BY-THREE CIRCUIT WAVEFORMS

Special Functions

5

TEXAS
INSTRUMENTS
POST OFFICE BOX 225012 • DALLAS, TEXAS 75265

TYPICAL APPLICATION DATA

pulse-width modulation

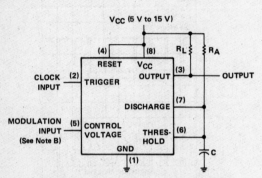

NOTE B: The modulating signal may be direct or capacitively coupled to the control voltage terminal. For direct coupling, the effects of modulation source voltage and impedance on the bias of the timer should be considered.

FIGURE 19—CIRCUIT FOR PULSE-WIDTH MODULATION

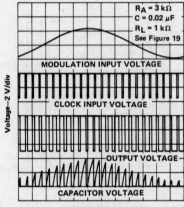

FIGURE 20—PULSE-WIDTH MODULATION WAVEFORMS

The operation of the timer may be modified by modulating the internal threshold and trigger voltages. This is accomplished by applying an external voltage (or current) to the control voltage pin. Figure 19 is a circuit for pulse-width modulation. The monostable circuit is triggered by a continuous input pulse train and the threshold voltage is modulated by a control signal. The resultant effect is a modulation of the output pulse width, as shown in Figure 20. A sine-wave modulation signal is illustrated, but any wave-shape could be used.

TEXAS
INSTRUMENTS

POST OFFICE BOX 225012 • DALLAS, TEXAS 75265

TYPICAL APPLICATION DATA

pulse-position modulation

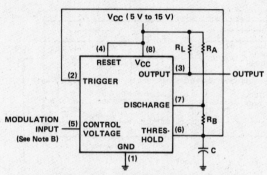

NOTE B: The modulating signal may be direct or capacitively coupled to the control voltage terminal. For direct coupling, the effects of modulation source voltage and impedance on the bias of the timer should be considered.

FIGURE 21—CIRCUIT FOR PULSE-POSITION MODULATION

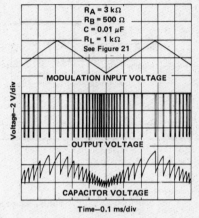

FIGURE 22—PULSE POSITION-MODULATION WAVEFORMS

Any of these timers may be used as a pulse-position modulator as shown in Figure 21. In this application, the threshold voltage, and thereby the time delay, of a free-running oscillator is modulated. Figure 22 shows such a circuit, with a triangular-wave modulation signal, however, any modulating wave-shape could be used.

Special Functions

5

TYPICAL APPLICATION DATA

sequential timer

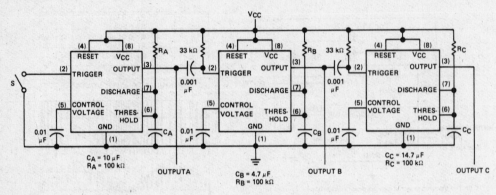

S closes momentarily at t = 0.

FIGURE 23—SEQUENTIAL TIMER CIRCUIT

Many applications, such as computers, require signals for initializing conditions during start-up. Other applications such as test equipment require activation of test signals in sequence. These timing circuits may be connected to provide such sequential control. The timers may be used in various combinations of astable or monostable circuit connections, with or without modulation, for extremely flexible waveform control. Figure 23 illustrates a sequencer circuit with possible applications in many systems and Figure 24 shows the output waveforms.

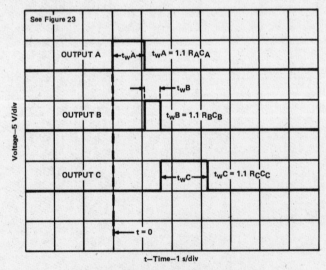

FIGURE 24—SEQUENTIAL TIMER WAVEFORMS

TEXAS
INSTRUMENTS
POST OFFICE BOX 225012 • DALLAS, TEXAS 75265

Special Functions

5

**LINEAR
INTEGRATED
CIRCUITS**

**TYPES SE556, SE556C, SA556, NE556
DUAL PRECISION TIMERS**

D2440, APRIL 1978—REVISED OCTOBER 1983

- Two Precision Timing Circuits per Package
- Astable or Monstable Operation
- TTL-Compatible Output can Sink or Source up to 150 mA
- Active Pull-Up or Pull-Down
- Designed to be Interchangeable with Signetics SE556, SE556C, SA556, NE556

APPLICATIONS

Precision Timer from
   Microseconds to Hours
Sequential Timer
Pulse-Shaping Circuit
Pulse Generator
Missing-Pulse Detector
Tone-Burst Generator
Pulse-Width Modulator
Time-Delay Circuit
Frequency Divider
Pulse-Position Modulator
Appliance Timer
Touch-Tone Encoder
Industrial Controls

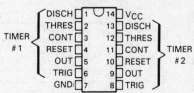

SE556 . . . J
SE556C, SA556, NE556 . . . D, J, OR N
DUAL-IN-LINE PACKAGE
(TOP VIEW)

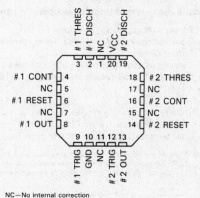

SE556 . . . FH OR FK PACKAGE
(TOP VIEW)

NC—No internal correction

Special Functions

5

description

These devices provide two monolithic, independent timing circuits of the SE555, SE555C, SA555, or NE555 type in each package. These circuits can be operated in the astable or the monostable mode with external resistor-capacitor timing control. The basic timing provided by the RC time constant may be actively controlled by modulating the bias of the control voltage input.

The threshold and trigger levels are normally two-thirds and one-third respectively of V_{CC}. These levels can be altered by use of the control voltage terminal. When the trigger input falls below trigger level, the flip-flop is set and the output goes high. If the trigger input is above the trigger level and the threshold input is above the threshold level, the flip-flop is reset and the output is low. The reset input can override all other inputs and can be used to initiate a new timing cycle. When the reset input goes low, the flip-flop is reset and the output goes low. Whenever the output is low, a low-impedance path is provided between the discharge terminal and ground.

functional block diagram (each timer)

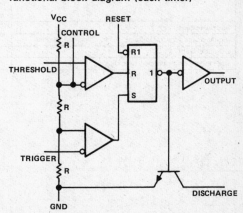

Reset can override Trigger, which can override Threshold.

**TEXAS
INSTRUMENTS**
POST OFFICE BOX 225012 • DALLAS, TEXAS 75265

The SE556 and SE556C are characterized for operation over the full military range of $-55\,°C$ to $125\,°C$. The SA556 is characterized for operation from $-40\,°C$ to $85\,°C$, and the NE556 is characterized for operation from $0\,°C$ to $70\,°C$.

FUNCTION TABLE

RESET	TRIGGER VOLTAGE†	THRESHOLD VOLTAGE†	OUTPUT	DISCHARGE SWITCH
Low	Irrelevant	Irrelevant	Low	On
High	$< 1/3\ V_{DD}$	Irrelevant	High	Off
High	$> 1/3\ V_{DD}$	$> 2/3\ V_{DD}$	Low	On
High	$> 1/3\ V_{DD}$	$< 2/3\ V_{DD}$	As previously established	

†Voltage levels shown are nominal.

absolute maximum ratings over operating free-air temperature range (unless otherwise noted)

Supply voltage, V_{CC} (see Note 1) . . . . 18 V
Input voltage (control voltage, reset, threshold, trigger) . . . . V_{CC}
Output current . . . . ±225 mA
Continuous total dissipation at (or below) 25 °C free-air temperature (see Note 2) . . . . 600 mW
Operating free-air temperature range: SE556, SE556C . . . . $-55\,°C$ to $125\,°C$
 SA556 . . . . $-40\,°C$ to $85\,°C$
 NE556 . . . . $0\,°C$ to $70\,°C$
Storage temperature range . . . . $-65\,°C$ to $150\,°C$
Lead temperature 1,6 mm (1/16 inch) from case for 60 seconds: FH, FK, or J package . . . . 300 °C
Lead temperature 1,6 mm (1/16 inch) from case for 10 seconds: D or N package . . . . 260 °C

NOTES: 1. All voltage values are with respect to network ground terminal.
2. For operation above 25 °C free-air temperature, refer to Dissipation Derating Curves, Section 2. In the J package, SE556 and SE556C chips are alloy mounted, SA556 and NE556 chips are glass mounted.

recommended operating conditions

	SE556 MIN	SE556 MAX	SE556C MIN	SE556C MAX	SA556 MIN	SA556 MAX	NE556 MIN	NE556 MAX	UNIT
Supply voltage, V_{CC}	4.5	18	4.5	16	4.5	16	4.5	16	V
Input voltage (control voltage, reset, threshold, trigger)		V_{CC}		V_{CC}		V_{CC}		V_{CC}	V
Output current		±200		±200		±200		±200	mA
Operating free-air temperature, T_A	-55	125	-55	125	-40	85	0	70	°C

TEXAS
INSTRUMENTS
POST OFFICE BOX 225012 • DALLAS, TEXAS 75265

electrical characteristics at 25 °C free-air temperature, V_{CC} = 5 V to 15 V (unless otherwise noted)

PARAMETER	TEST CONDITIONS	SE556			SE556C, SA556 NE556			UNIT
		MIN	TYP	MAX	MIN	TYP	MAX	
Threshold voltage level	V_{CC} = 15 V	9.4	10	10.6	8.8	10	11.2	V
	V_{CC} = 5 V	2.7	3.3	4	2.4	3.3	4.2	
Threshold current (see Note 1)			30	250		30	250	nA
Trigger voltage level	V_{CC} = 15 V	4.8	5	5.2	4.5	5	5.6	V
	V_{CC} = 5 V	1.45	1.67	1.9	1.1	1.67	2.2	
Trigger current	Trigger at 0 V		0.5	0.9		0.5	2	μA
Reset voltage level		0.4	0.7	1	0.4	0.7	1	V
Reset current	Reset at V_{CC}		0.1	0.4		0.1	0.4	mA
	Reset at 0 V		−0.4	−1		−0.4	−1	
Discharge switch off-state current			20	100		20	100	nA
Control voltage (open-circuit)	V_{CC} = 15 V	9.6	10	10.4	9	10	11	V
	V_{CC} = 5 V	2.9	3.3	3.8	2.6	3.3	4	
Low-level output voltage	V_{CC} = 15 V	I_{OL} = 10 mA		0.1	0.15		0.1	0.25	V
		I_{OL} = 50 mA		0.4	0.5		0.4	0.75	
		I_{OL} = 100 mA		2	2.25		2	3.2	
		I_{OL} = 200 mA		2.5			2.5		
	V_{CC} = 5 V	I_{OL} = 5 mA		0.05	0.15		0.05	0.25	
		I_{OL} = 8 mA		0.1	0.2		0.25	0.3	
High-level output voltage	V_{CC} = 15 V	I_{OH} = −100 mA	13	13.3		12.75	13.3		V
		I_{OH} = −200 mA		12.5			12.5		
	V_{CC} = 5 V	I_{OH} = −100 mA	3	3.3		2.75	3.3		
Supply current	Output low, No load	V_{CC} = 15 V		20	24		20	30	mA
		V_{CC} = 5 V		6	10		6	12	
	Output high, No load	V_{CC} = 15 V		18	20		18	26	
		V_{CC} = 5 V		4	8		4	10	

NOTE 1: This parameter influences the maximum value of the timing resistors R_A and R_B in the circuit of Figure 1. For example, when V_{CC} = 5 V the maximum value is R = R_A + R_B = 3.4 MΩ and for V_{CC} = 15 V the maximum value for R_A + R_B = 10 MΩ.

operating characteristics, V_{CC} = 5 V and 15 V

PARAMETER		TEST CONDITIONS[†]	SE556			SE556C, SA556 NE556			UNIT
			MIN	TYP	MAX	MIN	TYP	MAX	
Initial error of timing interval[‡]	Each timer, monostable[§]	T_A = 25 °C		0.5	1.5		1	3	%
	Each timer, astable[¶]			1.5			2.25		
	Timer 1 − Timer 2			±0.5			±1		
Temperature coefficient of timing interval	Each timer, monostable[§]	T_A = MIN to MAX		30	100		50		ppm/°C
	Each timer, astable[¶]			90			150		
	Timer 1 − Timer 2			±10			±10		
Supply voltage sensitivity of timing interval	Each timer, monostab;e[§]	T_A = 25 °C		0.05	0.2		0.1	0.5	%/V
	Each timer, astable[¶]			0.15			0.3		
	Timer 1− Timer 2			±0.1			±0.2		
Output pulse rise time		C_L = 15 pF, T_A = 25 °C		100	200		100	300	ns
Output pulse fall time				100	200		100	300	ns

[†]For conditions shown as MIN or MAX, use the appropriate value specified under recommended operating conditions.
[‡]Timing interval error is defined as the difference between the measured value and the nominal value computed by the formula: t_w = 1.1 $R_A C$.
[§]Values specified are for a device in a monostable circuit similar to Figure 2, with component values as follow: R_A = 2 kΩ to 100 kΩ, C = 0.1 μF.
[¶]Values specified are for a device in an astable circuit similar to Figure 1, with component values as follow: R_A = 1 kΩ to 100 kΩ, C = 0.1 μF.

Special Functions

5

TEXAS
INSTRUMENTS
POST OFFICE BOX 225012 • DALLAS, TEXAS 75265

TYPICAL APPLICATION DATA

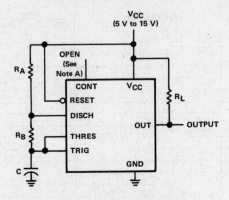

NOTE A: Bypassing the control voltage input to ground with a capacitor may improve operation. This should be evaluated for individual applications.

FIGURE 1—CIRCUIT FOR ASTABLE OPERATION

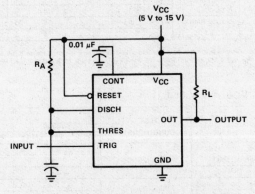

FIGURE 2—CIRCUIT FOR MONOSTABLE OPERATION

TEXAS
INSTRUMENTS
POST OFFICE BOX 225012 • DALLAS, TEXAS 75265

**LINEAR
INTEGRATED
CIRCUITS**

**TYPES SE592, NE592, NE592A
DIFFERENTIAL VIDEO AMPLIFIERS**

D2667, FEBRUARY 1984

- 90-MHz Bandwidth
- Adjustable Gain to 400
- No Frequency Compensation Required
- Adjustable Passband
- Designed to be Interchangeable with Signetics SE592 and NE592

J OR N DUAL-IN-LINE PACKAGE
(TOP VIEW)

IN +	1	14	IN −
NC	2	13	NC
GAIN ADJ 2A	3	12	GAIN ADJ 2B
GAIN ADJ 1A	4	11	GAIN ADJ 1B
$V_{CC}-$	5	10	$V_{CC}+$
NC	6	9	NC
OUT +	7	8	OUT −

NC—No internal connection

Special Functions

5

description

These devices are monolithic two-stage amplifiers with differential inputs and differential outputs.

Internal series-shunt feedback provides wide bandwidth, low phase distortion, and excellent gain stability. Emitter-follower outputs enable the devices to drive capacitive loads, and all stages are current-source biased to obtain high common-mode and supply-voltage rejection ratios.

Fixed differential amplification of 100 or 400 may be selected without external components; or amplification may be adjusted from 0 to 400 by the use of a single external resistor connected between the gain-adjustment pins 1A and 1B. External frequency-compensating components are not required for any gain option.

The devices are particularly useful in magnetic-tape or disc-file systems using phase or NRZ encoding and in high-speed thin-film or plated-wire memories. Other applications include general purpose video and pulse amplifiers where wide bandwidth, low phase shift, and excellent gain stability are required.

The SE592 is characterized for operation over the full military temperature range of −55°C to 125°C. The NE592 and NE592A are characterized for operation from 0°C to 70°C.

schematic

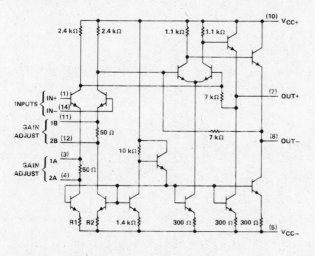

All resistor values shown are in ohms and nominal.
In NE592 or SE592, R1 = 500 Ω, R2 = 500 Ω.
In NE592A, R1 = 600 Ω, R2 = 600 Ω.

symbol

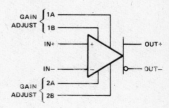

absolute maximum ratings over operating free-air temperature range (unless otherwise noted)

Supply voltage V_{CC+} (see Note 1) .... 8 V
Supply voltage V_{CC-} (see Note 1) .... −8 V
Differential input voltage .... ±5 V
Common-mode input voltage .... ±6 V
Output current .... 10 mA
Continuous total power dissipation at (or below) 25 °C free-air temperature (see Note 2) .... 500 mW
Operating free-air temperature range .... 0 °C to 70 °C
Storage temperature range .... −65 °C to 150 °C
Lead temperature 1,6 mm (1/16 inch) from case for 60 seconds: J package .... 300 °C
Lead temperature 1,6 mm (1/16 inch) from case for 10 seconds: N package .... 260 °C

NOTES: 1. All voltage values except differential input voltages are with respect to the midpoint between V_{CC+} and V_{CC-}.
2. For operation above 25 °C free-air temperature, refer to Dissipation Derating Curves in Section 2. In the J package, SE592 chips are alloy mounted, NE592 and NE592A chips are glass mounted.

recommended operating conditions

	SE592			NE592 NE592A			UNIT
	MIN	NOM	MAX	MIN	NOM	MAX	
Supply voltage, V_{CC+}	3	6	8	3	6	8	V
Supply voltage, V_{CC-}	−3	−6	−8	−3	−6	−8	V
Operating free-air temperature, T_A	−55		125	0		70	°C

Texas
Instruments
POST OFFICE BOX 225012 • DALLAS, TEXAS 75265

electrical characteristics at 25°C operating free-air temperature, $V_{CC+} = 6$ V, $V_{CC-} = -6$ V (unless otherwise noted)

PARAMETER		TEST FIGURE	TEST CONDITIONS	GAIN OPTION[†]	SE592			UNIT
					MIN	TYP	MAX	
A_{VD}	Large-signal differential voltage amplification	1	$V_{OPP} = 3$ V, $R_L = 2$ kΩ	1	300	400	500	V/V
				2	90	100	110	
BW	Bandwidth (−3 dB)	2	$V_{OPP} = 1$ V	1		40		MHz
				2		90		
I_{IO}	Input offset current			1, 2, or 3		0.4	3	μA
I_{IB}	Input bias current			1, 2, or 3		9	20	μA
V_{ICR}	Common-mode input voltage range	3		1, 2, or 3	±1			V
V_{OC}	Common-mode output voltage	1	$R_L = \infty$	1, 2, or 3	2.4	2.9	3.4	V
V_{OO}	Output offset voltage	1	$V_{IO} = 0$, $R_L = \infty$	1			1.5	V
				2			1	
				3		0.35	0.75	
V_{OPP}	Maximum peak-to-peak output voltage swing	1	$R_L = 2$ kΩ	1, 2, or 3	3	4		V
r_i	Input resistance			1		4		kΩ
				2	20	30		
r_o	Output resistance					20		Ω
C_i	Input capacitance					2		pF
CMRR	Common-mode rejection ratio	3	$V_{IC} = ±1$ V, $f = 100$ kHz	2	60	86		dB
		3	$V_{IC} = ±1$ V, $f = 5$ MHz	2		60		
k_{SVR}	Supply-voltage rejection ratio ($\Delta V_{CC}/\Delta V_{IO}$)	4	$\Delta V_{CC+} = ±0.5$ V, $\Delta V_{CC-} = ±0.5$ V	2	50	70		dB
V_n	Broadband equivalent noise voltage	4	BW = 1 kHz to 10 MHz	1, 2, or 3		12		μV
t_{pd}	Propagation delay time	2	$\Delta V_O = 1$ V	1		7.5		ns
				2		6	10	
t_r	Rise time	2	$\Delta V_O = 1$ V	1		10.5		ns
				2		4.5	10	
$I_{sink(max)}$	Maximum output sink current[‡]			1, 2, or 3	3	4		mA
I_{CC}	Supply current		No load, No signal	1, 2, or 3		18	24	mA

[†] The gain option is selected as follows:
  Gain Option 1 . . Gain Adjust pin 1A is connected to pin 1B, pins 2A and 2B are open.
  Gain Option 2 . . Gain Adjust pin 2A is connected to pin 2B, pins 1A and 1B are open.
  Gain Option 3 . . All Gain Adjust pins are open.
[‡] For interchangeability considerations it should be kept in mind that this parameter is not guaranteed by all major manufacturers of SE592 as of the publication of this data sheet.

Special Functions

5

TYPE SE592
DIFFERENTIAL VIDEO AMPLIFIER

electrical characteristics over recommended operating free-air temperature range, V_{CC+} = 6 V, V_{CC-} = −6 V (unless otherwise noted)

PARAMETER		TEST FIGURE	TEST CONDITIONS		GAIN OPTION[†]	SE592			UNIT
						MIN	TYP	MAX	
A_{VD}	Large-signal differential voltage amplification	1	V_{OPP} = 3 V		1	200		600	V/V
					2	80		120	
I_{IO}	Input offset current				1 or 2			5	µA
I_{IB}	Input bias current				1 or 2			40	µA
V_{ICR}	Common-mode input voltage range	3			1 or 2	±1			V
V_{OO}	Output offset voltage	1	V_{ID} = 0,	$R_L = \infty$	1			1.5	V
					2			1.2	
					3			1	
V_{OPP}	Maximum output voltage peak-to-peak swing	1	R_L = 2 kΩ		1 or 2	2.5			V
r_i	Input resistance				2		8		kΩ
CMRR	Common-mode rejection ratio	3	V_{IC} = ±1 V,	f = 100 kHz	2		50		dB
k_{SVR}	Supply voltage rejection ratio ($\Delta V_{CC}/\Delta V_{IO}$)	4	ΔV_{CC+} = ±0.5 V, ΔV_{CC-} = ±0.5 V		2		50		dB
$I_{sink(max)}$	Maximum output sink current				1, 2, or 3	2.5			mA
I_{CC}	Supply current	1	No load,	No signal	1, 2, or 3			27	mA

[†] The gain option is selected as follows:
 Gain Option 1 . . Gain Adjust pin 1A is connected to pin 1B; pins 2A and 2B are open.
 Gain Option 2 . . Gain Adjust pin 2A is connected to pin 2B; pins 1A and 1B are open.
 Gain Option 3 . . All Gain Adjust pins are open.

Special Functions

5

TEXAS INSTRUMENTS
POST OFFICE BOX 225012 ● DALLAS, TEXAS 75265

electrical characteristics at 25 °C operating free-air temperature, V_{CC+} = 6 V, V_{CC-} = −6 V (unless otherwise noted)

PARAMETER		TEST FIGURE	TEST CONDITIONS	GAIN OPTION†	NE592 MIN	NE592 TYP	NE592 MAX	NE592A MIN	NE592A TYP	NE592A MAX	UNIT
A_{VD}	Large signal differential voltage amplification	1	V_{OPP} = 3 V, R_L = 2 kΩ	1	250	400	600	400	440	600	V/V
				2	80	100	120	80	100	120	
BW	Bandwidth (−3 dB)	2	V_{OPP} = 1 V	1		40			40		MHz
				2		90			90		
I_{IO}	Input offset current			1, 2, or 3		0.4	5		0.4	5	µA
I_{IB}	Input bias current			1, 2, or 3		9	30		10	30	µA
V_{ICR}	Common-mode input voltage range	3		1, 2, or 3	±1			±1			V
V_{OC}	Common-mode output voltage	1	R_L = ∞	1, 2, or 3	2.4	2.9	3.4	2.4	2.9	3.4	V
V_{OO}	Output offset voltage	1	V_{ID} = 0, R_L = ∞	1 or 2			1.5			1.5	V
				3		0.35	0.75		0.35	0.75	
V_{OPP}	Maximum peak-to-peak output voltage swing	1	R_L = 2 kΩ	1, 2, or 3	3	4		3	4		V
r_i	Input resistance			1		4			4		kΩ
				2	10	30		10	30		
r_o	Output resistance			2		20			20		Ω
C_i	Input capacitance			2		2			2		pF
CMRR	Common-mode rejection ratio	3	V_{IC} = ±1 V, f = 100 kHz	2	60	86		60	86		dB
		3	V_{IC} = ±1 V, f = 5 MHz	2		60			60		
k_{SVR}	Supply-voltage rejection ratio ($\Delta V_{CC}/\Delta V_{IO}$)	4	ΔV_{CC+} = ±0.5 V, ΔV_{CC-} = ±0.5 V	2	50	70		50	70		dB
V_n	Broadband equivalent noise voltage	4	BW = 1 kHz to 10 MHz	1, 2, or 3		12			12		µV
t_{pd}	Propagation delay time	2	ΔV_O = 1 V	1		7.5			7.5		ns
				2		6	10		6	10	
t_r	Rise time	2	ΔV_O = 1 V	1		10.5			10.5		ns
				2		4.5	12		4.5	12	
$I_{sink(max)}$	Maximum output sink current‡			1, 2, or 3	3	4		3	4		mA
I_{CC}	Supply current		No load, No signal	1, 2, or 3		18	24		19	24	mA

† The gain option is selected as follows:
Gain Option 1 . . Gain Adjust pin 1A is connected to pin 1B, pins 2A and 2B are open.
Gain Option 2 . . Gain Adjust pin 2A is connected to pin 2B, pins 1A and 1B are open.
Gain Option 3 . . All Gain Adjust pins are open.
‡ For interchangeability considerations it should be kept in mind that this parameter is not guaranteed by all major manufacturers of NE592 as of the publication of this data sheet.

Special Functions

5

Special Functions

5

electrical characteristics over recommended operating free-air temperature range, V_{CC+} = 6 V, V_{CC-} = −6 V (unless otherwise noted)

	PARAMETER	TEST FIGURE	TEST CONDITIONS	GAIN OPTION †	NE592 MIN	NE592 TYP	NE592 MAX	NE592A MIN	NE592A TYP	NE592A MAX	UNIT
A_{VD}	Large-signal differential voltage amplification	1	V_{OPP} = 3 V	1	250		600	400		600	V/V
				2	80		120	80		120	V/V
I_{IO}	Input offset current			1 or 2			6			6	µA
I_{IB}	Input bias current			1 or 2			40			40	µA
V_{ICR}	Common-mode input voltage range	3		1 or 2	±1			±1			V
V_{OO}	Output offset voltage	1	V_{ID} = 0, $\quad R_L$ = ∞	1 or 2			1.5			1.5	V
				3			1			1	V
V_{OPP}	Maximum output voltage peak-to-peak swing	1	R_L = 2 kΩ	1 or 2	2.8			2.8			V
r_i	Input resistance			2	8			8			kΩ
CMRR	Common-mode rejection ratio	3	V_{IC} = ±1 V, $\quad$ f = 100 kHz	2	50			50			dB
k_{SVR}	Supply voltage rejection ratio ($\Delta V_{CC}/\Delta V_{IO}$)	4	ΔV_{CC+} = ±0.5 V, ΔV_{CC-} = ±0.5 V	2	50			50			dB
$I_{sink(max)}$	Maximum output sink current		No load, $\quad$ No signal	1, 2, or 3	2.8			2.8	4		mA
I_{CC}	Supply current	1	No signal	1, 2, or 3			27			27	mA

† The gain option is selected as follows:
Gain Option 1 . . Gain Adjust pin 1A is connected to pin 1B; pins 2A and 2B are open.
Gain Option 2 . . Gain Adjust pin 2A is connected to pin 2B; pins 1A and 1B are open.
Gain Option 3 . . All Gain Adjust pins are open.

TEXAS INSTRUMENTS
POST OFFICE BOX 225012 • DALLAS, TEXAS 75265

PARAMETER MEASUREMENT INFORMATION

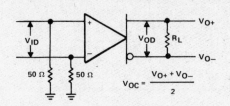

FIGURE 1

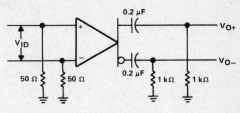

FIGURE 2

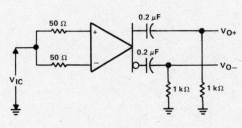

FIGURE 3

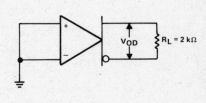

FIGURE 4

Special Functions

5

TEXAS
INSTRUMENTS
POST OFFICE BOX 225012 • DALLAS, TEXAS 75265

LINEAR
INTEGRATED
CIRCUITS

TYPE SN28827
SONAR RANGING MODULE

D2780, OCTOBER 1983

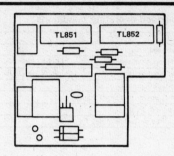

- Accurate Sonar Ranging from 6 Inches to 35 Feet

- Drives 50-kHz Electrostatic Transducer with No Additional Interface

- Operates from Single Supply

- Accurate Clock Output Provided for External Use

- Selective Echo Exclusion

- TTL-Compatible

- Multiple Measurement Capability

- Uses TL851 and TL852 Sonar Ranging Integrated Circuits

description

The SN28827 is an economical sonar ranging module that can drive a 50-kilohertz, 300-volt electrostatic transducer with no additional interface. This module, with a simple interface, is able to measure distances ranging from 6 inches to 35 feet. The typical absolute accuracy is ±2% at one foot or greater.

This module has an external blanking input that allows selective echo exclusion for operation in a multiple-echo mode. The module is able to differentiate echos from objects that are only three inches apart. The digitally controlled-gain, variable-bandwidth amplifier minimizes noise and side-lobe detection in sonar applications.

The module has an accurate ceramic-resonator-controlled 420-kilohertz time-base generator. An output based on the 420-kilohertz time base is provided for external use. The sonar transmit output is 16 cycles at a frequency of 49.4 kilohertz.

The SN28827 operates over a supply voltage range of from 4.5 volts to 6.8 volts and is characterized for operation from 0°C to 40°C.

absolute maximum ratings

Voltage from any pin to ground (see Note 1)	7 V
Voltage from any pin except XDCR to V_{CC} (see Note 1)	−7 V to 0.5 V
Operating free-air temperature range	0°C to 40°C
Storage temperature range	−40°C to 85°C

NOTE 1: The XDCR pin may be driven from −1 volt to 300 volts typical with respect to ground.

Special Functions

5

Copyright © 1983 by Texas Instruments Incorporated

TEXAS
INSTRUMENTS
POST OFFICE BOX 225012 • DALLAS, TEXAS 75265

recommended operating conditions

		MIN	MAX	UNIT
Supply voltage, V_{CC}		4.5	6.8	V
High-level input voltage, V_{IH}	BLNK, BINH, INIT	2.1		V
Low-level input voltage, V_{IL}	BLNK, BINH, INIT		0.6	V
ECHO and OSC output voltage			6.8	V
Delay time, power up to INIT high		5		ms
Recycle period		80		ms
Operating free-air temperature, T_A		0	40	°C

electrical characteristics over recommended ranges of supply voltage and operating free-air temperature (unless otherwise noted)

PARAMETER		TEST CONDITIONS	MIN	TYP	MAX	UNIT
Input current	BLNK, BINH, INIT	V_I = 2.1 V			1	mA
High-level output current, I_{OH}	ECHO, OSC	V_{OH} = 5.5 V			100	µA
Low-level output voltage, V_{OL}	ECHO, OSC	I_{OL} = 1.6 mA			0.4	V
Transducer bias voltage		T_A = 25 °C		150		V
Transducer output voltage (peak-to-peak)		T_A = 25 °C		300		V
Number of cycles for XDCR output to reach 300 V		C = 500 pF			7	
Internal blanking interval				2.38[†]		ms
Frequency during 16-pulse transmit period	OSC output			49.4[†]		kHz
	XMIT output			49.4[†]		
Frequency after 16-pulse transmit period	OSC output			93.3[†]		kHz
	XMIT output			0		
Supply current, I_{CC}	During transmit period				2000	mA
	After transmit period				100	

[†]These typical values apply for a 420-kHz ceramic resonator.

TEXAS
INSTRUMENTS
POST OFFICE BOX 225012 ● DALLAS, TEXAS 75265

operation with Polaroid electrostatic transducer

There are two basic modes of operation for the SN28827 Sonar ranging module: single-echo mode and multiple-echo mode. The application of power (V_{CC}), the activation of the Initiate (INIT) input, and the resulting transmit output, and the use of the Blanking Inhibit (BINH) input are basically the same for either mode of operation. After applying power (V_{CC}), a minimum of 5 milliseconds must elapse before the INIT input can be taken high. During this time, all internal circuitry is reset and the internal oscillator stabilizes. When INIT is taken high, drive to the Transducer XDCR output occurs. Sixteen pulses at 49.4 kilohertz with 300-volt amplitude will excite the transducer as transmission occurs. At the end of the 16 transmit pulses, a dc bias of 150 volts will remain on the transducer as recommended for optimum operation by the transducer manufacturer.

In order to eliminate ringing of the transducer from being detected as a return signal, the Receive (REC) input of the ranging control IC is inhibited by internal blanking for 2.38 milliseconds after the initiate signal. If a reduced blanking time is desired, then the BINH input can be taken high to end the blanking of the Receive input anytime prior to internal blanking. This may be desired to detect objects closer than 1.33 feet corresponding to 2.38 milliseconds and may be done if transducer damping is sufficient that ringing is not detected as a return signal.

In the single-echo mode of operation (Figure 1), all that must be done next is to wait for the return of the transmitted signal, traveling at approximately 0.9 milliseconds per foot out and back. The returning signal is amplified and appears as a high-logic-level echo output. The time between INIT going high and the Echo (ECHO) output going high is proportional to the distance of the target from the transducer. If desired, the cycle can now be repeated by returning INIT to a low-logic level and then taking it high when the next transmission is desired.

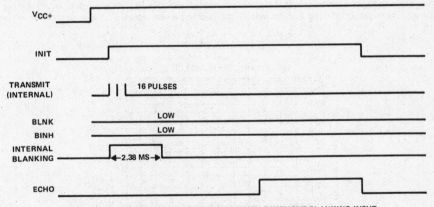

FIGURE 1—EXAMPLE OF A SINGLE-ECHO-MODE CYCLE WITHOUT BLANKING INPUT

If there is more than one target and multiple echos are to be detected from a single transmission, then the cycle is slightly different (Figure 2). After receiving the first return signal, which causes the ECHO output to go high, the Blanking (BLNK) input must be taken high then back low to reset the ECHO output for the next return signal. The blanking signal must be at least 0.44 milliseconds in duration to account for all 16 returning pulses from the first target and allow for internal delay times. This corresponds to the two targets being 3 inches apart.

Special Functions

5

TYPE SN28827
SONAR RANGING MODULE

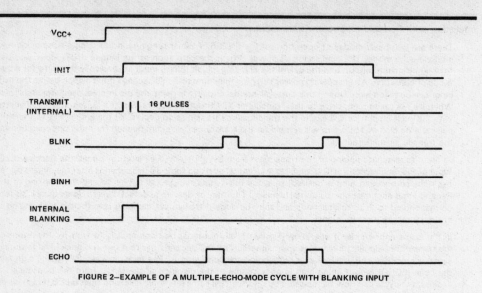

FIGURE 2—EXAMPLE OF A MULTIPLE-ECHO-MODE CYCLE WITH BLANKING INPUT

During a cycle starting with INIT going high, the receiver amplifier gain is incremented higher at discrete times (Figure 3) since the transmitted signal is attenuated with distance. At approximately 38 milliseconds, the maximum gain is attained. For this reason, sufficient gain may not be available for objects greater than

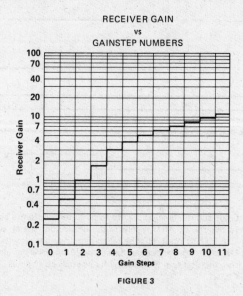

FIGURE 3

TEXAS
INSTRUMENTS

POST OFFICE BOX 225012 • DALLAS, TEXAS 75265

35 feet away. Although gain can be increased by varying R1 (Figure 4), there is a limit to which the gain can be increased for reliable module operation. This will vary from application to application. The modules are "kitted" prior to their final test during manufacture. This is necessary because the desired gain distribution is much narrower than the module gain distribution if all were kitted with one value resistor. As kitted, these modules will perform satisfactorily in most applications. As a rule of thumb, the gain can be increased by up to a factor of 4, if required, by increasing R1 correspondingly. Gain is directly proportional to R1.

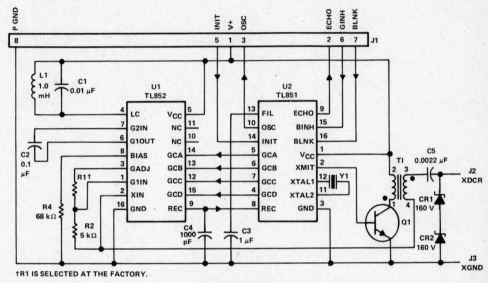

†R1 IS SELECTED AT THE FACTORY.

FIGURE 4—SCHEMATIC

Special Functions

5

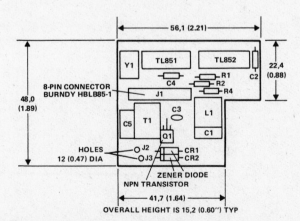

FIGURE 5 – COMPONENT LAYOUT AND DIMENSIONS OF MODULE

NOTE: All dimensions are in millimeters and parenthetically in inches.

**LINEAR
INTEGRATED
CIRCUITS**

**TYPES TL010I, TL010C
ADJUSTABLE-RATIO CURRENT MIRRORS**

D2738, SEPTEMBER 1983

- 33 Distinct Input-to-Output Emitter Ratios from 3:1 to 1:15
- Wide Input Current Range: 1 µA to 3 mA
- 35-Volt Output Capability
- High Output Impedance

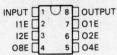

**P DUAL-IN-LINE PACKAGE
(TOP VIEW)**

INPUT	1		8	OUTPUT
I1E	2		7	O1E
I2E	3		6	O2E
O8E	4		5	O4E

description

The TL010 is a Wilson current mirror that provides output current in a selectable fixed ratio to the input current. The ratio is substantially independent of changes in load, voltages, and temperature. Selecting the ratio consists of connecting appropriate input emitter pins and output emitter pins to ground as shown in Figure 1.

The TL010 is designed to operate with up to 3 milliamperes input current if all three input emitter pins are used. It will also operate at voltages up to 35 volts.

The TL010I is characterized for operation from −40°C to 85°C. The TL010C is characterized for operation from 0°C to 70°C.

Special Functions

5

typical values of current ratio at $T_A = 25°C$

EMITTER RATIO m:n[†]	CURRENT RATIO $h_F = I_O/I_I$	EMITTER RATIO m:n[†]	CURRENT RATIO $h_F = I_O/I_I$	EMITTER RATIO m:n[†]	CURRENT RATIO $h_F = I_O/I_I$
1:15	14.1	1:6	5.78	3:8	2.61
1:14	13.2	2:11	5.34	2:5	2.43
1:13	12.3	1:5	4.82	3:7	2.26
1:12	11.4	3:14	4.53	1:2	1.98
1:11	10.5	2:9	4.38	3:5	1.64
1:10	9.55	3:13	4.21	2:3	1.45
1:9	8.62	1:4	3.89	3:4	1.32
1:8	7.72	3:11	3.57	1:1	0.99
2:15	7.23	2:7	3:40	3:2	0.663
1:7	6:71	3:10	3:25	2:1	0.50
2:13	6.29	1:3	2.90	3:1	0.332

[†]m is the number of input emitters used, n is the number of output emitters used.

schematic

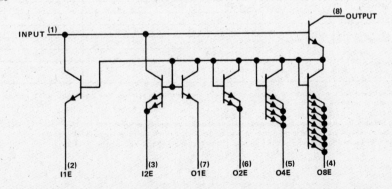

TYPES TL010I, TL010C
ADJUSTABLE-RATIO CURRENT MIRRORS

absolute maximum ratings over operating free-air temperature range (unless otherwise noted)

Output voltage (see Note 1) . . . . 45 V
Input current . . . . 5 mA
Continuous total dissipation at (or below) 25 °C free-air temperature (see Note 2) . . . . 725 mW
Operating free-air temperature range: TL010I . . . . −40 °C to 85 °C
                               TL010C . . . . 0 °C to 70 °C
Storage temperature range . . . . −65 °C to 150 °C
Lead temperature 1,6 mm (1/16 inch) from case for 10 seconds . . . . 260 °C

NOTES: 1. Input and output voltages are with respect to the common terminal. Neither voltage should be more negative than −0.3 V.
       2. For operation above 25 °C free-air temperature, derate linearly at the rate of 5.8 mW/°C.

recommended operating conditions

	TL010I		TL010C		UNIT
	MIN	MAX	MIN	MAX	
Output voltage, V_O	5	35	5	35	V
Input voltage, V_I	0.6	1.7	0.65	1.6	V
Input current per input emitter, I_I	0.001	1	0.001	1	mA
Operating free-air temperature, T_A	−40	85	. 0	70	°C

electrical characteristics over recommended ranges of operating free-air temperature and output voltage (unless otherwise noted)

PARAMETER		TEST CONDITIONS[†]		TL010I			TL010C			UNIT
				MIN	TYP[‡]	MAX	MIN	TYP[‡]	MAX	
V_I	Input voltage	$I_I = m \times 1\ \mu A$			1			1		V
		$I_I = m \times 10\ \mu A$			1.1			1.1		
		$I_I = m \times 100\ \mu A$			1.25			1.25		
		$I_I = m \times 1\ mA$			1.4			1.4		
h_F	Current ratio (I_O/I_I)	$I_I = $ MIN to MAX	m:n = 1:8	6.97	7.72	8.13	7.05	7.72	8.13	
			m:n = 1:4	3.61	3.89	4.05	3.64	3.89	4.05	
			m:n = 1:2	1.84	1.98	2.07	1.88	1.98	2.07	
			m:n = 1:1	0.89	0.99	1.08	0.94	0.99	1.04	
			m:n = 2:1	0.46	0.50	0.56	0.475	0.50	0.525	
α_{hF}	Temperature coefficient of current ratio	$I_I = $ MIN to MAX				300			300	ppm/°C
	Output-to-input isolation	$I_I = $ MIN to MAX, f = 1 kHz		60			60			dB
$V_{O(th)}$	Output threshold voltage§	$I_I = $ MIN to MAX	$T_A = $ MIN			1.1			1.05	V
			$T_A = $ 25 °C			1			1	
r_o	Output resistance¶	F = 1 kHz	$I_I = m \times 10\ \mu A$		200 m/n			200 m/n		MΩ
			$I_I = m \times 100\ \mu A$		20 m/n			20 m/n		
			$I_I = m \times 1\ mA$		2 m/n			2 m/n		
f_{max}	Maximum operating frequency #	$I_I = m \times 1\ mA$, $R_L = 500\ \Omega$		10			10			MHz

[†] m is the number of input emitters, n is the number of output emitters. For conditions shown as MIN or MAX, use the appropriate value specified under recommended operating conditions.
[‡] All typical values are at $T_A = $ 25 °C.
§ Output threshold voltage is the voltage at which the current ratio is equal to 90% of its value at $V_O = $ 15 V.
¶ The output resistance is directly proportional to the number of input emitters divided by the number of output emitters (m/n).
# Maximum operating frequency is the frequency at which the output current is down 3 dB from its low-frequency value.

Special Functions

5

TEXAS INSTRUMENTS
POST OFFICE BOX 225012 • DALLAS, TEXAS 75265

108

TYPICAL APPLICATION INFORMATION

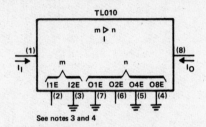

FIGURE 1—CURRENT MIRROR SET FOR A CURRENT RATIO OF 2:13

NOTES: 3. Selected emitters must be grounded as close as possible to the package to avoid unstable device behavior.

Using the fixed-Beta model, the current ratio for a current mirror of m input emitters and n output emitters may be calculated as

$$\frac{I_O}{I_I} = \frac{\beta^2 n + \beta (n+m)}{\beta^2 m + (\beta+1)(m+n)}$$

Second-order effects, such as on-chip self-heating, may slightly perturb the observed ratio from the calculated value.

4. At high current levels a small capacitor (270 pF) may be required between the input and output terminals to improve stability.

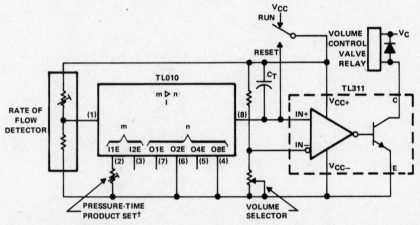

†Adjust for a mirror of 11.9

In this application of the TL010, the problem is to measure a precise volume of liquid flowing through a line and shut off the flow with a relay when the limit is reached. For the particular volume to be measured and the pressure detector used, a current gain of 11.9 is required. By setting the TL010 for a gain of 10 with the emitter selection, the exact gain of 11.9 may be obtained by adjusting the pressure-time product control.

TEXAS INSTRUMENTS
POST OFFICE BOX 225012 ● DALLAS, TEXAS 75265

Special Functions

5

LINEAR
INTEGRATED
CIRCUITS

SERIES TL011, TL012, TL014, TL021
FIXED-RATIO N-P-N CURRENT MIRRORS

D2614, NOVEMBER 1983

- **Wide Input Current Range:**
  1 µA to 1 mA

- **35-Volt Output Capability**

- **High Output Impedance**

- **Guaranteed Current-Ratio Tolerances over
  Full Temperature Range;**
  ±8% for I Suffix
  ±7% for C Suffix

- **Typically Less Than ±1% Error at 25°C**

LP PACKAGE
(TOP VIEW)

	INPUT
	COMMON
	OUTPUT

TEMPERATURE	INPUT-TO-OUTPUT CURRENT RATIO			
RANGE	1:1	1:2	1:4	2:1
−40°C to 85°C	TL011I	TL012I	TL014I	TL021I
0°C to 70°C	TL011C	TL012C	TL014C	TL021C

description

The TL011, TL012, TL014, and TL021 are Wilson current mirrors with output currents in fixed proportion to the input currents and substantially independent of changes in voltage, load, and temperature. These devices make use of the tight matching properties of identical bipolar transistors on a monolithic integrated circuit chip to achieve current-ratio accuracy typically better than 98%.

Current mirrors are used extensively in linear integrated circuit designs as active loads for operational-amplifier stages and as current sources for other stages. The TL011 family gives the designer this same capability with no sacrifice in accuracy or stability.

The TL011, TL012, and TL014 are designed to operate with input currents up to 1 milliampere and output voltage up to 35 volts. The TL021 is designed for 2 milliamperes and 35 volts.

schematics

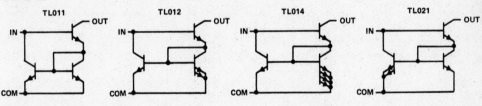

symbols

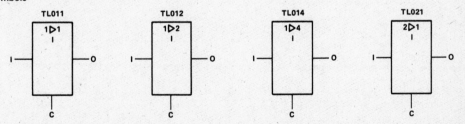

Special Functions

5

TEXAS
INSTRUMENTS

POST OFFICE BOX 225012 • DALLAS, TEXAS 75265

absolute maximum ratings over operating free-air temperature range (unless otherwise noted)

Output voltage (see Note 1)... 45 V

Input current .... 5 mA

Continuous total dissipation at (or below) 25°C free-air temperature (see Note 2) .... 775 mW

Operating free-air temperature range: TL011I, TL012I, TL014I, TL021I................. −40°C to 85°C

TL011C, TL012C, TL014C, TL021C .... 0°C to 70°C

Storage temperature range .... −65°C to 150°C

Lead temperature 1,6 mm (1/16 inch) from case for 10 seconds .... 260°C

NOTES: 1. Input and output voltages are with respect to the common terminal. Neither voltage should be more negative than −0.3 V.
2. For operation above 25°C free-air temperature, derate linearly at the rate of 6.2 mW/°C.

recommended operating conditions

		TL0_ _I		TL0_ _C		
		MIN	MAX	MIN	MAX	UNIT
Output voltage, V_O		5	35	5	35	V
Input current, I_O	TL021	0.002	2	0.002	2	mA
	All others	0.001	1	0.001	1	
Operating free-air temperature, T_A		−40	85	0	70	°C

Special Functions

5

TEXAS
INSTRUMENTS

POST OFFICE BOX 225012 ● DALLAS, TEXAS 75265

electrical characteristics over recommended ranges of operating free-air temperature and output voltage (unless otherwise noted)

PARAMETER	TEST CONDITIONS	TL011			TL012			TL014			TL021			UNIT
		MIN	TYP†	MAX	MIN	TYP†	MAX	MIN	TYP†	MAX	MIN	TYP†	MAX	
V_I Input voltage	$I_I = 1\ \mu A$		1			1			1			1		V
	$I_I = 2\ \mu A$													
	$I_I = 10\ \mu A$		1.1			1.1			1.1			1.1		
	$I_I = 20\ \mu A$													
	$I_I = 100\ \mu A$		1.25			1.25			1.25			1.25		
	$I_I = 200\ \mu A$													
	$I_I = 1\ mA$													
	$I_I = 2\ mA$		1.4			1.4			1.4			1.4		
h_F Current ratio (I_O/I_I) TLO_I	I_I = MIN to MAX‡	0.92	1	1.08	1.84	2	2.16	3.68	4	4.32	0.46	0.5	0.54	
h_F Current ratio (I_O/I_I) TLO_C	I_I = MIN to MAX	0.93	1	1.07	1.86	2	2.14	3.72	4	4.28	0.465	0.5	0.535	
α_{hF} Temperature coefficient of current ratio	I_I = MIN to MAX		50			100			200			200		ppm/°C
Output-to-input isolation	I_I = MIN to MAX, f = 1 kHz	80			80			80			80			dB
$V_{O(th)}$ Output threshold voltage § TLO_I	$T_A = -40°C$			1.35			1.35			1.35			1.35	V
$V_{O(th)}$ Output threshold voltage § TLO_C	$T_A = 0°C$			1.25			1.25			1.25			1.25	
$V_{O(th)}$ Output threshold voltage § All	$T_A = 25°C$			1.2			1.2			1.2			1.2	
r_O Output resistance	$I_I = 10\ \mu A$, f = 1 kHz		200			100			50			200		MΩ
	$I_I = 100\ \mu A$		20			10			5			20		
	$I_I = 1\ mA$		2			1			0.5			2		
f_{max} Maximum operating frequency¶	I_I = MAX, $R_L = 500\ \Omega$		10			10			10			10		MHz

† All typical values are at $T_A = 25°C$.

‡ For test conditions shown as MIN or MAX, use the appropriate value specified under recommended operating conditions.

§ Output threshold voltage is the voltage at which the current ratio is equal to 90% of its value at $V_O = 15$ V.

¶ Maximum operating frequency is the frequency at which the output current is down 3 dB from its low frequency value.

5

Special Functions

TEXAS INSTRUMENTS
POST OFFICE BOX 225012 • DALLAS, TEXAS 75265

TYPICAL CHARACTERISTICS

TL011
CURRENT RATIO
vs
FREE-AIR TEMPERATURE

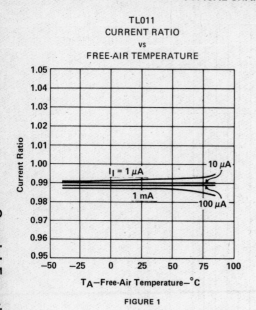

FIGURE 1

TL012
CURRENT RATIO
vs
FREE-AIR TEMPERATURE

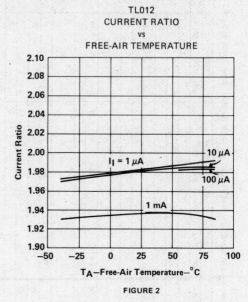

FIGURE 2

TL014
CURRENT RATIO
vs
FREE-AIR TEMPERATURE

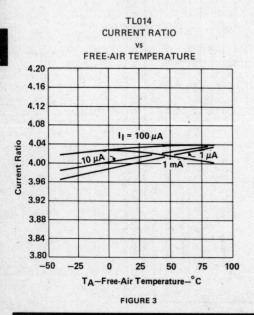

FIGURE 3

TL021
CURRENT RATIO
vs
FREE-AIR TEMPERATURE

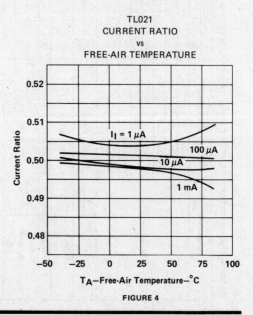

FIGURE 4

Special Functions

5

TEXAS
INSTRUMENTS
POST OFFICE BOX 225012 • DALLAS, TEXAS 75265

TYPICAL APPLICATIONS INFORMATION

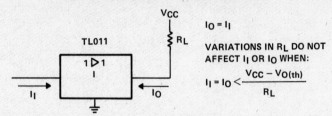

$I_O = I_I$

VARIATIONS IN R_L DO NOT
AFFECT I_I OR I_O WHEN:

$$I_I = I_O < \frac{V_{CC} - V_{O(th)}}{R_L}$$

FIGURE 5—BASIC CURRENT BUFFER

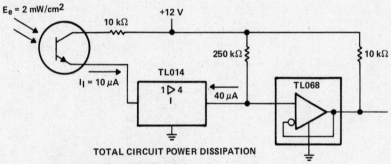

TOTAL CIRCUIT POWER DISSIPATION

Idle condition: P_D = 1.5 mW typical
On condition: P_D = 12.5 mW typical
10 μA from phototransistor provides a V_O swing of 10 V at 1 mA.

FIGURE 6—PHOTOTRANSISTOR PREAMPLIFIER

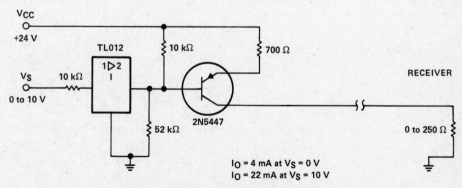

I_O = 4 mA at V_S = 0 V
I_O = 22 mA at V_S = 10 V

FIGURE 7—TWO-WIRE LINEAR CURRENT-MODE TRANSMITTER

Special Functions

5

LINEAR
INTEGRATED
CIRCUITS

TYPE TL026
AGC VIDEO AMPLIFIER

D2790, OCTOBER 1983

- Low Output Common-Mode Sensitivity to AGC Voltages
- Input and Output Impedances Independent of AGC Voltage
- Maximum Gain of 100 Typ
- Wide AGC Range
- 3-dB Bandwidth at 60 MHz
- Other Characteristics Similar to NE592 and uA733

JG OR P
DUAL-IN-LINE PACKAGE
(TOP VIEW)

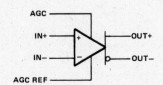

symbol

description

This device is a monolithic two-stage video amplifier with differential inputs and outputs.

Internal feedback provides wide bandwidth, low phase distortion, and excellent gain stability. Variable gain based on signal summation provides large AGC control over a wide bandwidth with low harmonic distortion. Emitter-follower outputs enable the device to drive capacitive loads. All stages are current-source biased to obtain high common-mode and supply-voltage rejection ratios. The gain of 100 may be electronically attenuated as much as 50 dB at 60 MHz by applying a control voltage to the AGC pins. No external frequency compensation components are required.

This device is particularly useful in TV and Radio IF and RF AGC circuits, as well as magnetic-tape and disc-file systems where AGC is needed. Other applications include video and pulse amplifiers where a large AGC range, wide bandwidth, low phase shift, and excellent gain stability are required.

Special Functions

5

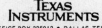

TEXAS
INSTRUMENTS

POST OFFICE BOX 225012 • DALLAS, TEXAS 75265

- Operates from ±6-V Supplies or 12-V Supply
- Head Read-Select-Diode Array
- Dual Write Current Inputs
- Read Amplifier Gain Adjustable with Single External Resistor and Capacitor
- Wide Bandwidth, Low Phase Distortion, and Excellent Gain Stability
- High Common-Mode Rejection

N DUAL-IN-LINE PACKAGE
(TOP VIEW)

WRITE A	1	16	HEAD 1A
WRITE B	2	15	HEAD 1B
GAIN SELECT GA	3	14	HEAD 2A
GAIN SELECT GB	4	13	HEAD 2B
V_{EE}	5	12	HEAD 3A
OUTPUT A	6	11	HEAD 3B
OUTPUT B	7	10	HEAD 4A
V_{CC}	8	9	HEAD 4B

description

The TL030 is a monolithic high-speed disk-memory read amplifier fabricated with bipolar Schottky process technology. The device consists of a diode selection matrix comprised of write, read, and head diodes preceding a video amplifier. The head read diode array may be externally biased to select one of four-disk memory heads. The resultant analog signal is then amplified by the read amplifier and presented as differential emitter-follower output voltages. The TL030 is characterized for operation from 0°C to 70°C.

Special Functions

5

functional block diagram

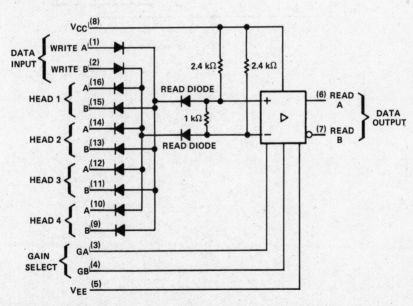

Copyright © 1983 by Texas Instruments Incorporated

TEXAS
INSTRUMENTS

POST OFFICE BOX 225012 • DALLAS, TEXAS 75265

TYPICAL APPLICATION DATA

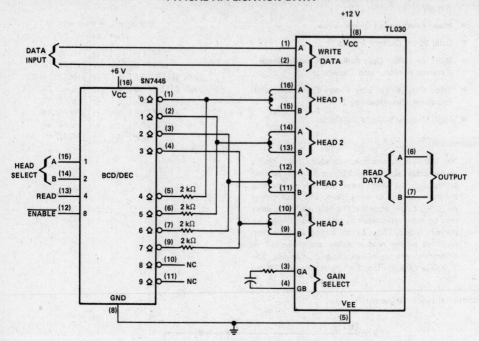

TEXAS
INSTRUMENTS
POST OFFICE BOX 225012 • DALLAS, TEXAS 75265

Special Functions

5

- Magnetic-Field-Sensing Hall-Effect Input
- Activated with Small, Commercially Available Permanent Magnets
- Special Switching Thresholds are Easily Programmable at the Factory
- Voltage Range: 4 V to 30 V
- Output Compatible with All Digital Logic Families
- This Series will be Available for Pin-for-Pin Compatibility with Sprague's UGN3013, UGN3019, UGN3020, UGN3030, and UGN3040

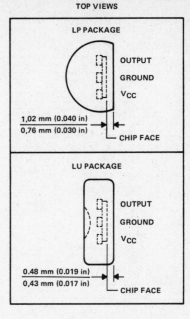

TOP VIEWS

LP PACKAGE

OUTPUT
GROUND
V_{CC}

1,02 mm (0.040 in)
0,76 mm (0.030 in)
CHIP FACE

LU PACKAGE

OUTPUT
GROUND
V_{CC}

0.48 mm (0.019 in)
0,43 mm (0.017 in)
CHIP FACE

Special Functions

5

description

The TL160 series is a complete line of Hall-Effect switches. Each device consists of a voltage regulator, a Hall sensing element, amplifier, Schmitt trigger, and an open-collector output stage integrated on a single monolithic silicon chip. Operate and release points will be independently programmable at the factory. These switching points will be in the range of 0 to 50 milliteslas (500 gauss).

absolute maximum ratings over operating free-air temperature range (unless otherwise noted)

Supply voltage, V_{CC} .... 30 V
Output voltage .... 30 V
Output current .... 30 mA
Storage temperature range .... −65 °C to 150 °C
Magnetic flux density, B .... unlimited

Copyright © 1983 by Texas Instruments Incorporated

TEXAS INSTRUMENTS

POST OFFICE BOX 225012 • DALLAS, TEXAS 75265

- Magnetic-Field Sensing Hall-Effect Input
- On-Off Hysteresis
- Small Size
- Solid-State Technology
- Open-Collector Output

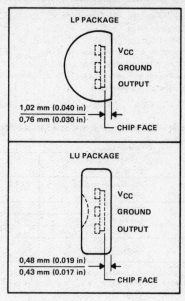

TOP VIEWS

LP PACKAGE

VCC
GROUND
OUTPUT

1,02 mm (0.040 in)
0,76 mm (0.030 in)

CHIP FACE

LU PACKAGE

VCC
GROUND
OUTPUT

0,48 mm (0.019 in)
0,43 mm (0.017 in)

CHIP FACE

description

The TL170C is a low-cost magnetically operated electronic switch that utilizes the Hall Effect to sense steady-state magnetic fields. Each circuit consists of a Hall-Effect sensor, signal conditioning and hysteresis functions, and an output transistor integrated into a monolithic chip. The outputs of these circuits can be directly connected to many different types of electronic components.

The TL170C is characterized for operation over the temperature range of 0 °C to 70 °C.

FUNCTION TABLE (T_A = 25 °C)

FLUX DENSITY	OUTPUT
≤ -25 mT	Off
-25 mT $< B < 25$ mT	Undefined
≥ 25 mT	On

functional block diagram

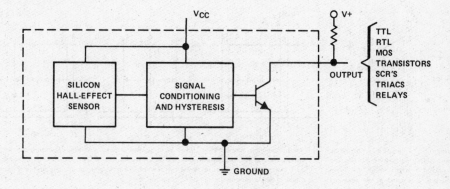

VCC

V+

SILICON HALL-EFFECT SENSOR

SIGNAL CONDITIONING AND HYSTERESIS

OUTPUT

TTL
RTL
MOS
TRANSISTORS
SCR'S
TRIACS
RELAYS

GROUND

Special Functions

5

TEXAS INSTRUMENTS
POST OFFICE BOX 225012 • DALLAS, TEXAS 75265

absolute maximum ratings over operating free-air temperature range (unless otherwise noted)

Supply voltage, V_{CC} (see Note 1) . . . . 7 V
Output voltage . . . . 30 V
Output current . . . . 20 mA
Operating free-air temperature range . . . . 0 °C to 70 °C
Storage temperature range . . . . −65 °C to 150 °C
Lead temperature 1,6 mm (1/16 inch) from case for 10 seconds . . . . 260 °C
Magnetic flux density . . . . unlimited

NOTE 1. Voltage values are with respect to network ground terminal.

electrical characteristics at specified free-air temperature, V_{CC} = 5 V ± 5% (unless otherwise noted)

	PARAMETER	TEST CONDITIONS		MIN	TYP	MAX	UNIT
B_{T+}	Threshold of positive-going		25 °C			25	mT §
	magnetic flux density†		0 °C to 70 °C			35	
B_{T-}	Threshold of negative-going		25 °C	−25¶			mT §
	magnetic flux density†		0 °C to 70 °C	−35¶			
$B_{T+} - B_{T-}$	Hysteresis		0 °C to 70 °C		20		mT §
I_{OH}	High-level output current	V_{OH} = 20 V	0 °C to 70 °C			100	µA
V_{OL}	Low-level output voltage	V_{CC} = 4.75 V, I_{OL} = 16 mA	0 °C to 70 °C			0.4	V
I_{CC}	Supply current	V_{CC} = 5.25 V — Output low	0 °C to 70 °C			6	mA
		Output high				4	

†Threshold values are those levels of magnetic flux denisity at which the output changes state. For the TL170C, a level more positive than B_{T+} causes
 the output to go to a low level and a level more negative than B_{T-} causes the output to go to a high level. See Figures 1 and 2.

§The unit of magnetic flux density in the International System of Units (SI) is the tesla (T). The tesla is equal to one weber per square meter. Values expressed
 in milliteslas may be converted to gauss by multiplying by ten.

¶The algebraic convention, where the most negative limit is designated as minimum, is used in this data sheet for flux-density threshold levels only.

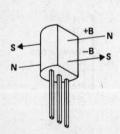

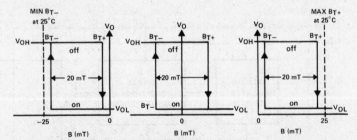

The north pole of a magnet is the pole that is attracted by the geographical north pole. The north pole of a magnet repels the north-seeking pole of a compass. By accepted magnetic convention, lines of flux emanate from the north pole of a magnet and enter the south pole.

**FIGURE 1—DEFINITION OF
MAGNETIC FLUX POLARITY**

The positive-going threshold (B_{T+}) may be a negative or positive B level at which a positive-going (decreasing negative or increasing positive) flux density results in the TL170 output turning on. The negative-going threshold is a positive or negative B level at which a negative-going (decreasing positive or increasing negative) flux density results in the TL170 turning off.

FIGURE 2—REPRESENTATIVE CURVES OF V_O vs B

**TEXAS
INSTRUMENTS**
POST OFFICE BOX 225012 • DALLAS, TEXAS 75265

LINEAR
INTEGRATED
CIRCUITS

TYPE TL172C
NORMALLY OFF SILICON HALL-EFFECT SWITCH

D2490, AUGUST 1977—REVISED NOVEMBER 1983

- ● Magnetic-Field Sensing Hall-Effect Input
- ● On-Off Hysteresis
- ● Small Size
- ● Solid-State Technology
- ● Open-Collector Output
- ● Normally Off Switch

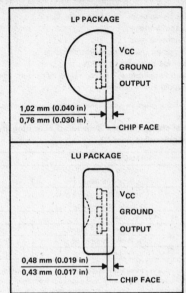

TOP VIEWS

LP PACKAGE

VCC
GROUND
OUTPUT

1,02 mm (0.040 in)
0,76 mm (0.030 in)
CHIP FACE

LU PACKAGE

VCC
GROUND
OUTPUT

0,48 mm (0.019 in)
0,43 mm (0.017 in)
CHIP FACE

Special Functions

5

description

The TL172C is a low-cost magnetically operated normally off electronic switch that utilizes the Hall Effect to sense the presence of a magnetic field. Each circuit consists of a Hall-Effect sensor, signal conditioning and hysteresis functions, and an output transistor integrated into a monolithic chip. A magnetic field of sufficient strength in the positive direction will cause the TL172C output to be in a low-impedance state. Otherwise the output will present a high impedance. The output of this circuitry can be connected to many different types of electronic components.

The TL172C is characterized for operation over the temperature range of 0°C to 70°C.

FUNCTION TABLE

FLUX DENSITY	OUTPUT
≤ 10 mT	Off
10 mT < B < 60 mT	Undefined
≥ 60 mT	On

functional block diagram

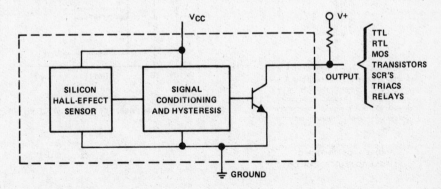

VCC

V+

TTL
RTL
MOS
TRANSISTORS
SCR'S
TRIACS
RELAYS

OUTPUT

SILICON
HALL-EFFECT
SENSOR

SIGNAL
CONDITIONING
AND HYSTERESIS

GROUND

1183

TYPE TL172C
NORMALLY OFF SILICON HALL-EFFECT SWITCH

absolute maximum ratings over operating free-air temperature range (unless otherwise noted)

Supply voltage, V_{CC} (see Note 1) .... 7 V
Output voltage .... 30 V
Output current .... 20 mA
Operating free-air temperature range .... 0°C to 70°C
Storage temperature range .... −65°C to 150°C
Lead temperature 1,6 mm (1/16 inch) from case for 10 seconds .... 260°C
Magnetic flux density .... unlimited

NOTE 1: Voltage values are with respect to network ground terminal.

electrical characteristics over rated operating free-air temperature range, V_{CC} = 5 V ± 5% (unless otherwise noted)

	PARAMETER	TEST CONDITIONS	MIN	TYP	MAX	UNIT
B_{T+}	Threshold of positive-going magnetic flux density [†]				60	mT [§]
B_{T-}	Threshold of negative-going magnetic flux density [†]		10			mT [§]
$B_{T+} - B_{T-}$	Hysteresis				23	mT [§]
I_{OH}	High-level output current	V_{OH} = 20 V			100	μA
V_{OL}	Low-level output voltage	V_{CC} = 4.75 V, I_{OL} = 16 mA			0.4	V
I_{CC}	Supply current	V_{CC} = 5.25 V			6	mA

[†]Threshold values are those levels of magnetic flux density at which the output changes state. For the TL172C, a level more positive than B_{T+} causes the output to go to a low level, and a level more negative than B_{T-} causes the output to go to a high level. See Figures 1 and 2.
[§]The unit of magnetic flux density in the International System of Units (SI) is the tesla (T). The tesla is equal to one weber per square meter. Values expressed in milliteslas may be converted to gauss by multiplying by ten.

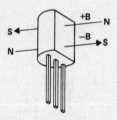

The north pole of a magnet is the pole that is attracted by the geographical north pole. The north pole of a magnet repels the north-seeking pole of a compass. By accepted magnetic convention, lines of flux emanate from the north pole of a magnet and enter the south pole.

FIGURE 1—DEFINITION OF MAGNETIC FLUX POLARITY

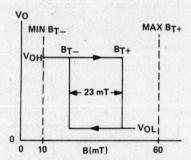

FIGURE 2—REPRESENTATIVE CURVE OF V_O vs B

TEXAS
INSTRUMENTS
POST OFFICE BOX 225012 • DALLAS, TEXAS 75265

1183

- Output Voltage Linear with Applied Magnetic Field
- Sensitivity Constant Over Wide Operating Temperature Range
- Solid-State Technology
- Three-Terminal Device
- Senses Static or Dynamic Magnetic Fields

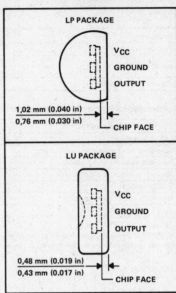

TOP VIEWS

LP PACKAGE

V_{CC}
GROUND
OUTPUT

1,02 mm (0.040 in)
0,76 mm (0.030 in)
CHIP FACE

LU PACKAGE

V_{CC}
GROUND
OUTPUT

0,48 mm (0.019 in)
0,43 mm (0.017 in)
CHIP FACE

description

The TL173I and TL173C are low-cost magnetic-field sensors designed to provide a linear output voltage proportional to the magnetic field they sense. These monolithic circuits incorporate a Hall element as the primary sensor along with a voltage reference and a precision amplifier. Temperature stabilization and internal trimming circuitry yield a device that features high overall sensitivity accuracy with less than 5% error over its operating temperature range.

The TL173I is characterized for operation from −20°C to 85°C. The TL173C is characterized for operation from 0°C to 70°C.

functional block diagram

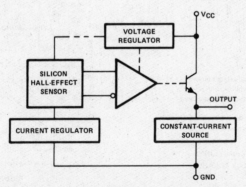

V_{CC}

VOLTAGE REGULATOR

SILICON HALL-EFFECT SENSOR

OUTPUT

CURRENT REGULATOR

CONSTANT-CURRENT SOURCE

GND

absolute maximum ratings over operating free-air temperature range (unless otherwise noted)

Supply voltage, V_{CC} (see Note 1) .... 25 V
Continuous total dissipation at (or below) 25°C free-air temperature (see Note 2) .... 775 mW
Operating free-air temperature range: TL173I .... −20°C to 85°C
                                    TL173C .... 0°C to 70°C
Storage temperature range .... −65°C to 150°C
Lead temperature 1,6 mm (1/16 inch) from case for 10 seconds .... 260°C
Magnetic flux density .... unlimited

NOTES: 1. Voltage values are with respect to network ground terminal.
       2. For operation above 25°C free-air temperature, derate linearly at the rate of 6.2 mW/°C.

Special Functions

5

recommended operating conditions

		TL173I MIN	TL173I NOM	TL173I MAX	TL173C MIN	TL173C NOM	TL173C MAX	UNIT
Supply voltage, V_{CC}		10.8	12	13.2	10.8	12	13.2	V
Magnetic flux density, B				±50			±50	mT
Output current, I_O	Sink			0.5			0.5	mV
	Source			−2			−2	
Operating free-air temperature, T_A		−20		85	0		70	°C

electrical characteristics over full range of recommended operating conditions (unless otherwise noted)

	PARAMETER	TEST CONDITIONS[†]	MIN	TYP[‡]	MAX	UNIT
V_O	Output voltage	$I_O = -2$ mA to 0.5 mA,	5.8	6	6.2	V
k_{SVS}	Supply voltage sensitivity ($\Delta V_{IO}/\Delta V_{CC}$)	B = 0 mT§, $\quad$ $T_A = 25°C$		18		mV/V
S	Magnetic sensitivity ($\Delta V_O/\Delta B$)	B = −50 to 50 mT§, $T_A = 25°C$	13.5	15	18	V/T§
ΔS	Magnetic sensitivity change with temperature	$\Delta T_A = 25°C$ to MIN or MAX			±5	%
I_{CC}	Supply current	B = 0 mT§, $\quad$ $I_O = 0$		8	12	mA
f_{max}	Maximum operating frequency			100		kHz

[†]For conditions shown as MIN or MAX, use the appropriate value specified under recommended operating conditions.
[‡]Typical values are at $V_{CC} = 12$ V and $T_A = 25°C$.
§The unit of magnetic flux density in the International System of Units (SI) is the tesla (T). The tesla is equal to one weber per square meter. Values expressed in milliteslas may be converted to gauss by multiplying by ten, e.g., 50 millitesla = 500 gauss.

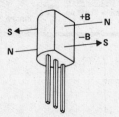

The north pole of a magnet is the pole that is attracted by the geographical north pole. The north pole of a magnet repels the north-seeking pole of a compass. By accepted magnetic convention, lines of flux emanate from the north pole of a magnet and enter the south pole.

**FIGURE 1—DEFINITION OF
MAGNETIC FLUX POLARITY**

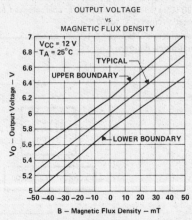

OUTPUT VOLTAGE
vs
MAGNETIC FLUX DENSITY

FIGURE 2

TEXAS
INSTRUMENTS
POST OFFICE BOX 225012 • DALLAS, TEXAS 75265

TYPICAL APPLICATION DATA

The circuit in Figure 3 may be used to set the output voltage at zero field strength to exactly 6 V (using R1), and to set the sensitivity to exactly −15 V/T (using R2), as depicted in Figure 4.

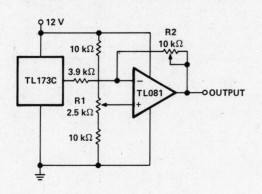

FIGURE 3—COMPENSATION CIRCUIT

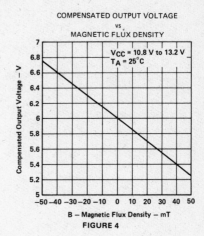

COMPENSATED OUTPUT VOLTAGE
vs
MAGNETIC FLUX DENSITY

FIGURE 4

Special Functions

5

TEXAS
INSTRUMENTS
POST OFFICE BOX 225012 • DALLAS, TEXAS 75265

LINEAR
INTEGRATED
CIRCUITS

TYPES TL592, TL592A
DIFFERENTIAL VIDEO AMPLIFIERS

D2668, NOVEMBER 1983

- 8-Pin Version of NE592 . . . Saves Printed Circuit Board Space
- Adjustable Gain to 400
- No Frequency Compensation Required
- Adjustable Passband

P DUAL-IN-LINE PACKAGE
(TOP VIEW)

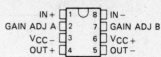

description

This device is a monolithic two-stage video amplifier with differential inputs and differential outputs.

Internal series-shunt feedback provides wide bandwidth, low phase distortion, and excellent gain stability. Emitter-follower outputs enable the device to drive capacitive loads. All stages are current-source biased to obtain high common-mode and supply-voltage rejection ratios.

Fixed differential amplification of nominally 400 may be selected without external components, or amplification may be adjusted from 0 to approximately 400 by the use of a single external resistor connected between the gain-adjustment pins A and B. No external frequency-compensating components are required for any gain option.

The device is particularly useful in magnetic-tape or disc-file systems using phase or NRZ encoding and in high-speed thin-film or plated-wire memories. Other applications include general-purpose video and pulse amplifiers where wide bandwidth, low phase shift, and excellent gain stability are required.

The TL592 and TL592A are characterized for operation from 0°C to 70°C.

symbol

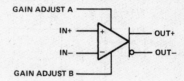

Special Functions

5

TEXAS
INSTRUMENTS
POST OFFICE BOX 225012 • DALLAS, TEXAS 75265

schematic

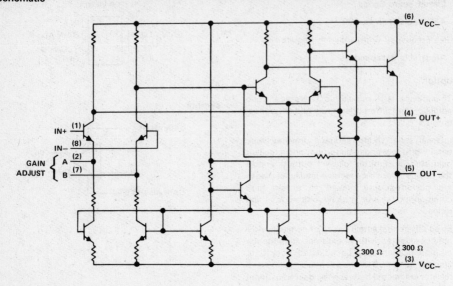

absolute maximum ratings over operating free-air temperature (unless otherwise noted)

Supply voltage, V_{CC+} (see Note 1) . . . . 8 V
Supply voltage, V_{CC-} (see Note 1) . . . . −8 V
Differential input voltage . . . . ±5 V
Voltage range, any input . . . . V_{CC+} to V_{CC-}
Output current . . . . 10 mA
Continuous total power dissipation . . . . 500 mW
Operating free-air temperature range . . . . 0°C to 70°C
Storage temperature range . . . . −65°C to 150°C
Lead temperature 1,6 mm (1/16 inch) from case for 10 seconds . . . . 260°C

NOTE 1: All voltage values except differential input voltages are with respect to the midpoint between V_{CC+} and V_{CC-}.

recommended operating conditions

	MIN	NOM	MAX	UNIT
Supply voltage, V_{CC+}	3	6	8	V
Supply voltage, V_{CC-}	−3	−6	−8	V
Operating free-air temperature, T_A	0		70	°C

TEXAS
INSTRUMENTS
POST OFFICE BOX 225012 • DALLAS, TEXAS 75265

electrical characteristics at specified free-air temperature, V_{CC+} = 6 V, V_{CC-} = −6 V, R_L = 2 kΩ (unless otherwise noted)

PARAMETER	TEST FIGURE	TEST CONDITIONS		GAIN OPTION†	TL592 MIN	TL592 TYP	TL592 MAX	TL592A MIN	TL592A TYP	TL592A MAX	UNIT
A_{VD} Large-signal differential voltage amplification	1	V_{OPP} = 3 V, R_L = 2 kΩ	25°C	1	250	400	600	400	440	600	V/V
			0°C to 70°C		250	400	600	400		600	
B_W Bandwidth (−3 dB)	2	V_{OPP} = 1 V	25°C	1		50			50		MHz
I_{IO} Input offset current			25°C	1 or 2		0.4	5		0.4	5	µA
			0°C to 70°C				6			6	
I_{IB} Input bias current			25°C	1 or 2		9	30		10	30	µA
			0°C to 70°C				40			40	
V_{ICR} Common-mode input voltage range	3		25°C	1 or 2	±1			±1			V
			0°C to 70°C		±1			±1			
V_{OC} Common-mode output voltage	1	R_L = ∞	25°C	1 or 2	2.4	2.9	3.4	2.4	2.9	3.4	V
V_{OO} Output offset voltage	1	V_{ID} = 0, R_L = ∞	25°C	2		0.35	0.75		0.35	0.75	V
			0°C to 70°C				1.5			1.5	
V_{OPP} Peak-to-peak output voltage swing	1	R_L = 2 kΩ	25°C	1	3	4		3	4		V
			0°C to 70°C		2.8			2.8			
z_i Input impedance	1	V_{OD} = 1 V, f = 1 kHz to 10 MHz, f = 100 kHz	25°C	1		4			3.6		kΩ
		f = 5 MHz				3.6			3.3		
CMRR Common-mode rejection ratio	3	V_{IC} = ±1 V, f = 100 kHz	25°C		60	86		60	86		dB
		f = 5 MHz			50	60		50	60		
k_{SVR} Supply voltage rejection ratio ($\Delta V_{CC}/\Delta V_{IO}$)	4	ΔV_{CC+} = ±0.5 V, ΔV_{CC-} = ±0.5 V	25°C	1	50	70		50	70		dB
			0°C to 70°C		50			50			
V_n Broadband equivalent input noise voltage	4	BW = 1 kHz to 10 MHz	25°C	1 or 2		12			12		µV
t_{pd} Propagation delay time	2	ΔV_O = 1 V	25°C	2		7.5			7.5		ns
t_r Rise time	2	ΔV_O = 1 V	25°C	2		10.5			10.5		ns
$I_{sink(max)}$ Maximum output sink current		No load,	25°C	1, 2, or 3	3	4		3	4		ma
I_{CC} Supply current		No signal	0°C to 70°C	1 or 2		18	24		19	24	mA
							27			27	

†The gain option is selected as follows:
Gain Option 1 . . . Gain adjust pin A is connected to pin B.
Gain Option 2 . . . Gain adjust pins A and B are open.

TYPES TL592, TL592A
DIFFERENTIAL VIDEO AMPLIFIERS

PARAMETER MEASUREMENT INFORMATION

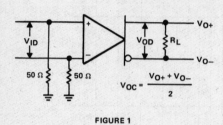

$$V_{OC} = \frac{V_{O+} + V_{O-}}{2}$$

FIGURE 1

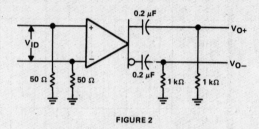

FIGURE 2

FIGURE 3

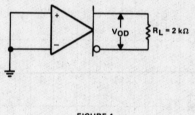

FIGURE 4

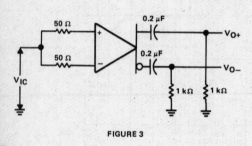

Special Functions

5

TEXAS
INSTRUMENTS
POST OFFICE BOX 225012 • DALLAS, TEXAS 75265

1183

LINEAR
INTEGRATED
CIRCUITS

TYPE TL592B
DIFFERENTIAL VIDEO AMPLIFIER

D2668, OCTOBER 1983

- Adjustable Gain to 400
- No Frequency Compensation Required
- Adjustable Passband
- Lower Noise . . . Less than 3 μV

description

This device is a monolithic two-stage video amplifier with differential inputs and differential outputs. It features internal series-shunt feedback that provides wide bandwidth, low phase distortion, and excellent gain stability. Emitter-follower outputs enable the device to drive capacitive loads. All stages are current-source biased to obtain high common-mode and supply-voltage rejection ratios.

Fixed differential amplification of 400 may be selected without external components, or amplification may be adjusted from 0 to 400 by the use of a single external resistor connected between the gain-adjustment pins A and B. No external frequency-compensating components are required for any gain option.

The device is particularly useful in magnetic-tape or disc-file systems that use phase or NRZ encoding and in high-speed thin-film or plated-wire memories. Other applications include general-purpose video and pulse amplifiers.

The device achieves lower equivalent noise through special processing and a new circuit layout incorporating input transistors with low base resistance.

The TL592B will be characterized for operation from 0 °C to 70 °C.

N DUAL-IN-LINE PACKAGE
(TOP VIEW)

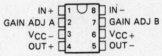

IN+	1	14 IN−
NC	2	13 NC
NC	3	12 NC
GAIN ADJ A	4	11 GAIN ADJ B
$V_{CC}-$	5	10 $V_{CC}+$
NC	6	9 NC
OUT+	7	8 OUT−

P DUAL-IN-LINE PACKAGE
(TOP VIEW)

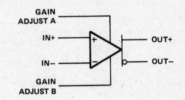

IN+	1	8 IN−
GAIN ADJ A	2	7 GAIN ADJ B
$V_{CC}-$	3	6 $V_{CC}+$
OUT+	4	5 OUT−

NC — No internal connection

symbol

GAIN
ADJUST A

IN+

IN−

GAIN
ADJUST B

OUT+

OUT−

Special Functions

5

TEXAS
INSTRUMENTS

POST OFFICE BOX 225012 • DALLAS, TEXAS 75265

LINEAR
INTEGRATED
CIRCUITS

TYPE TL851
SONAR RANGING CONTROL

D2760, SEPTEMBER 1983

- Designed for Use with the TL852 in Sonar Ranging Modules Like the SN28827

- Operates with Single Supply

- Accurate Clock Output for External Use

- Synchronous 4-Bit Gain Control Output

- Internal 1.2-V Level Detector for Receive

- TTL-Compatible

- Interfaces to Electrostatic or Piezoelectric Transducers

N DUAL-IN-LINE PACKAGE
(TOP VIEW)

V_{CC}	1	16	BLNK
XMIT	2	15	BINH
GND	3	14	INIT
GCD	4	13	FILT
GCA	5	12	XTAL2
GCB	6	11	XTAL1
GCC	7	10	OSC
REC	8	9	ECHO

description

The TL851 is an economical digital I^2L ranging control integrated circuit designed for use with the Texas Instruments TL852 Sonar ranging receiver integrated circuit.

The TL851 is designed for distance measurement from six inches to 35 feet. The device has an internal oscillator that uses a low-cost external ceramic resonator. With a simple interface and a 420-kHz ceramic resonator, the device will drive a 50-kHz electrostatic transducer.

The device cycle begins when Initiate (INIT) is taken to the high logic level. There must be at least 5 ms from initial power up (V_{CC}) to the first initiate signal in order for all the device internal latches to reset and for the ceramic-resonator-controlled oscillator to stabilize. The device will transmit a burst of 16 pulses each time INIT is taken high.

The oscillator output (OSC) is enabled by INIT. The oscillator frequency is the ceramic resonator frequency divided by 8.5 for the first 16 cycles (during transmit) and then the oscillator frequency changes to the ceramic resonator frequency divided by 4.5 for the remainder of the device cycle.

When used with an external 420-kilohertz ceramic resonator, the device internal blanking disables the receive input (REC) for 2.38 ms after initiate to exclude false receive inputs that may be caused by transducer ringing. The internal blanking feature also eliminates echos from objects closer than 1.3 feet from the transducer. If it is necessary to detect objects closer than 1.3 feet, then the internal blanking may be shortened by taking the blanking inhibit (BINH) high, enabling the receive input. The blanking input (BLNK) may be used to disable the receive input and reset ECHO to a low logic level at any time during the device cycle for selective echo exclusion or for a multiple-echo mode of operation.

The device provides a synchronous 4-bit gain control output (12 steps) designed to control the gain of the TL852 sonar ranging receiver integrated circuit. The digital gain control waveforms are shown in Figure 2 with the nominal transition times from INIT listed in the Gain Control Output Table.

The threshold of the internal receive level detector is 1.2 volts. The TL851 operates over a supply voltage range of 4.5 volts to 6.8 volts and is characterized for operation from 0°C to 40°C.

Special Functions

5

Copyright © 1983 by Texas Instruments Incorporated

TEXAS
INSTRUMENTS
POST OFFICE BOX 225012 • DALLAS, TEXAS 75265

TYPE TL851
SONAR RANGING CONTROL

GAIN CONTROL OUTPUT TABLE

STEP NUMBER	GCD	GCC	GCB	GCA	TIME (ms) FROM INITIATE↑[†]
0	L	L	L	L	2.38 ms
1	L	L	L	H	5.12 ms
2	L	L	H	L	7.87 ms
3	L	L	H	H	10.61 ms
4	L	H	L	L	13.35 ms
5	L	H	L	H	16.09 ms
6	L	H	H	L	18.84 ms
7	L	H	H	H	21.58 ms
8	H	L	L	L	27.07 ms
9	H	L	L	H	32.55 ms
10	H	L	H	L	38.04 ms
11	H	L	H	H	INIT ↓

[†] This is the time to the end of the indicated step and assumes a nominal 420-kHz ceramic resonator.

functional block diagram

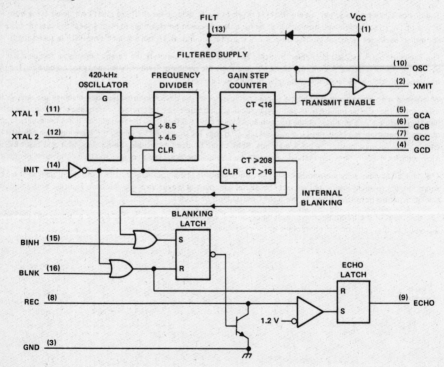

TEXAS
INSTRUMENTS
POST OFFICE BOX 225012 • DALLAS, TEXAS 75265

983

schematics of inputs and outputs

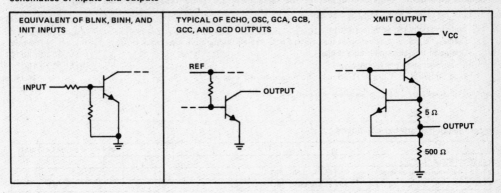

| EQUIVALENT OF BLNK, BINH, AND INIT INPUTS | TYPICAL OF ECHO, OSC, GCA, GCB, GCC, AND GCD OUTPUTS | XMIT OUTPUT |

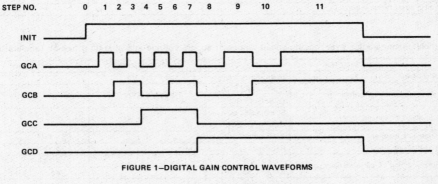

FIGURE 1—DIGITAL GAIN CONTROL WAVEFORMS

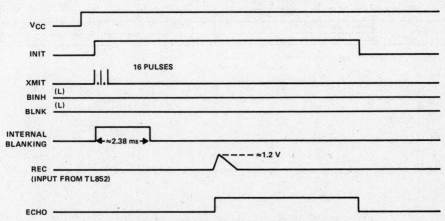

FIGURE 2—EXAMPLE OF SINGLE-ECHO-MODE CYCLE WHEN USED WITH THE
TL852 RECEIVER AND 420-kHz CERAMIC RESONATOR

983

TYPE TL851
SONAR RANGING CONTROL

absolute maximum ratings over operating free-air temperature range (unless otherwise noted)

Voltage at any pin with respect to GND .... −0.5 V to 7 V
Voltage at any pin with respect to V_{CC} .... −7 V to 0.5 V
Continuous power dissipation at (or below) 25°C free-air temperature (see Note 1) .... 1150 mW
Operating free-air temperature range .... 0°C to 70°C
Storage temperature range .... −65°C to 150°C
Lead temperature 1,6 mm (1/16 inch) from case for 10 seconds .... 260°C

NOTE 1: For operation above 25°C, derate linearly at the rate of 9.2 mW/°C.

recommended operating conditions

		MIN	MAX	UNIT
Supply voltage, V_{CC}		4.5	6.8	V
High-level input voltage, V_{IH}	BLNK, BINH, INIT	2.1		V
Low-level input voltage, V_{IL}	BLNK, BINH, INIT		0.6	V
Delay time, power up to INIT high		5		ms
Operating free-air temperature, T_A		0	40	°C

electrical characteristics over recommended ranges of supply voltage and operating free-air temperature

PARAMETER		TEST CONDITIONS	MIN	TYP	MAX	UNIT
Input current	BLNK, BINH, INIT	V_I = 2.1 V			1	mA
High-level output current, I_{OH}	ECHO, OSC, GCA, GCB, GCC, GCD	V_{OH} = 5.5 V			100	µA
Low-level output voltage, V_{OL}	ECHO, OSC, GCA, GCB, GCC, GCD	I_{OL} = 1.6 mA			0.4	V
On-state output current	XMIT output	V_O = 1 V		−140		mA
Internal blanking interval	REC input			2.38[†]		ms
Frequency during 16-pulse transmit period	OSC output			49.4[†]		kHz
	XMIT output			49.4[†]		
Frequency after 16-pulse transmit period	OSC output			93.3[†]		kHz
	XMIT output			0		
Supply current, I_{CC}	During transmit period				260	mA
	After transmit period				55	

[†]These typical values apply for a 420-kHz ceramic resonator.

TEXAS
INSTRUMENTS
POST OFFICE BOX 225012 • DALLAS, TEXAS 75265

LINEAR
INTEGRATED
CIRCUITS

TYPE TL852
SONAR RANGING RECEIVER

D2779, NOVEMBER 1983

- Designed for Use with the TL851 in Sonar Ranging Modules Like the SN28827

- Digitally Controlled Variable-Gain Variable-Bandwidth Amplifier

- Operational Frequency Range of 20 kHz to 90 kHz

- TTL-Compatible

- Operates from Power Sources of 4.5 V to 6.8 V

- Interfaces to Electrostatic or Piezoelectric Transducers

- Overall Gain Adjustable with One External Resistor

N DUAL-IN-LINE PACKAGE
(TOP VIEW)

G1IN	1	16 GND
XIN	2	15 GCD
GADJ	3	14 GCA
LC	4	13 GCB
V_{CC}	5	12 GCC
G1OUT	6	11 NC
G2IN	7	10 NC
BIAS	8	9 REC

NC—No internal connection

description

The TL852 is an economical sonar ranging receiver integrated circuit for use with the TL851 control integrated circuit. A minimum of external components is required for operation, and this amplifier easily interfaces to Polaroid's 50-kilohertz electrostatic transducer. An external 68-kilohm ±5% resistor from pin 8 (Bias) to pin 16 (GND) provides the internal biasing reference. Amplifier gain can be set with a resistor from pin 1 (G1IN) to pin 3 (GADJ). Required amplifier gain will vary for different applications. Using the detect-level measurement circuit of Figure 1, a nominal peak-to-peak value of 230 millivolts input during gain step 2 is recommended for most applications. For reliable operation, a level no lower than 50 millivolts should be used. The recommended detect level of 230 millivolts can be obtained for most amplifiers with an R1 value between 5 kilohms and 20 kilohms.

Digital control of amplifier gain is provided with gain control inputs on pins 12 through 15. These inputs must be driven synchronously (all inputs stable within 0.1 microsecond) to avoid false receive output signals due to invalid logic counts. This can be done easily with the TL851 control IC. A plot showing relative gain for the various gain steps versus time can be seen in Figure 2. To dampen ringing of the 50-kilohertz electrostatic transducer, a 5-kilohm resistor from pin 1 (GAIN) to pin 2 (XIN) is recommended.

An external parallel combination of inductance and capacitance between pin 4 (LC) and pin 5 (V_{CC}) provides an amplifier with an externally controlled gain and Q. This not only allows control of gain to compensate for attenuation of signal with distance, but also maximizes noise and sidelobe rejection. Care must be taken to accurately tune the L-C combination at operating frequency or gain and Q will be greatly reduced at higher gain steps.

AC coupling between stages of the amplifier is accomplished with a 0.01-microfarad capacitor for proper biasing.

The receive output is normally held at a low level by an internal 1-microampere current source. When an input of sufficient amplitude is received, the output is driven alternately by the 1-microampere discharge current and a 50-microampere charging current. A 1000-picofarad capacitor is required from the receive output (pin 9) to ground (pin 16) to integrate the received signal so that one or two noise pulses will not be recognized.

Pin 2 (XIN) provides clamping for the transformer secondary when used for transducer transmit drive as shown in Figure 4 of the SN28827 data sheet.

Special Functions

5

TEXAS
INSTRUMENTS
POST OFFICE BOX 225012 • DALLAS, TEXAS 75265

TYPE TL852
SONAR RANGING RECEIVER

functional block diagram

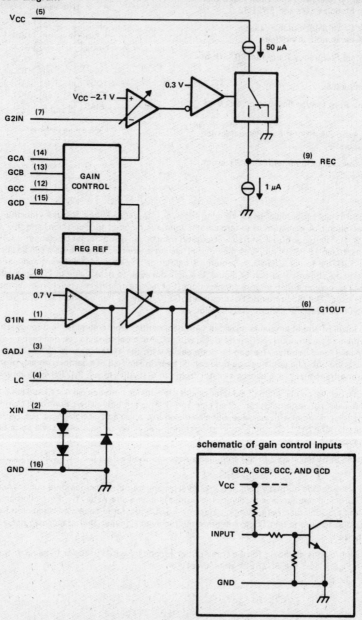

schematic of gain control inputs

GCA, GCB, GCC, AND GCD

TEXAS
INSTRUMENTS
POST OFFICE BOX 225012 • DALLAS, TEXAS 75265

absolute maximum ratings over operating free-air temperature range (unless otherwise noted)

Voltage at any pin with respect to GND . . . . −0.5 V to 7 V
Voltage at any pin with respect to V_{CC} . . . . −7 V to 0.5 V
XIN input current (50% duty cycle) . . . . ±60 mA
Continuous power dissipation at (or below) 25°C free-air temperature (see Note 1) . . . . . 1150 mW
Operating free-air temperature range . . . . −40°C to 85°C
Storage temperature range . . . . −65°C to 150°C
Lead temperature 1,6 mm (1/16 inch) from case for 10 seconds . . . . 260°C

NOTE 1: For operation above 25°C, derate linearly at the rate of 9.2 mW/°C.

recommended operating conditions

		MIN	MAX	UNIT
Supply voltage, V_{CC}		4.5	6.8	V
High-level input voltage, V_{IH}	GCA, GCB, GCC, GCD	2.1		V
Low-level input voltage, V_{IL}			0.6	V
Bias resistor between pins 8 and 16		64	72	kΩ
Operation free-air temperature, T_A		0	40	°C

electrical characteristics over recommended ranges of supply voltage and operating free-air temperature (T_A = 0°C to 40°C)

PARAMETER	TEST CONDITIONS		MIN	TYP	MAX	UNIT
Input clamp voltage at XIN	I_I = 40 mA				2.5	V
	I_I = −40 mA				−1.5	
Open-circuit input voltage at GCA, GCB, GCC, GCD	V_{CC} = 5 V,	I_I = 0		2.5		V
High-level input current, I_{IH}, into GCA, GCB, GCC, GCD	V_{CC} = 5 V,	V_{IH} = 2 V		−0.5		mA
Low-level input current, I_{IL}, into GCA, GCB, GCC, GCD	V_{CC} = 5 V,	V_{IL} = 0			−3	mA
Receive output current	I_{G2IN} = −100 μA,	V_O = 0.3 V		1		μA
	I_{G2IN} = 100 μA,	V_O = 0.1 V		−50		
Supply current, I_{CC}					45	mA

Special Functions

5

TYPE TL852
SONAR RANGING RECEIVER

TYPICAL APPLICATION INFORMATION

detect level vs gain step

Detect level is measured by applying a 15-cycle burst of 49.4 kilohertz square wave just after the beginning of the gain step to be tested. The least burst amplitude that makes the REC pin reach the trip level is defined to be the detect level. System gain is then inversely proportional to detect level. See the test circuit in Figure 1.

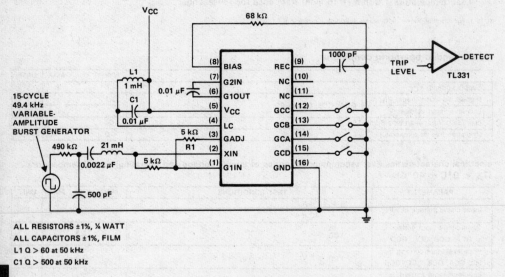

ALL RESISTORS ±1%, ¼ WATT
ALL CAPACITORS ±1%, FILM
L1 Q > 60 at 50 kHz
C1 Q > 500 at 50 kHz

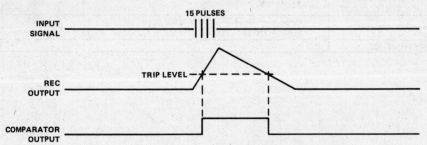

FIGURE 1–DETECT-LEVEL MEASUREMENT CIRCUIT AND WAVEFORMS

Special Functions

5

TEXAS
INSTRUMENTS
POST OFFICE BOX 225012 • DALLAS, TEXAS 75265

1183

TYPICAL APPLICATION INFORMATION

GAIN STEP TABLE

GCD	GCC	GCB	GCA	STEP NUMBER
L	L	L	L	0
L	L	L	H	1
L	L	H	L	2
L	L	H	H	3
L	H	L	L	4
L	H	L	H	5
L	H	H	L	6
L	H	H	H	7
H	L	L	L	8
H	L	L	H	9
H	L	H	L	10
H	L	H	H	11

RECEIVER GAIN
vs
GAIN STEP NUMBERS

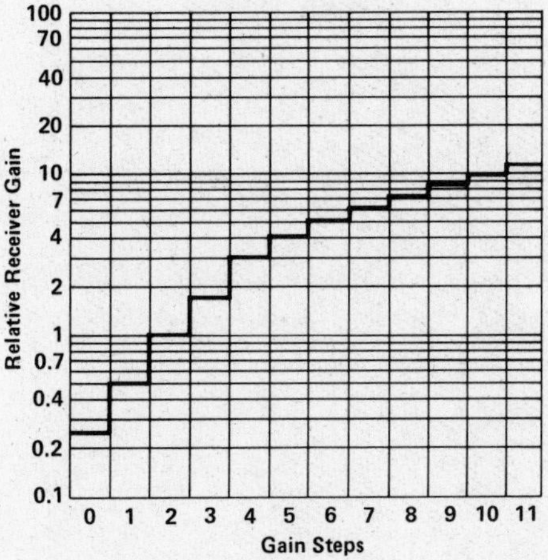

FIGURE 2

Special Functions

5

**LINEAR
INTEGRATED
CIRCUITS**

**TYPE TLC551C
LinCMOS™ TIMER**

D2791, FEBRUARY 1984

- Very Low Power Consumption . . . 1 mW
  Typ at V_{DD} = 5 V
- Capable of Very-High-Speed Operation
  Typically 2 MHz in Astable Mode
- Complementary MOS Output Capable of
  Swinging Rail-to-Rail
- High Output-Current Capability
  Sink 100 mA Typ
  Source 10 mA Typ
- Output Fully CMOS-, TTL-, and
  MOS-Compatible
- Low Supply Current Reduces Spikes During
  Output Transitions
- High Impedance Inputs . . . 10^{12} Ω Typ
- Single-Supply Operation from 1 to 18 Volts
- Functionally Interchangeable with the
  Signetics NE555; has Same Pinout

**D, JG, OR P
DUAL-IN-LINE PACKAGE
(TOP VIEW)**

```
GND  [ 1  U  8 ]  VDD
TRIG [ 2     7 ]  DSCH
OUT  [ 3     6 ]  THRES
RESET[ 4     5 ]  CONT
```

description

The TLC551 is a monolithic timing circuit fabricated using TI's LinCMOS™ process. Due to its high-impedance inputs (typically 10^{12} Ω), it is capable of producing accurate time delays and oscillations while using less-expensive, smaller timing capacitors than the NE555. Like the NE555 , the TLC551 achieves both monostable (using one resistor and one capacitor) and astable (using two resistors and one capacitor) operation. In addition, 50% duty cycle astable operation is possible using only a single resistor and one capacitor. The LinCMOS™ process allows the TLC551 to operate at frequencies up to 2 MHz and be fully compatible with CMOS, TTL, and MOS logic. It also provides very low power consumption (typically 1 mW at V_{DD} = 5 V) over a wide range of supply voltages ranging from 1 volt to 18 volts.

functional block diagram

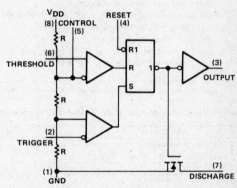

Reset can override Trigger, which can override Threshold.

Special Functions

5

The threshold and trigger levels are normally two-thirds and one-third respectively of V_{DD}. These levels can be altered by use of the control voltage terminal. When the trigger input falls below the trigger level, the flip-flop is set and the output goes high. If the trigger input is above the trigger level and the threshold input is above the threshold level, the flip-flop is reset and the output is low. The reset input can override all other inputs and can be used to initiate a new timing cycle. When the reset input goes low, the flip-flop is reset and the output goes low. Whenever the output is low, a low-impedance path is provided between the discharge terminal and ground.

While the CMOS output is capable of sinking over 100 mA and sourcing over 10 mA, the TLC551 exhibits greatly reduced supply current spikes during output transitions. This minimizes the need for the large decoupling capacitors required by the NE555.

The TLC551C is characterized for operation from 0°C to 70°C.

Copyright © 1984 by Texas Instruments Incorporated

**TEXAS
INSTRUMENTS**
POST OFFICE BOX 225012 • DALLAS, TEXAS 75265

FUNCTION TABLE

RESET	TRIGGER VOLTAGE†	THRESHOLD VOLTAGE†	OUTPUT	DISCHARGE SWITCH
Low	Irrelevant	Irrelevant	Low	On
High	< 1/3 V_{DD}	Irrelevant	High	Off
High	> 1/3 V_{DD}	> 2/3 V_{DD}	Low	On
High	> 1/3 V_{DD}	< 2/3 V_{DD}	As previously established	

†Voltages levels shown are nominal.

absolute maximum ratings over operating free-air temperature range (unless otherwise noted)

Supply voltage, V_{DD} (see Note 1) . . . . 18 V
Input voltage range (any input) . . . . − 0.3 V to 18 V
Continuous total dissipation at (or below) 25 °C free-air temperature (see Note 2) . . . . 600 mW
Operating free-air temperature range . . . . 0 °C to 70 °C
Storage temperature range . . . . − 65 °C to 150 °C
Lead temperature 1,6 mm (1/16 inch) from case for 60 seconds: JG package . . . . 300 °C
Lead temperature 1,6 mm (1/16 inch) from case for 10 seconds: D or P package . . . . 260 °C

NOTES: 1. All voltage values are with respect to network ground terminal.
2. For operation above 25 °C free-air temperature, refer to Dissipation Derating Curves, Section 2.

electrical characteristics at 25 °C free-air temperature, V_{DD} = 1 V to 15 V (unless otherwise noted)

PARAMETER	TEST CONDITIONS		MIN	TYP	MAX	UNIT
Threshold voltage level as a percentage of supply voltage				66.7%		
Threshold current	V_{DD} = 5 V			10		pA
Trigger voltage level as a percentage of supply voltage				33.3%		
Trigger current	V_{DD} = 5 V			10		pA
Reset voltage level				0.7		V
Reset current	V_{DD} = 5 V			± 10		pA
Control voltage (open-circuit) as a percentage of supply voltage				66.7%		
Low-level output voltage	V_{DD} = 15 V	I_{OL} = 10 mA		0.1		V
		I_{OL} = 50 mA		0.5		
		I_{OL} = 100 mA		1		
	V_{DD} = 5 V	I_{OL} = 5 mA		0.1		
		I_{OL} = 8 mA		0.16		
High-level output voltage	V_{DD} = 15 V	I_{OH} = − 1 mA		14.8		V
		I_{OH} = − 5 mA		14		
		I_{OH} = − 10 mA		12.7		
	V_{DD} = 5 V	I_{OH} = − 2 mA		4		
		I_{OH} = − 1 mA		4.5		
Supply current	V_{DD} = 15 V			360		µA
	V_{DD} = 5 V			170		

TEXAS
INSTRUMENTS
POST OFFICE BOX 225012 • DALLAS, TEXAS 75265

operating characteristics, V_{DD} = 1.2 V, T_A = 25°C (unless otherwise noted)

PARAMETER	TEST CONDITIONS	MIN	TYP	MAX	UNIT
Initial error of timing interval	V_{DD} = 1 V to 15 V, R_A = R_B = 1 kΩ to 100 kΩ, C_T = 0.1 μF, See Note 1		1%		
Supply voltage sensitivity of timing interval			0.1		%/V
Output pulse rise time	R_L = 10 MΩ, C_L = 10 pF		20		ns
Output pulse fall time			20		
Maximum frequency in astable mode	R_A = 470 Ω, R_B = 200 Ω, C_T = 200 pF, See Note 1		2.1		MHz

NOTE 1: R_A, R_B, and C_T are as defined in Figure 1.

TYPICAL APPLICATION DATA

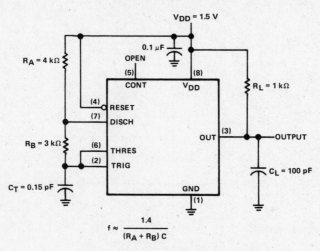

$$f \approx \frac{1.4}{(R_A + R_B) C}$$

FIGURE 1—CIRCUIT FOR ASTABLE OPERATION

Special Functions

5

LINEAR
INTEGRATED
CIRCUITS

TYPES TLC552C, TLC556M, TLC556C
DUAL LinCMOS™ TIMERS

D2796, FEBRUARY 1984

- **Very Low Power Consumption**
  . . . 2 mW Typ at V_{DD} = 5 V

- **Capable of Very High-Speed Operation**
  . . . Typically 2 MHz in Astable Mode

- **Complementary MOS output Capable of**
  Swinging Rail-to-Rail

- **High Output-Current Capability**
  . . . Sink 100 mA Typ
  . . . Source 10 mA Typ

- **Output Fully CMOS-, TTL-, and**
  MOS-Compatible

- **Low Supply Current Reduces Spikes During**
  Output Transitions

- **High Impedance Inputs . . . 10^{12} Ω Typ**

- **Single-Supply Operation**
  TLC552 . . . from 1 to 18 Volts
  TLC556 . . . from 2 to 18 Volts

- **Functionally Interchangeable with the**
  Signetics NE556 and SE556; Has Same
  Pinout

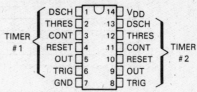

TLC556M . . . J DUAL-IN-LINE PACKAGE
TLC552C, TLC556C . . . D OR N DUAL-IN-LINE PACKAGE
(TOP VIEW)

```
              DSCH  [ 1   14 ]  VDD
             THRES  [ 2   13 ]  DSCH
 TIMER        CONT  [ 3   12 ]  THRES
  #1         RESET  [ 4   11 ]  CONT     TIMER
               OUT  [ 5   10 ]  RESET     #2
              TRIG  [ 6    9 ]  OUT
               GND  [ 7    8 ]  TRIG
```

description

The TLC552 and TLC556 are dual monolithic timing circuits fabricated using TI's LinCMOS™ process. Due to their high-impedance inputs (typically 10^{12} Ω), they are capable of producing accurate time delays and oscillations while using less expensive, smaller timing capacitors than the NE556. Like the NE556, the TLC552 and TLC556 achieve both monostable (using one resistor and one capacitor) and astable (using two resistors and one capacitor) operation. In addition, 50% duty cycle astable operation is possible using only a single resistor and one capacitor. The LinCMOS™ process allows the TLC552 and TLC556

functional block diagram (each timer)

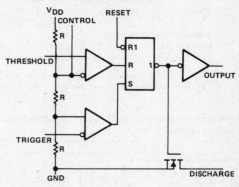

Reset can override Trigger and Threshold.
Trigger can override Threshold.

to operate at frequencies up to 2 MHz and be fully compatible with CMOS, TTL, and MOS logic. It also provides very low power consumption (typically 2 mW at V_{DD} = 5 V) over a wide range of supply voltages ranging from 1 volt to 18 volts for the TLC552 and 2 volts to 18 volts for the TLC556.

The threshold and trigger levels are normally two-thirds and one-third respectively of V_{DD}. These levels can be altered by use of the control voltage terminal. When the trigger input falls below trigger level, the flip-flop is set and the output goes high. If the trigger input is above the trigger level and the threshold input is above the threshold level, the flip-flop is reset and the output is low. The reset input can override all other inputs and can be used to initiate a new timing cycle. When the reset input goes low, the flip-flop is reset and the output goes low. Whenever the output is low, a low impedance path is provided between the discharge terminal and ground.

While the CMOS output is capable of sinking over 100 mA and sourcing over 10 mA, the TLC552 and TLC556 exhibit greatly reduced supply current spikes during output transitions. This minimizes the need for the large decoupling capacitors required by the NE556.

The TLC556M will be characterized for operation over the full military temperature range of −55°C to 125°C. The TLC552CM and TLC556C are characterized for operation from 0°C to 70°C.

Special Functions

5

Copyright © 1984, Texas Instruments Incorporated

TEXAS
INSTRUMENTS
POST OFFICE BOX 225012 • DALLAS, TEXAS 75265

FUNCTION TABLE

RESET	TRIGGER VOLTAGE†	THRESHOLD VOLTAGE†	OUTPUT	DISCHARGE SWITCH
Low	Irrelevant	Irrelevant	Low	On
High	< 1/3 V_{DD}	Irrelevant	High	Off
High	> 1/3 V_{DD}	> 2/3 V_{DD}	Low	On
High	> 1/3 V_{DD}	< 2/3 V_{DD}	As previously established	

†Voltages levels shown are nominal.

absolute maximum ratings over operating free-air temperature range (unless otherwise noted)

Supply voltage, V_{DD} (see Note 1) . . . . 18 V
Input voltage range (any input) . . . . −0.3 V to 18 V
Continuous total dissipation at (or below 25 °C free-air temperature (see Note 2) . . . . 950 mW
Operating free-air temperature range: TLC556M . . . . −55°C to 125°C
                                  TLC552C, TLC556C . . . . 0°C to 70°C
Storage temperature range . . . . −65°C to 150°C
Lead temperature 1,6 mm (1/16 inch) from case for 60 seconds: J package . . . . 300°
Lead temperature 1,6 mm (1/16 inch) from case for 10 seconds: D or N package . . . . 260°C

NOTES: 1. All voltage values are with respect to network ground terminal.
        2. For operation above 25 °C free-air temperature, refer to Dissipation Derating Curves, Section 2.

electrical characteristics at 25 °C free-air temperature, V_{DD} = 1 V to 15 V for TLC552 or 2 V to 15 V for TLC556 (unless otherwise noted)

PARAMETER	TEST CONDITIONS		MIN	TYP	MAX	UNIT
Threshold voltage level as a percentage of supply voltage				66.7%		
Threshold current	V_{DD} = 5 V			10		pA
Trigger voltage level as a percentage of supply voltage				33.3%		
Trigger current	V_{DD} = 5 V			10		pA
Reset voltage level				0.7		V
Reset current	V_{DD} = 5 V			±10		pA
Control voltage (open-circuit) as a percentage of supply voltage				66.7%		
Low-level output voltage	V_{DD} = 15 V	I_{OL} = 10 mA		0.1		V
		I_{OL} = 50 mA		0.5		
		I_{OL} = 100 mA		1		
	V_{DD} = 5 V	I_{OL} = 5 mA		0.1		
		I_{OL} = 8 mA		0.16		
High-level output voltage	V_{DD} = 15 V	I_{OH} = −1 mA		14.8		V
		I_{OH} = −5 mA		14		
		I_{OH} = −10 mA		12.7		
	V_{DD} = 5 V	I_{OH} = −2 mA		4		
		I_{OH} = −1 mA		4.5		
Supply current	V_{DD} = 15 V			360		μA
	V_{DD} = 5 V			170		

TEXAS
INSTRUMENTS
POST OFFICE BOX 225012 • DALLAS, TEXAS 75265

Special Functions

5

operating characteristics, V$_{DD}$ = 1.2 V for TLC552 or 2 V for TLC556, T$_A$ = 25°C (unless otherwise noted)

PARAMETER	TEST CONDITIONS		MIN	TYP	MAX	UNIT
Initial error of timing interval	V$_{DD}$ = 1 V to 15 V for TLC552, or 2 V to 15 V for TLC556 R$_A$ = R$_B$ = 1 kΩ to 100 kΩ,			1%		
Supply voltage sensitivity of timing interval	C$_T$ = 0.1 μF, See Figure 1			0.1		%/V
Output pulse rise time	V$_{DD}$ = 5 V,	R$_L$ = 10 MΩ,		20		ns
Output pulse fall time	C$_L$ = 10 pF			20		
Maximum frequency in astable mode	R$_A$ = 470 Ω, R$_B$ = 200 Ω, C$_T$ = 200 pF			2.1		MHz

TYPICAL APPLICATION DATA

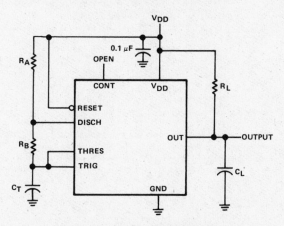

FIGURE 1—CIRCUIT FOR ASTABLE OPERATION

Special Functions

5

LINEAR
INTEGRATED
CIRCUITS

TYPES TLC555M, TLC555C
LinCMOS™ TIMERS

D2784, SEPTEMBER 1983

- Very Low Power Consumption . . . 1 mW Typ at V_{DD} = 5 V

- Capable of Very-High-Speed Operation . . . Typically 2 MHz in Astable Mode

- Complementary CMOS output Capable of Swinging Rail-to-Rail

- High Output-Current Capability
  . . . Sink 100 mA Typ
  . . . Source 10 mA Typ

- Output Fully CMOS-, TTL-, and MOS-Compatible

- Low Supply Current Reduces Spikes During Output Transitions

- High Impedance Inputs . . . 10^{12} Ω Typ

- Single-Supply Operation from 2 to 18 V

- Functionally Interchangeable with the Signetics NE555; has Same Pinout

TLC555M . . . JG PACKAGE
TLC555C . . . D, JG, or P PACKAGE
(TOP VIEW)

GND	1	8	V_{DD}
TRIG	2	7	DSCH
OUT	3	6	THRES
RESET	4	5	CONT

functional block diagram

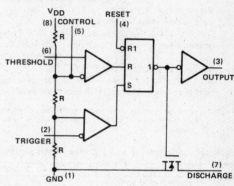

Reset can override Trigger, which can override Threshold.

Special Functions

5

description

The TLC555 is a monolithic timing circuit fabricated using TI's LinCMOS™ process. Due to its high-impedance inputs (typically 10^{12} Ω), it is capable of producing accurate time delays and oscillations while using less expensive, smaller timing capacitors than the NE555. Like the NE555, the TLC555 achieves both monostable (using one resistor and one capacitor) and astable (using two resistors and one capacitor) operation. In addition, 50% duty cycle astable operation is possible using only a single resistor and one capacitor. The LinCMOS™ process allows the TLC555 to operate at frequencies up to 2 MHz and be fully compatible with CMOS, TTL, and MOS logic. It also provides very low power consumption (typically 1 mW at V_{DD} = 5 V) over a wide range of supply voltages ranging from 2 volts to 18 volts.

Like the NE555, the threshold and trigger levels are normally two-thirds and one-third respectively of V_{DD}. These levels can be altered by use of the control voltage terminal. When the trigger input falls below trigger level, the flip-flop is set and the output goes high. If the trigger input is above the trigger level and the threshold input is above the threshold level, the flip-flop is reset and the output is low. The reset input can override all other inputs and can be used to initiate a new timing cycle. When the reset input goes low, the flip-flop is reset and the output goes low. Whenever the output is low, a low-impedance path is provided between the discharge terminal and ground.

While the complementary CMOS output is capable of sinking over 100 mA and sourcing over 10 mA, the TLC555 exhibits greatly reduced supply current spikes during output transitions. This minimizes the need for the large decoupling capacitors required by the NE555.

These devices have internal electrostatic discharge (ESD) protection circuits that will prevent catastrophic failures at voltage up to 2000 volts as tested under MIL-STD-883B, Method 3015.1. However, care should be exercised in handling these devices as exposure to ESD may result in a degradation of the device parametric performance.

All unused inputs should be tied to an appropriate logic level to prevent false triggering.

The TLC555M is characterized for operation over the full military temperature range of −55 °C to 125 °C; the TLC555C is characterized for operation from 0 °C to 70 °C.

Copyright © 1983 by Texas Instruments Incorporated

TEXAS
INSTRUMENTS

POST OFFICE BOX 225012 • DALLAS, TEXAS 75265

TYPES TLC555M, TLC555C
LinCMOS™ TIMERS

FUNCTION TABLE

RESET	TRIGGER VOLTAGE†	THRESHOLD VOLTAGE†	OUTPUT	DISCHARGE SWITCH
Low	Irrelevant	Irrelevant	Low	On
High	< 1/3 V_{DD}	Irrelevant	High	Off
High	> 1/3 V_{DD}	> 2/3 V_{DD}	Low	On
High	> 1/3 V_{DD}	< 2/3 V_{DD}	As previously established	

†Voltages levels shown are nominal.

absolute maximum ratings over operating free-air temperature range (unless otherwise noted)

Supply voltage, V_{DD} (see Note 1) . . . . 18 V
Input voltage range (any input) . . . . −0.3 V to 18 V
Continuous total dissipation at (or below) 25 °C free-air temperature (see Note 2). . . . . 600 mW
Operating free-air temperature range: TLC555M . . . . −55 °C to 125 °C
                                                            TLC555C . . . . 0 °C to 70 °C
Storage temperature range . . . . −65 °C to 150 °C
Lead temperature 1,6 mm (1/16 inch) from case for 60 seconds: JG package . . . . 300 °C
Lead temperature 1,6 mm (1/16 inch) from case for 10 seconds: D or P package . . . . 260 °C

NOTES: 1. All voltage values are with respect to network ground terminal.
2. For operation above 25 °C free-air temperature, refer to Dissipation Derating Curves, Section 2. In the JG package, TLC555M chips are alloy-mounted.

electrical characteristics at 25 °C free-air temperature, V_{DD} = 5 V to 15 V (unless otherwise noted)

PARAMETER	TEST CONDITIONS		MIN	TYP	MAX	UNIT
Threshold voltage level as a percentage of supply voltage				66.7%		
Threshold current	V_{DD} = 5 V			10		pA
Trigger voltage level as a percentage of supply voltage				33.3%		
Trigger current	V_{DD} = 5 V			10		pA
Reset voltage level				0.7		V
Reset current	V_{DD} = 5 V			±10		pA
Control voltage (open-circuit) as a percentage of supply voltage				66.7%		
Low-level output voltage	V_{DD} = 15 V	I_{OL} = 10 mA		0.1		V
		I_{OL} = 50 mA		0.5		
		I_{OL} = 100 mA		1		
	V_{DD} = 5 V	I_{OL} = 5 mA		0.1		
		I_{OL} = 8 mA		0.16		
High-level output voltage	V_{DD} = 15 V	I_{OH} = −1 mA		14.8		V
		I_{OH} = −5 mA		14		
		I_{OH} = −10 mA		12.7		
	V_{DD} = 5 V	I_{OH} = −2 mA		4		
		I_{OH} = −1 mA		4.5		
Supply current	V_{DD} = 15 V			360		µA
	V_{DD} = 5 V			170		

Special Functions

5

983

TEXAS
INSTRUMENTS
POST OFFICE BOX 225012 • DALLAS, TEXAS 75265

operating characteristics, V_{DD} = 5 V, T_A = 25°C (unless otherwise noted)

PARAMETER	TEST CONDITIONS		MIN	TYP	MAX	UNIT
Initial error of timing interval	V_{DD} = 5 V to 15 V, R_A = R_B = 1 kΩ to 100 kΩ,			1%		
Supply voltage sensitivity of timing interval	C_T = 0.1 μF, See Figure 1			0.1		%/V
Output pulse rise time	V_{DD} = 5 V,	R_L = 10 MΩ,		20		ns
Output pulse fall time	C_L = 10 pF			20		
Maximum frequency in astable mode	R_A = 470 Ω, C_T = 200 pF	R_B = 200 Ω,		2.1		MHz

TYPICAL APPLICATION DATA

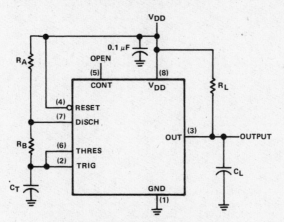

FIGURE 1—CIRCUIT FOR ASTABLE OPERATION

Special Functions

5

**TEXAS
INSTRUMENTS**
POST OFFICE BOX 225012 • DALLAS, TEXAS 75265

**LINEAR
INTEGRATED
CIRCUITS**

**TYPES uA733M, uA733C
DIFFERENTIAL VIDEO AMPLIFIERS**

D922, NOVEMBER 1970—REVISED OCTOBER 1979

- **200 MHz Bandwidth**
- **250 kΩ Input Resistance**
- **Selectable Nominal Amplification of 10, 100, or 400**
- **No Frequency Compensation Required**
- **Designed to be Interchangeable with Fairchild µA733M and µA733C**

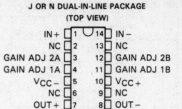

**J OR N DUAL-IN-LINE PACKAGE
(TOP VIEW)**

NC—No internal connection

**uA733M . . . U FLAT PACKAGE
(TOP VIEW)**

description

The uA733 is a monolithic two-stage video amplifier with differential inputs and differential outputs.

Internal series-shunt feedback provides wide bandwidth, low phase distortion, and excellent gain stability. Emitter-follower outputs enable the device to drive capacitive loads and all stages are current-source biased to obtain high common-mode and supply-voltage rejection ratios.

Fixed differential amplification of 10, 100, or 400 may be selected without external components, or amplification may be adjusted from 10 to 400 by the use of a single external resistor connected between 1A and 1B. No external frequency-compensating components are required for any gain option.

The device is particularly useful in magnetic-tape or disc-file systems using phase or NRZ encoding and in high-speed thin-film or plated-wire memories. Other applications include general purpose video and pulse amplifiers where wide bandwidth, low phase shift, and excellent gain stability are required.

The uA733M is characterized for operation over the full military temperature range of -55°C to 125°C; the uA733C is characterized for operation from 0°C to 70°C.

symbol

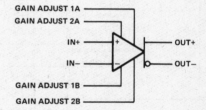

Special Functions

5

absolute maximum ratings over operating free-air temperature range (unless otherwise noted)

		uA733M	uA733C	UNIT
Supply voltage V_{CC+} (See Note 1)		8	8	V
Supply voltage V_{CC-} (See Note 1)		-8	-8	V
Differential input voltage		±5	±5	V
Common-mode input voltage		±6	±6	V
Output current		10	10	mA
Continuous total power dissipation at (or below) 25°C free-air temperature (see Note 2)		500	500	mW
Operating free-air temperature range		−55 to 125	0 to 70	°C
Storage temperature range		−65 to 150	−65 to 150	°C
Lead temperature 1,6 mm (1/16 inch) from case for 60 seconds	J or U package	300	300	°C
Lead temperature 1,6 mm (1/16 inch) from case for 10 seconds	N package		260	°C

NOTES: 1. All voltage values, except differential input voltages, are with respect to the midpoint between V_{CC+} and V_{CC-}.

2. For operation above 25°C free-air temperature, refer to Dissipation Derating Curves, Section 2. In the J package, uA733M chips are alloy mounted; uA733C chips are glass mounted.

electrical characteristics, V_{CC+} = 6 V, V_{CC-} = −6 V, T_A = 25°C

PARAMETER		TEST FIGURE	TEST CONDITIONS	GAIN OPTION†	uA733M MIN	TYP	MAX	uA733C MIN	TYP	MAX	UNIT
A_{VD}	Large-signal differential voltage amplification	1	V_{OD} = 1 V	1	300	400	500	250	400	600	
				2	90	100	110	80	100	120	
				3	9	10	11	8	10	12	
BW	Bandwidth	2	R_S = 50 Ω	1		50			50		MHz
				2		90			90		
				3		200			200		
I_{IO}	Input offset current			Any		0.4	3		0.4	5	µA
I_{IB}	Input bias current			Any		9	20		9	30	µA
V_{ICR}	Common-mode input voltage range	1		Any	±1			±1			V
V_{OC}	Common-mode output voltage	1		Any	2.4	2.9	3.4	2.4	2.9	3.4	V
V_{OO}	Output offset voltage	1		1		0.6	1.5		0.6	1.5	V
				2 & 3		0.35	1		0.35	1.5	
V_{OPP}	Maximum peak-to-peak output voltage swing	1		Any	3	4.7		3	4.7		V
r_i	Input resistance	3	$V_{OD} \leqslant 1$ V	1		4			4		kΩ
				2	20	24		10	24		
				3		250			250		
r_o	Output resistance					20			20		Ω
C_i	Input capacitance	3	$V_{OD} \leqslant 1$ V	2		2			2		pF
CMRR	Common-mode rejection ratio	4	V_{IC} = ±1 V, f ≤ 100 kHz	2	60	86		60	86		dB
			V_{IC} = ±1 V, f = 5 MHz	2		70			70		
k_{SVR}	Supply voltage rejection ratio ($\Delta V_{CC}/\Delta V_{IO}$)	1	ΔV_{CC+} = ± 0.5 V, ΔV_{CC-} = ± 0.5 V	2	50	70		50	70		dB
V_n	Broadband equivalent input noise voltage	5	BW = 1 kHz to 10 MHz	Any		12			12		µV
t_{pd}	Propagation delay time	2	R_S = 50 Ω, Output voltage step = 1 V	1		7.5			7.5		ns
				2		6.0	10		6.0	10	
				3		3.6			3.6		
t_r	Rise time	2	R_S = 50 Ω, Output voltage step = 1 V	1		10.5			10.5		ns
				2		4.5	10		4.5	12	
				3		2.5			2.5		
$I_{sink(max)}$	Maximum output sink current			Any	2.5	3.6		2.5	3.6		mA
I_{CC}	Supply current		No load, no signal	Any		16	24		16	24	mA

†The gain option is selected as follows:
Gain Option 1 . . . Gain-adjust pin 1A is connected to pin 1B, and pins 2A and 2B are open.
Gain Option 2 . . . Gain-adjust pin 1A and pin 1B are open, pin 2A is connected to pin 2B.
Gain Option 3 . . . All four gain-adjust pins are open.

1183

TEXAS
INSTRUMENTS
POST OFFICE BOX 225012 ● DALLAS, TEXAS 75265

electrical characteristics (continued), V_{CC+} = 6 V, V_{CC-} = −6 V
T_A = −55°C to 125°C for uA733M, 0°C to 70°C for uA733C

PARAMETER		TEST FIGURE	TEST CONDITIONS	GAIN OPTION†	uA733M MIN	uA733M MAX	uA733C MIN	uA733C MAX	UNIT
A_{VD}	Large-signal differential voltage amplification	1	V_{OD} = 1 V	1	200	600	250	600	
				2	80	120	80	120	
				3	8	12	8	12	
I_{IO}	Input offset current			Any		5		6	µA
I_{IB}	Input bias current			Any		40		40	µA
V_{ICR}	Common-mode input voltage range	1		Any	±1		±1		V
V_{OO}	Output offset voltage	1		1		1.5		1.5	V
				2 & 3		1.2		1.5	
V_{OPP}	Maximum peak-to-peak output voltage swing	1		Any	2.5		2.8		V
r_i	Input resistance	3	V_{OD} ≤ 1 V	2	8		8		kΩ
CMRR	Common-mode rejection ratio	4	V_{IC} = ±1 V, f ≤ 100 kHz	2	50		50		dB
			V_{IC} = ±1 V, f = 5 MHz	2					
k_{SVR}	Supply voltage rejection ratio ($\Delta V_{CC}/\Delta V_{IO}$)	1	ΔV_{CC+} = ±0.5 V, ΔV_{CC-} = ±0.5 V	2	50		50		dB
$I_{sink(max)}$	Maximum output sink current			Any	2.2		2.5		mA
I_{CC}	Supply current		No load, No signal	Any		27		27	mA

†The gain option is selected as follows:
Gain Option 1 . . . Gain-adjust pin 1A is connected to pin 1B, and pins 2A and 2B are open.
Gain Option 2 . . . Gain-adjust pin 1A and pin 1B are open, pin 2A is connected to pin 2B.
Gain Option 3 . . . All four gain-adjust pins are open.

schematic

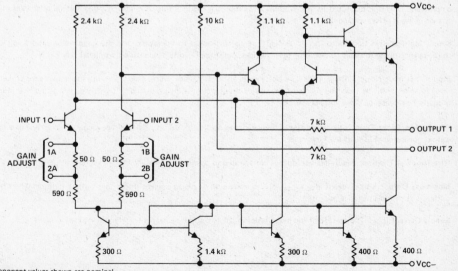

Component values shown are nominal.

TEXAS
INSTRUMENTS
POST OFFICE BOX 225012 • DALLAS, TEXAS 75265

Special Functions

5

DEFINITION OF TERMS

Large-Signal Differential Voltage Amplification (A_{VD}) The ratio of the change in voltage between the output terminals to the change in voltage between the input terminals producing it.

Bandwidth (BW) The range of frequencies within which the differential gain of the amplifier is not more than 3 dB below its low-frequency value.

Input Offset Current (I_{IO}) The difference between the currents into the two input terminals with the inputs grounded.

Input Bias Current (I_{IB}) The average of the currents into the two input terminals with the inputs grounded.

Input Voltage Range (V_I) The range of voltage that if exceeded at either input terminal will cause the amplifier to cease functioning properly.

Common-Mode Output Voltage (V_{OC}) The average of the d-c voltages at the two output terminals.

Output Offset Voltage (V_{OO}) The difference between the d-c voltages at the two output terminals when the input terminals are grounded.

Maximum Peak-to-Peak Output Voltage Swing (V_{OPP}) The maximum peak-to-peak output voltage swing that can be obtained without clipping. This includes the unbalance caused by output offset voltage.

Input Resistance (r_i) The resistance between the input terminals with either input grounded.

Output Resistance (r_o) The resistance between either output terminal and ground.

Input Capacitance (C_i) The capacitance between the input terminals with either input grounded.

Common-Mode Rejection Ratio (CMRR) The ratio of differential voltage amplification to common-mode voltage amplification. This is measured by determining the ratio of a change in input common-mode voltage to the resulting change in input offset voltage.

Supply Voltage Rejection Ratio ($\Delta V_{CC}/\Delta V_{IO}$) The absolute value of the ratio of the change in power supply voltages to the change in input offset voltage. For these devices, both supply voltages are varied symmetrically.

Equivalent Input Noise Voltage (V_n) The voltage of an ideal voltage source (having an internal impedance equal to zero) in series with the input terminals of the device that represents the part of the internally generated noise that can properly be represented by a voltage source.

Propagation Delay Time (t_{pd}) The interval between the application of an input voltage step and its arrival at either output, measured at 50% of the final value.

Rise Time (t_r) The time required for an output voltage step to change from 10% to 90% of its final value.

Maximum Output Sink Current ($I_{sink(max)}$) The maximum available current into either output terminal when that output is at its most negative potential.

Supply Current (I_{CC}) The average of the magnitudes of the two supply currents I_{CC1} and I_{CC2}.

TEXAS
INSTRUMENTS
POST OFFICE BOX 225012 • DALLAS, TEXAS 75265

PARAMETER MEASUREMENT INFORMATION

test circuits

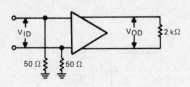

FIGURE 1

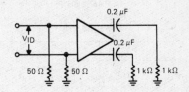

FIGURE 2

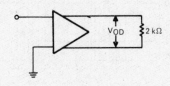

FIGURE 3

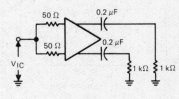

FIGURE 4

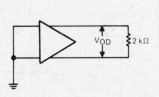

FIGURE 5

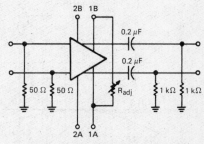

VOLTAGE AMPLIFICATION ADJUSTMENT

FIGURE 6

Special Functions

5

TYPES uA733M, uA733C
DIFFERENTIAL VIDEO AMPLIFIERS

TYPICAL CHARACTERISTICS

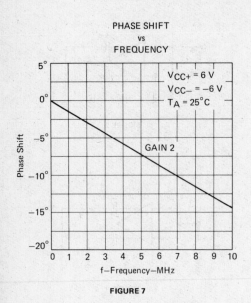

PHASE SHIFT vs FREQUENCY

FIGURE 7

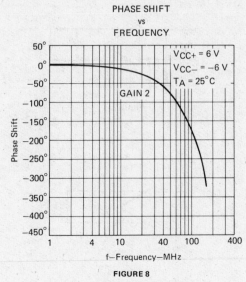

PHASE SHIFT vs FREQUENCY

FIGURE 8

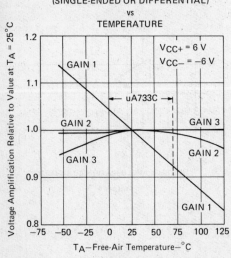

VOLTAGE AMPLIFICATION
(SINGLE-ENDED OR DIFFERENTIAL)
vs
TEMPERATURE

FIGURE 9

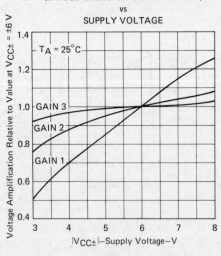

VOLTAGE AMPLIFICATION
(SINGLE-ENDED OR DIFFERENTIAL)
vs
SUPPLY VOLTAGE

FIGURE 10

5

TEXAS INSTRUMENTS
POST OFFICE BOX 225012 • DALLAS, TEXAS 75265

1183

TYPICAL CHARACTERISTICS

Special Functions

5

DIFFERENTIAL VOLTAGE AMPLIFICATION
vs
RESISTANCE BETWEEN G1A AND G1B

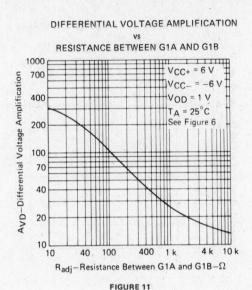

FIGURE 11

SINGLE-ENDED VOLTAGE AMPLIFICATION
vs
FREQUENCY

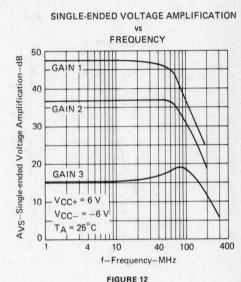

FIGURE 12

SUPPLY CURRENT
vs
FREE-AIR TEMPERATURE

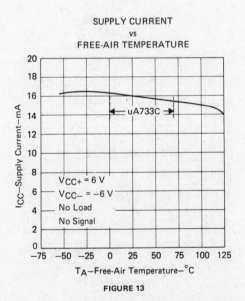

FIGURE 13

SUPPLY CURRENT
vs
SUPPLY VOLTAGE

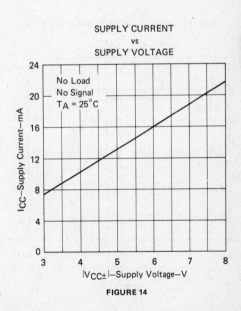

FIGURE 14

TEXAS
INSTRUMENTS
POST OFFICE BOX 225012 ● DALLAS, TEXAS 75265

Special Functions

5

TYPICAL CHARACTERISTICS

MAXIMUM PEAK-TO-PEAK OUTPUT VOLTAGE
vs
LOAD RESISTANCE

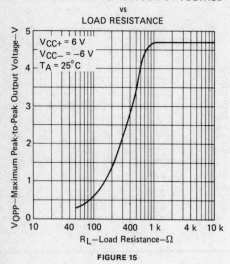

FIGURE 15

MAXIMUM PEAK-TO-PEAK OUTPUT VOLTAGE
vs
SUPPLY VOLTAGE

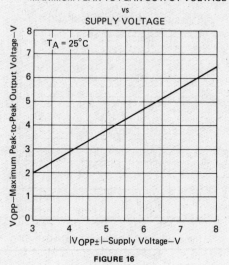

FIGURE 16

MAXIMUM PEAK-TO-PEAK OUTPUT VOLTAGE
vs
FREQUENCY

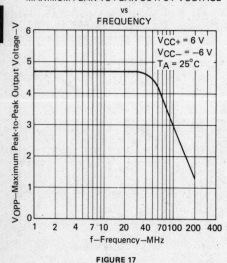

FIGURE 17

INPUT RESISTANCE
vs
FREE-AIR TEMPERATURE

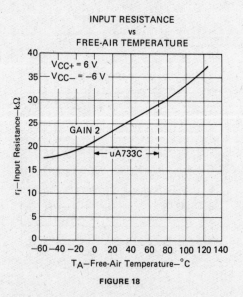

FIGURE 18

TEXAS
INSTRUMENTS
POST OFFICE BOX 225012 • DALLAS, TEXAS 75265

676

**LINEAR
INTEGRATED
CIRCUITS**

**TYPE uA2240C
PROGRAMMABLE TIMER/COUNTER**

D2442, JUNE 1978—REVISED FEBRUARY 1984

- **Accurate Timing from Microseconds to Days**

- **Programmable Delays from 1 Time Constant to 255 Time Constants**

- **Outputs Compatible with TTL and CMOS**

- **Wide Supply-Voltage Range**

- **External Sync and Modulation Capability**

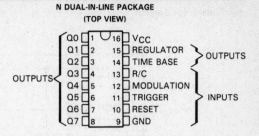

N DUAL-IN-LINE PACKAGE
(TOP VIEW)

description

These circuits consist of a time-base oscillator, an eight-bit counter, a control flip-flop, and a voltage regulator. The frequency of the time-base oscillator is set by the time constant of an external resistor and capacitor at pin 13 and can be synchronized or modulatd by signals applied to the modulation input. The output of the time-base section is applied directly to the input of the counter section and also appears at pin 14 (time base). The time-base pin may be used to monitor the frequency of the oscillator, to provide an output pulse to other circuitry, or (with the time-base section disabled) to drive the counter input from an external source. The counter input is activated on a negative-going transition. The reset input stops the time-base oscillator and sets each binary output, Q0 through Q7, and the time-base output to a TTL high level. After resetting, the trigger input starts the oscillator and all Q outputs go low. Once triggered, the uA2240C will ignore any signals at the trigger input until it is reset.

The uA2240C timer/counter may be operated in the free-running mode or with output-signal feeedback to the reset input for automatic reset. Two or more binary outputs may be connected together to generate complex pulse patterns, or each output may be used separately to provide eight output frequencies. Using two circuits in cascade can provide precise time delays of up to three years.

The uA2240C is characterized for operation from 0 °C to 70 °C.

Special Functions

5

284

**TEXAS
INSTRUMENTS**
POST OFFICE BOX 225012 • DALLAS, TEXAS 75265

functional block diagram

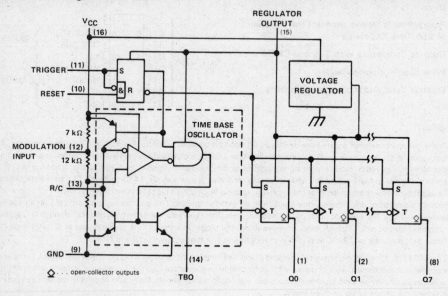

$\Diamond$ ... open-collector outputs

absolute maximum ratings

Supply voltage, V_{CC} (see Note 1)	18 V
Output voltage: Q0 thru Q7	18 V
Output current: Q0 thru Q7	10 mA
Regulator output current	−5 mA
Continuous dissipation at (or below) 25 °C free-air temperature (see Note 2):	650 mW
Operating free-air temperature range	0 °C to 70 °C
Lead temperature 1,6 mm (1/16 inch) from case for 10 seconds	260 °C

NOTES: 1. Voltage values are with respect to the network ground terminal.
2. For operation above 25 °C, see the Dissipation Derating Curves, Section 2.

recommended operating conditions

	uA2240C			UNIT
	MIN	NOM	MAX	
Supply voltage, V_{CC} (see Note 3)	4		14	V
Timing resistor	0.001		10	MΩ
Timing capacitor	0.01		1000	μF
Counter input frequency (Pin 14)			1.5	MHz
Pull-up resistor, time-base output			20	kΩ
Trigger and reset input pulse voltage	2	3		V
Trigger and reset input pulse duration	2			μs
External clock input pulse voltage	3			V
External clock input pulse duration	1			μs

NOTE 3: For operation with $V_{CC} \leq 4.5$ V, short regulator output to V_{CC}.

Texas Instruments
POST OFFICE BOX 225012 • DALLAS, TEXAS 75265

electrical characteristics at 25 °C free-air temperature

PARAMETER	TEST CIRCUIT	TEST CONDITIONS		MIN	TYP	MAX	UNIT
Regulator output voltage	1	$V_{CC} = 5$ V,	Trigger and reset open or grounded	3.9	4.4		V
	2	$V_{CC} = 15$ V,	Trigger and reset open or grounded	5.8	6.3	6.8	
Modulation input open circuit voltage	1	$V_{CC} = 5$ V,	Trigger and reset open or grounded	2.8	3.5	4.2	V
		$V_{CC} = 15$ V,	Trigger and reset open or grounded		10.5		
Trigger threshold voltage	1	$V_{CC} = 5$ V,	Reset at 0 V		1.4	2	V
High-level trigger current	1	$V_{CC} = 5$ V,	Trigger at 2 V, Reset at 0 V		10		μA
Reset threshold voltage	1	$V_{CC} = 5$ V,	Trigger at 0 V		1.4	2	V
High-level reset current	1	$V_{CC} = 5$ V,	Trigger at 0 V		10		μA
Couonter input (time base) threshold voltage	2	$V_{CC} = 5$ V,	Trigger and reset open or grounded	1	1.4		V
Low-level output current, Q0 thru Q7	2	$V_{CC} = 5$ V, Trigger at 2 V, Reset at 0 V, $V_{OL} < 0.4$ V		2	4		mA
High-level output current, Q0 thru Q7	2	$V_{OH} = 15$ V, Reset at 2 V, Trigger at 0 V			0.01	15	μA
Supply current	1	$V_{CC} = 5$ V,	Trigger at 0 V, Reset at 5 V		4	7	mA
	1	$V_{CC} = 15$ V,	Trigger at 0 V, Reset at 5 V		13	18	
	3	V+ = 4 V			1.5		

operating characteristics at 25 °C free-air temperature (unless otherwise noted)

PARAMETER	TEST CIRCUIT	TEST CONDITIONS[†]		MIN	TYP	MAX	UNIT
Initial error of time base[‡]	1	$V_{CC} = 5$ V, Trigger at 5 V, Reset at 0 V			±0.5	±5	%
Temperature coefficient of time-base period	1	$T_A = 0$ °C to 70 °C	$V_{CC} = 5$ V		−200		ppm/°C
			$V_{CC} = 15$ V		−80		
Supply voltage sensitivity of time-base period	1	$V_{CC} \geq 8$ V			−0.08	−0.3	%/V
Time-base output frequency	1	$V_{CC} = 5$ V, R = MIN, C = MIN			130		kHz
Propagation delay time		see Note 4	From trigger input		1		μs
			From reset input		0.8		
Output rise time	2	$R_L = 3$ kΩ, $C_L = 10$ pF	Q0 thru Q7		180		ns
Output fall time					180		

[†]For conditions shown as MIN or MAX, use the appropriate value specified under recommended operating conditions.

[‡]This is the time-base period error due only to the uA2240C and expressed as a percentage of nominal (1.00 RC).

NOTE 4: Propagation delay time is measured from the 50% point on the leading edge of an input pulse to the 50% point on the leading edge of the resulting change of state at Q0.

Special Functions

5

TEXAS INSTRUMENTS
POST OFFICE BOX 225012 • DALLAS, TEXAS 75265

Special Functions

5

PARAMETER MEASUREMENT INFORMATION

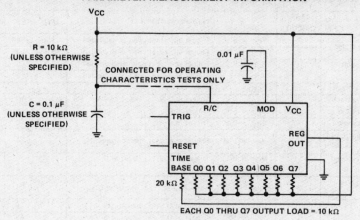

FIGURE 1—GENERAL TEST CIRCUIT

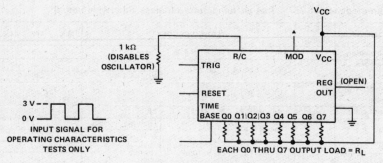

FIGURE 2—COUNTER TEST CIRCUIT

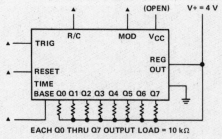

FIGURE 3—REDUCED-POWER TEST CIRCUIT
(TIME BASE DISABLED)

▲These connections may be open or grounded for this test.

TEXAS
INSTRUMENTS
POST OFFICE BOX 225012 • DALLAS, TEXAS 75265

TYPICAL CHARACTERISTICS

NORMALIZED TIME-BASE PERIOD
vs
MODULATION INPUT VOLTAGE

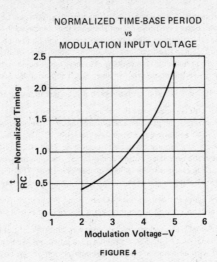

FIGURE 4

Special Functions

5

TYPICAL APPLICATION INFORMATION

Figure 5 shows voltage waveforms for typical operation of the uA2240C. If both reset and trigger inputs are low during power-up, the timer/counter will be in a reset state with all binary (Q) outputs high and the oscillator stopped. In this state, a high level on the trigger input starts the time-base oscillator. The initial negative-going pulse from the oscillator sets the Q outputs to low logic levels at the beginning of the first time-base period. The uA2240C will ignore any further signals at the trigger input until after a reset signal is applied to the reset input. With the trigger input low, a high level at the reset input will set Q outputs high and stop the time-base oscillator. If the reset signal occurs while the trigger input is high, the reset is ignored. If the reset input remains high when the trigger input goes low, the uA2240C will reset.

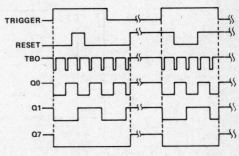

FIGURE 5—TIMING DIAGRAM OF OUTPUT WAVEFORMS

TEXAS INSTRUMENTS
POST OFFICE BOX 225012 • DALLAS, TEXAS 75265

TYPICAL APPLICATION INFORMATION

In monostable applications of the uA2240C one or more of the binary outputs will be connected to the reset terminal as shown in Figure 6. The binary outputs are open-collector stages that can be connected together to a common pull-up resistor to provide a "wired-OR" function. The combined output will be low as long as any one of the outputs is low. This type of arrangement can be used for time delays that are integer multiples of the time-base period. For example, if Q5 ($2^5 = 32$) only is connected to the reset input, every trigger pulse will generate a 32-period active-low output. Similarly, if Q0, Q4, and Q5 are connected to reset, each trigger pulse creates a 49-period delay.

In astable operation, the uA2240C will free-run from the time it is triggered until it receives an external reset signal.

The period of the time-base oscillator is equal to the RC time constant of an external resistor and capacitor connected as shown in Figure 6 when the modulation input is open (approximately 3.5 volts internal, see Figure 4). Under conditions of high supply voltage ($V_{CC} > 7$ V) and low value of timing capacitor (C < 0.1 μF), the pulse duration of the time-base oscillator may be too short to properly trigger the counters. This situation can be corrected by adding a 300-picofarad capacitor between the time-base output and ground. The time-base output (TBO) is an open-collector output that requires a 20-kΩ pull-up resistor to Pin 15 for proper operation. The time-base pin may also be used as an input to the counters for an external time-base or as an active-low inhibit input to interrupt counting without resetting.

The modulation input varies the ratio of the time-base period to the RC time constant as a function of the dc bias voltage (see Figure 4). It can also be used to synchronize the timer/counter to an external clock or sync signal.

The regulator output is used internally to drive the binary counters and the control logic. This terminal can also be used to supply voltage to additional uA2240C devices to minimize power dissipation when several timer circuits are cascaded. For circuit operation with an external clock, the regulator output can be used as the V_{CC} input terminal to power down the internal time base and reduce power dissipation. When supply voltages less than 4.5 volts are used with the internal time base, Pin 15 should be shorted to Pin 16.

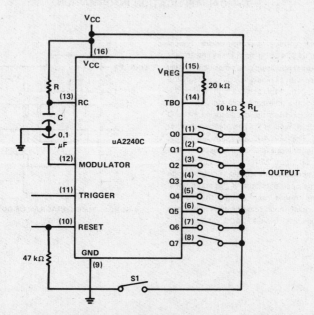

FIGURE 6—BASIC CONNECTIONS FOR TIMING APPLICATIONS

TEXAS
INSTRUMENTS
POST OFFICE BOX 225012 ● DALLAS, TEXAS 75265

General Information | 1

Thermal Information | 2

Operational Amplifiers | 3

Voltage Comparators | 4

Special Functions | 5

Voltage Regulators | 6

Data Acquisition | 7

Appendix | A

Voltage Regulators

6

FIXED-OUTPUT VOLTAGE REGULATORS

positive-voltage regulators

DEVICE SERIES	OUTPUT VOLTAGE TOLERANCE	MINIMUM DIFFERENTIAL VOLTAGE	OUTPUT CURRENT RATING	PACKAGES	PAGE
LM2930-0	±10%	0.6 V	150 mA	KC	6-45
LM2931-0	±10%	0.6 V	150 mA	KC	6-51
LM330-0	±4%	0.6 V	150 mA	KC	6-27
LM340-00	+4%	2 V	1.5 A	KC	6-33
TL780-00C	±1%	2 V	1.5 A	KC	6-137
uA7800C	+4%	2 V − 3 V	1.5 A	KC	6-175
uA78L00AC	+5%	2 V	100 mA	LP	6-183
uA78L00C	±10%	2 V − 2.5 V	100 mA	LP	6-183
uA78M00C	±5%	2 V − 3 V	500 mA	KC	6-189
uA78M00M	±5%	2 V − 3 V	500 mA	KC	6-189

negative-voltage regulators

DEVICE SERIES	OUTPUT VOLTAGE TOLERANCE	MINIMUM DIFFERENTIAL VOLTAGE	OUTPUT CURRENT RATING	PACKAGES	PAGE
LM320-00	±4%	2 V	1.5 A	KC	6-21
MC79L00AC	±5%	1.7 V	100 mA	LP	6-57
MC79L00C	±10%	1.7 V	100 mA	LP	6-57
uA7900C	±5%	2 V − 3 V	1.5 A	KC	6-201
uA79M00C	±5%	2 V − 3 V	1.5 A	KC	6-207
uA79M00M	±5%	2 V − 3 V	1.5 A	KC	6-207

available output voltage for above regulator series

DEVICE SERIES	2.6	5.0	5.2	6.0	6.2	8.0	8.5	9.0	10.0	12.0	15.0	18.0	20.0	24.0
LM2930-0		X				X								
LM2931-0		X												
LM320-00		X								X	X			
LM330-0		X												
LM340-00		X								X	X			
MC7900AC		X								X	X			
MC79L00C		X								X	X			
TL780-00C		X								X	X			
uA7800C		X		X		X	X		X	X	X	X		X
uA78L00AC	X	X			X	X		X	X	X	X			
UA78L00C	X	X			X	X		X	X	X	X			
uA78M00C		X		X		X			X	X	X		X	X
uA78M00M		X		X		X			X	X	X			
uA7900C		X	X	X		X				X	X	X		X
uA7900M		X	X	X		X				X	X	X		X
uA79M00C		X		X		X				X	X		X	X
uA79M00M		X		X		X				X	X			

Voltage Regulators

6

VARIABLE-OUTPUT VOLTAGE REGULTORS

positive-voltage series regulators

DEVICE NUMBER	OUTPUT VOLTAGE		MAXIMUM DIFFERENTIAL VOLTAGE	OUTPUT CURRENT RATING	PACKAGES	PAGE
	MIN	MAX				
LM217	1.2 V	37 V	$V_I - 1.2$ V	1.5 A	KC	6-11
LM317	1.2 V	37 V	$V_I - 1.2$ V	1.5 A	KC	6-11
LM350	1.2 V	33 V	$V_I - 1.2$ V	3 A	KC	6-41
TL317C	1.2 V	32 V	$V_I - 1.2$ V	100 mA	LP	6-91
TL317M	1.2 V	32 V	$V_I - 1.2$ V	100 mA	LP	6-91
TL783AC	5 V	200 V	200 V	700 mA	KC	6-141
TL783C	10 V	125 V	37 V	700 mA	KC	6-141
uA723C	3 V	38 V	37 V	25 mA	J, N	6-169
uA723M	3 V	38 V	37 V	25 mA	J, N	6-169

negative-voltage series regulators

DEVICE NUMBER	OUTPUT VOLTAGE		MAXIMUM DIFFERENTIAL VOLTAGE	OUTPUT CURRENT RATING	PACKAGES	PAGE
	MIN	MAX				
LM237	1.2 V	37 V	$V_I - 1.2$ V	1.5 A	KC	6-17
LM337	1.2 V	37 V	$V_I - 1.2$ V	1.5 A	KC	6-17

positive-shunt regulators

DEVICE NUMBER	SHUNT VOLTAGE		SHUNT CURRENT		TEMP COEFFICIENT RATING	PACKAGES	PAGE
	MIN	MAX	MIN	MAX			
TL430C	3 V	30 V	2 mA	100 mA	200 ppm/°C	LP	6-95
TL430I	3 V	30 V	2 mA	100 mA	200 ppm/°C	LP, P	6-95
TL431C	3 V	30 V	0.5 mA	100 mA	100 ppm/°C	LP, P	6-99
TL431I	2.55 V	36 V	1 mA	100 mA	100 ppm/°C	LP	6-99
TL431M	2.55 V	36 V	1 mA	100 mA	100 ppn/°C	JG	6-99

PROTECTION CIRCUITS

undervoltage protection circuits

DEVICE NUMBER	FEATURES	TEMP RANGE	PACKAGES	PAGE
TL7700C	Power-up and voltage reset generator specifically for microcomputer control supervision. These devices operate over a wide supply voltage range (3 V to 18 V) and have externally adjustable pulse duration to ensure system reset.	0°C to 70°C	D, P	6-163
TL7700I		−25°C to 85°C	D, P	
TL7702 TL7705 TL7709 TL7712 TL7715	Power-up and voltage reset generator specifically for microcomputer control supervision. These devices operate over a wide supply voltage range (3 V to 18 V) and have externally adjustable pulse duration to ensure system reset.	0°C to 70°C	D, P	6-165

Voltage Regulators

6

Texas Instruments
POST OFFICE BOX 225012 • DALLAS, TEXAS 75265

overvoltage protection circuit

DEVICE NUMBER	FEATURES	TEMP RANGE	PACKAGES	PAGE
MC3423	Separate outputs for "crowbar" and logic circuitry, progrmmable time delay, TTL-level activation isolated from voltage-sensing inputs	0°C to 70°C	JG, P	6-55

SWITCHING VOLTAGE REGULATOR/CONTROLLERS

general-purpose switching regulators/controllers

		BASE DEVICE NUMBERS					
	MC35060 MC34060	SG3524 SG2524 SG1524	SG3525A SG2525A SG1525A	SG3527A SG2527A SG1527A	TL3525A TL2525A TL1525A	TL3527A TL2537A TL1527A	TL497A
PAGES	6-61	6-69	6-81	6-81	6-153	6-153	6-119
FEATURES							
General Features							
Fixed On Time	—	—	—	—	—	—	X
Fixed Frequency PWM	X	X	X	X	X	X	X
Expandable	X	X	X	X	X	X	—
Control Features							
On Chip Reference	X	X	X	X	X	X	X
Precision On Chip Reference	—	—	X	X	X	X	—
Dead Time Adjust	X	—	X	X	X	X	—
Current Sense Amplifier	—	—	—	—	—	—	X
Error Amplifier	2	2	1	1	1	1	1
Operates to 40 V	X	35 V	35 V	35 V	35 V	35 V	—
Operates above 40 V	—	—	—	—	—	—	—
Protection Features							
On Chip Regulator	X	—	—	—	—	—	—
Internal Soft Start	—	X	X	X	X	X	—
Under Voltage Lockout	—	X	X	X	X	X	—
Inhibit Control	—	X	X	X	X	X	X
Double Pulse Protection	—	X	X	X	X	X	—
Output Features							
Single-ended Output	X	—	—	—	—	—	—
Double-ended Outputs	—	X	X	X	X	X	—
Totem-Pole Outputs	—	—	X	X	X	X	—
Parallelable Outputs	—	—	—	—	—	—	—
External Output Trigger	—	—	—	—	—	—	—
AVAILABILITY							
Commercial Temp Range							
Plastic (N Package)	X	X	X	X	X	X	X
Ceramic (J Package)	X	X	X	X	X	X	X
Industrial Temp Range							
Plastic (N Package)	—	X	X	X	X	X	X
Ceramic (J Package)	—	X	X	X	X	X	X
Military Temp Range							
Ceramic (J Package)	X	X	X	X	X	X	X

Voltage Regulators

6

SWITCHING VOLTAGE REGULATOR/CONTROLLERS

general-purpose switching regulators/controllers

	BASE DEVICE NUMBERS						
	TL493	TL494	TL495	TL593	TL594	TL595	TL1451
PAGES	6-107	6-107	6-107	6-127	6-127	6-127	6-151
FEATURES							
General Features							
Fixed On time	—	—	—	—	—	—	—
Fixed Frequency PWM	X	X	X	X	X	X	X
Low Bias Current Requirements	—	—	—	—	—	—	—
Expandable	X	X	X	X	X	X	X
Control Features							
On Chip Reference	X	X	X	X	X	X	X
Precision On Chip Reference	—	—	—	X	X	X	—
Dead Time Adjust	X	X	X	X	X	X	X
Current Sense Amplifier	1	—	—	1	—	—	—
Error Amplifier	1	2	2	1	2	2	2
Operates to 40 V	X	X	X	X	X	X	X
Operates above 40 V	—	—	X	—	—	X	—
Protection Features							
On Chip Regulator	—	—	X	—	—	X	—
Internal Soft Start	—	—	—	—	—	—	—
Under Voltage Lockout	—	—	—	X	X	—	—
Inhibit Control	X	X	X	X	X	X	—
Double Pulse Protection	X	X	X	X	X	X	—
Output Features							
Single-ended Output	—	—	—	—	—	—	2
Double-ended Outputs	X	X	X	X	X	X	—
Totem-pole Outputs	—	—	—	—	—	—	—
Parallelable Outputs	X	X	X	X	X	X	—
External Output Trigger	—	—	X	—	—	X	—
AVAILABILITY							
Commercial Temp Range							
Plastic (N Package)	X	X	X	X	X	X	X
Ceramic (J Package)	—	X	—	—	—	—	X
Industrial Temp Range							
Plastic (N Package)	—	X	—	—	X	—	—
Ceramic (J Package)	—	X	—	—	X	—	—
Military Temp Range							
Ceramic (J Package)	—	X	—	X	X	—	—

Voltage Regulators

6

TEXAS
INSTRUMENTS
POST OFFICE BOX 225012 • DALLAS, TEXAS 75265

special-purpose regulators and controllers

DEVICE NUMBER	FUNCTION	FEATURES	PAGE
TL580C	Micropower Dual Switching Regulator	High-efficiency, low bias current, two control channels.	6-125
RM4193 RC4193	Micropower Switching Regulator	High-efficiency, low bias current adjustable output voltage, good for battery-backup circuit.	6-67
UC3846 UC3847	Curent-mode PWM Controller	Pulse-by-pulse programmable current limiting, self start, under-voltage lockout, and shutdown.	6-217
TL499	Wide-Range Power Supply Controller	Adjustable regulator that switches over to battery-backup when line voltage is low.	6-124
TL496	9-Volt Power Supply Controller	Operates from a variety of sources including 1- and 2-cell batteries and step-down ac line voltage.	6-115

Voltage Regulators

6

GLOSSARY

SERIES REGULATORS

Input Regulation

The change in output voltage, often expressed as a percentage of output voltage, for a change in input voltage from one level to another level.
NOTE: Sometimes this characteristic is normalized with respect to the input voltage change.

Ripple Rejection

The ratio of the peak-to-peak input ripple voltage to the peak-to-peak output ripple voltage.
NOTE: This is the reciprocal of ripple sensitivity.

Ripple Sensitivity

The ratio of the peak-to-peak output ripple voltage, sometimes expressed as a percentage of output voltage, to the peak-to-peak input ripple voltage.
NOTE: This is the reciprocal of ripple rejection.

Output Regulation

The change in output voltage, often expressed as a percentage of output voltage, for a change in load current from one level to another level.

Output Resistance

The output resistance under small-signal conditions.

Temperature Coefficient of Output Voltage (α_{VO})

The ratio of the change in output voltage, usually expressed as a percentage of output voltage, to the change in temperature. This is the average value for the total temperature change.

$$\alpha_{VO} = \pm \left[\frac{V_O \text{ at } T_2 - V_O \text{ at } T_1}{V_O \text{ at } 25\,^{\circ}C} \right] \left[\frac{100\%}{T_2 - T_1} \right]$$

Output Voltage Change with Temperature

The percentage change in the output voltage for a change in temperature. This is the net change over the total temperature range.

Output Voltage Long-Term Drift

The change in output voltage over a long period of time.

Output Noise Voltage

The rms value of the ac component of the output voltage, sometimes expressed as a percentage of the dc output voltage, with constant load and no input ripple.

Current-Limit Sense Voltage

The current-sense voltage at which current limiting occurs.

Voltage Regulators

6

TEXAS INSTRUMENTS
POST OFFICE BOX 225012 • DALLAS, TEXAS 75265

Current-Sense Voltage

The voltage that is a function of the load current and is normally used for control of the current-limiting circuitry.

Dropout Voltage

The low input-to-output differential voltage at which the circuit ceases to regulate against further reductions in input voltage.

Feedback Sense Voltage

The voltage that is a function of the output voltage and is used for feedback control of the regulator.

Reference Voltage

The voltage that is compared with the feedback sense voltage to control the regulator.

Bias Current

The difference between input and output currents.
NOTE: This is sometimes referred to as quiescent current.

Standby Current

The input current drawn by the regulator with no output load and no reference voltage load.

Short-Circuit Output Current

The output current of the regulator with the output shorted to ground.

Peak Output Current

The maximum output current that can be obtained from the regulator due to limiting circuitry within the regulator.

Overvoltage Shutdown Voltage

The input voltage applied to a regulator having overvoltage shutdown protection that will cause the output voltage to go nearly to zero.

Junction Temperature, Virtual Junction Temperature

A temperature representing the temperature of the junction(s), field-effect transistor channel(s), or other internal point(s) of heat generation calculated on the basis of a simplified model of the thermal and electrical behavior of the semiconductor device.

SHUNT REGULATORS

NOTE: These terms and symbols are based on JEDEC and IEC standards for voltage regulator diodes.

Shunt Regulator

A device having a voltage-current characteristic similar to that of a voltage-regulator diode; normally biased to operate in a region of low differential resistance (corresponding to the breakdown region of a regulator diode) to develop across its terminals an essentially constant voltage throughout a specified current range.

Voltage Regulators

6

GLOSSARY

Anode

The electrode to which the regulator current flows within the regulator when it is biased for regulation.

Cathode

The electrode from which the regulator current flows within the regulator when it is biased for regulation.

Reference Input Voltage (V_{ref}) (of an adjustable shunt regulator)

The voltage at the reference input terminal with respect to the anode terminal.

Temperature Coefficient of Reference Voltge (αV_{ref})

The ratio of the change in reference voltage to the change in temperature. This is the average value for the total temperature change.
To obtain a value in ppm/°C:

$$\alpha V_{ref} = \left[\frac{V_{ref} \text{ at } T_2 - V_{ref} \text{ at } T_1}{V_{ref} \text{ at } 25\,^{\circ}C} \right] \left[\frac{10^6}{T_2 - T_1} \right]$$

Regulator Voltage (V_Z)

The dc voltage across the regulator when it is biased for regulation.

Regulator Current (I_Z)

The dc current through the regulator when it is biased for regulation.

Regulator Current near Lower Knee of Regulation Range (I_{ZK})

The regulator current near the lower limit of the region within which regulation occurs; this corresponds to the breakdown knee of a regulator diode.

Regulator Current at Maximum Limit of Regulation Range (I_{ZM})

The regulator current above which the differential resistance of the regulator significantly increases.

Differential Regulator Resistance (r_z)

The quotient of a change in voltage across the regulator and the corresponding change in current through the regulator when it is biased for regulation.

Noise Voltage (V_{nz})

The rms value of the ac component of the voltage across the regulator with the regulator biased for regulation and with no input ripple.

Texas
Instruments
POST OFFICE BOX 225012 • DALLAS, TEXAS 75265

**LINEAR
INTEGRATED
CIRCUITS**

**TYPES LM217, LM317
3-TERMINAL ADJUSTABLE REGULATORS**

D2212, SEPTEMBER 1977 — REVISED DECEMBER 1982

- Output Voltage Range Adjustable from 1.2 V to 37 V

- Guaranteed Output Current Capability of 1.5 A

- Input Regulation Typically 0.01% Per Input-Volt Change

- Output Regulation Typically 0.1%

- Peak Output Current Constant Over Temperature Range of Regulator

- Popular 3-Lead TO-220AB Package

- Ripple Rejection Typically 80 dB

- Direct Replacement for National LM217 and LM317

terminal assignments

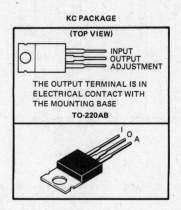

KC PACKAGE

(TOP VIEW)

INPUT
OUTPUT
ADJUSTMENT

THE OUTPUT TERMINAL IS IN
ELECTRICAL CONTACT WITH
THE MOUNTING BASE

TO-220AB

Voltage Regulators

6

description

The LM217, and LM317 are adjustable 3-terminal positive-voltage regulators capable of supplying 1.5 amperes over a differential voltage range of 3 volts to 40 volts. They are exceptionally easy to use and require only two external resistors to set the output voltage. Both input and output regulation are better than standard fixed regulators. The devices are packaged in a standard transistor package that is easily mounted and handled.

In addition to higher performance than fixed regulators, these regulators offer full overload protection available only in integrated circuits. Included on the chip are current limit, thermal overload protection, and safe-area protection. All overload protection circuitry remains fully functional even if the adjustment terminal is disconnected. Normally, no capacitors are needed unless the device is situated far from the input filter capacitors in which case an input bypass is needed. An optional output capacitor can be added to improve transient response. The adjustment terminal can be bypassed to achieve very high ripple rejection, which is difficult to achieve with standard 3-terminal regulators.

Besides replacing fixed regulators, these regulators are useful in a wide variety of other applications. The primary applications of each of these regulators is that of a programmable output regulator, but by connecting a fixed resistor between the adjustment terminal and the output terminal, each device can be used as a precision current regulator. Even though the regulator is floating and sees only the input-to-output differential voltage, use of these devices to regulate output voltages that would cause the maximum-rated differential voltage to be exceeded if the output became shorted to ground is not recommended. The TL783 or TL783A is recommended for output voltages exceeding 37 volts. Supplies with electronic shutdown can be achieved by clamping the adjustment terminal to ground, which programs the output to 1.2 volts where most loads draw little current.

The LM217 and LM317 are characterized for operation from $-25\,°C$ to $150\,°C$ and from $0\,°C$ to $125\,°C$, respectively.

**TEXAS
INSTRUMENTS**

POST OFFICE BOX 225012 • DALLAS, TEXAS 75265

schematic

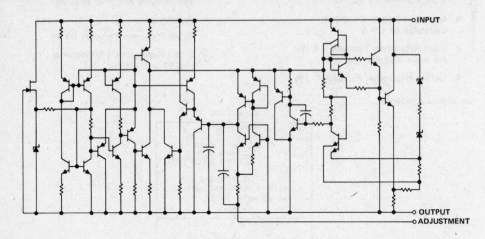

absolute maximum ratings over operation temperature range (unless otherwise noted)

	LM217	LM317	UNIT
Input-to-output differential voltage, $V_I - V_O$	40	40	V
Continuous total dissipation at 25°C free-air temperature (see Note 1)	2000	2000	mW
Continuous total dissipation at (or below) 25°C case temperature (see Note 1)	20	20	W
Operating free-air, case, or virtual junction temperature range	−25 to 150	0 to 125	°C
Storage temperature range	−65 to 150	−65 to 150	°C
Lead temperature 1,6 mm (1/16 inch) from case for 10 seconds	260	260	°C

NOTE 1: For operation above 25°C free-air or case temperature, refer to Figures 15 and 16. To avoid exceeding the design maximum virtual junction temperature, these ratings should not be exceeded. Due to variations in individual device electrical characteristics and thermal resistance, the built-in thermal overload protection may be activated at power levels slightly above or below the rated dissipation.

recommended operating conditions

	LM217		LM317		UNIT
	MIN	MAX	MIN	MAX	
Output current, I_O	5	1500	10	1500	mA
Operating virtual junction temperature, T_J	−25	150	0	125	°C

TEXAS
INSTRUMENTS
POST OFFICE BOX 225012 • DALLAS, TEXAS 75265

electrical characteristics over recommended ranges of operating virtual junction temperature (unless otherwise noted)

PARAMETER	TEST CONDITIONS†		LM217 MIN	LM217 TYP	LM217 MAX	LM317 MIN	LM317 TYP	LM317 MAX	UNIT
Input regulation (See Note 2)	$V_I - V_O$ = 3 V to 40 V, See Note 3	T_J = 25°C		0.01	0.02		0.01	0.04	%/V
		I_O = 10 mA to 1.5 A		0.02	0.05		0.02	0.07	
Ripple rejection	V_O = 10 V,	f = 120 Hz		65			65		dB
	V_O = 10 V, f = 120 Hz 10-μF capacitor between ADJ and ground		66	80		66	80		
Output regulation	I_O = 10 mA to 1.5 A, T_J = 25°C, See Note 3	$V_O \leq$ 5 V		5	15		5	25	mV
		$V_O >$ 5 V		0.1	0.3		0.1	0.5	%
	I_O = 10 mA to 1.5 A, See Note 3	$V_O \leq$ 5 V		20	50		20	70	mV
		$V_O >$ 5 V		0.3	1		0.3	1.5	%
Output voltage change with temperature	T_J = MIN to MAX			1			1		%
Output voltage long-term drift (see Note 4)	After 1000 h at T_J = MAX and $V_I - V_O$ = 40 V			0.3	1		0.3	1	%
Output noise voltage	f = 10 Hz to 10 kHz, T_J = 25°C			0.003			0.003		%
Minimum output current to maintain regulation	$V_I - V_O$ = 40 V			3.5	5		3.5	10	mA
Peak output current	$V_I - V_O \leq$ 15 V		1.5	2.2		1.5	2.2		A
	$V_I - V_O \leq$ 40 V			0.4			0.4		
Adjustment-terminal current				50	100		50	100	μA
Change in adjustment-terminal current	$V_I - V_O$ = 2.5 V to 40 V, I_O = 10 mA to 1.5 A			0.2	5		0.2	5	μA
Reference voltage (output to ADJ)	$V_I - V_O$ = 3 V to 40 V, I_O = 10 mA to 1.5 A, P $\leq$ 20 W		1.2	1.25	1.3	1.2	1.25	1.3	V

† Unless otherwise noted, these specifications apply for the following test conditions; $V_I - V_O$ = 5 V and I_O = 0.5 A. For conditions shown as MIN or MAX, use the appropriate value specified under recommended operating conditions.

NOTES: 2. Input regulation is expressed here as the percentage change in output voltage per 1-volt change at the input.

3. Input regulation and output regulation are measured using pulse techniques ($t_w \leq$ 10 μs, duty cycle $\leq$ 5%) to limit changes in average internal dissipation. Output voltage changes due to large changes in internal dissipation must be taken into account separately.

4. Since long-term drift cannot be measured on the individual devices prior to shipment, this specification is not intended to be a guarantee or warranty. It is an engineering estimate of the average drift to be expected from lot to lot.

Voltage Regulators

6

TEXAS
INSTRUMENTS
POST OFFICE BOX 225012 • DALLAS, TEXAS 75265

TYPICAL APPLICATION DATA

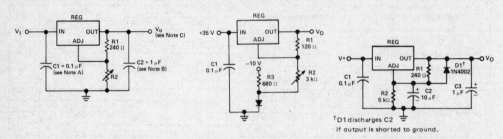

FIGURE 1—ADJUSTABLE VOLTAGE REGULATOR

FIGURE 2—0-V to 30-V REGULATOR CIRCUIT

FIGURE 3—ADJUSTABLE REGULATOR CIRCUIT WITH IMPROVED RIPPLE REJECTION

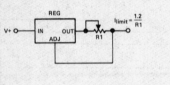

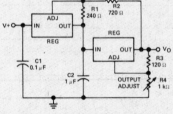

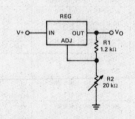

FIGURE 4—PRECISION CURRENT LIMITER CIRCUIT

FIGURE 5—TRACKING PREREGULATOR CIRCUIT

FIGURE 6—1.2 to 20-V REGULATOR CIRCUIT WITH MINIMUM PROGRAM CURRENT

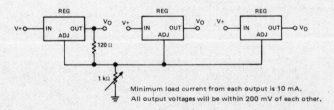

Minimum load current from each output is 10 mA.
All output voltages will be within 200 mV of each other.

FIGURE 7—ADJUSTING MULTIPLE ON-CARD REGULATORS WITH A SINGLE CONTROL

NOTES: A. Use of an input bypass capacitor is recommended if regulator is far from filter capacitors.

B. Use of an output capacitor improves transient response but is optional.

C. Output voltage is calculated from the equation: $V_O = V_{ref}\left(1 + \dfrac{R2}{R1}\right)$

V_{ref} equals the difference between the output and adjustment terminal voltages.

TEXAS INSTRUMENTS
POST OFFICE BOX 225012 • DALLAS, TEXAS 75265

Voltage Regulators

6

TYPICAL APPLICATIONS

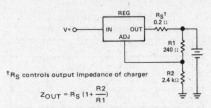

†R_S controls output impedance of charger

$$z_{OUT} = R_S \left(1 + \frac{R2}{R1}\right)$$

The use of R_S allows low charging rates with a fully-charged battery.

FIGURE 8—BATTERY CHARGER CIRCUIT

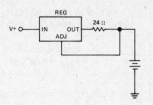

FIGURE 9—50-mA CONSTANT-CURRENT BATTERY CHARGER CIRCUIT

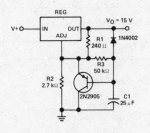

FIGURE 10—SLOW-TURN-ON 15-V REGULATOR CIRCUIT

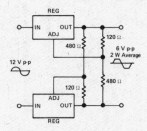

FIGURE 11—A-C VOLTAGE REGULATOR CIRCUIT

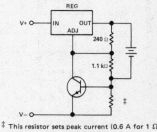

‡ This resistor sets peak current (0.6 A for 1 Ω)

FIGURE 12—CURRENT-LIMITED 6-V CHARGER

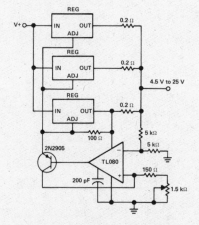

FIGURE 13—ADJUSTABLE 4-A REGULATOR

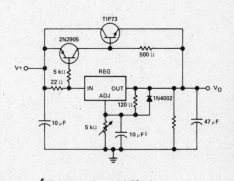

¶ Minimum load current is 30 mA.
§ Optional capacitor improves ripple rejection

FIGURE 14—HIGH-CURRENT ADJUSTABLE REGULATOR

Voltage Regulators

6

282

THERMAL INFORMATION

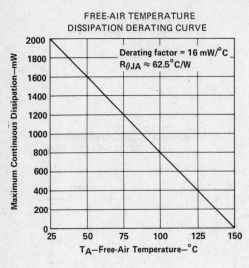

FREE-AIR TEMPERATURE
DISSIPATION DERATING CURVE

Derating factor = 16 mW/°C
$R_{\theta JA} \approx 62.5$°C/W

FIGURE 15

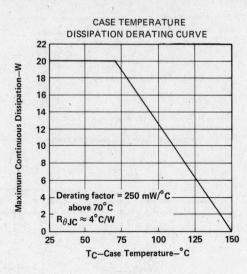

CASE TEMPERATURE
DISSIPATION DERATING CURVE

Derating factor = 250 mW/°C
above 70°C
$R_{\theta JC} \approx 4$°C/W

FIGURE 16

Voltage Regulators

6

TEXAS
INSTRUMENTS
POST OFFICE BOX 225012 • DALLAS, TEXAS 75265

1282

**LINEAR
INTEGRATED
CIRCUITS**

**TYPES LM237, LM337
3-TERMINAL ADJUSTABLE REGULATORS**

D2640, NOVEMBER 1981

- Output Voltage Range Adjustable from −1.2 V to −37 V

- Guaranteed I_O Capability of 1.5 A

- Input Regulation Typically 0.01% per Input-Volt Change

- Output Regulation Typically 0.3%

- Peak Output Current Constant Over Temperature Range of Regulator

- Ripple Rejection Typically 77 dB

- Direct Replacement for National Semiconductor LM237, LM337

LM237, LM337 . . . KC PACKAGE

(TOP VIEW)

OUTPUT
INPUT
ADJUSTMENT

THE INPUT TERMINAL IS IN
ELECTRICAL CONTACT WITH
THE MOUNTING BASE

TO-220AB

description

The LM237 and LM337 are adjustable 3-terminal negative-voltage regulators capable of supplying in excess of −1.5 A over an output voltage range of −1.2 V to −37 V. They are exceptionally easy to use, requiring only two external resistors to set the output voltage and one output capacitor for frequency compensation. The current design has been optimized for excellent regulation and low thermal transients. In addition the LM237 and LM337 feature internal current limiting, thermal shutdown, and safe-area compensation, making them virtually immune to blowout by overloads.

The LM237 and LM337 serve a wide variety of applications including local on-card regulation, programmable output voltage regulation, or precision current regulation. They are ideal complements to the LM217 and LM317 adjustable positive-voltage regulators.

schematic diagram

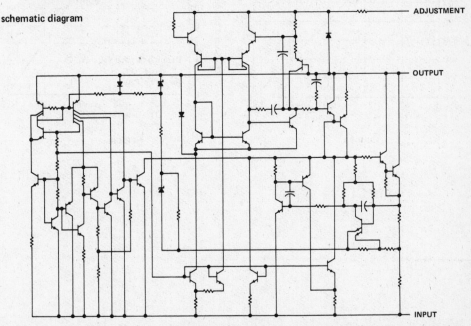

Voltage Regulators

6

TEXAS
INSTRUMENTS

POST OFFICE BOX 225012 • DALLAS, TEXAS 75265

Voltage Regulators

absolute maximum ratings over operating temperature range (unless otherwise noted)

Input-to-output differential voltage, $V_I - V_O$ .... −40 V

Continuous total dissipation at 25°C free-air temperature (see Note 1) .... 2 W

Continuous total dissipation at (or below) 25°C case temperature (see Note 1) .... 20 W

Operating free-air, case, or virtual junction temperature range: LM237 .... −25°C to 150°C

LM337 .... 0°C to 125°C

Storage temperature range .... −65°C to 150°C

Lead temperature 1/16 inch from case for 10 seconds .... 260°C

NOTE 1: For operation above 25 °C free-air or case temperature, refer to Figures 1 and 2. To avoid exceeding the design maximum virtual junction temperature, these ratings should not be exceeded. Due to variations in individual device electrical characteristics and thermal resistance, the built-in thermal overload protection may be activated at power levels slightly above or below the rated dissipation.

FREE-AIR TEMPERATURE
DISSIPATION DERATING CURVE

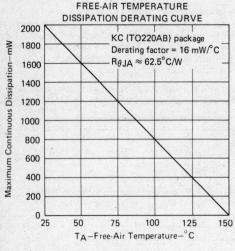

KC (TO220AB) package
Derating factor = 16 mW/°C
$R_{\theta JA} \approx 62.5°C/W$

FIGURE 1

CASE TEMPERATURE
DISSIPATION DERATING CURVE

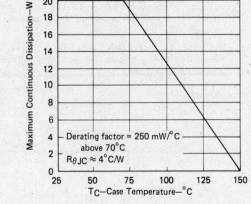

Derating factor = 250 mW/°C
above 70°C
$R_{\theta JC} \approx 4°C/W$

FIGURE 2

6

recommended operating conditions

		LM237		LM337		UNIT
		MIN	MAX	MIN	MAX	
Output current, I_O	$|V_I - V_O| \leqslant 40$ V, $P \leqslant 15$ W	10	1500	10	1500	mA
	$|V_I - V_O| \leqslant 10$ V, $P \leqslant 15$ W	6	1500	6	1500	
Operating virtual junction temperature, T_J		−25	150	0	125	°C

electrical characteristics over recommended ranges of operating virtual junction temperature (unless otherwise noted)

PARAMETER	TEST CONDITIONS†		LM237			LM337			UNIT
			MIN	TYP	MAX	MIN	TYP	MAX	
Input regulation‡	$V_I - V_O = -3$ V to −40 V, See Note 2	$T_J = 25°$C		0.01	0.02		0.01	0.04	%/V
		T_J = MIN to MAX		0.02	0.05		0.02	0.07	
Ripple rejection	$V_O = -10$ V,	f = 120 Hz		60			60		dB
	$V_O = -10$ V, $C_{ADJ} = 10 \mu$F	f = 120 Hz	66	77		66	77		
Output regulation	I_O = 10 mA to 1.5 A, $T_J = 25°$C, See Note 2	$|V_O| \leqslant 5$ V			25			50	mV
		$|V_O| \geqslant 5$ V			0.5			1	%
	I_O = 10 mA to 1.5 A, See Note 2	$|V_O| \leqslant 5$ V			50			70	mV
		$|V_O| \geqslant 5$ V			1			1.5	%
Output voltage change with temperature	T_J = MIN to MAX			0.6			0.6		%
Output voltage long-term drift (see Note 3)	After 1000 h at T_J = MAX and $V_I - V_O = -40$ V			0.3	1		0.3	1	%
Output noise voltage	f = 10 Hz to 10 kHz,	$T_J = 25°$C		0.003			0.003		%
Minimum output current to maintain regulation	$|V_I - V_O| \leqslant 40$ V			2.5	5		2.5	10	mA
	$|V_I - V_O| \leqslant 10$ V			1.2	3		1.5	6	
Peak output current	$|V_I - V_O| \leqslant 15$ V		1.5	2.2		1.5	2.2		A
	$|V_I - V_O| \leqslant 40$ V,	$T_J = 25°$C	0.24	0.4		0.15	0.4		
Adjustment- terminal current				65	100		65	100	μA
Change in adjustment terminal current	$V_I - V_O = -2.5$ V to −40 V, I_O = 10 mA to MAX,	$T_J = 25°$C		2	5		2	5	μA
Reference voltage (output to ADJ)	$V_I - V_O = -3$ to −40 V, I_O = 10 mA to 1.5 A, $P \leqslant$ rated dissipation	$T_J = 25°$C	−1.225	−1.250	−1.275	−1.213	−1.25	−1.287	V
		T_J = MIN to MAX	−1.2	−1.25	−1.3	−1.2	−1.25	−1.3	
Thermal regulation	Initial $T_J = 25°$C,	10-ms pulse		0.002	0.02		0.003	0.04	%/W

†Unless otherwise noted, these specifications apply for the following test conditions $|V_I - V_O| = 5$ V and I_O = 0.5 A. For conditions shown as MIN or MAX, use the appropriate value specified under recommended operating conditions.

‡Input regulation is expressed here as the percentage change in output voltage per 1 volt change at the input.

NOTES: 2. Input regulation and output regulation are measured using pulse techniques ($t_w \leqslant 10 \mu$s, duty cycle $\leqslant$ 5%) to limit changes in average internal dissipation. Output voltage changes due to large changes in internal dissipation must be taken into account separately.

3. Since long-term drift cannot be measured on the individual devices prior to shipment, this specification is not intended to be a guarantee or warranty. It is an engineering estimate of the average drift to be expected from lot to lot.

Voltage Regulators

6

TEXAS
INSTRUMENTS
POST OFFICE BOX 225012 • DALLAS, TEXAS 75265

TYPICAL APPLICATION DATA

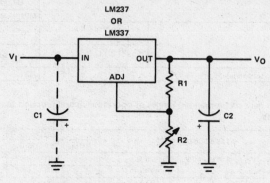

R1   R1 is typically 120 Ω.

$$R2 = R1\left(\frac{-V_O}{-1.25} -1\right) \text{ where } V_O \text{ is the output in volts.}$$

C1 is a 1-μF solid tantalum required only if the regulator is more than 10 cm (4 in.) from the power supply filter capacitor.

C2 is a 1-μF solid tantalum or 10-μF aluminum electrolytic required for stability.

FIGURE 3 — ADJUSTABLE NEGATIVE-VOLTAGE REGULATOR

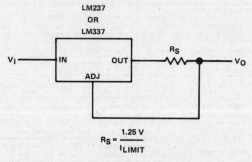

$$R_S = \frac{1.25 \text{ V}}{I_{LIMIT}}$$

FIGURE 4—CURRENT-LIMITING CIRCUIT

TEXAS
INSTRUMENTS
POST OFFICE BOX 225012 • DALLAS, TEXAS 75265

118

- 3-Terminal Regulators
- Internal Thermal Overload Protection
- Internal Short-Circuit Current Limiting
- Easily Adjustable to Higher Output Voltage
- Interchangeable with National Semiconductor LM320 Series

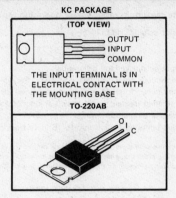

KC PACKAGE
(TOP VIEW)

OUTPUT
INPUT
COMMON

THE INPUT TERMINAL IS IN
ELECTRICAL CONTACT WITH
THE MOUNTING BASE

TO-220AB

NOMINAL OUTPUT VOLTAGE	MAXIMUM OUTPUT CURRENT	REGULATOR
−5 V	1.5 A	LM320-5
−12 V	1 A	LM320-12
−15 V	1A	LM320-15

description

The LM320 series of three-terminal, fixed-negative-voltage monolithic integrated circuit voltage regulators are designed to provide a fixed output voltage of −5 volts, −12 volts, and −15 volts with up to 1.5 amperes of output current. Each is designed for a wide range of applications which includes on-card regulation for elimination of noise and distribution problems associated with single-point regulation.

The internal current limiting and thermal shutdown features of these regulators make them essentially immune to overload. The LM320, when used as a fixed-voltage regulator, needs only one external component: a compensation capacitor at the output terminal. In addition, these devices can be used with external components to obtain adjustable output voltages and currents or as the power-pass element in precision regulators.

schematic diagram

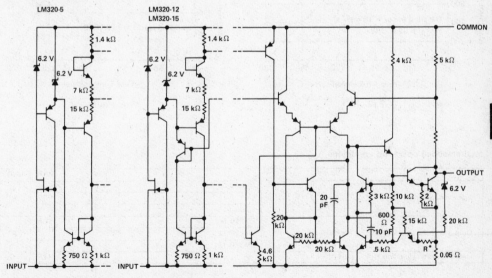

For LM320-5, R* = 50 Ω. For LM320-12 and LM320-15, R* = 150 Ω
All component values are nominal.

Voltage Regulators

6

Voltage Regulators

6

absolute maximum ratings over operating temperature range (unless otherwise noted)

Input voltage: LM320-5 .... -25 V

　　　　　　　 LM320-12 .... -35 V

　　　　　　　 LM320-15 .... -35 V

Input-output voltage differential .... 25 V

Continuous total dissipation at 25 °C free-air temperature (see Note 1) .... 2 W

Continuous total dissipation at (or below) 25 °C case temperature (see Note 1) .... 15 W

Operating free-air, case, or virtual junction temperature range .... -55 °C to 150 °C

Storage temperature range .... -65 °C to 150 °C

Lead temperature 1,6 mm (1/16 inch) from case for 10 seconds .... 260 °C

NOTE 1: For operation above 25 °C free-air or case temperature, refer to Figures 1 and 2. To avoid exceeding the design maximum virtual junction temperature, these ratings should not be exceeded. Due to variations in individual device electrical characteristics and thermal resistance, the built-in thermal overload protection may be activated at power levels slightly above or below the rated dissipation.

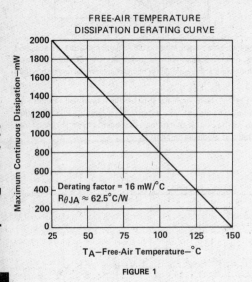

FIGURE 1

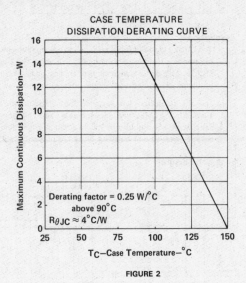

FIGURE 2

recommended operating conditions

		MIN	MAX	UNIT
Input voltage, V_I	LM320-5	-7.5	-25	V
	LM320-12	-14.5	-32	
	LM320-15	-17.5	-35	
Output current, I_O	LM320-5		1.5	A
	LM320-12		1	
	LM320-15		1	
Operating virtual junction temperature, T_J		0	125	°C

TEXAS
INSTRUMENTS
POST OFFICE BOX 225012 • DALLAS, TEXAS 75265

LM320-5 electrical characteristics at specified virtual junction temperature, I_O = 5 mA, V_I = −10 V, (unless otherwise noted)

PARAMETER	TEST CONDITIONS†		MIN	TYP	MAX	UNIT
Output voltage		T_J = 25°C	−4.8		−5.2	V
	V_I = −7.5 V to −25 V, P ≤ 15 W,	I_O = 5 mA to 1.5 A, T_J = 0°C to 125°C	−4.75		−5.25	
Input regulation	V_I = −7.5 V to −25 V,	T_J = 25°C		10	40	mV
Ripple rejection	f = 120 Hz,	T_J = 0°C to 125°C	54	64		dB
Output regulation	I_O = 5 mA to 1.5 A,	T_J = 25°C		50	100	mV
Output noise voltage	C_L = 1 μF, f = 10 Hz to 100 kHz,	T_J = 25°C		150		μV
Output voltage long-term drift (see Note 2)	After 1000 h at T_J = 125°,	T_J = 25°C		10		mV
Bias current	V_I = −7.5 V to −25 V,	T_J = 0°C to 125°C		1	2	mA
Bias current change	V_I = −7.5 V to −25 V	T_J = 25°C		0.1	0.4	mA
	I_O = 5 mA to 1.5 A			0.1	0.4	

LM320-12 electrical characteristics at specified virtual junction temperature, I_O = 5 mA, V_I = −17 V, (unless otherwise noted)

PARAMETER	TEST CONDITIONS†		MIN	TYP	MAX	UNIT
Output voltage		T_J = 25°C	−11.6	−12	−12.4	V
	V_I = −14.5 V to −32 V, P ≤ 15 W,	I_O = 5 mA to 1 A, T_J = 0°C to 125°C	−11.4		−12.6	
Input regulation	V_I = −14.5 V to −32 V,	T_J = 25°C		4	20	mV
Ripple rejection	f = 120 Hz,	T_J = 0°C to 125°C	56	80		dB
Output regulation	I_O = 5 mA to 1 A,	T_J = 25°C		30	80	mV
Output noise voltage	C_L = 1 μF, f = 10 Hz to 100 kHz,	T_J = 25°C		400		μV
Output voltage long-term drift (see Note 2)	After 1000 h at T_J = 125°C,	T_J = 25°C		24		mV
Bias current	V_I = −14.5 V to −32 V,	T_J = 0°C to 125°C		2	4	mA
Bias current change	V_I = −14.5 V to −32 V	T_J = 25°C		0.1	0.4	mA
	I_O = 5 mA to 1 A			0.1	0.4	

LM320-15 electrical characteristics at specified virtual junction temperature, I_O = 5 mA, V_I = −20 V, (unless otherwise noted)

PARAMETER	TEST CONDITIONS†		MIN	TYP	MAX	UNIT
Output voltage		T_J = 25°C	−14.5	−15	−15.5	V
	V_I = −17.5 V to −35 V, P ≤ 15 W,	I_O = 5 mA to 1 A, T_J = 0°C to 125°C	−14.3		−15.7	
Input regulation	V_I = −17.5 V to −35 V,	T_J = 25°C		5	20	mV
Ripple rejection	f = 120 Hz,	T_J = 0°C to 125°C	56	80		dB
Output regulation	I_O = 5 mA to 1 A	T_J = 25°C		30	80	mV
Output noise voltage	C_L = 1 μF, f = 10 Hz to 100 kHz,	T_J = 25°C		400		μV
Output voltage long-term drift (see Note 2)	After 1000 h at T_J = 125°C,	T_J = 25°C		30		mV
Bias current	V_I = −17.5 V to −35 V,	T_J = 0°C to 125°C		2	4	mA
Bias current change	V_I = −17.5 V to −35 V,	T_J = 25°C		0.1	0.4	mA
	I_O = 5 mA to 1 A			0.1	0.4	

†All characteristics are measured with a 1-μF capacitor across the input and a 2-μF solid-tantalum capacitor across the output. All characteristics except ripple rejection and output noise voltage are measured using pulse techniques (t_w ≤ 10 ms, duty cycle ≤ 5%). Output voltage changes due to changes in internal temperature must be taken into account separately.

NOTE 2: Since long-term drift cannot be measured on the individual devices prior to shipment, this specification is not intended to be a guarantee or warranty. It is an engineering estimate of the average drift to be expected from lot to lot.

Voltage Regulators

6

TYPICAL CHARACTERISTICS

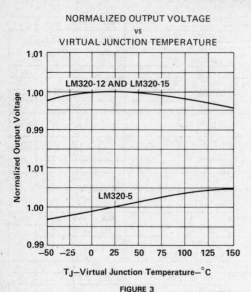

NORMALIZED OUTPUT VOLTAGE
vs
VIRTUAL JUNCTION TEMPERATURE

FIGURE 3

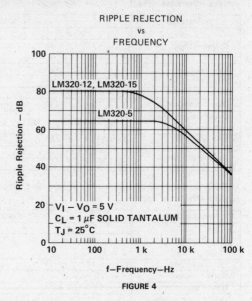

RIPPLE REJECTION
vs
FREQUENCY

FIGURE 4

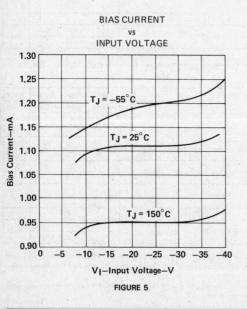

BIAS CURRENT
vs
INPUT VOLTAGE

FIGURE 5

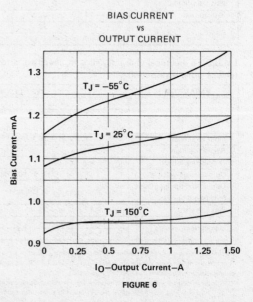

BIAS CURRENT
vs
OUTPUT CURRENT

FIGURE 6

TEXAS
INSTRUMENTS
POST OFFICE BOX 225012 • DALLAS, TEXAS 75265

483

TYPICAL CHARACTERISTICS

SHORT-CIRCUIT OUTPUT CURRENT
vs
INPUT-OUTPUT VOLTAGE DIFFERENTIAL

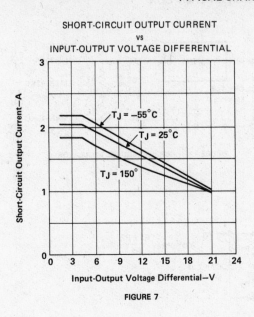

FIGURE 7

DROPOUT VOLTAGE
vs
OUTPUT CURRENT

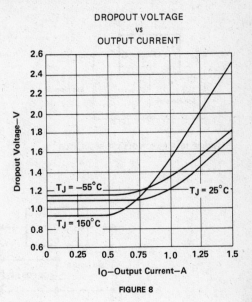

FIGURE 8

OUTPUT IMPEDANCE
vs
FREQUENCY

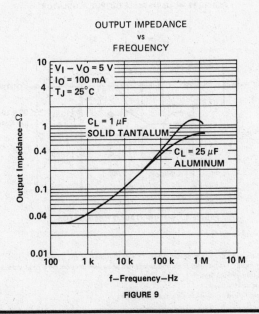

FIGURE 9

Voltage Regulators

6

483

TEXAS
INSTRUMENTS
POST OFFICE BOX 225012 • DALLAS, TEXAS 75265

TYPE SERIES LM320
3-TERMINAL NEGATIVE-VOLTAGE REGULATORS

TYPICAL APPLICATION INFORMATION

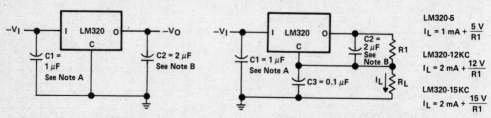

FIGURE 10 — FIXED-VOLTAGE REGULATOR

FIGURE 11 — CURRENT SOURCE REGULATOR

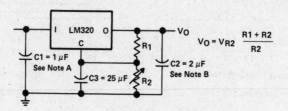

$$V_O = V_{R2} \frac{R1 + R2}{R2}$$

FIGURE 12 — ADJUSTABLE OUTPUT REGULATOR

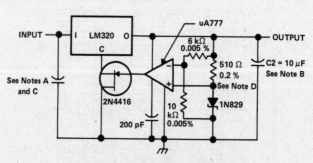

FIGURE 13 — HIGH-STABILITY REGULATOR

NOTES: A. Capacitor C1 is required if the regulator is not located within 75 mm (3 inches) of the power supply filter.
   B. Capacitor C2 is required for stability. For the value given, the capacitor must be solid tantalum but a 25-μF aluminum electrolytic may be substituted. Values given may be increased without limit.
   C. In Figure 13 capacitor C1 is solid tantalum.
   D. This resistor determines zener current. Adjust to minimize thermal drift.

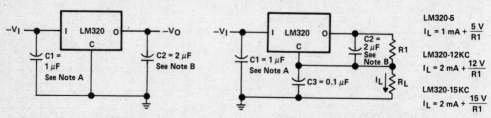

TEXAS INSTRUMENTS
POST OFFICE BOX 225012 • DALLAS, TEXAS 75265

Voltage Regulators

6

LINEAR
INTEGRATED
CIRCUITS

TYPE LM330
3-TERMINAL POSITIVE REGULATOR

D2700, APRIL 1983

- Input-Output Differential Less than 0.6 V
- Output Current of 150 mA
- Reverse Polarity Protection
- Line Transient Protection
- Internal Short-Circuit Current Limiting
- Internal Thermal Overload Protection
- Mirror-Image Insertion Protection
- Direct Replacement for National LM330T-5.0

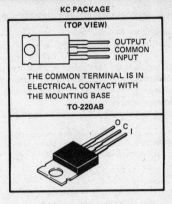

KC PACKAGE
(TOP VIEW)

OUTPUT
COMMON
INPUT

THE COMMON TERMINAL IS IN
ELECTRICAL CONTACT WITH
THE MOUNTING BASE

TO-220AB

description

The LM330 3-terminal positive regulator features an ability to source 150 milliamperes of output current with an input-output differential of 0.6 volt or less. Familar regulator features such as current limit and thermal overload protection are also provided.

The LM330 has low dropout voltage making it useful for certain battery applications. For example, since the low dropout voltage allows a longer battery discharge before the output falls out of regulation, a battery supplying the regulator input voltage may discharge to 5.6 volts and still properly regulate the system and load voltage. The LM330 protects both itself and the regulated system from reverse installation of batteries.

Other protection features include line transient protection above 40 volts, where the output actually shuts down to avoid damaging internal and external circuits. The LM330 regulator cannot be harmed by temporary mirror-image insertion.

schematic diagram

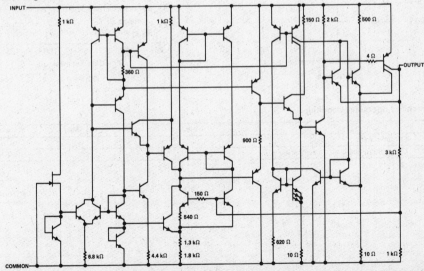

INPUT

1 kΩ   1 kΩ   150 Ω 2 kΩ   500 Ω

360 Ω   4 Ω   OUTPUT

900 Ω   3 kΩ

150 Ω
540 Ω

1.3 kΩ   620 Ω
6.8 kΩ   4.4 kΩ   1.8 kΩ   10 Ω   10 Ω   1 kΩ

COMMON

Voltage Regulators

6

TYPE LM330
3-TERMINAL POSITIVE REGULATOR

absolute maximum ratings over operating free-air temperature range (unless otherwise noted)

Continuous input voltage	26 V
Transient input voltage t = 1 s	50 V
t = 100 ms	60 V
Continuous total dissipation at 25 °C free-air temperature (see Note 1)	2 W
Continuous total dissipation at (or below) case temperature (see Note 1)	20 W
Operating free-air, case, or virtual junction temperature	−55°C to 150°C
Storage temperature	−65°C to 150°C
Lead temperature 1,6 mm (1/16 inch) from case for 10 seconds	260°C

NOTE 1: For operation above 25 °C free-air or case temperature, refer to Figures 1 and 2. To avoid exceeding the design maximum virtual junction temperature, these ratings should not be exceeded. Due to variations in individual device electrical characteristics and thermal resistance, the built-in thermal overload protection may be activated at power levels slightly above or below the rated dissipation.

FREE-AIR TEMPERATURE
DISSIPATION DERATING CURVE

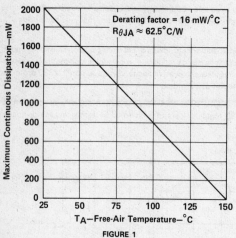

FIGURE 1

CASE TEMPERATURE
DISSIPATION DERATING CURVE

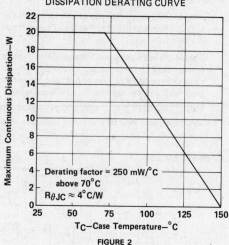

FIGURE 2

recommended operating conditions

		MIN	MAX	UNIT
I_O	Output current	5	150	mA
T_A	Operating virtual junction temperature	0	100	°C

Voltage Regulators

6

TEXAS
INSTRUMENTS
POST OFFICE BOX 225012 • DALLAS, TEXAS 75265

electrical characteristics at 25 °C virtual junction temperature, $V_I = 14$ V, $I_O = 150$ mA, (unless otherwise noted)

PARAMETERS	TEST CONDITIONS[†]		MIN	TYP	MAX	UNIT
Output voltage	$V_I = 6$ V to 26 V, $\quad I_O = 5$ mA to 150 mA,		4.8	5	5.2	V
	$T_J = 0$ °C to 100 °C		4.75		5.25	
Input regulation	$I_O = 5$ mA	$V_I = 9$ V to 16 V		7	25	mV
		$V_I = 6$ V to 26 V		30	60	
Ripple rejection	$f = 120$ Hz			56		dB
Output regulation	$I_O = 5$ mA to 150 mA			14	50	mV
Output voltage long-term drift[‡]	After 1000 h at $T_J = 100$ °C			20		mV
Dropout voltage	$I_O = 150$ mA			0.32	0.6	V
Output noise voltage	$f = 10$ Hz to 100 kHz			50		μV
Output voltage with input polarity reversed	$R_L = 100\ \Omega$	$V_I = -30$ V, $t = 100$ ms	> -0.3			V
		$V_I = -12$ V, DC	> -0.3			
Output voltage with input transient	$V_I = 60$ V,	$t = 100$ ms	< 5.5			V
	$V_I = 50$ V,	$t = 1$ s	< 5.5			
Bias current with input transient	$R_L = 100\ \Omega$	$V_I = 40$ V, $t = 1$ s		14		mA
		$V_I = -6$ V, $t = 1$ s		-80		
Overvoltage shutdown voltage			26	45		V
Output impedance	$I_O = 100$ mA, $I_O = 10$ mA (rms), $f = 100$ Hz to 10 kHz			200		mΩ
Bias current	$I_O = 10$ mA			3.5	7	mA
	$I_O = 50$ mA			5	11	
	$I_O = 150$ mA			18	40	
Bias current change	$V_I = 6$ V to 26 V			10		%
Peak output current			150	420	700	mA

[†]Unless otherwise specified, all characteristics except ripple rejection and noise voltage measurements are measured using pulse techniques ($t_w \leq 10$ ms, duty cycle $\leq 5\%$) with a capacitor of 0.1 μF across the input and a capacitor of 10 μF across the output. Output voltage changes due to changes in internal temperature must be taken into account separately.

[‡]Since long-term drift cannot be measured on the individual devices prior to shipment, this specification is not intended to be a guarantee or warranty. It is an engineering estimate of the average drift to be expected from lot to lot.

Voltage Regulators

6

TEXAS
INSTRUMENTS
POST OFFICE BOX 225012 ● DALLAS, TEXAS 75265

TYPICAL CHARACTERISTICS

OUTPUT VOLTAGE
vs
VIRTUAL JUNCTION TEMPERATURE

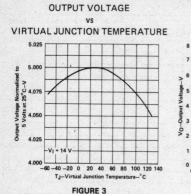

FIGURE 3

OUTPUT VOLTAGE
vs
INPUT VOLTAGE

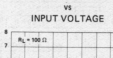

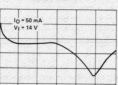
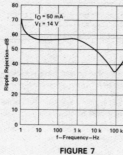

FIGURE 4

OUTPUT VOLTAGE
vs
INPUT VOLTAGE

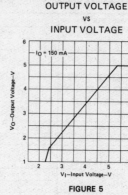

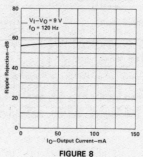

FIGURE 5

PEAK OUTPUT CURRENT
vs
INPUT VOLTAGE

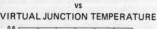

FIGURE 6

RIPPLE REJECTION
vs
FREQUENCY

FIGURE 7

RIPPLE REJECTION
vs
OUTPUT CURRENT

FIGURE 8

DROPOUT VOLTAGE
vs
VIRTUAL JUNCTION TEMPERATURE

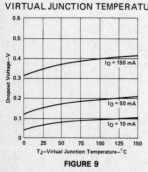

FIGURE 9

DROPOUT VOLTAGE
vs
OUTPUT CURRENT

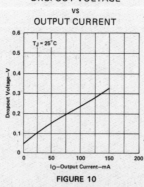

FIGURE 10

OUTPUT IMPEDANCE
vs
FREQUENCY

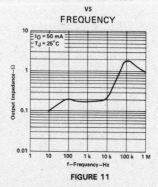

FIGURE 11

Voltage Regulators

6

TEXAS
INSTRUMENTS
POST OFFICE BOX 225012 • DALLAS, TEXAS 75265

TYPICAL CHARACTERISTICS

INPUT CURRENT
vs
INPUT VOLTAGE

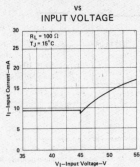

FIGURE 12

LINE TRANSIENT RESPONSE

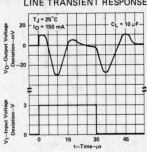

FIGURE 13

INPUT CURRENT
vs
REVERSE INPUT VOLTAGE

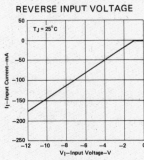

FIGURE 14

OUTPUT VOLTAGE
vs
REVERSE INPUT VOLTAGE

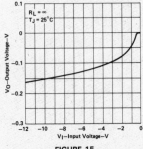

FIGURE 15

LOAD TRANSIENT RESPONSE

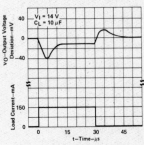

FIGURE 16

BIAS CURRENT
vs
OUTPUT CURRENT

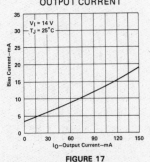

FIGURE 17

BIAS CURRENT
vs
VIRTUAL JUNCTION TEMPERATURE

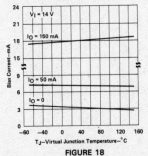

FIGURE 18

BIAS CURRENT
vs
INPUT VOLTAGE

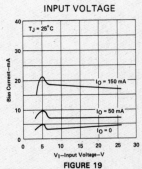

FIGURE 19

Voltage Regulators

6

TEXAS
INSTRUMENTS
POST OFFICE BOX 225012 • DALLAS, TEXAS 75265

TYPICAL APPLICATION DATA

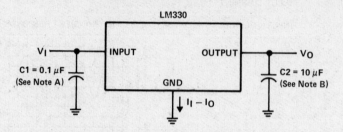

NOTES: A. Use of C1 is required if the regulator is not located in close proximity to the supply filter.

B. Capacitor C2 must be located as close as possible to the regulator and may be an aluminum or tantalum type capacitor. The minimum capacitance that will provide stability is 10 μF. The capacitor must be rated for operation at -40°C to guarantee stability to that extreme.

FIGURE 20

TEXAS
INSTRUMENTS
POST OFFICE BOX 225012 • DALLAS, TEXAS 75265

- 3-Terminal Regulators
- Output Current up to 1.5 A
- No External Components
- Internal Thermal Overload
- High Power Dissipation Capability
- Internal Short-Circuit Current Limiting
- Output Transistor Safe-Area Compensation
- Output Load Regulation . . . 0.3% Typ
- Direct Replacements for National LM340 Series

NOMINAL OUTPUT VOLTAGE	REGULATOR
5 V	LM340-5
12 V	LM340-12
15 V	LM340-15

description

This series of fixed-voltage monolithic integrated-circuit voltage regulators is designed for a wide range of applications. These applications include on-card regulation for elimination of noise and distribution problems associated with single-point regulation. Any of these regulators can deliver up to 1.5 amperes of output current. The internal current limiting and thermal shutdown features of these regulators make them essentially immune to overload. In addition to use as fixed-voltage regulators, these devices can be used with external components to obtain adjustable output voltages and currents and also as the power-pass element in precision regulators.

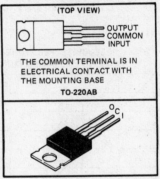

KC PACKAGE

(TOP VIEW)

OUTPUT
COMMON
INPUT

THE COMMON TERMINAL IS IN ELECTRICAL CONTACT WITH THE MOUNTING BASE

TO-220AB

Voltage Regulators

6

schematic

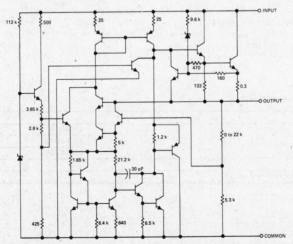

Resistor values shown are nominal and in ohms.

TEXAS INSTRUMENTS

POST OFFICE BOX 225012 • DALLAS, TEXAS 75265

absolute maximum ratings over operating temperature range (unless otherwise noted)

Input voltage . . . . 35 V
Continuous total dissipation at 25 °C free-air temperature (see Note 1) . . . . 2 W
Continuous total dissipation at (or below) 25 °C case temperature (see Note 1) . . . . 15 W
Operating free-air, case, or virtual junction temperature range . . . . −55 °C to 150 °C
Storage temperature range . . . . −65 °C to 150 °C
Lead temperature 1,6 mm (1/16 inch) from case for 10 seconds . . . . 260 °C

NOTE 1: For operation above 25 °C free-air or case temperature, refer to Figures 1 and 2. To avoid exceeding the design maximum virtual junction temperature, these ratings should not be exceeded. Due to variations in individual device electrical characteristics and thermal resistance, the built-in thermal overload protection may be activated at power levels slightly above or below the rated dissipation.

FREE-AIR TEMPERATURE
DISSIPATION DERATING CURVE

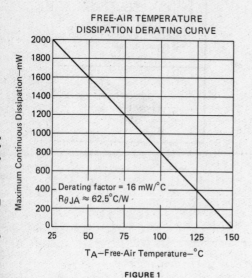

Derating factor = 16 mW/°C
$R_{\theta JA} \approx 62.5$ °C/W

FIGURE 1

CASE TEMPERATURE
DISSIPATION DERATING CURVE

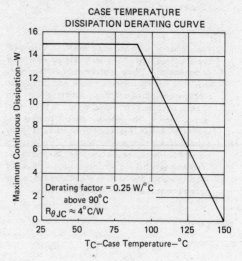

Derating factor = 0.25 W/°C
above 90°C
$R_{\theta JC} \approx 4$ °C/W

FIGURE 2

recommended operating conditions

		MIN	MAX	UNIT
Input voltage, V_I	LM340-5	7	25	V
	LM340-12	14.5	30	
	LM340-15	17.5	30	
Output current, I_O			1.5	A
Operating virtual junction temperature, T_J		0	125	°C

Voltage Regulators

6

LM340-5 electrical characteristics at specified virtual junction temperature, V_I = 10 V, I_O = 1 A (unless otherwise noted)

PARAMETER	TEST CONDITIONS[†]			MIN	TYP	MAX	UNIT
Output voltage	I_O = 5 mA to 1 A		25°C	4.8	5	5.2	V
	V_I = 7 V to 20 V, $\quad$ I_O = 5 mA to 1 A, $\quad$ P ≤ 15 W		0°C to 125°C	4.75		5.25	
Input regulation	I_O = 500 mA	V_I = 7 V to 25 V	25°C		3	50	mV
		V_I = 8 V to 20 V	0°C to 125°C			50	
	I_O = 1 A	V_I = 7.3 V to 20 V	25°C			50	
		V_I = 8 V to 12 V	0°C to 125°C			25	
Ripple rejection	V_I = 8 V to 18 V, $\quad$ I_O ≤ 1 A		25°C	62	80		dB
	f = 120 Hz $\qquad$ I_O ≤ 500 mA		0°C to 125°C	62			
Output regulation	I_O = 250 mA to 750 mA		25°C			25	mV
	I_O = 5 mA to 1.5 A				10	50	
	I_O = 5 mA to 1 A		0°C to 125°C			50	
Output noise voltage	f = 10 Hz to 100 kHz		25°C		40		µV
Dropout voltage	I_O = 1 A		25°C		2		V
Temperature coefficient of output voltage	I_O = 5 mA		0°C to 125°C		−0.6		mV/°C
Output impedance	f = 1 kHz		25°C		8		mΩ
Bias current	I_O ≤ 1 A		25°C			8	mA
			0°C to 125°C			8.5	
Bias current change	V_I = 7.5 V to 20 V, $\quad$ I_O ≤ 1 A		25°C			1	mA
	V_I = 7 V to 25 V, $\quad$ I_O ≤ 500 mA		0°C to 125°C			1	
	I_O = 5 mA to 1 A					0.5	
Peak output current			25°C		2.4		A
Short-circuit current			25°C		2.1		A

† All characteristics are measured with a capacitor across the input of 0.22 µF and a capacitor across the output of 0.1 µF. All characteristics except noise voltage rejection ratio are measured using pulse techniques (t_w ≤ 10 ms, duty cycle ≤ 5%). Output voltage changes due to changes in internal temperature must be taken into account separately.

Voltage Regulators

6

Texas Instruments
POST OFFICE BOX 225012 • DALLAS, TEXAS 75265

LM340-12 electrical characteristics at specified virtual junction temperature, V_I = 19 V, I_O = 1 A (unless otherwise noted)

PARAMETER	TEST CONDITIONS[†]			MIN	TYP	MAX	UNIT
Output voltage	I_O = 5 mA to 1 A		25 °C	11.5	12	12.5	V
	V_I = 14.5 V to 27 V, P ≤ 15 W	I_O = 5 mA to 1 A,	0 °C to 125 °C	11.4		12.6	
Input regulation	I_O = 500 mA	V_I = 14.5 V to 30 V	25 °C		4	120	mV
		V_I = 15 V to 27 V	0 °C to 125 °C			120	
	I_O = 1 A	V_I = 14.6 V to 27 V	25 °C			120	
		V_I = 16 V to 22 V	0 °C to 125 °C			120	
Ripple rejection	V_I = 15 V to 25 V, f = 120 Hz	I_O ≤ 1 A	25 °C	55	72		dB
		I_O ≤ 500 mA	0 °C to 125 °C	55			
Output regulation	I_O = 250 mA to 750 mA		25 °C			60	mV
	I_O = 5 mA to 1.5 A				12	120	
	I_O = 5 mA to 1 A		0 °C to 125 °C			120	
Output noise voltage	f = 10 Hz to 100 kHz		25 °C		75		µV
Dropout voltage	I_O = 1 A		25 °C		2		V
Temperature coefficient of output voltage	I_O = 5 mA		0 °C to 125 °C		−1.5		mV/°C
Output impedance	f = 1 kHz		25 °C		18		mΩ
Bias current	I_O ≤ 1 A		25 °C			8	mA
			0 °C to 125 °C			8.5	
Bias current change	V_I = 14.8 V to 27 V, I_O ≤ 1 A		25 °C			1	mA
	V_I = 14.5 V to 30 V, I_O ≤ 500 mA		0 °C to 125 °C			1	
	I_O = 5 mA to 1 A					0.5	
Peak output current			25 °C		2.4		A
Short-circuit current			25 °C		1.5		A

† All characteristics are measured with a capacitor across the input of 0.22 µF and a capacitor across the output of 0.1 µF. All characteristics except noise voltage and ripple rejection ratio are measured using pulse techniques (t_w ≤ 10 ms, duty cycle ≤ 5%). Output voltage changes due to changes in internal temperature must be taken into account separately.

Voltage Regulators

6

TEXAS
INSTRUMENTS
POST OFFICE BOX 225012 ● DALLAS, TEXAS 75265

LM340-15 electrical characteristics at specified virtual junction temperature, V_I = 23 V, I_O = 1 A (unless otherwise noted)

PARAMETER	TEST CONDITIONS[†]			MIN	TYP	MAX	UNIT
Output voltage	I_O = 5 mA to 1 A		25°C	14.4	15	15.6	V
	V_I = 17.5 V to 30 V, I_O = 5 mA to 1 A, P ≤ 15 W		0°C to 125°C	14.25		15.75	
Input regulation	I_O = 500 mA	V_I = 17.5 V to 30 V	25°C		4	150	mV
		V_I = 18.5 V to 30 V	0°C to 125°C			150	
	I_O = 1 A	V_I = 17.7 V to 30 V	25°C			150	
		V_I = 20 V to 26 V	0°C to 125°C			75	
Ripple rejection	V_I = 18.5 V to 28.5 V, f = 120 Hz	I_O ≤ 1 A	25°C	54	70		dB
		I_O ≤ 500 mA	0°C to 125°C	54			
Output regulation	I_O = 250 mA to 750 mA		25°C			75	mV
	I_O = 5 mA to 1.5 A				12	150	
	I_O = 5 mA to 1 A		0°C to 125°C			150	
Output noise voltage	f = 10 Hz to 100 kHz		25°C		90		µV
Dropout voltage	I_O = 1 A		25°C		2		V
Temperature coefficient of output voltage	I_O = 5 mA		0°C to 125°C		−1.8		mV/°C
Output impedance	f = 1 kHz		25°C		19		mΩ
Bias current	I_O ≤ 1 A		25°C			8	mA
			0°C to 125°C			8.5	
Bias current change	V_I = 17.9 V to 30 V, I_O ≤ 1 A		25°C			1	mA
	V_I = 17.5 V to 30 V, I_O ≤ 500 mA		0°C to 125°C			1	
	I_O = 5 mA to 1 A					0.5	
Peak output current			25°C		2.4		A
Short-circuit current			25°C		1.2		A

† All characteristics are measured with a capacitor across the input of 0.22 µF and a capacitor across the output of 0.1 µF. All characteristics except noise voltage and ripple rejection ratio are measured using pulse techniques (t_w ≤ 10 ms, duty cycle ≤ 5%). Output voltage changes due to changes in internal temperature must be taken into account separately.

Voltage Regulators

6

TEXAS
INSTRUMENTS
POST OFFICE BOX 225012 ● DALLAS, TEXAS 75265

Voltage Regulators

6

TYPICAL CHARACTERISTICS

NORMALIZED OUTPUT VOLTAGE
vs
VIRTUAL JUNCTION TEMPERATURE

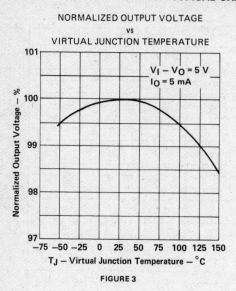

FIGURE 3

RIPPLE REJECTION
vs
FREQUENCY

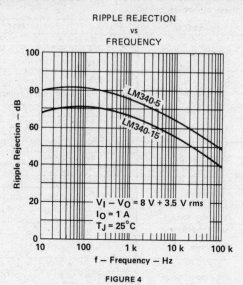

FIGURE 4

DROPOUT VOLTAGE
vs
VIRTUAL JUNCTION TEMPERATURE

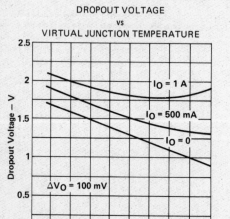

FIGURE 5

LM340-5
DROPOUT VOLTAGE
vs
INPUT VOLTAGE

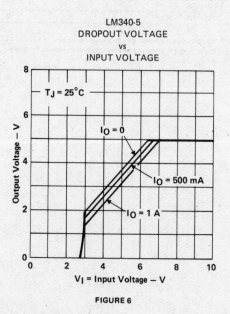

FIGURE 6

TEXAS
INSTRUMENTS
POST OFFICE BOX 225012 • DALLAS, TEXAS 75265

TYPICAL CHARACTERISTICS

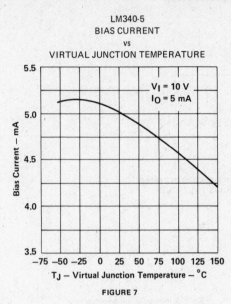

LM340-5
BIAS CURRENT
vs
VIRTUAL JUNCTION TEMPERATURE

$V_I = 10$ V
$I_O = 5$ mA

Bias Current — mA

T_J — Virtual Junction Temperature — °C

FIGURE 7

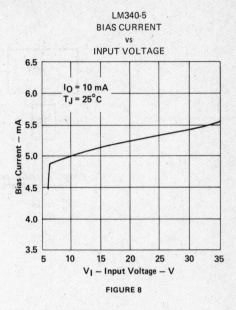

LM340-5
BIAS CURRENT
vs
INPUT VOLTAGE

$I_O = 10$ mA
$T_J = 25°C$

Bias Current — mA

V_I — Input Voltage — V

FIGURE 8

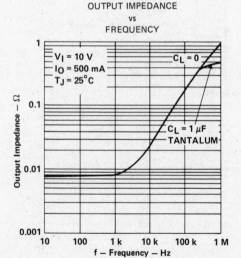

LM340-5
OUTPUT IMPEDANCE
vs
FREQUENCY

$V_I = 10$ V
$I_O = 500$ mA
$T_J = 25°C$

$C_L = 0$

$C_L = 1 \mu$F
TANTALUM

Output Impedance — Ω

f — Frequency — Hz

FIGURE 9

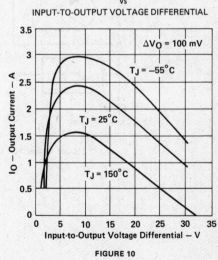

PEAK OUTPUT CURRENT
vs
INPUT-TO-OUTPUT VOLTAGE DIFFERENTIAL

$\Delta V_O = 100$ mV

$T_J = -55°C$

$T_J = 25°C$

$T_J = 150°C$

I_O — Output Current — A

Input-to-Output Voltage Differential — V

FIGURE 10

Voltage Regulators

6

TEXAS
INSTRUMENTS
POST OFFICE BOX 225012 ● DALLAS, TEXAS 75265

TYPICAL APPLICATION DATA

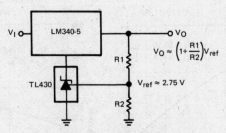

$$V_O \approx \left(1 + \frac{R1}{R2}\right)V_{ref}$$

$$V_{ref} \approx 2.75 \text{ V}$$

FIGURE 11—ADJUSTABLE SUPPLY WITH STABLE OUTPUT FROM 8 VOLTS TO 35 VOLTS

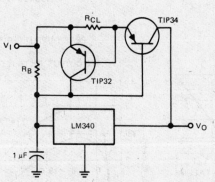

The boost circuit takes over at a level determined by R_B.

$$R_B \approx \frac{0.6 \text{ V}}{I_B}$$

where I_B is the LM340 operating level.

Maximum current limit I_{CL} is determined by R_{CL}.

$$R_{CL} \approx \frac{0.6 \text{ V}}{I_{CL}}$$

Example: If I_B is selected to be 0.5 A, then
$R_B = 1.2 \, \Omega$.
If I_{CL} is 3 A, then
$R_{CL} = 0.2 \, \Omega$.

FIGURE 12—OUTPUT CURRENT BOOST CIRCUIT

TEXAS INSTRUMENTS

POST OFFICE BOX 225012 • DALLAS, TEXAS 75265

- Adjustable Output . . . 1.2 V to 33 V
- 3-A Output Current Capability
- Line Regulation . . . 0.005%/V Typ
- Load Regulation . . . 0.1% Typ
- Current Limit Constant with Temperature
- Guaranteed Thermal Regulation
- Direct Replacement for National Semiconductor LM350

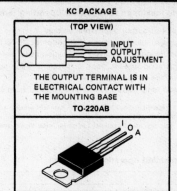

KC PACKAGE

(TOP VIEW)

INPUT
OUTPUT
ADJUSTMENT

THE OUTPUT TERMINAL IS IN
ELECTRICAL CONTACT WITH
THE MOUNTING BASE

TO-220AB

description

The LM350 is an adjustable 3-terminal positive-voltage regulator capable of supplying 3 amperes over an output voltage range of 1.2 volts to 33 volts. The device is easy to use and requires only two external resistors to set the output voltage. Both input and output regulation are better than standard fixed regulators.

In addition to higher performance than fixed regulators, the LM350 offers full overload protection available only in integrated circuits. Included on the chip current limit, thermal overload protection, and safe-area protection. All overload protection circuitry remains fully functional even if the adjustment terminal is disconnected. Normally, no capacitors are needed unless the device is situated far from the input filter capacitors in which case an input bypass is needed. An optional output capacitor can be added to improve transient response. The adjustment terminal can be bypassed to achieve very high ripple rejection, which is difficult to achieve with standard 3-terminal regulators.

Besides replacing fixed regulators, the LM350 is useful in a wide variety of other applications. Even though the regulator is floating and sees only the input-to-output differential voltage, use of these devices to regulate voltages that would cause the maximum-rated differential voltage to be exceeded if the output became shorted to ground is not recommended. The TL783 or TL783A is recommended for output voltages exceeding 33 volts. The primary application of the LM350 is that of a programmable output regulator, but by connecting a fixed resistor between the adjustment terminal and the output terminal, this device can be used as a precision current regulator. Supplies with electronic shutdown can be achieve by clamping the adjustment terminal to ground, which programs the output to 1.2 volts where most loads draw little current.

The LM350 is characterized for operation from 0 °C to 125 °C.

Voltage Regulators

6

TEXAS INSTRUMENTS

POST OFFICE BOX 225012 • DALLAS, TEXAS 75265

Voltage Regulators

6

absolute maximum ratings over operating temperature range (unless otherwise noted)

Input-to-output differential voltage .... 35 V
Continuous total power dissipation at 25 °C free-air temperature (see Note 1) .... 2 W
Continuous total power dissipation at (or below) 25 °C case temperature (see Note 1) .... 30 W
Operating free-air, case, or virtual junction temperature range .... −55 °C to 150 °C
Storage temperature range .... −65 °C to 150 °C
Lead temperature 1,6 mm (1/16 inch) from case for 60 seconds .... 260 °C

NOTE 1: For operation above 25 °C free-air or case temperature, refer to Figures 1 and 2. To avoid exceeding the design maximum virtual junction temperature, these ratings should not be exceeded. Due to variations in individual device electrical characteristics and thermal resistance, the built-in thermal overload protection may be activated at power levels slightly above or below the rated dissipation.

recommended operating conditions

	MIN	MAX	UNIT
Output current, I_O		3	A
Operating virtual junction temperature, T_J	0	125	°C

electrical characteristics over recommended ranges of operating virtual junction temperature, $V_I - V_O = 5$ V, $I_O = 1.5$ A (unless otherwise noted)

PARAMETER	TEST CONDITIONS		MIN	TYP	MAX	UNIT
Input regulation (see Note 2)	$V_I - V_O = 3$ V to 35 V	$T_J = 25$ °C		0.005	0.03	%/V
	See Note 3	$T_J = 0$ °C to 125 °C		0.02	0.07	
Ripple rejection	$V_O = 10$ V,	$f = 120$ Hz		65		dB
	$V_O = 10$ V, 10-μF capacitor between ADJ and ground	$f = 120$ Hz,	66			
Output regulation	$I_O = 10$ mA to 3 A, $T_J = 25$ °C, See Note 3	$V_O \leq 5$ V		5	25	mV
		$V_O > 5$ V		0.1	0.5	%
	$I_O = 10$ mA to 3 A, See Note 3	$V_O \leq 5$ V		20	70	mV
		$V_O > 5$ V		0.3	1.5	%
Output voltage change with temperature	$T_J = 0$ °C to 125 °C			1		%
Thermal regulation	$t_w = 20$ ms			0.002	0.03	%/W
Output voltage long-term drift (see Note 4)	After 1000 h at $T_J = 125$ °C			0.3	1	%
Output noise voltage	$f = 10$ Hz to 10 kHz,	$T_J = 25$ °C		0.003		%
Minimum output current to maintain regulation	$V_I - V_O = 35$ V			3.5	10	mA
Peak output current	$V_I - V_O \leq 10$ V		3	4.5		A
	$V_I - V_O = 30$ V,	$T_J = 25$ °C	0.25	1		
Adjustment-terminal current				50	100	μA
Change in adjustment-terminal current	$V_I - V_O = 3$ V to 35 V, $I_O = 10$ mA to 3 A			0.2	5	μA
Reference voltage (output to ADJ)	$V_I - V_O = 3$ V to 35 V, $I_O = 10$ mA to 3 A,	$P \leq 30$ W	1.2	1.25	1.3	V

NOTES: 2. Input regulation is expressed as the percentage change in output voltage per 1-volt change at the input.
3. Input regulation and output regulation are measured using pulse techniques ($t_w \leq 10$ μs, duty cycle ≤ 5%) to limit changes in average internal dissipation. Output voltage changes due to large changes in internal dissipation must be taken into account separately.
4. Since long-term drift cannot be measured on the individual devices prior to shipment, this specification is not intended to be a guarantee or warranty. It is an engineering estimate of the average drift to be expected from lot to lot.

TEXAS INSTRUMENTS
POST OFFICE BOX 225012 ● DALLAS, TEXAS 75265

1183

Voltage Regulators

6

THERMAL INFORMATION

FREE-AIR TEMPERATURE
DISSIPATION DERATING CURVE

FIGURE 1

CASE TEMPERATURE
DISSIPATION DERATING CURVE

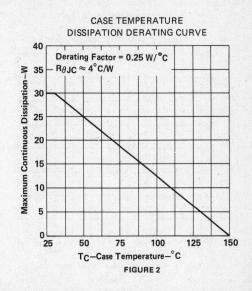

FIGURE 2

TYPICAL CHARACTERISTICS

OUTPUT VOLTAGE DEVIATION
vs
VIRTUAL JUNCTION TEMPERATURE

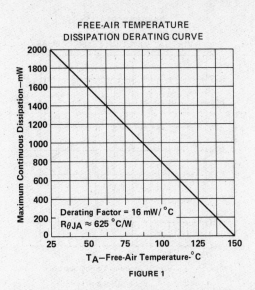

FIGURE 3

OUTPUT CURRENT
vs
INPUT-OUTPUT DIFFERENTIAL VOLTAGE

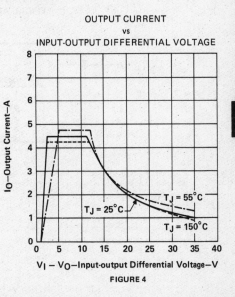

FIGURE 4

Voltage Regulators

TYPICAL CHARACTERISTICS

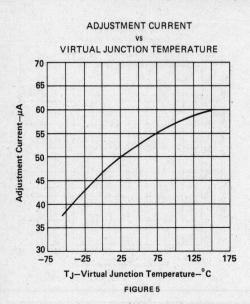

ADJUSTMENT CURRENT
vs
VIRTUAL JUNCTION TEMPERATURE

FIGURE 5

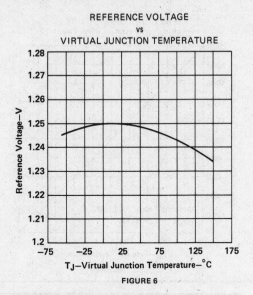

REFERENCE VOLTAGE
vs
VIRTUAL JUNCTION TEMPERATURE

FIGURE 6

TYPICAL APPLICATION DATA

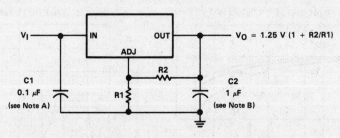

$V_O = 1.25\ V\ (1 + R2/R1)$

NOTES: A. Capacitor C1 is required if regulator is not located in close proximity to the power supply amplifier.
       B. Capacitor C2 may be used to improve transient response.

TEXAS
INSTRUMENTS
POST OFFICE BOX 225012 • DALLAS, TEXAS 75265

6

LINEAR INTEGRATED CIRCUITS

TYPES LM2930-5, LM2930-8
3-TERMINAL POSITIVE REGULATORS

- Input-Output Differential Less than 0.6 V
- Output Current of 150 mA
- Reverse Battery Protection
- Line Transient Protection
- 40-Volt Load-Dump Protection
- Internal Short-Circuit Current Limiting
- Internal Thermal Overload Protection
- Mirror-Image Insertion Protection
- Direct Replacement for National LM2930 Series

KC PACKAGE

(TOP VIEW)

OUTPUT
COMMON
INPUT

THE COMMON TERMINAL IS IN
ELECTRICAL CONTACT WITH
THE MOUNTING BASE

TO-220AB

description

The LM2930-5 and LM2930-8 are 3-terminal positive regulators that provide fixed 5-volt and 8-volt regulated outputs. Each features the ability to source 150 milliamperes of output current with an input-output differential of 0.6 volt or less. Familiar regulator features such as current limit and thermal overload protection are also provided.

The LM2930 series has low voltage dropout making it useful for certain battery applications. For example, the low voltage dropout feature allows a longer battery discharge before the output falls out of regulation; the battery supplying the regulator input voltage may discharge to 5.6 volts and still properly regulate the system and load voltage. Supporting this feature, the LM2930 series protects both itself and the regulated system from reverse battery installation or two-battery jumps.

Other protection features include line transient protection for load-dump of up to 40 volts. In this case the regulator shuts down to avoid damaging internal and external circuits. The LM2930 series regulator cannot be harmed by temporary mirror-image insertion.

schematic diagram

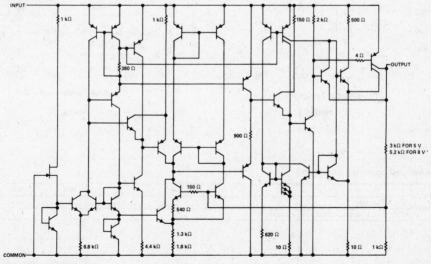

All component values are nominal.

TEXAS INSTRUMENTS

POST OFFICE BOX 225012 • DALLAS, TEXAS 75265

Voltage Regulators

6

absolute maximum ratings over operating free-air temperature range (unless otherwise noted)

Continuous input voltage . . . . 26 V
Transient input voltage: t = 1 s . . . . 40 V
Continuous reverse input voltage . . . . −6 V
Transient reverse input voltage: t = 100 ms . . . . −12 V
Continuous total dissipation at 25°C free-air temperature (see Note 1) . . . . 2 W
Continuous total dissipation at (or below) 25°C case-temperature (see Note 1) . . . . 20 W
Operating free-air, case, or virtual junction temperature . . . . −40°C to 150°C
Storage temperature range . . . . −65°C to 150°C
Lead temperature 1,6 mm (1/16 inch) from case to 10 seconds . . . . 260°C

NOTE 1: For operation above 25°C free-air or case temperature, refer to Figures 1 and 2. To avoid exceeding the design maximum virtual junction temperature, these ratings should not be exceeded. Due to variation in individual device elecrical characteristics and thermal resistance, the bult-in thermal overload protection may be activated at power levels slightly above or below the rated dissipation.

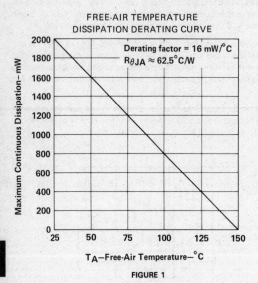

FREE-AIR TEMPERATURE
DISSIPATION DERATING CURVE

Derating factor = 16 mW/°C
$R_{\theta}JA \approx 62.5°C/W$

FIGURE 1

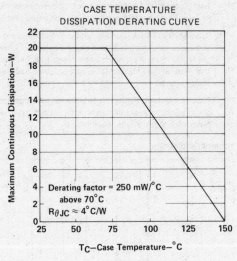

CASE TEMPERATURE
DISSIPATION DERATING CURVE

Derating factor = 250 mW/°C
above 70°C
$R_{\theta}JC \approx 4°C/W$

FIGURE 2

recommended operating conditions

		MIN	MAX	UNIT
I_O	Output current		150	mA
T_J	Operating virtual junction temperature	−40	125	°C

TEXAS
INSTRUMENTS
POST OFFICE BOX 225012 • DALLAS, TEXAS 75265

LM2930-5 electrical characteristics at 25 °C virtual junction temperature, V_I = 14 V, I_O = 150 mA, (unless otherwise noted)

PARAMETER	TEST CONDITIONS[†]		MIN	TYP	MAX	UNIT
Output voltage	V_I = 6 V to 26 V, T_J = −40 °C to 125 °C	I_O = 5 mA to 150 mA,	4.5	5	5.5	V
Input regulation	I_O = 5 mA	V_I = 9 V to 16 V		7	25	mV
		V_I = 6 V to 26 V		30	80	
Ripple rejection	f = 120 Hz			56		dB
Output regulation	I_O = 5 mA to 150 mA			14	50	mV
Output voltage long-term drift[‡]	After 1000 h at T_J = 125 °C			20		mV
Dropout voltage	I_O = 150 mA			0.32	0.6	V
Output noise voltage	f = 10 Hz to 100 kHz			60		µV
Output voltage during line transients	V_I = −12 V to 40 V, R_L = 100 Ω		−0.3		5.5	V
Output impedance	I_O = 100 mA, I_o = 10 mA (rms), f = 100 Hz to 10 kHz			200		mΩ
Bias current	I_O = 10 mA			4	7	mA
	I_O = 150 mA			18	40	
Peak output current			150	300	700	mA

LM2930-8 electrical characteristics at 25 °C virtual junction temperature, V_I = 14 V, I_O = 150 mA, (unless otherwise noted)

PARAMETER	TEST CONDITIONS[†]		MIN	TYP	MAX	UNIT
Output voltage	V_I = 9.4 V to 26 V, T_J = −40 °C to 125 °C	I_O = 5 mA to 150 mA,	7.2	8	8.8	V
Input regulation	I_O = 5 mA	V_I = 9.4 V to 16 V		12	50	V
		V_I = 9.4 V to 26 V		50	100	
Ripple rejection	f = 120 Hz			52		dB
Output regulation	I_O = 5 mA to 150 mA			25	50	mV
Output voltage long-term drift[‡]	After 1000 h at T_J = 125 °C			30		mV
Dropout voltage	I_O = 150 mA			0.32	0.6	V
Output noise voltage	f = 10 Hz to 100 kHz			90		µV
Output voltage during line transients	V_I = −12 V to 40 V, R_L = 100 Ω		−0.3		8.8	V
Output impedance	I_O = 100 mA, I_o = 10 mA (rms), f = 100 Hz to 10 kHz			300		mΩ
Bias current	I_O = 10 mA			4	7	mA
	I_O = 150 mA			18	40	
Peak output current			150	300	700	mA

[†] Unless otherwise specified, all characteristics, except ripple rejection and noise voltage measurements, are measured using pulse techniques (t_w ≤ 10 ms, duty cycle ≤ 5%) with a capacitor of 0.1 µF across the input and a capacitor of 10 µF across the output. Output voltage changes due to changes in internal temperature must be taken into account separately.

[‡] Since long-term drift cannot be measured on the individual devices prior to shipment, this specification is not intended to be a guarantee or warranty. It is an engineering estimate of the average drift to be expected from lot to lot.

Voltage Regulators

6

TYPICAL CHARACTERISTICS

Voltage Regulators

6

NORMALIZED OUTPUT VOLTAGE
vs
VIRTUAL JUNCTION TEMPERATURE

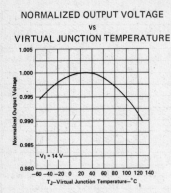

FIGURE 3

LM2930-5
OUTPUT VOLTAGE
vs
INPUT VOLTAGE

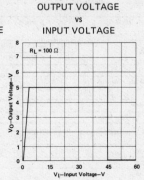

FIGURE 4

LM2930-5
OUTPUT VOLTAGE
vs
INPUT VOLTAGE

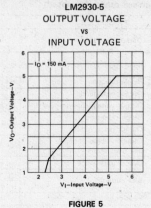

FIGURE 5

RIPPLE REJECTION
vs
FREQUENCY

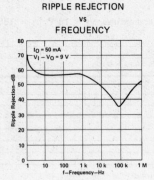

FIGURE 6

RIPPLE REJECTION
vs
OUTPUT CURRENT

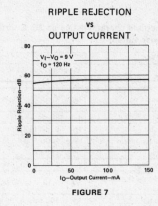

FIGURE 7

DROPOUT VOLTAGE
vs
VIRTUAL JUNCTION TEMPERATURE

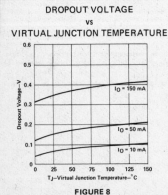

FIGURE 8

DROPOUT VOLTAGE
vs
OUTPUT CURRENT

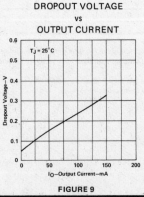

FIGURE 9

OUTPUT IMPEDANCE
vs
FREQUENCY

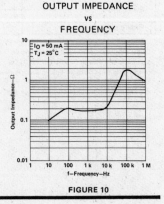

FIGURE 10

TEXAS INSTRUMENTS

POST OFFICE BOX 225012 • DALLAS, TEXAS 75265

TYPICAL CHARACTERISTICS

INPUT CURRENT
vs
INPUT VOLTAGE

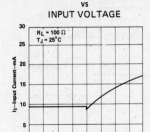

FIGURE 11

LINE TRANSIENT RESPONSE

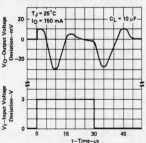

FIGURE 12

INPUT CURRENT
vs
REVERSE INPUT VOLTAGE

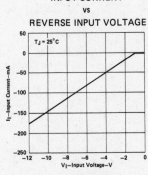

FIGURE 13

OUTPUT VOLTAGE
vs
REVERSE INPUT VOLTAGE

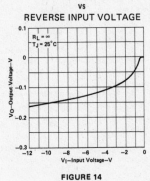

FIGURE 14

LOAD TRANSIENT RESPONSE

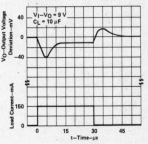

FIGURE 15
LM2905-5

BIAS CURRENT
vs
OUTPUT CURRENT

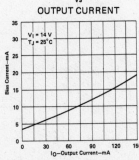

FIGURE 16

BIAS CURRENT
vs
VIRTUAL JUNCTION TEMPERATURE

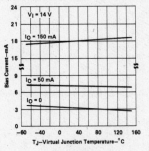

FIGURE 17

BIAS CURRENT
vs
INPUT VOLTAGE

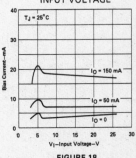

FIGURE 18

Voltage Regulators

6

483

TEXAS
INSTRUMENTS
POST OFFICE BOX 225012 • DALLAS, TEXAS 75265

TYPICAL APPLICATION DATA

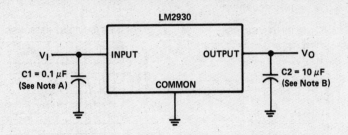

NOTES: A. Use of C1 is required if the regulator is not located in close proximity to the supply filter.
       B. Capacitor C2 must be located as close as possible to the regulator and may be an aluminum or tantalum type capacitor. The minimum value required for stability is 10 μF. The capacitor must be rated for operation at $-40\,°C$ to guarantee stability to that extreme.

FIGURE 19

TEXAS
INSTRUMENTS
POST OFFICE BOX 225012 ● DALLAS, TEXAS 75265

LINEAR
INTEGRATED
CIRCUITS

TYPE LM2931-5
3-TERMINAL POSITIVE VOLTAGE REGULATOR

D2828, FEBRUARY 1984

- Input-Output Differential Less than 0.6 V
- Output Current of 150 mA
- Reverse Battery Protection
- Very Low Quiescent Current
- 60-Volt Load-Dump Protection
- Internal Short-Circuit Current Limiting
- Internal Thermal Overload Protection
- Mirror-Image Insertion Protection
- Reverse Transient Protection to −50 V
- Direct Replacement for National LM2931-5

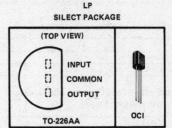

KC PACKAGE

(TOP VIEW)

OUTPUT
COMMON
INPUT

THE COMMON TERMINAL IS IN
ELECTRICAL CONTACT WITH
THE MOUNTING BASE

TO-220AB

LP
SILECT PACKAGE

(TOP VIEW)

[] INPUT
[] COMMON
[] OUTPUT

OCI

TO-226AA

description

The LM2931-5 is a 3-terminal positive voltage regulator that provides a 5-volt regulated output. It features the ability to source 150 milliamperes of output current with an input-output differential of 0.6 volt or less. Familiar regulator features such as current limit and thermal overload protection are also provided.

The LM2931-5 has low voltage dropout making it useful for certain battery applications. For example, the low-voltage-dropout feature allows a longer battery discharge before the output falls out of regulation; the battery supplying the regulator input voltage may discharge to 5.6 volts and still properly regulate the system and load voltage. Supporting this feature, the LM2931-5 protects both itself and the regulated system from reverse battery installation or two-battery jumps. The very low quiescent current feature is especially useful in battery powered applications.

Other protection features include line transient protection for load-dump of up to 60 volts. In this case the regulator shuts down to avoid damaging internal and external circuits. The LM2931-5 regulator is virtually immune to temporary mirror-image insertion.

The LM2931-5 is characterized for operation from −40°C to 150°C.

Voltage Regulators

6

TEXAS
INSTRUMENTS

POST OFFICE BOX 225012 • DALLAS, TEXAS 75265

TYPE LM2931-5
3-TERMINAL POSITIVE VOLTAGE REGULATOR

absolute maximum ratings over operating free-air temperature range (unless otherwise noted)

Continuous input voltage	26 V
Transient input voltage: t = 1 s	60 V
Continuous reverse input voltage	−15 V
Transient reverse input voltage: t = 100 ms	−50 V
Continuous total dissipation at (or below) 25 °C free-air temperature (see Note 1)	
LP package	775 mW
KC package	2 W
Continuous total dissipation at (or below) 25 °C case temperature (see Note 1)	
LP package	1.6 W
KC package	4 W
Operating free-air, case, or virtual junction temperature	−40 °C to 150 °C
Storage temperature range	−65 °C to 150 °C
Lead temperature 1,6 mm (1/16 inch) from case for 10 seconds	260 °C

NOTE 1: For operation above 25 °C free-air temperature, refer to Figures 1, 2, 3, and 4. To avoid exceeding the design maximum virtual junction temperature, these ratings should not be exceeded. Due to variation in individual device electrical characteristics and thermal resistance, the built-in thermal overload protection may be activated at power levels slightly above or below the rated dissipation.

recommended operating conditions

	MIN	TYP	MAX	UNIT
Output current, I_O			150	mA
Operating virtual junction temperature, T_J	−40		125	°C

electrical characteristics at 25 °C virtual junction temperature, V_I = 14 V, I_O = 10 mA, (unless otherwise noted)

PARAMETER	TEST CONDITIONS[†]		MIN	TYP	MAX	UNIT
Output voltage	V_I = 6 V to 26 V, I_O ≤ 150 mA, T_J = −40 °C to 125 °C		4.75	5	5.25	V
Input regulation	I_O = 10 mA	V_I = 9 V to 16 V		2	10	mV
		V_I = 6 V to 26 V		4	30	
Ripple rejection	f = 120 Hz			80		dB
Output regulation	I_O = 5 mA to 150 mA			14	50	mV
Output voltage long-term drift[‡]	After 100 h at T_J = 125 °C			20		mV
Dropout voltage	I_O = 10 mA			0.05	0.2	V
	I_O = 150 mA			0.3	0.6	
Output noise voltage	f = 10 Hz to 100 kHz			500		µV rms
Bias current	V_I = 6 V to 26 V, I_O = 10 mA, T_J = −40 °C to 125 °C			0.4	1	mA
	V_I = 14 V, I_O = 150 mA, T_J = 25 °C			15		

[†]Unless otherwise specified, all characteristics, except ripple rejection and noise voltage measurements, are measured using pulse techniques (t_w ≤ 10 ms, duty cycle ≤ 5%) with a capacitor of 0.1 µF across the input and a capacitor of 100 µF across the output. Output voltage changes due to changes in internal temperature must be taken into account separately.

[‡]Since long-term drift cannot be measured on the individual devices prior to shipment, this specification is not intended to be a guarantee or warranty. It is an engineering estimate of the average drift to be expected from lot to lot.

TEXAS INSTRUMENTS
POST OFFICE BOX 225012 • DALLAS, TEXAS 75265

THERMAL INFORMATION

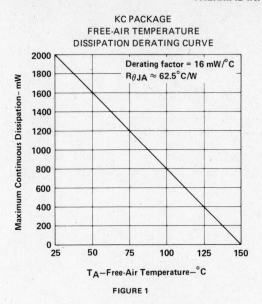

KC PACKAGE
FREE-AIR TEMPERATURE
DISSIPATION DERATING CURVE

Derating factor = 16 mW/°C
$R_{\theta JA} \approx 62.5°C/W$

T_A—Free-Air Temperature—°C

FIGURE 1

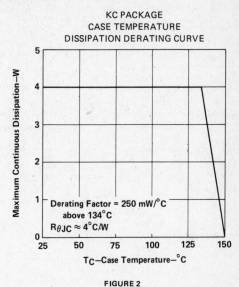

KC PACKAGE
CASE TEMPERATURE
DISSIPATION DERATING CURVE

Derating Factor = 250 mW/°C
above 134°C
$R_{\theta JC} \approx 4°C/W$

T_C—Case Temperature—°C

FIGURE 2

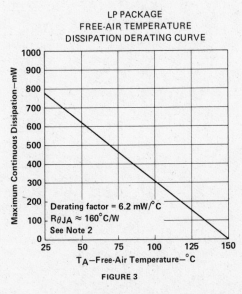

LP PACKAGE
FREE-AIR TEMPERATURE
DISSIPATION DERATING CURVE

Derating factor = 6.2 mW/°C
$R_{\theta JA} \approx 160°C/W$
See Note 2

T_A—Free-Air Temperature—°C

FIGURE 3

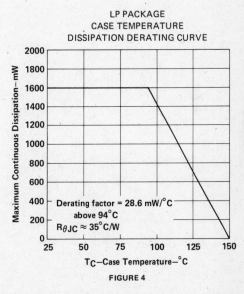

LP PACKAGE
CASE TEMPERATURE
DISSIPATION DERATING CURVE

Derating factor = 28.6 mW/°C
above 94°C
$R_{\theta JC} \approx 35°C/W$

T_C—Case Temperature—°C

FIGURE 4

NOTE 2: This curve for the LP package is based on thermal resistance, $R_{\theta JA}$, measured in still air with the device mounted in an Augat socket. The bottom of the package was 3/8 inch above the socket.

Voltage Regulators

6

TEXAS INSTRUMENTS
POST OFFICE BOX 225012 ● DALLAS, TEXAS 75265

Voltage Regulators

6

- Separate Outputs for "Crowbar" and Logic Circuitry
- Programmable Time Delay to Eliminate Noise Triggering
- TTL-Level Activation Isolated from Voltage-Sensing Inputs
- 2.6-Volt Internal Voltage Reference with Temperature Coefficient Typically 0.08%/°C

MC3423 . . . . . JG OR P
DUAL-IN-LINE PACKAGE
(TOP VIEW)

V$_{CC}$	1	8	OUT
SENSE 1	2	7	V$_{EE}$
SENSE 2	3	6	IND OUT
CURR SOURCE	4	5	REMOTE ACTIVATE

Voltage Regulators

6

description

The MC3423 overvoltage-sensing circuit is designed to protect sensitive electronic circuitry by monitoring the supply rail and triggering an external "crowbar" SCR in the event of a voltage transient or loss of regulation. The protective mechanism may be activated by an overvoltage condition at the Sense 2 input or by application of a TTL high level to the Remote Activate terminal. Separate outputs are available to trigger the crowbar circuit and to provide a logic pulse to indicator or power supply control circuitry. The Sense 2 input provides a direct control of the output circuitry. The Sense 1 input controls an internal current source that may be utilized to implement a delayed trigger by connecting its output to an external capacitor and the Sense 2 input. This protects against false triggering due to noise at the Sense 1 input.

The MC3423 is characterized for operation from 0 °C to 70 °C.

functional block diagram

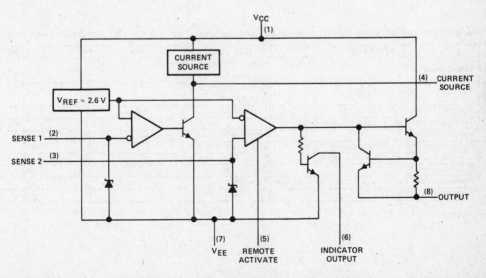

TEXAS INSTRUMENTS
POST OFFICE BOX 225012 • DALLAS, TEXAS 75265

TYPE MC3423
OVERVOLTAGE-SENSING CIRCUIT

absolute maximum ratings

Supply voltage, V_{CC} (see Note 1) . . . . 40 V
Sense 1 voltage . . . . 6.5 V
Sense 2 voltage . . . . 6.5 V
Remote activate input voltage . . . . 7 V
Output current, I_O . . . . 300 mA
Continuous dissipation at (or below) 25°C free-air temperature (see Note 2): JG package . . . . 825 mW
P package . . . . 1000 mW
Operating free-air temperature range . . . . 0°C to 70°C
Storage temperature range . . . . −65°C to 150°C

NOTES: 1. Voltage values are measured with respect to the V_{EE} terminal.
2. For operating above 25°C free-air temperature, refer to the Dissipation Derating Table. In the JG package, MC3423 chips are glass-mounted.

DISSIPATION DERATING TABLE

PACKAGE	POWER RATING	DERATING FACTOR	ABOVE T_A
JG (Glass-Mounted Chip)	825 mW	6.6 mW/°C	25°C
P	1000 mW	8 mW/°C	25°C

recommended operating conditions

	MIN	MAX	UNIT
Supply voltage, V_{CC}	4.5	40	V
High-level input voltage, remote activate input	2		V
Low-level input voltage, remote activate input		0.5	V

electrical characteristics over operating free-air temperature range, V_{CC} = 5 V to 36 V (unless otherwise noted)

PARAMETER	TEST CONDITIONS	MIN	TYP	MAX	UNIT
Output voltage	Remote Activate at 2 V, I_O = 100 mA	$V_{CC} - 2.2$	$V_{CC} - 1.8$		V
Indicator low-level output voltage	Remote Activate at 2 V, I_O = 1.6 mA		0.1	0.4	V
Threshold voltage of either sense input	T_A = 25°C	2.45	2.6	2.75	V
Temperature coefficient of input threshold voltage			0.06		%/°C
Source current (pin 4)	Sense 1 at 3 V, Pin 4 at 1.3 V	0.1	0.22	0.3	mA
High-level input current, Remote Activate input	V_{CC} = 5 V, V_I = 2 V		5	40	µA
Low-level input current, Remote Activate input	T_{CC} = 5 V, V_I = 0.8 V		−120	−180	µA
Supply current	Outputs open		6	10	mA
Propagation delay time, Remote Activate input to Output	T_A = 25°C		0.5		µs
Output current rate of rise	T_A = 25°C		400		mA/µs

TEXAS
INSTRUMENTS
POST OFFICE BOX 225012 • DALLAS, TEXAS 75265

Voltage Regulators

6

- 3-Terminal Regulators
- Output Current up to 100 mA
- No External Components Required
- Internal Thermal Overload Protection
- Internal Short Circuit Current Limiting
- Direct Replacement for Motorola MC79L00 Series
- Available in 5% or 10% Selections

NOMINAL OUTPUT VOLTAGE	5% OUTPUT VOLTAGE TOLERANCE	10% OUTPUT VOLTAGE TOLERANCE
−5 V	MC79L05AC	MC79L05C
−12 V	MC79L12AC	MC79L12C
−15 V	MC79L15AC	MC79L15C

description

This series of fixed-voltage monolithic integrated-circuit voltage regulators is designed for a wide range of applications. These include on-card regulation for elimination of noise and distribution problems associated with single-point regulation. In addition, they can be used to control series pass elements to make high-current voltage-regulator circuits. One of these regulators can deliver up to 100 mA of output current. The internal current-limiting and thermal-shutdown features make them essentially immune to overload. When used as a replacement for a Zener-diode and resistor combination, these devices can provide an effective improvement in output impedance of two orders of magnitude and lower bias current.

LP SILECT PACKAGE

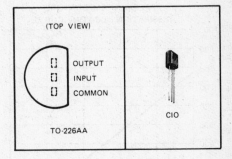

(TOP VIEW)

[] OUTPUT
[] INPUT
[] COMMON

TO-226AA

CIO

schematic

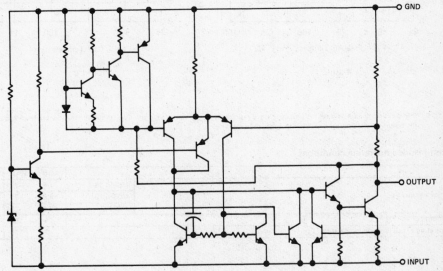

Voltage Regulators

6

absolute maximum ratings over operating temperature range (unless otherwise noted)

	MC79L05	MC79L12 MC79L15	UNIT
Input voltage	−30	−35	V
Continuous total dissipation at 25°C free-air temperature (see Note 1)	775	775	mW
Continuous total dissipation at (or below) 25°C case temperature (see Note 1)	1600	1600	mW
Operating free-air, case, or virtual junction temperature range	0 to 150	0 to 150	°C
Storage temperature range	−65 to 150	−65 to 150	°C
Lead temperature 1/16 inch (1,6 mm) from case for 10 seconds	260	260	°C

NOTE 1: For operation above 25°C free-air temperature, refer to Dissipation Derating Curves, Figure 1 and Figure 2.

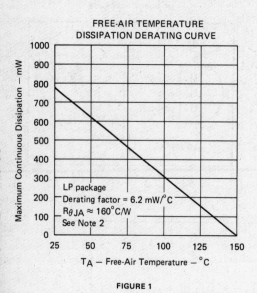

FREE-AIR TEMPERATURE
DISSIPATION DERATING CURVE

FIGURE 1

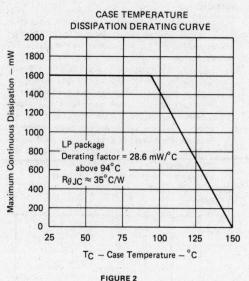

CASE TEMPERATURE
DISSIPATION DERATING CURVE

FIGURE 2

NOTE 2: This curve for the LP package is based on thermal resistance, $R_{\theta JA}$, measured in still air with the device mounted in an Augat socket. The bottom of the package was 3/8 inch above the socket.

recommended operating conditions

		MIN	MAX	UNIT
Input voltage, V_I	MC79L05	−7	−20	V
	MC79L12	−14.5	−27	
	MC79L15	−17.5	−30	
Output current, I_O			100	mA
Operating virtual junction temperature, T_J		0	125	°C

Voltage Regulators

6

TEXAS
INSTRUMENTS
POST OFFICE BOX 225012 ● DALLAS, TEXAS 75265

1082

MC79L05 electrical characteristics at specified virtual junction temperature,
V_I = −10 V, I_O = 40 mA (unless otherwise noted)

PARAMETER	TEST CONDITIONS[†]		MC79L05C			MC79L05AC			UNIT
			MIN	TYP	MAX	MIN	TYP	MAX	
Output voltage		25°C	−4.6	−5	−5.4	−4.8	−5	−5.2	V
	V_I = −7 V to −20 V, I_O = 1 mA to 40 mA	0°C to 125°C	−4.5		−5.5	−4.75		−5.25	
	V_I = −10 V, I_O = 1 mA to 70 mA	0°C to 125°C	−4.5		−5.5	−4.75		−5.25	
Input regulation	V_I = −7 V to −20 V	25°C			200			150	mV
	V_I = −8 V to −20 V				150			100	
Ripple rejection	V_I = −8 V to −18 V, f = 120 Hz	25°C	40	49		41	49		dB
Output regulation	I_O = 1 mA to 100 mA	25°C			60			60	mV
	I_O = 1 mA to 40 mA				30			30	
Output noise voltage	f = 10 Hz to 100 kHz	25°C		40			40		µV
Dropout voltage	I_O = 40 mA	25°C		1.7			1.7		V
Bias current		25°C			6			6	mA
		125°C			5.5			5.5	
Bias current change	V_I = −8 V to −20 V	0°C to 125°C			1.5			1.5	mA
	I_O = 1 mA to 40 mA				0.2			0.1	

MC79L12 electrical characteristics at specified virtual junction temperature,
V_I = −19, I_O = 40 mA (unless otherwise noted)

PARAMETER	TEST CONDITIONS[†]		MC79L12C			MC79L12AC			UNIT
			MIN	TYP	MAX	MIN	TYP	MAX	
Output voltage		25°C	−11.1	−12	−12.9	−11.5	−12	−12.5	V
	V_I = −14.5 to −27 V, I_O = 1 mA to 40 mA	0°C to 125°C	−10.8		−13.2	−11.4		−12.6	
	V_I = −19 V, I_O = 1 mA to 70 mA	0°C to 125°C	−10.8		−13.2	−11.4		−12.6	
Input regulation	V_I = −14.5 to −27 V	25°C			250			250	mV
	V_I = −16 V to −27 V				200			200	
Ripple rejection	V_I = −15 V to −25 V, f = 120 Hz	25°C	36	42		37	42		dB
Output regulation	I_O = 1 mA to 100 mA	25°C			100			100	mV
	I_O = 1 mA to 40 mA				50			50	
Output noise voltage	f = 10 Hz to 100 kHz	25°C		80			80		µV
Dropout voltage	I_O = 40 mA	25°C		1.7			1.7		V
Bias current		25°C			6.5			6.5	mA
		125°C			6			6	
Bias current change	V_I = −16 V to −27 V	0°C to 125°C			1.5			1.5	mA
	I_O = 1 mA to 40 mA				0.2			0.1	

[†] All characteristics are measured with a 0.33-µF capacitor across the input and a 0.1-µF capacitor across the output. All characteristics except noise voltage and ripple rejection ratio are measured using pulse techniques ($t_w \leqslant$ 10 ms, duty cycle $\leqslant$ 5%). Output voltage changes due to changes in internal temperature must be taken into account separately.

Voltage Regulators

6

TEXAS
INSTRUMENTS
POST OFFICE BOX 225012 • DALLAS, TEXAS 75265

MC79L15 electrical characteristics at specified virtual junction temperature,
$V_I = -23$ V, $I_O = 40$ mA (unless otherwise noted)

PARAMETER	TEST CONDITIONS†		MC79L15C			MC79L15AC			UNIT
			MIN	TYP	MAX	MIN	TYP	MAX	
Output voltage		25°C	-13.8	-15	-16.2	-14.4	-15	-15.6	V
	$V_I = -17.5$ V to -30 V, $I_O = 1$ mA to 40 mA	0°C to 125°C	-13.5		-16.5	-14.25		-15.75	
	$V_I = -23$ V, $I_O = 1$ mA to 70 mA	0°C to 125°C	-13.5		-16.5	-14.25		-15.75	
Input regulation	$V_I = -17.5$ V to -30 V	25°C			300			300	mV
	$V_I = -20$ V to -30 V				250			250	
Ripple rejection	$V_I = -18.5$ V to -28.5 V, f = 120 Hz	25°C	33	39		34	39		dB
Output regulation	$I_O = 1$ mA to 100 mA	25°C			150			150	mV
	$I_O = 1$ mA to 40 mA				75			75	
Output noise voltage	f = 10 Hz to 100 kHz	25°C		90			90		µV
Dropout voltage	$I_O = 40$ mA	25°C		1.7			1.7		V
Bias current		25°C			6.5			6.5	mA
		125°C			6			6	
Bias current change	$V_I = -20$ V to -30 V	0°C to 125°C			1.5			1.5	mA
	$I_O = 1$ mA to 40 mA				0.2			0.1	

†All characteristics are measured with a 0.33-µF capacitor across the input and a 0.1-µF capacitor across the output. All characteristics except noise voltage and ripple rejection ratio are measured using pulse techniques ($t_w \leqslant 10$ ms, duty cycle $\leqslant 5\%$). Output voltage changes due to changes in internal temperature must be taken into account separately.

TEXAS
INSTRUMENTS
POST OFFICE BOX 225012 • DALLAS, TEXAS 75265

1082

LINEAR INTEGRATED CIRCUITS

TYPES MC35060, MC34060
PULSE-WIDTH-MODULATION CONTROL CIRCUITS

D2714, MARCH 1983

- Complete PWM Power Control Circuitry
- Uncommitted Output for 200-mA Sink or Source Current
- Variable Dead-Time Provides Control Over Total Range
- Internal Regulator Provides a Stable 5-V Reference Supply
- Circuit Architecture Provides Easy Synchronization
- Direct Replacements for Motorola MC35060 and MC34060

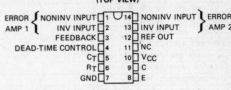

J OR N
DUAL-IN-LINE PACKAGE
(TOP VIEW)

ERROR AMP 1 { NONINV INPUT [1 14] NONINV INPUT } ERROR AMP 2
        { INV INPUT [2 13] INV INPUT }
        FEEDBACK [3 12] REF OUT
DEAD-TIME CONTROL [4 11] NC
        C_T [5 10] V_{CC}
        R_T [6 9] C
        GND [7 8] E

NC—No internal connections

description

The MC35060 and MC34060 incorporate on a single monolithic chip all the functions required in the construction of a pulse-width-modulation control circuit. Designed primarily for power supply control, each of the devices contains an on-chip 5-volt regulator, two error amplifiers, an adjustable oscillator, and a dead-time control comparator. The uncommitted output transistor provides either common-emitter or emitter-follower output capability. The internal amplifiers exhibit a common-mode voltage range from -0.3 volt to $V_{CC} - 2$ volts. The dead-time control comparator has a fixed offset that provides approximately 5% dead time unless externally altered. The on-chip oscillator may be bypassed by terminating R_T (pin 6) to the reference output and providing a sawtooth input to C_T (pin 5), or it may be used to drive the common MC35060 or MC34060 circuitry and provide a sawtooth input for associated control circuitry in multiple rail power supplies.

The MC35060 is characterized for operation over the full military temperature range of $-55\,^{\circ}C$ to $125\,^{\circ}C$. The MC34060 is characterized for operation from $0\,^{\circ}C$ to $70\,^{\circ}C$.

functional block diagram

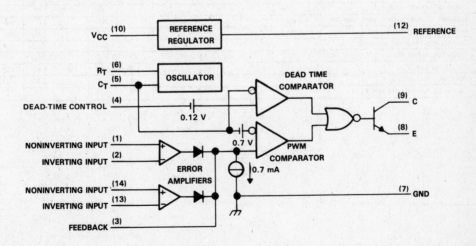

All voltage and current values shown are nominal.

383

Voltage Regulators

6

TEXAS INSTRUMENTS
POST OFFICE BOX 225012 • DALLAS, TEXAS 75265

absolute maximum ratings over operation temperature range (unless otherwise noted)

	MC35060	MC34060	UNIT
Supply voltage, V_{CC} (see Note 1)	42	42	V
Amplifier input voltages	$V_{CC} + 0.3$	$V_{CC} + 0.3$	V
Collector output voltage	42	42	V
Collector output current	250	250	mA
Continuous total dissipation at (or below) 25°C free-air temperature (see Note 2)	1000	1000	mW
Operating free-air temperature range	−55 to 125	0 to 70	°C
Storage temperature range	−65 to 150	−65 to 150	°C
Lead temperature 1,6 mm (1/16 inch) from case for 60 seconds: J package	300	300	°C
Lead temperature 1,6 mm (1/16 inch) from case for 10 seconds: N package		260	°C

NOTES: 1. All voltage values, except differential voltages, are with respect to the network ground terminal.
2. For operation above 25°C free-air temperature, refer to Dissipation Derating Table. In the J package, MC35060 chips are alloy-mounted and MC34060 chips are glass-mounted.

DISSIPATION DERATING TABLE

PACKAGE	POWER RATING	DERATING FACTOR	ABOVE T_A
J (Alloy-Mounted Chip)	1000 mW	11.0 mW/°C	59°C
J (Glass-Mounted Chip)	1000 mW	8.2 mW/°C	28°C
N	1000 mW	9.2 mW	41°C

recommended operating conditions

	MC35060		MC34060		UNIT
	MIN	MAX	MIN	MAX	
Supply voltage, V_{CC}	7	40	7	40	V
Amplifier input voltages, V_I	−0.3	$V_{CC} - 2$	−0.3	$V_{CC} - 2$	V
Collector output voltage, V_O		40		40	V
Collector output current (each transistor)		200		200	mA
Reference output current		10		10	mA
Current into feedback terminal		0.3		0.3	mA
Timing capacitor, C_T	0.47	10 000	0.47	10 000	nF
Timing resistor, R_T	1.8	500	1.8	500	kΩ
Oscillator frequency	1	200	1	200	kHz
Operating free-air temperature, T_A	−55	125	0	70	°C

Voltage Regulators

6

TEXAS
INSTRUMENTS
POST OFFICE BOX 225012 • DALLAS, TEXAS 75265

electrical characteristics over recommended operating free-air temperature range, $V_{CC} = 15$ V, f = 10 kHz (unless otherwise noted)

reference section

PARAMETER	TEST CONDITIONS[†]	MC35060			MC34060			UNIT
		MIN	TYP[‡]	MAX	MIN	TYP[‡]	MAX	
Output voltage (V_{ref})	$I_O = 1$ mA	4.75	5	5.25	4.75	5	5.25	V
Input regulation	$V_{CC} = 7$ V to 40 V, $T_A = 25°C$		2	25		2	25	mV
Output regulation	$I_{IO} = 1$ to 10 mA, $T_A = 25°C$		1	15		1	15	mV
Output voltage change with temperature	ΔT_A = MIN to MAX		0.2	2		0.2	2.6	%
Short-circuit output current[§]	$V_{ref} = 0$	10	35	50		35		mA

oscillator section

PARAMETER	TEST CONDITIONS[†]	MC35060			MC34060			UNIT
		MIN	TYP[‡]	MAX	MIN	TYP[‡]	MAX	
Frequency	$C_T = 0.001$ μF, $R_T = 47$ kΩ		25			25		kHz
Standard deviation of frequency[¶]	$C_T = 0.001$ μF, $R_T = 47$ kΩ		3			3		%
Frequency change with voltage	$V_{CC} = 7$ V to 40 V, $T_A = 25°C$		0.1			0.1		%
Frequency change with temperature	$C_T = 0.001$ μF, $R_T = 47$ kΩ, ΔT_A = MIN to MAX			4			2	%

dead-time control-section (see figure 1)

PARAMETER	TEST CONDITIONS	MIN	TYP[†]	MAX	UNIT
Input bias current (pin 4)	$V_I = 0$ to 5.25 V		−2	−10	μA
Maximum duty cycle	V_I (pin 4) = 0 $C_T = 0.1$ μF, $R_T = 12$ kΩ	90	96	100	%
	$C_T = 0.1$ μF, $R_T = 47$ kΩ		92	100	
Input threshold voltage (pin 4)	Zero duty cycle		3	3.3	V
	Maximum duty cycle	0			

error-amplifier sections

PARAMETER	TEST CONDITIONS	MIN	TYP[†]	MAX	UNIT
Input offset voltage	V_O (pin 3) = 2.5 V		2	10	mV
Input offset current	V_O (pin 3) = 2.5 V		25	250	nA
Input bias current	V_O (pin 3) = 2.5 V		0.2	1	μA
Common-mode input voltage range	$V_{CC} = 7$ V to 40 V	−0.3 to $V_{CC} - 2$			V
Open-loop voltage amplification	$\Delta V_O = 3$ V, $R_L = 2$ kΩ, $V_O = 0.5$ V to 3.5 V	70	95		dB
Unit-gain bandwidth			800		kHz
Common-mode rejection ratio	$V_{CC} = 40$ V	65	80		dB
Output sink current (pin 3)	$V_{ID} = -15$ mV to −5 V, $V_{(pin 3)} = 0.5$ V	0.3	0.7		mA
Output source current (pin 3)	$V_{ID} = 15$ mV to 5 V, $V_{(pin 3)} = 3.5$ V		−2		mA

output section

PARAMETER		TEST CONDITIONS	MC35060			MC34060			UNIT
			MIN	TYP[‡]	MAX	MIN	TYP[‡]	MAX	
Collector off-state current		$V_{CE} = 40$ V, $V_{CC} = 40$ V		2	100		2	100	μA
Emitter off-state current		$V_{CC} = V_C = 40$ V, $V_E = 0$			−150			−100	μA
Collector-emitter saturation voltage	Common-emitter	$V_E = 0$, $I_C = 200$ mA		1.1	1.5		1.1	1.3	V
	Emitter follower	$V_C = 15$ V, $I_E = -200$ mA		1.5	2.5		1.5	2.5	

[†]For conditions shown as MIN or MAX, use the appropriate value specified under recommended operation conditions.
[‡]All typical values except for temperature coefficients are at $T_A = 25°C$.
[§]Duration of the short-circuit should not exceed one second.
[¶]Standard deviation is a measure of the statistical distribution about the mean as derived from the formula $\sigma = \sqrt{\dfrac{\sum\limits_{n=1}^{N}(x_n - \bar{x})^2}{N-1}}$

Voltage Regulators

6

Texas
INSTRUMENTS
POST OFFICE BOX 225012 • DALLAS, TEXAS 75265

Voltage Regulators

6

electrical characteristics over recommended operating free-air temperature range, V_{CC} = 15 V, f = 10 kHz (unless otherwise noted)

pwm comparator section (see figure 1)

PARAMETER	TEST CONDITIONS	MIN	TYP‡	MAX	UNIT
Input threshold voltage (pin 3)	Zero duty cycle		4	4.5	V
Input sink current (pin 3)	$V_{(pin\ 3)}$ = 0.7 V	0.3	0.7		mA

total device

PARAMETER	TEST CONDITIONS		MIN	TYP‡	MAX	UNIT
Standby supply current	Pin 6 at V_{ref}, All other inputs and outputs open	V_{CC} = 15 V		6	10	mA
		V_{CC} = 40 V		9	15	
Average supply current	$V_{(pin\ 4)}$ = 2 V,	See Figure 1		7.5		mA

switching characteristics, T_A = 25°C

PARAMETER	TEST CONDITIONS	MIN	TYP‡	MAX	UNIT
Output voltage rise time	Common-emitter configuration, See Figure 3		100	200	ns
Output voltage fall time			25	100	ns
Output voltage rise time	Emitter-follower configuration, See Figure 4		100	200	ns
Output voltage fall time			40	100	ns

‡All typical values except for temperature coefficients are at T_A = 25°C.

PARAMETER MEASUREMENT INFORMATION

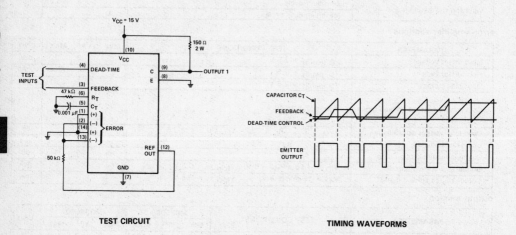

TEST CIRCUIT

TIMING WAVEFORMS

FIGURE 1 — DEAD-TIME AND FEEDBACK CONTROL

TEXAS
INSTRUMENTS
POST OFFICE BOX 225012 • DALLAS, TEXAS 75265

PARAMETER MEASUREMENT INFORMATION

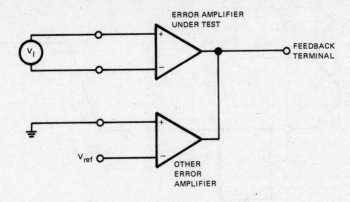

FIGURE 2 — ERROR-AMPLIFIER CHARACTERISTICS

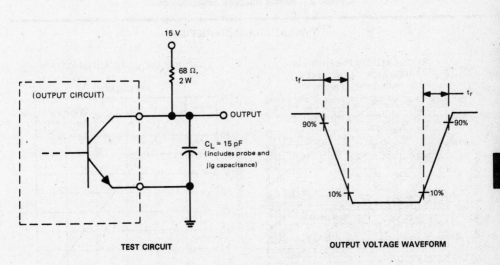

TEST CIRCUIT

OUTPUT VOLTAGE WAVEFORM

FIGURE 3 — COMMON-EMITTER CONFIGURATION

Voltage Regulators

6

TEXAS
INSTRUMENTS
POST OFFICE BOX 225012 • DALLAS, TEXAS 75265

PARAMETER MEASUREMENT INFORMATION

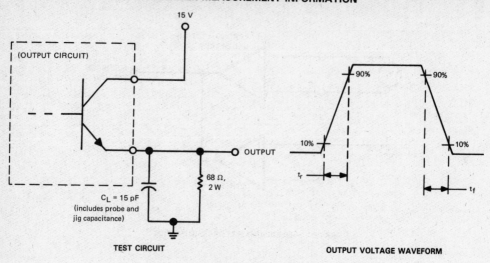

TEST CIRCUIT

OUTPUT VOLTAGE WAVEFORM

FIGURE 4 — EMITTER-FOLLOWER CONFIGURATION

TYPICAL CHARACTERISTICS

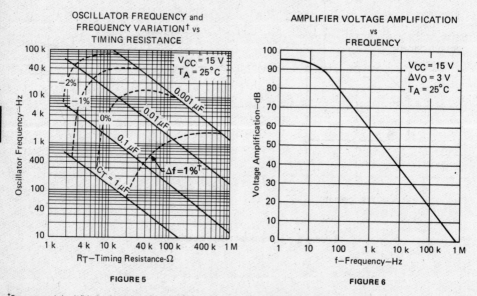

OSCILLATOR FREQUENCY and
FREQUENCY VARIATION† vs
TIMING RESISTANCE

FIGURE 5

AMPLIFIER VOLTAGE AMPLIFICATION
vs
FREQUENCY

FIGURE 6

†Frequency variation (Δf) is the change in oscillator frequency that occurs over the full temperature range.

TEXAS
INSTRUMENTS
POST OFFICE BOX 225012 • DALLAS, TEXAS 75265

LINEAR
INTEGRATED
CIRCUITS

TYPES RM4193, RC4193
MICROPOWER SWITCHING REGULATOR

D2718, SEPTEMBER 1983

- High Efficiency . . . 80% Typ
- Low Bias Current . . . 135 μA
- Adjustable Output . . . 2.5 V to 24 V
- Output Current . . . 150 mA
- Internal Reference . . . 1.3 V ±5%
- Remote Shutdown Capabilities
- Interchangeable with Raytheon RM4193 and RC4193

RM4193 . . . JG
RC4193 . . . JG OR P
DUAL-IN-LINE PACKAGE
(TOP VIEW)

```
LBR [ 1    8 ] LBD
 CX [ 2    7 ] VFB
 LX [ 3    6 ] IC
GND [ 4    5 ] VCC
```

description

The RM4193 and RC4193 are monolithic micropower switching regulators designed to provide all the functions required to make a complete low-power switching regulator primarily for battery operated instruments. The RM4193 and RC4193 offer the system designer the flexibility of tailoring the circuit to the application. Typical applications include step-up switching regulation, step-down switching regulation, and inverting switch regulation. The devices each contain a 1.3-volt temperature-compensated band-gap reference, an adjustable free-running oscillator, voltage comparator, low battery detection circuitry, and a 150-milliampere output-switch transistor.

FUNCTION TABLE

PIN	FUNCTION	DESCRIPTION
1	LBR	Low battery resistor
2	CX	External capacitor
3	LX	External inductor
4	GND	Ground
5	VCC	Supply voltage
6	IC	Reference set control
7	VFB	Feedback voltage
8	LBD	Low battery detector

For most applications, these regulators can achieve up to 80% efficiency while operating over a wide supply voltage range from 2.4 volts to 24 volts at an ultra-low bias current drain of 135 microamperes. The RM4193 and RC4193 have an adjustable 100-hertz to 160-kilohertz free-running oscillator that provides the drive circuitry for the on-chip 150-milliampere output-switch transistor. An external capacitor on pin 2 determines the oscillator frequency.

The low-battery detection circuitry contains an open-collector output transistor that can be used to activate a liquid crystal display whenever the battery voltage drops below a programmed level. This programmed level is determined by the selection of external resistors connected to pin 1.

The regulator will shut off when pin 6 (IC) is below 0.5 volt. The shut-off feature is useful in battery-backup applications requiring operation only when the line power is removed. Another use of this feature is connecting a zener diode between pin 6 and the battery line to shut down the regulator whenever the battery voltage drops below a predetermined level.

The RM4193 will be characterized for operation over the full military temperature range of −55°C to 125°C. The RC4193 will be characterized for operation from 0°C to 70°C.

functional block diagram

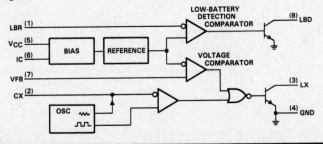

Voltage Regulators

6

Copyright © 1983 by Texas Instruments Incorporated

TEXAS
INSTRUMENTS

POST OFFICE BOX 225012 • DALLAS, TEXAS 75265

Voltage Regulators

6

TYPES SG1524, SG2524, SG3524
REGULATING PULSE WIDTH MODULATORS

D2294, APRIL 1977 — REVISED DECEMBER 1982

- Complete PWM Power Control Circuitry
- Uncommitted Outputs for Single-Ended or Push-Pull Applications
- Low Standby Current . . . 8 mA Typ
- Interchangeable With Silicon General SG1524, SG2524, and SG3524

description

The SG1524, SG2524, and SG3524 incorporate on single monolithic chips all the functions required in the construction of a regulating power supply, inverter, or switching regulator. They can also be used as the control element for high-power-output applications. The SG1524 family was designed for switching regulators of either polarity, transformer-coupled dc-to-dc converters, transformerless voltage doublers, and polarity converter applications employing fixed-frequency, pulse-width-modulation techniques. The complementary output allows either single-ended or push-pull application. Each device includes an on-chip regulator, error amplifier, programmable oscillator, pulse-steering flip-flop, two uncommitted pass transistors, a high-gain comparator, and current-limiting and shut-down circuitry.

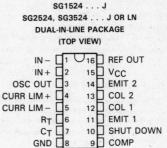

SG1524 . . . J
SG2524, SG3524 . . . J OR LN
DUAL-IN-LINE PACKAGE
(TOP VIEW)

IN −	1	16 REF OUT
IN +	2	15 V_{CC}
OSC OUT	3	14 EMIT 2
CURR LIM +	4	13 COL 2
CURR LIM −	5	12 COL 1
R_T	6	11 EMIT 1
C_T	7	10 SHUT DOWN
GND	8	9 COMP

The SG1524 is characterized for operation over the full military temperature range of −55°C to 125°C. The SG2524 is characterized for operation from −25°C to 85°C, and the SG3524 is characterized for operation from 0°C to 70°C.

functional block diagram

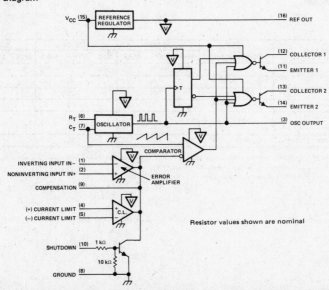

Resistor values shown are nominal

6

Voltage Regulators

TEXAS
INSTRUMENTS
POST OFFICE BOX 225012 ● DALLAS, TEXAS 75265

TYPES SG1524, SG2524, SG3524
REGULATING PULSE WIDTH MODULATORS

absolute maximum ratings over operating free-air temperature range (unless otherwise noted)

Supply Voltage, V_{CC} (see Notes 1 and 2) .... 40 V
Collector Output Current .... 100 mA
Reference Output Current .... 50 mA
Current Through C_T Terminal .... −5 mA
Continuous Total Dissipation at (or below) 25°C Free-Air Temperature (See Note 3) .... 1000 mW
Operating Free-Air Temperature Range: SG1524 .... −55°C to 125°C
                                   SG2524 .... −25°C to 85°C
                                   SG3524 .... 0°C to 70°C
Storage Temperature Range .... −65°C to 150°C

NOTES: 1. All voltage values are with respect to network ground terminal.
        2. The reference regulator may be bypassed for operation from a fixed 5-volt supply by connecting the V_{CC} and reference output pins both to the supply voltage. In this configuration the maximum supply voltage is 6 volts.
        3. For operation above 25°C free-air temperature refer to Figures 16 and 17. In the J package, SG1524 chips are alloy mounted; SG2524 and SG3524 chips are glass mounted.

recommended operating conditions

	SG1524		SG2524		SG3524		UNIT
	MIN	MAX	MIN	MAX	MIN	MAX	
Supply voltage, V_{CC}	8	40	8	40	8	40	V
Reference output current	0	50	0	50	0	50	mA
Current thru C_T terminal	−0.03	−2	−0.03	−2	−0.03	−2	mA
Timing resistor, R_T	1.8	100	1.8	100	1.8	100	kΩ
Timing capacitor, C_T	0.001	0.1	0.001	0.1	0.001	0.1	µF
Operating free-air temperature	−55	125	−25	85	0	70	°C

electrical characteristics over recommended operating free-air temperature range, V_{CC} = 20 V, f = 20 kHz (unless otherwise noted)

reference section

PARAMETER	TEST CONDITIONS[†]	SG1524			SG2524			SG3524			UNIT
		MIN	TYP[‡]	MAX	MIN	TYP[‡]	MAX	MIN	TYP[‡]	MAX	
Output voltage		4.8	5	5.2	4.8	5	5.2	4.6	5	5.4	V
Input regulation	V_{CC} = 8 to 40 V		10	20		10	20		10	30	mV
Ripple rejection	f = 120 Hz		66			66			66		dB
Output regulation	I_O = 0 to 20 mA		20	50		20	50		20	50	mV
Output voltage change with temperature	T_A = MIN to MAX		0.6	2		0.3	1		0.3	1	%
Short-circuit output current §	V_{ref} = 0		100			100			100		mA

[†] For conditions shown as MIN or MAX, use the appropriate value specified under recommended operating conditions.
[‡] All typical values except output voltage change with temperature are at T_A = 25°C.
§ Duration of the short-circuit should not exceed one second.

TEXAS INSTRUMENTS
POST OFFICE BOX 225012 • DALLAS, TEXAS 75265

electrical characteristics over recommended operating free-air temperature range, V_{CC} = 20 V, f = 20 kHz (unless otherwise noted)

oscillator section

PARAMETER	TEST CONDITIONS†	MIN	TYP‡	MAX	UNIT
Frequency	C_T = 0.001 μF, R_T = 2 kΩ		450		kHz
Standard deviation of frequency §	All values of voltage, temperature, resistance, and capacitance constant		5		%
Frequency change with voltage	V_{CC} = 8 to 40 V, T_A = 25°C			1	%
Frequency change with temperature	T_A = MIN to MAX			2	%
Output amplitude at pin 3			3.5		V
Output pulse width at pin 3	C_T = 0.01 μF		0.5		μs

error amplifier section

PARAMETER	TEST CONDITIONS	SG1524, SG2524			SG3524			UNIT
		MIN	TYP‡	MAX	MIN	TYP‡	MAX	
Input offset voltage	V_{IC} = 2.5 V		0.5	5		2	10	mV
Input bias current	V_{IC} = 2.5 V		2	10		2	10	μA
Open-loop voltage amplification		72	80		60	80		dB
Common-mode input voltage range	T_A = 25°C	1.8 to 3.4			1.8 to 3.4			V
Common-mode rejection ratio			70			70		dB
Unity-gain bandwidth			3			3		MHz
Output swing	T_A = 25°C	0.5		3.8	0.5		3.8	V

output section

PARAMETER	TEST CONDITIONS	MIN	TYP‡	MAX	UNIT
Collector-emitter breakdown voltage		40			V
Collector off-state current	V_{CE} = 40 V		0.01	50	μA
Collector-emitter saturation voltage	I_C = 50 mA		1	2	V
Emitter output voltage	V_C = 20 V, I_E = −250 μA	17	18		V
Turn-off voltage rise time	R_C = 2 kΩ		0.2		μs
Turn-on voltage fall time	R_C = 2 kΩ		0.1		μs

comparator section

PARAMETER	TEST CONDITIONS	MIN	TYP‡	MAX	UNIT
Maximum duty cycle, each output			45		%
Input threshold voltage at pin 9	Zero duty cycle		1		V
	Maximum duty cycle		3.5		
Input bias current			−1		μA

†For conditions shown as MIN or MAX, use the appropriate value specified under recommended operating conditions.

‡All typical values except for temperature coefficients are at T_A = 25°C.

§Standard deviation is a measure of the statistical distribution about the mean as derived from the formula $\sigma = \sqrt{\sum\limits_{n=1}^{N} \frac{(X_n - \bar{X})^2}{N-1}}$

Voltage Regulators

6

Texas Instruments
POST OFFICE BOX 225012 • DALLAS, TEXAS 75265

electrical characteristics over recommended operating free-air temperature range, V_{CC} = 20 V, f = 20 kHz (unless otherwise noted)

current limiting section

PARAMETER	TEST CONDITIONS	SG1524, SG2524			SG3524			UNIT
		MIN	TYP‡	MAX	MIN	TYP‡	MAX	
Input voltage range (either input)		−1 to +1			−1 to +1			V
Sense voltage at $T_A = 25°C$	$V_{(pin\ 2)} - V_{(pin\ 1)} \geqslant 50$ mV,	190	200	210	180	200	220	mV
Temperature coefficient of sense voltage	$V_{(pin\ 9)} = 2$ V		0.2			0.2		mV/°C

total device

PARAMETER	TEST CONDITIONS		MIN	TYP‡	MAX	UNIT
Standby current	$V_{CC} = 40$ V, Pin 2 at 2 V,	Pins 1,4,7,8,9,11,14 grounded, All other inputs and outputs open		8	10	mA

‡All typical values except for temperature coefficients are at $T_A = 25°C$.

PARAMETER MEASUREMENT INFORMATION

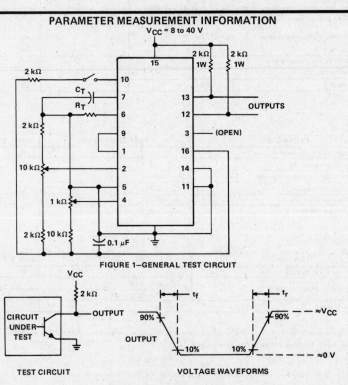

FIGURE 1—GENERAL TEST CIRCUIT

FIGURE 2—SWITCHING TIMES

Texas Instruments

POST OFFICE BOX 225012 ● DALLAS, TEXAS 75265

TYPICAL CHARACTERISTICS

OPEN-LOOP VOLTAGE AMPLIFICATION
OF ERROR AMPLIFIER
vs
FREQUENCY

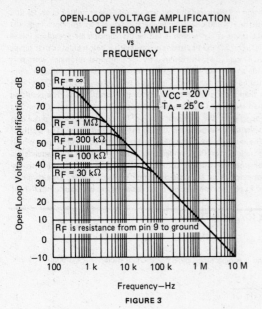

FIGURE 3

OSCILLATOR FREQUENCY
vs
TIMING RESISTANCE

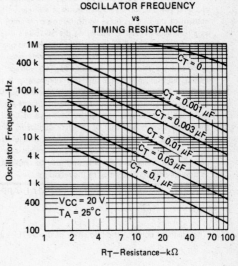

FIGURE 4

OUTPUT DEAD TIME
vs
TIMING CAPACITANCE VALUE

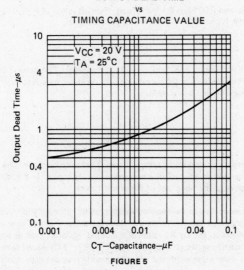

FIGURE 5

Voltage Regulators

6

TEXAS
INSTRUMENTS
POST OFFICE BOX 225012 • DALLAS, TEXAS 75265

PRINCIPLES OF OPERATION

The SG1524[†] is a fixed-frequency pulse-width-modulation voltage-regulator control circuit. The regulator operates at a fixed frequency that is programmed by one timing resistor R_T and one timing capacitor C_T. R_T establishes a constant charging current for C_T. This results in a linear voltage ramp at C_T, which is fed to the comparator providing linear control of the output pulse width by the error amplifier. The SG1524 contains an on-board 5-volt regulator that serves as a reference as well as supplying the SG1524's internal regulator control circuitry. The internal reference voltage is divided externally by a resistor ladder network to provide a reference within the common-mode range of the error amplifier as shown in Figure 6, or an external reference may be used. The output is sensed by a second resistor divider network and the error signal is amplified. This voltage is then compared to the linear voltage ramp at C_T. The resulting modulated pulse out of the high-gain comparator is then steered to the appropriate output pass transistor (Q1 or Q2) by the pulse-steering flip-flop, which is synchronously toggled by the oscillator output. The oscillator output pulse also serves as a blanking pulse to assure both outputs are never on simultaneously during the transition times. The width of the blanking pulse is controlled by the value of C_T. The outputs may be applied in a push-pull configuration in which their frequency is half that of the base oscillator, or paralleled for single-ended applications in which the frequency is equal to that of the oscillator. The output of the error amplifier shares a common input to the comparator with the current-limiting and shut-down circuitry and can be overridden by signals from either of these inputs. This common point is also available externally and may be employed to control the gain of, or to compensate, the error amplifier, or to provide additional control to the regulator.

TYPICAL APPLICATION DATA

oscillator

The oscillator controls the frequency of the SG1524 and is programmed by R_T and C_T as shown in Figure 4.

$$f \approx \frac{1.15}{R_T C_T}$$

where R_T is in kilohms
C_T is in microfarads
f is in kilohertz

Practical values of C_T fall between 0.001 and 0.1 microfarad. Practical values of R_T fall between 1.8 and 100 kilohms. This results in a frequency range typically from 140 hertz to 500 kilohertz.

blanking

The output pulse of the oscillator is used as a blanking pulse at the output. This pulse width is controlled by the value of C_T as shown in Figure 5. If small values of C_T are required, the oscillator output pulse width may still be maintained by applying a shunt capacitance from pin 3 to ground.

synchronous operation

When an external clock is desired, a clock pulse of approximately 3 volts can be applied directly to the oscillator output terminal. The impedance to ground at this point is approximately 2 kilohms. In this configuration R_T C_T must be selected for a clock period slightly greater than that of the external clock.

If two or more SG1524 regulators are to be operated synchronously, all oscillator output terminals should be tied together. The oscillator programmed for the minimum clock period will be the master from which all the other SG1524's operate. In this application, the C_T R_T values of the slaved regulators must be set for a period approximately 10% longer than that of the master regulator. In addition, C_T (master) = 2 C_T (slave) to ensure that the master output pulse, which occurs first, has a wider pulse width and will subsequently reset the slave regulators.

[†]Throughout these discussions, references to SG1524 apply also to SG2524 and SG3524.

TEXAS
INSTRUMENTS
POST OFFICE BOX 225012 • DALLAS, TEXAS 75265

TYPICAL APPLICATION DATA

voltage reference

The 5-volt internal reference may be employed by use of an external resistor divider network to establish a reference within the error amplifiers common-mode voltage range (1.8 to 3.4 volts) as shown in Figure 6, or an external reference may be applied directly to the error amplifier. For operation from a fixed 5-volt supply, the internal reference may be bypassed by applying the input voltage to both the V_{CC} and V_{REF} terminals. In this configuration, however, the input voltage is limited to a maximum of 6 volts.

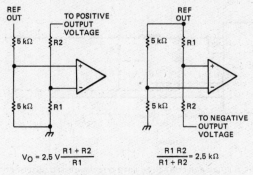

$$V_O = 2.5\ V \frac{R1 + R2}{R1} \qquad \frac{R1\ R2}{R1 + R2} = 2.5\ k\Omega$$

FIGURE 6—ERROR AMPLIFIER BIAS CIRCUITS

error amplifier

The error amplifier is a differential-input transconductance amplifier. The output is available for dc gain control or ac phase compensation. The compensation node (pin 9) is a high-impedance node (R_L = 5 megohms). The gain of the amplifier is $A_V = (0.002\ \Omega^{-1})\ R_L$ and can easily be reduced from a nominal 10,000 by an external shunt resistance from pin 9 to ground. Refer to Figure 3 for data.

compensation

Pin 9, as discussed above, is made available for compensation. Since most output filters will introduce one or more additional poles at frequencies below 200 hertz, which is the pole of the uncompensated amplifier, introduction of a zero to cancel one of the output filter poles is desirable. This can best be accomplished with a series RC circuit from pin 9 to ground in the range of 50 kilohms and 0.001 microfarads. Other frequencies can be canceled by use of the formula $f \approx 1/RC$.

shut down circuitry

Pin 9 can also be employed to introduce external control of the SG1524. Any circuit that can sink 200 microamperes can pull the compensation terminal to ground and thus disable the SG1524.

In addition to constant-current limiting, pins 4 and 5 may also be used in transformer-coupled circuits to sense primary current and shorten an output pulse should transformer saturation occur. Pin 5 may also be grounded to convert pin 4 into an additional shutdown terminal.

Voltage Regulators

6

TYPICAL APPLICATION DATA

current limiting

A current-limiting sense amplifier is provided in the SG1524. The current-limiting sense amplifier exhibits a threshold of 200 millivolts and must be applied in the ground line since the voltage range of the inputs is limited to +1 volt to −1 volt. Caution should be taken to ensure the −1-volt limit is not exceeded by either input, otherwise damage to the device may result.

Fold-back current limiting can be provided with the network shown in Figure 7. The current-limit schematic is shown in Figure 8.

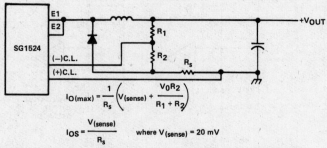

$$I_{O(max)} = \frac{1}{R_s}\left(V_{(sense)} + \frac{V_0 R_2}{R_1 + R_2}\right)$$

$$I_{OS} = \frac{V_{(sense)}}{R_s} \qquad \text{where } V_{(sense)} = 20 \text{ mV}$$

FIGURE 7—FOLDBACK CURRENT LIMITING FOR SHORTED OUTPUT CONDITIONS

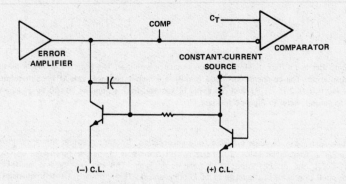

FIGURE 8—CURRENT-LIMIT SCHEMATIC

output circuitry

The SG1524 contains two identical n-p-n transistors the collectors and emitters of which are uncommitted. Each transistor has antisaturation circuitry that limits the current through that transistor to a maximum of 100 milliamperes for fast response.

TEXAS
INSTRUMENTS
POST OFFICE BOX 225012 • DALLAS, TEXAS 75265

TYPICAL APPLICATION DATA

general

There are a wide variety of output configurations possible when considering the application of the SG1524 as a voltage regulator control circuit. They can be segregated into three basic categories:

1. Capacitor-diode-coupled voltage multipliers

2. Inductor-capacitor-implemented single-ended circuits

3. Transformer-coupled circuits

Examples of these categories are shown in Figures 9, 10 and 11, respectively. Detailed diagrams of specific applications are shown in Figures 12 through 15.

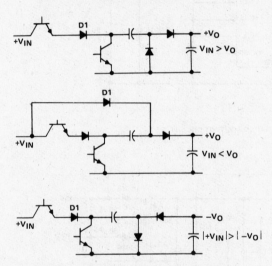

FIGURE 9—CAPACITOR-DIODE-COUPLED VOLTAGE-MULTIPLIER OUTPUT STAGES

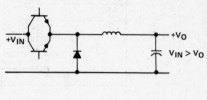
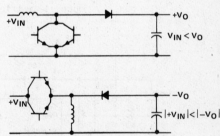
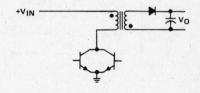

FIGURE 10—SINGLE-ENDED INDUCTOR CIRCUIT

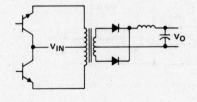

PUSH PULL

FLYBACK

FIGURE 11—TRANSFORMER-COUPLED OUTPUTS

Voltage Regulators

6

TEXAS
INSTRUMENTS
POST OFFICE BOX 225012 • DALLAS, TEXAS 75265

TYPICAL APPLICATION DATA

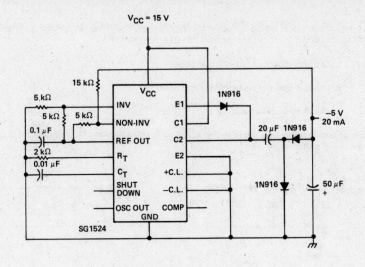

FIGURE 12—CAPACITOR-DIODE OUTPUT CIRCUIT

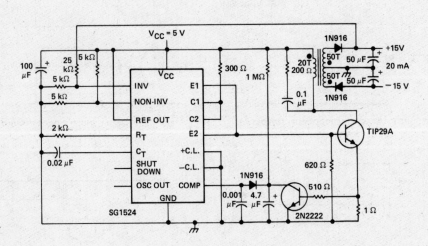

FIGURE 13 – FLYBACK CONVERTER CIRCUIT

TEXAS INSTRUMENTS
POST OFFICE BOX 225012 ● DALLAS, TEXAS 75265

TYPICAL APPLICATION DATA

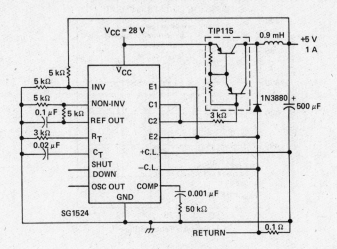

FIGURE 14—SINGLE-ENDED LC CIRCUIT

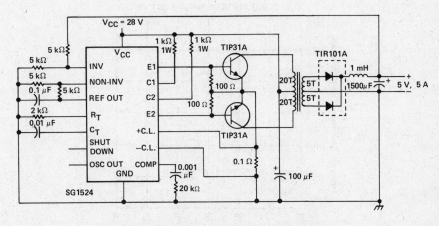

FIGURE 15—PUSH-PULL TRANSFORMER-COUPLED CIRCUIT

TEXAS
INSTRUMENTS
POST OFFICE BOX 225012 • DALLAS, TEXAS 75265

Voltage Regulators

6

THERMAL INFORMATION

J PACKAGE FREE-AIR TEMPERATURE DISSIPATION DERATING CURVE

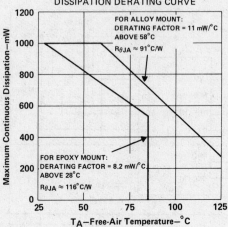

FOR ALLOY MOUNT:
DERATING FACTOR = 11 mW/°C
ABOVE 58°C
$R_{\theta}JA \approx 91°C/W$

FOR EPOXY MOUNT:
DERATING FACTOR = 8.2 mW/°C
ABOVE 28°C
$R_{\theta}JA \approx 116°C/W$

Maximum Continuous Dissipation—mW

T_A—Free-Air Temperature—°C

FIGURE 16

N PACKAGE FREE-AIR TEMPERATURE DISSIPATION DERATING CURVE

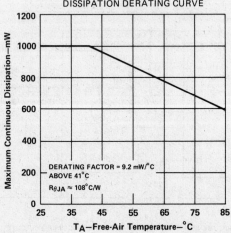

DERATING FACTOR = 9.2 mW/°C
ABOVE 41°C
$R_{\theta}JA \approx 108°C/W$

Maximum Continuous Dissipation—mW

T_A—Free-Air Temperature—°C

FIGURE 17

TEXAS
INSTRUMENTS
POST OFFICE BOX 225012 • DALLAS, TEXAS 75265

- Complete PWM Power Control Circuitry
- 8-Volt to 35-Volt Operation
- 5.1-Volt Reference Trimmed to ±1%
- Frequency Range . . . 100 Hz to 500 Hz
- Adjustable Deadtime Control
- Under-Voltage Lockout for Low V_{CC} Conditions
- Latched PWM Prevents Multiple Pulses
- Dual Sink or Source Output Drivers
- Direct Replacements for Silicon General SG1525A/SG1527A Series

SG1525A, SG1527A . . . J
SG2525A, SG2527A . . . J OR N
SG3525A, SG3527A . . . J OR N

DUAL-IN-LINE PACKAGE
(TOP VIEW)

INVERTING INPUT	1	16	REFERENCE
NONINVERTING INPUT	2	15	V_{CC} (V_I)
SYNC	3	14	OUTPUT B
OSCILLATOR OUT	4	13	V_C
C_T	5	12	GND
R_T	6	11	OUTPUT A
DISCHARGE	7	10	SHUTDOWN
SOFT-START	8	9	COMPENSATION

output logic

SG1525A, SG2525A, SG3525A . . . NOR
SG1527A, SG2572A, SG3527A . . . OR

description

The SG1525A/SG1527A series of pulse-width modulation integrated circuits are designed to offer improved performance and lower external parts count when used to implement various types of switching power supplies. Each device includes an on-chip 5.1-volt reference, error amplifier, programmable oscillator, pulse-steering flip-flop, a latched comparator under-voltage lockout, shutdown circuitry, and complementary source or sink outputs. The on-chip 5.1-volt reference is trimmed to ±1% initial accuracy, serves as a reference output, and supplies the internal regulator control circuitry. The input common-mode range of the error amplifier includes the reference voltage, which eliminates the need for external divider resistors.

The oscillator operates at a fixed frequency determined by one timing resistor R_T and one timing capacitor C_T. The timing resistor establishes the constant charging current for C_T, resulting in a linear voltage ramp at C_T, which is fed to the PWM comparator providing linear control of the output pulse duration by the error amplifier. A Sync input to the oscillator allows for external synchronization or for multiple units to be slaved together. A single external resistor between the C_T pin and the Discharge pin provides a wide range of dead-time adjustment. These devices also feature built-in soft-start circuitry that requires only an external timing capacitor. The Shutdown pin controls both the soft-start and the output drivers, and provides instantaneous turn-off with soft-start recycle for slow turn-on. The soft-start and output driver circuitry are also controlled by the under-voltage lockout circuit, which, during low-input supply voltage of less than that required for normal operation, keeps the soft-start capacitor discharged and the output drivers off.

Another unique feature is the S-R latch following the PWM comparator. This feature enables the output drivers to be turned off any time the PWM pulse is terminated. The latch is reset with each clock pulse. However, the PWM outputs will remain turned off for the duration of the period if the PWM comparator output is in a low-level state. The SG1525A, SG2525A, and SG3525A output stages feature NOR logic, resulting in a low output for an off-state. The SG1527A, SG2527A, and SG3527A output stages feature OR logic, resulting in a high-level output for an off-state. The output stages are totem-pole designs capable of sourcing or sinking 200 milliamperes of output current.

The SG1525A and SG1527A are characterized for operation over the full military temperature range of −55°C to 125°C. The SG2525A and SG2527A are characterized for operation from −25°C to 85°C. The SG3525A and SG3527A are characterized for operation for 0°C to 70°C.

Voltage Regulators

6

TYPES SG1525A, SG1527A, SG2525A, SG2527A, SG3525A, SG3527A
PULSE-WIDTH MODULATION CONTROLLERS

functional block diagram (positive logic)

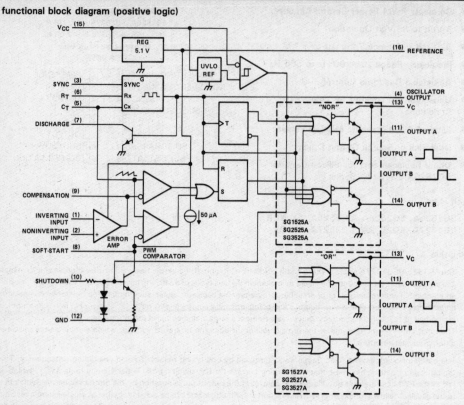

absolute maximum ratings over operating free-air temperature range (unless otherwise noted)

Supply voltage, V_{CC} (see Note 1) .... 40 V
Collector voltage, V_C .... 40 V
Logic input voltage range sync and shutdown .... −0.3 V to 5.5 V
Analog input voltage range error amplifier inputs .... −0.3 V to V_{CC}
Output current, I_O .... 500 mA
Reference output current, I_{REF} .... 50 mA
Current through C_T terminal .... −5 mA
Continuous total dissipation at (or below) 25°C free-air temperature (see Note 2) .... 1000 mW
Operating free-air temperature range: SG1525A, SG1527A .... −55°C to 125°C
                                      SG2525A, SG2527A .... −25°C to 85°C
                                      SG3525A, SG3527A .... 0°C to 70°C
Operating virtual junction temperature range .... 0°C to 150°C
Storage temperature range .... −65°C to 150°C
Lead temperature 1,6 mm (1/16 inch) from case for 60 seconds: J Package .... 300°C
Lead temperature 1,6 mm (1/16 inch) from case for 10 seconds: N Package .... 260°C

NOTES: 1. All voltage values are with respect to network ground terminal.
         2. For operating above 25°C free-air temperature, see Dissipation Derating Curves, Figures 1 and 2. In the J package, SG1525A and SG1527A chips are alloy-mounted; SG2525A, SG2527A, SG3525A, and SG3527A chips are epoxy mounted.

TEXAS INSTRUMENTS
POST OFFICE BOX 225012 • DALLAS, TEXAS 75265

Voltage Regulators

6

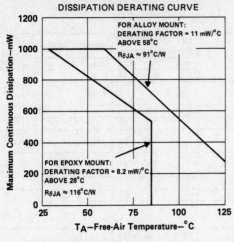

J PACKAGE FREE-AIR TEMPERATURE DISSIPATION DERATING CURVE

FOR ALLOY MOUNT:
DERATING FACTOR = 11 mW/°C
ABOVE 58°C
$R_{\theta JA} \approx 91°C/W$

FOR EPOXY MOUNT:
DERATING FACTOR = 8.2 mW/°C
ABOVE 28°C
$R_{\theta JA} \approx 116°C/W$

Maximum Continuous Dissipation—mW

T_A—Free-Air Temperature—°C

FIGURE 1

N PACKAGE FREE-AIR TEMPERATURE DISSIPATION DERATING CURVE

DERATING FACTOR = 9.2 mW/°C
ABOVE 41°C
$R_{\theta JA} \approx 108°C/W$

Maximum Continuous Dissipation—mW

T_A—Free-Air Temperature—°C

FIGURE 2

recommended operating conditions

PARAMETER		SG1525A, SG1527A		SG2525A, SG2527A		SG3525A, SG3527A		UNIT
		MIN	MAX	MIN	MAX	MIN	MAX	
Supply voltage, V_{CC}		8	35	8	35	8	35	V
Collector voltage, V_C		4.5	35	4.5	35	4.5	35	V
Output current, I_O	Steady state	0	±100	0	±100	0	±100	mA
	Peak	0	±400	0	±400	0	±400	
Reference output current, I_{REF}		0	20	0	20	0	20	mA
Oscillator frequency range		100	500	100	500	100	500	kHz
Timing resistor, R_T		2	150	2	150	2	150	kΩ
Timing capacitor, C_T		0.001	0.1	0.001	0.1	0.001	0.1	µF
Dead-time resistor, R_D		0	500	0	500	0	500	Ω
Operating free-air temperature range, T_A		−55	125	−25	85	0	70	°C

Voltage Regulators

6

TEXAS INSTRUMENTS
POST OFFICE BOX 225012 • DALLAS, TEXAS 75265

electrical characteristics over recommended operating free-air temperature range, V_{CC} = 20 V (unless otherwise noted)

reference section

PARAMETER	TEST CONDITIONS	SG1525A, SG1527A SG2525A, SG2527A			SG3525A, SG3527A			UNIT
		MIN	TYP	MAX	MIN	TYP	MAX	
Output voltage	T_J = 25°C	5.05	5.1	5.15	5	5.1	5.2	V
	V_{CC} = 8 V to 35 V, I_O = 0 to 20 mA	5		5.2	4.95		5.25	
Input regulation	V_{CC} = 8 V to 35 V		14	20		14	20	mV
Output regulation	I_O = 0 to 20 mA		5	50		5	50	mV
Output voltage change with temperature			24	50		24	50	mV
Output voltage long-term drift (see Note 3)	After 1000 h at T_J = 125°C		25	50		25	50	mV
Output noise voltage (RMS)	f = 10 Hz to 10 kHz, T_J = 25°C		40	200		40	200	μV
Short-circuit output current	V_O = 0 V, T_J = 25°C		80	100		80	100	mA

oscillator section

PARAMETER	TEST CONDITIONS	SG1525A, SG1527A SG2525A, SG2527A			SG3525A, SG3527A			UNIT
		MIN	TYP	MAX	MIN	TYP	MAX	
Maximum frequency	R_T = 2 kΩ, C_T = 1 nF	400			400			kHz
Minimum frequency	R_T = 150 kΩ, C_T = 0.1 μF			100			100	Hz
Initial frequency error	R_T = 3.6 kΩ, R_D = 0 Ω, C_T = 0.1 μF, f = 40 kHz, T_A = 25°C		±2%	±6%		±2%	±6%	
Frequency change with supply voltage	V_{CC} = 8 V to 35 V		±0.3%	±1%		±1%	±2%	
Frequency change with temperature	T_A = MIN to MAX		±3%	±6%		±3%	±6%	
Output amplitude at Pin 4	R_T = 3.6 kΩ, R_D = 0 Ω, C_T = 0.1 μF, f = 40 kHz	3	3.5		3	3.5		V
Output pulse duration at Pin 4	R_T = 3.5 kΩ, R_D = 0 Ω C_T = 0.1 μF, f = 40 kHz, T_J = 25°C	0.3	0.5	1	0.3	0.6	1	μs
Input threshold voltage at Pin 3		1.2	2	2.8	1.2	2	2.8	V
Input current at Pin 3	$V_{I(Pin3)}$ = 3.5 V		1.6	2.5		1.6	2.5	mA
Current through Pin 5 due to internal current mirror	Current through Pin 6 = 6 mA	1.7	2	2.2	1.7	2	2.2	mA

NOTE   3: Since long-term drift cannot be measured on the individual devices prior to shipment, this specification is not intended to be a guarantee or warranty. It is an engineering estimate of the average drift to be expected from lot to lot.

Voltage Regulators

6

TEXAS
INSTRUMENTS
POST OFFICE BOX 225012 ● DALLAS, TEXAS 75265

983

electrical characteristics over recommended operating free-air temperature range, V_{CC} = 20 V (unless otherwise noted)

error amplifier section

PARAMETER	TEST CONDITIONS	SG1525A, SG1527A SG2525A, SG2527A			SG3525A, SG3527A			UNIT
		MIN	TYP	MAX	MIN	TYP	MAX	
High-level output voltage		3.8	5.6		3.8	5.6		V
Low-level output voltage			0.2	0.5		0.2	0.5	V
Input offset voltage			0.5	5		2	10	mV
Input bias current			1	10		1	10	µA
Input offset current				1			1	µA
Open-loop voltage amplification	$R_L \geq$ 10 M	60	75		60	75		dB
Common-mode rejection ratio	V_{IC} = 1.5 V to 5.2 V	60	75		60	75		dB
Supply voltage rejection ratio	V_{CC} = 8 V to 35 V	50	60		50	60		dB
Gain-bandwidth product	A_V = 0 dB, T_J = 25°C	1	2		1	2		MHz

comparator section

PARAMETER	TEST CONDITIONS		MIN	TYP	MAX	UNIT
Input threshold voltage	R_T = 3.6 kΩ, R_D = 0 Ω, C_T = 10 nF, f = 40 kHz	Duty cycle = 0%	0.6	0.9		V
		Duty cycle = MAX		3.3	3.6	
Input bias current				0.5	1	µA

soft-start section

PARAMETER	TEST CONDITIONS	MIN	TYP	MAX	UNIT
Soft-start voltage	V_I at Pin 10 = 2 V		0.4	0.6	V
Soft-start current	V_I at Pin 10 = 0 V	25	50	80	µA
Input current, Shutdown	V_I at Pin 10 = 2.5 V		0.4	1	mA

output section

PARAMETER	TEST CONDITIONS	MIN	TYP	MAX	UNIT
High-level output voltage	I_{OH} = −20 mA	18	19		V
	I_{OH} = −100 mA	17	18		
Low-level output voltage	I_{OL} = 20 mA		0.2	0.4	V
	I_{OL} = 100 mA		1	2	
Under-voltage lockout voltage	V_I at Pins 8 and 9 = high	6	7	8	V
Collector cutoff current (see Note 4)	V_C = 35 V, I_O = 100 mA			200	µA
Output pulse rise time	C_L = 1 nF, T_J = 25°C		100	600	ns
Output pulse fall time	C_L = 1 nF, T_J = 25°C		50	300	ns
Shutdown delay time	V_I at pin 10 = 3 V, capacitance at pin 8 = 0, T_J = 25°C		0.2	0.5	µs

NOTE 4: Collector cutoff current specifications apply only for the SG1525A, SG2525A, and SG3525A devices.

total device

PARAMETER	TEST CONDITIONS	MIN	TYP	MAX	UNIT
Minimum duty cycle				0%	
Maximum duty cycle		45%	49%		
Standby current	V_{CC} = 35 V		14	20	mA

Voltage Regulators

6

TEXAS
INSTRUMENTS
POST OFFICE BOX 225012 • DALLAS, TEXAS 75265

TYPES SG1525A, SG1527A, SG2525A, SG2527A, SG3525A, SG3527A
PULSE-WIDTH MODULATION CONTROLLERS

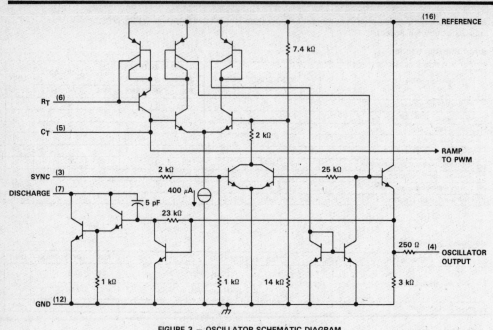

FIGURE 3 — OSCILLATOR SCHEMATIC DIAGRAM

TYPICAL CHARACTERISTICS

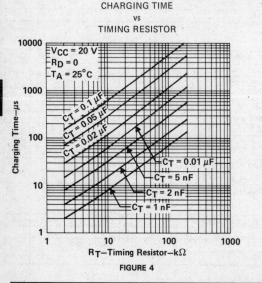

CHARGING TIME
vs
TIMING RESISTOR

FIGURE 4

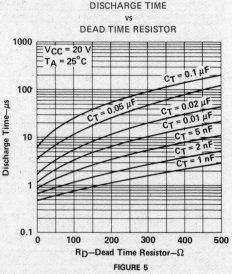

DISCHARGE TIME
vs
DEAD TIME RESISTOR

FIGURE 5

TEXAS INSTRUMENTS
POST OFFICE BOX 225012 • DALLAS, TEXAS 75265

983

TYPICAL CHARACTERISTICS

ERROR AMPLIFIER OPEN-LOOP
FREQUENCY RESPONSE

SG1525A OUTPUT SATURATION VOLTAGE
vs
OUTPUT CURRENT

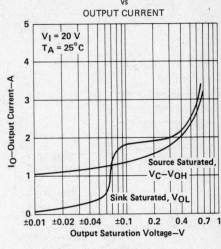

FIGURE 6

FIGURE 7

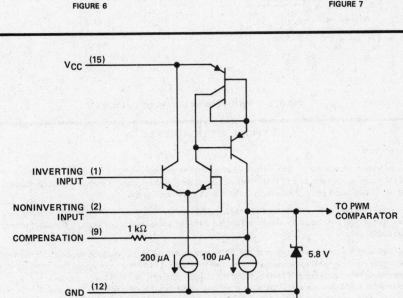

FIGURE 8 — ERROR AMPLIFIER SCHEMATIC DIAGRAM

TEXAS
INSTRUMENTS
POST OFFICE BOX 225012 • DALLAS, TEXAS 75265

Voltage Regulators

6

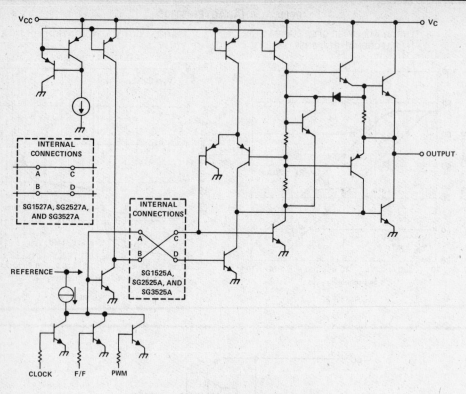

FIGURE 9 — OUTPUT CIRCUIT SCHEMATIC DIAGRAM

TYPICAL APPLICATION DATA

shutdown options

1. Use an external transistor or open-collector comparator to pull down on the Compensation terminal (Pin 9). This will set the PWM latch and turn off both driver outputs. If the shutdown signal is momentary, pulse-by-pulse protection will be accomplished as the PWM latch is reset with each clock pulse.

2. The same results may be accomplished by pulling down on the Soft-Start terminal (Pin 8) with the only difference being that on this pin shutdown will not affect the amplifier compensation network, but must discharge any soft-start capacitance.

3. Application of a positive-going signal to the Shutdown terminal (Pin 10) will provide the most rapid shutdown of the driver outputs but will not immediately set the PWM latch if there is a capacitor at the Soft-Start terminal. The capacitor will discharge but at a current twice the charging current. The PWM latch can be set on a pulse-by-pulse basis by the shutdown terminal if there is no external capacitance on the Soft-start terminal (Pin 8). Slow turn-on may still be accomplished by connecting an external capacitor, blocking diode, and charging resistor to the Compensation terminal (Pin 9).

TEXAS
INSTRUMENTS

POST OFFICE BOX 225012 • DALLAS, TEXAS 75265

TYPICAL APPLICATION DATA

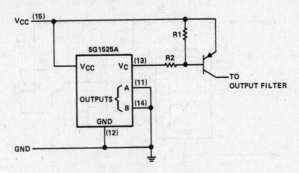

For single-ended supplies, the driver outputs are grounded. The V$_C$ terminal is switched to ground by the totem-pole source transistors on the alternate oscillator cycles.

FIGURE 10 — SINGLE-ENDED CIRCUIT

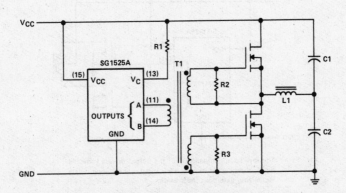

Low-power transformers can be directly driven by the SG1525A. Automatic reset occurs during deadtime when both ends of the primary winding are switched to ground.

FIGURE 11 — TRANSFORMER-COUPLED CIRCUIT

Voltage Regulators

6

TEXAS INSTRUMENTS
POST OFFICE BOX 225012 • DALLAS, TEXAS 75265

TYPICAL APPLICATION DATA

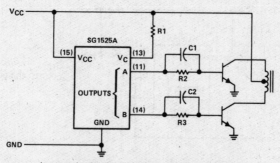

In conventional push-pull bipolar designs, forward base drive is controlled by $R_1 - R_3$. Rapid turn-off times for the power devices are achieved with speed-up capacitors C_1 and C_2.

FIGURE 12 — BIPOLAR PUSH-PULL CIRCUIT

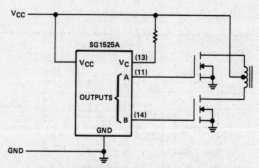

The low source impedance of the output drivers provides rapid charging of power FET input capacitance while minimizing external components.

**FIGURE 13 — LOW-IMPEDANCE BIPOLAR-DRIVE
PUSH-PULL CIRCUIT**

TEXAS
INSTRUMENTS
POST OFFICE BOX 225012 • DALLAS, TEXAS 75265

LINEAR
INTEGRATED
CIRCUITS

TYPE TL317M, TL317C
3-TERMINAL ADJUSTABLE REGULATOR

D2527, APRIL 1979–REVISED JANUARY 1983

- Output Voltage Range Adjustable from 1.2 V to 32 V
- Guaranteed Output Current Capability of 100 mA

- Input Regulation Typically 0.01% Per Input-Volt Change
- Output Regulation Typically 0.5%
- Ripple Rejection Typically 80 dB

terminal assignments

TL317M . . . JG PACKAGE

(TOP VIEW)

INPUT	1	8	OUTPUT
NC	2	7	NC
NC	3	6	ADJUSTMENT
NC	4	5	NC

TL317C . . . LP SILECT PACKAGE

(TOP VIEW)

INPUT
OUTPUT
ADJUSTMENT

AO

NC—No internal connection

description

The TL317 is an adjustable 3-terminal positive-voltage regulator capable of supplying 100 milliamperes over an output-voltage range of 1.2 volts to 32 volts. It is exceptionally easy to use and requires only two external resistors to set the output voltage. Both input and output regulation are better than standard fixed regulators. The device is packaged in standard packages that are easily mounted and handled.

In addition to higher performance than fixed regulators, this regulator offers full overload protection available only in integrated circuits. Included on the chip are current limit and thermal overload protection. All overload protection circuitry remains fully functional even if the adjustment terminal is disconnected. Normally, no capacitors are needed unless the device is situated far from the input filter capacitors in which case an input bypass is needed. An optional output capacitor can be added to improve transient response. The adjustment terminal can be bypassed to achieve very high ripple rejection, which is difficult to achieve with standard 3-terminal regulators.

Besides replacing fixed regulators, the regulator is useful in a wide variety of other applications. Since the regulator is floating and sees only the input-to-output differential voltage, supplies of several hundred volts can be regulated as long as the maximum input-to-output differential is not exceeded. Its primary application is that of a programmable output regulator, but by connecting a fixed resistor between the adjustment terminal and the output terminal, this device can be used as a precision current regulator. Supplies with electronic shutdown can be achieved by clamping the adjustment terminal to ground, which programs the output to 1.2 volts where most loads draw little current.

The TL317M is characterized for operation over the full military temperature range from −55°C to 125°C. The TL317C is characterized for operation from 0°C to 125°C.

Voltage Regulators

6

TEXAS
INSTRUMENTS

POST OFFICE BOX 225012 • DALLAS, TEXAS 75265

schematic

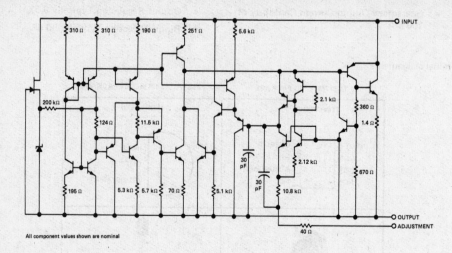

All component values shown are nominal

absolute maximum ratings over operation temperature range (unless otherwise noted)

Input-to-output differential voltage, $V_I - V_O$ .... 35 V
Continuous total dissipation at (or below) 25°C free-air temperature (see Note 1): JG package ..... 1050 mW
                  LP package ...... 775 mW
Continuous total dissipation at (or below) 25°C case temperature (see Note 1) .... 1600 mW
Operating free-air, case, or virtual junction temperature range: TL317M .... −55°C to 150°C
                       TL317C .... 0°C to 150°C
Storage temperature range .... −65°C to 150°C
Lead temperature 1,6 mm (1/16 inch) from case for 60 seconds, JG package .... 300°C
Lead temperature 1,6 mm (1/16 inch) from case for 10 seconds, LP package .... 260°C

NOTE 1: For operation above 25°C free-air or case temperature, refer to Dissipation Derating Table.

DISSIPATION DERATING TABLE

PACKAGE	REFERENCE POINT	POWER RATING	DERATING FACTOR	ABOVE (T_A OR T_C)
JG	Free-air	1050 mW	8.4 mW/°C	25°C
	Case	1600 mW	38.4 mW/°C	108°C
LP	Free-air	775 mW	6.2 mW/°C	25°C
	Case	1600 mW	28.6 mW/°C	94°C

recommended operating conditions

	TL317M		TL317C		UNIT
	MIN	MAX	MIN	MAX	
Output current, I_O	2.5	100	2.5	100	mA
Operating virtual junction temperature, T_J	−55	125	0	125	°C

TEXAS
INSTRUMENTS
POST OFFICE BOX 225012 • DALLAS, TEXAS 75265

electrical characteristics over recommended ranges of operating virtual junction temperature (unless otherwise noted)

PARAMETER	TEST CONDITIONS[†]		MIN	TYP	MAX	UNIT
Input regulation (see Note 2)	$V_I - V_O$ = 3 V to 35 V, See Note 3	$T_J = 25°C$		0.01	0.02	%/V
		I_O = 2.5 mA to 100 mA		0.02	0.05	
Ripple rejection	V_O = 10 V,	f = 120 Hz		65		dB
	V_O = 10 V, 10-µF capacitor between ADJ and ground	f = 120 Hz,	66	80		
Output regulation	I_O = 2.5 mA to 100 mA, $T_J = 25°C$, See Note 3	$V_O ≤ 5$ V		25		mV
		$V_O ≥ 5$ V		0.5		%
	I_O = 2.5 mA to 100 mA, See Note 3	$V_O ≤ 5$ V		50		mV
		$V_O ≥ 5$ V		1		%
Output voltage change with temperature	$T_J = 0°C$ to 125°C			1		%
Output voltage long-term drift (see Note 4)	After 1000 h at $T_J = 125°C$ and $V_I - V_O$ = 35 V			0.3	1	%
Output noise voltage	f = 10 Hz to 10 kHz, $T_J = 25°C$			0.003		%
Minimum output current to maintain regulation	$V_I - V_O$ = 35 V			1.5	2.5	mA
Peak output current	$V_I - V_O ≤ 35$ V		100	200		mA
Adjustment-terminal current				50	100	µA
Change in adjustment-terminal current	$V_I - V_O$ = 2.5 V to 35 V, I_O = 2.5 mA to 100 mA			0.2	5	µA
Reference voltage (output to ADJ)	$V_I - V_O$ = 3 V to 35 V, I_O = 2.5 mA to 100 mA, P ≤ rated dissipation		1.2	1.25	1.3	V

[†]Unless otherwise noted, these specifications apply for the following test conditions: $V_I - V_O$ = 5 V and I_O = 2.5 mA.

NOTES: 2. Input regulation is expressed here as the percentage change in output voltage per 1-volt change at the input.
    3. Input regulation and output regulation are measured using pulse techniques ($t_w ≤ 10$ µs, duty cycle ≤ 5%) to limit changes in average internal dissipation. Output voltage changes due to large changes in internal dissipation must be taken into account separately.
    4. Since long-term drift cannot be measured on the individual devices prior to shipment, this specification is not intended to be a guarantee or warranty. It is an engineering estimate of the average drift to be expected from lot to lot.

TYPICAL APPLICATION DATA

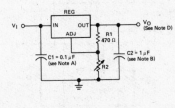

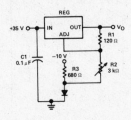

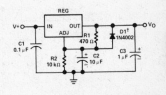

[†]D1 discharges C2 if output is shorted to ground.

FIGURE 1—ADJUSTABLE VOLTAGE REGULATOR

FIGURE 2—0-V to 30-V REGULATOR CIRCUIT

FIGURE 3—ADJUSTABLE REGULATOR CIRCUIT WITH IMPROVED RIPPLE REJECTION

NOTES: A. Use of an input bypass capacitor is recommended if regulator is far from filter capacitors.
    B. Use of an output capacitor improves transient response but is optional.
    C. V_{ref} equals the difference between the output and adjustment terminal voltages.
    D. Output voltage is calculated from the equation: $V_O = V_{ref}\left(1 + \dfrac{R2}{R1}\right)$

Voltage Regulators

6

183

TEXAS
INSTRUMENTS
POST OFFICE BOX 225012 • DALLAS, TEXAS 75265

TYPICAL APPLICATION DATA

Voltage Regulators

6

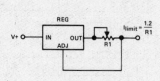

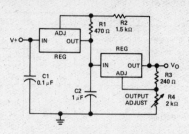

FIGURE 4—PRECISION CURRENT
LIMITER CIRCUIT

FIGURE 5—TRACKING PREREGULATOR
CIRCUIT

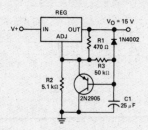

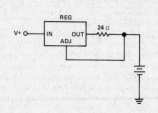

FIGURE 6—SLOW-TURN-ON 15-V
REGULATOR CIRCUIT

FIGURE 7—50-mA CONSTANT-CURRENT
BATTERY CHARGER CIRCUIT

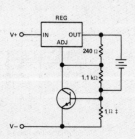

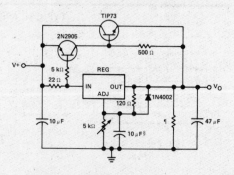

‡This resistor sets peak current (100 mA for 6 Ω).

¶Minimum load current is 30 mA.
§Optional capacitor improves ripple rejection

FIGURE 8—CURRENT-LIMITED
6-V CHARGER

FIGURE 9—HIGH-CURRENT ADJUSTABLE REGULATOR

**TEXAS
INSTRUMENTS**

POST OFFICE BOX 225012 • DALLAS, TEXAS 75265

**LINEAR
INTEGRATED
CIRCUITS**

**TYPES TL430I, TL430C
ADJUSTABLE SHUNT REGULATORS**

D2165, JUNE 1976—REVISED DECEMBER 1982

- Temperature Compensated
- Programmable Output Voltage
- Low Output Resistance
- Low Output Noise
- Sink Capability to 100 mA

**LP
SILECT PACKAGE**

(TOP VIEW)

▯ CATHODE

▯ ANODE

▯ REF

RAK

description

The TL430 is a three-terminal adjustable shunt regulator featuring excellent temperature stability, wide operating current range, and low output noise. The output voltage may be set by two external resistors to any desired value between 3 volts and 30 volts. The TL430 can replace zener diodes in many applications providing improved performance.

The TL430I is characterized for operation from −25 °C to 85 °C, and the TL430C is characterized for operating from 0 °C to 70 °C.

functional block diagram

absolute maximum ratings over operating free-air temperature range (unless otherwise noted)

Regulator voltage (see Note 1) . . . . 30 V
Continuous regulator current . . . . 150 mA
Continuous dissipation at (or below) 25 °C free-air temperature (see Note 2) . . . . 775 mW
Operating free-air temperature range: TL430I . . . . −40 °C to 85 °C
                           TL430C . . . . 0 °C to 70 °C
Storage temperature range . . . . −65 °C to 150 °C
Lead temperature 1,6 mm (1/16 inch) from case for 10 seconds . . . . 260 °C

recommended operating conditions

	MIN	MAX	UNIT
Regulator Voltage, V_Z	V_{ref}	30	V
Regulator current, I_Z	2	100	mA

NOTES: 1. All voltage values are with respect to the anode terminal.
         2. For operation above 25 °C free-air temperature, refer to Dissipation Derating Curves, Figure 5.

Voltage Regulators

6

1282

**TEXAS
INSTRUMENTS**

POST OFFICE BOX 225012 ● DALLAS, TEXAS 75265

TYPES TL430I, TL430C
ADJUSTABLE SHUNT REGULATORS

electrical characteristics at 25°C free-air temperature (unless otherwise noted)

	PARAMETER	TEST FIGURE	TEST CONDITIONS		TL430I MIN	TL430I TYP	TL430I MAX	TL430C MIN	TL430C TYP	TL430C MAX	UNIT
V_{ref}	Reference input voltage	1	$V_Z = V_{ref}$,	I_Z = 10 mA	2.6	2.75	2.9	2.5	2.75	3	V
αV_{ref}	Temperature coefficient of reference input voltage	1	$V_Z = V_{ref}$, $T_A = 0°C$ to $70°C$	I_Z = 10 mA,		+120	+200		+120		ppm/°C
I_{ref}	Reference input current	2	I_Z = 10 mA, $R2 = \infty$	R1 = 10 kΩ,		3	10		3	10	µA
I_{ZK}	Regulator current near lower knee of regulation range	1	$V_Z = V_{ref}$			0.5	2		0.5	2	mA
I_{ZM}	Regulator current at maximum limit of regulation range	1	$V_Z = V_{ref}$		50			50			mA
		2	V_Z = 5 V to 30 V,	See Note 3	100			100			
r_z	Differential regulator resistance (see Note 4)	1	$V_Z = V_{ref}$, $\Delta I_Z = (52-2)$ mA			1.5	3		1.5	3	Ω
V_{nz}	Noise voltage	2	f = 0.1 Hz to 10 Hz V_Z = 3 V			50			50		µV
			V_Z = 12 V			200			200		
			V_Z = 30 V			650			650		

NOTES: 3. The average power dissipation, $V_Z \cdot I_Z \cdot$ duty cycle, must not exceed the maximum continuous rating in any 10-ms interval.
4. The regulator resistance for $V_Z > V_{ref}$, r_z', is given by:

$$r_z' = r_z \left(1 + \frac{R1}{R2}\right)$$

PARAMETER MEASUREMENT INFORMATION

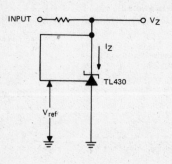

FIGURE 1—TEST CIRCUIT FOR $V_Z = V_{ref}$

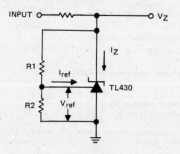

$$V_Z = V_{ref} \left(1 + \frac{R1}{R2}\right) + I_{ref} \bullet R1$$

FIGURE 2—TEST CIRCUIT FOR $V_Z > V_{ref}$

97

TEXAS
INSTRUMENTS
POST OFFICE BOX 225012 • DALLAS, TEXAS 75265

TYPICAL CHARACTERISTICS

SMALL-SIGNAL REGULATOR IMPEDANCE
vs
FREQUENCY

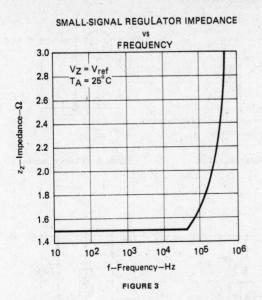

FIGURE 3

CURRENT
vs
VOLTAGE

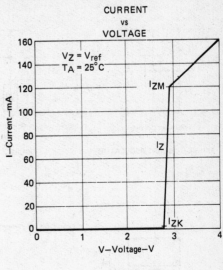

FIGURE 4

THERMAL INFORMATION

LP PACKAGE
DISSIPATION DERATING CURVE

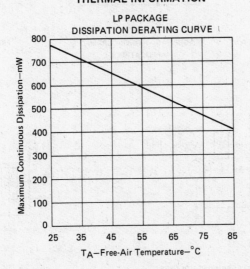

FIGURE 5

Voltage Regulators

6

TEXAS
INSTRUMENTS
POST OFFICE BOX 225012 • DALLAS, TEXAS 75265

TYPICAL APPLICATION DATA

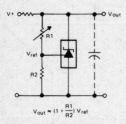

$$V_{out} \approx (1 + \frac{R1}{R2})\, V_{ref}$$

FIGURE 6—SHUNT REGULATOR

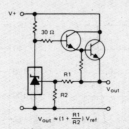

$$V_{out} \approx (1 + \frac{R1}{R2})\, V_{ref}$$

FIGURE 7—SERIES REGULATOR

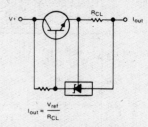

$$I_{out} = \frac{V_{ref}}{R_{CL}}$$

FIGURE 8—CURRENT LIMITER

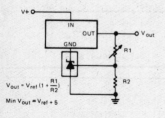

$$V_{out} = V_{ref} (1 + \frac{R1}{R2})$$
$$\text{Min } V_{out} \approx V_{ref} + 5$$

FIGURE 9—OUTPUT CONTROL OF A
THREE-THERMAL
FIXED REGULATOR

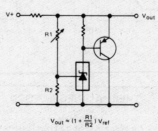

$$V_{out} \approx (1 + \frac{R1}{R2})\, V_{ref}$$

FIGURE 10—HIGHER-CURRENT
APPLICATIONS

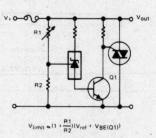

$$V_{limit} \approx (1 + \frac{R1}{R2})(V_{ref} + V_{BE(Q1)})$$

FIGURE 11—CROW BAR

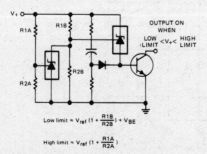

OUTPUT ON
WHEN
LOW LIMIT $<V+<$ HIGH LIMIT

$$\text{Low limit} \approx V_{ref} (1 + \frac{R1B}{R2B}) + V_{BE}$$

$$\text{High limit} \approx V_{ref} (1 + \frac{R1A}{R2A})$$

FIGURE 12—OVER-VOLTAGE/UNDER-VOLTAGE
PROTECTION CIRCUIT

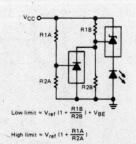

$$\text{Low limit} \approx V_{ref} (1 + \frac{R1B}{R2B}) + V_{BE}$$

$$\text{High limit} \approx V_{ref} (1 + \frac{R1A}{R2A})$$

FIGURE 13—V_{CC} MONITOR

Voltage Regulators

6

TEXAS
INSTRUMENTS
POST OFFICE BOX 225012 • DALLAS, TEXAS 75265

- Equivalent Full-Range Temperature Coefficient . . . 30 ppm/°C Typ
- Temperature Compensated for Operation Over Full Rated Operating Temperature Range
- Adjustable Output Voltage
- Fast Turn-On Response
- Sink Current Capability . . . 1 mA to 100 mA
- Low (0.2-Ω Typ) Dynamic Output Impedance
- Low Output Noise Voltage

description

The TL431 is a three-terminal adjustable regulator series with guaranteed thermal stability over applicable temperature ranges. The output voltage may be set to any value between V_{ref} (approximately 2.5 volts) and 36 volts with two external resistors (see Figure 16). These devices have a typical dynamic output impedance of 0.2 Ω. Active output circuitry provides a very sharp turn-on characteristic, making these devices excellent replacements for zener diodes in many applications.

The TL431M is characterized for operation over the full military temperature range of −55°C to 125°C. The TL431I is characterized for operation from −40°C to 85°C, and the TL431C from 0°C to 70°C.

terminal assignments

TL431M . . JG DUAL-IN-LINE PACKAGE (TOP VIEW)	TL431I, TL431C . . . LP SILECT PACKAGE (TOP VIEW)	TL431I, TL431C . . . P DUAL-IN-LINE PACKAGE (TOP VIEW)
CATHODE 1 — 8 REF NC 2 — 7 NC NC 3 — 6 ANODE NC 4 — 5 NC	CATHODE ANODE REF	CATHODE 1 — 8 REF NC 2 — 7 NC NC 3 — 6 ANODE NC 4 — 5 NC
	RAK	

NC—No internal connection

schematic

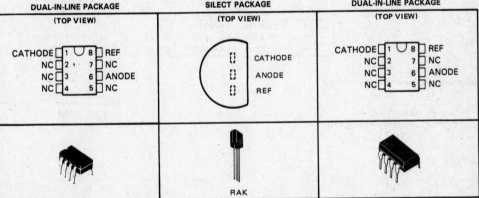

Component values are nominal.

functional block diagram

REFERENCE (R)

ANODE (A)   CATHODE (K)

Voltage Regulators

6

TEXAS INSTRUMENTS
POST OFFICE BOX 225012 • DALLAS, TEXAS 75265

absolute maximum ratings over operating free-air temperature range (unless otherwise noted)

Cathode voltage (see Note 1)	37 V
Continuous cathode current range	−100 mA to 150 mA
Reference input current range	−50 μA to 10 mA
Continuous power dissipation at (or below) 25°C free-air temperature (see Note 2): JG package . . . . . . 1050 mW
LP package . . . . . 775 mW
P package . . . . . 1000 mW
Operating free-air temperature range: TL431C . . . . 0°C to 70°C
TL431I . . . . −40°C to 85°C
TL431M . . . . −55°C to 125°C
Storage temperature range . . . . −65°C to 150°C
Lead temperature 1,6 mm (1/16 inch) from case for 60 seconds: JG package . . . . 300°C
Lead temperature 1,6 mm (1/16 inch) from case for 10 seconds: LP or P package . . . . 260°C

NOTES: 1. Voltage values are with respect to the anode terminal unless otherwise noted.

2. For operation above 25°C free-air temperature, refer to the Dissipation Derating Table.

DISSIPATION DERATING TABLE

PACKAGE	POWER RATING	DERATING FACTOR	ABOVE T_A
JG	1050 mW	8.4 mW/°C	25°C
LP	775 mW	6.2 mW/°C	25°C
P	1000 mW	8.0 mW/°C	25°C

recommended operating conditions

		MIN	MAX	UNIT
Cathode voltage, V_{KA}		V_{ref}	36	V
Cathode current, I_K, (for regulation)		1	100	mA

TEXAS
INSTRUMENTS
POST OFFICE BOX 225012 • DALLAS, TEXAS 75265

electrical characteristics at 25°C free-air temperature (unless otherwise noted)

PARAMETER		TEST CIRCUIT	TEST CONDITIONS		TL431M			TL431I			TL431C			UNIT
					MIN	TYP	MAX	MIN	TYP	MAX	MIN	TYP	MAX	
V_{ref}	Reference input voltage	1	$V_{KA} = V_{ref}$, $I_K = 10$ mA		2440	2495	2550	2440	2495	2550	2440	2495	2550	mV
$V_{ref(dev)}$	Deviation of reference input voltage over full temperature range‡	1	$V_{KA} = V_{ref}$, $I_K = 10$ mA, T_A = full range†			22	44		15	30		8	17	mV
$\dfrac{\Delta V_{ref}}{\Delta V_{KA}}$	Ratio of change in reference input voltage to the change in cathode voltage	2	$I_K = 10$ mA	$\Delta V_{KA} = 10$ V $- V_{ref}$		−1.4	−2.7		−1.4	−2.7		−1.4	−2.7	mV/V
				$\Delta V_{KA} = 36$ V $- 10$ V		−1	−2		−1	−2		−1	−2	V
I_{ref}	Reference input current	2	$I_K = 10$ mA, R1 = 10 kΩ, R2 = ∞			2	4		2	4		2	4	µA
$I_{ref(dev)}$	Deviation of reference input current over full temperature range‡	2	$I_K = 10$ mA, R1 = 10 kΩ, R2 = ∞, T_A = full range†			1	3		0.8	2.5		0.4	1.2	µA
I_{min}	Minimum cathode current for regulation	1	$V_{KA} = V_{ref}$			0.4	1		0.4	1		0.4	1	mA
I_{off}	Off-state cathode current	3	$V_{KA} = 36$ V, $V_{ref} = 0$			0.1	1		0.1	1		0.1	1	µA
$\lvert z_{ka} \rvert$	Dynamic impedance§	1	$V_{KA} = V_{ref}$, $I_K = 1$ mA to 100 mA, $f \leqslant 1$ kHz			0.2	0.5		0.2	0.5		0.2	0.5	Ω

†Full temperature range is −55°C to 125°C for the TL431M, −40°C to 85°C for the TL431I, and 0°C to 70°C for the TL431C.

‡The deviation parameters $V_{ref(dev)}$ and $I_{ref(dev)}$ are defined as the differences between the maximum and minimum values obtained over the rated temperature range. The equivalent full-range temperature coefficient of the reference input voltage, αV_{ref}, is defined as:

$$\left| \alpha V_{ref} \left(\frac{ppm}{°C} \right) \right| = \frac{\left(\dfrac{V_{ref(dev)}}{V_{ref} @ 25°C} \right) \times 10^6}{\Delta T_A}$$

where ΔT_A is the rated operating free-air temperature range of the device,

αV_{ref} can be positive or negative depending on whether minimum V_{ref} or maximum V_{ref}, respectively, occurs at the lower temperature (see Figure 8).

Example: Max V_{ref} = 2500 mV @ 30°C, Min V_{ref} = 2492 mV @ 0°C, V_{ref} = 2495 mV @ 25°C, $\Delta T_A = 70°C$ for TL431C

$$\left| \alpha V_{ref} \right| = \frac{\left(\dfrac{8 \text{ mV}}{2495 \text{ mV}} \right) \times 10^6}{70°C} = 46 \, ppm/°C$$

Because minimum V_{ref} occurs at the lower temperature, the coefficient is positive.

§The dynamic impedance is defined as:

$$\left| z_{ka} \right| = \frac{\Delta V_{KA}}{\Delta I_K}$$

When the device is operated with two external resistors (see Figure 2), the total dynamic impedance of the circuit is given by:

$$\left| z' \right| = \frac{\Delta V}{\Delta I} \approx \left| z_{ka} \right| \left(1 + \frac{R1}{R2} \right)$$

Voltage Regulators

6

TEXAS INSTRUMENTS
POST OFFICE BOX 225012 • DALLAS, TEXAS 75265

TYPES TL431M, TL431I, TL431C
ADJUSTABLE PRECISION SHUNT REGULATORS

PARAMETER MEASUREMENT INFORMATION

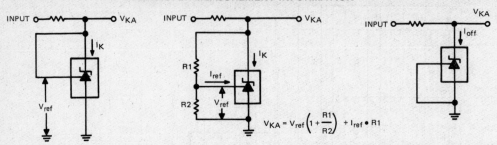

$$V_{KA} = V_{ref}\left(1 + \frac{R1}{R2}\right) + I_{ref} \cdot R1$$

FIGURE 1—TEST CIRCUIT FOR $V_{KA} = V_{ref}$ FIGURE 2—TEST CIRCUIT FOR $V_{KA} > V_{ref}$ FIGURE 3—TEST CIRCUIT FOR I_{off}

TYPICAL CHARACTERISTICS

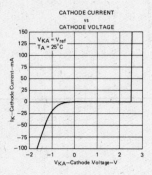

FIGURE 4

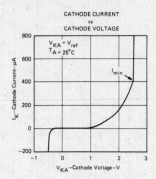

FIGURE 5

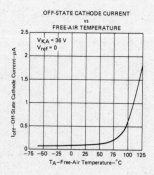

FIGURE 6

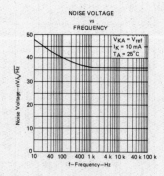

FIGURE 7

Texas
Instruments
POST OFFICE BOX 225012 • DALLAS, TEXAS 75265

TYPICAL CHARACTERISTICS

REFERENCE INPUT VOLTAGE
vs
FREE-AIR TEMPERATURE

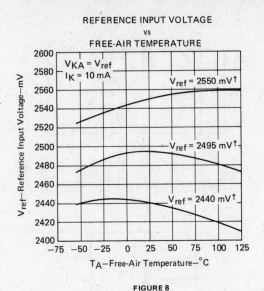

FIGURE 8

REFERENCE INPUT CURRENT
vs
FREE-AIR TEMPERATURE

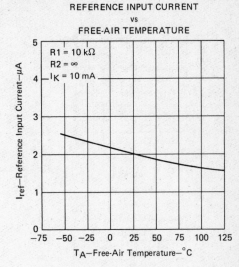

FIGURE 9

CHANGE IN REFERENCE INPUT VOLTAGE
vs
CATHODE VOLTAGE

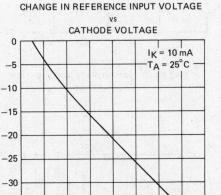

FIGURE 10

DYNAMIC IMPEDANCE
vs
FREE-AIR TEMPERATURE

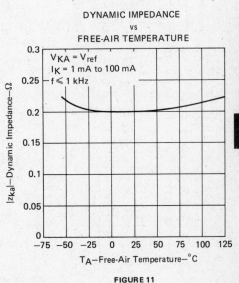

FIGURE 11

†Data is for devices having the indicated value of V_{ref} at I_K = 10 mA, T_A = 25°C.

Voltage Regulators

6

TEXAS INSTRUMENTS
POST OFFICE BOX 225012 • DALLAS, TEXAS 75265

TYPICAL CHARACTERISTICS

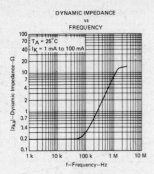

DYNAMIC IMPEDANCE
vs
FREQUENCY

FIGURE 12

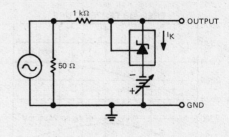

TEST CIRCUIT FOR DYNAMIC IMPEDANCE

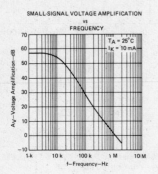

SMALL-SIGNAL VOLTAGE AMPLIFICATION
vs
FREQUENCY

FIGURE 13

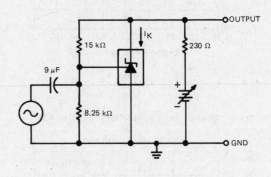

TEST CIRCUIT FOR VOLTAGE AMPLIFICATION

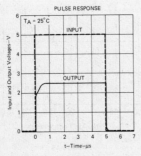

PULSE RESPONSE

FIGURE 14

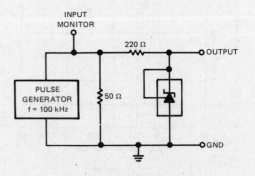

TEST CIRCUIT FOR PULSE RESPONSE

Voltage Regulators

6

TEXAS
INSTRUMENTS
POST OFFICE BOX 225012 • DALLAS, TEXAS 75265

TYPICAL CHARACTERISTICS

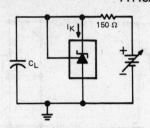

TEST CIRCUIT FOR CURVE A BELOW

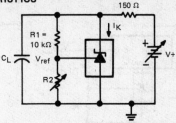

TEST CIRCUIT FOR CURVES B, C, AND D BELOW

STABILITY BOUNDARY CONDITIONS

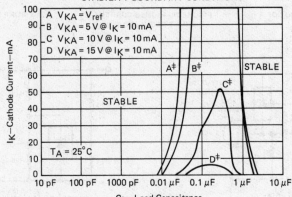

A $V_{KA} = V_{ref}$
B $V_{KA} = 5\,V @ I_K = 10\,mA$
C $V_{KA} = 10\,V @ I_K = 10\,mA$
D $V_{KA} = 15\,V @ I_K = 10\,mA$

$T_A = 25°C$

FIGURE 15

‡The areas under the curves represent conditions that may cause the device to oscillate. For curves B, C, and D, R2 and V+ were adjusted to establish the initial V_{KA} and I_K conditions with $C_L = 0$. V+ and C_L were then adjusted to determine the ranges of stability.

TYPICAL APPLICATIONS

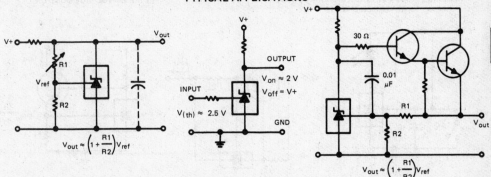

$$V_{out} \approx \left(1 + \frac{R1}{R2}\right)V_{ref}$$

OUTPUT
$V_{on} \approx 2\,V$
$V_{off} = V+$
INPUT
$V_{(th)} \approx 2.5\,V$
GND

$$V_{out} \approx \left(1 + \frac{R1}{R2}\right)V_{ref}$$

FIGURE 16—SHUNT REGULATOR   **FIGURE 17—SINGLE-SUPPLY COMPARATOR WITH TEMPERATURE-COMPENSATED THRESHOLD**   **FIGURE 18—SERIES REGULATOR**

Voltage Regulators

6

TYPICAL APPLICATIONS

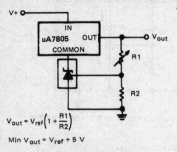

$$V_{out} = V_{ref}\left(1 + \frac{R1}{R2}\right)$$

Min $V_{out} = V_{ref} + 5$ V

FIGURE 19— OUTPUT CONTROL OF A THREE-TERMINAL FIXED REGULATOR

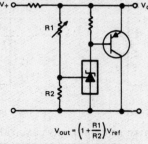

$$V_{out} = \left(1 + \frac{R1}{R2}\right)V_{ref}$$

FIGURE 20—HIGHER-CURRENT SHUNT REGULATOR

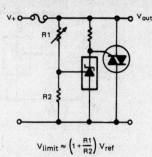

$$V_{limit} \approx \left(1 + \frac{R1}{R2}\right)V_{ref}$$

FIGURE 21—CROW BAR

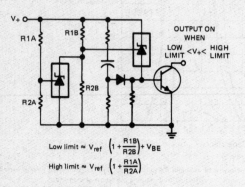

OUTPUT ON WHEN

LOW LIMIT $< V+ <$ HIGH LIMIT

Low limit $\approx V_{ref}\left(1 + \frac{R1B}{R2B}\right) + V_{BE}$

High limit $\approx V_{ref}\left(1 + \frac{R1A}{R2A}\right)$

FIGURE 22—OVER-VOLTAGE/UNDER-VOLTAGE PROTECTION CIRCUIT

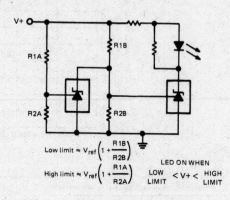

Low limit $\approx V_{ref}\left(1 + \frac{R1B}{R2B}\right)$

High limit $\approx V_{ref}\left(1 + \frac{R1A}{R2A}\right)$

LED ON WHEN

LOW LIMIT $< V+ <$ HIGH LIMIT

FIGURE 23—VOLTAGE MONITOR

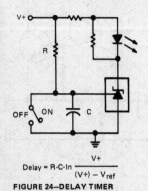

$$\text{Delay} = R \cdot C \cdot \ln \frac{V+}{(V+) - V_{ref}}$$

FIGURE 24—DELAY TIMER

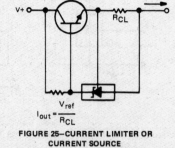

$$I_{out} = \frac{V_{ref}}{R_{CL}}$$

FIGURE 25—CURRENT LIMITER OR CURRENT SOURCE

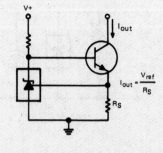

$$I_{out} = \frac{V_{ref}}{R_S}$$

FIGURE 26—CONSTANT-CURRENT SINK

Voltage Regulators

6

TEXAS INSTRUMENTS
POST OFFICE BOX 225012 • DALLAS, TEXAS 75265

LINEAR
INTEGRATED
CIRCUITS

TYPES TL493, TL494, TL495
PULSE-WIDTH-MODULATION CONTROL CIRCUITS

D2535, JANUARY 1983—REVISED SEPTEMBER 1983

- Complete PWM Power Control Circuitry
- Uncommitted Outputs for 200-mA Sink or Source Current
- Output Control Selects Single-Ended or Push-Pull Operation
- Internal Circuitry Prohibits Double Pulse at Either Output
- Variable Dead-Time Provides Control Over Total Range
- Internal Regulator Provides a Stable 5-V Reference Supply Trimmed to 1%
- Circuit Architecture Allows Easy Synchronization
- TL493 Has Output Current-Limit Sensing
- TL495 Has On-Chip 39-V Zener and External Control of Output Steering

description

The TL493, TL494, and TL495 each incorporate on a single monolithic chip all the functions required in the construction of a pulse-width-modulation control circuit. Designed primarily for power supply control, these devices offer the systems engineer the flexibility to tailor the power supply control circuitry to his application.

The TL493 contains an error amplifier, current-limiting amplifier, an on-chip adjustable oscillator, a dead-time control comparator, pulse-steering control flip-flop, a 5-volt, 1%-precision regulator, and output-control circuits.

The error amplifier exhibits a common-mode voltage range from -0.3 volts to $V_{CC} - 2$ volts. The current-limit amplifier exhibits a common-mode voltage range from -0.3 volts to 3 volts with an offset voltage of approximately 80 millivolts in series with the inverting input to ease circuit design requirements. The dead-time control comparator has a fixed offset that provides approximately 5% dead time when externally altered. The on-chip oscillator may be bypassed by terminating R_T (pin 6) to the reference output and providing a sawtooth input to C_T (pin 5), or it may be used to drive the common circuits in synchronous multiple-rail power supplies.

TL493C . . . N
DUAL-IN-LINE PACKAGE (TOP VIEW)

```
ERROR  { NONINV INPUT [ 1   16 ] NONINV INPUT } CURRENT
AMP 1  {   INV INPUT [ 2   15 ] INV INPUT     } LIMIT AMP
            FEEDBACK [ 3   14 ] REF OUT
   DEAD-TIME CONTROL [ 4   13 ] OUTPUT CONTROL
                  CT [ 5   12 ] VCC
                  RT [ 6   11 ] C2
                 GND [ 7   10 ] E2
                  C1 [ 8    9 ] E1
```

TL494M . . . J
TL494I, TL494C . . . J OR N
DUAL-IN-LINE PACKAGE (TOP VIEW)

```
ERROR  { NONINV INPUT [ 1   16 ] NONINV INPUT } ERROR
AMP 1  {   INV INPUT [ 2   15 ] INV INPUT     } AMP 2
            FEEDBACK [ 3   14 ] REF OUT
   DEAD-TIME CONTROL [ 4   13 ] OUTPUT CONTROL
                  CT [ 5   12 ] VCC
                  RT [ 6   11 ] C2
                 GND [ 7   10 ] E2
                  C1 [ 8    9 ] E1
```

TL495C . . . N
DUAL-IN-LINE PACKAGE (TOP VIEW)

```
ERROR  { NONINV INPUT [ 1   18 ] NONINV INPUT } ERROR
AMP 1  {   INV INPUT [ 2   17 ] INV INPUT     } AMP 2
            FEEDBACK [ 3   16 ] REF OUT
   DEAD-TIME CONTROL [ 4   15 ] VZ
                  CT [ 5   14 ] OUTPUT CONTROL
                  RT [ 6   13 ] STEERING INPUT
                 GND [ 7   12 ] VCC
                  C1 [ 8   11 ] C2
                  E1 [ 9   10 ] E2
```

DEVICE TYPES, SUFFIX VERSIONS, AND PACKAGES

	TL493	TL494	TL495
TL49–M	*	J	*
TL49–I	*	J,N	*
TL49–C	N	J,N	N

*These combinations are not defined by this data sheet.

FUNCTION TABLE

INPUTS		OUTPUT FUNCTION
OUTPUT CONTROL	STEERING INPUT (TL495 only)	
$V_I \leq 0.4$ V	Open	Single-ended or parallel output
$V_I \geq 2.4$ V	Open	Normal push-pull operation
$V_I \geq 2.4$ V	$V_I \leq 0.4$ V	PWM Output at Q1
$V_I \geq 2.4$ V	$V_I \geq 2.4$ V	PWM Output at Q2

Voltage Regulators

6

TEXAS
INSTRUMENTS
POST OFFICE BOX 225012 • DALLAS, TEXAS 75265

description (continued)

The uncommitted output transistors provide either common-emitter or emitter-follower output capability. Each device provides for push-pull or single-ended output operation, which may be selected through the output-control function. The architecture of these devices prohibits the possibility of either output being pulsed twice during push-pull operation.

The TL493 and TL494 are similar except that an additional error amplifier is included in the TL494 instead of a current-limiting amplifier. The TL495 provides the identical functions found in the TL494. In addition, it contains an on-chip 39-volt zener diode for high-voltage applications where V_{CC} is greater than 40 volts, and an output-steering control that overrides the internal control of the pulse-steering flip-flop.

The TL494M is characterized for operation over the full military temperature range from $-55\,°C$ to $125\,°C$. The TL494I is characterized for operation from $-25\,°C$ to $85\,°C$. The TL493C, TL494C, and TL495C are characterized for operation from $0\,°C$ to $70\,°C$.

functional block diagram

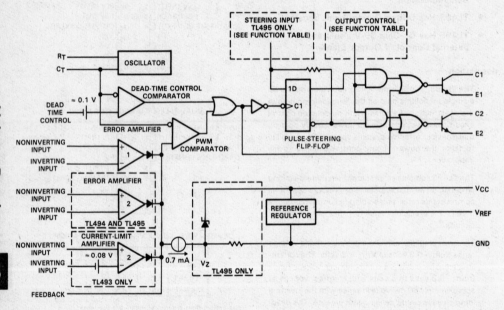

TEXAS INSTRUMENTS

POST OFFICE BOX 225012 • DALLAS, TEXAS 75265

absolute maximum ratings over operating free-air temperature range (unless otherwise noted)

	TL494M	TL494I	TL493C TL494C TL495C	UNIT
Supply voltage, V_{CC} (see Note 1)	41	41	41	V
Amplifier input voltages	$V_{CC}+0.3$	$V_{CC}+0.3$	$V_{CC}+0.3$	V
Collector output voltage	41	41	41	V
Collector output current	250	250	250	mA
Continuous total dissipation at (or below) 25°C free-air temperature (see Note 2)	1000	1000	1000	mW
Operating free-air temperature range	−55 to 125	−25 to 85	0 to 70	°C
Storage temperature range	−65 to 150	−65 to 150	−65 to 150	°C
Lead temperature 1,6 mm (1/16 inch) from case for 60 seconds: J package	300	300	300	°C
Lead temperature 1,6 mm (1/16 inch) from case for 10 seconds: N package		260	260	°C

NOTES: 1. All voltage values, except differential voltages, are with respect to the network ground terminal.
2. For operation above 25°C free-air temperature, refer to Dissipation Derating Table. In the J package, TL494M chips are alloy-mounted; TL494I and TL494C chips are glass mounted.

DISSIPATION DERATING TABLE

PACKAGE	POWER RATING	DERATING FACTOR	ABOVE T_A
J (Alloy-Mounted Chip)	1000 mW	11.0 mW/°C	59°C
J (Glass-Mounted Chip)	1000 mW	8.2 mW/°C	28°C
N	1000 mW	9.2 mW/°C	41°C

recommended operating conditions

	TL494M		TL494I		TL493C TL494C TL495C		UNIT
	MIN	MAX	MIN	MAX	MIN	MAX	
Supply voltage, V_{CC}	7	40	7	40	7	40	V
Amplifier input voltages, V_I	−0.3	$V_{CC}-2$	−0.3	$V_{CC}-2$	−0.3	$V_{CC}-2$	V
Collector output voltage, V_O		40		40		40	V
Collector output current (each transistor)		200		200		200	mA
Current into feedback terminal		0.3		0.3		0.3	mA
Timing capacitor, C_T	0.47	10 000	0.47	10 000	0.47	10 000	nF
Timing resistor, R_T	1.8	500	1.8	500	1.8	500	kΩ
Oscillator frequency	1	300	1	300	1	300	kHz
Operating free-air temperature, T_A	−55	125	−25	85	0	70	°C

TEXAS INSTRUMENTS
POST OFFICE BOX 225012 • DALLAS, TEXAS 75265

electrical characteristics over recommended operating free-air temperature range, V_{CC} = 15 V, f = 10 kHz (unless otherwise noted)

reference section

PARAMETER	TEST CONDITIONS[†]	TL494M			TL493C TL494I, TL494C TL495C			UNIT
		MIN	TYP[‡]	MAX	MIN	TYP[‡]	MAX	
Output voltage (V_{ref})	I_O = 1 mA	4.75	5	5.25	4.75	5	5.25	V
Input regulation	V_{CC} = 7 V to 40 V		2	25		2	25	mV
Output regulation	I_O = 1 to 10 mA		1	15		1	15	mV
Output voltage change with temperature	ΔT_A = MIN to MAX		0.2	1		0.2	1	%
Short-circuit output current[§]	V_{ref} = 0	10	35	50		35		mA

oscillator section (see Figure 1)

PARAMETER	TEST CONDITIONS[†]	TL494M			TL493C TL494I, TL494C TL495C			UNIT
		MIN	TYP[‡]	MAX	MIN	TYP[‡]	MAX	
Frequency	C_T = 0.01 μF, R_T = 12 kΩ		10			10		kHz
Standard deviation of frequency[¶]	All values of V_{CC}, C_T, R_T, T_A constant		10			10		%
Frequency change with voltage	V_{CC} = 7 V to 40 V, T_A = 25°C		0.1			0.1		%
Frequency change with temperature	C_T = 0.01 μF, R_T = 12 kΩ, ΔT_A = MIN to MAX			12			12	%

amplifier sections (see Figure 2)

PARAMETER		TEST CONDITIONS	MIN	TYP[‡]	MAX	UNIT
Input offset voltage	Error	V_O (pin 3) = 2.5 V		2	10	mV
	current-limit (TL493 only)			80		
Input offset current		V_O (pin 3) = 2.5 V		25	250	nA
Input bias current		V_O (pin 3) = 2.5 V		0.2	1	μA
Common-mode input voltage range	Error	V_{CC} = 7 V to 40 V	−0.3 to V_{CC}−2			V
	Current-limit (TL493 only)		−0.3 to 3			
Open-loop voltage amplification	Error	ΔV_O = 3 V, V_O = 0.5 V to 3.5 V	70	95		dB
	Current-limit (TL493 only)			90		
Unity-gain bandwidth				800		kHz
Common-mode rejection ratio	Error	V_{CC} = 40 V, T_A = 25°C	65	80		dB
	Current-limit (TL493 only)			70		
Output sink current (pin 3)		V_{ID} = −15 mV to −5 V, $V_{(pin 3)}$ = 0.5 V	0.3	0.7		mA
Output source current (pin 3)		V_{ID} = 15 mV to 5 V, $V_{(pin 3)}$ = 3.5 V		−2		mA

[†]For conditions shown as MIN or MAX, use the appropriate value specified under recommended operating conditions.

[‡]All typical values except for parameter changes with temperature are at T_A = 25°C.

[§]Duration of the short-circuit should not exceed one second.

[¶]Standard deviation is a measure of the statistical distribution about the mean as derived from the formula $\sigma = \sqrt{\dfrac{\sum\limits_{n=1}^{N}(x_n - \bar{x})^2}{N-1}}$

Voltage Regulators

6

TEXAS INSTRUMENTS
POST OFFICE BOX 225012 • DALLAS, TEXAS 75265

electrical characteristics over recommended operating free-air temperature range, V_{CC} = 15 V, f = 10 kHz (unless otherwise noted)

output section

PARAMETER		TEST CONDITIONS		TL494M			TL493C TL494I, TL494C TL495C			UNIT
				MIN	TYP‡	MAX	MIN	TYP†	MAX	
Collector off-state current		V_{CE} = 40 V, V_{CC} = 40 V			2	100		2	100	μA
Emitter off-state current		V_{CC} = V_C = 40 V, V_E = 0				−150			−100	μA
Collector-emitter saturation voltage	Common-emitter	V_E = 0, I_C = 200 mA		1.1	1.5		1.1	1.3		V
	Emitter-follower	V_C = 15 V, I_E = −200 mA		1.5	2.5		1.5	2.5		
Output control input current		V_I = V_{ref}				3.5			3.5	mA

dead-time control-section (see Figure 1)

PARAMETER	TEST CONDITIONS	MIN	TYP†	MAX	UNIT
Input bias current (pin 4)	V_I = 0 to 5.25 V		−2	−10	μA
Maximum duty cycle, each output	V_I (pin 4) = 0	45			%
Input threshold voltage (pin 4)	Zero duty cycle		3	3.3	V
	Maximum duty cycle	0			

pwm comparator section (see Figure 1)

PARAMETER	TEST CONDITIONS	MIN	TYP†	MAX	UNIT
Input threshold voltage (pin 3)	Zero duty cycle		4	4.5	V
Input sink current (pin 3)	$V_{(pin 3)}$ = 0.7 V	0.3	0.7		mA

steering control (TL495 only)

PARAMETER	TEST CONDITIONS	MIN	MAX	UNIT
Input current	V_I = 0.4 V		−200	μA
	V_I = 2.4 V		200	

zener-diode circuit (TL495 only)

PARAMETER	TEST CONDITIONS	MIN	TYP†	MAX	UNIT
Breakdown voltage	V_{CC} = 41 V, I_Z = 2 mA		39		V
Sink current	$V_{I(pin 15)}$ = 1 V		0.3		mA

total device (see Figure 1)

PARAMETER	TEST CONDITIONS		MIN	TYP†	MAX	UNIT
Standby supply current	Pin 6 at V_{ref}, All other inputs and outputs open	V_{CC} = 15 V		6	10	mA
		V_{CC} = 40 V		9	15	
Average supply current	$V_{(pin 4)}$ = 2 V			7.5		mA

switching characteristics, T_A = 25°C

PARAMETER	TEST CONDITIONS	MIN	TYP†	MAX	UNIT
Output voltage rise time	Common-emitter configuration,		100	200	ns
Output voltage fall time	See Figure 3		25	100	ns
Output voltage rise time	Emitter-follower configuration,		100	200	ns
Output voltage fall time	See Figure 4		40	100	ns

†All typical values except for temperature coefficient are at T_A = 25°C.

Voltage Regulators

6

PARAMETER MEASUREMENT INFORMATION

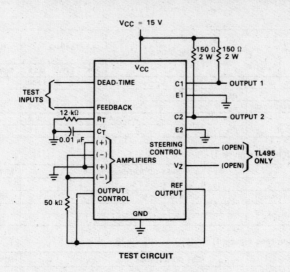

TEST CIRCUIT

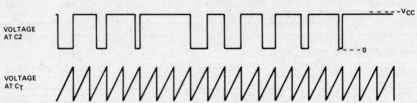

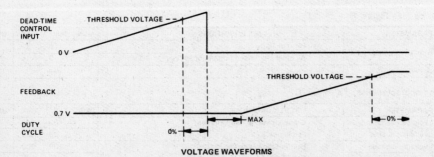

VOLTAGE WAVEFORMS

FIGURE 1—OPERATIONAL TEST CIRCUIT AND WAVEFORMS

TEXAS
INSTRUMENTS
POST OFFICE BOX 225012 • DALLAS, TEXAS 75265

Voltage Regulators

6

983

PARAMETER MEASUREMENT INFORMATION

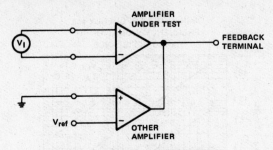

FIGURE 2 — AMPLIFIER CHARACTERISTICS

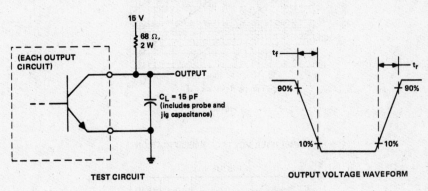

FIGURE 3 — COMMON-EMITTER CONFIGURATION

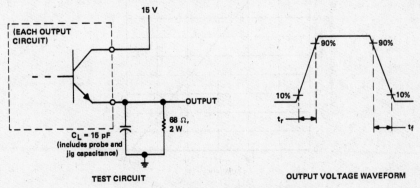

FIGURE 4 — EMITTER-FOLLOWER CONFIGURATION

Voltage Regulators

6

TEXAS INSTRUMENTS
POST OFFICE BOX 225012 • DALLAS, TEXAS 75265

TYPICAL CHARACTERISTICS

OSCILLATOR FREQUENCY and
FREQUENCY VARIATION† vs
TIMING RESISTANCE

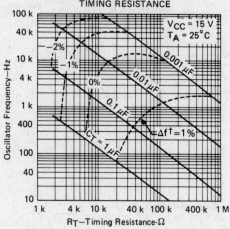

FIGURE 5

AMPLIFIER VOLTAGE AMPLIFICATION
vs
FREQUENCY

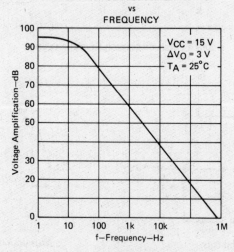

FIGURE 6

†Frequency variation (Δf) is the change in oscillator frequency that occurs over the full temperature range.

**TEXAS
INSTRUMENTS**

POST OFFICE BOX 225012 • DALLAS, TEXAS 75265

- **Internal Step-Up Switching Regulator**
- **Fixed 9-Volt Output**
- **Charges Battery Source During Transformer-Coupled-Input Operation**
- **Minimum External Components Required (1 Inductor, 1 Capacitor, 1 Diode)**
- **1- or 2-Cell-Input Operation**

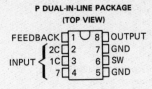

P DUAL-IN-LINE PACKAGE
(TOP VIEW)

FEEDBACK	1	8	OUTPUT
2C	2	7	GND
INPUT { 1C	3	6	SW
7	4	5	GND

Pins 5 and 7 are connected together internally.

description

The TL496 power supply control circuit is designed to provide a 9-volt regulated supply from a variety of input sources. Operable from a 1- or 2-cell-battery input, the TL496 performs as a switching regulator with the addition of a single inductor and filter capacitor. When ac coupled with a step-down transformer, the TL496 operates as a series regulator to maintain the regulated output voltage and, with the addition of a single catch diode, time shares to recharge the input batteries.

The design of the TL496 allows minimal supply current drain during stand-by operation (125 μA typical). With most battery sources this allows a constant bias to be maintained on the power supply. This makes power instantly available to the system thus eliminating power-up sequencing problems.

functional block diagram

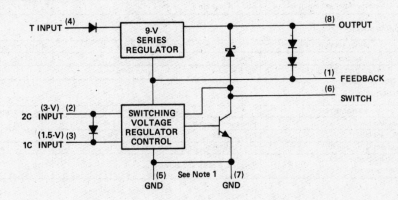

NOTE 1: Pins 5 and 7, though connected together internally, must both be terminated to ground to ensure proper circuit operation.

Voltage Regulators

6

TEXAS
INSTRUMENTS
POST OFFICE BOX 225012 • DALLAS, TEXAS 75265

absolute maximum ratings

Input voltage:
Pin 2 . . . . 3.5 V
Pin 3 . . . . 2.5 V
Pin 4 . . . . 20 V
Output voltage (Pin 6) . . . . 12 V
Diode reverse voltage (Pin 8) . . . . 12 V
Switch current (Pin 6) . . . . 1.2 A
Diode current (Pin 8) . . . . 1.2 A
Operating free-air temperature range . . . . 0°C to 70°C
Storage temperature range . . . . −65°C to 150°C
Lead temperature 1,6 mm (1/16 inch) from case for 10 seconds . . . . 260°C

electrical characteristics at 25°C free-air temperature

series regulator section (input is pin 4)

PARAMETER	TEST CONDITIONS			MIN	TYP	MAX	UNIT
Dropout voltage	V_I = 5 V,	I_O = −50 mA			1.5	2	V
Regulated output voltage	V_I = 20 V		I_O = −50 μA	9.5	10.1	11.2	V
			I_O = −80 mA	9.0	10.0	11.0	
	V_I = 20 V,		I_O = −50 μA	8.5	9.0	9.7	
	Pin 1 shorted to pin 8		I_O = −80 mA	6.7	8.6	9.5	
Standby current (pin 4)	V_I = 20 V,	Pin 8 at 12 V				400	μA
Reverse current thru pin 4	V_I = −1.5 V,	1 mA into Pin 8				−25	μA

output switch

PARAMETER		TEST CONDITIONS		MIN	TYP	MAX	UNIT
$V_{CE(sat)}$	Collector-emitter saturation voltage	800 mA into Pin 6,	Pin 2 at 2.25 V		0.35	0.6	V

diode (pin 6 to pin 8)

PARAMETER		TEST CONDITIONS		MIN	TYP	MAX	UNIT
V_F	Forward voltage	I_F = 1.5 A			1.6	2.5	V
I_R	Reverse current thru pin 6	Pin 6 at 0 V,	1 mA into Pin 8			−20	μA

control section

PARAMETER	TEST CONDITIONS		MIN	TYP	MAX	UNIT
On-state current (pin 2)	Pins 1 and 8 at 0 V,	Pin 2 at 3 V		60	100	mA
Standby current (pin 1)	Pin 1 at 8.65 V,	Pins 2 and 6 at 3 V			40	μA
Standby current (pin 2 and 6)	Pin 1 at 8.65 V,	Pins 2 and 6 at 3 V			400	μA
Start-up current (current into pin 6 to initiate cycle)	Pins 1, 2, 6 and 8 at 2.25 V		16			mA

TEXAS
INSTRUMENTS

POST OFFICE BOX 225012 • DALLAS, TEXAS 75265

Voltage Regulators

6

128

TYPICAL APPLICATION DATA

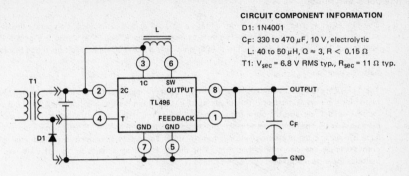

CIRCUIT COMPONENT INFORMATION
D1: 1N4001
C_F: 330 to 470 μF, 10 V, electrolytic
L: 40 to 50 μH, Q ≈ 3, R < 0.15 Ω
T1: V_{sec} = 6.8 V RMS typ., R_{sec} = 11 Ω typ.

FIGURE 1—ONE-CELL OPERATION

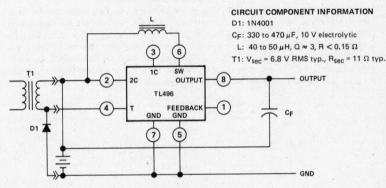

CIRCUIT COMPONENT INFORMATION
D1: 1N4001
C_F: 330 to 470 μF, 10 V electrolytic
L: 40 to 50 μH, Q ≈ 3, R < 0.15 Ω
T1: V_{sec} = 6.8 V RMS typ., R_{sec} = 11 Ω typ.

FIGURE 2—TWO-CELL OPERATION

Voltage Regulators

6

recommended operating conditions

	MIN	MAX	UNIT
Input voltage, one-cell operation (pins 2 and 3 to ground)	1.1	1.5	V
Input voltage, two-cell operation (pin 2 to ground)	2.3	3	V
Input voltage, one-cell or two-cell operation (pin 4 to ground)	V_O+2	20	V

typical electrical characteristics for circuits above

PARAMETER		ONE-CELL OPERATION (FIGURE 1)	TWO-CELL OPERATION (FIGURE 2)
Input current	No load	125 uA	125 uA
	R_L = 120 Ω	525 mA	405 mA
Output voltage	Without T1	7.2 V	8.6 V
	With T1	8.6 V	10 V
Output current capability		40 mA	80 mA
Efficiency		66%	66%
Battery life (AA NiCad) no load		60 days	166 days

TEXAS
INSTRUMENTS
POST OFFICE BOX 225012 • DALLAS, TEXAS 75265

functional description

The TL496 is designed to operate from either a single-cell or two-cell source. To operate the device from a single cell (1.1 V to 1.5 V) the source must be connected to both inputs 1C and 2C as shown in Figure 1. For two-cell operation (2.3 V to 3.0 V), the input is applied to the 2C input only and the 1C input is left open (see Figure 2).

battery operation

The TL496 operates as a switching regulator from a battery input. The cycle is initiated when a low voltage condition is sensed by the internal feedback (the thresholds at pin 1 and pin 8 are approximately 7.2 and 8.6 volts respectively). An internal latch is set and the output transistor is turned "on." This causes the current in the external inductor (L) to increase linearly until it reaches a peak value of approximately 1 ampere. When the peak current is sensed the internal latch is reset and the output transistor is turned "off." The energy developed in the inductor is then delivered to the output storage capacitor through the blocking diode. The latch remains in the off state until the feedback signal indicates the output voltage is again deficient.

transformer-coupled operation

The TL496 operates on alternate half cycles of the ac input during transformer-coupled operation to, first, sustain the output voltage and, second, recharge the batteries. The TL496 performs like a series regulator to supply charge to the output filter/storage capacitor during the first half cycle. The output voltage of the series regulator is slightly higher voltage than that created by the switching circuit; this maintains the feedback voltage above the switching regulator control circuit threshold. This effectively inhibits the switching control circuitry. During the second half cycle an external diode (1N4001) is used to clamp the negative going end of the transformer secondary to ground thus allowing the positive-going end (end connected to V+ side of battery) to pump charge into the stand-by batteries.

TEXAS
INSTRUMENTS
POST OFFICE BOX 225012 • DALLAS, TEXAS 75265

- All Monolithic
- High Efficiency . . . 60% or Greater
- Output Current . . . 500 mA
- Input Current Limit Protection
- TTL Compatible Inhibit
- Adjustable Output Voltage
- Input Regulation . . . 0.2% Typ
- Output Regulation . . . 0.4% Typ
- Soft Start-up Capability

TL497AM . . . J
TL497AI, TL497AC . . . J OR N
DUAL-IN-LINE PACKAGE
(TOP VIEW)

COMP INPUT	1	14	V$_{CC}$
INHIBIT	2	13	CUR LIM SENS
FREQ CONTROL	3	12	BASE DRIVE[†]
SUBSTRATE	4	11	BASE[†]
GND	5	10	COL OUT
CATHODE	6	9	NC
ANODE	7	8	EMIT OUT

NC—No internal connection

[†] The Base pin (#11) and Base Drive pin (#12) are used for device testing only. They are not normally used in circuit applications of the device.

description

The TLC497A incorporates on a single monolithic chip all the active functions required in the construction of a switching voltage regulator. It can also be used as the control element to drive external components for high-power-output applications. The TL497A was designed for ease of use in step-up, step-down, or voltage inversion applications requiring high efficiency.

The TL497A is a fixed-on-time variable-frequency switching voltage regulator control circuit. The on-time is programmed by a single external capacitor connected between the frequency control pin and ground. This capacitor, C_T, is charged by an internal constant-current generator to a predetermined threshold. The charging current and the threshold vary proportionally with V_{CC}, thus the one time remains constant over the specified range of input voltage (5 to 12 volts). Typical on-times for various values of C_T are as follows:

TIMING CAPACITOR, C_T (pF)	200	250	350	400	500	750	1000	1500	2000
ON-TIME (μs)	19	22	26	32	44	56	80	120	180

The output voltage is controlled by an external resistor ladder network (R1 and R2 in Figures 1, 2, and 3) that provides a feedback voltage to the comparator input. This feedback voltage is compared to the reference voltage of 1.2 volts (relative to the substrate pin) by the high-gain comparator. When the output voltage decays below the value required to maintain 1.2 V at the comparator input, the comparator enables the oscillator circuit, which charges and discharges C_T as described above. The internal pass transistor is driven on during the charging of C_T. The internal transistor may be used directly for switching currents up to 500 milliamperes. Its collector and emitter are uncommitted and it is current driven to allow operation from the positive supply voltage or ground. An internal Schottky diode matched to the current characteristics of the internal transistor is also available for blocking or commutating purposes. The TL497A also has on-chip current-limit circuitry that senses the peak currents in the switching regulator and protects the inductor against saturation and the pass transistor against overstress. The current limit is adjustable and is programmed by a single sense resistor, R_{CL}, connected between pin 14 and pin 13. The current-limit circuitry is activated when 0.7 volt is developed across R_{CL}. External gating is provided by the inhibit input. When the inhibit input is high, the output is turned off.

Simplicity of design is a primary feature of the TL497A. With only six external components (three resistors, two capacitors, and one inductor), the TL497A will operate in numerous voltage conversion applications (step-up, step-down, invert) with as much as 85% of the source power delivered to the load. The TL497A replaces the TL497 in all applications.

The TL497AM is characterized for operation over the full military temperature range of −55°C to 125°C, the TL497AI is characterized for operation from −25°C to 85°C, and the TL497AC from 0°C to 70°C.

Voltage Regulators

6

functional block diagram

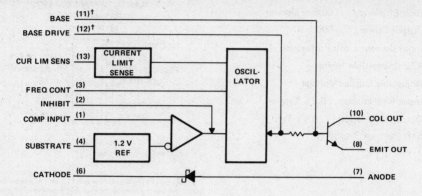

†The Base pin (#11) and Base Drive pin (#12) are used for device testing only. They are not normally used in circuit applications of the device.

absolute maximum ratings over operating free-air temperature range (unless otherwise noted)

Input voltage, V_{CC} (see Note 1)	15 V
Output voltage	35 V
Comparator input voltage	5 V
Inhibit input voltage	5 V
Diode reverse voltage	35 V
Power switch current	750 mA
Diode forward current	750 mA
Continuous total dissipation at (or below) 25°C free-air temperature (see Note 2)	1000 mW
Operating free-air temperature range: TL497AM	−55°C to 125°C
TL497AI	−25°C to 85°C
TL497AC	0°C to 70°C
Storage temperature range	−65°C to 150°C
Lead temperature 1,6 mm (1/16 inch) from case for 60 seconds: J package	300°C
Lead temperature 1,6 mm (1/16 inch) from case for 10 seconds: N package	260°C

NOTES: 1. All voltage values except diode voltages are with respect to network ground terminal.
2. Above 28°C free-air temperature, derate the N package at the rate of 9.2 mW/°C. Above 41°C free-air temperature, derate the J glass-mounted package at the rate of 8.2 mW/°C. Above 59°C free-air temperature, derate the J alloy-mounted package at the rate of 11.0 mW/°C. In the J package, TL4974AM chips are alloy mounted, TL4974AC chips are glass mounted.

recommended operating conditions

	MIN	MAX	UNIT
Input voltage, V_I	4.5	12	V
Output voltage: step-up configuration (see Figure 1)	$V_I + 2$	30	V
step-down configuration (see Figure 2)	V_{ref}	$V_I - 1$	V
inverting regulator (see Figure 3)	$-V_{ref}$	−25	V
Power switch current		500	mA
Diode forward current		500	mA

983

TEXAS
INSTRUMENTS
POST OFFICE BOX 225012 • DALLAS, TEXAS 75265

electrical characteristics at specified free-air temperature, V_I = 6 V (unless otherwise noted)

PARAMETER	TEST CONDITIONS†		TL497AM, TL497AI			TL497AC			UNIT
			MIN	TYP‡	MAX	MIN	TYP‡	MAX	
High-level inhibit input voltage		25°C	2.5			2.5			V
Low-level inhibit input voltage		25°C			0.8			0.8	V
High-level inhibit input current	$V_{I(I)}$ = 5 V	Full range		0.8	1.5		0.8	1.5	mA
Low-level inhibit input current	$V_{I(I)}$ = 0 V	Full range		5	20		5	10	μA
Comparator reference voltage	V_I = 4.5 V to 6 V	Full range	1.14	1.20	1.26	1.08	1.20	1.32	V
Comparator input bias current	V_I = 6 V	Full range		40	100		40	100	μA
Switch on-state voltage	V_I = 4.5 V	I_O = 100 mA	25°C		0.13	0.2		0.13	0.2	V
		I_O = 500 mA	Full range			1			0.85	
Switch off-state current	V_I = 4.5 V, V_O = 30 V	25°C		10	50		10	50	μA
		Full range			500			200	
Current-limit sense voltage	V_I = 6 V	25°C	0.45		1	0.45		1	V
Diode forward voltage	I_O = 10 mA	Full range		0.75	0.95		0.75	0.85	V
	I_O = 100 mA	Full range		0.9	1.1		0.9	1	
	I_O = 500 mA	Full range		1.33	1.75		1.33	1.55	
Diode reverse voltage	I_O = 500 μA	Full range	30						V
	I_O = 200 μA	Full range				30			
On-state supply current		25°C		11	14		11	14	mA
		Full range			16			15	
Off-state supply current		25°C		6	9		6	9	mA
		Full range			11			10	

† Full range for TL497AM is −55°C to 125°C, for TL497AI is −25°C to 85°C, and for TL497AC is 0°C to 70°C.
‡ All typical values are at T_A = 25°C.

Voltage Regulators

6

TYPES TL497AM, TL497AI, TL497AC, SWITCHING VOLTAGE REGULATORS

TYPICAL APPLICATION DATA

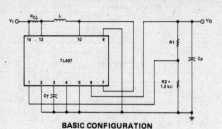

BASIC CONFIGURATION
(I$_{PK}$ < 500 mA)

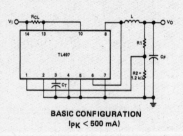

EXTENDED POWER CONFIGURATION
(USING EXTERNAL TRANSISTOR)

DESIGN EQUATIONS

- $I_{PK} = 2 I_O \, max \left[\dfrac{V_O}{V_I} \right]$

- $L \, (\mu H) = \dfrac{V_I}{I_{PK}} \, t_{on}(\mu s)$

Choose L (50 to 500 μH), calculate t_{on} (25 to 150 μs)

- $C_T (pF) \approx 12 \, t_{on}(\mu s)$

- $R1 = (V_O - 1.2) \, k\Omega$

- $R_{CL} = \dfrac{0.5 \, V}{I_{PK}}$

- $C_F \, (\mu F) \approx t_{on}(\mu s) \dfrac{\left[\dfrac{V_I}{V_O} \, I_{PK} + I_O \right]}{V_{ripple \, (PK)}}$

FIGURE 1—POSITIVE REGULATOR, STEP-UP CONFIGURATIONS

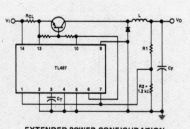

BASIC CONFIGURATION
I$_{PK}$ < 500 mA)

EXTENDED POWER CONFIGURATION
(USING EXTERNAL TRANSISTOR)

DESIGN EQUATIONS

- $I_{PK} = 2 I_O \, max$

- $L \, (\mu H) = \dfrac{V_I - V_O}{I_{PK}} \, t_{on}(\mu s)$

Choose L (50 to 500 μH), calculate t_{on} (10 to 150 μs)

- $C_T (pF) \approx 12 \, t_{on}(\mu s)$

- $R1 = (V_O - 1.2) \, k\Omega$

- $R_{CL} = \dfrac{0.5 \, V}{I_{PK}}$

- $C_F(\mu F) \approx t_{on}(\mu s) \dfrac{\left[\dfrac{V_I}{V_O} \, I_{PK} + I_O \right]}{V_{ripple \, (PK)}}$

FIGURE 2—POSITIVE REGULATOR, STEP-DOWN CONFIGURATIONS

TEXAS INSTRUMENTS
POST OFFICE BOX 225012 • DALLAS, TEXAS 75265

TYPICAL APPLICATION DATA

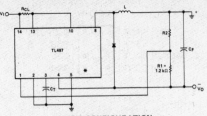

BASIC CONFIGURATION

($I_{PK} < 500$ mA)

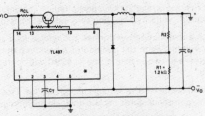

EXTENDED POWER CONFIGURATION

(USING EXTERNAL TRANSISTOR)

FIGURE 3—INVERTING APPLICATIONS

- $I_{PK} = 2 I_O \text{ max} \left[1 + \dfrac{|V_O|}{V_I}\right]$

- $L\,(\mu H) = \dfrac{V_I}{I_{PK}}\, t_{on}(\mu s)$

Choose L (50 to 500 μH), calculate t_{on} (25 to 150 μs)

- $C_T(pF) \approx 12\, t_{on}(\mu s)$

- $R2 = (V_O - 1.2)\ k\Omega$

- $R_{CL} = \dfrac{0.5\ V}{I_{PK}}$

- $C_F\,(\mu F) \approx t_{on}(\mu s)\ \dfrac{\left[\dfrac{V_I}{V_O}\, I_{PK} + I_O\right]}{V_{ripple\,(PK)}}$

*Use external catch-diode, e.g., 1N4001, when building an inverting supply with the TL497A.

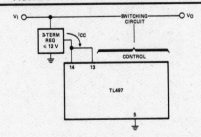

EXTENDED INPUT CONFIGURATION WITHOUT CURRENT LIMIT

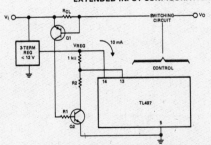

DESIGN EQUATIONS

$R_{CL} = \dfrac{V_{BE(Q1)}}{I_{limit\,(PK)}}$

$R1 = \dfrac{V_I}{I_{B(Q2)}}$

$R2 = (V_{reg} - 1)\ 10\ k\Omega$

CURRENT LIMIT FOR EXTENDED INPUT CONFIGURATION

FIGURE 4—EXTENDED INPUT VOLTAGE RANGE ($V_I > 15$ V)

TEXAS INSTRUMENTS

POST OFFICE BOX 225012 • DALLAS, TEXAS 75265

Voltage Regulators

6

**LINEAR
INTEGRATED
CIRCUITS**

**TYPE TL499C
WIDE-RANGE POWER SUPPLY CONTROLLER**

D2762, JANUARY 1984

- ● Internal Step-Up Switching Regulator
- ● Adjustable Output Voltage
- ● 1.1-Volt to 25-Volt Input Switching Operation
- ● Thermal Protection During Switching Operation
- ● Externally Controlled Switching Current
- ● No External Rectifier Required

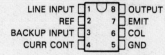

**JG OR P
DUAL-IN-LINE PACKAGE
(TOP VIEW)**

LINE INPUT	1	8	OUTPUT
REF	2	7	EMIT
BACKUP INPUT	3	6	COL
CURR CONT	4	5	GND

description

The TL499C is a monolithic integrated circuit designed to provide regulated supply voltages. The regulated voltage can be set to any value between 2.9 volts and 30 volts by adjusting two external resistors.

When the TL499C is ac coupled to line power through a step-down transformer, it operates as a series dc voltage regulator to maintain the regulated output voltage. With the addition of a backup battery of from 1.1 volts to 25 volts, an inductor, a filter capacitor, and two resistors, the TL499C will operate as a step-up switching regulator during an ac line failure.

The adjustable regulated output voltage makes the TL499C useful for a wide range of applications. Providing backup power during an ac line failure makes the TL499 extremely useful as backup power in microprocessor memory applications.

The TL499C is characterized for operation from −20°C to 85°C.

Copyright © 1984 by Texas Instruments Incorporated

**Texas
Instruments**

POST OFFICE BOX 225012 ● DALLAS, TEXAS 75265

**LINEAR
INTEGRATED
CIRCUITS**

**TYPES TL580C
MICROPOWER DUAL SWITCHING REGULATOR**

D2723, MARCH 1983

- High Efficiency . . . 80% Typ

- Low Bias Current . . . 140 μA

- Two Channels, Each with Output Voltage Adjustment
  Channel A: Output Voltage 2.5 V to 24 V
               Output Current 100 mA
  Channel B: Output Voltage 2.5 V to 24 V
               Output Current 1.8 mA

- Special Multifunctional Operation-Select Pin

**JG OR P
DUAL-IN-LINE PACKAGE
(TOP VIEW)**

INPUT B	1	8 OUTPUT B
C_T	2	7 INPUT A
OUTPUT A	3	6 SYNC B
GND	4	5 V_{CC}

description

The TL580 is a monolithic, micropower, dual-switching regulator designed for use in battery applications. The output voltage of each channel is adjustable. Floating the special pin, SYNC B, causes Channel B to be synchronized to the oscillator in the same manner as Channel A. Shorting SYNC B to ground blocks the oscillator from Channel B, then Channel B becomes a single-input comparator for low-battery indicator detection.

Both Channel A and Channel B are referenced to a band-gap generator. An external capacitor on the C_T input (Pin 2) sets the oscillator frequency between 100 hertz and 160 kilohertz.

The TL580C can attain up to 80-percent efficiency while operating over a supply voltage range of 2.4 volts to 30 volts at an ultralow bias current of 140 microamperes.

The TL580C is characterized for operation from 0°C to 70°C.

functional block diagram (positive logic)

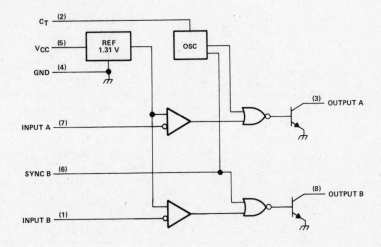

Voltage Regulators

6

Copyright © 1983 by Texas Instruments Incorporated

**TEXAS
INSTRUMENTS**
POST OFFICE BOX 225012 • DALLAS, TEXAS 75265

**LINEAR
INTEGRATED
CIRCUITS**

**TYPES TL593, TL594, TL595
PULSE-WIDTH-MODULATION CONTROL CIRCUITS**

D2712, APRIL 1983—REVISED DECEMBER 1983

- ● Complete PWM Power Control Circuitry
- ● Uncommitted Outputs for 200-mA Sink or Source Current
- ● Output Control Selects Single-Ended or Push-Pull Operation
- ● Internal Circuitry Prohibits Double Pulse at Either Output
- ● Variable Dead-Time Provides Control Over Total Range
- ● Internal Regulator Provides a Stable 5-V Reference Supply Trimmed to 1%
- ● Circuit Architecture Allows Easy Synchronization
- ● Under-Voltage Lockout for Low V_{CC} Conditions
- ● TL593 has Output Current-Limit Sensing
- ● TL595 has On-Chip 39-V Zener and External Control of Output Steering
- ● Improved Direct Replacements for TL493, TL494, and TL495

description

The TL593, TL594, and TL595 devices, each incorporate on a single monolithic chip all the functions required in the construction of a pulse-width-modulation control circuit. Designed primarily for power supply control, these devices offer the systems engineer the flexibility to tailor the power supply control circuitry to his application. The TL593, TL594, and TL595 are improved direct replacements for the TL493, TL494, and TL495.

The TL593 contains an error amplifier, current-limiting amplifier, an on-chip adjustable oscillator, a dead-time control comparator, pulse-steering control flip-flop, 5-volt regulator with a precision of 1%, an under-voltage lockout control circuit, and output control circuitry.

The error amplifier exhibits a common-mode voltage range from −0.3 volts to V_{CC} −2 volts. The current-limit amplifier exhibits a common-mode voltage range from −0.3 volts to V_{CC} −6 volts with an offset voltage of approximately 80 millivolts in series with the inverting input to ease circuit design requirements. The dead-time control comparator has a fixed offset that provides approximately 5% dead time when externally altered. The on-chip oscillator may be bypassed by terminating R_T (pin 6) to the reference output and providing a sawtooth input to C_T (pin 5), or it may be used to drive the common circuitry in synchronous multiple-rail power supplies.

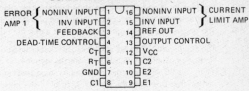

TL593M . . . J
TL593C . . . N
DUAL-IN-LINE PACKAGE (TOP VIEW)

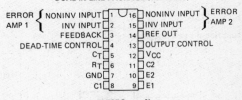

TL594M . . . J
TL594I, TL594C . . . J OR N
DUAL-IN-LINE PACKAGE (TOP VIEW)

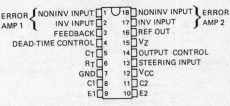

TL595C . . . N
DUAL-IN-LINE PACKAGE (TOP VIEW)

Voltage Regulators

6

DEVICE TYPES, SUFFIX VERSIONS, AND PACKAGES

	TL593	TL594	TL595
TL59—M	J	J	*
TL59—I	*	J,N	*
TL59—C	N	J,N	N

*These combinations are not defined by this data sheet.

FUNCTION TABLE

INPUTS		OUTPUT FUNCTION
OUTPUT CONTROL	STEERING INPUT (TL595 only)	
$V_I < 0.4$ V	Open	Single ended or parallel output
$V_I > 2.4$ V	Open	Normal push-pull operation
$V_I > 2.4$ V	$V_I < 0.4$ V	PWM Output at Q1
$V_I > 2.4$ V	$V_I > 2.4$ V	PWM Output at Q2

1283

TEXAS INSTRUMENTS
POST OFFICE BOX 225012 ● DALLAS, TEXAS 75265

TYPES TL593, TL594, TL595
PULSE-WIDTH-MODULATION CONTROL CIRCUITS

description (continued)

The uncommitted output transistors provide either common-emitter or emitter-follower output capability. Each device provides for push-pull or single-ended output operation with selection by means of the output-control function. The architecture of these devices prohibits the possibility of either output being pulsed twice during push-pull operation. The under-voltage lockout control circuit locks the outputs off until the internal circuitry is operational.

The TL593 and TL594 are similar except that an additional error amplifier is included in the TL594 instead of a current-limiting amplifier. The TL595 provides the identical functions found in the TL594. In addition, the TL595 also contains an on-chip 39-volt zener diode for high-voltage applications where V_{CC} is greater than 40 volts, and an output steering control that overrides the internal control of the pulse-steering flip-flop.

The TL593M and TL594M are characterized for operation over the full military temperature range from $-55\,^{\circ}$C to $125\,^{\circ}$C. The TL594I is characterized for operation from $-25\,^{\circ}$C to $85\,^{\circ}$C. The TL593C, TL594C, and TL595C are characterized for operation from $0\,^{\circ}$C to $70\,^{\circ}$C.

functional block diagram

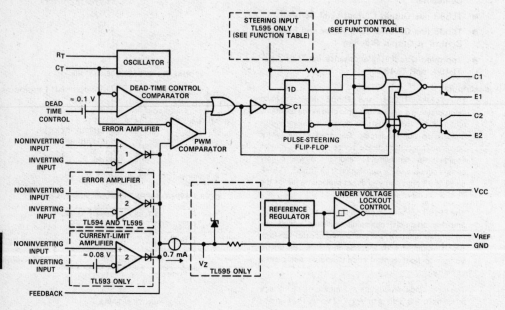

TEXAS INSTRUMENTS
POST OFFICE BOX 225012 • DALLAS, TEXAS 75265

Voltage Regulators

6

absolute maximum ratings over operating free-air temperature range (unless otherwise noted)

	TL593M TL594M	TL594I	TL593C TL594C TL595C	UNIT
Supply voltage, V_{CC} (see Note 1)	41	41	41	V
Amplifier input voltages	$V_{CC}+0.3$	$V_{CC}+0.3$	$V_{CC}+0.3$	V
Collector output voltage	41	41	41	V
Collector output current	250	250	250	mA
Continuous total dissipation at (or below) 25 °C free-air temperature (see Note 2)	1000	1000	1000	mW
Operating free-air temperature range	−55 to 125	−25 to 85	0 to 70	°C
Storage temperature range	−65 to 150	−65 to 150	−65 to 150	°C
Lead temperature 1,6 mm (1/16 inch) from case for 60 seconds: J package	300	300	300	°C
Lead temperature 1,6 mm (1/16 inch) from case for 10 seconds: N package		260	260	°C

NOTES: 1. All voltage values, except differential voltages, are with respect to the network ground terminal.
2. For operation above 25 °C free-air temperature, refer to Dissipation Derating Table. In the J package, the TL593M and TL594M chips are alloy mounted; TL594I and TL594C chips are glass mounted.

DISSIPATION DERATING TABLE

PACKAGE	POWER RATING	DERATING FACTOR	ABOVE T_A
J (Alloy-Mounted Chip)	1000 mW	11.0 mW/°C	59 °C
J (Glass-Mounted Chip)	1000 mW	8.2 mW/°C	28 °C
N	1000 mW	9.2 mW	41 °C

recommended operating conditions

	TL593M TL594M		TL594I		TL593C TL594C TL595C		UNIT
	MIN	MAX	MIN	MAX	MIN	MAX	
Supply voltage, V_{CC}	7	40	7	40	7	40	V
Amplifier input voltages, V_I	−0.3	$V_{CC}-2$	−0.3	$V_{CC}-2$	−0.3	$V_{CC}-2$	V
Collector output voltage, V_O		40		40		40	V
Collector output current (each transistor)		200		200		200	mA
Current into feedback terminal		0.3		0.3		0.3	mA
Timing capacitor, C_T	0.47	10 000	0.47	10 000	0.47	10 000	nF
Timing resistor, R_T	1.8	500	1.8	500	1.8	500	kΩ
Oscillator frequency	1	300	1	300	1	300	kHz
Operating free-air temperature, T_A	−55	125	−25	85	0	70	°C

Voltage Regulators

6

TEXAS
INSTRUMENTS
POST OFFICE BOX 225012 • DALLAS, TEXAS 75265

Voltage Regulators

electrical characteristics over recommended operating free-air temperature range, V_{CC} = 15 V, f = 10 kHz (unless otherwise noted)

reference section

PARAMETER	TEST CONDITIONS[†]	TL593M TL594M			TL593C TL594I, TL594C TL595C			UNIT
		MIN	TYP[‡]	MAX	MIN	TYP[‡]	MAX	
Output voltage (V_{ref})	I_O = 1 mA, T_A = 25 °C	4.95	5	5.05	4.95	5	5.05	V
Input regulation	V_{CC} = 7 V to 40 V, T_A = 25 °C		2	25		2	25	mV
Output regulation	I_O = 1 to 10 mA, T_A = 25 °C		14	35		14	35	mV
Output voltage change with temperature	ΔT_A = MIN to MAX		0.2	1		0.2	1	%
Short-circuit output current[§]	V_{ref} = 0	10	35	60	10	35	50	mA

oscillator section (see Figure 2)

PARAMETER	TEST CONDITIONS[†]	TL593M TL594M			TL593C TL594I, TL594C TL595C			UNIT
		MIN	TYP[‡]	MAX	MIN	TYP[‡]	MAX	
Frequency			10			10		kHz
Standard deviation of frequency[¶]	All values of V_{CC}, C_T, R_T, T_A constant		10			10		%
Frequency change with voltage	V_{CC} = 7 V to 40 V, T_A = 25 °C		0.1			0.1		%
Frequency change with temperature	ΔT_A = MIN to MAX			12			12	%

amplifier sections (see Figure 1)

PARAMETER		TEST CONDITIONS	MIN	TYP[‡]	MAX	UNIT
Input offset voltage	Error	Feedback pin at 2.5 V		2	10	mV
	current-limit (TL593 only)			80		
Input offset current		Feedback control at 2.5 V		25	250	nA
Input bias current		Feedback control at 2.5 V		0.2	1	µA
Common-mode input voltage range	Error	V_{CC} = 7 V to 40 V		−0.3 to $V_{CC}-2$		V
	Current-limit (TL593 only)			−0.3 to $V_{CC}-6$		
Open-loop voltage amplification	Error	ΔV_O = 3 V, V_O = 0.5 V to 3.5 V	70	95		dB
	Current-limit (TL593 only)			90		
Unity-gain bandwidth				800		kHz
Common-mode rejection ratio	Error	V_{CC} = 40 V, T_A = 25 °C	65	80		dB
	Current-limit (TL593 only)			70		
Output sink current (pin 3)		V_{ID} = −15 mV to −5 mV, Feedback control at 0.5 V	0.3	0.7		mA
Output source current (pin 3)		V_{ID} = 15 mV to 5 mV, Feedback at 3.5 V	−2			mA

[†]For conditions shown as MIN or MAX, use the appropriate value specified under recommended operating conditions.
[‡]All typical values except for parameter changes with temperature are at T_A = 25 °C.
[§]Duration of the short-circuit should not exceed one second.
[¶]Standard deviation is a measure of the statistical distribution about the mean as derived from the formula

$$\sigma = \sqrt{\dfrac{\sum\limits_{n=1}^{N}(x_n - \bar{x})^2}{N-1}}$$

TEXAS
INSTRUMENTS
POST OFFICE BOX 225012 • DALLAS, TEXAS 75265

electrical characteristics over recommended operating free-air temperature range, V_{CC} = 15 V, f = 10 kHz (unless otherwise noted)

dead-time control section (see Figure 2)

PARAMETER	TEST CONDITIONS	MIN	TYP‡	MAX	UNIT
Input bias current (pin 4)	V_I = 0 to 5.25 V		−2	−10	µA
Maximum duty cycle, each output	Dead-time control at 0 V	45			%
Input threshold voltage (pin 4)	Zero duty cycle		3	3.3	V
	Maximum duty cycle	0			

output section

PARAMETER	TEST CONDITIONS	TL593M TL594M			TL593C TL594I, TL594C TL595C			UNIT
		MIN	TYP‡	MAX	MIN	TYP‡	MAX	
Collector off-state current	V_{CE} = 40 V, V_{CC} = 40 V		2	100		2	100	µA
	V_C = 15 V, V_E = 0 V, V_{CC} = 1 to 3 V, Dead-time and output control pins at 0 V		4	200		4	200	
Emitter off-state current	V_{CC} = V_C = 40 V, V_E = 0			−150			−100	µA
Collector-emitter saturation voltage — Common-emitter	V_E = 0, I_C = 200 mA	1.1	1.5		1.1	1.3		V
Collector-emitter saturation voltage — Emitter-follower	V_C = 15 V, I_E = −200 mA	1.5	2.5		1.5	2.5		
Output control input current	V_I = V_{ref}		3.5			3.5		mA

pwm comparator section (see Figure 2)

PARAMETER	TEST CONDITIONS	MIN	TYP‡	MAX	UNIT
Input threshold voltage (pin 3)	Zero duty cycle		4	4.5	V
Input sink current (pin 3)	$V_{(pin\ 3)}$ = 0.5 V	0.3	0.7		mA

under-voltage lockout section (see Figure 2)

PARAMETER	TEST CONDITIONS†	TL593M TL594M			TL593C TL594I, TL594C TL595C			UNIT
		MIN	TYP	MAX	MIN	TYP	MAX	
Threshold voltage	T_A = 25 °C			6			6	V
	ΔT_A = MIN to MAX	3		6.9	3.5		6.9	
Hysteresis|		30			100			mV

total device (see Figure 2)

PARAMETER	TEST CONDITIONS		MIN	TYP‡	MAX	UNIT
Standby supply current	Pin 6 at V_{ref}, All other inputs and outputs open	V_{CC} = 15 V		9	15	mA
		V_{CC} = 40 V		11	18	
Average supply current	Dead-time Control at 2 V, See Figure 2			12.4		mA

†For conditions shown as MIN or MAX, use the appropriate value specified under recommended operating conditions.
‡All typical values except for parameter changes with temperature are at T_A = 25 °C.
|Hysteresis is the difference between the positive-going input threshold voltage and the negative-going input threshold voltage.

TEXAS INSTRUMENTS
POST OFFICE BOX 225012 ● DALLAS, TEXAS 75265

switching characteristics, T_A = 25°C

PARAMETER	TEST CONDITIONS	MIN	TYP‡	MAX	UNIT
Output voltage rise time	Common-emitter configuration,		100	200	ns
Output voltage fall time	See Figure 3		30	100	
Output voltage rise time	Emitter-follower configuration,		200	400	ns
Output voltage fall time	See Figure 4		45	100	

‡All typical values are at T_A = 25°C.

PARAMETER MEASUREMENT INFORMATION

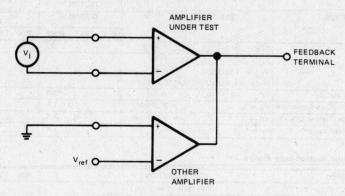

FIGURE 1 — AMPLIFIER CHARACTERISTICS

TEXAS
INSTRUMENTS
POST OFFICE BOX 225012 • DALLAS, TEXAS 75265

Voltage Regulators

6

PARAMETER MEASUREMENT INFORMATION

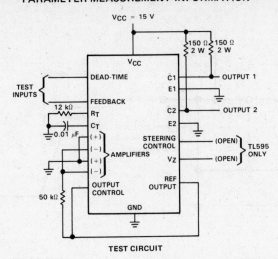

TEST CIRCUIT

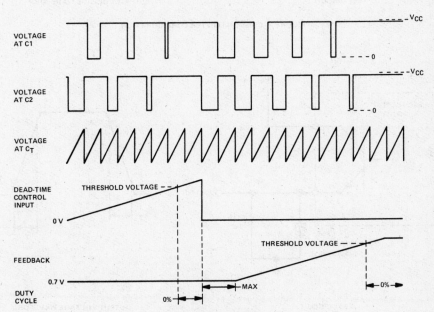

VOLTAGE WAVEFORMS

FIGURE 2—OPERATIONAL TEST CIRCUIT AND WAVEFORMS

Voltage Regulators

6

TEXAS
INSTRUMENTS
POST OFFICE BOX 225012 • DALLAS, TEXAS 75265

PARAMETER MEASUREMENT INFORMATION

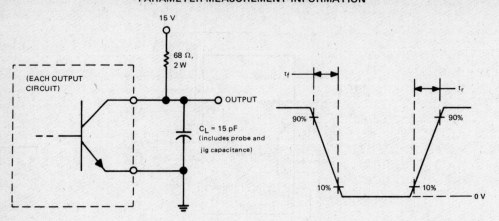

TEST CIRCUIT

OUTPUT VOLTAGE WAVEFORM

FIGURE 3—COMMON-EMITTER CONFIGURATION

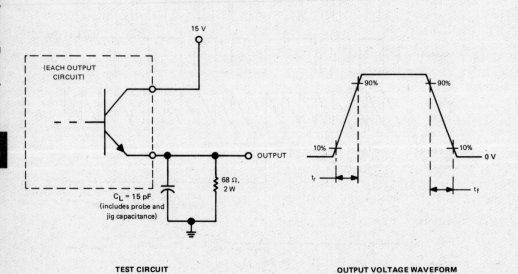

TEST CIRCUIT

OUTPUT VOLTAGE WAVEFORM

FIGURE 4—EMITTER-FOLLOWER CONFIGURATION

Voltage Regulators

6

TEXAS
INSTRUMENTS
POST OFFICE BOX 225012 • DALLAS, TEXAS 75265

483

TYPICAL CHARACTERISTICS

OSCILLATOR FREQUENCY and
FREQUENCY VARIATION† vs
TIMING RESISTANCE

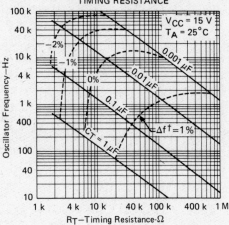

FIGURE 5

AMPLIFIER VOLTAGE AMPLIFICATION
vs
FREQUENCY

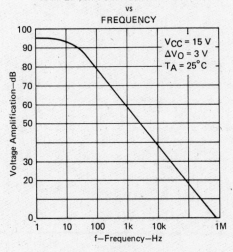

FIGURE 6

†Frequency variation (Δf) is the change in oscillator frequency that occurs over the full temperature range.

Voltage Regulators

6

**LINEAR
INTEGRATED
CIRCUITS**

**SERIES TL780
POSITIVE VOLTAGE REGULATORS**

D2643, APRIL 1981

- ±1% Output tolerance at 25°C
- ±2% Output Tolerance Over Full Operating Range
- Thermal Shutdown
- Internal Short-Circuit Current Limiting
- Pinout Identical to uA7800 Series
- Improved Version of uA7800 Series

NOMINAL OUTPUT VOLTAGE	REGULATOR
5 V	TL780-05C
12 V	TL780-12C
15 V	TL780-15C

KC PACKAGE

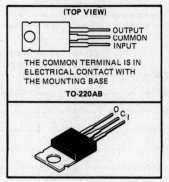

(TOP VIEW)

OUTPUT
COMMON
INPUT

THE COMMON TERMINAL IS IN
ELECTRICAL CONTACT WITH
THE MOUNTING BASE

TO-220AB

description

Each fixed-voltage precision regulator in this series is capable of supplying 1.5 amperes of load current. A unique temperature-compensation technique coupled with an internally trimmed bandgap reference has resulted in improved accuracy when compared to other three-terminal regulators. Advanced layout techniques provide excellent line, load, and thermal regulation. The internal current limiting and thermal shutdown features make the devices essentially immune to overload.

schematic

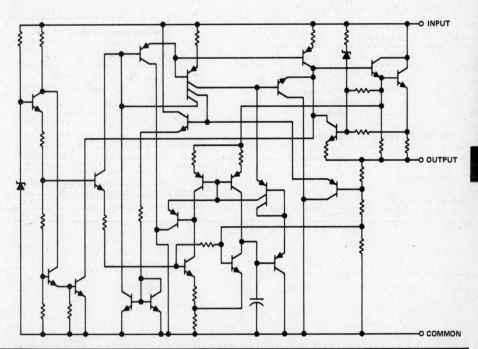

INPUT

OUTPUT

COMMON

Voltage Regulators

6

**TEXAS
INSTRUMENTS**

POST OFFICE BOX 225012 • DALLAS, TEXAS 75265

absolute maximum ratings over operating temperature range (unless otherwise noted)

Input voltage .... 35 V
Continuous total dissipation at 25°C free-air temperature (see Note 1) .... 2 W
Continuous total dissipation at (or below) 25°C case temperature (see Note 1) .... 15 W
Operating free-air, case, or virtual junction temperature range .... 0 to 150°C
Storage temperature range .... −65 to 150°C
Lead temperature 1.6 mm (1/16 inch) from case for 10 seconds .... 260°C

NOTE 1: For operation above 25 °C free-air or case temperature, refer to Figures 1 and 2. To avoid exceeding the design maximum virtual junction temperature, these ratings should not be exceeded. Due to variations in individual device electrical characteristics and thermal resistance, the built-in thermal overload protection may be activated at power levels slightly above or below the rated dissipation.

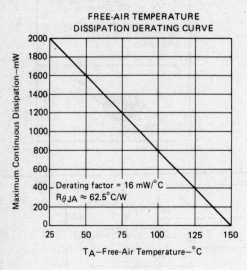

FREE-AIR TEMPERATURE
DISSIPATION DERATING CURVE

Derating factor = 16 mW/°C
$R_{\theta JA} \approx 62.5°C/W$

FIGURE 1

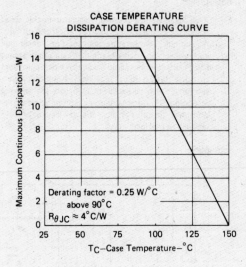

CASE TEMPERATURE
DISSIPATION DERATING CURVE

Derating factor = 0.25 W/°C
above 90°C
$R_{\theta JC} \approx 4°C/W$

FIGURE 2

recommended operating conditions

		MIN	MAX	UNIT
Input voltage, V_I	TL780-05C	7	25	V
	TL780-12C	14.5	30	
	TL780-15C	17.5	30	
Output current, I_O			1.5	A
Operating virtual junction temperature, T_J		0	125	°C

Voltage Regulators

6

TEXAS
INSTRUMENTS
POST OFFICE BOX 225012 • DALLAS, TEXAS 75265

883

TL780-05C electrical characteristics at specified virtual junction temperature,
V_I = 10 V, I_O = 500 mA (unless otherwise noted)

PARAMETER	TEST CONDITIONS†		MIN	TYP	MAX	UNIT
Output voltage	I_O = 5 mA to 1 A, P ≤ 15 W V_I = 7 V to 20 V	25°C	4.95	5	5.05	V
		0°C to 125°C	4.9		5.1	
Input regulation	V_I = 7 V to 25 V	25°C		0.5	5	mV
	V_I = 8 V to 12 V			0.5	5	
Ripple rejection	V_I = 8 V to 18 V, f = 120 Hz	0°C to 125°C	70	85		dB
Output regulation	I_O = 5 mA to 1.5 A	25°C		4	25	mV
	I_O = 250 mA to 750 mA			1.5	15	
Output resistance	f = 1 kHz	0°C to 125°C		0.0035		Ω
Temperature coefficient of output voltage	I_O = 5 mA	0°C to 125°C		0.25		mV/°C
Output noise voltage	f = 10 Hz to 100 kHz	25°C		75		µV
Dropout voltage	I_O = 1 A	25°C		2		V
Bias current		25°C		5	8	mA
Bias current change	V_I = 7 V to 25 V	0°C to 125°C		0.7	1.3	mA
	I_O = 5 mA to 1 A			0.03	0.5	
Short-circuit output current	V_I = 35 V	25°C		750		mA
Peak output current		25°C		2.2		A

TL780-12C electrical characteristics at specified virtual junction temperature,
V_I = 19 V, I_O = 500 mA (unless otherwise noted)

PARAMETER	TEST CONDITIONS†		MIN	TYP	MAX	UNIT
Output voltage	I_O = 5 mA to 1 A, P ≤ 15 W V_I = 14.5 V to 27 V	25°C	11.88	12	12.12	V
		0°C to 125°C	11.76		12.24	
Input regulation	V_I = 14.5 V to 30 V	25°C		1.2	12	mV
	V_I = 16 V to 22 V			1.2	12	
Ripple rejection	V_I = 15 V to 25 V, f = 120 Hz	0°C to 125°C	65	80		dB
Output regulation	I_O = 5 mA to 1.5 A	25°C		6.5	60	mV
	I_O = 250 mA to 750 mA			2.5	36	
Output resistance	f = 1 kHz	0°C to 125°C		0.0035		Ω
Temperature coefficient of output voltage	I_O = 5 mA	0°C to 125°C		0.6		mV/°C
Output noise voltage	f = 10 Hz to 100 kHz	25°C		180		µV
Dropout voltage	I_O = 1 A	25°C		2		V
Bias current		25°C		5.5	8	mA
Bias current change	V_I = 14.5 V to 30 V	0°C to 125°C		0.4	1.3	mA
	I_O = 5 mA to 1 A			0.03	0.5	
Short-circuit output current	V_I = 35 V	25°C		350		mA
Peak output current		25°C		2.2		A

†All characteristics are measured with a capacitor across the input of 0.33 µF and a capacitor across the output of 0.22 µF. All characteristics except noise voltage and ripple rejection ratio are measured using pulse techniques (t_w ≤ 10 ms, duty cycles ≤ 5%). Output voltage changes due to changes in internal temperature must be taken into account separately.

Voltage Regulators

6

TL780-15C electrical characteristics at specified virtual junction temperature,
V_I = 23 V, I_O = 500 mA (unless otherwise noted)

PARAMETER	TEST CONDITIONS[†]		MIN	TYP	MAX	UNIT
Output voltage	I_O = 5 mA to 1 A, P ≤ 15 W V_I = 17.5 V to 30 V	25°C	14.85	15	15.15	V
		0°C to 125°C	14.7		15.3	
Input regulation	V_I = 17.5 V to 30 V	25°C		1.5	15	mV
	V_I = 20 V to 26 V			1.5	15	
Ripple rejection	V_I = 18.5 V to 28.5 V f = 120 Hz	0°C to 125°C	60	75		dB
Output regulation	I_O = 5 mA to 1.5 A	25°C		7	75	mV
	I_O = 250 mA to 750 mA			2.5	45	
Output resistance	f = 1 kHz	0°C to 125°C		0.0035		Ω
Temperature coefficient of output voltage	I_O = 5 mA	0°C to 125°C		0.62		mV/°C
Output noise voltage	f = 10 Hz to 100 kHz	25°C		225		µV
Dropout voltage	I_O = 1 A	25°C		2		V
Bias current		25°C		5.5	8	mA
Bias current change	V_I = 17.5 V to 30 V	0°C to 125°C		0.4	1.3	mA
	I_O = 5 mA to 1 A			0.02	0.5	
Short-circuit output current	V_I = 35 V	25°C		230		mA
Peak output current		25°C		2.2		A

[†]All characteristics are measured with a capacitor across the input of 0.33 µF and a capacitor across the output of 0.22 µF. All characteristics except noise voltage and ripple rejection ratio are measured using pulse techniques (t_W ≤ 10 ms, duty cycle ≤ 5%). Output voltage changes due to changes in internal temperature must be taken into account separately.

TYPICAL APPLICATION DATA

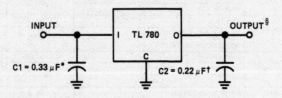

*C1 required if regulator is far from power supply filter.
[†]C2 not required for stability, however transient response is improved
[§]Permanent damage can occur if output is pulled below ground.

TEXAS
INSTRUMENTS
POST OFFICE BOX 225012 • DALLAS, TEXAS 75265

**LINEAR
INTEGRATED
CIRCUITS**

**TYPE TL783C
HIGH-VOLTAGE ADJUSTABLE REGULATOR**

D2659, SEPTEMBER 1981—REVISED JANUARY 1983

- Output Adjustable From 1.25 V To 125-Volt

- 700 mA Output Current

- Full Short-Circuit, Safe-Operating-Area, and Thermal Shutdown Protection

- 0.001 %/V Typical Input Regulation

- 0.15% Typical Output Regulation

- 76 dB Typical Ripple Rejection

- Standard TO-220AB Package

KC PACKAGE

(TOP VIEW)

INPUT
OUTPUT
ADJUSTMENT

THE OUTPUT TERMINAL IS IN
ELECTRICAL CONTACT WITH
THE MOUNTING BASE

TO-220AB

description

The TL783 is an adjustable 3-terminal positive-voltage regulator with an output range of 1.25 volts to 125 volts and a DMOS output transistor capable of sourcing more than 700 milliamperes. It is designed for use in high-voltage applications where standard bipolar regulators cannot be used. Excellent performance specifications . . . superior to those of most bipolar regulators . . . are achieved through circuit design and advanced layout techniques.

As a state-of-the-art regulator, the TL783 combines standard bipolar circuitry with high-voltage double-diffused MOS transistors on one chip to yield a device capable of withstanding voltages far higher than standard bipolar integrated circuits. Because of its lack of secondary breakdown and thermal runaway characteristics usually associated with bipolar outputs, the TL783 maintains full overload protection while operating at up to 125 volts from input to output. Other features of the device include current limiting, safe-operating-area (SOA) protection, and thermal shutdown. Even if the adjustment pin is inadvertently disconnected, the protection circuitry remains functional.

Only two external resistors are required to program the output voltage. An input bypass capacitor is necessary only when the regulator is situated far from the input filter. An output capacitor, although not required, will improve transient response and protection from instantaneous output short-circuits. Excellent ripple rejection can be achieved without a bypass capacitor at the adjustment terminal.

functional block diagram

$$V_O \approx V_{REF} \left(1 + \frac{R_2}{R_1} \right)$$

Voltage Regulators

6

**TEXAS
INSTRUMENTS**

POST OFFICE BOX 225012 • DALLAS, TEXAS 75265

TYPE TL783C
HIGH-VOLTAGE ADJUSTABLE REGULATOR

absolute maximum ratings over operating temperature range (unless otherwise noted)

Input-to-output differential voltage, $V_I - V_O$	125 V
Continuous total dissipation at (or below) 25°C free-air temperature (see Note 1)	2 W
Continuous total dissipation at (or below) 25°C case temperature (see Note 1)	20 W
Operating free-air, case, or virtual junction temperature range	0°C to 150°C
Lead temperature 1/16 inch (1,6 mm) from case for 10 seconds	260°C

NOTE 1: For operation above 25 °C free-air or case temperature, refer to Figures 1 and 2. To avoid exceeding the design maximum virtual junction temperature, these ratings should not be exceeded. Due to variations in individual dvice electrical characteristics and thermal resistance, the built-in thermal overload protection may be activated at power levels slightly above or below the rated dissipation.

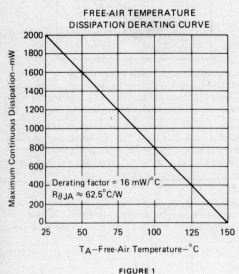

FREE-AIR TEMPERATURE
DISSIPATION DERATING CURVE

Derating factor = 16 mW/°C
$R_{\theta JA} \approx 62.5°C/W$

T_A — Free-Air Temperature—°C

FIGURE 1

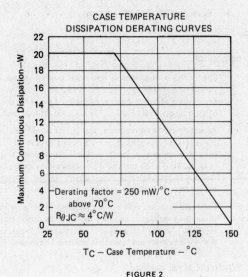

CASE TEMPERATURE
DISSIPATION DERATING CURVES

Derating factor = 250 mW/°C
above 70°C
$R_{\theta JC} \approx 4°C/W$

T_C — Case Temperature — °C

FIGURE 2

recommended operating conditions

	MIN	MAX	UNIT
Input-to-output voltage differential, $V_I - V_O$		125	V
Output current, I_O	15	700	mA
Operating virtual junction temperature, T_J	0	125	°C

TEXAS INSTRUMENTS
POST OFFICE BOX 225012 ● DALLAS, TEXAS 75265

electrical characteristics at $V_I - V_O = 25$ V, $I_O = 0.5$ A, $T_J = 0°C$ to $125°C$ (unless otherwise noted)

PARAMETER	TEST CONDITIONS[†]		MIN	TYP	MAX	UNIT
Input regulation[‡]	$V_I - V_O = 20$ V to 125 V	$T_J = 25°C$		0.001	0.01	%/V
		$T_J = 0°C$ to $125°C$		0.004	0.02	
Ripple rejection	$\Delta V_{I(p-p)} = 10$ V, $\qquad V_O = 10$ V, $\qquad$ f = 120 Hz		66	76		dB
Output regulation	$I_O = 15$ mA to 700 mA, $\quad T_J = 25°C$	$V_O \leqslant 5$ V		7.5	25	mV
		$V_O \geqslant 5$ V		0.15	0.5	%
	$I_O = 15$ mA to 700 mA	$V_O \leqslant 5$ V		20	70	mV
		$V_O \geqslant 5$ V		0.3	1.5	%
Output voltage change with temperature				0.4		%
Output voltage long-term drift	1000 h at $T_J = 125°C$, $\qquad V_I - V_O = 125$ V, See Note 2			0.2		%
Output noise voltage	f = 10 Hz to 10 kHz, $\qquad T_J = 25°C$			0.003		%
Minimum output current to maintain regulation	$V_I - V_O = 125$ V				15	mA
Peak output current	$V_I - V_O = 25$ V, $\qquad$ t = 1 ms			1100		mA
	$V_I - V_O = 15$ V, $\qquad$ t = 30 ms			715		
	$V_I - V_O = 25$ V, $\qquad$ t = 30 ms		700	900		
	$V_I - V_O = 125$ V, $\qquad$ t = 30 ms		100	250		
Adjustment-terminal current				83	110	µA
Change in adjustment-terminal current	$V_I - V_O = 15$ V to 125 V, $\quad I_O = 15$ mA to 700 mA, $\quad$ P ≤ rated dissipation			0.5	5	µA
Reference voltage (output to ADJ)	$V_I - V_O = 10$ V to 125 V, $\quad I_O = 15$ mA to 700 mA, $\quad$ P ≤ rated dissipation		1.2	1.27	1.3	V

[†] All characteristics except noise voltage and ripple rejection are measured using pulse techniques ($t_w \leqslant 10$ ms, duty cycle ≤ 5%) to limit changes in average internal dissipation. Output voltage changes due to large changes in internal dissipation must be taken into account separately.

[‡] Input regulation is expressed here as the percentage change in output voltage per 1-volt change at the input.

NOTE 2: Since long-term drift cannot be measured on the individual devices prior to shipment, this specification is not intended to be a guarantee or warranty. It is an engineering estimate of the average drift to be expected from lot to lot.

Voltage Regulators

6

TEXAS
INSTRUMENTS
POST OFFICE BOX 225012 • DALLAS, TEXAS 75265

Voltage Regulators

6

TYPICAL CHARACTERISTICS

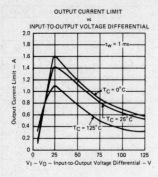

FIGURE 3

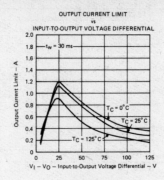

FIGURE 4

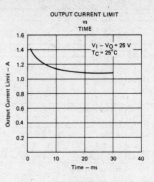

FIGURE 5

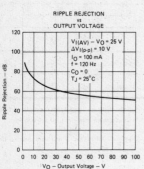

FIGURE 6

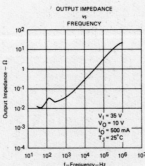

FIGURE 7

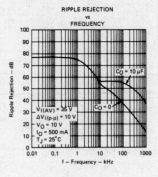

FIGURE 8

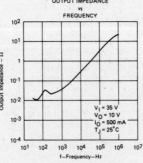

FIGURE 9

Texas
Instruments
POST OFFICE BOX 225012 • DALLAS, TEXAS 75265

TYPICAL CHARACTERISTICS

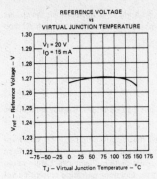

REFERENCE VOLTAGE
vs
VIRTUAL JUNCTION TEMPERATURE

FIGURE 10

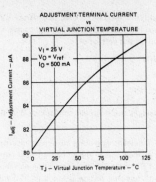

ADJUSTMENT-TERMINAL CURRENT
vs
VIRTUAL JUNCTION TEMPERATURE

FIGURE 11

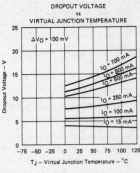

DROPOUT VOLTAGE
vs
VIRTUAL JUNCTION TEMPERATURE

FIGURE 12

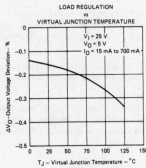

LOAD REGULATION
vs
VIRTUAL JUNCTION TEMPERATURE

FIGURE 13

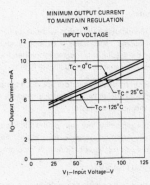

MINIMUM OUTPUT CURRENT
TO MAINTAIN REGULATION
vs
INPUT VOLTAGE

FIGURE 14

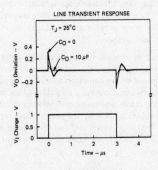

LINE TRANSIENT RESPONSE

FIGURE 15

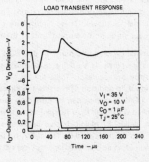

LOAD TRANSIENT RESPONSE

FIGURE 16

Voltage Regulators

6

DESIGN CONSIDERATIONS

The internal reference (see functional block diagram) is used to generate 1.25 volts nominal (V_{ref}) between the output and adjustment terminals. This voltage is developed across R1 and causes a constant current to flow through R1 and the programming resistor R2, giving an output voltage of:

$$V_O = V_{ref} (1 + R2/R1) + I_{adj} (R2)$$

or

$$V_O \approx V_{ref} (1 + R2/R1).$$

The TL783 was designed to minimize I_{adj} and maintain consistency over line and load variations, thereby minimizing the I_{adj} (R2) error term.

To maintain I_{adj} at a low level, all quiescent operating current is returned to the output terminal. This quiescent current must be sunk by the external load and is the minimum load current necessary to prevent the output from rising. The recommended R1 value of 82 ohms will provide a minimum load current of 15 milliamperes. Larger values may be used if the input-to-output differential voltage is less than 125 volts (see minimum operating current curve) or if the load will sink some portion of the minimum current.

Bypass capacitors

The TL783 regulator is stable without bypass capacitors; however, any regulator will become unstable with certain values of output capacitance if an input capacitor is not used. Therefore, the use of input bypassing is recommended whenever the regulator is located more than four inches from the power-supply filter capacitor. A 1-microfarad tantalum or electrolytic capacitor is usually sufficient.

Adjustment-terminal capacitors are not recommended for use on the TL783 because they can seriously degrade load transient response as well as create a need for extra protection circuitry. Excellent ripple rejection is presently achieved without this added capacitor.

Due to the relatively low gain of the MOS output stage, output voltage drop-out may occur under large load transient conditions. Addition of an output bypass capacitor will greatly enhance load transient response as well as prevent drop-out. For most applications it is recommended that an output bypass capacitor be used with a minimum value of:

$$C_O \ (\mu f) = 15/V_O$$

Larger values will provide proportionally better transient response characteristics.

Protection circuitry

The TL783 regulator includes built-in protection circuitry capable of guarding the device against most overload conditions encountered in normal operation. These protective features are current limiting, safe-operating-area protection, and thermal shutdown. These circuits are meant to protect the device under occasional fault conditions only. Continuous operation in the current limit or thermal shutdown mode is not recommended.

The internal protection circuits of the TL783 will protect the device up to maximum rated V_I as long as certain precautions are taken. If V_I is instantaneously switched on, transients exceeding maximum input ratings may occur, which can destroy the regulator. These are usually caused by lead inductance and bypass capacitors causing a ringing voltage on the input. In addition, if rise times in excess of 10 V/ns are applied to the input, a parasitic n-p-n transistor in parallel with the DMOS output can be turned on causing the device to fail. If the device is operated over 50 volts and the input is switched on rather than ramped on, a low-Q capacitor, such as a tantalum or electrolytic should be used rather than ceramic, paper, or plastic bypass capacitors. A dissipation factor of 0.015 or greater will usually provide adequate damping to suppress ringing. Normally, no problems will occur if the input voltage is allowed to ramp upward through the action of an ac line rectifier and filter network.

TEXAS
INSTRUMENTS
POST OFFICE BOX 225012 ● DALLAS, TEXAS 75265

Similarly, if an instantaneous short circuit is applied to the outputs, both ringing and excessive fall times can result. A tantalum or electrolytic bypass capacitor is recommended to eliminate this problem. However, if a large output capacitor is used and the input is shorted, addition of a protection diode may be necessary to prevent capacitor discharge through the regulator. The amount of discharge current delivered is dependent on output voltage, size of capacitor, and fall time of V_I. A protective diode (see Figure 17) is required only for capacitance values greater than

$$C_O \ (\mu f) = 3 \times 10^4/(V_O)^2.$$

Care should always be taken to prevent insertion of regulators into a socket with power on. Power should be turned off before removing or inserting regulators.

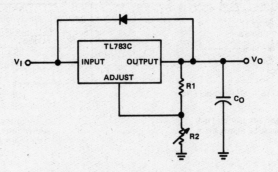

FIGURE 17— REGULATOR WITH PROTECTIVE DIODE

Load regulation

The current set resistor (R1) should be located close to the regulator output terminal rather than near the load. This eliminates long line drops from being amplified through the action of R1 and R2 to degrade load regulation. To provide remote ground sensing, R2 should be near the load ground.

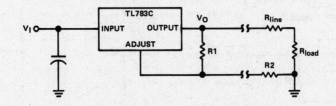

FIGURE 18—REGULATOR WITH CURRENT-SET RESISTOR

Voltage Regulators

6

TEXAS
INSTRUMENTS
POST OFFICE BOX 225012 • DALLAS, TEXAS 75265

TYPE TL783C
HIGH-VOLTAGE ADJUSTABLE REGUALTOR

TYPICAL APPLICATION DATA

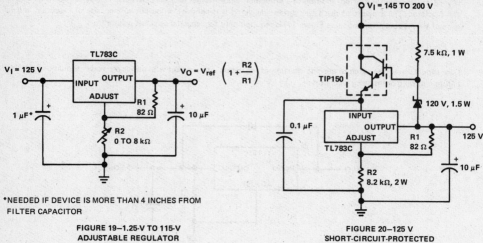

*NEEDED IF DEVICE IS MORE THAN 4 INCHES FROM
 FILTER CAPACITOR

FIGURE 19—1.25-V TO 115-V
ADJUSTABLE REGULATOR

FIGURE 20—125 V
SHORT-CIRCUIT-PROTECTED
OFF-LINE REGULATOR

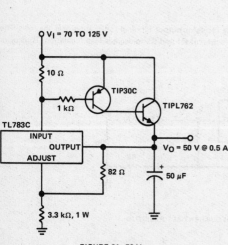

FIGURE 21—50-V
REGULATOR WITH CURRENT BOOST

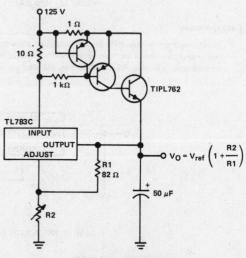

FIGURE 22—ADJUSTABLE
REGULATOR WITH CURRENT BOOST
AND CURRENT LIMIT

Texas
INSTRUMENTS
POST OFFICE BOX 225012 • DALLAS, TEXAS 75265

981

TYPICAL APPLICATION DATA

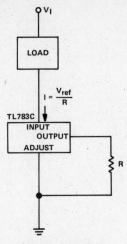

FIGURE 23—CURRENT-
SINKING REGULATOR

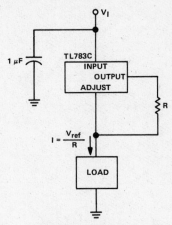

FIGURE 24—CURRENT-
SOURCING REGULATOR

Voltage Regulators

6

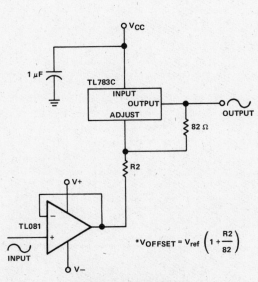

$$*V_{OFFSET} = V_{ref}\left(1 + \frac{R2}{82}\right)$$

FIGURE 25—HIGH-VOLTAGE
UNITY-GAIN OFFSET AMPLIFIER

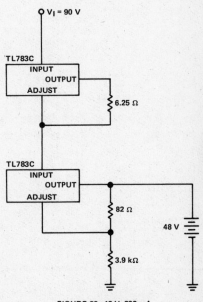

FIGURE 26—48-V, 200-mA
FLOAT CHARGER

TYPE TL1451C
DUAL PULSE-WIDTH-MODULATION CONTROL CIRCUIT

D2730, FEBRUARY 1983

- Complete PWM Power Control Circuitry
- Completely Synchronized Operation
- Internal Under-Voltage Lockout Protection
- Wide Supply Voltage Range
- Internal Short-Circuit Protection
- Oscillator Frequency . . . 500 kHz Max
- Variable Dead Time Provides Control Over Total Range
- Internal Regulator Provides A Stable 2.5-V Reference Supply

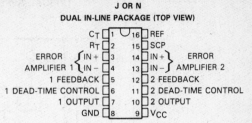

J OR N
DUAL IN-LINE PACKAGE (TOP VIEW)

C_T	1	16	REF
R_T	2	15	SCP
ERROR AMPLIFIER 1 { IN +	3	14	IN + } ERROR AMPLIFIER 2
IN −	4	13	IN −
1 FEEDBACK	5	12	2 FEEDBACK
1 DEAD-TIME CONTROL	6	11	2 DEAD-TIME CONTROL
1 OUTPUT	7	10	2 OUTPUT
GND	8	9	V_{CC}

description

The TL1451 incorporates on a single monolithic chip all the functions required in the construction of two pulse-width-modulation control circuits. Designed primarily for power supply control, the TL1451 contains an on-chip 2.5-volt regulator, two error amplifiers, an adjustable oscillator, two dead-time comparators, under-voltage lockout circuitry, and dual common-emitter output transistor circuits.

The uncommitted output transistors provide common-emitter output capability for each controller. The internal amplifiers exhibit a common-mode voltage range from 0.4 volts to 1.5 volts. The dead-time control comparator has no offset unless externally altered and may be used to provide 0% to 100% dead time. The on-chip oscillator may be operated by terminating R_T (pin 2) and C_T (pin 1). During low V_{CC} conditions, the under-voltage lockout control circuit feature locks the outputs off until the internal circuitry is operational.

The TL1451 is characterized for operation from −20°C to 85°C.

recommended operating conditions

	MIN	NOM	MAX	UNIT
Supply voltage, V_{CC}	3.6		40	V
High-level output voltage, V_{OH}			40	V
High-level output current, I_{OH}			20	mA
Error amplifier common-mode input voltage, V_{IC}	0.4		1.5	V
Input voltage range at dead-time terminal	1.4		2.05	V
Input current at feedback terminal			−50	μA
Timing capacitor, C_T	0.15		15	μF
Timing resistor, R_T	5		50	kΩ
Oscillator frequency, f_{osc}	1		500	kHz
Operating free-air temperature, T_A	−20		85	°C

Voltage Regulators

6

PRODUCT PREVIEW

This document contains information on a product under development. Texas Instruments reserves the right to change or discontinue this product without notice.

TEXAS
INSTRUMENTS

POST OFFICE BOX 225012 • DALLAS, TEXAS 75265

TYPE TL1451C
DUAL PULSE-WIDTH-MODULATION CONTROL CIRCUIT

TYPICAL APPLICATION DATA

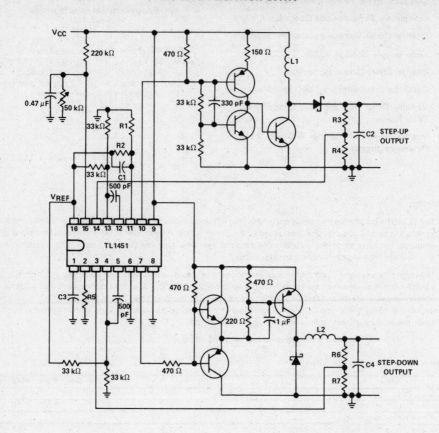

Values for R1 through R7, C1 through C4, and L1 and L2 depend upon individual application.

Voltage Regulators

6

TEXAS
INSTRUMENTS

POST OFFICE BOX 225012 • DALLAS, TEXAS 75265

LINEAR
INTEGRATED
CIRCUITS

TYPES TL1525A, TL1527A, TL2525A, TL2527A, TL3525A, TL3527A
PULSE-WIDTH MODULATION CONTROLLERS

D2724, APRIL 1983

- Complete PWM Power Control Circuitry
- 8-Volt to 35-Volt Operation
- 5.1-Volt Reference Trimmed to ±1%
- Frequency Range . . . 100 Hz to 500 Hz
- Adjustable Deadtime Control
- Under-Voltage Lockout for Low V_{CC} Conditions
- Latched PWM Prevents Multiple Pulses
- Dual Sink or Source Output Drivers
- Improved Direct Replacements for Silicon General SG1525A/SG1527A Series

TL1525A, TL1527A . . . J
TL2525A, TL2527A . . . J OR N
TL3525A, TL3527A . . . J OR N
DUAL-IN-LINE PACKAGE
(TOP VIEW)

INVERTING INPUT	1		16	REFERENCE
NONINVERTING INPUT	2		15	V_{CC} (V_I)
SYNC	3		14	OUTPUT B
OSCILLATOR OUT	4		13	V_C
C_T	5		12	GND
R_T	6		11	OUTPUT A
DISCHARGE	7		10	SHUTDOWN
SOFT-START	8		9	COMPENSATION

output logic

TL1525A, TL2525A, TL3525A . . . OR
TL1527A, TL2572A, TL3527A . . . NOR

description

The TL1525A/TL1527A series of pulse-width modulation integrated circuits are designed to offer improved performance and lower external parts count when used to implement various types of switching power supplies. Each device includes an on-chip 5.1-volt reference, error amplifier, programmable oscillator, pulse-steering flip-flop, a latched comparator under-voltage lockout, shutdown circuitry, and complementary source or sink outputs. The on-chip 5.1-volt reference is trimmed to ±1% initial accuracy, serves as a reference output, and supplies the internal regulator control circuitry. The input common-mode range of the error amplifier includes the reference voltage, which eliminates the need for external divider resistors.

The oscillator operates at a fixed frequency determined by one timing resistor R_T and one timing capacitor C_T. The timing resistor establishes the constant charging current for C_T, resulting in a linear voltage ramp at C_T, which is fed to the PWM comparator providing linear control of the output pulse duration by the error amplifier. A Sync input to the oscillator allows for external synchronization or for multiple units to be slaved together. A single external resistor between the C_T pin and the Discharge pin provides a wide range of dead-time adjustment. These devices also feature built-in soft-start circuitry that requires only an external timing capacitor. The Shutdown pin controls both the soft-start and the output drivers, and provides instantaneous turn-off with soft-start recycle for slow turn-on. The soft-start and output driver circuitry are also controlled by the under-voltage lockout circuit, which, during low-input supply voltage of less than that required for normal operation, keeps the soft-start capacitor discharged and the output drivers off.

Another unique feature is the S-R latch following the PWM comparator. This feature enables the output drivers to be turned off any time the PWM pulse is terminated. The latch is reset with each clock pulse. However, the PWM outputs will remain turned off for the duration of the period if the PWM comparator output is in a low-level state. The TL1525A, TL2525A, and TL3525A output stages feature NOR logic resulting in a low output for an off-state. The TL1527A, TL2527A, and TL3527A output stages feature OR logic resulting in a high-level output for an off-state. The output stages are totem-pole designs capable of sourcing or sinking 200 milliamperes of output current.

The TL1525A and TL1527A are characterized for operation over the full military temperature range of −55°C to 125°C. The TL2525A and TL2527A are characterized for operation from −25°C to 85°C. The TL3525A and TL3527A are characterized for operation for 0°C to 70°C.

Voltage Regulators

6

functional block diagram (positive logic)

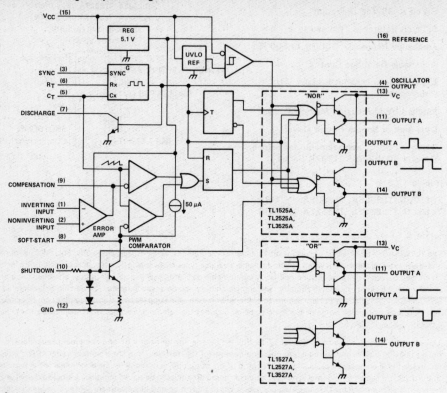

absolute maximum ratings over operating free-air temperature range (unless otherwise noted)

Supply voltage, V_{CC} (see Note 1)	40 V
Collector voltage, V_C	40 V
Logic input voltage range sync and shutdown	−0.3 V to 5.5 V
Analog input voltage range error amplifier inputs	−0.3 V to V_{CC}
Output current, I_O	500 mA
Reference output current, I_{REF}	50 mA
Current through C_T terminal	−5 mA
Continuous total dissipation at (or below) 25°C free-air temperature (see Note 2)	1000 mW
Operating free-air temperature range: TL1525A, TL1527A	−55°C to 125°C
TL2525A, TL2527A	−25°C to 85°C
TL3525A, TL3527A	0°C to 70°C
Operating virtual junction temperature range	0°C to 150°C
Storage temperature range	−65°C to 150°C
Lead temperature 1,6 mm (1/16 inch) from case for 60 seconds: J Package	300°C
Lead temperature 1,6 mm (1/16 inch) from case for 10 seconds: N Package	260°C

NOTES: 1. All voltage values are with respect to network ground terminal.

2. For operating above 25°C free-air temperature, see Dissipation Derating Curves, Figures 1 and 2. In the J package, TL1525A and TL1527A chips are alloy-mounted; TL2525A, TL2527A, TL3525A, and TL3527A chips are epoxy mounted.

TEXAS INSTRUMENTS

POST OFFICE BOX 225012 • DALLAS, TEXAS 75265

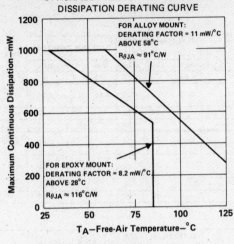

FIGURE 1

N PACKAGE FREE-AIR TEMPERATURE
DISSIPATION DERATING CURVE

FIGURE 2

recommended operating conditions

PARAMETER		TL1525A, TL1527A		TL2525A, TL2527A		TL3525A, TL3527A		UNIT
		MIN	MAX	MIN	MAX	MIN	MAX	
Supply voltage, V_{CC}		8	35	8	35	8	35	V
Collector voltage, V_C		4.5	35	4.5	35	4.5	35	V
Output current, I_O	Steady state	0	±100	0	±100	0	±100	mA
	Peak	0	±400	0	±400	0	±400	
Reference output current, I_{REF}		0	20	0	20	0	20	mA
Oscillator frequency range		100	500	100	500	100	500	kHz
Timing resistor, R_T		2	150	2	150	2	150	kΩ
Timing capacitor, C_T		0.001	0.1	0.001	0.1	0.001	0.1	μF
Dead-time resistor, R_D		0	500	0	500	0	500	Ω
Operating free-air temperature range, T_A		−55	125	−25	85	0	70	°C

Voltage Regulators

6

TEXAS
INSTRUMENTS
POST OFFICE BOX 225012 • DALLAS, TEXAS 75265

electrical characteristics over recommended operating free-air temperature range, $V_{CC} = 20$ V (unless otherwise noted)

reference section

PARAMETER	TEST CONDITIONS	TL1525A, TL1527A TL2525A, TL2527A			TL3525A, TL3527A			UNIT
		MIN	TYP	MAX	MIN	TYP	MAX	
Output voltage	$T_J = 25°C$	5.05	5.1	5.15	5	5.1	5.2	V
	$V_{CC} = 8$ V to 35 V, $I_O = 0$ to 20 mA	5		5.2	4.95		5.25	
Input regulation	$V_{CC} = 8$ V to 35 V		14	20		14	20	mV
Output regulation	$I_O = 0$ to 20 mA		5	50		5	50	mV
Output voltage change with temperature			24	50		24	50	mV
Output voltage long-term drift (see Note 3)	After 1000 h at $T_J = 125°C$		25	50		25	50	mV
Output noise voltage (RMS)	$f = 10$ Hz to 10 kHz, $T_J = 25°C$		40	200		40	200	μV
Short-circuit output current	$V_O = 0$ V, $T_J = 25°C$		80	100		80	100	mA

oscillator section

PARAMETER	TEST CONDITIONS	TL1525A, TL1527A TL2525A, TL2527A			TL3525A, TL3527A			UNIT
		MIN	TYP	MAX	MIN	TYP	MAX	
Maximum frequency	$R_T = 2$ kΩ, $\qquad C_T = 1$ nF	400			400			kHz
Minimum frequency	$R_T = 150$ kΩ, $\quad C_T = 0.1$ μF			100			100	Hz
Initial frequency error	$R_T = 3.6$ kΩ, $\qquad R_D = 0$ Ω, $C_T = 0.1$ μF, $\quad f = 40$ kHz, $T_A = 25°C$		$\pm 2\%$	$\pm 6\%$		$\pm 2\%$	$\pm 6\%$	
Frequency change with supply voltage	$V_{CC} = 8$ V to 35 V		$\pm 0.3\%$	$\pm 1\%$		$\pm 1\%$	$\pm 2\%$	
Frequency change with temperature	$T_A = $ MIN to MAX		$\pm 3\%$	$\pm 6\%$		$\pm 3\%$	$\pm 6\%$	
Output amplitude at Pin 4	$R_T = 3.6$ kΩ, $\qquad R_D = 0$ Ω, $C_T = 0.1$ μF, $\quad f = 40$ kHz	3	3.5		3	3.5		V
Output pulse duration at Pin 4	$R_T = 3.5$ kΩ, $\qquad R_D = 0$ Ω $C_T = 0.1$ μF, $\quad f = 40$ kHz, $T_J = 25°C$	0.3	0.5	1	0.3	0.6	1	μs
Input threshold voltage at Pin 3		1.2	2	2.8	1.2	2	2.8	V
Input current at Pin 3	$V_{I(Pin3)} = 3.5$ V		1.6	2.5		1.6	2.5	mA
Current through Pin 5 due to internal current mirror	Current through Pin 6 = 6 mA	1.7	2	2.2	1.7	2	2.2	mA

NOTE 3: Since long-term drift cannot be measured on the individual devices prior to shipment, this specification is not intended to be a guarantee or warranty. It is an engineering estimate of the average drift to be expected from lot to lot.

Voltage Regulators

6

483

TEXAS INSTRUMENTS
POST OFFICE BOX 225012 • DALLAS, TEXAS 75265

electrical characteristics over recommended operating free-air temperature range, V_{CC} = 20 V (unless otherwise noted)

error amplifier section

PARAMETER	TEST CONDITIONS	TL1525A, TL1527A TL2525A, TL2527A			TL3525A, TL3527A			UNIT
		MIN	TYP	MAX	MIN	TYP	MAX	
High-level output voltage		3.8	5.6		3.8	5.6		V
Low-level output voltage			0.2	0.5		0.2	0.5	V
Input offset voltage			0.5	5		2	10	mV
Input bias current			1	10		1	10	µA
Input offset current				1			1	µA
Open-loop voltage amplification	$R_L \geq$ 10 M	60	75		60	75		dB
Common-mode rejection ratio	V_{IC} = 1.5 V to 5.2 V	60	75		60	75		dB
Supply voltage rejection ratio	V_{CC} = 8 V to 35 V	50	60		50	60		dB
Gain-bandwidth product	A_V = 0 dB, T_J = 25 °C	1	2		1	2		MHz

comparator section

PARAMETER	TEST CONDITIONS		MIN	TYP	MAX	UNIT
Input threshold voltage	R_T = 3.6 kΩ, R_D = 0 Ω, C_T = 10 nF, f = 40 kHz	Duty cycle = 0%	0.6	0.9		V
		Duty cycle = MAX		3.3	3.6	
Input bias current				0.5	1	µA

soft-start section

PARAMETER	TEST CONDITIONS	MIN	TYP	MAX	UNIT
Soft-start voltage	V_I at Pin 10 = 2 V		0.4	0.6	V
Soft-start current	V_I at Pin 10 = 0 V	25	50	80	µA
Input current, Shutdown	V_I at Pin 10 = 2.5 V		0.4	1	mA

output section

PARAMETER	TEST CONDITIONS	MIN	TYP	MAX	UNIT
High-level output voltage	I_{OH} = −20 mA	18	19		V
	I_{OH} = −100 mA	17	18		
Low-level output voltage	I_{OL} = 20 mA		0.2	0.4	V
	I_{OL} = 100 mA		1	2	
Under-voltage lockout voltage	V_I at Pins 8 and 9 = high	6	7	8	V
Collector cutoff current (see Note 4)	V_C = 35 V, I_O = 100 mA			200	µA
Output pulse rise time	C_L = 1 nF, T_J = 25 °C		100	600	ns
Output pulse fall time	C_L = 1 nF, T_J = 25 °C		50	300	ns
Shutdown delay time	V_I at Pin 10 = 3 V, capacitance at pin 8 = 0, T_J = 25 °C		0.2	0.5	µs

NOTE 4: Collector cutoff current specifications apply only for the TL1525A, TL2525A, and TL3525A devices.

total device

PARAMETER	TEST CONDITIONS	MIN	TYP	MAX	UNIT
Minimum duty cycle				0%	
Maximum duty cycle		45%	49%		
Standby current	V_{CC} = 35 V		14	20	mA

Voltage Regulators

6

TEXAS
INSTRUMENTS
POST OFFICE BOX 225012 • DALLAS, TEXAS 75265

Voltage Regulators

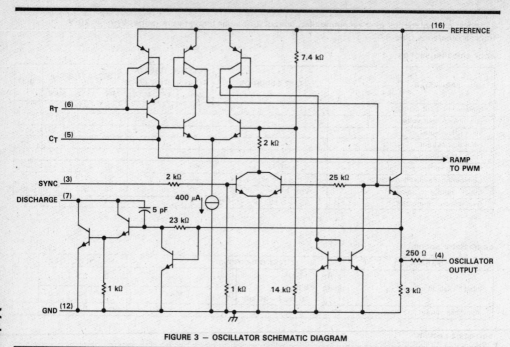

FIGURE 3 — OSCILLATOR SCHEMATIC DIAGRAM

TYPICAL CHARACTERISTICS

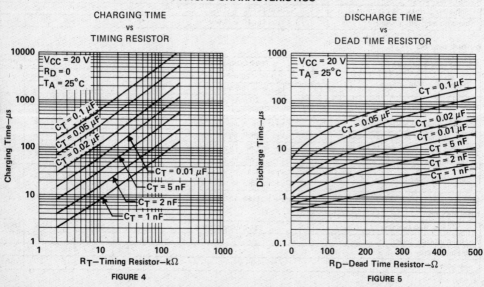

CHARGING TIME
vs
TIMING RESISTOR

FIGURE 4

DISCHARGE TIME
vs
DEAD TIME RESISTOR

FIGURE 5

TYPICAL CHARACTERISTICS

ERROR AMPLIFIER OPEN-LOOP
FREQUENCY RESPONSE

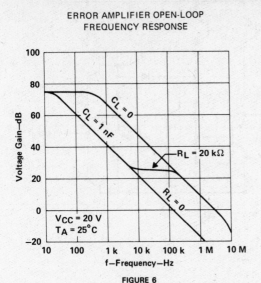

FIGURE 6

TL1525A OUTPUT SATURATION VOLTAGE
vs
OUTPUT CURRENT

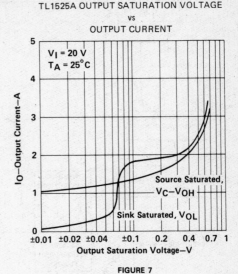

FIGURE 7

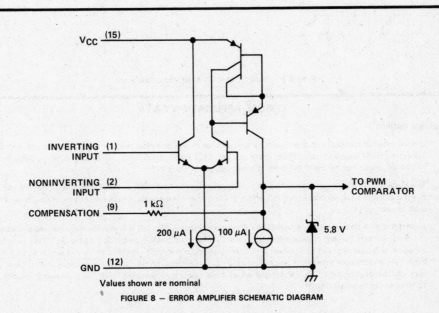

FIGURE 8 — ERROR AMPLIFIER SCHEMATIC DIAGRAM

Voltage Regulators

6

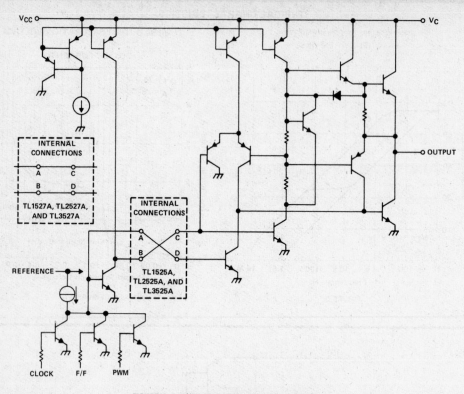

FIGURE 9 — OUTPUT CIRCUIT SCHEMATIC DIAGRAM

TYPICAL APPLICATION DATA

shutdown options

1. Use an external transistor or open-collector comparator to pull down on the Compensation terminal (Pin 9). This will set the PWM latch and turn off both driver outputs. If the shutdown signal is momentary, pulse-by-pulse protection will be accomplished as the PWM latch is reset with each clock pulse.

2. The same results may be accomplished by pulling down on the Soft-Start terminal (Pin 8) with the only difference being that on this pin shutdown will not affect the amplifier compensation network, but must discharge any soft-start capacitance.

3. Application of a positive-going signal to the Shutdown terminal (Pin 10) will provide the most rapid shutdown of the driver outputs but will not immediately set the PWM latch if there is a capacitor at the Soft-Start terminal. The capacitor will discharge but at a current twice the charging current. The PWM latch can be set on a pulse-by-pulse basis by the shutdown terminal if there is no external capacitance on the Soft-start terminal (Pin 8). Slow turn-on may still be accomplished by connecting an external capacitor, blocking diode, and charging resistor to the Compensation terminal (Pin 9).

TEXAS INSTRUMENTS
POST OFFICE BOX 225012 • DALLAS, TEXAS 75265

TYPICAL APPLICATION DATA

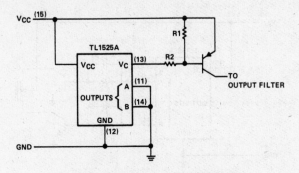

For single-ended supplies, the driver outputs are grounded. The V_C terminal is switched to ground by the totem-pole source transistors on the alternate oscillator cycles.

FIGURE 10 — SINGLE-ENDED CIRCUIT

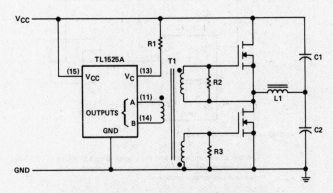

Low-power transformers can be directly driven by the TL1525A. Automatic reset occurs during deadtime when both ends of the primary winding are switched to ground.

FIGURE 11 — TRANSFORMER-COUPLED CIRCUIT

Voltage Regulators

6

TYPICAL APPLICATION DATA

Voltage Regulators

6

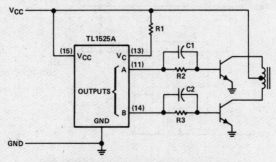

In conventional push-pull bipolar designs, forward base drive is controlled by $R_1 - R_3$. Rapid turn-off times for the power devices are achieved with speed-up capacitors C_1 and C_2.

FIGURE 12 — BIPOLAR PUSH-PULL CIRCUIT

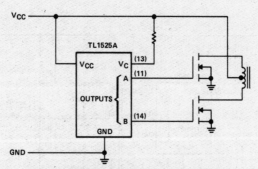

The low source impedance of the output drivers provides rapid charging of power FET input capacitance while minimizing external components.

**FIGURE 13 — LOW-IMPEDANCE BIPOLAR-DRIVE
PUSH-PULL CIRCUIT**

TEXAS INSTRUMENTS
POST OFFICE BOX 225012 • DALLAS, TEXAS 75265

LINEAR
INTEGRATED
CIRCUITS

TYPES TL7700I, TL7700C
SUPPLY VOLTAGE SUPERVISORS

D2812, DECEMBER 1983

- Power-On Reset Generator
- Automatic Reset Generation After Voltage Drop
- Wide Supply Voltage Range . . . 1.8 V to 40 V
- Precision Voltage Sensor
- Temperature-Compensated Voltage Reference
- Externally-Adjustable Pulse Duration
- Programmable Sense Voltage
- Programmable Hysteresis

D OR P DUAL-IN-LINE PACKAGE
(TOP VIEW)

```
        C_T  [ 1  U  8 ]  RESET OUT
SENSE INPUT  [ 2     7 ]  NC
        NC   [ 3     6 ]  NC
        GND  [ 4     5 ]  V_CC
```

NC—No internal connection

description

The TL7700 is a monolithic integrated circuit supply voltage supervisor specifically designed for use as a reset controller in microcomputer and microprocessor systems. During power-up the device tests the supply voltage and keeps the RESET output active as long as the supply voltage has not reached its nominal voltage value. The device internal time delay is determined by an external capacitor connected to the C_T input (pin 1).

$$t_d = 10^5 \times C_T$$
Where: C_T is in farads (F),
t_d is in seconds (s)

The TL7700I is characterized for operation from $-25\,°C$ to $85\,°C$; the TL7700C is characterized from $0\,°C$ to $70\,°C$.

absolute maximum ratings over operating free-air temperature (unless otherwise noted)

Supply voltage, V_{CC} (see Note 1) . . . . 41 V
Input voltage at SENSE . . . . -0.3 to 41 V
Output current . . . . 5 mA
Operating free-air temperature range: TL7700I . . . . $-25\,°C$ to $85\,°C$
                                     TL7700C . . . . $0\,°C$ to $70\,°C$
Storage temperature range . . . . $-65\,°C$ to $150\,°C$

NOTE 1: All voltage values are with respect to the network ground terminal.

recommended operating conditions

		MIN	NOM	MAX	UNIT
Supply voltage, V_{CC}		1.8		40	V
Low-level output current, I_{OL}				3	mA
Operating free-air temperature, T_A	TL7700I	-25		85	°C
	TL7700C	0		70	

Voltage Regulators

6

Copyright © 1983 by Texas Instruments Incorporated

TEXAS INSTRUMENTS
POST OFFICE BOX 225012 ● DALLAS, TEXAS 75265

Voltage Regulators

6

LINEAR
INTEGRATED
CIRCUITS

TYPES TL7702A, TL7705A, TL7709A, TL7712A, TL7715A
SUPPLY VOLTAGE SUPERVISORS

D2722, APRIL 1983—REVISED FEBRUARY 1984

- Power-On Reset Generator
- Automatic Reset Generation After Voltage Drop
- Wide Supply Voltage Range . . . 3 V to 18 V
- Precision Voltage Sensor
- Temperature-Compensated Voltage Reference
- True and Complement Reset Outputs
- Externally Adjustable Pulse Width

D OR P DUAL-IN-LINE PACKAGE
(TOP VIEW)

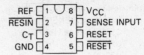

REF	1	8	V_CC
$\overline{\text{RESIN}}$	2	7	SENSE INPUT
C_T	3	6	RESET
GND	4	5	$\overline{\text{RESET}}$

description

The TL7702A series are monolithic integrated circuit supply voltage supervisors specifically designed for use as reset controllers in microcomputer and microprocessor systems. During power-up the device tests the supply voltage and keeps the RESET and $\overline{\text{RESET}}$ outputs active (high and low, respectively) as long as the supply voltage has not reached its nominal voltage value. Taking $\overline{\text{RESIN}}$ low has the same effect. To ensure that the microcomputer system has reset, the TL7702A then initiates an internal time delay that delays the return of the reset outputs to their inactive states. Since the time delay for most microcomputers and microprocessors is in the order of several machine cycles, the device internal time delay is determined by an external capacitor connected to the C_T input (pin 3).

$$t_d = 1.3 \times 10^4 \times C_T$$
Where: C_T is in farads (F) and t_d is in seconds(s)

In addition, when the supply voltage drops below the nominal value, the outputs will be active until the supply voltage returns to the nominal value. An external capacitor (typically 0.1 μF) must be connected to the REF output (pin 1) to reduce the influence of fast transients in the supply voltage.

The TL7702AI series is characterized for operation from −25°C to 85°C; the TL7702AC series is characterized from 0°C to 70°C.

functional block diagram

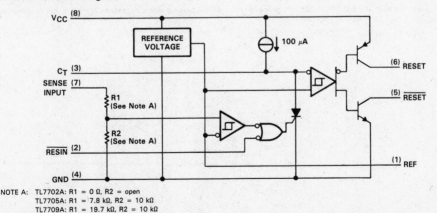

NOTE A: TL7702A: R1 = 0 Ω, R2 = open
TL7705A: R1 = 7.8 kΩ, R2 = 10 kΩ
TL7709A: R1 = 19.7 kΩ, R2 = 10 kΩ
TL7712A: R1 = 32.7 kΩ, R2 = 10 kΩ
TL7715A: R1 = 43.4 kΩ, R2 = 10 kΩ

Voltage Regulators

6

Copyright © 1983 by Texas Instruments Incorporated

TEXAS
INSTRUMENTS

POST OFFICE BOX 225012 • DALLAS, TEXAS 75265

Voltage Regulators

6

absolute maximum ratings over operating free-air temperature (unless otherwise noted)

Supply voltage, V_{CC} (see Note 1) . . . . 20 V

Input voltage range at $\overline{RESIN}$ . . . . −0.3 V to 20 V

Input voltage at SENSE:

TL7702A (see Note 2) . . . . −0.3 V to 6 V

TL7705A . . . . −0.3 V to 10 V

TL7709A . . . . −0.3 V to 15 V

TL7712A . . . . −0.3 V to 20 V

TL7715A . . . . −0.3 V to 20 V

High-level output current at $\overline{RESET}$ . . . . −30 mA

Low-level output current at $\overline{RESET}$ . . . . 30 mA

Operating free-air temperature range:

TL77_I . . . . −25°C to 85°C

TL77_C . . . . 0°C to 70°C

Storage temperature range . . . . −65°C to 150°C

NOTES: 1. All voltage values are with respect to the network ground terminal.
2. For the TL7702A, the voltage applied to the SENSE terminal must never exceed V_{CC}.

recommended operating conditions

		MIN	NOM	MAX	UNIT
Supply voltage, V_{CC}		3.6		18	V
High-level input voltage at $\overline{RESIN}$, V_{IH}		2			V
Low-level input voltage at $\overline{RESIN}$, V_{IL}				0.6	V
Voltage at sense input, V_I:	TL7702A	0		See Note 3	V
	TL7705A	0		10	
	TL7709A	0		15	
	TL7712A	0		20	
	TL7715A	0		20	
High-level output current at $\overline{RESET}$, I_{OH}				−16	mA
Low-level output current at $\overline{RESET}$, I_{OL}				16	mA
Operating free-air temperature range, T_A:	TL77_I	−25		85	°C
	TL77_C	0		70	

NOTE 3: For proper operation of the TL7702A, the voltage applied to the SENSE terminal should not exceed V_{CC} − 1 V or 6 V, whichever is less.

Texas Instruments

POST OFFICE BOX 225012 ● DALLAS, TEXAS 75265

electrical characteristics over recommended ranges of supply voltage, input voltage, output current, and free-air temperature (unless otherwise noted)

	PARAMETER		TEST CONDITIONS[†]	MIN	TYP	MAX	UNIT
V_{OH}	High-level output voltage at RESET		$I_{OH} = -16$ mA	$V_{CC} - 1.5$			V
V_{OL}	Low-level output voltage at $\overline{RESET}$		$I_{OL} = 16$ mA			0.4	V
V_{ref}	Reference voltage		$T_A = 25°C$	2.48	2.53	2.58	V
V_T	Threshold voltage at SENSE input	TL7702A	$V_{CC} = 3.6$ V to 18 V $T_A = 25°C$	2.48	2.53	2.58	V
		TL7705A		4.5	4.55	4.6	
		TL7709A		7.5	7.6	7.7	
		TL7712A		10.6	10.8	11	
		TL7715A		13.2	13.5	13.8	
$V_{T+} - V_{T-}$	Hysteresis[‡] at SENSE input	TL7702A	$V_{CC} = 3.6$ V to 18 V, $T_A = 25°C$		10		mV
		TL7705A			15		
		TL7709A			20		
		TL7712A			35		
		TL7715A			45		
I_I	Input current at $\overline{RESIN}$ input		$V_I = 2.4$ V to V_{CC}			20	μA
			$V_I = 0.4$ V			-100	
I_I	Input current at SENSE input	TL7702A	$V_{ref} < V_I < V_{CC} - 1.5$ V		0.5	2	μA
I_{OH}	High-level output current at $\overline{RESET}$		$V_O = 18$ V			50	μA
I_{OL}	Low-level output current at RESET		$V_O = 0$			-50	μA
I_{CC}	Supply current		All inputs and outputs open		1.8	3	mA

[†]All characteristics are measured with C = 0.1 μF from Pin 1 to GND, and with C = 0.1 μF from Pin 3 to GND.
[‡]Hysteresis is the difference between the positive-going input threshold voltage, V_{T+}, and the negative-going input threshold voltage, V_{T-}.

TYPICAL CHARACTERISTICS

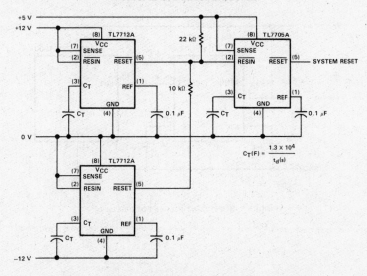

$$C_T(F) = \frac{1.3 \times 10^4}{t_d(s)}$$

FIGURE 1—MULTIPLE POWER SUPPLY SYSTEM RESET GENERATION

Voltage Regulators

6

TEXAS INSTRUMENTS
POST OFFICE BOX 225012 • DALLAS, TEXAS 75265

TYPES TL7702A, TL7705A, TL7709A, TL7712A, TL7715A
SUPPLY VOLTAGE SUPERVISORS

TYPICAL APPLICATION DATA

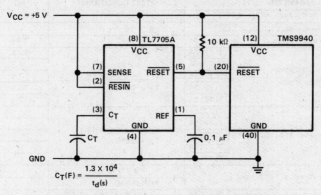

$$C_T(F) = \frac{1.3 \times 10^4}{t_d(s)}$$

FIGURE 2−RESET CONTROLLER FOR TMS9940 SYSTEM

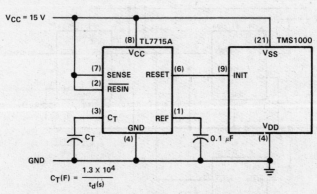

$$C_T(F) = \frac{1.3 \times 10^4}{t_d(s)}$$

FIGURE 3−RESET CONTROLLER FOR TMS1000

Voltage Regulators

6

Texas
Instruments
POST OFFICE BOX 225012 ● DALLAS, TEXAS 75265

LINEAR
INTEGRATED
CIRCUITS

TYPES uA723M, uA723C
PRECISION VOLTAGE REGULATORS

D1063, AUGUST 1972—REVISED DECEMBER 1982

- 150-mA Load Current Without External Power Transistor
- Typically 0.02% Input Regulation and 0.03% Load Regulation (uA723M)
- Adjustable Current Limiting Capability
- Input Voltages to 40 Volts
- Output Adjustable from 2 to 37 Volts
- Direct Replacement for Fairchild μA723M and μA723C

uA723M . . . J PACKAGE
uA723C . . . J OR N PACKAGE
(TOP VIEW)

NC	1	14 NC
CURR LIM	2	13 FREQ COMP
CURR SENS	3	12 $V_{CC}+$
IN−	4	11 V_C
IN+	5	10 OUTPUT
$V_{(ref)}$	6	9 V_Z
$V_{CC}−$	7	8 NC

uA723M . . . U PACKAGE
(TOP VIEW)

CURR SENS	1	10 CURR LIM
IN−	2	9 FREQ COMP
IN+	3	8 $V_{CC}+$
$V_{(ref)}$	4	7 V_C
$V_{CC}−$	5	6 OUTPUT

NC—No internal connection

description

The uA723M and uA723C are monolithic integrated circuit voltage regulators featuring high ripple rejection, excellent input and load regulation, excellent temperature stability, and low standby current. The circuit consists of a temperature-compensated reference voltage amplifier, an error amplifier, a 150-milliampere output transistor, and an adjustable output current limiter.

The uA723M and uA723C are designed for use in positive or negative power supplies as a series, shunt, switching, or floating regulator. For output currents exceeding 150 mA, additional pass elements may be connected as shown in Figure 4 and 5.

The uA723M is characterized for operation over the full military temperature range of −55 °C to 125 °C. The uA723C is characterized for operation from 0 °C to 70 °C.

functional block diagram

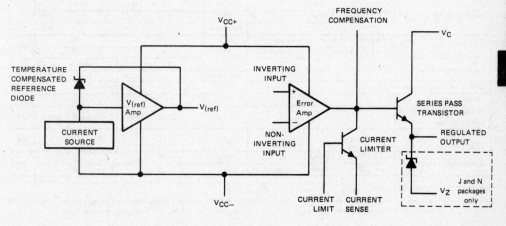

Voltage Regulators

6

TEXAS
INSTRUMENTS
POST OFFICE BOX 225012 • DALLAS, TEXAS 75265

absolute maximum ratings over operating free-air temperature range (unless otherwise noted)

Peak voltage from V_{CC+} to V_{CC-} ($t_W \leqslant 50$ ms)	50 V
Continuous voltage from V_{CC+} to V_{CC-}	40 V
Input-to-output voltage differential	40 V
Differential input voltage to error amplifier	±5 V
Voltage between noninverting input and V_{CC-}	. 8 V
Current from V_Z	25 mA
Current from $V_{(ref)}$	15 mA
Continuous total dissipation at (or below) 25°C free-air temperature (see Note 1):	
J or N package	1000 mW
U package	. 675 mW
Operating free-air temperature range: uA723M Circuits	−55°C to 125°C
uA723C Circuits	0°C to 70°C
Storage temperature range	−65°C to 150°C
Lead temperature 1,6 mm (1/16 inch) from case for 60 seconds, J or U package	300°C
Lead temperature 1,6 mm (1/16 inch) from case for 10 seconds, N package	260°C

NOTE 1: Power dissipation = [I(standby) + I(ref)] V_{CC} + [V_C − V_O] I_O. For operation at elevated temperature, refer to Dissipation Derating Table. In the J package, uA723M chips are alloy-mounted; uA723C chips are glass-mounted.

recommended operating conditions

	MIN	MAX	UNIT
Input voltage, V_I	9.5	40	V
Output voltage, V_O	2	37	V
Input-to-output voltage differential, $V_C - V_O$	3	38	V
Output current, I_O		150	mA

electrical characteristics at specified free-air temperature (see Note 2)

PARAMETER	TEST CONDITIONS†		uA723M			uA723C			UNIT
			MIN	TYP	MAX	MIN	TYP	MAX	
Input regulation	V_I = 12 V to V_I = 15 V	25°C		0.01%	0.1%		0.01%	0.1%	
	V_I = 12 V to V_I = 40 V	25°C		0.02%	0.2%		0.1%	0.5%	
	V_I = 12 V to V_I = 15 V	Full range			0.3%			0.3%	
Ripple rejection	f = 50 Hz to 10 kHz, C(ref) = 0	25°C		74			74		dB
	f = 50 Hz to 10 kHz, C(ref) = 5 μF	25°C		86			86		
Output regulation	I_O = 1 mA to I_O = 50 mA	25°C		−0.03%	−0.15%		−0.03%	−0.2%	
		Full range			−0.6%			−0.6%	
Reference voltage, $V_{(ref)}$		25°C	6.95	7.15	7.35	6.8	7.15	7.5	V
Standby current	V_I = 30 V, I_O = 0	25°C		2.3	3.5		2.3	4	mA
Temperature coefficient of output voltage		Full range		0.002	0.015		0.003	0.015	%/°C
Short-circuit output current	R_{SC} = 10 Ω, V_O = 0	25°C		65			65		mA
Output noise voltage	BW = 100 Hz to 10 kHz, C(ref) = 0	25°C		20			20		μV
	BW = 100 Hz to 10 kHz, C(ref) = 5 μF	25°C		2.5			2.5		

†Full range for uA723M is −55°C to 125°C and for uA723C is 0°C to 70°C.

NOTE 2: For all values in this table the device is connected as shown in Figure 1 with the divider resistance as seen by the error amplifier ≤ 10 kΩ. Unless otherwise specified, V_I = V_{CC+} = V_C = 12 V, V_{CC-} = 0, V_O = 5 V, I_O = 1 mA, R_{SC} = 0, and C(ref) = 0.

schematic

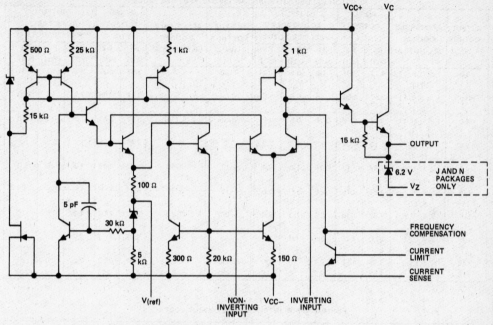

RESISTOR AND CAPACITOR VALUES SHOWN ARE NOMINAL.

DISSIPATION DERATING TABLE

POWER	POWER RATING	DERATING FACTOR	ABOVE T_A
J (Alloy-Mounted Chip)	1000 mW	11.0 mW/°C	59°C
J (Glass-Mounted Chip)	1000 mW	8.2 mW/°C	28°C
N	1000 mW	9.2 mW/°C	41°C
U	675 mW	5.4 mW/°C	25°C

Voltage Regulators

6

TYPICAL APPLICATION DATA

TABLE I
RESISTOR VALUES (kΩ) FOR STANDARD OUTPUT VOLTAGES

OUTPUT VOLTAGE (V)	APPLICABLE FIGURES (SEE NOTE 3)	FIXED OUTPUT ± 5%		OUTPUT ADJUSTABLE ±10% (SEE NOTE 4)			OUTPUT VOLTAGE (V)	APPLICABLE FIGURES (SEE NOTE 3)	FIXED OUTPUT ± 5%		OUTPUT ADJUSTABLE ±10% (SEE NOTE 4)		
		R1 (kΩ)	R2 (kΩ)	R1 (kΩ)	P1 (kΩ)	R2 (kΩ)			R1 (kΩ)	R2 (kΩ)	R1 (kΩ)	P1 (kΩ)	R2 (kΩ)
+3.0	1, 5, 6, 9, 11, 12 (4)	4.12	3.01	1.8	0.5	1.2	+100	7	3.57	105	2.2	10	91
+3.6	1, 5, 6, 9, 11, 12 (4)	3.57	3.65	1.5	0.5	1.5	+250	7	3.57	255	2.2	10	240
+5.0	1, 5, 6, 9, 11, 12 (4)	2.15	4.99	0.75	0.5	2.2	−6 (Note 5)	3, (10)	3.57	2.43	1.2	0.5	0.75
+6.0	1, 5, 6, 9, 11, 12 (4)	1.15	6.04	0.5	0.5	2.7	−9	3, 10	3.48	5.36	1.2	0.5	2.0
+9.0	2, 4, (5, 6, 9, 12)	1.87	7.15	0.75	1.0	2.7	−12	3, 10	3.57	8.45	1.2	0.5	3.3
+12	2, 4, (5, 6, 9, 12)	4.87	7.15	2.0	1.0	3.0	−15	3, 10	3.57	11.5	1.2	0.5	4.3
+15	2, 4, (5, 6, 9, 12)	7.87	7.15	3.3	1.0	3.0	−28	3, 10	3.57	24.3	1.2	0.5	10
+28	2, 4, (5, 6, 9, 12)	21.0	7.15	5.6	1.0	2.0	−45	8	3.57	41.2	2.2	10	33
+45	7	3.57	48.7	2.2	10	39	−100	8	3.57	95.3	2.2	10	91
+75	7	3.57	78.7	2.2	10	68	−250	8	3.57	249	2.2	10	240

TABLE II
FORMULAS FOR INTERMEDIATE OUTPUT VOLTAGES

Outputs from +2 to +7 volts [Figures 1, 5, 6, 9, 11, 12, (4)] $$V_O = V_{(ref)} \times \frac{R2}{R1 + R2}$$	Outputs from +4 to +250 volts [Figure 7] $$V_O = \frac{V_{(ref)}}{2} \times \frac{R2 - R1}{R1};$$ R3 = R4	Current Limiting $$I_{(limit)} \approx \frac{0.65\ V}{R_{sc}}$$
Outputs from +7 to +37 volts [Figures 2, 4, (5, 6, 9, 11, 12)] $$V_O = V_{(ref)} \times \frac{R1 + R2}{R2}$$	Outputs from −6 to −250 volts [Figures 3, 8, 10] $$V_O = -\frac{V_{(ref)}}{2} \times \frac{R1 + R2}{R1};$$ R3 = R4	Foldback Current Limiting [Figure 6] $$I_{(knee)} \approx \frac{V_O R3 + (R3 + R4)\ 0.65\ V}{R_{sc} R4}$$ $$I_{OS} \approx \frac{0.65\ V}{R_{sc}} \times \frac{R3 + R4}{R4}$$

NOTES: 3. The R1/R2 divider may be across either V_O or $V_{(ref)}$. If the divider is across $V_{(ref)}$ and uses figures without parentheses, use figures with parentheses when the divider is across V_O.
4. To make the voltage adjustable, the R1/R2 divider shown in the figures must be replaced by the divider shown at the right.
5. The device requires a minimum of 9 V between V_{CC+} and V_{CC-} when V_O is equal to or more positive than −9 V.

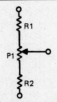

ADJUSTABLE OUTPUT CIRCUITS

TEXAS INSTRUMENTS
POST OFFICE BOX 225012 • DALLAS, TEXAS 75265

Voltage Regulators

6

TYPICAL APPLICATION DATA

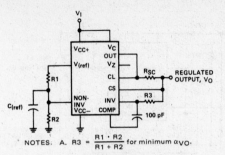

NOTES: A. $R3 = \dfrac{R1 \cdot R2}{R1 + R2}$ for minimum α_{VO}.

B. R3 may be eliminated for minimum component count. Use direct connection (i.e., $R_3 = 0$).

FIGURE 1—BASIC LOW-VOLTAGE REGULATOR
(V_O = 2 TO 7 VOLTS)

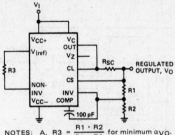

NOTES: A. $R3 = \dfrac{R1 \cdot R2}{R1 + R2}$ for minimum α_{VO}.

B. R3 may be eliminated for minimum component count. Use direct connection (i.e., $R_3 = 0$).

FIGURE 2—BASIC HIGH-VOLTAGE REGULATOR
(V_O = 7 TO 37 VOLTS)

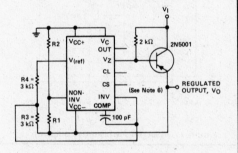

FIGURE 3—NEGATIVE-VOLTAGE REGULATOR

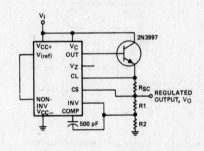

FIGURE 4—POSITIVE-VOLTAGE REGULATOR
(EXTERNAL N-P-N- PASS TRANSISTOR)

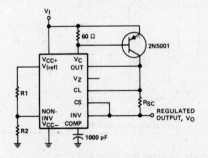

FIGURE 5—POSITIVE-VOLTAGE REGULATOR
(EXTERNAL P-N-P PASS TRANSISTOR)

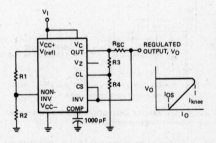

FIGURE 6—FOLDBACK CURRENT LIMITING

Voltage Regulators

6

TYPICAL APPLICATION DATA

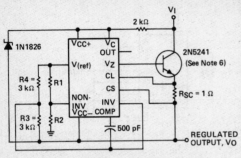

FIGURE 7—POSITIVE FLOATING REGULATOR

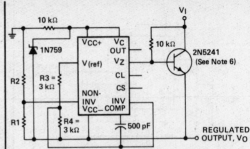

FIGURE 8—NEGATIVE FLOATING REGULATOR

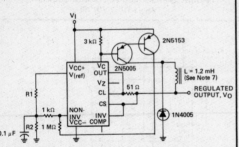

FIGURE 9—POSITIVE SWITCHING REGULATOR

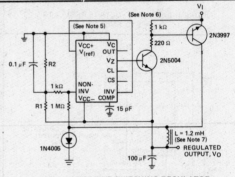

FIGURE 10—NEGATIVE SWITCHING REGULATOR

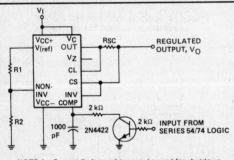

NOTE A: Current limit transistor may be used for shutdown
if current limiting is not required.

FIGURE 11—REMOTE SHUTDOWN REGULATOR WITH
CURRENT LIMITING

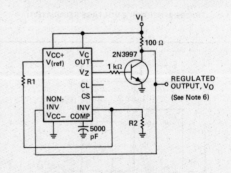

FIGURE 12—SHUNT REGULATOR

NOTES: 5. The device requires a minimum of 9 V between V_{CC+} and V_{CC-} when V_O is equal to or more positive than -9 V.
6. When 10-lead uA723 devices are used in applications requiring V_Z, an external 6.2-V regulator diode must be connected in series with the V_O terminal.
7. L is 40 turns of No. 20 enameled copper wire wound on Ferroxcube P36/22-3B7 potted core, or equivalent, with 0.009-inch air gap.

Voltage Regulators

6

TEXAS
INSTRUMENTS
POST OFFICE BOX 225012 • DALLAS, TEXAS 75265

- 3-Terminal Regulators
- Output Current up to 1.5 A
- No External Components
- Internal Thermal Overload Protection
- High Power Dissipation Capability
- Internal Short-Circuit Current Limiting
- Output Transistor Safe-Area Compensation
- Direct replacements for Fairchild μA7800 Series

NOMINAL OUTPUT VOLTAGE	REGULATOR
5 V	uA7805C
6 V	uA7806C
8 V	uA7808C
8.5 V	uA7885C
10 V	uA7810C
12 V	uA7812C
15 V	uA7815C
18 V	uA7818C
24 V	uA7824C

description

This series of fixed-voltage monolithic integrated-circuit voltage regulators is designed for a wide range of applications. These applications include on-card regulation for elimination of noise and distribution problems associated with single-point regulation. Each of these regulators can deliver up to 1.5 amperes of output current. The internal current limiting and thermal shutdown features of these regulators make them essentially immune to overload. In addition to use as fixed-voltage regulators, these devices can be used with external components to obtain adjustable output voltages and currents and also as the power-pass element in precision regulators.

KC PACKAGE

(TOP VIEW)

OUTPUT
COMMON
INPUT

THE COMMON TERMINAL IS IN ELECTRICAL CONTACT WITH THE MOUNTING BASE

TO-220AB

O C
I

Voltage Regulators

6

schematic

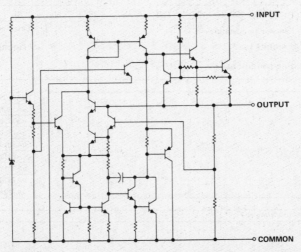

Resistor values shown are nominal and in ohms.

282

TEXAS INSTRUMENTS

POST OFFICE BOX 225012 • DALLAS, TEXAS 75265

Voltage Regulators

absolute maximum ratings over operating temperature range (unless otherwise noted)

		uA78__C	UNIT
Input voltage	uA7824C	40	V
	All others	35	
Continuous total dissipation at 25°C free-air temperature (see Note 1)		2	W
Continuous total dissipation at (or below) 25°C case temperature (see Note 1)		15	W
Operating free-air, case, or virtual junction temperature range		0 to 150	°C
Storage temperature range		−65 to 150	°C
Lead temperature 1,6 mm (1/16 inch) from case for 10 seconds		260	°C

NOTE 1: For operation above 25°C free-air or case temperature, refer to Figures 1 and 2. To avoid exceeding the design maximum virtual junction temperature, these ratings should not be exceeded. Due to variations in individual device electrical characteristics and thermal resistance, the built-in thermal overload protection may be activated at power levels slightly above or below the rated dissipation.

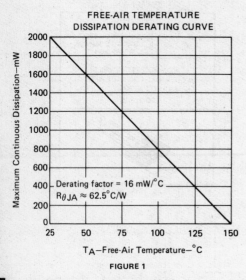

FREE-AIR TEMPERATURE
DISSIPATION DERATING CURVE

Derating factor = 16 mW/°C
$R_{\theta JA} \approx 62.5°$C/W

T_A−Free-Air Temperature−°C

FIGURE 1

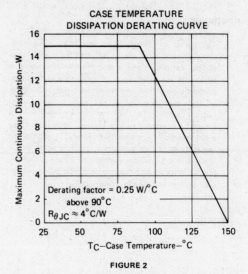

CASE TEMPERATURE
DISSIPATION DERATING CURVE

Derating factor = 0.25 W/°C
above 90°C
$R_{\theta JC} \approx 4°$C/W

T_C−Case Temperature−°C

FIGURE 2

recommended operating conditions

		MIN	MAX	UNIT
Input voltage, V_I	uA7805C	7	25	V
	uA7806C	8	25	
	uA7808C	10,5	25	
	uA7885C	10,5	25	
	uA7810C	12,5	28	
	uA7812C	14,5	30	
	uA7815C	17,5	30	
	uA7818C	21	33	
	uA7824C	27	38	
Output current, I_O			1,5	A
Operating virtual junction temperature, T_J		0	125	°C

Texas Instruments

POST OFFICE BOX 225012 • DALLAS, TEXAS 75265

6

uA7805C electrical characteristics at specified virtual junction temperature, V_I = 10 V, I_O = 500 mA (unless otherwise noted)

PARAMETER	TEST CONDITIONS[†]		uA7805C			UNIT
			MIN	TYP	MAX	
Output voltage	I_O = 5 mA to 1 A, V_I = 7 V to 20 V, P ≤ 15 W	25°C	4.8	5	5.2	V
		0°C to 125°C	4.75		5.25	
Input regulation	V_I = 7 V to 25 V	25°C		3	100	mV
	V_I = 8 V to 12 V			1	50	
Ripple rejection	V_I = 8 V to 18 V, f = 120 Hz	0°C to 125°C	62	78		dB
Output regulation	I_O = 5 mA to 1.5 A	25°C		15	100	mV
	I_O = 250 mA to 750 mA			5	50	
Output resistance	f = 1 kHz	0°C to 125°C		0.017		Ω
Temperature coefficient of output voltage	I_O = 5 mA	0°C to 125°C		−1.1		mV/°C
Output noise voltage	f = 10 Hz to 100 kHz	25°C		40		µV
Dropout voltage	I_O = 1 A	25°C		2.0		V
Bias current		25°C		4.2	8	mA
Bias current change	V_I = 7 V to 25 V	0°C to 125°C			1.3	mA
	I_O = 5 mA to 1 A				0.5	
Short-circuit output current		25°C		750		mA
Peak output current		25°C		2.2		A

uA7806C electrical characteristics at specified virtual junction temperature, V_I = 11 V, I_O = 500 mA (unless otherwise noted)

PARAMETER	TEST CONDITIONS[†]		uA7806C			UNIT
			MIN	TYP	MAX	
Output voltage	I_O = 5 mA to 1 A, V_I = 8 V to 21 V, P ≤ 15 W	25°C	5.75	6	6.25	V
		0°C to 125°C	5.7		6.3	
Input regulation	V_I = 8 V to 25 V	25°C		5	120	mV
	V_I = 9 V to 13 V			1.5	60	
Ripple rejection	V_I = 9 V to 19 V, f = 120 Hz	0°C to 125°C	59	75		dB
Output regulation	I_O = 5 mA to 1.5 A	25°C		14	120	mV
	I_O = 250 mA to 750 mA			4	60	
Output resistance	f = 1 kHz	0°C to 125°C		0.019		Ω
Temperature coefficient of output voltage	I_O = 5 mA	0°C to 125°C		−0.8		mV/°C
Output noise voltage	f = 10 Hz to 100 kHz	25°C		45		µV
Dropout voltage	I_O = 1 A	25°C		2.0		V
Bias current		25°C		4.3	8	mA
Bias current change	V_I = 8 V to 25 V	0°C to 125°C			1.3	mA
	I_O = 5 mA to 1 A				0.5	
Short-circuit output current		25°C		550		mA
Peak output current		25°C		2.2		A

[†]All characteristics are measured with a capacitor across the input of 0.33 µF and a capacitor across the output of 0.1 µF. All characteristics except noise voltage and ripple rejection ratio are measured using pulse techniques (t_W ≤ 10 ms, duty cycle ≤ 5%). Output voltage changes due to changes in internal temperature must be taken into account separately.

Voltage Regulators

6

TEXAS
INSTRUMENTS
POST OFFICE BOX 225012 • DALLAS, TEXAS 75265

uA7808C electrical characteristics at specified virtual junction temperature,
V_I = 14 V, I_O = 500 mA (unless otherwise noted)

PARAMETER	TEST CONDITIONS†		uA7808C			UNIT
			MIN	TYP	MAX	
Output voltage	I_O = 5 mA to 1 A, V_I = 10.5 V to 23 V, P ⩽ 15 W	25°C	7.7	8	8.3	V
		0°C to 125°C	7.6		8.4	
Input regulation	V_I = 10.5 V to 25 V	25°C		6	160	mV
	V_I = 11 V to 17 V			2	80	
Ripple rejection	V_I = 11.5 V to 21.5 V, f = 120 Hz	0°C to 125°C	56	72		dB
Output regulation	I_O = 5 mA to 1.5 A	25°C		12	160	mV
	I_O = 250 mA to 750 mA			4	80	
Output resistance	f = 1 kHz	0°C to 125°C		0.016		Ω
Temperature coefficient of output voltage	I_O = 5 mA	0°C to 125°C		−0.8		mV/°C
Output noise voltage	f = 10 Hz to 100 kHz	25°C		52		µV
Dropout voltage	I_O = 1 A	25°C		2.0		V
Bias current		25°C		4.3	8	mA
Bias current change	V_I = 10.5 V to 25 V	0°C to 125°C			1	mA
	I_O = 5 mA to 1 A				0.5	
Short-circuit output current		25°C		450		mA
Peak output current		25°C		2.2		A

uA7885C electrical characteristics at specified virtual junction temperature,
V_I = 15 V, I_O = 500 mA (unless otherwise noted)

PARAMETER	TEST CONDITIONS†		uA7885C			UNIT
			MIN	TYP	MAX	
Output voltage	I_O = 5 mA to 1 A, V_I = 11 V to 23.5 V, P ⩽ 15 W	25°C	8.15	8.5	8.85	V
		0°C to 125°C	8.1		8.9	
Input regulation	V_I = 10.5 V to 25 V	25°C		6	170	mV
	V_I = 11 V to 17 V			2	85	
Ripple rejection	V_I = 11.5 V to 21.5 V, f = 120 Hz	0°C to 125°C	54	70		dB
Output regulation	I_O = 5 mA to 1.5 A	25°C		12	170	mV
	I_O = 250 mA to 750 mA			4	85	
Output resistance	f = 1 kHz	0°C to 125°C		0.016		Ω
Temperature coefficient of output voltage	I_O = 5 mA	0°C to 125°C		−0.8		mV/°C
Output noise voltage	f = 10 Hz to 100 kHz	25°C		55		µV
Dropout voltage	I_O = 1 A	25°C		2.0		V
Bias current		25°C		4.3	8	mA
Bias current change	V_I = 10.5 V to 25 V	0°C to 125°C			1	mA
	I_O = 5 mA to 1 A				0.5	
Short-circuit output current		25°C		450		mA
Peak output current		25°C		2.2		A

†All characteristics are measured with a capacitor across the input of 0.33 µF and a capacitor across the output of 0.1 µF. All characteristics except noise voltage and ripple rejection ratio are measured using pulse techniques (t_w ⩽ 10 ms, duty cycle ⩽ 5%). Output voltage changes due to changes in internal temperature must be taken into account separately.

Voltage Regulators

6

1283

uA7810C electrical characteristics at specified virtual junction temperature, V_I = 17 V, I_O = 500 mA (unless otherwise noted)

PARAMETER	TEST CONDITIONS†		uA7810C			UNIT
			MIN	TYP	MAX	
Output voltage	I_O = 5 mA to 1 A, V_I = 12.5 V to 25 V, P ≤ 15 W	25°C	9.6	10	10.4	V
		0°C to 125°C	9.5	10	10.5	
Input regulation	V_I = 12.5 V to 28 V	25°C		7	200	mV
	V_I = 14 V to 20 V			2	100	
Ripple rejection	V_I = 13 V to 23 V, f = 120 Hz	0°C to 125°C	55	71		dB
Output regulation	I_O = 5 mA to 1.5 A	25°C		12	200	mV
	I_O = 250 mA to 750 mA			4	100	
Output resistance	f = 1 kHz	0°C to 125°C		0.018		Ω
Temperature coefficient of output voltage	I_O = 5 mA	0°C to 125°C		−1.0		mV/°C
Output noise voltage	f = 10 Hz to 100 kHz	25°C		70		µV
Dropout voltage	I_O = 1 A	25°C		2.0		V
Bias current		25°C		4.3	8	mA
Bias current change	V_I = 12.5 V to 28 V	0°C to 125°C			1	mA
	I_O = 5 mA to 1 A				0.5	
Short-circuit output current		25°C		400		mA
Peak output current		25°C		2.2		A

uA7812C electrical characteristics at specified virtual junction temperature, V_I = 19 V, I_O = 500 mA (unless otherwise noted)

PARAMETER	TEST CONDITIONS†		uA7812C			UNIT
			MIN	TYP	MAX	
Output voltage	I_O = 5 mA to 1 A, V_I = 14.5 V to 27 V, P ≤ 15 W	25°C	11.5	12	12.5	V
		0°C to 125°C	11.4		12.6	
Input regulation	V_I = 14.5 V to 30 V	25°C		10	240	mV
	V_I = 16 V to 22 V			3	120	
Ripple rejection	V_I = 15 V to 25 V, f = 120 Hz	0°C to 125°C	55	71		dB
Output regulation	I_O = 5 mA to 1.5 A	25°C		12	240	mV
	I_O = 250 mA to 750 mA			4	120	
Output resistance	f = 1 kHz	0°C to 125°C		0.018		Ω
Temperature coefficient of output voltage	I_O = 5 mA	0°C to 125°C		−1.0		mV/°C
Output noise voltage	f = 10 Hz to 100 kHz	25°C		75		µV
Dropout voltage	I_O = 1 A	25°C		2.0		V
Bias current		25°C		4.3	8	mA
Bias current change	V_I = 14.5 V to 30 V	0°C to 125°C			1	mA
	I_O = 5 mA to 1 A				0.5	
Short-circuit output current		25°C		350		mA
Peak output current		25°C		2.2		A

†All characteristics are measured with a capacitor across the input of 0.33 µF and a capacitor across the output of 0.1 µF. All characteristics except noise voltage and ripple rejection ratio are measured using pulse techniques (t_w ≤ 10 ms, duty cycle ≤ 5%). Output voltage changes due to changes in internal temperature must be taken into account separately.

Voltage Regulators

6

TEXAS INSTRUMENTS
POST OFFICE BOX 225012 ● DALLAS, TEXAS 75265

uA7815C electrical characteristics at specified virtual junction temperature, V_I = 23 V, I_O = 500 mA (unless otherwise noted)

PARAMETER	TEST CONDITIONS[†]		uA7815C			UNIT
			MIN	TYP	MAX	
Output voltage	I_O = 5 mA to 1 A, $\quad$ V_I = 17.5 V to 30 V, P ≤ 15 W	25°C	14.4	15	15.6	V
		0°C to 125°C	14.25		15.75	
Input regulation	V_I = 17.5 V to 30 V	25°C		11	300	mV
	V_I = 20 V to 26 V			3	150	
Ripple rejection	V_I = 18.5 V to 28.5 V, $\quad$ f = 120 Hz	0°C to 125°C	54	70		dB
Output regulation	I_O = 5 mA to 1.5 A	25°C		12	300	mV
	I_O = 250 mA to 750 mA			4	150	
Output resistance	f = 1 kHz	0°C to 125°C		0.019		Ω
Temperature coefficient of output voltage	I_O = 5 mA	0°C to 125°C		−1.0		mV/°C
Output noise voltage	f = 10 Hz to 100 kHz	25°C		90		µV
Dropout voltage	I_O = 1 A	25°C		2.0		V
Bias current		25°C		4.4	8	mA
Bias current change	V_I = 17.5 V to 30 V	0°C to 125°C			1	mA
	I_O = 5 mA to 1 A				0.5	
Short-circuit output current		25°C		230		mA
Peak output current		25°C		2.1		A

uA7818C electrical characteristics at specified virtual junction temperature, V_I = 27 V, I_O = 500 mA (unless otherwise noted)

PARAMETER	TEST CONDITIONS[†]		uA7818C			UNIT
			MIN	TYP	MAX	
Output voltage	I_O = 5 mA to 1 A, $\quad$ V_I = 21 V to 33 V, P ≤ 15 W	25°C	17.3	18	18.7	V
		0°C to 125°C	17.1		18.9	
Input regulation	V_I = 21 V to 33 V	25°C		15	360	mV
	V_I = 24 V to 30 V			5	180	
Ripple rejection	V_I = 22 V to 32 V, $\quad$ f = 120 Hz	0°C to 125°C	53	69		dB
Output regulation	I_O = 5 mA to 1.5 A	25°C		12	360	mV
	I_O = 250 mA to 750 mA			4	180	
Output resistance	f = 1 kHz	0°C to 125°C		0.022		Ω
Temperature coefficient of output voltage	I_O = 5 mA	0°C to 125°C		−1.0		mV/°C
Output noise voltage	f = 10 Hz to 100 kHz	25°C		110		µV
Dropout voltage	I_O = 1 A	25°C		2.0		V
Bias current		25°C		4.5	8	mA
Bias current change	V_I = 21 V to 33 V	0°C to 125°C			1	mA
	I_O = 5 mA to 1 A				0.5	
Short-circuit output current		25°C		200		mA
Peak output current		25°C		2.1		A

[†]All characteristics are measured with a capacitor across the input of 0.33 µF and a capacitor across the output of 0.1 µF. All characteristics except noise voltage and ripple rejection ratio are measured using pulse techniques (t_w ≤ 10 ms, duty cycle ≤ 5%). Output voltage changes due to changes in internal temperature must be taken into account separately.

Voltage Regulators

6

TEXAS INSTRUMENTS
POST OFFICE BOX 225012 ● DALLAS, TEXAS 75265

128

uA7824C electrical characteristics at specified virtual junction temperature,
V_I = 33 V, I_O = 500 mA (unless otherwise noted)

PARAMETER	TEST CONDITIONS†		uA7824C			UNIT
			MIN	TYP	MAX	
Output voltage	I_O = 5 mA to 1 A, V_I = 27 V to 38 V, P ≤ 15 W	25°C	23	24	25	V
		0°C to 125°C	22.8		25.2	
Input regulation	V_I = 27 V to 38 V	25°C		18	480	mV
	V_I = 30 V to 36 V			6	240	
Ripple rejection	V_I = 28 V to 38 V, f = 120 Hz	0°C to 125°C	50	66		dB
Output regulation	I_O = 5 mA to 1.5 A	25°C		12	480	mV
	I_O = 250 mA to 750 mA			4	240	
Output resistance	f = 1 kHz	0°C to 125°C		0.028		Ω
Temperature coefficient of output voltage	I_O = 5 mA	0°C to 125°C		−1.5		mV/°C
Output noise voltage	f = 10 Hz to 100 kHz	25°C		170		µV
Dropout voltage	I_O = 1 A	25°C		2.0		V
Bias current		25°C		4.6	8	mA
Bias current change	V_I = 27 V to 38 V	0°C to 125°C			1	mA
	I_O = 5 mA to 1 A				0.5	
Short-circuit output current		25°C		150		mA
Peak output current		25°C		2.1		A

†All characteristics are measured with a capacitor across the input of 0.33 µF and a capacitor across the output of 0.1 µF. All characteristics except noise voltage and ripple rejection ratio are measured using pulse techniques (t_w ≤ 10 ms, duty cycle ≤ 5%). Output voltage changes due to changes in internal temperature must be taken into account separately.

Voltage Regulators

6

TEXAS INSTRUMENTS
POST OFFICE BOX 225012 ● DALLAS, TEXAS 75265

LINEAR
INTEGRATED
CIRCUITS

SERIES uA78L00
POSITIVE-VOLTAGE REGULATORS
D2203, JANUARY 1976—REVISED AUGUST 1983

- 3-Terminal Regulators
- Output Current up to 100 mA
- No External Components
- Internal Thermal Overload Protection
- Unusually High Power Dissipation Capability
- Internal Short-Circuit Current Limiting
- Direct Replacement for Fairchild μA78L00 Series

NOMINAL OUTPUT VOLTAGE	5% OUTPUT VOLTAGE TOLERANCE	10% OUTPUT VOLTAGE TOLERANCE
2.6 V	uA78L02AC	uA78L02C
5 V	uA78L05AC	uA78L05C
6.2 V	UA78L06AC	uA78L06C
8 V	uA78L08AC	uA78L08C
9 V	uA78L09AC	uA78L09C
10 V	uA78L10AC	uA78L10C
12 V	uA78L12AC	uA78L12C
15 V	uA78L15AC	uA78L15C

description

This series of fixed-voltage monolithic integrated-circuit voltage regulators is designed for a wide range of applications. These applications include on-card regulation for elimination of noise and distribution problems associated with single-point regulation. In addition, they can be used with power-pass elements to make high-current voltage regulators. One of these regulators can deliver up to 100 mA of output current. The internal current limiting and thermal shutdown features of these regulators make them essentially immune to overload. When used as a replacement for a Zener-diode—resistor combination, an effective improvement in output impedance of typically two orders of magnitude can be obtained together with lower-bias current.

JG
DUAL-IN-LINE PACKAGE

(TOP VIEW)

```
INPUT   1   8   OUTPUT
NC      2   7   NC
NC      3   6   COMMON
NC      4   5   NC
```

LP
SILECT PACKAGE

(TOP VIEW)

INPUT
COMMON
OUTPUT

TO-226AA

NC—No internal connection

schematic

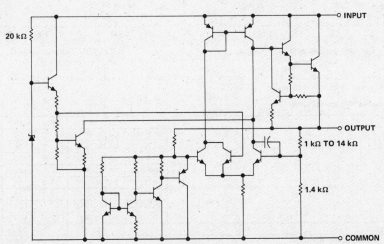

20 kΩ

INPUT

OUTPUT

1 kΩ TO 14 kΩ

1.4 kΩ

COMMON

Resistor values shown are nominal

883

Voltage Regulators

6

TEXAS
INSTRUMENTS
POST OFFICE BOX 225012 • DALLAS, TEXAS 75265

Voltage Regulators

6

absolute maximum ratings over operating temperature range (unless otherwise noted)

		uA78L02AC, uA78L02C THRU uA78L10AC, uA78L10C	uA78L12AC, uA78L12C uA78L15AC, uA78L15C	UNIT
Input voltage		30	35	V
Continuous total dissipation at 25 °C free-air temperature (see Note 1)	JG package	825	825	mW
	LP package	775	775	
Continuous total dissipation at (or below) 25 °C case temperature (see Note 1)		1600	1600	mW
Operating free-air, case, or virtual junction temperature range		0 to 150	0 to 150	°C
Storage temperature range		−65 to 150	−65 to 150	°C
Lead temperature 1,6 mm (1/16 inch) from case for 10 seconds		260	260	°C

NOTE 1: For operation above 25 °C free-air temperature, refer to Figures 1 and 2. To avoid exceeding the design maximum virtual junction temperature, these ratings should not be exceeded. Due to variations in individual device electrical characteristics and thermal resistance, the built-in thermal overload protection may be activated at power levels slightly above or below the rated dissipation.

FREE-AIR TEMPERATURE DISSIPATION DERATING CURVE

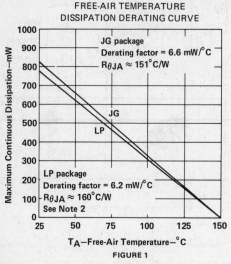

FIGURE 1

CASE TEMPERATURE DISSIPATION DERATING CURVE

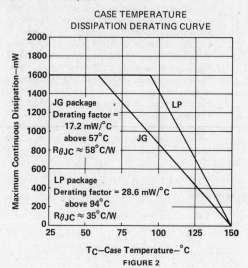

FIGURE 2

NOTE 2: This curve for the LP package is based on thermal resistance, $R_{\theta JA}$, measured in still air with the device mounted in an Augat socket. The bottom of the package was 3/8 inch above the socket.

recommended operating conditions

		MIN	MAX	UNIT
Input voltage, V_I	uA78L02C, uA78L02AC	4.75	20	V
	uA78L05C, uA78L05AC	7	20	
	uA78L06C, uA78L06AC	8.5	20	
	uA78L08C, uA78L08AC	10.5	23	
	uA78L09C, uA78L09AC	11.5	24	
	uA78L10C, uA78L10AC	12.5	25	
	uA78L12C, uA78L12AC	14.5	27	
	uA78L15C, uA78L15AC	17.5	30	
Output current, I_O			100	mA
Operating virtual junction temperature, T_J		0	125	°C

883

TEXAS INSTRUMENTS

POST OFFICE BOX 225012 • DALLAS, TEXAS 75265

uA78L02AC, uA78L02C electrical characteristics at specified virtual junction temperature, V_I = 9 V, I_O = 40 mA (unless otherwise noted)

PARAMETER	TEST CONDITIONS†		uA78L02AC			uA78L02C			UNIT
			MIN	TYP	MAX	MIN	TYP	MAX	
Output voltage		25°C	2.5	2.6	2.7	2.4	2.6	2.8	V
	V_I = 4.75 V to 20 V, I_O = 1 mA to 40 mA	0°C to 125°C	2.45		2.75	2.35		2.85	
	I_O = 1 mA to 70 mA		2.45		2.75	2.35		2.85	
Input regulation	V_I = 4.75 V to 20 V	25°C		20	100		20	125	mV
	V_I = 5 V to 20 V			16	75		16	100	
Ripple rejection	V_I = 6 V to 16 V, f = 120 Hz	25°C	43	51		42	51		dB
Output regulation	I_O = 1 mA to 100 mA	25°C		12	50		12	50	mV
	I_O = 1 mA to 40 mA			6	25		6	25	
Output noise voltage	f = 10 Hz to 100 kHz	25°C		30			30		µV
Dropout voltage		25°C		1.7			1.7		V
Bias current		25°C		3.6	6		3.6	6	mA
		125°C			5.5			5.5	
Bias current change	V_I = 5 V to 20 V	0°C to 125°C			2.5			2.5	mA
	I_O = 1 mA to 40 mA				0.1			0.2	

uA78L05AC, uA78L05C electrical characteristics at specified virtual junction temperature, V_I = 10 V, I_O = 40 mA (unless otherwise noted)

PARAMETER	TEST CONDITIONS†		uA78L05AC			uA78L05C			UNIT
			MIN	TYP	MAX	MIN	TYP	MAX	
Output voltage		25°C	4.8	5	5.2	4.6	5	5.4	V
	V_I = 7 V to 20 V, I_O = 1 mA to 40 mA	0°C to 125°C	4.75		5.25	4.5		5.5	
	I_O = 1 mA to 70 mA		4.75		5.25	4.5		5.5	
Input regulation	V_I = 7 V to 20 V	25°C		32	150		32	200	mV
	V_I = 8 V to 20 V			26	100		26	150	
Ripple rejection	V_I = 8 V to 18 V, f = 120 Hz	25°C	41	49		40	49		dB
Output regulation	I_O = 1 mA to 100 mA	25°C		15	60		15	60	mV
	I_O = 1 mA to 40 mA			8	30		8	30	
Output noise voltage	f = 10 Hz to 100 kHz	25°C		42			42		µV
Dropout voltage		25°C		1.7			1.7		V
Bias current		25°C		3.8	6		3.8	6	mA
		125°C			5.5			5.5	
Bias current change	V_I = 8 V to 20 V	0°C to 125°C			1.5			1.5	mA
	I_O = 1 mA to 40 mA				0.1			0.2	

† All characteristics are measured with a capacitor across the input of 0.33 µF and a capacitor across the output of 0.1 µF. All characteristics except noise voltage and ripple rejection ratio are measured using pulse tchniques (t_w ≤ 10 ms, duty cycle ≤ 5%). Output voltage changes due to changes in internal temperature must be taken into account separately.

Voltage Regulators

6

983

TEXAS
INSTRUMENTS
POST OFFICE BOX 225012 • DALLAS, TEXAS 75265

uA78L06AC, uA78L06C electrical characteristics at specified virtual junction temperature,
V_I = 12 V, I_O = 40 mA (unless otherwise noted)

PARAMETER	TEST CONDITIONS[†]		uA78L06AC			uA78L06C			UNIT
			MIN	TYP	MAX	MIN	TYP	MAX	
Output voltage	V_I = 8.5 V to 20 V, I_O = 1 mA to 40 mA	25°C	5.95	6.2	6.45	5.7	6.2	6.7	V
		0°C to 125°C	5.9		6.5	5.6		6.8	
	I_O = 1 mA to 70 mA		5.9		6.5	5.6		6.8	
Input regulation	V_I = 8.5 V to 20 V	25°C		35	175		35	200	mV
	V_I = 9 V to 20 V			29	125		29	150	
Ripple rejection	V_I = 10 V to 20 V, f = 120 Hz	25°C	40	48		39	48		dB
Output regulation	I_O = 1 mA to 100 mA	25°C		16	80		16	80	mV
	I_O = 1 mA to 40 mA			9	40		9	40	
Output noise voltage	f = 10 Hz to 100 kHz	25°C		46			46		μV
Dropout voltage		25°C		1.7			1.7		V
Bias current		25°C		3.9	6		3.9	6	mA
		125°C			5.5			5.5	
Bias current change	V_I = 9 V to 20 V	0°C to 125°C			1.5			1.5	mA
	I_O = 1 mA to 40 mA				0.1			0.2	

uA78L08AC, uA78L08C electrical characteristics at specified virtual junction temperature,
V_I = 14 V, I_O = 40 mA (unless otherwise noted)

PARAMETER	TEST CONDITIONS[†]		uA78L08AC			uA78L08C			UNIT
			MIN	TYP	MAX	MIN	TYP	MAX	
Output voltage	V_I = 10.5 V to 23 V, I_O = 1 mA to 40 mA	25°C	7.7	8	8.3	7.36	8	8.64	V
		0°C to 125°C	7.6		8.4	7.2		8.8	
	I_O = 1 mA to 70 mA		7.6		8.4	7.2		8.8	
Input regulation	V_I = 10.5 V to 23 V	25°C		42	175		42	200	mV
	V_I = 11 V to 23 V			36	125		36	150	
Ripple rejection	V_I = 13 V to 23 V, f = 120 Hz	25°C	37	46		36	46		dB
Output regulation	I_O = 1 mA to 100 mA	25°C		18	80		18	80	mV
	I_O = 1 mA to 40 mA			10	40		10	40	
Output noise voltage	f = 10 Hz to 100 kHz	25°C		54			54		μV
Dropout voltage		25°C		1.7			1.7		V
Bias current		25°C		4	6		4	6	mA
		125°C			5.5			5.5	
Bias current change	V_I = 11 V to 23 V	0°C to 125°C			1.5			1.5	mA
	I_O = 1 mA to 40 mA				0.1			0.2	

[†] All characteristics are measured with a capacitor across the input of 0.33 μF and a capacitor across the output of 0.1 μF. All characteristics except noise voltage and ripple rejection ratio are measured using pulse techniques (t_W ≤ 10 ms, duty cycle ≤ 5%). Output voltage changes due to changes in internal temperature must be taken into account separately.

Voltage Regulators

6

Texas
Instruments
POST OFFICE BOX 225012 • DALLAS, TEXAS 75265

883

uA78L09AC, uA78L09C electrical characteristics at specified virtual junction temperature,
V_I = 16 V, I_O = 40 mA (unless otherwise noted)

PARAMETER	TEST CONDITIONS[†]		uA78L09AC			uA78L09C			UNIT
			MIN	TYP	MAX	MIN	TYP	MAX	
Output voltage	V_I = 12 V to 24 V, I_O = 1 mA to 40 mA	25°C	8.6	9	9.4	8.3	9	9.7	V
		0°C to 125°C	8.55		9.45	8.1		9.9	
	I_O = 1 mA to 70 mA		8.55		9.45	8.1		9.9	
Input regulation	V_I = 12 V to 24 V	25°C		45	175		45	225	mV
	V_I = 13 V to 24 V			40	125		40	175	
Ripple rejection	V_I = 13 V to 24 V, f = 120 Hz	25°C	37	45		36	45		dB
Output regulation	I_O = 1 mA to 100 mA	25°C		19	90		19	90	mV
	I_O = 1 mA to 40 mA	25°C		11	40		11	40	
Output noise voltage	f = 10 Hz to 100 kHz	25°C		58			58		µV
Dropout voltage		25°C		1.7			1.7		V
Bias current		25°C		4.1	6		4.1	6	mA
		125°C			5.5			5.5	
Bias current change	V_I = 13 V to 24 V	0°C to 125°C			1.5			1.5	mA
	I_O = 1 mA to 40 mA				0.1			0.2	

uA78L10AC, uA78L10C electrical characteristics at specified virtual junction temperature,
V_I = 17 V, I_O = 40 mA (unless otherwise noted)

PARAMETER	TEST CONDITIONS[†]		uA78L10AC			uA78L10C			UNIT
			MIN	TYP	MAX	MIN	TYP	MAX	
Output voltage	V_I = 13 V to 25 V, I_O = 1 mA to 40 mA	25°C	9.6	10	10.4	9.2	10	10.8	V
		0°C to 125°C	9.5		10.5	9		10	
	I_O = 1 mA to 70 mA		9.5		10.5	9		10	
Input regulation	V_I = 13 V to 25 V	25°C		51	175		51	225	mV
	V_I = 14 V to 25 V			42	125		42	175	
Ripple rejection	V_I = 14 V to 25 V, f = 120 Hz	25°C	37	44		36	44		dB
Output regulation	I_O = 1 mA to 100 mA	25°C		20	90		20	90	mV
	I_O = 1 mA to 40 mA	25°C		11	40		11	40	
Output noise voltage	f = 10 Hz to 100 kHz	25°C		62			62		µV
Dropout voltage		25°C		1.7			1.7		V
Bias current		25°C		4.2	6		4.2	6	mA
		125°C			5.5			5.5	
Bias current change	V_I = 14 V to 25 V	0°C to 125°C			1.5			1.5	mA
	I_O = 1 mA to 40 mA				0.1			0.2	

† All characteristics are measured with a capacitor across the input of 0.33 µF and a capacitor across the output of 0.1 µF. All characteristics except noise voltage and ripple rejection ratio are measured using pulse techniques (t_w ≤ 10 ms, duty cycle ≤ 5%). Output voltage changes due to changes in internal temperature must be taken into account separately.

Voltage Regulators

6

TEXAS
INSTRUMENTS
POST OFFICE BOX 225012 • DALLAS, TEXAS 75265

uA78L12AC, uA78L12C electrical characteristics at specified virtual junction temperature,
$V_I = 19$ V, $I_O = 40$ mA (unless otherwise noted)

PARAMETER	TEST CONDITIONS[†]		uA78L12AC			uA78L12C			UNIT
			MIN	TYP	MAX	MIN	TYP	MAX	
Output voltage		25°C	11.5	12	12.5	11.1	12	12.9	V
	V_I = 14.5 V to 27 V, I_O = 1 mA to 40 mA	0°C to 125°C	11.4		12.6	10.8		13.2	
	I_O = 1 mA to 70 mA		11.4		12.6	10.8		13.2	
Input regulation	V_I = 14.5 V to 27 V	25°C		55	250		55	250	mV
	V_I = 16 V to 27 V			49	200		49	200	
Ripple rejection	V_I = 15 V to 25 V, f = 120 Hz	25°C	37	42		36	42		dB
Output regulation	I_O = 1 mA to 100 mA	25°C		22	100		22	100	mV
	I_O = 1 mA to 40 mA			13	50		13	50	
Output noise voltage	f = 10 Hz to 100 kHz	25°C		70			70		μV
Dropout voltage		25°C		1.7			1.7		V
Bias current		25°C		4.3	6.5		4.3	6.5	mA
		125°C			6			6	
Bias current change	V_I = 16 V to 27 V	0°C to 125°C			1.5			1.5	mA
	I_O = 1 mA to 40 mA				0.1			0.2	

uA78L15AC, uA78L15C electrical characteristics at specified virtual junction temperature,
$V_I = 23$ V, $I_O = 40$ mA (unless otherwise noted)

PARAMETER	TEST CONDITIONS[†]		uA78L15AC			uA78L15C			UNIT
			MIN	TYP	MAX	MIN	TYP	MAX	
Output voltage		25°C	14.4	15	15.6	13.8	15	16.2	V
	V_I = 17.5 V to 30 V, I_O = 1 mA to 40 mA	0°C to 125°C	14.25		15.75	13.5		16.5	
	I_O = 1 mA to 70 mA		14.25		15.75	13.5		16.5	
Input regulation	V_I = 17.5 V to 30 V	25°C		65	300		65	300	mV
	V_I = 20 V to 30 V			58	250		58	250	
Ripple rejection	V_I = 18.5 to 28.5 V, f = 120 Hz	25°C	34	39		33	39		dB
Output regulation	I_O = 1 mA to 100 mA	25°C		25	150		25	150	mV
	I_O = 1 mA to 40 mA			15	75		15	75	
Output noise voltage	f = 10 Hz to 100 kHz	25°C		82			82		μV
Dropout voltage		25°C		1.7			1.7		V
Bias current		25°C		4.6	6.5		4.6	6.5	mA
		125°C			6			6	
Bias current change	V_I = 10 V to 30 V	0°C to 125°C			1.5			1.5	mA
	I_O = 1 mA to 40 mA				0.1			0.2	

[†] All characteristics are measured with a capacitor across the input of 0.33 μF and a capacitor across the output of 0.1 μF. All characteristics except noise voltage and ripple rejection ratio are measured using pulse techniques ($t_w \leq$ 10 ms, duty cycle $\leq$ 5%). Output voltage changes due to changes in internal temperature must be taken into account separately.

TEXAS
INSTRUMENTS

POST OFFICE BOX 225012 • DALLAS, TEXAS 75265

LINEAR INTEGRATED CIRCUITS

- 3-Terminal Regulators
- Output Current up to 500 mA
- No External Components
- Internal Thermal Overload Protection
- High Power Dissipation Capability
- Internal Short-Circuit Current Limiting
- Output Transistor Safe-Area Compensation
- Direct Replacements for Fairchild μA78M00 Series and National LM78MXX and LM341 Series

NOMINAL OUTPUT VOLTAGE	−55°C to 150°C OPERATING TEMPERATURE RANGE	0°C to 125°C OPERATING TEMPERATURE RANGE
5 V	uA78M05M	uA78M05C
6 V	uA78M06M	uA78M06C
8V	uA78M08M	uA78M08C
10 V	uA78M10M	uA78M10C
12 V	uA78M12M	uA78M12C
15 V	uA78M15M	uA78M15C
20 V		uA78M20C
24 V		uA78M24C
PACKAGES	JG	KC

description

This series of fixed-voltage monolithic integrated-circuit voltage regulators is designed for a wide range of applications. These applications include on-card regulation for elimination of noise and distribution problems associated with single-point regulation. Each of these regulators can deliver up to 500 milliamperes of output current. The internal current limiting and thermal shutdown features of these regulators make them essentially immune to overload. In addition to use as fixed-voltage regulators, these devices can be used with external components to obtain adjustable output voltages and currents and also as the power pass element in precision regulators.

schematic

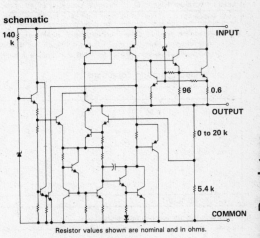

Resistor values shown are nominal and in ohms.

terminal assignments

uA78M_M ... JG PACKAGE

(TOP VIEW)

COMMON 1 8 NC
NC 2 7 NC
NC 3 6 OUTPUT
INPUT 4 5 NC

NC — No internal connection.

uA78M_C ... KC PACKAGE

(TOP VIEW)

OUTPUT
COMMON
INPUT

THE COMMON TERMINAL IS IN ELECTRICAL CONTACT WITH THE MOUNTING BASE

TO-220AB

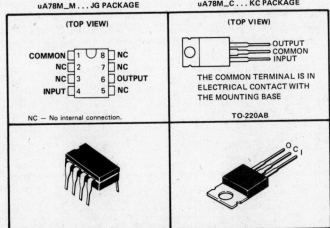

Voltage Regulators

6

TEXAS INSTRUMENTS

POST OFFICE BOX 225012 • DALLAS, TEXAS 75265

absolute maximum ratings over operating temperature range (unless otherwise noted)

		uA78M05M THRU uA78M24M	uA78M05C THRU uA78M24C	UNIT
Input voltage	uA78M20 thru uA78M24		40	V
	All others	35	35	
Continuous total dissipation at 25°C free-air temperature (see Note 1)	JG package	1.05		W
	KC (TO-220AB) package		2	
Continuous total dissipation at (or below) 25°C case temperature (see Note 1)	KC package		7.5	W
Operating free-air, case, or virtual junction temperature range		−55 to 150	0 to 150	°C
Storage temperature range		−65 to 150	−65 to 150	°C
Lead temperature 1,6 mm (1/16 inch) from case for 60 seconds	JG package	300		°C
Lead temperature 1,6 mm (1/16 inch) from case for 10 seconds	KC package		260	°C

NOTE 1: For operation above 25°C free-air or case temperature, refer to Figures 1 through 3. To avoid exceeding the design maximum virtual junction temperature, these ratings should not be exceeded. Due to variations in individual device electrical characteristics and thermal resistance, the built-in thermal overload protection may be activated at power levels slightly above or below the rated dissipation.

recommended operating conditions

		MIN	MAX	UNIT
Input voltage, V_I	uA78M05M, uA78M05C	7	25	V
	uA78M06M, uA78M06C	8	25	
	uA78M08M, uA78M08C	10.5	25	
	uA78M10M, uA78M10C	12.5	28	
	uA78M12M, uA78M12C	14.5	30	
	uA78M15M, uA78M15C	17.5	30	
	uA78M20C	23	35	
	uA78M24C	27	38	
Output current, I_O	All devices		500	mA
Operating virtual junction temperature, T_J	uA78M05M thru uA78M15M	−55	150	°C
	uA78M05C thru uA78M24C	0	125	

983

TEXAS
INSTRUMENTS
POST OFFICE BOX 225012 • DALLAS, TEXAS 75265

uA78M05M, uA78M05C electrical characteristics at specified virtual junction temperature,
V_I = 10 V, I_O = 350 mA (unless otherwise noted)

PARAMETER	TEST CONDITIONS		uA78M05M			uA78M05C			UNIT
			MIN	TYP	MAX	MIN	TYP	MAX	
Output voltage	I_O = 5 mA to 350 mA	25°C	4.8	5	5.2	4.8	5	5.2	V
	V_I = 8 V to 20 V	−55°C to 150°C	4.7		5.3				
	V_I = 7 V to 20 V	0°C to 125°C				4.75		5.25	
Input regulation	I_O = 200 mA	V_I = 7 V to 25 V	25°C		3	50		3	100	mV
		V_I = 8 V to 20 V	25°C		1	25				
		V_I = 8 V to 25 V	25°C					1	50	
Ripple rejection	I_O = 100 mA, V_I = 8 V to 18 V, f = 120 Hz	−55°C to 150°C	62						dB
	I_O = 300 mA	0°C to 125°C				62			
		25°C	62	80		62	80		
Output regulation	I_O = 5 mA to 500 mA	25°C		20	50		20	100	mV
	I_O = 5 mA to 200 mA			10	25		10	50	
Temperature coefficient of output voltage	I_O = 5 mA	−55°C to 25°C			−2				mV/°C
		25°C to 150°C			−1.5				
		0°C to 125°C					−1		
Output noise voltage	f = 10 Hz to 100 kHz	25°C		40	200		40	200	µV
Dropout voltage		25°C		2	2.5		2	2.5	V
Bias current		25°C		4.5	7		4.5	6	mA
Bias current change	I_O = 200 mA, V_I = 8 V to 25 V	−55°C to 150°C			0.8			0.8	mA
		0°C to 125°C							
	I_O = 5 mA to 350 mA	−55°C to 150°C			0.5			0.5	
		0°C to 125°C							
Short-circuit output current	V_I = 35 V	25°C		300			300		mA
Peak output current		25°C	0.5	0.7	1.4		0.7		A

†All characteristics are measured with a capacitor across the input of 0.33 µF and a capacitor across the output of 0.1 µF. All characteristics except noise voltage and ripple rejection ratio are measured using pulse techniques (t_w ≤ 10 ms, duty cycle ≤ 5%). Output voltage changes due to changes in internal temperature must be taken into account separately.

6

Voltage Regulators

TEXAS
INSTRUMENTS
POST OFFICE BOX 225012 • DALLAS, TEXAS 75265

uA78M06M, uA78M06C electrical characteristics at specified virtual junction temperature, V_I = 11 V, I_O = 350 mA (unless otherwise noted)

PARAMETER	TEST CONDITIONS†	uA78M06M			uA78M06C			UNIT
		MIN	TYP	MAX	MIN	TYP	MAX	
Output voltage	I_O = 5 mA to 350 mA, 25°C	5.75	6	6.25	5.75	6	6.25	V
	V_I = 9 V to 21 V, −55°C to 150°C	5.7		6.3				
	V_I = 8 V to 21 V, 0°C to 125°C				5.7		6.3	
Input regulation	I_O = 200 mA, V_I = 8 V to 25 V, 25°C		5	60		5	100	mV
	V_I = 9 V to 20 V, 25°C		1.5	30				
	V_I = 9 V to 25 V, 25°C					1.5	50	
Ripple rejection	V_I = 9 V to 19 V, f = 120 Hz, I_O = 100 mA, −55°C to 150°C	59						dB
	I_O = 100 mA, 0°C to 125°C				59			
	I_O = 300 mA, 25°C	59			59			
Output regulation	I_O = 5 mA to 500 mA, 25°C		20	60		20	120	mV
	I_O = 5 mA to 200 mA, 25°C		10	30		10	60	
Temperature coefficient of output voltage	I_O = 5 mA, −55°C to 25°C			−2.4				mV/°C
	25°C to 150°C			−1.8				
	0°C to 125°C					−1		
Output noise voltage	f = 10 Hz to 100 kHz, 25°C		45			45		µV
Dropout voltage	25°C		2	2.5		2		V
Bias current	25°C		4.5	7		4.5	6	mA
Bias current change	I_O = 200 mA, V_I = 9 V to 25 V, −55°C to 150°C			0.8				mA
	0°C to 125°C						0.8	
	I_O = 5 mA to 350 mA, −55°C to 150°C			0.5				
	0°C to 125°C						0.5	
Short-circuit output current	V_I = 35 V, 25°C		270	600		270		mA
Peak output current	25°C	0.5	0.7	1.4		0.7		A

†All characteristics are measured with a capacitor across the input of 0.33 µF and a capacitor across the output of 0.1 µF. All characteristics except noise voltage and ripple rejection ratio are measured using pulse techniques ($t_w \leq$ 10 ms, duty cycle $\leq$ 5%). Output voltage changes due to changes in internal temperature must be taken into account separately.

TEXAS INSTRUMENTS
POST OFFICE BOX 225012 • DALLAS, TEXAS 75265

uA78M08M, uA78M08C electrical characteristics at specified virtual junction temperature,
V_I = 14 V, I_O = 350 mA (unless otherwise noted)

PARAMETER	TEST CONDITIONS†	uA78M08M MIN	uA78M08M TYP	uA78M08M MAX	uA78M08C MIN	uA78M08C TYP	uA78M08C MAX	UNIT
Output voltage	I_O = 5 mA to 350 mA, 25°C	7.7	8	8.3	7.7	8	8.3	V
	V_I = 11.5 V to 23 V, −55°C to 150°C	7.6		8.4				
	V_I = 10.5 V to 23 V, 0°C to 125°C				7.6		8.4	
Input regulation	V_I = 10.5 V to 25 V, I_O = 200 mA, 25°C		6	60		6	100	mV
	V_I = 11 V to 20 V (V_I = 11 V to 25 V), 25°C		2	30		2	50	
Ripple rejection	V_I = 11.5 V to 21.5 V, I_O = 100 mA, f = 120 Hz, −55°C to 150°C	56						dB
	0°C to 125°C				56	80		
	25°C	56	80		56	80		
Output regulation	I_O = 5 mA to 500 mA, 25°C		25	80		25	160	mV
	I_O = 5 mA to 200 mA, 25°C		10	40		10	80	
Temperature coefficient of output voltage	I_O = 5 mA, −55°C to 25°C			−3.2				mV/°C
	25°C to 150°C			−2.4				
	0°C to 125°C					−1		
Output noise voltage	f = 10 Hz to 100 kHz, 25°C		52	320		52		µV
Dropout voltage	25°C		2	2.5		2		V
Bias current	25°C		4.6	7		4.6	6	mA
Bias current change	I_O = 200 mA, V_I = 11.5 V to 25 V, −55°C to 150°C			0.8				mA
	V_I = 10.5 V to 25 V, 0°C to 125°C						0.8	
	I_O = 5 mA to 350 mA, −55°C to 150°C			0.5				
	0°C to 125°C						0.5	
Short-circuit output current	V_I = 35 V, 25°C		250	600		250		mA
Peak output current	25°C	0.5	0.7	1.4		0.7		A

†All characteristics are measured with a capacitor across the input of 0.33 µF and a capacitor across the output of 0.1 µF. All characteristics except noise voltage and ripple rejection ratio are measured using pulse techniques (t_W ≤ 10 ms, duty cycle ≤ 5%). Output voltage changes due to changes in internal temperature must be taken into account separately.

6 Voltage Regulators

TEXAS
INSTRUMENTS
POST OFFICE BOX 225012 • DALLAS, TEXAS 75265

uA78M10M, uA78M10C electrical characteristics at specified virtual junction temperature, V_I = 17 V, I_O = 350 mA (unless otherwise noted)

PARAMETER	TEST CONDITIONS[†]		uA78M10M MIN	TYP	MAX	uA78M10C MIN	TYP	MAX	UNIT
Output voltage	I_O = 5 mA to 350 mA	25°C	9.6	10	10.4	9.6	10	10.4	V
	V_I = 13.5 V to 25 V (M) V_I = 12.5 V to 25 V (C)	−55°C to 150°C (M) 0°C to 125°C (C)	9.5		10.5	9.5		10.5	
Input regulation	I_O = 200 mA	V_I = 12.5 V to 28 V, 25°C		7	60		7	100	mV
		V_I = 14 V to 20 V (M) / V_I = 14 V to 28 V (C), 25°C		2	30		2	50	
Ripple rejection	V_I = 15 V to 25 V, f = 120 Hz	I_O = 100 mA, 25°C	55	80		55	80		dB
Output regulation	I_O = 5 mA to 500 mA	25°C		25	100		25	200	mV
	I_O = 5 mA to 200 mA	25°C		10	50		10	100	
Temperature coefficient of output voltage	I_O = 5 mA	−55°C to 25°C			−4		−1		mV/°C
		25°C to 150°C (M) / 0°C to 125°C (C)			−3				
Output noise voltage	f = 10 Hz to 100 kHz	25°C		64			64		µV
Dropout voltage		25°C		2	2.5		2		V
Bias current		25°C		4.7	6		4.7	6	mA
Bias current change	I_O = 200 mA, V_I = 13.5 V to 28 V (M) / V_I = 12.5 V to 28 V (C)	−55°C to 150°C (M) / 0°C to 125°C (C)			0.8			0.8	mA
	I_O = 5 mA to 350 mA	−55°C to 150°C (M) / 0°C to 125°C (C)			0.5			0.5	
Short-circuit output current	V_I = 35 V	25°C		245	600		245		mA
Peak output current		25°C	0.5	0.7	1.4		0.7		A

[†]All characteristics are measured with a capacitor across the input of 0.33 µF and a capacitor across the output of 0.1 µF. All characteristics except noise voltage and ripple rejection ratio are measured using pulse techniques ($t_w \leq$ 10 ms, duty cycle $\leq$ 5%). Output voltage changes due to changes in internal temperature must be taken into account separately.

TEXAS INSTRUMENTS

POST OFFICE BOX 225012 • DALLAS, TEXAS 75265

uA78M12M, uA78M12C electrical characteristics at specified virtual junction temperature,
V_I = 19 V, I_O = 350 mA (unless otherwise noted)

PARAMETER	TEST CONDITIONS†	uA78M12M			uA78M12C			UNIT
		MIN	TYP	MAX	MIN	TYP	MAX	
Output voltage	I_O = 5 mA to 350 mA, 25°C	11.5	12	12.5	11.5	12	12.5	V
	V_I = 15.5 V to 27 V, −55°C to 150°C	11.4		12.6				
	V_I = 14.5 V to 27 V, 0°C to 125°C				11.4		12.6	
Input regulation	V_I = 16 V to 25 V, 25°C		8	60		8	100	mV
	V_I = 16 V to 30 V, 25°C		2	30		2	50	
Ripple rejection	V_I = 15 V to 25 V, I_O = 100 mA, f = 120 Hz, −55°C to 150°C	55						dB
	I_O = 100 mA, 0°C to 125°C				55			
	I_O = 300 mA, 25°C	55	80		55	80		
Output regulation	I_O = 5 mA to 500 mA, 25°C		25	120		25	240	mV
	I_O = 5 mA to 200 mA, 25°C		10	60		10	120	
Temperature coefficient of output voltage	I_O = 5 mA, −55°C to 25°C			−4.8				mV/°C
	25°C to 150°C			−3.6				
	0°C to 125°C					−1		
Output noise voltage	f = 10 Hz to 100 kHz, 25°C		75	480		75		µV
Dropout voltage	25°C		2	2.5		2		V
Bias current	25°C		4.8	7		4.8	6	mA
Bias current change	I_O = 200 mA, −55°C to 150°C			0.8				mA
	I_O = 200 mA, 0°C to 125°C						0.8	
	V_I = 15 V to 30 V, −55°C to 150°C			0.5				
	V_I = 14.5 V to 30 V, 0°C to 125°C						0.5	
Short-circuit output current	V_I = 35 V, 25°C		240	600		240		mA
Peak output current	25°C	0.5	0.7	1.4		0.7		A

†All characteristics are measured with a capacitor across the input of 0.33 µF and a capacitor across the output of 0.1 µF. All characteristics except noise voltage and ripple rejection ratio are measured using pulse techniques (t_w ≤ 10 ms, duty cycle ≤ 5%). Output voltage changes due to changes in internal temperature must be taken into account separately.

Voltage Regulators

6

Voltage Regulators

6

uA78M15M, uA78M15C electrical characteristics at specified virtual junction temperature, VI = 23 V, IO = 350 mA (unless otherwise noted)

PARAMETER	TEST CONDITIONS†			uA78M15M			uA78M15C			UNIT
				MIN	TYP	MAX	MIN	TYP	MAX	
Output voltage	IO = 5 mA to 350 mA	VI = 18.5 V to 30 V	25°C	14.4	15	15.6	14.4	15	15.6	V
		VI = 17.5 V to 30 V	-55°C to 150°C 0°C to 125°C	14.25		15.75	14.25		15.75	
Input regulation	IO = 200 mA	VI = 17.5 V to 30 V	25°C		10	60		10	100	mV
		VI = 20 V to 30 V	25°C		3	30		3	50	
Ripple rejection	VI = 18.5 V to 28.5 V, f = 120 Hz		-55°C to 150°C 0°C to 125°C	54			54			dB
			25°C	54	70		54	70		
Output regulation	IO = 5 mA to 500 mA		25°C		25	150		25	300	mV
	IO = 5 mA to 200 mA		25°C		10	75		10	150	
Temperature coefficient of output voltage	IO = 5 mA		-55°C to 25°C			-6				mV/°C
			25°C to 150°C			-4.5				
			0°C to 125°C					-1		
Output noise voltage	f = 10 Hz to 100 kHz		25°C		90	600		90		µV
Dropout voltage	IO = 200 mA		25°C		2	2.5		2		V
Bias current			25°C		4.8	7		4.8	6	mA
Bias current change	IO = 200 mA, VI = 18.5 V to 30 V		-55°C to 150°C 0°C to 125°C			0.8			0.8	mA
	IO = 5 mA to 350 mA	VI = 17.5 V to 30 V	-55°C to 150°C 0°C to 125°C			0.5			0.5	
Short-circuit output current	VI = 35 V		25°C		240	600		240		mA
Peak output current			25°C	0.5	0.7	1.4		0.7	0.5	A

†All characteristics are measured with a capacitor across the input of 0.33 µF and a capacitor across the output of 0.1 µF. All characteristics except noise voltage and ripple rejection ratio are measured using pulse techniques (t_w ≤ 10 ms, duty cycle ≤ 5%). Output voltage changes due to changes in internal temperature must be taken into account separately.

TEXAS INSTRUMENTS

POST OFFICE BOX 225012 • DALLAS, TEXAS 75265

383

uA78M20C electrical characteristics at specified virtual junction temperature,
V_I = 29 V, I_O = 350 mA (unless otherwise noted)

PARAMETER	TEST CONDITIONS†			uA78M20C MIN	TYP	MAX	UNIT
Output voltage	I_O = 5 mA to 350 mA,		25°C	19.2	20	20.8	V
	V_I = 23 V to 35 V		0°C to 125°C	19		21	
Input regulation	V_I = 23 V to 35 V	I_O = 200 mA	25°C		10	100	mV
	V_I = 24 V to 35 V		25°C		5	50	
Ripple rejection	V_I = 24 V to 34 V, f = 120 Hz	I_O = 100 mA	0°C to 125°C	53	70		dB
		I_O = 300 mA	25°C	53			
Output regulation	I_O = 5 mA to 500 mA		25°C		30	400	mV
	I_O = 5 mA to 200 mA				10	200	
Temperature coefficient of output voltage	I_O = 5 mA		0°C to 125°C		-1.1		mV/°C
Output noise voltage	f = 10 Hz to 100 kHz		25°C		110		µV
Dropout voltage			25°C		2		V
Bias current			25°C		4.9	6	mA
Bias current change	I_O = 200 mA,	V_I = 23 V to 35 V	0°C to 125°C			0.8	mA
	I_O = 5 mA to 350 mA		0°C to 125°C			0.5	
Short-circuit output current	V_I = 35 V		25°C		240		mA
Peak output current			25°C		0.7		A

†All characteristics are measured with a capacitor across the input of 0.33 µF and a capacitor across the output of 0.1 µF. All characteristics except noise voltage and ripple rejection ratio are measured using pulse techniques $t_w \le$ 10 ms, duty cycle $\le$ 5%). Output voltage changes due to changes in internal temperature must be taken into account separately.

Voltage Regulators

6

Texas
Instruments
POST OFFICE BOX 225012 ● DALLAS, TEXAS 75265

Voltage Regulators

6

uA78M24C electrical characteristics at specified virtual junction temperature,
V_I = 33 V, I_O = 350 mA (unless otherwise noted)

PARAMETER	TEST CONDITIONS†		uA78M24C MIN	TYP	MAX	UNIT
Output voltage	I_O = 5 mA to 350 mA, V_I = 27 V to 38 V	25°C	23	24	25	V
		0°C to 125°C	22.8		25.2	V
Input regulation	I_O = 200 mA, V_I = 27 V to 38 V	25°C		10	100	mV
	V_I = 28 V to 38 V	25°C		5	50	mV
Ripple rejection	V_I = 28 V to 38 V, f = 120 Hz, I_O = 100 mA	−55°C to 150°C	50			dB
	I_O = 300 mA	0°C to 125°C	50	70		dB
Output regulation	I_O = 5 mA to 500 mA	25°C		30	480	mV
	I_O = 5 mA to 200 mA	25°C		10	240	mV
Temperature coefficient of output voltage	I_O = 5 mA	0°C to 125°C		−1.2		mV/°C
Output noise voltage	f = 10 Hz to 100 kHz	25°C		170		µV
Dropout voltage		25°C		2		V
Bias current		25°C		5	6	mA
Bias current change	I_O = 200 mA, V_I = 27 V to 38 V	0°C to 125°C			0.8	mA
	I_O = 5 mA to 350 mA	0°C to 125°C			0.5	mA
Short-circuit output current	V_I = 35 V	25°C		240		mA
Peak output current		25°C		0.7		A

†All characteristics are measured with a capacitor across the input of 0.33 µF and a capacitor across the output of 0.1 µF. All characteristics except noise voltage and ripple rejection ratio are measured using pulse techniques ($t_w \leq$ 10 ms, duty cycle $\leq$ 5%). Output voltage changes due to changes in internal temperature must be taken into account separately.

TEXAS
INSTRUMENTS
POST OFFICE BOX 225012 ● DALLAS, TEXAS 75265

THERMAL INFORMATION

KC PACKAGE
FREE-AIR TEMPERATURE
DISSIPATION DERATING CURVE

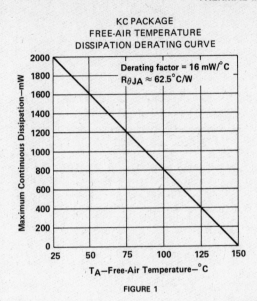

Derating factor = 16 mW/°C
$R_{\theta}JA \approx 62.5°C/W$

FIGURE 1

KC PACKAGE
CASE TEMPERATURE
DISSIPATION DERATING CURVE

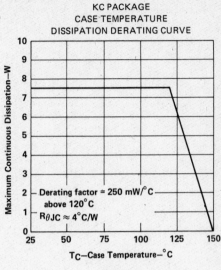

Derating factor = 250 mW/°C
above 120°C
$R_{\theta}JC \approx 4°C/W$

FIGURE 2

JG PACKAGE
CASE TEMPERATURE
DISSIPATION DERATING CURVE

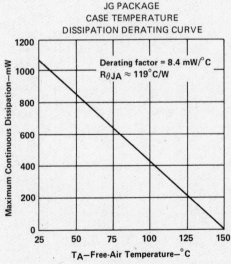

Derating factor = 8.4 mW/°C
$R_{\theta}JA \approx 119°C/W$

FIGURE 3

Voltage Regulators

6

Voltage Regulators

6

- 3-Terminal Regulators
- Output Current up to 1.5 A
- No External Components
- Internal Thermal Overload Protection
- High Power Dissipation Capability
- Internal Short-Circuit Current Limiting
- Output Transistor Safe-Area Compensation
- Essentially Equivalent to National LM320 Series
- Direct Replacements for Fairchild μA7900 Series and National LM79XX Series

NOMINAL OUTPUT VOLTAGE	REGULATOR
−5 V	uA7905C
−5.2 V	uA7952C
−6 V	uA7906C
−8 V	uA7908C
−12 V	uA7912C
−15 V	uA7915C
−18 V	uA7918C
−24 V	uA7924C

description

This series of fixed-negative-voltage monolithic integrated-circuit voltage regulators is designed to complement Series uA7800 in a wide range of applications. These applications include on-card regulation for elimination of noise and distribution problems associated with single-point regulation. Each of these regulators can deliver up to 1.5 amperes of output current. The internal current limiting and thermal shutdown features of these regulators make them essentially immune to overload. In addition to use as fixed-voltage regulators, these devices can be used with external components to obtain adjustable output voltages and currents and also as the power pass element in precision regulators.

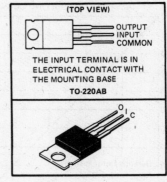

KC PACKAGE

(TOP VIEW)

OUTPUT
INPUT
COMMON

THE INPUT TERMINAL IS IN ELECTRICAL CONTACT WITH THE MOUNTING BASE

TO-220AB

Voltage Regulators

6

schematic

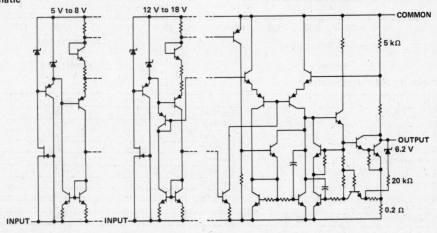

5 V to 8 V

12 V to 18 V

COMMON

5 kΩ

OUTPUT
6.2 V

20 kΩ

0.2 Ω

INPUT — — INPUT

All component values are nominal.

TEXAS INSTRUMENTS
POST OFFICE BOX 225012 • DALLAS, TEXAS 75265

absolute maximum ratings over operating temperature range (unless otherwise noted)

		uA7905C THRU uA7924C	UNIT
Input voltage	uA7924C	−40	V
	All others	−35	
Continuous total dissipation at 25°C free-air temperature (see Note 1)		2	W
Continuous total dissipation at (or below) 25°C case temperature (see Note 1)		15	W
Operating free-air, case, or virtual junction temperature range		0 to 150	°C
Storage temperature range		−65 to 150	°C
Lead temperature 3,2 mm (1/8 inch) from case for 10 seconds		260	°C

NOTE 1: For operation above 25 °C free-air or case temperature, refer to Figures 1 and 2. To avoid exceeding the design maximum virtual junction temperature, these ratings should not be exceeded. Due to variations in individual device electrical characteristics and thermal resistance, the built-in thermal overload protection may be activated at power levels slightly above or below the rated dissipation.

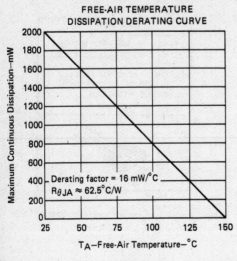

FREE-AIR TEMPERATURE
DISSIPATION DERATING CURVE

Derating factor = 16 mW/°C
$R_{\theta JA} \approx 62.5°C/W$

FIGURE 1

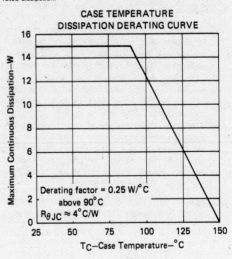

CASE TEMPERATURE
DISSIPATION DERATING CURVE

Derating factor = 0.25 W/°C above 90°C
$R_{\theta JC} \approx 4°C/W$

FIGURE 2

recommended operating conditions

		MIN	MAX	UNIT
Input voltage, V_I	uA7905C	−7	−25	V
	uA7952C	−7.2	−25	
	uA7906C	−8	−25	
	uA7908C	−10.5	−25	
	uA7912C	−14.5	−30	
	uA7915C	−17.5	−30	
	uA7918C	−21	−33	
	uA7924C	−27	−38	
Output current, I_O			1.5	A
Operating virtual junction temperature, T_J		0	125	°C

**TEXAS
INSTRUMENTS**
POST OFFICE BOX 225012 ● DALLAS, TEXAS 75265

uA7905C electrical characteristics at specified virtual junction temperature, $V_I = -10$ V, $I_O = 500$ mA (unless otherwise noted)

PARAMETER	TEST CONDITIONS†		uA7905C			UNIT
			MIN	TYP	MAX	
Output voltage	I_O = 5 mA to 1 A, $V_I = -7$ V to -20 V, P ⩽ 15 W	25°C	−4.8	−5	−5.2	V
		0°C to 125°C	−4.75		−5.25	
Input regulation	$V_I = -7$ V to -25 V	25°C		12.5	50	mV
	$V_I = -8$ V to -12 V			4	15	
Ripple rejection	$V_I = -8$ V to -18 V, f = 120 Hz	0°C to 125°C	54	60		dB
Output regulation	I_O = 5 mA to 1.5 A	25°C		15	100	mV
	I_O = 250 mA to 750 mA			5	50	
Temperature coefficient of output voltage	I_O = 5 mA	0°C to 125°C		−0.4		mV/°C
Output noise voltage	f = 10 Hz to 100 kHz	25°C		125		µV
Dropout voltage	I_O = 1 A	25°C		1.1		V
Bias current		25°C		1.5	2	mA
Bias current change	$V_I = -7$ V to -25 V	0°C to 125°C		0.15	0.5	mA
	I_O = 5 mA to 1 A			0.08	0.5	
Peak output current		25°C		2.1		A

uA7952C electrical characteristics at specified virtual junction temperature, $V_I = -10$ V, $I_O = 500$ mA (unless otherwise noted)

PARAMETER	TEST CONDITIONS†		uA7952C			UNIT
			MIN	TYP	MAX	
Output voltage	I_O = 5 mA to 1 A, $V_I = -7.2$ V to -20 V, P ⩽ 15 W	25°C	−5	−5.2	−5.4	V
		0°C to 125°C	−4.95		−5.45	
Input regulation	$V_I = -7.2$ V to -25 V	25°C		12.5	100	mV
	$V_I = -8.2$ V to -12 V			4	50	
Ripple rejection	$V_I = -8.2$ V to -18 V, f = 120 Hz	0°C to 125°C	54	60		dB
Output regulation	I_O = 5 mA to 1.5 A	25°C		15	100	mV
	I_O = 250 mA to 750 mA			5	50	
Temperature coefficient of output voltage	I_O = 5 mA	0°C to 125°C		−0.4		mV/°C
Output noise voltage	f = 10 Hz to 100 kHz	25°C		125		µV
Dropout voltage	I_O = 1 A	25°C		1.1		V
Bias current		25°C		1.5	2	mA
Bias current change	$V_I = -7.2$ V to -25 V	0°C to 125°C		0.15	1.3	mA
	I_O = 5 mA to 1 A			0.08	0.5	
Peak output current		25°C		2.1		A

†All characteristics are measured with a solid-tantalum capacitor across the input of 2 µF and a solid-tantalum capacitor across the output of 1 µF. All characteristics except noise voltage and ripple rejection ratio are measured using pulse techniques (t_w ⩽ 10 ms, duty cycle ⩽ 5%). Output voltage changes due to changes in internal temperature must be taken into account separately.

Voltage Regulators

6

TEXAS
INSTRUMENTS
POST OFFICE BOX 225012 • DALLAS, TEXAS 75265

TYPES uA7906C, uA7908C
NEGATIVE-VOLTAGE REGULATORS

uA7906C electrical characteristics at specified virtual junction temperature, $V_I = -11$ V, $I_O = 500$ mA (unless otherwise noted)

PARAMETER	TEST CONDITIONS[†]		uA7906C			UNIT
			MIN	TYP	MAX	
Output voltage	$I_O = 5$ mA to 1 A, $V_I = -8$ V to -21 V, P ≤ 15 W	25°C	−5.75	−6	−6.25	V
		0°C to 125°C	−5.7		−6.3	
Input regulation	$V_I = -8$ V to -25 V	25°C		12.5	120	mV
	$V_I = -9$ V to -13 V			4	60	
Ripple rejection	$V_I = -9$ V to -19 V, f = 120 Hz	0°C to 125°C	54	60		dB
Output regulation	$I_O = 5$ mA to 1.5 A	25°C		15	120	mV
	$I_O = 250$ mA to 750 mA			5	60	
Temperature coefficient of output voltage	$I_O = 5$ mA	0°C to 125°C		−0.4		mV/°C
Output noise voltage	f = 10 Hz to 100 kHz	25°C		150		μV
Dropout voltage	$I_O = 1$ A	25°C		1.1		V
Bias current		25°C		1.5	2	mA
Bias current change	$V_I = -8$ V to -25 V	0°C to 125°C		0.15	1.3	mA
	$I_O = 5$ mA to 1 A			0.08	0.5	
Peak output current		25°C		2.1		A

uA7908C electrical characteristics at specified virtual junction temperature, $V_I = -14$ V, $I_O = 500$ mA (unless otherwise noted)

PARAMETER	TEST CONDITIONS[†]		uA7908C			UNIT
			MIN	TYP	MAX	
Output voltage	$I_O = 5$ mA to 1 A, $V_I = -10.5$ V to -23 V, P ≤ 15 W	25°C	−7.7	−8	−8.3	V
		0°C to 125°C	−7.6		−8.4	
Input regulation	$V_I = -10.5$ V to -25 V	25°C		12.5	160	mV
	$V_I = -11$ V to -17 V			4	80	
Ripple rejection	$V_I = -11.5$ V to -21.5 V, f = 120Hz	0°C to 125°C	54	60		dB
Output regulation	$I_O = 5$ mA to 1.5 A	25°C		15	160	mV
	$I_O = 250$ mA to 750 mA			5	80	
Temperature coefficient of output voltage	$I_O = 5$ mA	0°C to 125°C		−0.6		mV/°C
Output noise voltage	f = 10 Hz to 100 kHz	25°C		200		μV
Dropout voltage	$I_O = 1$ A	25°C		1.1		V
Bias current		25°C		1.5	2	mA
Bias current change	$V_I = -10.5$ V to -25 V	0°C to 125°C		0.15	1	mA
	$I_O = 5$ mA to 1 A			0.08	0.5	
Peak output current		25°C		2.1		A

[†]All characteristics are measured with a solid-tantalum capacitor across the input of 2 μF and a solid-tantalum capacitor across the output of 1 μF. All characteristics except noise voltage and ripple rejection ratio are measured using pulse techniques (t_W ≤ 10 ms, duty cycle ≤ 5%). Output voltage changes due to changes in internal temperature must be taken into account separately.

TEXAS
INSTRUMENTS
POST OFFICE BOX 225012 • DALLAS, TEXAS 75265

uA7912C electrical characteristics at specified virtual junction temperature, $V_I = -19$ V, $I_O = 500$ mA (unless otherwise noted)

PARAMETER	TEST CONDITIONS[†]		uA7912C			UNIT
			MIN	TYP	MAX	
Output voltage	$I_O = 5$ mA to 1 A, $V_I = -14.5$ V to -27 V, $P \leqslant 15$ W	25°C	−11.5	−12	−12.5	V
		0°C to 125°C	−11.4		−12.6	
Input regulation	$V_I = -14.5$ V to -30 V	25°C		5	80	mV
	$V_I = -16$ V to -22 V			3	30	
Ripple rejection	$V_I = -15$ V to -25 V, f = 120 Hz	0°C to 125°C	54	60		dB
Output regulation	$I_O = 5$ mA to 1.5 A	25°C		15	200	mV
	$I_O = 250$ mA to 750 mA			5	75	
Temperature coefficient of output voltage	$I_O = 5$ mA	0°C to 125°C		−0.8		mV/°C
Output noise voltage	f = 10 Hz to 100 kHz	25°C		300		µV
Dropout voltage	$I_O = 1$ A	25°C		1.1		V
Bias current		25°C		2	3	mA
Bias current change	$V_I = -14.5$ V to -30 V	0°C to 125°C		0.04	0.5	mA
	$I_O = 5$ mA to 1 A			0.06	0.5	
Peak output current		25°C		2.1		A

uA7915C electrical characteristics at specified virtual junction temperature, $V_I = -23$ V, $I_O = 500$ mA (unless otherwise noted)

PARAMETER	TEST CONDITIONS[†]		uA7915C			UNIT
			MIN	TYP	MAX	
Output voltage	$I_O = 5$ mA to 1 A, $V_I = -17.5$ V to -30 V, $P \leqslant 15$ W	25°C	−14.4	−15	−15.6	V
		0°C to 125°C	−14.25		−15.75	
Input regulation	$V_I = -17.5$ V to -30 V	25°C		5	100	mV
	$V_I = -20$ V to -26 V			3	50	
Ripple rejection	$V_I = -18.5$ V to -28.5 V, f = 120 Hz	0°C to 125°C	54	60		dB
Output regulation	$I_O = 5$ mA to 1.5 A	25°C		15	200	mV
	$I_O = 250$ mA to 750 mA			5	75	
Temperature coefficient of output voltage	$I_O = 5$ mA	0°C to 125°C		−1		mV/°C
Output noise voltage	f = 10 Hz to 100 kHz	25°C		375		µV
Dropout voltage	$I_O = 1$ A	25°C		1.1		V
Bias current		25°C		2	3	mA
Bias current change	$V_I = -17.5$ V to -30 V	0°C to 125°C		0.04	0.5	mA
	$I_O = 5$ mA to 1 A			0.06	0.5	
Peak output current		25°C		2.1		A

[†]All characteristics are measured with a solid-tantalum capacitor across the input of 2 µF and a solid-tantalum capacitor across the output of 1 µF. All characteristics except noise voltage and ripple rejection ratio are measured using pulse techniques ($t_w \leq 10$ ms, duty cycle $\leq 5\%$). Output voltage changes due to changes in internal temperature must be taken into account separately.

Voltage Regulators

6

uA7918C electrical characteristics at specified virtual junction temperature, $V_I = -27$ V, $I_O = 500$ mA (unless otherwise noted)

PARAMETER	TEST CONDITIONS[†]		uA7918C			UNIT
			MIN	TYP	MAX	
Output voltage	$I_O = 5$ mA to 1 A, $V_I = -21$ V to -33 V, P ≤ 15 W	25°C	−17.3	−18	−18.7	V
		0°C to 125°C	−17.1		−18.9	
Input regulation	$V_I = -21$ V to -33 V	25°C		5	360	mV
	$V_I = -24$ V to -30 V			3	180	
Ripple rejection	$V_I = -22$ V to -32 V, f = 120 Hz	0°C to 125°C	54	60		dB
Output regulation	$I_O = 5$ mA to 1.5 A	25°C		30	360	mV
	$I_O = 250$ mA to 750 mA			10	180	
Temperature coefficient of output voltage	$I_O = 5$ mA	0°C to 125°C		−1		mV/°C
Output noise voltage	f = 10 Hz to 100 kHz	25°C		450		μV
Dropout voltage	$I_O = 1$ A	25°C		1.1		V
Bias current		25°C		2	3	mA
Bias current change	$V_I = -21$ V to -33 V	0°C to 125°C		0.04	1	mA
	$I_O = 5$ mA to 1 A			0.06	0.5	
Peak output current		25°C		2.1		A

uA7924C electrical characteristics at specified virtual junction temperature, $V_I = -33$ V, $I_O = 500$ mA (unless otherwise noted)

PARAMETER	TEST CONDITIONS[†]		uA7924C			UNIT
			MIN	TYP	MAX	
Output voltage	$I_O = 5$ mA to 1 A, $V_I = -27$ V to -38 V, P ≤ 15 W	25°C	−23	−24	−25	V
		0°C to 125°C	−22.8		−25.2	
Input regulation	$V_I = -27$ V to -38 V	25°C		5	480	mV
	$V_I = -30$ V to -36 V			3	240	
Ripple rejection	$V_I = -28$ V to -38 V, f = 120 Hz	0°C to 125°C	54	60		dB
Output regulation	$I_O = 5$ mA to 1.5 A	25°C		85	480	mV
	$I_O = 250$ mA to 750 mA			25	240	
Temperature coefficient of output voltage	$I_O = 5$ mA	0°C to 125°C		−1		mV/°C
Output noise voltage	f = 10 Hz to 100 kHz	25°C		600		μV
Dropout voltage	$I_O = 1$ A	25°C		1.1		V
Bias current		25°C		2	3	mA
Bias current change	$V_I = -27$ V to -38 V	0°C to 125°C		0.04	1	mA
	$I_O = 5$ mA to 1 A			0.06	0.5	
Peak output current		25°C		2.1		A

[†]All characteristics are measured with a solid-tantalum capacitor across the input of 2 μF and a solid-tantalum capacitor across the output of 1 μF. All characteristics except noise voltage and ripple rejection ratio are measured using pulse techniques (t_w ≤ 10 ms, duty cycle ≤ 5%). Output voltage changes due to changes in internal temperature must be taken into account separately.

Voltage Regulators

6

Texas
Instruments
POST OFFICE BOX 225012 • DALLAS, TEXAS 75265

LINEAR
INTEGRATED
CIRCUITS

SERIES uA79M00
NEGATIVE-VOLTAGE REGULATORS

D2216, JUNE 1976–REVISED AUGUST 1983

- 3-Terminal Regulators
- Output Current up to 500 mA
- No External Components
- High Power Dissipation Capability
- Internal Short-Circuit Current Limiting
- Output Transistor Safe-Area Compensation
- Direct Replacements for Fairchild µA79M00 Series

NOMINAL OUTPUT VOLTAGE	−55°C TO 150°C OPERATING TEMPERATURE RANGE	0°C TO 125°C OPERATING TEMPERATURE RANGE
−5 V	uA79M05M	uA79M05C
−6 V	uA79M06M	uA79M06C
−8 V	uA79M08M	uA79M08C
−12 V	uA79M12M	uA79M12C
−15 V	uA79M15M	uA79M15C
−20 V		uA79M20C
−24 V		uA79M24C
PACKAGE	JG	KC

description

This series of fixed-negative-voltage monolithic integrated-circuit voltage regulators is designed to complement Series uA78M00 in a wide range of applications. These applications include on-card regulation for elimination of noise and distribution problems associated with single-point regulation. Each of these regulators can deliver up to 500 milliamperes of output current. The internal current limiting and thermal shutdown features of these regulators make them essentially immune to overload. In addition to use as fixed-voltage regulators, these devices can be used with external components to obtain adjustable output voltages and currents and also as the power pass element in precision regulators.

schematic

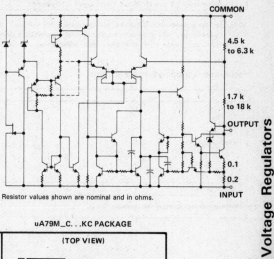

COMMON

4.5 k to 6.3 k

1.7 k to 18 k

OUTPUT

0.1

0.2

INPUT

Resistor values shown are nominal and in ohms.

terminal assignments

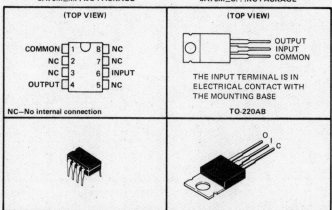

uA79M_M...JG PACKAGE

(TOP VIEW)

COMMON	1		8	NC
NC	2		7	NC
NC	3		6	INPUT
OUTPUT	4		5	NC

NC–No internal connection

uA79M_C...KC PACKAGE

(TOP VIEW)

OUTPUT
INPUT
COMMON

THE INPUT TERMINAL IS IN ELECTRICAL CONTACT WITH THE MOUNTING BASE

TO-220AB

Voltage Regulators

6

TEXAS
INSTRUMENTS
POST OFFICE BOX 225012 • DALLAS, TEXAS 75265

absolute maximum ratings over operating temperature range (unless otherwise noted)

		uA79M05M THRU uA79M15M	uA79M05C THRU uA79M24C	UNIT
Input voltage	uA79M20, uA79M24		−40	V
	All others	−35	−35	
Continuous total dissipation at 25°C free-air temperature (see Note 1)	JG package	1.05		W
	KC (TO-220AB) package		2	
Continuous total dissipation at (or below) 25°C case temperature (see Note 1)	KC package		7.5	W
Operating free-air, case or virtual junction temperature range		−55 to 150	0 to 150	°C
Storage temperature range		−65 to 150	−65 to 150	°C
Lead temperature 1,6 mm (1/16 inch) from case for 60 seconds	JG package	300		°C
Lead temperature 1,6 mm (1/16 inch) from case for 10 seconds	KC package		260	°C

NOTE 1: For operation above 25°C free-air or case temperature, refer to Figures 1 through 3. To avoid exceeding the design maximum virtual junction temperature, these ratings should not be exceeded. Due to variations in individual dvice electrical characteristics and thermal resistance, the built-in thermal overload protection may be activated at power levels slightly above or below the rated dissipation.

recommended operating conditions

		MIN	MAX	UNIT
Input voltage, V_I	uA79M05M, uA79M05C	−7	−25	V
	uA79M06M, uA79M06C	−8	−25	
	uA79M08M, uA79M08C	−10.5	−25	
	uA79M12M, uA79M12C	−14.5	−30	
	uA79M15M, uA79M15C	−17.5	−30	
	uA79M20C	−23	−35	
	uA79M24C	−27	−38	
Output current, I_O			500	mA
Operating virtual junction temperature, T_J	uA79M05M thru uA79M15M	−55	150	°C
	uA79M05C thru uA79M24C	0	125	

883

TEXAS
INSTRUMENTS
POST OFFICE BOX 225012 • DALLAS, TEXAS 75265

uA79M05M, uA79M05C electrical characteristics at specified virtual junction temperature, $V_I = -10$ V, $I_O = 350$ mA (unless otherwise noted)

PARAMETER	TEST CONDITIONS[†]		uA79M05M MIN	TYP	MAX	uA79M05C MIN	TYP	MAX	UNIT
Output voltage		25°C	−4.8	−5	−5.2	−4.8	−5	−5.2	V
	I_O = 5 mA to 350 mA, V_I = −7 V to −25 V	−55°C to 150°C	−4.75		−5.25				
		0°C to 125°C				−4.75		−5.25	
Input regulation	V_I = −7 V to −25 V	25°C		7	50		7	50	mV
	V_I = −8 V to −18 V			3	30		3	30	
Ripple rejection	V_I = −8 V to −18 V, f = 120 Hz, I_O = 100 mA	−55°C to 150°C	50						dB
		0°C to 125°C				50			
	I_O = 300 mA	25°C	54	60		54	60		
Output regulation	I_O = 5 mA to 500 mA	25°C		75	100		75	100	mV
	I_O = 5 mA to 350 mA			50			50		
Temperature coefficient of output voltage	I_O = 5 mA	−55°C to 150°C		−1.5					mV/°C
		0°C to 125°C					−0.4		
Output noise voltage	f = 10 Hz to 100 kHz	25°C		125	400		125		µV
Dropout voltage		25°C		1.1	2.3		1.1		V
Bias current		25°C		1	2		1	2	mA
Bias current change	V_I = −8 V to −25 V	−55°C to 150°C			0.4				mA
		0°C to 125°C						0.4	
	I_O = 5 mA to 350 mA	−55°C to 150°C			0.4				
		0°C to 125°C						0.4	
Short-circuit output current	V_I = −30 V	25°C			600		140		mA
Peak output current		25°C	0.5	0.65	1.4		0.65		A

[†]All characteristics are measured with a 2-µF capacitor across the input and a 1-µF capacitor across the output. All characteristics except noise voltage and ripple rejection ratio are measured using pulse techniques ($t_w \leqslant 10$ ms, duty cycle $\leqslant 5\%$). Output voltage changes due to changes in internal temperature must be taken into account separately.

Voltage Regulators

6

TEXAS
INSTRUMENTS
POST OFFICE BOX 225012 • DALLAS, TEXAS 75265

TYPES uA79M06M, uA79M06C
NEGATIVE-VOLTAGE REGULATORS

uA79M06M, uA79M06C electrical characteristics at specified virtual junction temperature,
$V_I = -11$ V, $I_O = 350$ mA (unless otherwise noted)

PARAMETER	TEST CONDITIONS†			uA79M06M			uA79M06C			UNIT
				MIN	TYP	MAX	MIN	TYP	MAX	
Output voltage			25°C	−5.75	−6	−6.25	−5.75	−6	−6.25	V
	I_O = 5 mA to 350 mA, V_I = −8 V to −25 V		−55°C to 150°C	−5.7		−6.3				
			0°C to 125°C				−5.7		−6.3	
Input regulation	V_I = −8 V to −25 V		25°C		7	60		7	60	mV
	V_I = −9 V to −19 V				3	40		3	40	
Ripple rejection	V_I = −9 V to −19 V, f = 120 Hz	I_O = 100 mA	−55°C to 150°C	50						dB
			0°C to 125°C				50			
		I_O = 300 mA	25°C	54	60		54	60		
Output regulation	I_O = 5 mA to 500 mA		25°C		80	120		80	120	mV
	I_O = 5 mA to 350 mA				55			55		
Temperature coefficient of output voltage	I_O = 5 mA		−55°C to 150°C			−1.5				mV/°C
			0°C to 125°C					−0.4		
Output noise voltage	f = 10 Hz to 100 kHz		25°C		150	480		150		µV
Dropout voltage			25°C		1.1	2.3		1.1		V
Bias current			25°C		1	2		1	2	mA
Bias current change	V_I = −9 V to −25 V		−55°C to 150°C			0.4				mA
			0°C to 125°C						0.4	
	I_O = 5 mA to 350 mA		−55°C to 150°C			0.4				
			0°C to 125°C						0.4	
Short-circuit output current	V_I = −30 V		25°C			600		140		mA
Peak output current			25°C	0.5	0.65	1.4		0.65		A

†All characteristics are measured with a 2-µF capacitor across the input and a 1-µF capacitor across the output. All characteristics except noise voltage and ripple rejection ratio are measured using pulse techniques ($t_w \leqslant 10$ ms, duty cycle ≤ 5%). Output voltage changes due to changes in internal temperature must be taken into account separately.

Voltage Regulators

6

TEXAS
INSTRUMENTS
POST OFFICE BOX 225012 • DALLAS, TEXAS 75265

uA79M08M, uA79M08C electrical characteristics at specified virtual junction temperature, $V_I = -19$ V, $I_O = 350$ mA (unless noted)

PARAMETER	TEST CONDITIONS†			uA79M08M			uA79M08C			UNIT
				MIN	TYP	MAX	MIN	TYP	MAX	
Output voltage	I_O = 5 mA to 350 mA, V_I = −10.5 V to −25 V		25°C	−7.7	−8	−8.3	−7.7	−8	−8.3	V
			−55°C to 150°C	−7.6		−8.4				
			0°C to 125°C				−7.6		−8.4	
Input regulation	V_I = −10.5 V to −25 V		25°C		8	80		8	80	mV
	V_I = −11 V to −21 V				4	50		4	50	
Ripple rejection	V_I = −11.5 V to −21.5 V, f = 120 Hz	I_O = 100 mA	−55°C to 150°C	50						dB
			0°C to 125°C				50			
		I_O = 300 mA	25°C	54	59		54	59		
Output regulation	I_O = 5 mA to 500 mA		25°C		90	160		90	160	mV
	I_O = 5 mA to 350 mA				60			60		
Temperature coefficient of output voltage	I_O = 5 mA		−55°C to 150°C		−2.4					mV/°C
			0°C to 125°C					−0.6		
Output noise voltage	f = 10 Hz to 100 kHz		25°C		200	640		200		µV
Dropout voltage			25°C		1.1	2.3		1.1		V
Bias current			25°C		1	2		1	2	mA
Bias current change	V_I = −10.5 V to −25 V		−55°C to 150°C		0.4					mA
			0°C to 125°C						0.4	
	I_O = 5 mA to 350 mA		−55°C to 150°C		0.4					
			0°C to 125°C						0.4	
Short-circuit output current	V_I = −30 V		25°C			600		140		mA
Peak output current			25°C	0.5	0.65	1.4		0.65		A

†All characteristics are measured with a 2-µF capacitor across the input and a 1-µF capacitor across the output. All characteristics except noise voltage and ripple rejection ratio are measured using pulse techniques ($t_w \leqslant$ 10 ms, duty cycle $\leqslant$ 5%). Output voltage changes due to changes in internal temperature must be taken into account separately.

Voltage Regulators

6

uA79M12M, uA79M12C electrical characteristics at specified virtual junction temperature,
$V_I = -19$ V, $I_O = 350$ mA (unless otherwise noted)

PARAMETER	TEST CONDITIONS†		uA79M12M MIN TYP MAX			uA79M12C MIN TYP MAX			UNIT
Output voltage	I_O = 5 mA to 350 mA, V_I = −14.5 V to −30 V	25°C	−11.5	−12	−12.5	−11.5	−12	−12.5	V
		−55°C to 150°C	−11.4		−12.6				
		0°C to 125°C				−11.4		−12.6	
Input regulation	V_I = −14.5 V to −30 V	25°C		9	80		9	80	mV
	V_I = −15 V to −25 V			5	50		5	50	
Ripple rejection	V_I = −15 V to −25 V, f = 120 Hz — I_O = 100 mA	−55°C to 150°C	50						dB
		0°C to 125°C				50			
	I_O = 300 mA	25°C	54	60		54	60		
Output regulation	I_O = 5 mA to 500 mA	25°C		65	240		65	240	mV
	I_O = 5 mA to 350 mA			45			45		
Temperature coefficient of output voltage	I_O = 5 mA	−55°C to 150°C			−3.6				mV/°C
		0°C to 125°C					−0.8		
Output noise voltage	f = 10 Hz to 100 kHz	25°C		300	960		300		µV
Dropout voltage		25°C		1.1	2.3		1.1		V
Bias current		25°C		1.5	3		1.5	3	mA
Bias current change	V_I = −14.5 V to −30 V	−55°C to 150°C			0.4				mA
		0°C to 125°C						0.4	
	I_O = 5 mA to 350 mA	−55°C to 150°C			0.4				
		0°C to 125°C						0.4	
Short-circuit output current	V_I = −30 V	25°C			600		140		mA
Peak output current		25°C	0.5	0.65	1.4		0.65		A

†All characteristics are measured with a 2-µF capacitor across the input and a 1-µF capacitor across the output. All characteristics except noise voltage and ripple rejection ratio are measured using pulse techniques ($t_w \leqslant 10$ ms, duty cycle $\leqslant 5\%$). Output voltage changes due to changes in internal temperature must be taken into account separately.

Voltage Regulators

6

TEXAS
INSTRUMENTS
POST OFFICE BOX 225012 • DALLAS, TEXAS 75265

uA79M15M, uA79M15C electrical characteristics at specified virtual junction temperature, $V_I = -23$ V, $I_O = 350$ mA (unless otherwise noted)

PARAMETER	TEST CONDITIONS†		uA79M15M MIN	TYP	MAX	uA79M15C MIN	TYP	MAX	UNIT
Output voltage		25°C	−14.4	−15	−15.6	−14.4	−15	−15.6	V
	I_O = 5 mA to 350 mA, V_I = −17.5 V to −30 V	−55°C to 150°C	−14.25		−15.75				
		0°C to 125°C				−14.25		−15.75	
Input regulation	V_I = −17.5 V to −30 V	25°C		9	80		9	80	mV
	V_I = −18 V to −28 V			7	50		7	50	
Ripple rejection	V_I = −18.5 V to −28.5 V, I_O = 100 mA f = 120 Hz	−55°C to 150°C	50						dB
		0°C to 125°C				50			
	I_O = 300 mA	25°C	54	59		54	59		
Output regulation	I_O = 5 mA to 500 mA	25°C		65	240		65	240	mV
	I_O = 5 mA to 350 mA			45			45		
Temperature coefficient of output voltage	I_O = 5 mA	−55°C to 150°C			−4.5				mV/°C
		0°C to 125°C					−1		
Output noise voltage	f = 10 Hz to 100 kHz	25°C		375	1200		375		µV
Dropout voltage		25°C		1.1	2.3		1.1		V
Bias current		25°C		1.5	3		1.5	3	mA
Bias current change	V_I = −17.5 V to −30 V	−55°C to 150°C			0.4				mA
		0°C to 125°C						0.4	
	I_O = 5 mA to 350 mA	−55°C to 150°C			0.4				
		0°C to 125°C						0.4	
Short-circuit output current	V_I = −30 V	25°C		600			140		mA
Peak output current		25°C	0.5	0.65			0.65		A

† All characteristics are measured with a 2-µF capacitor across the input and a 1-µF capacitor across the output. All characteristics except noise voltage and ripple rejection ratio are measured using pulse techniques ($t_w \leqslant$ 10 ms, duty cycle $\leqslant$ 5%). Output voltage changes due to changes in internal temperature must be taken into account separately.

Voltage Regulators

6

TEXAS
INSTRUMENTS
POST OFFICE BOX 225012 • DALLAS, TEXAS 75265

uA79M20C electrical characteristics at specified virtual junction temperature
$V_I = -29$ V, $I_O = 350$ mA (unless otherwise noted)

PARAMETER	TEST CONDITIONS[†]			uA79M20C			UNIT
				MIN	TYP	MAX	
Output voltage	$I_O = 5$ mA to 350 mA, $V_I = -23$ V to -35 V		25°C	−19.2	−20	−20.8	V
			0°C to 125°C	−19		−21	
Input regulation	$V_I = -23$ V to -35 V		25°C		12	80	mV
	$V_I = -24$ V to -34 V				10	70	
Ripple rejection	$V_I = -24$ V to -34 V,	$I_O = 100$ mA	0°C to 125°C	50			dB
	f = 120 Hz	$I_O = 300$ mA	25°C	54	58		
Output regulation	$I_O = 5$ mA to 500 mA		25°C		75	300	mV
	$I_O = 5$ mA to 350 mA				50		
Temperature coefficient of output voltage	$I_O = 5$ mA		0°C to 125°C		−1		mV/°C
Output noise voltage	f = 10 Hz to 100 kHz		25°C		500		µV
Dropout voltage			25°C		1.1		V
Bias current			25°C		1.5	3.5	mA
Bias current change	$V_I = -23$ V to -35 V		0°C to 125°C			0.4	mA
	$I_O = 5$ mA to 350 mA					0.4	
Short-circuit output current	$V_I = -30$ V		25°C		140		mA
Peak output current			25°C		650		A

[†]All characteristics are measured with a 2-µF capacitor across the input and a 1-µF capacitor across the output. All characteristics except noise voltage and ripple rejection ratio are measured using pulse techniques ($t_w \leqslant 10$ ms, duty cycle $\leqslant 5\%$). Output voltage changes due to changes in internal temperature must be taken into account separately.

Voltage Regulators

6

TEXAS
INSTRUMENTS
POST OFFICE BOX 225012 • DALLAS, TEXAS 75265

uA79M24C electrical characteristics at specified virtual junction temperature, $V_I = -33$ V, $I_O = 350$ mA (unless otherwise noted)

PARAMETER	TEST CONDITIONS†			uA79M24C			UNIT
				MIN	TYP	MAX	
Output voltage	I_O = 5 mA to 350 mA, V_I = −27 V to −38 V		25°C	−23	−24	−25	V
			0°C to 125°C	−22.8		−25.2	
Input regulation	V_I = −27 V to −38 V		25°C		12	80	mV
	V_I = −28 V to −38 V				12	70	
Ripple rejection	V_I = −28 V to −38 V, f = 120 Hz	I_O = 100 mA	0°C to 125°C	50			dB
		I_O = 300 mA	25°C	54	58		
Output regulation	I_O = 5 mA to 500 mA		25°C		75	300	mV
	I_O = 5 mA to 350 mA				50		
Temperature coefficient of output voltage	I_O = 5 mA		0°C to 125°C		−1		mV/°C
Output noise voltage	f = 10 Hz to 100 kHz		25°C		600		µV
Dropout voltage			25°C		1.1		V
Bias current			25°C		1.5	3.5	mA
Bias current change	V_I = −27 V to −38 V		0°C to 125°C			0.4	mA
	I_O = 5 mA to 350 mA					0.4	
Short-circuit output current	V_I = −30 V		25°C		140		mA
Peak output current			25°C		650		A

†All characteristics are measured with a 2-µF capacitor across the input and a 1-µF capacitor across the output. All characteristics except noise voltage and ripple rejection ratio are measured using pulse techniques ($t_w \leqslant 10$ ms, duty cycle $\leqslant 5\%$). Output voltage changes due to changes in internal temperature must be taken into account separately.

Voltage Regulators

6

TEXAS
INSTRUMENTS
POST OFFICE BOX 225012 • DALLAS, TEXAS 75265

SERIES uA79M00
NEGATIVE-VOLTAGE REGULATORS

THERMAL INFORMATION

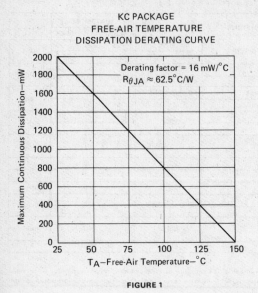

KC PACKAGE
FREE-AIR TEMPERATURE
DISSIPATION DERATING CURVE

Derating factor = 16 mW/°C
$R_{\theta JA} \approx 62.5°C/W$

FIGURE 1

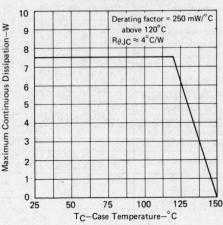

KC PACKAGE
CASE TEMPERATURE
DISSIPATION DERATING CURVE

Derating factor = 250 mW/°C
above 120°C
$R_{\theta JC} \approx 4°C/W$

FIGURE 2

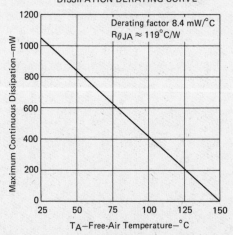

JG PACKAGE
FREE-AIR TEMPERATURE
DISSIPATION DERATING CURVE

Derating factor 8.4 mW/°C
$R_{\theta JA} \approx 119°C/W$

FIGURE 3

Voltage Regulators

6

TEXAS INSTRUMENTS
POST OFFICE BOX 225012 • DALLAS, TEXAS 75265

LINEAR
INTEGRATED
CIRCUITS

TYPES UC3846, UC3847
CURRENT-MODE PULSE-WIDTH-MODULATION CONTROLLERS

D2823, OCTOBER 1983

- Automatic Feed-Forward Compensation
- Programmable Pulse-by-Pulse Current Limiting
- Automatic Symmetry Correction in Push-Pull Configuration
- Parallel Operation Capability for Modular Power Systems
- Differential Current Sense Amplifier with Wide Common-Mode Voltage Range
- Double-Pulse Suppression
- 200-mA Totem-Pole Outputs
- ± 1% Bandgap Reference
- Under-Voltage Lockout
- Soft-Start Capability
- Shutdown Function
- 500-kHz Operation
- Direct Replacements for Unitrode UC3846 and UC3847

Output Logic:
UC3846 . . . NOR
UC3847 . . . OR

J OR N
DUAL-IN-LINE PACKAGE
(TOP VIEW)

```
              CLIM ADJ [ 1  U 16 ] SHTDWN
                 Vref [ 2    15 ] VCC
        CURR LIM { IN - [ 3   14 ] B OUT
        AMPL     { IN + [ 4   13 ] VC
        ERROR    { IN + [ 5   12 ] GND
        AMPL     { IN - [ 6   11 ] A OUT
                 COMP [ 7    10 ] SYNC
                   CT [ 8     9 ] RT
```

description

The UC3846 and UC3847 pulse-width-modulation controller integrated circuits are designed to offer improved performance and lower external parts count in fixed-frequency current-mode control systems. These devices provide improved line regulation, enhanced load response characteristics, and a simpler, easer-to-design control loop. Other advantages include inherent pulse-by-pulse current-limiting capability, automatic symmetry correction for push-pull converters, and the ability to parallel power modules while maintaining equal current sharing.

Protection circuitry includes a latched comparator under-voltage lockout, programmable current limiting, and self-start capability. A shutdown function is also available; this can either initiate a complete shutdown with automatic restart, or latch the supply off.

Other features include fully latched operation, double-pulse suppression, deadtime adjust capability, and a ± 1% trimmed bandgap reference.

Voltage Regulators

6

Copyright © 1983 by Texas Instruments Incorporated

TEXAS
INSTRUMENTS
POST OFFICE BOX 225012 • DALLAS, TEXAS 75265

General Information | 1

Thermal Information | 2

Operational Amplifiers | 3

Voltage Comparators | 4

Special Functions | 5

Voltage Regulators | 6

Data Acquisition | 7

Appendix | A

single- and dual-slope A/D converters

DEVICE NUMBER	FUNCTION	RESOLUTION	RECOMMENDED COMPLEMENT	TEMP[†] RANGE	PACKAGE	PAGE
TL500	Dual-slope analog procesors	13 Bits	TL502, TL505 or Microprocessor	C	N	7-43
TL501		10-12 Bits		C	N	7-43
TL505		8-10 Bits		C	N	7-57
TL502	Digital processors with seven-segment outputs	4 1/2 Digits	Microprocessor	C	N	7-43
TL503	Digital processors with BCD outputs	4 1/2 Digits	Microprocessor	C	N	7-43
TL507	Pulse-width modulator for single-slope converter	7 Bits	Microprocessor	C	N	7-63
TLC7126	Dual-slope A/D converter and LCD driver	3 1/2 Digits	Microprocessor	C	N	7-119

successive-approximation A/D converters

DEVICE NUMBER	SIGNAL INPUTS DEDICATED ANALOG	SIGNAL INPUTS ANALOG/ DIGITAL[‡]	ADDRESS AND DATA I/O FORMAT	CONVERSION SPEED[§] (μs)	UNADJUSTED ERROR ±LSB	POWER DISSIPATION (mW)	TEMP[†] RANGE	PACKAGE	PAGE
ADC0801	1	0	Parallel	100	—	29	I	N	7-5
ADC0802	1	0	Parallel	100	0.5	29	I	N	7-5
ADC0803	1	0	Parallel	100	—	29	I	N	7-5
ADC0805	1	0	Parallel	100	0.5	29	I	N	7-5
ADC0804C	1	0	Parallel	100	1	29	C	N	7-11
ADC0808	8	0	Parallel	100	0.75	12	I	N	7-17
ADC0809	8	0	Parallel	100	1.25	12	I	N	7-17
ADC0831	1	0	Serial	84	0.5	15	I	N	7-23
ADC0832	2	0	Serial	84	0.5	15	I	N	7-23
ADC0834	4	0	Serial	84	0.5	15	I	N	7-23
ADC0838	8	0	Serial	84	0.5	15	I	N	7-23
TL520	8	0	Parallel	70	0.75	2	I	N	7-67
TL521	8	0	Parallel	100	1	2	I	N	7-67
TL522	8	0	Parallel	208	0.5	0.3	I	N	7-67
TL530¶	8	6	Parallel	300	0.5	15	I	N	7-77
TL531¶	8	6	Parallel	300	1	15	I	N	7-77
TL532	5	6	Parallel	300	0.5	15	I	N	7-87
TLC532A	5	6	Parallel	15	0.5	6	M,I	FH,J,N	7-101
TL533	5	6	Parallel	300	1	15	I	N	7-87
TLC533A	5	6	Parallel	15	1	6	M,I	FH,J,N	7-101
TLC540	11	0	Serial	12	0.5	6	M,I	FK,FN,J,N	7-109
TLC541	11	0	Serial	34	1	6	M,I	FK,FN,J,N	7-109
TLC549	1	0	Serial	19	0.5	6	M,I	JG,P	7-115

[†] M ≡ −55°C to 125°C, I ≡ −40°C to 85°C, C ≡ 0°C to 70°C.
[‡] Analog/digital signal inputs can be used either as digital inputs for limiting sensing or digital data, or they can be used as analog inputs. For example: the TL530 can have 15 analog inputs and 6 digital outputs, 9 analog inputs and 12 digital inputs, or any combination in between.
[§] Includes access time.
[¶] The TL530 and TL531 devices also have 6 dedicated digital inputs.

Data Acquisition

7

SELECTION GUIDE

analog switches

DEVICE NUMBER	FUNCTION	TYPICAL IMPEDANCE (OHM)	VOLTAGE RANGE (V)	POWER SUPPLIES (V)	TEMP[†] RANGE	PACKAGE	PAGE
TL182	Twin SPDT	100	±10	±15, 5	M,I,C	J,N	7-37
TL185	Twin DPST	150	±10	±15, 5	M,I,C	J,N	7-37
TL188	Dual SPST	100	±10	±15, 5	M,I,C	J,N	7-37
TL191	Twin dual SPST	150	±10	±15, 5	M,I,C	J,N	7-37
TL601	SPDT	Not recommended for new designs			M,I,C	JG,P	7-95
TL604	Dual SPST				M,I,C	JG,P	7-95
TL607	SPDT with enable				M,I,C	JG,P	7-95
TL610	SPST with logic inputs				M,I,C	JG,P	7-95

[†]M ≡ −55°C to 125°C, I ≡ −40°C to 85°C, C ≡ 0°C to 70°C.

Data Acquisition

7

TEXAS
INSTRUMENTS
POST OFFICE BOX 225012 ● DALLAS, TEXAS 75265

**DATA
ACQUISITION
CIRCUITS**

**TYPES ADC0801, ADC0802, ADC0803, ADC0805
8-BIT ANALOG-TO-DIGITAL CONVERTERS
WITH DIFFERENTIAL INPUTS**

D2754, NOVEMBER 1983

- 8-Bit Resolution
- Ratiometric Conversion
- 100 μs Conversion Time
- 135 ns Access Time
- Guaranteed Monotonicity
- High Reference Ladder Impedance 8 kΩ Typical
- No Zero Adjust Requirement
- On-Chip Clock Generator
- Single 5-Volt Power Supply
- Operates With Microprocessor or as Stand-Alone
- Designed to be Interchangeable with National Semiconductor ADC0801, ADC0802, ADC0803, ADC0805

N DUAL-IN-LINE PACKAGE

(TOP VIEW)

$\overline{CS}$	1	V_{CC} (OR REF) 20
$\overline{RD}$	2	19 CLK OUT
$\overline{WR}$	3	18 DB0 (LSB)
CLK IN	4	17 DB1
$\overline{INTR}$	5	16 DB2
IN +	6	15 DB3
IN −	7	14 DB4
ANLG GND	8	13 DB5
REF/2	9	12 DB6
DGTL GND	10	11 DB7 (MSB)

DATA OUTPUTS (DB0–DB7)

description

The ADC0801, ADC0802, ADC0803, and ADC0805 are CMOS 8-bit successive-approximation analog-to-digital converters that use a modified potentiometric (256R) ladder. These devices are designed to operate from common microprocessor control buses, with the three-state output latches driving the data bus. The devices can be made to appear to the microprocessor as a memory location or an I/O port.

A differential analog voltage input allows increased common-mode rejection and offset of the zero-input analog voltage value. Although a reference input (REF/2) is available to allow 8-bit conversion over smaller analog voltage spans or to make use of an external reference, ratiometric conversion is possible with the REF/2 input open. Without an external reference, the conversion takes place over a span from V_{CC} to analog ground (ANLG GND). The devices can operate with an external clock signal or, with an additional resistor and capacitor, can operate using an on-chip clock generator.

The ADC0801I, ADC0802I, ADC0803I, and ADC0805I will be characterized for operation from −40°C to 85°C.

Data Acquisition

7

**TEXAS
INSTRUMENTS**

POST OFFICE BOX 225012 • DALLAS, TEXAS 75265

TYPES ADC0801, ADC0802, ADC0803, ADC0805
8-BIT ANALOG-TO-DIGITAL CONVERTERS
WITH DIFFERENTIAL INPUTS

functional block diagram (positive logic)

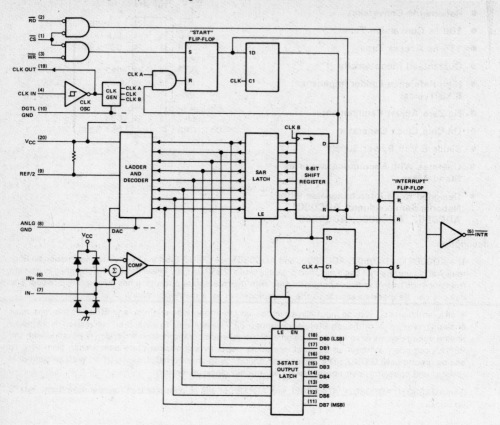

TEXAS
INSTRUMENTS

POST OFFICE BOX 225012 • DALLAS, TEXAS 75265

absolute maximum ratings over operating free-air temperature range (unless otherwise noted)

Supply voltage, V_{CC} (see Note 1) .... 6.5 V
Input voltage range $\overline{CS}$, $\overline{RD}$, $\overline{WR}$ .... −0.3 V to 18 V
Other inputs .... −0.3 V to V_{CC} +0.3 V
Output voltage range .... −0.3 V to V_{CC} +0.3 V
Continuous total power dissipation at 25 °C free-air temperature (see Note 2) .... 875 mW
Operating free-air temperature range .... −40 °C to 85 °C
Storage temperature range .... −65 °C to 150 °C
Lead temperature 1,6 mm (1/16 inch) from case for 10 seconds .... 260 °C

NOTES: 1. All voltage values are with respect to digital ground (DGTL GND) with DGTL GND and ANLG GND connected together (unless otherwise noted).
2. For operation above 25 °C free-air temperature, refer to Dissipation Derating Curves, section 2.

recommended operating conditions

		MIN	NOM	MAX	UNIT
V_{CC}	Supply voltage	4.5	5	6.3	V
$V_{REF/2}$	Voltage at REF/2 (see Note 3)	0.25	2.5		V
V_{IH}	High-level input voltage at $\overline{CS}$, $\overline{RD}$, or $\overline{WR}$	2		15	V
V_{IL}	Low-level input voltage at $\overline{CS}$, $\overline{RD}$, or $\overline{WR}$			0.8	V
	Analog ground voltage (see Note 4)	−0.05	0	1	V
	Analog input voltage (see Note 5)	GND −0.05		V_{CC} +0.05	V
f_{clock}	Clock input frequency (see Note 6)	100	640	1460	kHz
	Duty cycle above 640 kHz (see Note 6)	40		60	%
$t_{w(CLK)}$	Pulse duration clock input (high or low)	275	781		ns
$t_{w(WR)}$	Pulse duration, $\overline{WR}$ input low	100			ns
T_A	Operating free-air temperature	−40		85	°C

NOTES: 3. Proper operation is achieved over a differential input range of 0 V to V_{CC} when the REF/2 input is open.
4. These values are with respect to digital ground (pin 10).
5. When the positive analog input with respect to the negative analog input ($V_{in+} - V_{in-}$) is zero or negative, the output code is 0000 0000.
6. Total unadjusted error is guaranteed only at an f_{clock} of 640 kHz with a duty cycle of 40% to 60% (pulse duration 625 ns to 937 ns). For frequencies above this limit or pulse duration below 625 ns, error may increase. The duty cycle limits should be observed for an f_{clock} greater than 640 kHz. Below 640 kHz, this duty cycle limit can be exceeded provided $t_{w(CLK)}$ remains within limits.

Data Acquisition

7

TEXAS
INSTRUMENTS
POST OFFICE BOX 225012 • DALLAS, TEXAS 75265

electrical characteristics over recommended operating free-air temperature range, $V_{CC} = 5$ V, $f_{clock} = 640$ kHz, $V_{REF/2} = 2$ V (unless otherwise noted)

PARAMETER			TEST CONDITIONS		MIN	TYP[†]	MAX	UNIT
V_{OH}	High-level output voltage	All outputs	$V_{CC} = 4.75$ V,	$I_{OH} = -360\ \mu A$	2.4			V
		DB and $\overline{INTR}$	$V_{CC} = 4.75$ V,	$I_{OH} = -10\ \mu A$	4.5			
V_{OL}	Low-level output voltage	Data outputs	$V_{CC} = 4.75$ V,	$I_{OL} = 1.6$ mA			0.4	V
		$\overline{INTR}$ output	$V_{CC} = 4.75$ V,	$I_{OL} = 1$ mA			0.4	
		CLK OUT	$V_{CC} = 4.75$ V,	$I_{OL} = 360\ \mu A$			0.4	
V_{T+}	Clock positive-going threshold voltage				2.7	3.1	3.5	V
V_{T-}	Clock negative-going threshold voltage				1.5	1.8	2.1	V
$V_{T+} - V_{T-}$	Clock input hysteresis				0.6	1.3	2	V
I_{IH}	High-level input current					0.005	1	μA
I_{IL}	Low-level input current					-0.005	-1	μA
I_{OZ}	Off-state output current		$V_O = 0$				-3	μA
			$V_O = 5$ V				3	
I_{OHS}	Short-current output current	Output high	$V_O = 0$,	$T_A = 25°C$	-4.5	-6		mA
I_{OLS}	Short-circuit output current	Output low	$V_O = 5$ V,	$T_A = 25°C$	9	16		mA
I_{CC}	Supply current plus reference current		$V_{REF/2}$ = open, $\overline{CS}$ at 5 V	$T_A = 25°C$,		1.1	1.8	mA
$R_{REF/2}$	Input resistance to reference ladder		See Note 7		2.5	8		kΩ
C_i	Input capacitance (control)					5	7.5	pF
C_o	Output capacitance (DB)					5	7.5	pF

NOTE 7: Resistance is calculated from the current drawn from a 5-volt supply applied to pins 8 and 9.

operating characteristics over recommended operating free-air temperature, $V_{CC} = 5$ V, $V_{REF/2} = 2.5$ V, $f_{clock} = 640$ kHz (unless otherwise noted)

PARAMETER		TEST CONDITIONS		MIN	TYP[†]	MAX	UNIT
Supply-voltage-variation error		$V_{CC} = 4.5$ V to 5.5 V, See Note 8			±1/16	±1/8	LSB
Total adjusted error	ADC0801	With full-scale adjust, See Notes 8 and 9				±1/4	LSB
	ADC0803					±1/2	
Total unadjusted error	ADC0802	$V_{REF/2} = 2.5$ V, See Notes 8 and 9				±1/2	LSB
	ADC0805	$V_{REF/2}$ open, See Notes 8 and 9				±1	
DC common-mode error		See Note 8 and 9			±1/16	±1/8	LSB
t_{en}	Output enable time	$C_L = 100$ pF			135	200	ns
t_{dis}	Output disable time	$C_L = 10$ pF,	$R_L = 10$ kΩ		125	200	ns
$t_{d(INTR)}$	Delay time to reset $\overline{INTR}$				300	450	ns
t_{conv}	Conversion cycle time	$f_{clock} = 100$ kHz to 1.46 MHz, See Note 10		66		73	clock cycles
CR	Free-running conversion rate	$\overline{INTR}$ connected to $\overline{WR}$, $\overline{CS}$ at 0 V				8770	conv/s

[†]All typical values are at $T_A = 25°C$.

NOTES: 8. These parameters are guaranteed over the recommended analog input voltage range.
9. All errors are measured with reference to an ideal straight line through the end-points of the analog-to-digital transfer characteristic.
10. Although internal conversion is completed in 64 clock periods, a $\overline{CS}$ or $\overline{WR}$ low-to-high transition is followed by 1 to 8 clock periods before conversion starts. After conversion is completed, part of another clock period is required before a high-to-low transition of $\overline{INTR}$ completes the cycle.

Data Acquisition

7

TEXAS
INSTRUMENTS
POST OFFICE BOX 225012 • DALLAS, TEXAS 75265

PARAMETER MEASUREMENT INFORMATION

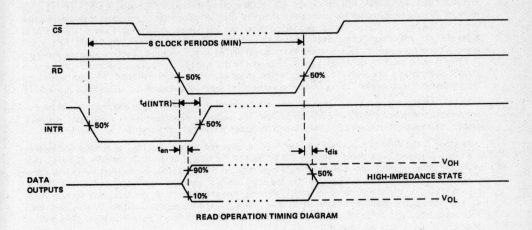

READ OPERATION TIMING DIAGRAM

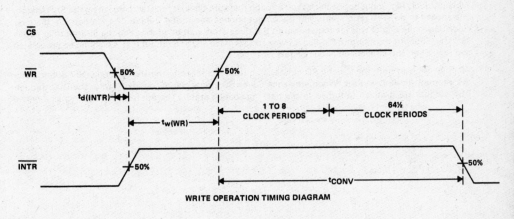

WRITE OPERATION TIMING DIAGRAM

Data Acquisition

7

PRINCIPLES OF OPERATION

The ADC0801, ADC0802, ADC0803, and ADC0805 each contain a circuit equivalent to a 256-resistor network. Analog switches are sequenced by successive approximation logic to match an analog differential input voltage ($V_{in+} - V_{in-}$) to a corresponding tap on the 256R network. The most-significant bit (MSB) is tested first. After eight comparisons (64 clock periods), an eight-bit binary code (1111 1111 = full scale) is transferred to an output latch and the interrupt ($\overline{INTR}$) output goes low. The device can be operated in a free-running mode by connecting the $\overline{INTR}$ output to the write ($\overline{WR}$) input and holding the conversion start ($\overline{CS}$) input at a low level. To ensure start-up under all conditions, a low-level $\overline{WR}$ input is required during the power-up cycle. Taking $\overline{CS}$ low anytime after that will interrupt a conversion in process.

When the $\overline{WR}$ input goes low, the internal successive approximation register (SAR) and eight-bit shift register are reset. As long as both $\overline{CS}$ and $\overline{WR}$ remain low, the analog-to-digital converter will remain in a reset state. One to eight clock periods after $\overline{CS}$ or $\overline{WR}$ makes a low-to-high transition, conversion starts.

When the $\overline{CS}$ and $\overline{WR}$ inputs are low, the start flip-flop is set and the interrupt flip-flop and eight-bit register are reset. The next clock pulse transfers a logic high to the output of the start flip-flop. The logic high is ANDed with the next clock pulse placing a logic high on the reset input of the start flip-flop. If either $\overline{CS}$ or $\overline{WR}$ have gone high, the set signal to the start flip-flop is removed causing it to be reset. A logic high is placed on the D input of the eight-bit shift register and the conversion process is started. If the $\overline{CS}$ and $\overline{WR}$ inputs are still low, the start flip-flop, the eight-bit shift register, and the SAR remain reset. This action allows for wide $\overline{CS}$ and $\overline{WR}$ inputs with conversion starting from one to eight clock periods after one of the inputs goes high.

When the logic high input has been clocked through the eight-bit shift register, completing the SAR search, it is applied to an AND gate controlling the output latches and to the D input of a flip-flop. On the next clock pulse, the digital word is transferred to the three-state output latches and the interrupt flip-flop is set. The output of the interrupt flip-flop is inverted to provide an $\overline{INTR}$ output that is high during conversion and low when the conversion is completed.

When a low is at both the $\overline{CS}$ and $\overline{RD}$ inputs, an output is applied to the DB0 through DB7 outputs and the interrupt flip-flop is reset. When either the $\overline{CS}$ or $\overline{RD}$ inputs return to a high state, the DB0 through DB7 outputs are disabled (returned to the high-impedance state). The interrupt flip-flop remains reset.

DATA
ACQUISITION
CIRCUITS

TYPE ADC0804C
8-BIT ANALOG-TO-DIGITAL CONVERTER
WITH DIFFERENTIAL INPUTS
D2755, OCTOBER 1983

- 8-Bit Resolution
- Ratiometric Conversion
- 100 μs Conversion Time
- 135 ns Access Time
- No Zero Adjust Requirement
- On-Chip Clock Generator
- Single 5-Volt Power Supply
- Operates With Microprocessor or as Stand-Alone
- Designed to be Interchangeable with National Semiconductor ADC0804LCN

N DUAL-IN-LINE PACKAGE
(TOP VIEW)

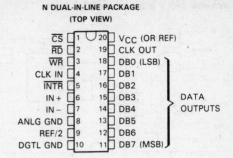

$\overline{CS}$	1	20	V_{CC} (OR REF)
$\overline{RD}$	2	19	CLK OUT
$\overline{WR}$	3	18	DB0 (LSB)
CLK IN	4	17	DB1
$\overline{INTR}$	5	16	DB2
IN +	6	15	DB3
IN −	7	14	DB4
ANLG GND	8	13	DB5
REF/2	9	12	DB6
DGTL GND	10	11	DB7 (MSB)

DATA OUTPUTS

description

The ADC0804C is a CMOS 8-bit successive-approximation analog-to-digital converter that uses a modified potentiometric (256R) ladder. The ADC0804 is designed to operate from common microprocessor control buses, with the three-state output latches driving the data bus. The ADC0804 can be made to appear to the microprocessor as a memory location or an I/O port.

A differential analog voltage input allows increased common-mode rejection and offset of the zero-input analog voltage value. Although a reference input (REF/2) is available to allow 8-bit conversion over smaller analog voltage spans or to make use of an external reference, ratiometric conversion is possible with the REF/2 input open. Without an external reference, the conversion takes place over a span from V_{CC} to analog ground (ANLG GND). The ADC0804 can operate with an external clock signal or, with an additional resistor and capacitor, can operate using an on-chip clock generator.

The ADC0804C is characterized for operation from 0°C to 70°C.

Data Acquisition

7

Copyright © 1983 by Texas Instruments Incorporated

TEXAS
INSTRUMENTS

POST OFFICE BOX 225012 • DALLAS, TEXAS 75265

functional block diagram (positive logic)

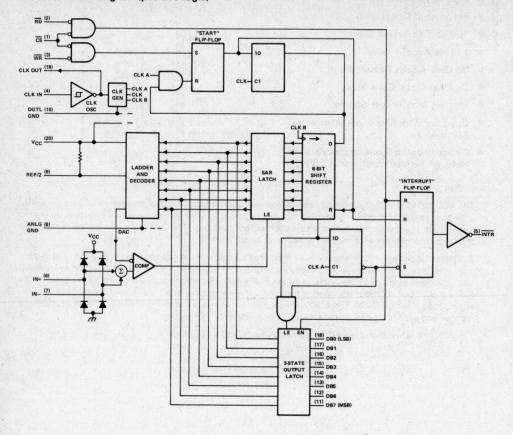

TEXAS
INSTRUMENTS
POST OFFICE BOX 225012 • DALLAS, TEXAS 75265

absolute maximum ratings over operating free-air temperature range (unless otherwise noted)

Supply voltage, V_{CC} (see Note 1) .... 6.5 V
Input voltage range $\overline{CS}$, $\overline{RD}$, $\overline{WR}$ .... −0.3 V to 18 V
          other inputs .... −0.3 V to V_{CC} + 0.3 V
Output voltage range .... −0.3 V to V_{CC} + 0.3 V
Continuous total power dissipation at 25 °C free-air temperature (see Note 2) .... 875 mW
Operating free-air temperature range .... 0 °C to 70 °C
Storage temperature range .... −65 °C to 150 °C
Lead temperature 1,6 mm (1/16 inch) from case for 10 seconds .... 260 °C

NOTES: 1. All voltage values are with respect to digital ground (DGTL GND) with DGTL GND and ANLG GND connected together (unless otherwise noted).
2. For operation above 25 °C free-air temperature, refer to Dissipation Derating Curves, Section 2.

recommended operating conditions

		MIN	NOM	MAX	UNIT
V_{CC}	Supply voltage	4.5	5	6.3	V
$V_{REF/2}$	Voltage at REF/2 (see Note 3)	0.25	2.5		V
V_{IH}	High-level input voltage at $\overline{CS}$, $\overline{RD}$, or $\overline{WR}$	2		15	V
V_{IL}	Low-level input voltage at $\overline{CS}$, $\overline{RD}$, or $\overline{WR}$			0.8	V
	Analog ground voltage (see Note 4)	−0.05	0	1	V
	Analog input voltage (see Note 5)	GND − 0.05		V_{CC} + 0.05	V
f_{clock}	Clock input frequency (see Note 6)	100	640	1460	kHz
	Duty cycle above 640 kHz (see Note 6)	40		60	%
$t_{w(CLK)}$	Pulse duration clock input (high or low) (see Note 6)	275	781		ns
$t_{w(WR)}$	Pulse duration, $\overline{WR}$ input low	100			ns
T_A	Operating free-air temperature	0		70	°C

NOTES: 3. Proper operation is achieved over a differential input range of 0 V to V_{CC} when the REF/2 input is open.
4. These values are with respect to digital ground (pin 10).
5. When the positive analog input with respect to the negative analog input ($V_{in +} - V_{in -}$) is zero or negative, the output code is 0000 0000.
6. Total unadjusted error is guaranteed only at an f_{clock} of 640 kHz with a duty cycle of 40% to 60% (pulse duration 625 ns to 937 ns). For frequencies above this limit or pulse duration below 625 ns, error may increase. The duty cycle limits should be observed for an f_{clock} greater than 640 kHz. Below 640 kHz, this duty cycle limit can be exceeded provided $t_{w(CLK)}$ remains within limits.

Data Acquisition

7

TEXAS
INSTRUMENTS
POST OFFICE BOX 225012 ● DALLAS, TEXAS 75265

electrical characteristics over recommended operating free-air temperature range,
V_{CC} = 5 V, f_{clock} = 640 kHz, REF/2 = 2.5 V (unless otherwise noted)

	PARAMETER		TEST CONDITIONS	MIN	TYP†	MAX	UNIT
V_{OH}	High-level output voltage	All outputs	V_{CC} = 4.75 V, I_{OH} = −360 μA	2.4			V
		DB and $\overline{INTR}$	V_{CC} = 4.75 V, I_{OH} = −10 μA	4.5			
V_{OL}	Low-level output voltage	Data outputs	V_{CC} = 4.75 V, I_{OL} = 1.6 mA			0.4	V
		$\overline{INTR}$ output	V_{CC} = 4.75 V, I_{OL} = 1 mA			0.4	
		CLK OUT	V_{CC} = 4.75 V, I_{OL} = 360 μA			0.4	
V_{T+}	Clock positive-going threshold voltage			2.7	3.1	3.5	V
V_{T-}	Clock negative-going threshold voltage			1.5	1.8	2.1	V
$V_{T+} - V_{T-}$	Clock input hysteresis			0.6	1.3	2	V
I_{IH}	High-level input current				0.005	1	μA
I_{IL}	Low-level input current				−0.005	−1	μA
I_{OZ}	Off-state output current		V_O = 0			−3	μA
			V_O = 5 V			3	
I_{OHS}	Short-circuit output current	Output high	V_O = 0, T_A = 25°C	−4.5	−6		mA
I_{OLS}	Short-circuit output current	Output low	V_O = 5 V, T_A = 25°C	9	16		mA
I_{CC}	Supply current plus reference current		REF/2 open, $\overline{CS}$ at 5 V, T_A = 25°C		1.9	2.5	mA
$R_{REF/2}$	Input resistance to reference ladder		See Note 7	1	1.3		kΩ
C_i	Input capacitance (control)				5	7.5	pF
C_o	Output capacitance (DB)				5	7.5	pF

†All typical values are at T_A = 25°C.
NOTE 7: The resistance is calculated from the current drawn from a 5-volt supply applied to pins 8 and 9.

operating characteristics over recommended operating free-air temperature,
V_{CC} = 5 V, $V_{REF/2}$ = 2.5 V, f_{clock} = 640 kHz (unless otherwise noted)

	PARAMETER	TEST CONDITIONS	MIN	TYP†	MAX	UNIT
	Supply-voltage-variation error	V_{CC} = 4.5 V to 5.5 V, See Note 8		±1/16	±1/8	LSB
	Total unadjusted error	See Notes 8 and 9			±1	LSB
	DC common-mode error	See Note 9		±1/16	±1/8	LSB
t_{en}	Output enable time	C_L = 100 pF		135	200	ns
t_{dis}	Output disable time	C_L = 10 pF, R_L = 10 kΩ		125	200	ns
$t_{d(INTR)}$	Delay time to reset $\overline{INTR}$			300	450	ns
t_{conv}	Conversion cycle time	f_{clock} = 100 kHz to 1.46 MHz, See Note 10	66		73	clock cycles
CR	Free-running conversion rate	$\overline{INTR}$ connected to $\overline{WR}$, $\overline{CS}$ at 0 V			8770	conv/s

†All typical values are at T_A = 25°C.
NOTES: 8. These parameters are guaranteed over the recommended analog input voltage range.
9. All errors are measured with reference to an ideal straight line through the end-points of the analog-to-digital transfer characteristic.
10. Although internal conversion is completed in 64 clock periods, a $\overline{CS}$ or $\overline{WR}$ low-to-high transition is followed by 1 to 8 clock periods before conversion starts. After conversion is completed, part of another clock period is required before a high-to-low transition of $\overline{INTR}$ completes the cycle.

Texas
Instruments
POST OFFICE BOX 225012 • DALLAS, TEXAS 75265

118

timing diagrams

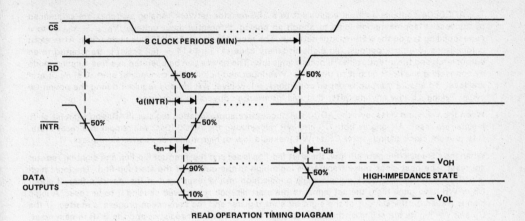

READ OPERATION TIMING DIAGRAM

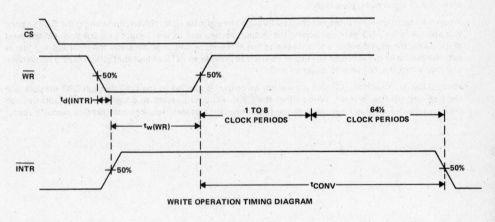

WRITE OPERATION TIMING DIAGRAM

Data Acquisition

7

**TEXAS
INSTRUMENTS**
POST OFFICE BOX 225012 • DALLAS, TEXAS 75265

PRINCIPLES OF OPERATION

The ADC0804 contains a circuit equivalent to a 256-resistor network. Analog switches are sequenced by successive approximation logic to match an analog differential input voltage ($V_{in+} - V_{in-}$) to a corresponding tap on the 256-resistor network. The most-significant bit (MSB) is tested first. After eight comparisons (64 clock periods), an eight-bit binary code (1111 1111 = full scale) is transferred to an output latch and the interrupt ($\overline{INTR}$) output goes low. The device can be operated in a free-running mode by connecting the $\overline{INTR}$ output to the write ($\overline{WR}$) input and holding the conversion start ($\overline{CS}$) input at a low level. To ensure start-up under all conditions, a low-level $\overline{WR}$ input is required during the power-up cycle. Taking $\overline{CS}$ low anytime after that will interrupt a conversion in process.

When the $\overline{WR}$ input goes low, the ADC0804 successive approximation register (SAR) and eight-bit shift register are reset. As long as both $\overline{CS}$ and $\overline{WR}$ remain low, the ADC0804C will remain in a reset state. One to eight clock periods after $\overline{CS}$ or $\overline{WR}$ makes a low-to-high transition, conversion starts.

When the $\overline{CS}$ and $\overline{WR}$ inputs are low, the start flip-flop is set and the interrupt flip-flop and eight-bit register are reset. The next clock pulse transfers a logic high to the output of the start flip-flop. The logic high is ANDed with the next clock pulse placing a logic high on the reset input of the start flip-flop. If either $\overline{CS}$ or $\overline{WR}$ have gone high, the set signal to the start flip-flop is removed causing it to be reset. A logic high is placed on the D input of the eight-bit shift register and the conversion process is started. If the $\overline{CS}$ and $\overline{WR}$ inputs are still low, the start flip-flop, the eight-bit shift register, and the SAR remain reset. This action allows for wide $\overline{CS}$ and $\overline{WR}$ inputs with conversion starting from one to eight clock periods after one of the inputs goes high.

When the logic high input has been clocked through the eight-bit shift register, completing the SAR search, it is applied to an AND gate controlling the output latches and to the D input of a flip-flop. On the next clock pulse, the digital word is transferred to the three-state output latches and the interrupt flip-flop is set. The output of the interrupt flip-flop is inverted to provide an $\overline{INTR}$ output that is high during conversion and low when the conversion is completed.

When a low is at both the $\overline{CS}$ and $\overline{RD}$ inputs, an output is applied to the DB0 through DB7 outputs and the interrupt flip-flop is reset. When either the $\overline{CS}$ or $\overline{RD}$ inputs return to a high state, the DB0 through DB7 outputs are disabled (returned to the high-impedance state). The interrupt flip-flop remains reset.

TYPES ADC0808, ADC0809
CMOS ANALOG-TO-DIGITAL CONVERTERS
WITH 8-CHANNEL MULTIPLEXERS

D2642, JUNE 1981–REVISED OCTOBER 1983

- Total Unadjusted Error . . . ±½ LSB Max for ADC0808 and ±1 LSB Max for ADC0809
- Resolution of 8 Bits
- 100 μs Conversion Time
- Ratiometric Conversion
- Guaranteed Monotonicity
- No Missing Codes
- Easy Interface with Microprocessors
- Latched 3-State Outputs
- Latched Address Inputs
- Single 5-Volt Supply
- Low Power Consumption
- Designed to be Interchangeable with National Semiconductor ADC0808, ADC0809

N
DUAL-IN-LINE PACKAGE
(TOP VIEW)

```
                       3 [ 1    28 ] 2
                       4 [ 2    27 ] 1  } INPUTS
             INPUTS <  5 [ 3    26 ] 0
                       6 [ 4    25 ] A
                       7 [ 5    24 ] B  } ADDRESS
    START OF CONVERSION [ 6    23 ] C
      END OF CONVERSION [ 7    22 ] ADDRESS LOAD CONTROL
                 2 − 5 [ 8    21 ] 2−1 (MSB)
         OUTPUT ENABLE [ 9    20 ] 2−2
                   CLK [ 10   19 ] 2−3
                   VCC [ 11   18 ] 2−4
                  REF+ [ 12   17 ] 2−8 (LSB)
                   GND [ 13   16 ] REF−
                 2 − 7 [ 14   15 ] 2−6
```

description

The ADC0808 and ADC0809 are monolithic CMOS devices with an 8-channel multiplexer, an 8-bit analog-to-digital (A/D) converter, and microprocessor-compatible control logic. The 8-channel multiplexer can be controlled by a microprocessor through a 3-bit address decoder with address load to select any one of eight single-ended analog switches connected directly to the comparator. The 8-bit A/D converter uses the successive-approximation conversion technique featuring a high-impedance threshold detector, a switched-capacitor array, a sample-and-hold, and a successive-approximation register (SAR).

The comparison and converting methods used eliminate the possibility of missing codes, nonmonotonicity, and the need for zero or full-scale adjustment. Also featured are latched 3-state outputs from the SAR and latched inputs to the multiplexer address decoder. The single 5-volt supply and low power requirements make the ADC0808 and ADC0809 especially useful for a wide variety of applications. Ratiometric conversion is made possible by access to the reference voltage input terminals.

The ADC0808 and ADC0809 are characterized for operation from -40°C to 85°C.

absolute maximum ratings over operating free-air temperature range (unless otherwise noted)

Supply voltage, V_{CC} (see Note 1) . . . . 6.5 V
Input voltage range: control inputs . . . . −0.3 V to 15 V
                    all other inputs . . . . −0.3 V to V_{CC} + 0.3 V
Continuous total dissipation at (or below) 25 °C free-air temperature (see Note 2) . . . . 875 mW
Operating free-air temperature range . . . . −40 °C to 85 °C
Storage temperature range . . . . −65 °C to 150 °C
Lead temperature 1,6 mm (1/16 inch) from case of 10 seconds . . . . 260 °C

NOTES: 1. All voltage values are with respect to network ground terminal.
        2. For operations above 25 °C free-air temperature, refer to Dissipation Derating Curves, Section 2.

Data Acquisition

7

TYPES ADC0808, ADC0809
CMOS ANALOG-TO-DIGITAL CONVERTERS
WITH 8-CHANNEL MULTIPLEXERS

functional block diagram (positive logic)

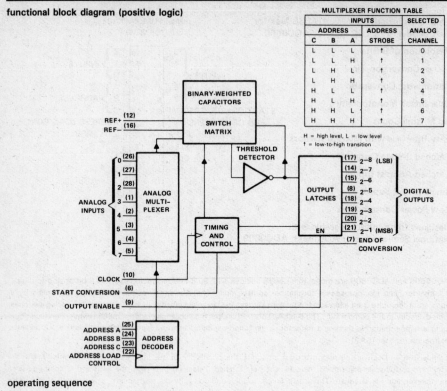

MULTIPLEXER FUNCTION TABLE				
INPUTS				SELECTED
ADDRESS			ADDRESS	ANALOG
C	B	A	STROBE	CHANNEL
L	L	L	↑	0
L	L	H	↑	1
L	H	L	↑	2
L	H	H	↑	3
H	L	L	↑	4
H	L	H	↑	5
H	H	L	↑	6
H	H	H	↑	7

H = high level, L = low level
↑ = low-to-high transition

operating sequence

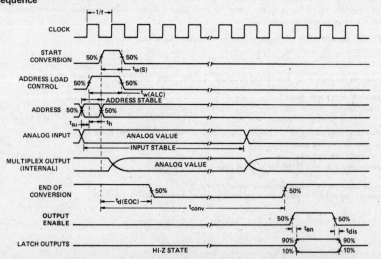

**TEXAS
INSTRUMENTS**

POST OFFICE BOX 225012 • DALLAS, TEXAS 75265

recommended operating conditions

	MIN	NOM	MAX	UNIT
Supply voltage, V_{CC}	4.5	5	6	V
Positive reference voltage, V_{ref+} (see Note 3)		V_{CC}	$V_{CC}+0.1$	V
Negative reference voltage, V_{ref-}		0	-0.1	V
Differential reference voltage, $V_{ref+} - V_{ref-}$		5		V
Start pulse duration $t_{w(S)}$	200			ns
Address load control pulse width, $t_{w(ALC)}$	200			ns
Address setup time, t_{su}	50			ns
Address hold time, t_h	50			ns
Clock frequency, f_{clock}	10	640	1280	kHz

NOTE 3: Care must be taken that this rating is observed even during power-up.

electrical characteristics over recommended operating free-air temperature range. V_{CC} = 4.75 V to 5.25 V (unless otherwise noted)

total device

PARAMETER		TEST CONDITIONS	MIN	TYP[†]	MAX	UNIT
V_{IH}	High-level input voltage, control inputs	V_{CC} = 5 V	$V_{CC}-1.5$			V
V_{IL}	Low-level input voltage, control inputs	V_{CC} = 5 V			1.5	V
V_{OH}	High-level output voltage	I_O = $-360\ \mu A$	$V_{CC}-0.4$			V
V_{OL}	Low-level output voltage — Data outputs	I_O = 1.6 mA			0.45	V
	Low-level output voltage — End of conversion	I_O = 1.2 mA			0.45	
I_{OZ}	Off-state (high-impedance-state) output current	V_O = 5 V			3	μA
		V_O = 0			-3	
I_I	Control input current at maximum input voltage	V_I = 15 V			1	μA
I_{IL}	Low-level control input current	V_I = 0			-1	μA
I_{CC}	Supply Current	f_{clock} = 640 kHz		0.3	3	mA
C_i	Input capacitance, control inputs	T_A = 25°C		10	15	pF
C_o	Output capacitance, data outputs	T_A = 25°C		10	15	pF
	Resistance from pin 12 to pin 16		1	1000		kΩ

analog multiplexer

PARAMETER		TEST CONDITIONS		MIN	TYP[†]	MAX	UNIT
I_{on}	Channel on-state current (see Note 4)	V_I = 5 V,	f_{clock} = 640 kHz			2	μA
		V_I = 0 V,	f_{clock} = 640 kHz			-2	
I_{off}	Channel off-state current	V_{CC} = 5 V, T_A = 25°C	V_I = 5 V		10	200	nA
			V_I = 0		-10	-200	
		V_{CC} = 5V	V_I = 5 V			1	μA
			V_I = 0			-1	

[†]Typical values are at V_{CC} = 5 V and T_A = 25°C.
NOTE 4: Channel on-state current is primarily due to the bias current into or out of the threshold detector, and it varies directly with clock frequency.

Data Acquisition

7

operating characteristics, $T_A = 25\,°C$, $V_{CC} = V_{REF+} = 5$ V, $V_{REF-} = 0$ V, $f_{clock} = 640$ kHz (unless otherwise noted)

	PARAMETER	TEST CONDITIONS	ADC0808			ADC0809			UNIT
			MIN	TYP†	MAX	MIN	TYP†	MAX	
k_{SVS}	Supply voltage sensitivity	$V_{CC} = V_{ref+} = 4.75$ V to 5.25 V, $T_A = -40\,°C$ to $85\,°C$, See Note 5		±0.05			±0.05		%/V
	Linearity error (see Note 6)			±0.25			±0.5		LSB
	Zero error (see Note 7)			±0.25			±0.25		LSB
	Total unadjusted error (See Note 8)	$T_A = 25\,°C$		±0.25	±0.5		±0.5		LSB
		$T_A = -40\,°C$ to $85\,°C$			±0.75			±1.25	
		$T_A = 0\,°C$ to $70\,°C$						±1	
t_{en}	Output enable time	$C_L = 50$ pF, $R_L = 10$ kΩ		80	250		80	250	ns
t_{dis}	Output disable time	$C_L = 10$ pF, $R_L = 10$ kΩ		105	250		105	250	ns
t_{conv}	Conversion time	See Note 10	90	100	116	90	100	116	µs
$t_{d(EOC)}$	Delay time, end of conversion output	See Notes 9 and 10	0		14.5	0		14.5	µs

†Typical values for all except supply voltage sensitivity are at $V_{CC} = 5$ V, and all are at T_A 25°C.

NOTES:
5. Supply voltage sensitivity relates to the ability of an analog-to-digital converter to maintain accuracy as the supply voltage varies. The supply and V_{ref+} are varied together and the change in accuracy is measured with respect to full-scale.
6. Linearity error is the maximum deviation from a straight line through the end points of the A/D transfer characteristic.
7. Zero error is the difference between the output of an ideal converter and the actual A/D converter for zero input voltage.
8. Total unadjusted error is the maximum sum of linearity error, zero error, and full-scale error.
9. For clock frequencies other than 640 kHz, $t_{d(EOC)}$ maximum is 8 clock periods plus 2 µs.
10. Refer to the operating sequence diagram.

Data Acquisition

7

TEXAS
INSTRUMENTS
POST OFFICE BOX 225012 ● DALLAS, TEXAS 75265

The ADC0808 and ADC0809 each consists of an analog signal multiplexer, an 8-bit successive-approximation converter, and related control and output circuitry.

multiplexer

The analog mutiplexer selects 1 of 8 single-ended input channels as determined by the address decoder. Address load control loads the address code into the decoder on a low-to-high transition.

converter

The CMOS threshold detector in the successive-approximation conversion system determines each bit by examining the charge on a series of binary-weighted capacitors (Figure 1). In the first phase of the conversion process, the analog input is sampled by closing switch S_C and all S_T switches, and by simultaneously charging all the capacitors to the input voltage.

In the next phase of the conversion process, all S_T and S_C switches are opened and the threshold detector begins identifying bits by identifying the charge (voltage) on each capacitor relative to the reference voltage. In the switching sequence, all eight capacitors are examined separately until all 8 bits are identified, and then the charge-convert sequence is repeated. In the first step of the conversion phase, the threshold detector looks at the first capacitor (weight = 128). Node 128 of this capacitor is switched to the reference voltage, and the equivalent nodes of all the other capacitors on the ladder are switched to REF −. If the voltage at the summing node is greater than the trip-point of the threshold detector (approximately one-half the V_{CC} voltage), a bit is placed in the output register, and the 128-weight capacitor is switched to REF −. If the voltage at the summing node is less than the trip point of the threshold detector, this 128-weight capacitor remains connected to REF + through the remainder of the capacitor-sampling (bit-counting) process. The process is repeated for the 64-weight capacitor, the 32-weight capacitor, and so forth down the line, until all bits are counted.

With each step of the capacitor-sampling process, the initial charge is redistributed among the capacitors. The conversion process is successive approximation, but relies on charge redistribution rather than a successive-approximation register (and reference D/A) to count and weigh the bits from MSB to LSB.

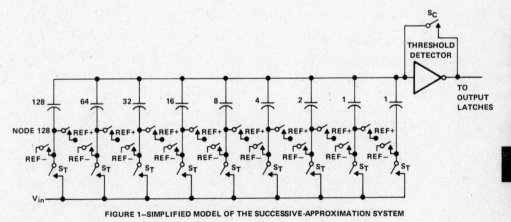

FIGURE 1—SIMPLIFIED MODEL OF THE SUCCESSIVE-APPROXIMATION SYSTEM

Data Acquisition

7

Texas Instruments
POST OFFICE BOX 225012 • DALLAS, TEXAS 75265

DATA
ACQUISITION
CIRCUITS

TYPES ADC0831, ADC0832, ADC0834, ADC0838
2-, 4-, 8-CHANNEL A/D PERIPHERALS WITH
SERIAL CONTROL AND MULTIPLEXER OPTIONS

D2995, OCTOBER 1983

- Easy Interface to Microprocessors or Stand-Alone Operation
- Operates Ratiometrically or With 5-V Reference
- No Full Scale or Zero Adjust
- 2-, 4-, or 8-Channel Multiplexer Options With Address Logic
- Shunt Regulator Allows Operation With High-Voltage Supplies
- 0-V to 5-V Input Range With Single 5-V Power Supply
- Remote Operation With Serial Data Link
- TTL/MOS Input/Output Compatible
- 8-, 14-, or 20-Pin DIP Package
- Designed to be Interchangeable with National Semiconductor ADC0831BC, ADC0832BC, ADC0834BC, and ADC0838BC Over Operating Temperature Range of −40°C to 85°C

P DUAL-IN-LINE PACKAGE
(TOP VIEW)

ADC0831

ADC0832

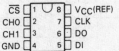

N DUAL-IN-LINE PACKAGE
(TOP VIEW)

ADC0834

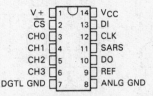

ADC0838

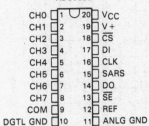

description

The ADC0831, ADC0832, ADC0834, and ADC0838 are 8-bit successive-approximation analog-to-digital converters each with a serial input/output and configurable input multiplexers with up to 8 channels. The serial input/output is configured to interface with standard shift registers or microprocessors.

The 2-, 4-, or 8-channel multiplexers are software configured for single-ended or differential inputs as well as channel assignment.

The differential analog voltage input allows increasing of the common-mode rejection and offsetting the analog zero input voltage value. In addition, the voltage reference input can be adjusted to allow encoding any smaller analog voltage span to the full 8 bits of resolution.

The ADC0831I, ADC0832I, ADC0834I, and ADC0838I are characterized for operation from −40°C to 85°C.

Data Acquisition

7

TEXAS
INSTRUMENTS

POST OFFICE BOX 225012 • DALLAS, TEXAS 75265

functional block diagram (ADC0838)

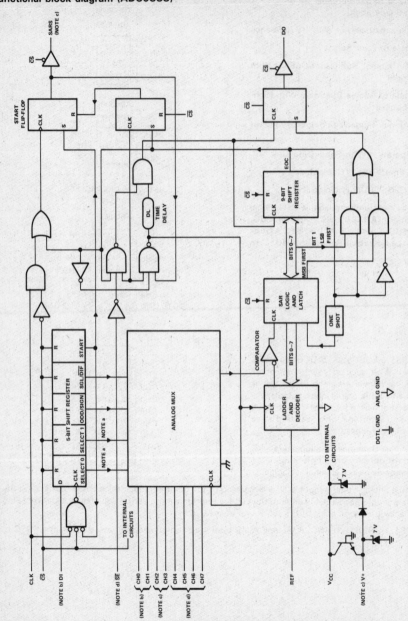

NOTES: a. For the ADC0834, DI is input directly to the D input of SELECT 1. SELECT 0 is forced to a high. For the ADC0832, DI is input directly to the D input of ODD/SIGN. SELECT 0 is forced to a low and SELECT 1 is forced to a high. (See multiplexer addressing tables.)
b. ADC0832, ADC0834, and ADC0838 only.
c. ADC0834 and ADC0838 only.
d. ADC0838 only.

Data Acquisition

7

TEXAS INSTRUMENTS
POST OFFICE BOX 225012 • DALLAS, TEXAS 75265

absolute maximum ratings over operating free-air temperature range (unless otherwise noted)

Supply voltage, V_{CC} (see Note 1) .... 6.5 V
Logic input voltage range .... −0.3 V to 15 V
Analog input voltage range .... −0.3 V to V_{CC} +0.3 V
Continuous total dissipation at (or below) 25 °C free-air temperature (see Note 2):
                                          P package .... 725 mW
                                          N package .... 800 mW
Operating free-air temperature range .... −40 °C to 85 °C
Storage temperature range .... −65 °C to 150 °C
Lead temperature 1,6 mm (1/16 inch) from case for 10 seconds .... 260 °C

NOTES: 1. All voltage values, except differential voltages, are with respect to the network ground terminal.
        2. For operation above 25 °C free-air temperature, refer to Dissipation Derating Curves, Section 2.

recommended operating conditions

		MIN	NOM	MAX	UNIT
V_{CC}	Supply voltage	4.5	5	6.3	V
V_{IH}	High-level input voltage	2		15	V
V_{IL}	Low-level input voltage			0.8	V
$t_{w(CSH)}$	Pulse duration, $\overline{CS}$ high	1·20			ns
t_{su}	Setup time, $\overline{CS}$ or $\overline{SE}$ low, or data valid before clock↑	250			ns
t_h	Hold time, data input valid after clock ↑	90			ns
f_{clock}	Clock frequency	10		400	kHz
	Clock duty cycle (see Note 3)	40		60	%
T_A	Operating free-air temperature	−40		85	°C

NOTE 3: The clock duty cycle range ensures proper operation at all clock frequencies. If a clock frequency is used outisde the recommended duty cycle range, the minimum pulse duration (high or low) is 1 μs.

TEXAS
INSTRUMENTS
POST OFFICE BOX 225012 ● DALLAS, TEXAS 75265

electrical characteristics over recommended operating free-air temperature range,
$V_{CC} = V+ = 5$ V (V+ applies to ADC0834 and ADC0838 only), $f_{clock} = 250$ kHz
(unless otherwise noted)

	PARAMETER	TEST CONDITIONS[†]		MIN	TYP[‡]	MAX	UNIT
V_{OH}	High-level output voltage	$V_{CC} = 4.75$ V,	$I_{OH} = -360\ \mu A$	2.4			V
		$V_{CC} = 4.75$ V,	$I_{OH} = -10\ \mu A$	4.5			
V_{OL}	Low-level output voltage	$V_{CC} = 4.75$ V,	$I_{OL} = 1.6$ mA			0.4	V
I_{IH}	High-level input current	$V_I = V_{CC}$			0.005	1	μA
I_{IL}	Low-level input current	$V_I = 0$			-0.005	-1	μA
I_{OZ}	High-impedance output current (DO, SARS)	$V_O = 0.4$ V,	$T_A = 25°C$		-0.1	-3	μA
		$V_O = 5$ V,	$T_A = 25°C$		0.1	3	
I_{OHS}	Source current	$V_O = 0$,	$T_A = 25°C$		-14		mA
I_{OLS}	Sink current	$V_O = V_{CC}$,	$T_A = 25°C$		16		mA
V_{ICR}	Common-mode input range (see Note 4)			-0.05 to $V_{CC}+0.05$			V
I_{on}	On-channel leakage current (see Note 8)	On-channel input at 0 V				-1	μA
		Off-channel inputs at 5 V, $T_A = 25°C$				-200	nA
		On-channel input at 0 V				1	μA
		Off-channel inputs at 5 V				200	nA
I_{off}	Off-channel leakage current (see Note 8)	On-channel input at 0 V				-1	μA
		Off-channel inputs at 5 V, $T_A = 25°C$				-50	nA
		On-channel input at 0 V				1	μA
		Off-channel inputs at 5 V				50	nA
R_{REF}	Input resistance to reference ladder			1.9	2.4		kΩ
C_i	Input capacitance (logic inputs)	$T_A = 25°C$			5		pF
C_O	Output capacitance	$T_A = 25°C$			5		pF
I_{CC}	Supply current (see Note 5)	ADC0832[§]			3	5.2	mA
		ADC0831, ADC0834, ADC0838			1	2.5	
$I+$	Current into V+ (see Note 5)					10	mA

[†]All parameters are measured under open-loop conditions with zero common-mode input voltage (unless otherwise specified).
[‡]All typical values are at $T_A = 25°C$.
[§]Includes ladder current.

NOTES: 4. For IN− more positive than IN+, the digital output code will be 0000 0000. Connected to each analog input are two on-chip diodes that will conduct forward current for analog input voltages one diode drop below ground or one diode drop above V_{CC}. Care must be taken during testing at low V_{CC} levels (4.5 V) because high-level analog input (5 V) can, especially at high temperatures, cause this input diode to conduct and cause errors for analog inputs that are near full-scale. As long as the analog input voltage does not exceed the supply voltage by more than 50 mV, the output code will be correct. To achieve an absolute 0-V to 5-V input voltage range requires a minimum V_{CC} of 4.950 volts for all variations of temperature and load.

5. An internal zener diode is connected from the V_{CC} input to ground and from V+ to ground. The breakdown voltage of each diode is approximately 7 V. The V+ diode is a shunt regulator and connects to V_{CC} via a diode. When the voltage regulator powers the converter, this diode ensures that the V_{CC} input is less than the zener breakdown voltage (6.4 V). A series resistor is recommended to limit the maximum current into the V+ input.

TEXAS INSTRUMENTS
POST OFFICE BOX 225012 • DALLAS, TEXAS 75265

operating characteristics over recommended operating free-air temperature range,
$V_{CC} = V+ = 5\ V$ (V + applies to ADC0834 and ADC0838 only), $f_{clock} = 250\ kHz$, $t_r = t_f = 20\ ns$
(unless otherwise noted)

PARAMETER		TEST CONDITIONS	MIN	TYP	MAX	UNIT
Supply-voltage variation error				± 1/16		LSB
Total unadjusted error (see Note 6)		V_{ref} forced to 5 V			± 1	LSB
Common-Mode error		Differential mode		± 1/16		LSB
t_{pd} Propagation delay time, clock ↓ to output data (see Note 7)	MSB first	$C_L = 100\ pF$, $T_A = 25°C$		650	1000	ns
	LSB first			250	600	
t_{dis} Output disable time from $\overline{CS}$ ↑		$C_L = 10\ pF$, $R_L = 10\ k\Omega$, $T_A = 25°C$		125	250	ns
t_{conv} Conversion time		Not including multiplexer addressing time, $T_A = 25°C$			8	clock period

NOTES: 6. Total unadjusted error includes offset, full scale, linearity, and multiplexer errors.
      7. If the MSB from the comparator is used first in the successive-approximation loop, then an additional built-in delay will allow for comparator response time.
      8. Leakage current is measured with the clock not switching.

PARAMETER MEASUREMENT INFORMATION

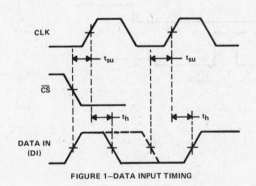

FIGURE 1–DATA INPUT TIMING

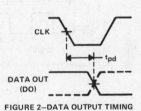

FIGURE 2–DATA OUTPUT TIMING

Data Acquisition

7

TEXAS
INSTRUMENTS
POST OFFICE BOX 225012 • DALLAS, TEXAS 75265

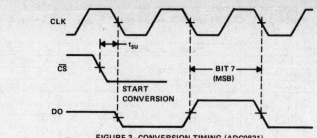

FIGURE 3—CONVERSION TIMING (ADC0831)

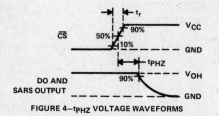

FIGURE 4—t$_{PHZ}$ VOLTAGE WAVEFORMS

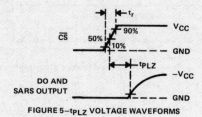

FIGURE 5—t$_{PLZ}$ VOLTAGE WAVEFORMS

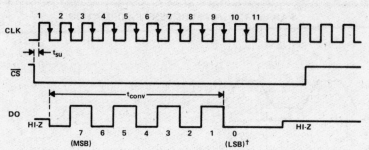

†LSB first output not available on ADC0831.

FIGURE 6—ADC0831 TIMING DIAGRAM

TEXAS
INSTRUMENTS
POST OFFICE BOX 225012 • DALLAS, TEXAS 75265

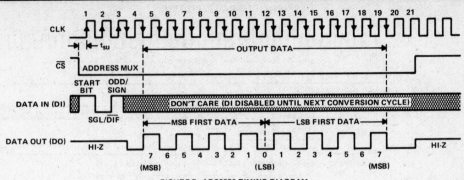

FIGURE 7—ADC0832 TIMING DIAGRAM

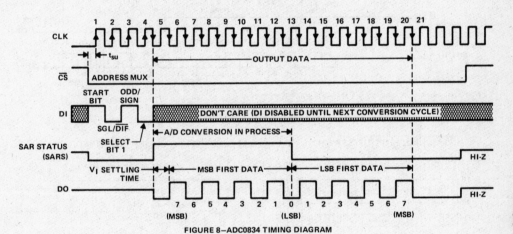

FIGURE 8—ADC0834 TIMING DIAGRAM

Data Acquisition

7

TEXAS
INSTRUMENTS

POST OFFICE BOX 225012 • DALLAS, TEXAS 75265

TYPICAL CHARACTERISTICS

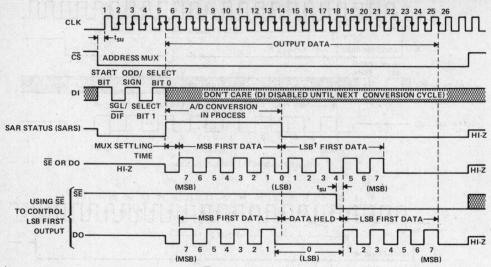

†Make sure clock edge of 18th clock, clocks in the LSB before $\overline{SE}$ is taken low.

FIGURE 9–ADC0838 TIMING DIAGRAM

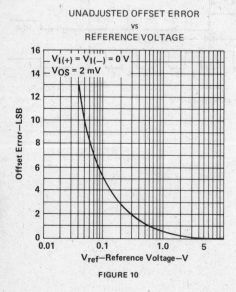

UNADJUSTED OFFSET ERROR
vs
REFERENCE VOLTAGE

$V_{I(+)} = V_{I(-)} = 0\ V$
$V_{OS} = 2\ mV$

FIGURE 10

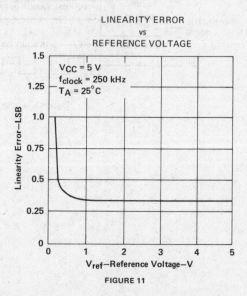

LINEARITY ERROR
vs
REFERENCE VOLTAGE

$V_{CC} = 5\ V$
$f_{clock} = 250\ kHz$
$T_A = 25°C$

FIGURE 11

Data Acquisition

7

TEXAS
INSTRUMENTS
POST OFFICE BOX 225012 • DALLAS, TEXAS 75265

118

TYPICAL CHARACTERISTICS

LINEARITY ERROR
vs
FREE-AIR TEMPERATURE

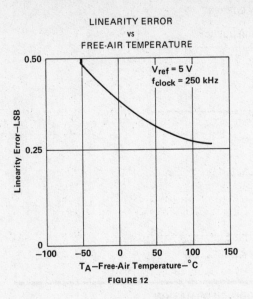

FIGURE 12

LINEARITY ERROR
vs
CLOCK FREQUENCY

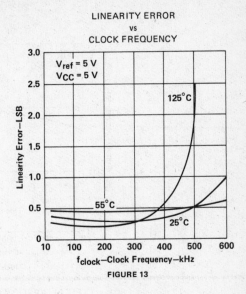

FIGURE 13

SUPPLY CURRENT†
vs
FREE-AIR TEMPERATURE

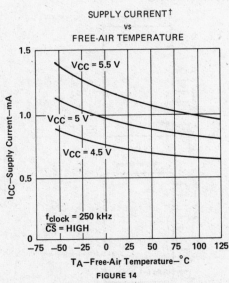

FIGURE 14

OUTPUT CURRENT
vs
FREE-AIR TEMPERATURE

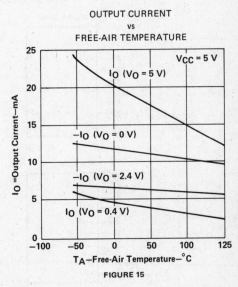

FIGURE 15

†For ADC0832, the ladder current I_{REF} should be added to I_{CC}.

Data Acquisition

7

TEXAS
INSTRUMENTS
POST OFFICE BOX 225012 • DALLAS, TEXAS 75265

TYPICAL CHARACTERISTICS

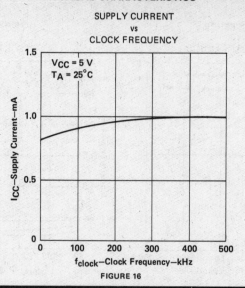

SUPPLY CURRENT
vs
CLOCK FREQUENCY

FIGURE 16

TYPICAL APPLICATION INFORMATION

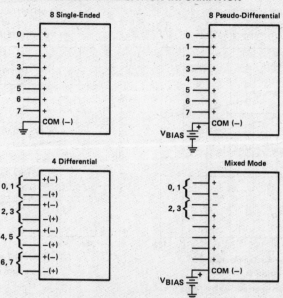

FIGURE 17—ANALOG INPUT MULTIPLEXER OPTIONS FOR THE ADC0838

TEXAS
INSTRUMENTS

POST OFFICE BOX 225012 • DALLAS, TEXAS 75265

ACD0832 MUX ADDRESSING (5-BIT SHIFT REGISTER) (See Note 9)

TABLE 1. SINGLE ENDED MUX MODE

MUX ADDRESS		CHANNEL NO.	
SGL/DIF	ODD/SIGN	0	1
1	0	+	
1	1		+

TABLE 2. DIFFERENTIAL MUX MODE

MUX ADDRESS		CHANNEL NO.	
SGL/DIF	ODD/SIGN	0	1
0	0	+	−
0	1	−	+

NOTE 9: Internally, Select 0 is low, Select 1 is high, COMMON is internally connected to ANLG GND.

ADC0834 MUX ADDRESSING (5-BIT SHIFT REGISTER) (See Note 10)

TABLE 3. SINGLE-ENDED MUX MODE

MUX ADDRESS			CHANNEL NO.			
SGL/DIF	ODD/SIGN	SELECT 1	0	1	2	3
1	0	0	+			
1	0	1		+		
1	1	0			+	
1	1	1				+

TABLE 4. DIFFERENTIAL MUX MODE

MUX ADDRESS			CHANNEL NO.			
SGL/DIF	ODD/SIGN	SELECT 1	0	1	2	3
0	0	0	+	−		
0	0	1			+	−
0	1	0	−	+		
0	1	1			−	+

NOTE 10: Internally, Select 0 is high, COMMON is internally connected to ANLG GND.

ADC0838 MUX ADDRESSING (5-BIT SHIFT REGISTER)

TABLE 5. SINGLE-ENDED MUX MODE

MUX ADDRESS		SELECT		ANALOG SINGLE-ENDED CHANNEL NO.								
SGL/DIF	ODD/SIGN	1	0	0	1	2	3	4	5	6	7	COM
1	0	0	0	+								−
1	0	0	1		+							−
1	0	1	0			+						−
1	0	1	1				+					−
1	1	0	0	+								−
1	1	0	1			+						−
1	1	1	0					+				−
1	1	1	1							+		−

TEXAS
INSTRUMENTS
POST OFFICE BOX 225012 • DALLAS, TEXAS 75265

Data Acquisition

7

1183

TYPES ADC0831, ADC0832, ADC0834, ADC0838
2-, 4-, 8-CHANNEL A/D PERIPHERALS WITH
SERIAL CONTROL AND MULTIPLEXER OPTIONS

TABLE 6. DIFFERENTIAL MUX MODE

MUX ADDRESS				ANALOG DIFFERENTIAL CHANNEL-PAIR NO.							
SGL/$\overline{\text{DIF}}$	ODD/SIGN	SELECT		0		1		2		3	
		1	0	0	1	2	3	4	5	6	7
0	0	0	0	+	−						
0	0	0	1			+	−				
0	0	1	0					+	−		
0	0	1	1							+	−
0	1	0	0	−	+						
0	1	0	1			−	+				
0	1	1	0					−	+		
0	1	1	1							−	+

PRINCIPLES OF OPERATION

The ADC0831, ADC0832, ADC0834, and ADC0838 use a sample data comparator structure that converts differential analog inputs by a successive-approximation routine. The input voltage to be converted is applied to a channel terminal and is compared to ground (single-ended), to an adjacent channel (differential), or to a common terminal (pseudo-differential) that can be an arbitrary voltage. The input terminals are assigned a positive (+) or negative (−) polarity. If the signal input applied to the assigned positive terminal is less than the signal on the negative terminal, the converter output is all zeroes.

Channel selection and input configuration are under software control using a serial data link from the controlling processor. A serial communication format allows more functions to be included in a converter package with no increase in size. In addition, it eliminates the transmission of low-level analog signals by locating the converter at the analog sensor. This process returns noise-free digital data to the processor.

A particular input configuration is assigned during the multiplexer addressing sequence. The multiplexer address is shifted into the converter through the data input (DI) line. The ADC0831 contains only one differential input channel having a fixed polarity assignment and not requiring addressing. The multiplexer address selects the analog inputs to be enabled and determines whether the input is single-ended or differential. When the input is differential, the polarity of the channel input is assigned. Differential inputs are assigned to adjacent channel pairs. For example, channel 0 and channel 1 may be selected as a differential pair. These channels can not act differentially with any other channel. In addition to selecting the differential mode, the polarity may also be selected. Either channel of the channel pair may be designated as negative or positive.

The common input on the ADC0838 can be used for a pseudo-differential input. In this mode, the voltage on the input is negative to any other channel. This voltage can be any reference potential common to all channel inputs. This feature is useful in single-supply applications where all analog circuits are biased to a potential other than ground.

Operation of the ADC0831, ADC0832, ADC0834, and ADC0838 is similar with the exception of multiplexer addressing. The ADC0838 has all the features of the other converts and is used for the functional block diagram.

A conversion is initiated by setting the chip select ($\overline{\text{CS}}$) input low. This enables all logic circuits. The $\overline{\text{CS}}$ input must be held low for the complete conversion process. A clock input is received from the processor. On each low-to-high transition of the clock input, the data on the DI input is clocked into the multiplexer address shift register. The first logic high on the input is the start bit. A 2- to 4-bit assignment word follows the start bit. On each successive low-to-high transition of the clock input, the start bit and assignment word are shifted through the shift register. When the start bit has been shifted into the start location of the multiplexer register, the input channel has been selected and conversion starts. The SAR status output (SARS) goes high to indicate that a conversion is in progress and the DI input to the multiplexer shift register is disabled for the duration of the conversion.

Data Acquisition

7

Data Acquisition

7-34

TEXAS INSTRUMENTS

POST OFFICE BOX 225012 • DALLAS, TEXAS 75265

An interval of one clock period is automatically inserted to allow for the selected multiplexer channel to settle. The data output DO comes out of the high-impedance state and provides a leading low for this one clock period of multiplexer settling time. The SAR comparator compares successive outputs from the resistive ladder with the incoming analog signal. The comparator output indicates whether the analog input is greater than or less than resistive ladder output. This data is parallel loaded into a 9-bit shift register which immediately outputs an 8-bit serial data word. This output is sent to the DO output with the most-significant bit (MSB) first. After eight clock periods the conversion is complete and the SAR status (SARS) output goes low. When $\overline{CS}$ goes high, all internal registers are cleared. At this time the output circuits go to three-state. If another conversion is desired, the $\overline{CS}$ line must make a high-to-low transition followed by address information.

In the ADC0831, only the MSB data is output first. The ADC0832 and ADC0834 output the LSB data first after the MSB first data stream is output. In the ADC0838, the programmer has the option of selecting MSB first or LSB first. To output LSB first, the shift enable ($\overline{SE}$) control input must go low. Data stored in the 9-bit shift register is now output with LSB first. The ADC0831 data is only output in MSB-first format.

The DI and DO pins can be tied together and controlled by a bi-directional processor I/O bit received on a single wire. This is possible because the DI input is only examined during the multiplexer addressing interval and the DO output is still in a high-impedance state.

Data Acquisition

7

TEXAS
INSTRUMENTS
POST OFFICE BOX 225012 • DALLAS, TEXAS 75265

Data Acquisition

7

- Functionally Interchangeble with Siliconix DG182, DG185, DG188, and DG191 with Same Terminal Assignments
- Monolithic Construction
- Adjustable Reference Voltage
- JFET Inputs

- Uniform On-State Resistance for Minimum Signal Distortion
- ± 10-V Analog Voltage Range
- TTL, MOS, and CMOS Logic Control Compatibility

description

The TL182, TL185, TL188, and TL191 are monolithic high-speed analog switches using BI-MOS technology. They comprise JFET-input buffers, level translators, and output JFET switches. The TL182 switches are SPST; the TL185 switches are SPDT. The TL188 is a pair of complementary SPST switches as is each half of the TL191.

A high level at a control input of the TL182 turns the associated switch off. A high level at a control input of the TL185 turns the associated switch on. For the TL188, a high level at the control input turns the associated switches S1 on and S2 off.

The threshold of the input buffer is determined by the voltage applied to the reference input (V_{ref}). The input threshold is related to the reference input by the equation $V_{th} = V_{ref} + 1.4$ V. Thus, for TTL compatibility, the V_{ref} input is connected to ground. The JFET input makes the device compatible with bipolar, MOS and CMOS logic families. Threshold compatibility may, again, be determined by $V_{th} = V_{ref} + 1.4$ V.

The output switches are junction field-effect transistors featuring low on-state resistance and high off-state resistance. The monolithic structure ensures uniform matching.

BI-MOS technology is a major breakthrough in linear integrated circuit processing. BI-MOS can have ion-implanted JFETs, p-channel MOS-FETs, plus the usual bipolar components all on the same chip. BI-MOS allows circuit designs that previously have been available only as expensive hybrids to be monolithic.

Devices with an "M" suffix are characterized for operation over the full military temperature range of −55°C to 125°C, those with an "I" suffix are characterized for operation from −25°C to 85°C, and those with a "C" suffix are characterized for operation from 0°C to 70°C.

TL182
J OR N DUAL-IN-LINE PACKAGE
(TOP VIEW)

1S	1	14	2S
1D	2	13	2D
NC	3	12	NC
NC	4	11	NC
1A	5	10	2A
V_{CC}	6	9	V_{EE}
V_{LL}	7	8	V_{ref}

TL185
J OR N DUAL-IN-LINE PACKAGE
(TOP VIEW)

1D1	1	16	1S1
NC	2	15	1A
1D2	3	14	V_{EE}
1S2	4	13	V_{ref}
2S1	5	12	V_{LL}
2D1	6	11	V_{CC}
NC	7	10	2A
2D2	8	9	2S2

TL188
J OR N DUAL-IN-LINE PACKAGE
(TOP VIEW)

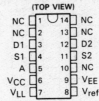

NC	1	14	NC
NC	2	13	NC
D1	3	12	D2
S1	4	11	S2
A	5	10	NC
V_{CC}	6	9	V_{EE}
V_{LL}	7	8	V_{ref}

TL191
J OR N DUAL-IN-LINE PACKAGE
(TOP VIEW)

1D1	1	16	1S1
NC	2	15	1A
1D2	3	14	V_{EE}
1S2	4	13	V_{ref}
2S2	5	12	V_{LL}
2D2	6	11	V_{CC}
NC	7	10	2A
2D1	8	9	2S1

NC—No internal connection

TEXAS INSTRUMENTS
POST OFFICE BOX 225012 • DALLAS, TEXAS 75265

TL182 TWIN SPST SWITCH

schematic (each channel)

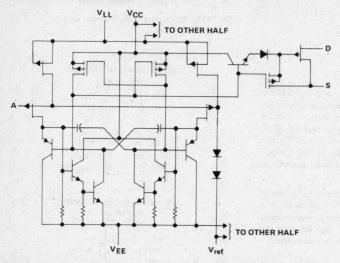

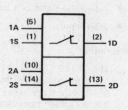

FUNCTION TABLE
(EACH HALF)

INPUT	SWITCH
A	S
L	ON (CLOSED)
H	OFF (OPEN)

TL185 TWIN DPST SWITCH

schematic (each channel)

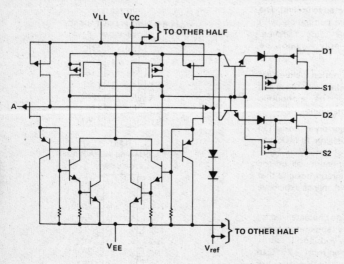

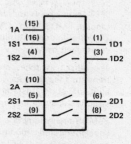

FUNCTION TABLE
(EACH HALF)

INPUT	SWITCHES
A	SW1 AND SW2
L	OFF (OPEN)
H	ON (CLOSED)

Data Acquisition

7

Texas
INSTRUMENTS
POST OFFICE BOX 225012 • DALLAS, TEXAS 75265

TL188 DUAL COMPLEMENTARY SPST SWITCH

schematic

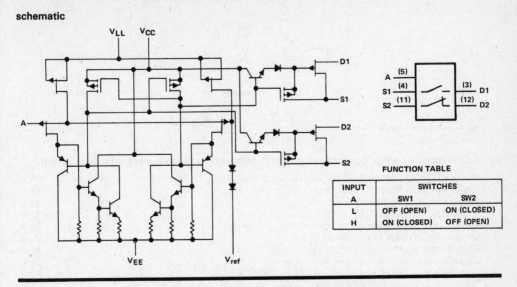

FUNCTION TABLE

INPUT	SWITCHES	
A	SW1	SW2
L	OFF (OPEN)	ON (CLOSED)
H	ON (CLOSED)	OFF (OPEN)

TL191 TWIN DUAL COMPLEMENTARY SPST SWITCH

schematic (each channel)

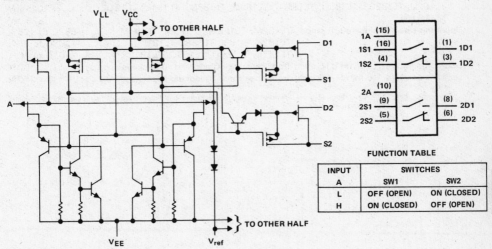

FUNCTION TABLE

INPUT	SWITCHES	
A	SW1	SW2
L	OFF (OPEN)	ON (CLOSED)
H	ON (CLOSED)	OFF (OPEN)

Data Acquisition

7

functional block diagram

See the preceding two pages for operation of the switches.

absolute maximum ratings over operating free-air temperature range (unless otherwise noted)

Positive supply to negative supply voltage, $V_{CC} - V_{EE}$ .... 36 V
Positive supply voltage to either drain, $V_{CC} - V_D$ .... 33 V
Drain to negative supply voltage, $V_D - V_{EE}$ .... 33 V
Drain to source voltage, $V_D - V_S$ .... ±22 V
Logic supply to negative supply voltage, $V_{LL} - V_{EE}$ .... 36 V
Logic supply to logic input voltage, $V_{LL} - V_I$ .... 33 V
Logic supply to reference voltage, $V_{LL} - V_{ref}$ .... 33 V
Logic input to reference voltage, $V_I - V_{ref}$ .... 33 V
Reference to negative supply voltage, $V_{ref} - V_{EE}$ .... 27 V
Reference to logic input voltage, $V_{ref} - V_I$ .... 2 V
Current (any terminal) .... 30 mA
Continuous dissipation at (or below) 25 °C free-air temperature (see Note 1):
    TL182MJ, TL185MJ, TL188MJ, TL191MJ .... 1375 mW
    TL182IJ, TL182CJ, TL185IJ, TL185CJ, TL188IJ, TL188CJ, TL191IJ, TL191CJ .... 1025 mW
    N package .... 875 mW
Operating free-air temperature range: TL182M, TL185M, TL188M, TL191M .... −55 °C to 125 °C
    TL182I, TL185I, TL188I, TL191I .... −25 °C to 85 °C
    TL182C, TL185C, TL188C, TL191C .... 0 °C to 70 °C
Lead temperature 1,6 mm (1/16 inch) from case for 60 seconds: J package .... 300 °C
Lead temperature 1,6 mm (1/16 inch) from case for 10 seconds: N package .... 260 °C

NOTE 1:   For operation above 25 °C free-air temperature, see Dissipation Derating Curves, Section 2. In the J package, "M" suffix chips are alloy mounted, "I" and "C" suffix chips are glass mounted.

Data Acquisition

7

TEXAS
INSTRUMENTS
POST OFFICE BOX 225012 • DALLAS, TEXAS 75265

electrical characteristics, V_{CC} = 15 V, V_{EE} = −15 V, V_{LL} = 5 V, V_{ref} = 0 V (unless otherwise noted)

PARAMETER		TEST CONDITIONS		TL1_M		TL1_I		TL1_C		UNIT
				MIN	MAX	MIN	MAX	MIN	MAX	
V_{IH}	High-level control input voltage		T_A = MIN TO MAX	V_{ref}+2		V_{ref}+2		V_{ref}+2		V
V_{IL}	Low-level control input voltage		T_A = MIN to MAX		V_{ref}+0.8		V_{ref}+0.8		V_{ref}+0.8	V
I_{IH}	High-level control input current	V_I = 5 V	T_A = 25°C		10		10		20	µA
			T_A = MAX		20		20		20	µA
I_{IL}	Low-level control input current	V_I = 0	T_A = MIN TO MAX		−250		−250		−250	µA
$I_{D(off)}$	Off-state drain current	V_D = 10 V, V_S = −10 V, V_{IL} = 0.8 V	T_A = 25°C		5		5		5	nA
		V_D = −10 V, V_S = 10 V, V_{IH} = 2 V	T_A = MAX		100		100		100	nA
$I_{S(off)}$	Off-state source current	V_D = −10 V, V_S = 10 V, V_{IH} = 2 V	T_A = 25°C		5		5		5	nA
			T_A = MAX		100		100		100	nA
$I_{D(on)}$ + $I_{S(on)}$	On-state channel leakage current	V_D = −10 V, V_S = −10 V, V_{IL} = 0.8 V	T_A = 25°C		−10		−10		−10	nA
		V_D = −10 V, V_S = −10 V, V_{IH} = 2 V	T_A = MAX		−200		−200		−200	nA
$r_{DS(on)}$	Drain-to-source on-state resistance	I_S = 1 mA, V_{IL} = 0.8 V	TL182, TL188 T_A = MIN to 25°C		75		100		100	Ω
			TL182, TL188 T_A = MAX		100		150		150	Ω
			TL185, TL191 T_A = MIN to 25°C		125		150		150	Ω
			TL185, TL191 T_A = MAX		250		300		300	Ω
I_{CC}	Supply current from V_{CC}	Both control inputs at 0 V	T_A = 25°C		1.5		1.5		1.5	mA
I_{EE}	Supply current from V_{EE}				−5		−5		−5	
I_{LL}	Supply current from V_{LL}				4.5		4.5		4.5	
I_{ref}	Reference current				−2		−2		−2	
I_{CC}	Supply current from V_{CC}	Both control inputs at 5 V	T_A = 25°C		1.5		1.5		1.5	mA
I_{EE}	Supply current from V_{EE}				−5		−5		−5	
I_{LL}	Supply current from V_{LL}				4.5		4.5		4.5	
I_{ref}	Reference current				−2		−2		−2	

switching characteristics, V_{CC} = 10 V, V_{EE} = −20 V, V_{LL} = 5 V, V_{ref} = 0 V, T_A = 25°C

PARAMETER		TEST CONDITIONS	TL1_M	TL1_I	TL1_C	UNIT
			TYP	TYP	TYP	
t_{on}	Turn-on time	R_L = 300 Ω, C_L = 30 pF, Figure 1	175	175	175	ns
t_{off}	Turn-off time		350	350	350	

Data Acquisition

7

PARAMETER MEASUREMENT INFORMATION

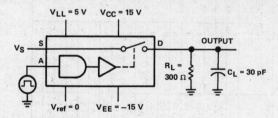

C_L includes probe and jig capacitance.

$V_S = 3$ V for t_{on} and -3 V for t_{off}.

$$V_O = V_S \frac{R_L}{R_L + r_{DS(on)}}$$

TEST CIRCUIT

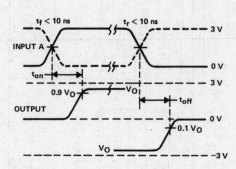

NOTE: A. The solid waveform applies for TL185 and SW1 of TL185 and TL191; the dashed waveform applies for TL182 and SW2 of TL185 and TL191.
B. V_O is the steady-state output with the switch on. Feed through via the gate capacitance may result in spikes (not shown) at the leading and trailing edges of the output waveform.

FIGURE 1—VOLTAGE WAVEFORMS

Data Acquisition

7

TEXAS
INSTRUMENTS
POST OFFICE BOX 225012 • DALLAS, TEXAS 75265

TL500C/TL501C ANALOG PROCESSORS

- True Differential Inputs
- Automatic Zero
- Automatic Polarity
- High Input Impedance . . . 10^9 Ohms Typically

TL500C CAPABILITIES

- Resolution . . . 14 Bits (with TL502C)
- Linearity Error . . . 0.001%
- 4 1/2-Digit Readout Accuracy with External Precision Reference

TL501C CAPABILITIES

- Resolution . . . 10-13 Bits (with TL502C)
- Linearity Error . . . 0.01%
- 3 1/2-Digit Readout Accuracy

TL502C/TL503C DIGITAL PROCESSORS

- Fast Display Scan Rates
- Internal Oscillator May Be Driven or Free-Running
- Interdigit Blanking
- Over-Range Blanking
- Display Test
- 4 1/2-Digit Display Circuitry
- High-Sink-Current Digit Driver for Large Displays

TL502C CAPABILITIES

- Compatible with Popular Seven-Segment Common-Anode Displays
- High-Sink-Current Segment Driver For Large Displays

TL503C CAPABILITIES

- Multiplexed BCD Outputs
- High-Sink-Current BCD Outputs

 Caution. These devices have limited built-in gate protection. The leads should be shorted together or the device placed in conductive foam during storage or handling to prevent electrostatic damage to the MOS gates.

The TL500C and TL501C analog processors and TL502C and TL503C digital processors provide the basic functions for a dual-slope-integrating analog-to-digital converter.

The TL500C and TL501C contain the necessary analog switches and decoding circuits, reference voltage generator, buffer, integrator, and comparator. These devices may be controlled by the TL502C, TL503C, by discrete logic, or by a software routine in a microprocessor.

The TL502C and TL503C each includes oscillator, counter, control logic, and digit enable circuits. The TL502C provides multiplexed outputs for seven-segment displays, while the TL503C has multiplexed BCD outputs.

When used in complementary fashion, these devices form a system that features automatic zero-offset compensation, true differential inputs, high input impedance, and capability for 4 1/2-digit accuracy. Applications include the conversion of analog data from high-impedance sensors of pressure, temperature, light, moisture, and position. Analog-to-digital-logic conversion provides display and control signals for weight scales, industrial controllers, thermometers, light-level indicators, and many other applications.

Data Acquisition

7

TEXAS INSTRUMENTS

POST OFFICE BOX 225012 • DALLAS, TEXAS 75265

principles of operation

The basic principle of dual-slope-integrating converters is relatively simple. A capacitor, C_X, is charged through the integrator from V_{CT} for a fixed period of time at a rate determined by the value of the unknown voltage input. Then the capacitor is discharged at a fixed rate (determined by the reference voltage) back to V_{CT} where the discharge time is measured precisely. The relationship of the charge and discharge values are shown below (see Figure 1).

$$V_{CX} = V_{CT} - \frac{V_I \, t_1}{R_X \, C_X} \qquad \text{Charge} \qquad (1)$$

$$V_{CT} = V_{CX} - \frac{V_{ref} \, t_2}{R_X \, C_X} \qquad \text{Discharge} \qquad (2)$$

Combining equations 1 and 2 results in:

$$\frac{V_I}{V_{ref}} = -\frac{t_2}{t_1} \qquad (3)$$

where:

V_{CT} = Comparator (offset) threshold voltage

V_{CX} = Voltage change across C_X during t_1 and during t_2 (equal in magnitude)

V_I = Average value of input voltage during t_1

t_1 = Time period over which unknown voltage is integrated

t_2 = Unknown time period over which a known reference voltage is integrated.

Equation (3) illustrates the major advantages of a dual-slop converter:
a. Accuracy is not dependent on absolute values of t_1 and t_2, but is dependent on their ratios. Long-term clock frequency variations will not affect the accuracy.
b. Offset values, V_{CT}, are not important.

The BCD counter in the digital processor (see Figure 2) and the control logic divide each measurement cycle into three phases. The BCD counter changes at a rate equal to one-half the oscillator frequency.

auto-zero phase

The cycle begins at the end of the integrate-reference phase when the digital processor applies low levels to inputs A and B of the analog processor. If the trigger input is at a high level, a free-running condition exists and continuous conversions are made. However, if the trigger input is low, the digital processor stops the counter at 20,000, entering a hold mode. In this mode, the processor samples the trigger input every 4000 oscillator pulses until a high level is detected. When this occurs, the counter is started again and is carried to completion at 30,000. The reference voltage is stored on reference capacitor C_{ref}, comparator offset voltage is stored on integration capacitor C_X, and the sum of the buffer and integrator offset voltages is stored on zero capacitor C_Z. During the auto-zero phase, the comparator output is characterized by an oscillation (limit cycle) of indeterminate waveform and frequency that is filtered and d-c shifted by the level shifter.

integrate-input phase

The auto-zero phase is completed at a BCD count of 30,000, and high levels are applied to both control inputs to initiate the integrate-input phase. The integrator charges C_X for a fixed time of 10,000 BCD counts at a rate determined by the input voltage. Note that during this phase, the analog inputs see only the high impedance of the noninverting operational amplifier input. Therefore, the integrator responds only to the difference between the analog input terminals, thus providing true differential inputs.

TEXAS
INSTRUMENTS
POST OFFICE BOX 225012 • DALLAS, TEXAS 75265

integrate-reference phase

At a BCD count of 39,999 + 1 = 40,000 or 0, the integrate-input phase is terminated and the integrate-reference phase is begun by sampling the comparator output. If the comparator output is low corresponding to a negative average analog input voltage, the digital processor applies a low and a high to inputs A and B, respectively, to apply the reference voltage stored on C_{ref} to the buffer. If the comparator output is high corresponding to a positive input, inputs A and B are made high and low, respectively, and the negative of the stored reference voltage is applied to the buffer. In either case, the processor automatically selects the proper logic state to cause the integrator to ramp back toward zero at a rate proportional to the reference voltage. The time required to return to zero is measured by the counter in the digital processor. The phase is terminated when the integrator output crosses zero and the counter contents are transferred to the register, or when the BCD counter reaches 20,000 and the over-range indication is activated. When activated, the over-range indication blands all but the most significant digit and sign.

Seventeen parallel bits (4-1/2 digits) of information are strobed into the buffer register at the end of the integration phase. Information for each digit is multiplexed out to the BCD outputs (TL503C) or the seven-segment drivers (TL502C) at a rate equal to the oscillator frequency divided by 400.

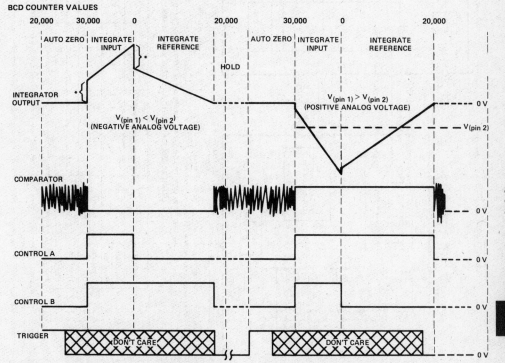

*This step is the voltage at pin 2 with respect to analog ground.

FIGURE 1—VOLTAGE WAVEFORMS AND TIMING DIAGRAM

Data Acquisition

7

TEXAS
INSTRUMENTS
POST OFFICE BOX 225012 • DALLAS, TEXAS 75265

Data Acquisition

7

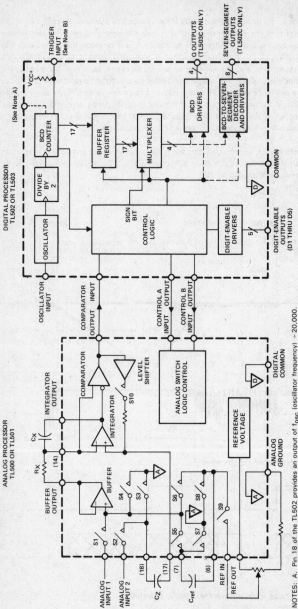

FIGURE 2— BLOCK DIAGRAM OF BASIC ANALOG-TO-DIGITAL CONVERTER
USING TL500C OR TL501C AND TL502C OR TL503C

NOTES: A. Pin 18 of the TL502 provides an output of f_{osc} (oscillator frequency) ÷ 20,000.
   B. The trigger input assumes a high level if not externally connected.

MODE	ANALOG INPUT	COMPARATOR	CONTROLS A AND B	ANALOG SWITCHES CLOSED
Auto Zero	X	Oscillation	L L	S3, S4, S7, S9, S10
Hold[†]				
Integrate Input	Positive	H	H H	S1, S2
	Negative	L	H L	
Integrate	X	H[‡]	H L	S3, S6, S7
Reference		L[‡]	L H	S3, S5, S8

H ≡ High, L ≡ Low, X ≡ Irrelevant

[†] If the trigger input is low at the beginning of the auto-zero cycle, the system will enter the hold mode. A high level (or open circuit) will signal the digital processor to continue or resume normal operation.

[‡] This is the state of the comparator output as determined by the polarity of the analog input during the integrate input phase.

TEXAS
INSTRUMENTS
POST OFFICE BOX 225012 • DALLAS, TEXAS 75265

108

description of analog processors

The TL500C and TL501C analog processors are designed to automatically compensate for internal zero offsets, integrate a differential voltage at the analog inputs, integrate a voltage at the reference input in the opposite direction, and provide an indication of zero-voltage crossing. The external control mechanism may be a microcomputer and software routine, discrete logic, or a TL502C or TL503C controller. The TL500C and TL501C are designed primarily for simple, cost-effective, dual-slope analog-to-digital converters. Both devices feature true differential analog inputs, high input impedance, and an internal reference-voltage source. The TL500C provides 4-1/2-digit readout accuracy when used with a precision external reference voltage. The TL501C provides 100-ppm linearity error and 3-1/2-digit accuracy capability. These devices are manufactured using TI's advanced technology to produce JFET, MOSFET, and bipolar devices on the same chip. The TL500C and TL501C are intended for operation over the temperature range of 0°C to 70°C.

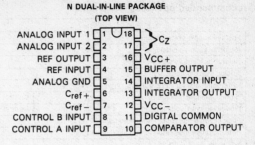

N DUAL-IN-LINE PACKAGE
(TOP VIEW)

ANALOG INPUT 1	1	18	C_Z
ANALOG INPUT 2	2	17	
REF OUTPUT	3	16	$V_{CC}+$
REF INPUT	4	15	BUFFER OUTPUT
ANALOG GND	5	14	INTEGRATOR INPUT
$C_{ref}+$	6	13	INTEGRATOR OUTPUT
$C_{ref}-$	7	12	$V_{CC}-$
CONTROL B INPUT	8	11	DIGITAL COMMON
CONTROL A INPUT	9	10	COMPARATOR OUTPUT

schematics of inputs and outputs

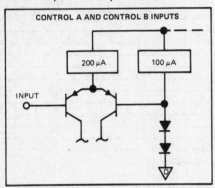

CONTROL A AND CONTROL B INPUTS

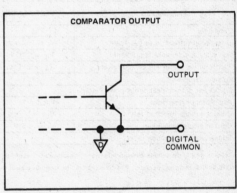

COMPARATOR OUTPUT

absolute maximum ratings over operating free-air temperature range (unless otherwise noted)

Positive supply voltage, $V_{CC}+$ (see Note 1) . . . . +18 V
Negative supply voltage, $V_{CC}-$ . . . . −18 V
Input voltage, V_I . . . . $\pm V_{CC}$
Comparator output voltage range (see Note 2) . . . . 0 V to $V_{CC}+$
Comparator output sink current (see Note 2) . . . . 20 mA
Buffer, reference, or integrator output source current (see Note 2) . . . . 10 mA
Total dissipation at (or below) 25°C free-air temperature (see Note 3) . . . . 875 mW
Operating free-air temperature range . . . . −0°C to 70°C
Storage temperature range . . . . −65°C to 125°C

NOTES: 1. Voltage values, except differential voltages, are with respect to the analog ground common pin tied together.
2. Buffer, integrator, and comparator outputs are not short-circuit protected.
3. For operation above 25°C free-air temperature, refer to Dissipation Derating Curves, Section 2.

Data Acquisition

7

TYPES TL500C, TL501C
ANALOG PROCESSORS

recommended operating conditions

		MIN	NOM	MAX	UNIT
Positive supply voltage, V_{CC+}		7	12	15	V
Negative supply voltage, V_{CC-}		−9	−12	−15	V
Reference input voltage, $V_{ref(I)}$		0.1		5	V
Analog input voltage, V_I				±5	V
Differential analog input voltage, V_{ID}				10	V
High-level input voltage, V_{IH}	Control inputs	2			V
Low-level input voltage, V_{IL}	Control inputs			0.8	V
Peak positive integrator output voltage, V_{OM+}		+9			V
Peak negative integrator output voltage, V_{OM-}		−5			V
Full scale input voltage				2 V_{ref}	
Autozero and reference capacitors, C_Z and C_{ref}		0.2			µF
Integrator capacitor, C_X		0.2			µF
Integrator resistor, R_X		15		100	kΩ
Integrator time constant, $R_X C_X$		See Note 4			
Free-air operating temperature, T_A		0		70	°C
Maximum conversion rate with TL502 or TL503			3	12.5	conv/sec

system electrical characteristics at $V_{CC} = ±12$ V, $V_{ref} = 1,000 ± 0.03$ mV, $T_A = 25°C$ (unless otherwise noted) (see Figure 3)

PARAMETER	TEST CONDITIONS	TL501C MIN	TL501C TYP	TL501C MAX	TL500C MIN	TL500C TYP	TL500C MAX	UNIT
Zero error			50	300		10	30	µV
Linearity error relative to full scale	$V_I = -2$ V to 2 V		0.005	0.05		0.001	0.005	%FS
Full scale temperature coefficient	$T_A = 0°C$ to 70°C		6			6		ppm/°C
Temperature coefficient of zero error	$T_A = 0°C$ to 70°C		4			1		µV/°C
Rollover error[†]			200	500		30	100	µV
Equivalent peak-to-peak input noise voltage			20			20		µV
Analog input resistance	Pin 1 or 2		10^9			10^9		Ω
Common-mode rejection ratio	$V_{IC} = -1$ V to +1 V		86			90		dB
Current into analog input	$V_I = ±5$ V		50			50		pA
Supply voltage rejection ratio			90			90		dB

[†] Rollover error is the voltage difference between the conversion results of the full-scale positive 2 volts and the full-scale negative 2 volts.

NOTE 4. The minimum integrator time constant may be found by use of the following formula:

$$\text{Minimum } R_X C_X = \frac{V_{ID}\,\text{(full scale)}\ t_1}{|V_{OM-} - V_I\text{(pin 2)}|}$$

where

V_{ID} = voltage at pin with respect to pin 2
V_I(pin 2) = voltage at pin 2 with respect to analog ground
t_1 = input integration time seconds

Data Acquisition

7

TEXAS
INSTRUMENTS
POST OFFICE BOX 225012 • DALLAS, TEXAS 75265

electrical characteristics at $V_{CC} = \pm 12$ V, $V_{ref} = 1$ V, $T_A = 25°C$ (see Figure 3)

integrator and buffer operational amplifiers

	PARAMETER	TEST CONDITIONS	MIN	TYP	MAX	UNITS
V_{IO}	Input offset voltage			15		mV
I_{IB}	Input bias current			50		pA
V_{OM+}	Positive output voltage swing		9	11		V
V_{OM-}	Negative output voltage swing		−5	−7		V
A_{VD}	Voltage amplification			110		dB
B_1	Unity-gain bandwidth			3		MHZ
CMRR	Common mode rejection	$V_{IC} = -1$ V to +1 V		100		dB
SR	Output slew rate			5		V/µs

comparator

	PARAMETER	TEST CONDITIONS	MIN	TYP	MAX	UNITS
V_{IO}	Input offset voltage			15		mV
I_{IB}	Input bias current			50		pA
A_{VD}	Voltage amplification			100		dB
V_{OL}	Low-level output voltage	$I_{OL} = 1.6$ mA		200	400	mV
I_{OH}	High-level output current	$V_{OH} = 3$ V		5	20	nA

voltage reference output

	PARAMETER	TEST CONDITIONS	MIN	TYP	MAX	UNITS
$V_{ref(0)}$	Reference voltage		1.12	1.22	1.32	V
α_{Vref}	Reference-voltage temperature coefficient	$T_A = 0°C$ to $70°C$		80		ppm/°C
r_o	Reference output resistance			3		Ω

logic control section

	PARAMETER	TEST CONDITIONS	MIN	TYP	MAX	UNITS
I_{IH}	High-level input current	$V_{IH} = 2$ V		1	10	µA
I_{IL}	Low-level input current	$V_{IL} = 0.8$ V		−40	−300	µA

total device

	PARAMETER	TEST CONDITIONS	MIN	TYP	MAX	UNITS
I_{CC+}	Positive supply current			15	20	mA
I_{CC-}	Negative supply current			12	18	mA

Data Acquisition

7

TEXAS
INSTRUMENTS
POST OFFICE BOX 225012 • DALLAS, TEXAS 75265

PARAMETER MEASUREMENT INFORMATION

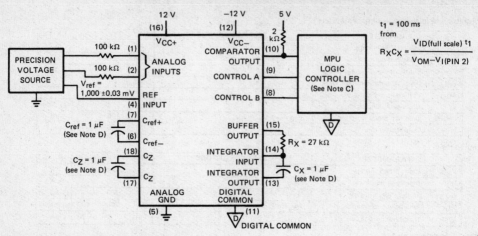

NOTES: C. Tests are started approximately 5 seconds after power-on.
D. Capacitors used are TRW's X363UW poly propylene or eqivalent for C_X, C_{ref}, and C_Z; however for C_{ref} and C_Z, film-dielectric capacitors may be substituted.

FIGURE 3–TEST CIRCUIT CONFIGURATION

external-component selection guide

The autozero capacitor C_Z and reference capacitor C_{ref} should be within the recommended range of operating conditions and should have low leakage characteristics. Most film-dielectric capacitors and some tantalum capacitors provide acceptable results. Ceramic and aluminum capacitors are not recommended because of their relatively high leakage characteristics.

The integrator capacitor C_X should also be within the recommended range and must have good voltage linearity and low dielectric absorbtion. A polypropylene-dielectric capacitor similar to TRW's X363UW is recommended for 4-1/2-digit accuracy. For 3-1/2-digit applications, polyster, polycarbonate, and other film dielectrics are usually suitable. Ceramic and electrolyic capacitors are not recommended.

Stray coupling from the comparator output to any analog pin (in order of importance 17, 18, 14, 7, 6, 13, 1, 2, 15) must be minimized to avoid oscillations. In addition, all power supply pins should be bypassed at the package, for example, by a 0.01- μF ceramic capacitor.

Analog and digital common are internally isolated and may be a different potentials. Digital common can be within 4 volts of positive or negative supply with the logic decode still functioning properly.

The time constant $R_X C_X$ should be kept as near the minimum value as possible and is given by the formula:

$$\text{Minimum } R_X C_X = \frac{V_{ID}\text{ (full scale) } t_1}{|V_{OM-}| - V_I(\text{pin 2})}$$

where:

$V_{ID}(\text{full scale})$ = Voltage on pin 1 with respect to pin 2

t_1 = Input integration time in seconds

$V_I(\text{pin 2})$ = Voltage on pin 2 with respect to analog ground

TEXAS
INSTRUMENTS
POST OFFICE BOX 225012 • DALLAS, TEXAS 75265

description of digital processors

The TL502C and TL503C are control logic devices designed to complement the TL500C and TL501C analog processors. They feature interdigit blanking, over-range blanking, an internal oscillator, and a fast display scan rate. The internal-oscillator input is a Schmitt trigger circuit that can be driven by an external clock pulse or provide its own time base with the addition of a capacitor. The typical oscillator frequency is 120 kHz with a 470-picofarad capacitor connected between the oscillator input and ground.

The TL502C provides seven-segment-display output drivers capable of sinking 100 milliamperes and compatible with popular common-anode displays. The TL503C has four BCD output drivers capable of 100-milliampere sink currents. The code (see next page and Figure 4) for each digit is multiplexed to the output drivers in phase with a pulse on the appropriate digit-enable line at a digit rate equal to f_{osc} divided by 200. Each digit-enable output is capable of sinking 20 milliamperes.

The comparator input of each device, in addition to monitoring the output of the zero-crossing detector in the analog processor, may be used in the display test mode to check for wiring and display faults. A high logic level (2 to 6.5 volts) at the trigger input with the comparator input at or below 6.5 volts starts the integrate-input phase. Voltage levels equal to or greater than 7.9 volts on both the trigger and comparator inputs clear the system and set the BCD counter to 20,000. When normal operation resumes, the conversion cycle is restarted at the auto zero phase.

These devices are manufactured using I^2L and bipolar techniques. The TL502C and TL503C are intended for operation from $0°C$ to $70°C$.

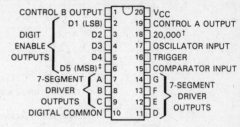

TL502 . . . N DUAL-IN-LINE PACKAGE
(TOP VIEW)

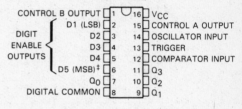

TL503 . . . N DUAL-IN-LINE PACKAGE
(TOP VIEW)

† Pin 18 of TL502 provides an output of f_{osc} (oscillator frequencies) ÷ 20,000.

‡ D5, the most significant bit, is also the sign bit.

Data Acquisition

7

TABLE OF SPECIAL FUNCTIONS
V_{CC} = 5 V ±10%

TRIGGER INPUT	COMPARATOR INPUT	FUNCTION
$V_I \leqslant 0.8$ V	$V_I \leqslant 6.5$ V	Hold at auto-zero cycle after completion of conversion
2 V $\leqslant V_I \leqslant 6.5$ V	$V_I \leqslant 6.5$ V	Normal operation (continuous conversion)
$V_I \leqslant 6.5$ V	$V_I \geqslant 7.9$ V	Display Test: All BCD outputs high
$V_I \geqslant 7.9$ V	$V_I \leqslant 6.5$ V	Internal Test
Both inputs go to $V_I \geqslant 7.9$ V simultaneously		System clear: Sets BCD counter to 20,000. When normal operation is resumed, cycle begins with Auto Zero.

TEXAS
INSTRUMENTS
POST OFFICE BOX 225012 • DALLAS, TEXAS 75265

DIGIT 5 (MOST SIGNIFICANT DIGIT) CHARACTER CODES

CHARACTER	TL502C SEVEN-SEGMENT LINES							TL503C BCD OUTPUT LINES			
	A	B	C	D	E	F	G	Q3 8	Q2 4	Q1 2	Q0 1
+	H	H	H	H	L	L	L	H	L	H	L
+1	H	L	L	H	L	L	L	H	H	H	L
−	L	H	H	L	H	H	L	H	L	H	H
−1	L	L	L	L	H	H	L	H	H	H	H

DIGITS 1 THRU 4 NUMERIC CODE (See Figure 4)

NUMBER	TL502C SEVEN-SEGMENT LINES							TL503C BCD OUTPUT LINES			
	A	B	C	D	E	F	G	Q3 8	Q2 4	Q1 2	Q0 1
0	L	L	L	L	L	L	H	L	L	L	L
1	H	L	L	H	H	H	H	L	L	L	H
2	L	L	H	L	L	H	L	L	L	H	L
3	L	L	L	L	H	H	L	L	L	H	H
4	H	L	L	H	H	L	L	L	H	L	L
5	L	H	L	L	H	L	L	L	H	L	H
6	L	H	L	L	L	L	L	L	H	H	L
7	L	L	L	H	H	H	H	L	H	H	H
8	L	L	L	L	L	L	L	H	L	L	L
9	L	L	L	L	H	L	L	H	L	L	H

H = high level, L = low level

schematics of inputs and outputs

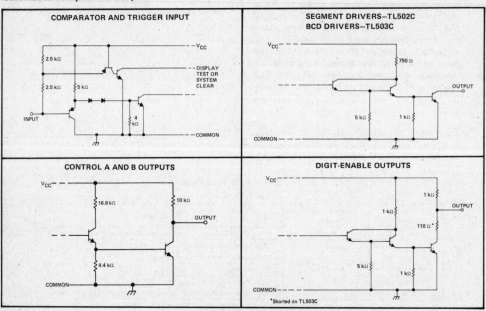

COMPARATOR AND TRIGGER INPUT

SEGMENT DRIVERS—TL502C
BCD DRIVERS—TL503C

CONTROL A AND B OUTPUTS

DIGIT-ENABLE OUTPUTS

*Shorted on TL503C

TEXAS INSTRUMENTS
INCORPORATED

POST OFFICE BOX 225012 ● DALLAS, TEXAS 75265

absolute maximum ratings

Supply voltage, V_{CC} (see Note 5)		7	V
Input voltage, V_I	Oscillator	5.5	V
	Comparator or Trigger	9	
Output current	BCD or Segment drivers	120	mA
	Digit-enable outputs	40	
	Pin 18 (TLC502 only)	20	
Total power dissipation at (or below) 30°C free-air temperature (see Note 6)		875	mW
Operating free-air temperature range		0 to 70	°C
Storage temperature range		−65 to 150	°C

NOTES: 5. Voltage values are with respect to the network ground terminal.
      6. For operation above 30°C free-air temperature, derate linearly at the rate of 9.2 mW/°C.

recommended operating conditions

		MIN	NOM	MAX	UNIT
Supply voltage, V_{CC}		4.5	5	5.5	V
High-level input voltage, V_{IH}	Comparator and trigger inputs	2			V
Low-level input voltage, V_{IL}	Comparator and trigger inputs			0.8	V
Operating free-air temperature		0		70	°C

Data Acquisition

7

Data Acquisition

7

electrical characteristics at 25 °C free-air temperature

PARAMETER		TERMINAL	TEST CONDITIONS	TL502C MIN	TL502C TYP	TL502C MAX	TL503C MIN	TL503C TYP	TL503C MAX	UNIT
V_{IK}	Input clamp voltage	All inputs	V_{CC} = 4.5 V, I_I = −12 mA		−0.8	−1.5		−0.8	−1.5	V
V_{T+}	Positive-going input threshold voltage	Oscillator	V_{CC} = 5 V		1.5			1.5		V
V_{T-}	Negative-going input threshold voltage	Oscillator	V_{CC} = 5 V		0.9			0.9		V
$V_{T+} - V_{T-}$	Hysteresis	Oscillator	V_{CC} = 5 V	0.4	0.6	0.8	0.4	0.6	0.8	
I_{T+}	Input current at positive-going input threshold voltage	Oscillator	V_{CC} = 5 V	−40	−94	−170	−40	−94	−170	µA
I_{T-}	Input current at negative-going input threshold voltage	Oscillator	V_{CC} = 5 V	40	117	170	40	117	170	µA
V_{OH}	High-level output voltage	Digit enable	V_{CC} = 4.5 V, I_{OH} = 0	4.15	4.4		4.15	4.4		V
		Control A and B		4.25	4.4		4.25	4.4		
V_{OL}	Low-level output voltage	Digit enable	I_{OL} = 20 mA		0.15	0.4		0.2	0.5	V
		Pin 18 (TL502C only)	I_{OL} = 10 mA							
		Control A and B	I_{OL} = 2 mA (V_{CC} = 4.5 V)		0.088	0.4		0.088	0.4	
		Segment drivers	I_{OL} = 100 mA		0.17	0.3		0.17	0.3	
		BCD drivers	I_{OL} = 100 mA							
I_I	Input current	Comparator, Trigger	V_{CC} = 5.5 V, V_I = 5.5 V		65	100		65	100	µA
		Oscillator				1			1	mA
I_{IH}	High-level input current	Comparator, Trigger	V_{CC} = 5.5 V, V_I = 2.4 V		−0.6	−1		−0.6	−1	mA
		Oscillator				0.5			0.5	
I_{IL}	Low-level input current	Comparator, Trigger	V_{CC} = 5.5 V, V_I = 0.4 V		−0.1	−0.17		−0.1	−0.17	mA
		Oscillator			−1	−1.6		−1	−1.6	
I_{OH}	High-level output current (Output transistor off)	Digit enable	V_O = 0.5 V	−2.5	−4		−2.5	−4		mA
		Pin 18 (TL502C only)	V_O = 0.5 V	−0.5	−0.9					
		Control A and B	V_O = 0.5 V (V_{CC} = 4.5 V)	−0.25	−0.4		−0.25	−0.4		
		Segment drivers	V_O = 5.5 V			0.25			0.25	
		BCD drivers	V_O = 5.5 V							
I_{OL}	Low-level output current (Output transistor on)	Digit enable	V_{CC} = 4.5 V, V_O = 3.55 V	18	23					mA
I_{CC}	Supply current	V_{CC}	V_{CC} = 5.5 V		73	110		73	110	mA

TEXAS
INSTRUMENTS
POST OFFICE BOX 225012 • DALLAS, TEXAS 75265

special functions† operating characteristics at 25 °C free-air temperature

PARAMETER		TEST CONDITIONS	MIN	TYP	MAX	UNIT
I_I	Input current into	V_{CC} = 5.5 V, V_I = 8.55 V		1.2	1.8	mA
	comparator or trigger inputs	V_{CC} = 5.5 V, V_I = 6.25 V			0.5	mA

†The comparator and trigger inputs may be used in the normal mode or to perform special functions. See the Table of Special Functions.

TYPICAL APPLICATION DATA

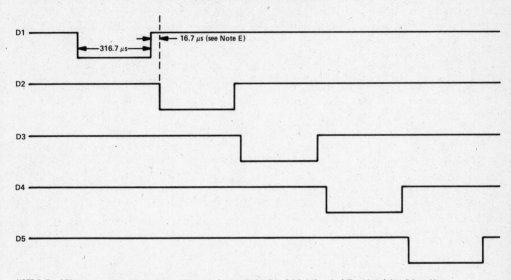

NOTE E. The BCD or seven-segment driver outputs are present for a particular digit slightly before the falling edge of that digit enable.

FIGURE 4—TL502C, TL503C DIGIT TIMING WITH 120-kHz CLOCK SIGNAL AT OSCILLATOR INPUT

Data Acquisition

7

Data Acquisition

7

- 3-Digit Accuracy (0.1%)
- 10-Bit Resolution
- Automatic Zero
- Internal Reference Voltage
- Single-Supply Operation
- High-Impedance MOS Input
- Designed for use with TMS 1000 Type Microprocessors for Cost-Effective High-Volume Applications
- BI-MOS Technology
- Only 40 mW Typical Power Consumption

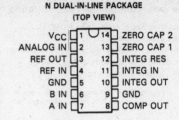

N DUAL-IN-LINE PACKAGE
(TOP VIEW)

VCC	1	14 ZERO CAP 2
ANALOG IN	2	13 ZERO CAP 1
REF OUT	3	12 INTEG RES
REF IN	4	11 INTEG IN
GND	5	10 INTEG OUT
B IN	6	9 GND
A IN	7	8 COMP OUT

 Caution. This device has limited built-in gate protection. The leads should be shorted together or the device placed in conductive foam during storage or handling to prevent electrostic damage to the MOS gates.

description

The TL505C is an analog-to-digital converter building block designed for use with TMS 1000 type microprocessors. It contains the analog elements (operational amplifier, comparator, voltage reference, analog switches, and switch drivers) necessary for a unipolar automatic-zeroing dual-slope converter. The logic for the dual-slope conversion can be performed by the associated MPU as a software routine or it can be implemented with other components such as the TL502 logic-control device.

The high-impedance MOS inputs permit the use of less expensive, lower value capacitors for the integration and offset capacitors and permit conversion speeds from 20 per second to 0.05 per second.

The TL505C is a product of TI's BI-MOS process, which incorporates bipolar and MOSFET transistors on the same monolithic integrated circuit. The TL505C is characterized for operation from 0°C to 70°C.

functional block diagram

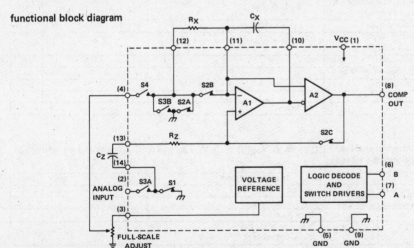

NOTE: Analog and digital GND are internally connected together.

Data Acquisition

7

TEXAS INSTRUMENTS
POST OFFICE BOX 225012 • DALLAS, TEXAS 75265

absolute maximum ratings over operating temperature range (unless otherwise noted)

Supply voltage, V_{CC} (see Note 1) . . . . 18 V
Input voltage, pins 2, 4, 6, and 7 . . . . V_{CC}
Continuous total dissipation at (or below) 25 °C free-air temperature (see Note 2) . . . . 875 mW
Operating free-air temperature range . . . . 0 °C to 70 °C
Storage temperature range . . . . −65 °C to 150 °C

NOTES: 1. Voltage values are with respect to the two ground terminals connected together.
2. For operation above 25 °C free-air temperature, refer to Dissipation Derating Curves, Section 2.

recommended operating conditions

	MIN	NOM	MAX	UNIT
Supply voltage, V_{CC}	7	9	15	V
Analog input voltage, V_I	0		4	V
Reference input voltage, $V_{ref(I)}$	0.5		3	V
High-level input voltage at A or B, V_{IH}	3.6		$V_{CC}+1$	V
Low-level input voltage at A or B, V_{IL}	0.2		1.8	V
Integrator capacitor, C_X	See "component selection"			
Integrator resistor, R_X	0.5		2	MΩ
Integration time, t_1	16.6		500	ms
Operating free-air temperature, T_A	0		70	°C

electrical characteristics, V_{CC} = 9 V, $V_{ref(I)}$ = 1 V, T_A = 25 °C, connected as shown in Figure 1 (unless otherwise noted)

	PARAMETER	TEST CONDITIONS	MIN	TYP	MAX	UNIT
V_{OH}	High-level output voltage at pin 8	I_{OH} = 0	7.5	8.5		V
I_{OH}	High-level output current at pin 8	V_{OH} = 7.5 V		−100		µA
V_{OL}	Low-level output voltage at pin 8	I_{OL} = 1.6 mA		200	400	mV
V_{OM}	Maximum peak output voltage swing at integrator output	$R_X \geq$ 500 kΩ	$V_{CC}-2$	$V_{CC}-1$		V
$V_{ref(O)}$	Reference output voltage	I_{ref} = −100 µA	1.15	1.22	1.35	V
αV_{ref}	Temperature coefficient of reference output voltage	T_A = 0 °C to 70 °C		±100		ppm/°C
I_{IH}	High-level input current into A or B	V_I = 9 V		1	10	µA
I_{IL}	Low-level input current into A or B	V_I = 1 V		10	200	µA
I_I	Current into analog input	V_I = 0 to 4 V, A input at 0 V		±10	±200	pA
I_{IB}	Total integrator input bias current			±10		pA
I_{CC}	Supply current	No load		4.5	8	mA

system electrical characteristics, V_{CC} = 9 V, $V_{ref(I)}$ = 1 V, T_A = 25 °C, connected as shown in Figure 1 (unless otherwise noted)

PARAMETER	TEST CONDITIONS	MIN	TYP	MAX	UNIT
Zero error	V_I = 0		0.1	0.4	mV
Linearity error	V_I = 0 to 4 V		0.02	0.1	%FS
Ratiometric reading	V_I = $V_{ref(I)}$ ≈ 1 V,	0.998	1.000	1.002	
Temperature coefficient of ratiometric reading	$V_{ref(I)}$ constant and ≈ 1 V, T_A = 0 °C to 70 °C		±10		ppm/°C

Data Acquisition

7

TEXAS
INSTRUMENTS
POST OFFICE BOX 225012 • DALLAS, TEXAS 75265

108

DEFINITION OF TERMS

Zero Error

The intercept (b) of the analog-to-digital converter system transfer function $y = mx + b$, where y is the digital output, x is the analog input, and m is the slope of the transfer function, which is approximated by the ratiometric reading.

Linearity Error

The maximum magnitude of the deviation from a straight line between the end points of the transfer function.

Ratiometric Reading

The ratio of negative integration time (t_2) to positive integration time (t_1).

PRINCIPLES OF OPERATION

A block diagram of an MPU system utilizing the TL505C is shown in Figure 1. The TL505C operates in a modified positive-integration three-step dual-slope conversion mode. The A/D converter waveforms during the conversion process are illustrated in Figure 2.

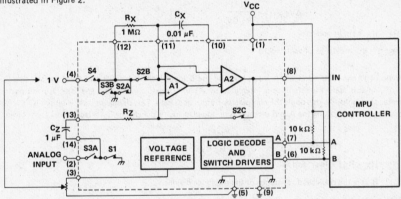

FIGURE 1—FUNCTIONAL BLOCK DIAGRAM OF TL505C INTERFACE WITH A MICROPROCESSOR SYSTEM

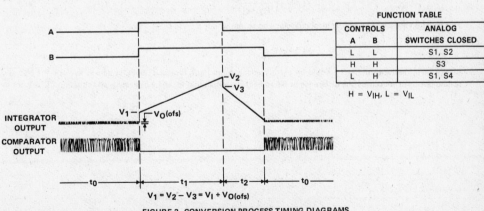

FUNCTION TABLE

CONTROLS		ANALOG
A	B	SWITCHES CLOSED
L	L	S1, S2
H	H	S3
L	H	S1, S4

$H = V_{IH}, L = V_{IL}$

$$V_1 = V_2 - V_3 = V_I + V_{O(ofs)}$$

FIGURE 2—CONVERSION PROCESS TIMING DIAGRAMS

TEXAS
INSTRUMENTS
POST OFFICE BOX 225012 • DALLAS, TEXAS 75265

PRINCIPLES OF OPERATION

The first step of the conversion cycle is the auto-zero period t_0 during which the integrator offset is stored in the auto-zero capacitor and the offset of the comparator is stored in the integrator capacitor. To accomplish this, the MPU takes the A and B inputs both low. This is decoded by the switch drivers, which close S1 and S2. The output of the comparator is connected to the input of the integrator through the low-pass filter consisting of R_Z and C_Z. The closed loop of A1 and A2 will seek a null condition where the offsets of the integrator and comparator are stored in C_Z and C_X, respectively. This null condition is characterized by a high-frequency oscillation at the output of the comparator. The purpose of S2B is to shorten the amount of time required to reach the null condition.

At the conclusion of t_0, the MPU takes the A and B inputs both high. This closes S3 and turns all other switches off. The input signal V_I is applied to the noninverting input of A1 through C_Z. V_I is then positively integrated by A1. Since the offset of A1 is stored in C_Z, the change in voltage across C_X will be due to only the input voltage. It should be noted that since the input is integrated in a positive integration during t_1, the output of A1 will be the sum of the input voltage, the integral of the input voltage, and the comparator offset, as shown in Figure 2. The change in voltage across capacitor C_X (V_{CX}) during t_1 is given by

$$\Delta V_{CX(1)} = \frac{V_I \, t_1}{R_1 \, C_X} \tag{1}$$

where $R_1 = R_X + R_{S3B}$ and

R_{S3B} is the resistance of switch S3B.

At the end of t_1 the MPU takes the A input low and the B input high. This turns on S1 and S4; all other switches are turned off. In this state the reference is integrated by A1 in a negative sense until the integrator output reaches the comparator threshold. At this point the comparator output goes high. This change in state is sensed by the MPU, which terminates t_2 by again taking the A and B inputs both low. During t_2 the change in voltage across C_X is given by

$$\Delta V_{CX(2)} = \frac{V_{ref} \, t_2}{R_2 \, C_X} \tag{2}$$

where $R_2 = R_X + R_{S4} + R_{ref}$ and

R_{ref} is the equivalent resistance of the reference divider.

Since $\Delta V_{CX1} = -\Delta V_{CX2}$, equations (1) and (2) can be combined to give

$$V_I = V_{ref} \, \frac{R_1 \cdot t_2}{R_2 \cdot t_1} \tag{3}$$

This equation is a variation on the ideal dual-slope equation, which is

$$V_I = V_{ref} \, \frac{t_2}{t_1} \tag{4}$$

Ideally then, the ratio of R_1/R_2 would be exactly equal to one. In a typical TL505C system where $R_X = 1 \text{ M}\Omega$, the scaling error introduced by the difference in R_1 and R_2 is so small that it can be neglected, and equation (3) reduces to (4).

Texas
Instruments
POST OFFICE BOX 225012 • DALLAS, TEXAS 75265

TYPICAL APPLICATION DATA

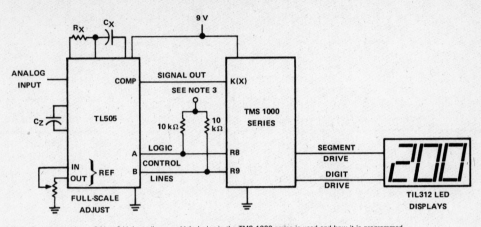

NOTE 3: Connect to either +9 V or 0 V depending on which device in the TMS 1000 series is used and how it is programmed.

FIGURE 5—TL505C IN CONJUNCTION WITH A TMS 1000 SERIES MICROPROCESSOR
FOR A 3-DIGIT DIGITAL PANEL METER APPLICATION

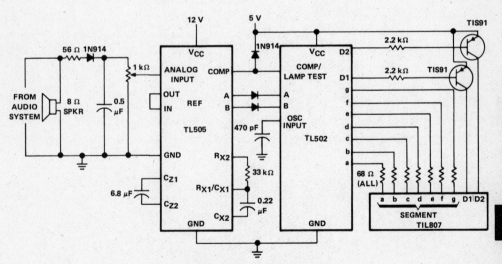

Data Acquisition

7

FIGURE 6—AUDIO PEAK POWER METER

Data Acquisition

7

- Low Cost
- 7-Bit Resolution
- Guaranteed Monotonicity
- Ratiometric Conversion
- Conversion Speed . . . approximately 1 ms
- Single-Supply Operation . . . Either Unregulated 8-V to 18-V (V_{CC2} Input), or Regulated 3.5-V to 6-V (V_{CC1} Input)
- I^2L Technology
- Power Consumption at 5 V . . . 25 mW Typ
- Regulated 5.5-V Output (≤ 1 mA)

P DUAL-IN-LINE PACKAGE
(TOP VIEW)

```
ENABLE [ 1  U  8 ] RESET
   CLK [ 2     7 ] VCC2
   GND [ 3     6 ] VCC1
OUTPUT [ 4     5 ] ANALOG INPUT
```

FUNCTION TABLE

ANALOG INPUT CONDITION	ENABLE	OUTPUT
X	L[†]	H
$V_I < 200$ mV	H	L
$V_{ramp} > V_I > 200$ mV	H	H
$V_I > V_{ramp}$	H	L

[†]Low level on enable also inhibits the reset function.

H = high level, L = low level, X = irrelevant

A high level on the reset pin clears the counter to zero, which sets the internal ramp to 0.75 V_{CC}. Internal pull-down resistors keep the reset and enable pins low when not connected.

description

The TL507 is a low-cost single-slope analog-to-digital converter designed to convert analog input voltages between 0.25 V_{CC1} and 0.75 V_{CC1} into a pulse-width-modulated output code. It contains a 7-bit synchronous counter, a binary-weighted resistor ladder network, an operational amplifier, two comparators, a buffer amplifier, an internal regulator, and necessary logic circuitry. Integrated-injection logic (I^2L) technology makes it possible to offer this complex circuit at low cost in a small dual-in-line 8-pin package.

In continuous operation, it is possible to obtain conversion speeds up to 1000 per second. The TL507 requires external signals for clock, reset, and enable. Versatility and simplicity of operation coupled with low cost, makes this converter especially useful for a wide variety of applications.

The TL507C is characterized for operation from 0°C to 70°C, and the TL507I is characterized for operation from −40°C to 85°C.

functional block diagram

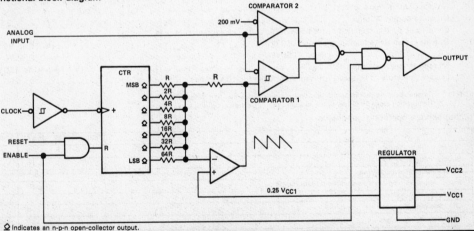

☒ Indicates an n-p-n open-collector output.

Data Acquisition

7

Data Acquisition

7

schematics of inputs and outputs

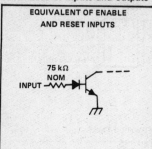

EQUIVALENT OF ENABLE
AND RESET INPUTS

75 kΩ
NOM

INPUT

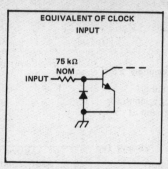

EQUIVALENT OF CLOCK
INPUT

75 kΩ
NOM

INPUT

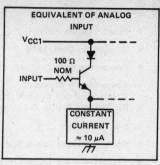

EQUIVALENT OF ANALOG
INPUT

V_{CC1}

100 Ω
NOM

INPUT

CONSTANT
CURRENT
≈ 10 µA

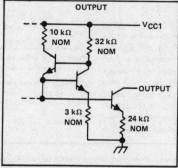

OUTPUT

10 kΩ
NOM

V_{CC1}

32 kΩ
NOM

OUTPUT

3 kΩ
NOM

24 kΩ
NOM

absolute maximum ratings over operating free-air temperature range (unless otherwise noted)

Supply voltage, V_{CC1} (see Note 1) .... 6.5 V
Supply voltage, V_{CC2} .... 20 V
Input voltage at analog input .... 6.5 V
Input voltage at enable, clock, and reset inputs .... ±20 V
On-state output voltage .... 6 V
Off-state output voltage .... 20 V
Continuous total dissipation at (or below) 25 °C free-air temperature (see Note 2) ..... 725 mW
Operating free-air temperature range: TL507I .... −40 °C to 85 °C
                  TL507C .... −0 °C to 70 °C
Storage temperature range .... −65 °C to 150 °C
Lead temperature 1/16 inch (1,6 mm) from case for 10 seconds .... 260 °C

NOTES: 1. Voltage values are with respect to network ground terminal unless otherwise noted.
       2. For operation above 25 °C free-air temperature, refer to Dissipation Derating Curves, Section 2.

recommended operating conditions

	MIN	NOM	MAX	UNIT
Supply voltage, V_{CC1}	3.5	5	6	V
Supply voltage, V_{CC2}	8	15	18	V
Input voltage at analog input	0		5.5	V
Input voltage at chip enable, clock, and reset inputs			±18	V
On-state output voltage			5.5	V
Off-state output voltage			18	V
Clock frequency, f_{clock}		125	150	kHz

**TEXAS
INSTRUMENTS**
POST OFFICE BOX 225012 • DALLAS, TEXAS 75265

electrical characteristics over recommended operating free-air temperature range, $V_{CC1} = V_{CC2} = 5$ V (unless otherwise noted)

regulator section

PARAMETER		TEST CONDITIONS		MIN	TYP‡	MAX	UNIT
V_{CC1}	Supply voltage (output)	$V_{CC2} = 10$ to 18 V,	$I_{CC1} = 0$ to -1 mA	5	5.5	6	V
I_{CC1}	Supply current	$V_{CC1} = 5$ V,	V_{CC2} open		5	8	mA
I_{CC2}	Supply current	$V_{CC2} = 15$ V,	V_{CC1} open		7	10	mA

inputs

PARAMETER			TEST CONDITIONS	MIN	TYP‡	MAX	UNIT
V_{IH}	High-level input voltage	Reset and		2			V
V_{IL}	Low-level input voltage	Enable				0.8	V
V_{T+}	Positive-going threshold voltage§	Clock Input		4.5			V
V_{T-}	Negative-going threshold voltage§					0.4	V
$V_{T+} - V_{T-}$	Hysteresis			2	2.6	4	V
I_{IH}	High-level input current	Reset, Enable, and Clock	$V_I = 2.4$ V		17	35	μA
			$V_I = 18$ V	130	220	320	
I_{IL}	Low-level input current		$V_I = 0$			±10	μA
I_I	Analog input current		$V_I = 4$ V		10	300	nA

output section

PARAMETER		TEST CONDITIONS	MIN	TYP‡	MAX	UNIT
I_{OH}	High-level output current	$V_{OH} = 18$ V		0.1	100	μA
I_{OL}	Low-level output current	$V_{OL} = 5.5$ V	5	10	15	mA
V_{OL}	Low-level output voltage	$I_{OL} = 1.6$ mA		80	400	mV

operating characteristics over recommended operating free-air temperature range, $V_{CC1} = V_{CC2} = 5.12$ V

PARAMETER	TEST CONDITIONS	MIN	TYP‡	MAX	UNIT
Overall error				±80	mV
Differential nonlinearity	See Figure 1			±1	LSB
Zero error§	Binary count = 0	1.20	1.28	1.36	V
Scale error	Binary count = 127			±80	mV
Full scale input voltage§	Binary count = 127	3.74	3.82	3.9	V
Propagation delay time from reset or enable				2	μs

‡All typical values are at $T_A = 25°C$.
§These parameters are linear functions of V_{CC1}.

definitions

zero error

The intercept (b) of the analog-to-digital converter-system transfer function $y = mx + b$, where y is the digital output, x is the analog input, and m is the slope of the transfer function.

overall error

The magnitude of the deviation from a straight line between the endpoints of the transfer function.

differential nonlinearity

Maximum deviation of an analog-value change that is asociated with a 1-bit code change (1 clock pulse) from its theroetical value of 1 LSB.

Data Acquisition

7

284

PARAMETER MEASUREMENT INFORMATION

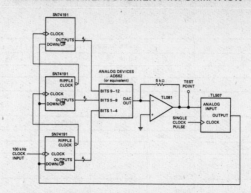

FIGURE 1—MONOTONICITY AND NONLINEARITY TEST CIRCUIT

PRINCIPLES OF OPERATION

The TL507 is a single-slope analog-to-digital converter. All single-slope converters are basically voltage-time or current-to-time converters. A study of the functional block diagram shows the versatility of the TL507.

An external clock signal is applied through a buffer to a negative-edge-triggered synchronous counter. Binary-weighted resistors from the counter are connected to an operational amplifier used as an adder. The operational amplifier generates a signal that ramps from $0.75 \cdot V_{CC1}$ down to $0.25 \cdot V_{CC1}$. Comparator 1 compares the ramp signal to the analog input signal. Comparator 2 functions as a fault defector. With the analog input voltage in the range $0.25 \cdot V_{CC1}$ to $0.75 \cdot V_{CC1}$, the duty cycle of the output signal is determined by the unknown analog input as shown in Figure 2 and the Function Table.

For illustration assume $V_{CC1} = 5.12$ V,

$$0.25 \cdot V_{CC1} = 1.28 \text{ V}$$

$$1 \text{ binary count} = \frac{(0.75 - 0.25) \cdot V_{CC1}}{128} = 20 \text{ mV}$$

$$0.75 \cdot V_{CC1} - 1 \text{ count} = 3.82 \text{ V}$$

The output is an open-collector n-p-n transistor capable of withstanding up to 18 volts in the off state. The output is current limited to the 8- to 12-milliampere range; however, care must be taken to ensure that the output does not exceed 5.5 volts in the on state.

The voltage regulator section allows operation from either an unregulated 8- to 18-volt V_{CC2} source or a regulated 3.5- to 6-volt V_{CC1} source. Regardless of which external power source is used, the internal circuitry operates at V_{CC1}. When operating from a V_{CC1} source, V_{CC2} may be connected to V_{CC1} or left open. When operating from a V_{CC2} source, V_{CC1} can be used as a reference voltage output.

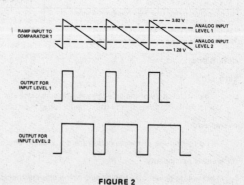

FIGURE 2

TEXAS INSTRUMENTS

POST OFFICE BOX 225012 • DALLAS, TEXAS 75265

DATA ACQUISITION CIRCUITS

TYPES TL520, TL521, TL522
CMOS ANALOG-TO-DIGITAL CONVERTERS
WITH 8-CHANNEL MULTIPLEXERS
D2666, SEPTEMBER 1982—REVISED SEPTEMBER 1983

- **Total Unadjusted Error at 85°C:**
  TL520 . . . ±3/4 LSB MAX
  TL521 . . . ±1 LSB MAX
  TL522 . . . ±1/2 LSB MAX

- **8-Bit Resolution**

- **Built-in 8-Input Analog Multiplexer**

- **Minimum Conversion Time:**
  TL520 . . . 70 μs
  TL521 . . . 100 μs
  TL522 . . . 200 μs

- **Ratiometric Conversion**

- **Guaranteed Monotonicity**

- **No Missing Codes**

- **Easy Interface with Microprocessors**

- **Latched 3-State Outputs**

- **Latched Address Inputs**

- **Single-Supply Operation**
  TL520, TL521 . . . 5 V
  TL522 . . . 3 V

- **Low Power Consumption**
  TL520, TL521 . . . 2.5 mW Typical
  TL522 . . . 0.3 mW Typical

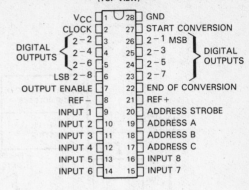

N DUAL-IN-LINE PACKAGE
(TOP VIEW)

V_{CC} [1	28] GND
CLOCK [2	27] START CONVERSION
2^{-2} [3	26] 2^{-1} MSB
2^{-4} [4	25] 2^{-3}
2^{-6} [5	24] 2^{-5}
LSB 2^{-8} [6	23] 2^{-7}
OUTPUT ENABLE [7	22] END OF CONVERSION
REF– [8	21] REF+
INPUT 1 [9	20] ADDRESS STROBE
INPUT 2 [10	19] ADDRESS A
INPUT 3 [11	18] ADDRESS B
INPUT 4 [12	17] ADDRESS C
INPUT 5 [13	16] INPUT 8
INPUT 6 [14	15] INPUT 7

DIGITAL OUTPUTS (pins 3, 4, 5) — DIGITAL OUTPUTS (pins 26, 25, 24, 23)

description

The TL520, TL521, and TL522 are monolithic CMOS devices each with an 8-channel multiplexer, and 8-bit analog-to-digital (A/D) converter, and microprocessor-compatible control logic. The 8-channel multiplexer can be controlled by a microprocessor through a 3-bit address decoder with address load to select any one of eight single-ended analog switches connected directly to a comparator. The 8-bit A/D converter uses a binary-weighted capacitor array to implement the high-speed, successive-approximation conversion technique.

The comparison and conversion methods used eliminate the possibility of missing codes, nonmonotonicity, and the need for zero or full-scale adjustment. Also featured are latched 3-state outputs and latched inputs to the multiplexer address decoder. The single 5-volt supply and low power requirements make the TL520 and TL521 especially useful for a wide variety of applications. The 3-volt and low power requirements make the TL522 especially useful for battery and LCD applications. Ratiometric conversion is made possible by access to the reference voltage input terminals.

The TL520, TL521, and TL522 are characterized for operation from −40°C to 85°C.

absolute maximum ratings over operating free-air temperature range (unless otherwise noted)

Supply voltage, V_{CC} (see Note 1)	6.5 V
Positive reference input voltage range, V_{REF+}	V_{REF-} to V_{CC} + 0.3 V
Negative reference input voltage range, V_{REF-} (see Note 1)	−0.3 V to V_{REF+}
Input voltage range: all other inputs	−0.3 V to V_{CC} + 0.3 V
Continuous total dissipation at 25°C free-air temperature (see Note 2)	1250 mW
Operating free-air temperature range	−40°C to 85°C
Storage temperature range	−65°C to 150°C
Lead temperature 1,6 mm (1/16 inch) from case for 10 seconds	260°C

NOTES: 1. All voltage values are with respect to network ground terminal.
2. For operation above 25°C free-air temperature, refer to Dissipation Derating Curves, Section 2.

functional block diagram (positive logic)

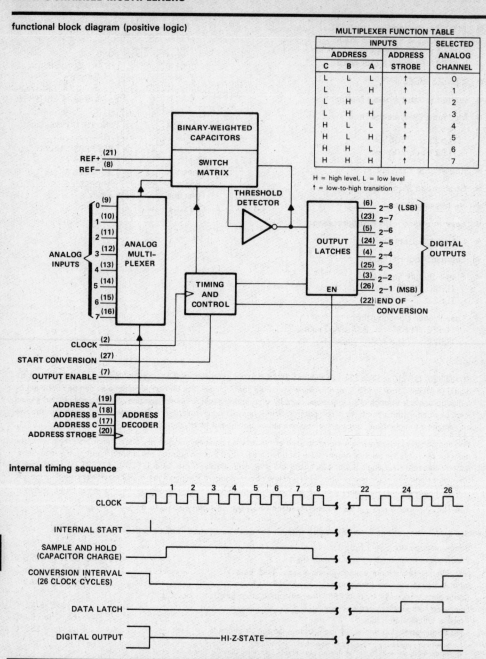

MULTIPLEXER FUNCTION TABLE

INPUTS				SELECTED
ADDRESS			ADDRESS	ANALOG
C	B	A	STROBE	CHANNEL
L	L	L	↑	0
L	L	H	↑	1
L	H	L	↑	2
L	H	H	↑	3
H	L	L	↑	4
H	L	H	↑	5
H	H	L	↑	6
H	H	H	↑	7

H = high level, L = low level
↑ = low-to-high transition

internal timing sequence

Data Acquisition

7

TEXAS
INSTRUMENTS
POST OFFICE BOX 225012 • DALLAS, TEXAS 75265

1083

TL520, TL521 recommended operating conditions

		TL520 MIN	NOM	MAX	TL521 MIN	NOM	MAX	UNIT
Supply voltage, V_{CC}		3	5	5.5	3	5	5.5	V
Positive reference voltage, V_{REF+}		3		V_{CC}	3		V_{CC}	V
Negative reference voltage, V_{REF-}		0		0.3	0		0.3	V
Supply voltage relative to V_{REF+} ($V_{CC} - V_{REF+}$)		0		1	0		1	V
Analog input voltage (see Note 3)		V_{REF-}		V_{REF+}	V_{REF-}		V_{REF+}	V
High-level control input voltage, V_{IH}	$V_{CC} \geq 4.75$ V	$V_{CC}-1.5$			$V_{CC}-1.5$			V
Low-level control input voltage, V_{IL}	$V_{CC} \geq 4.75$ V			1.5			1.5	V
Clock frequency, f_{clock}	$V_{REF+} = 5$ V		260	370		200	260	kHz
	$V_{REF+} = 3$ V		100			100		
Conversion time, t_{conv}	$V_{CC} = V_{REF+} = 5$ V	70			100			µs
Duration of start pulse, $t_{w(S)}$		100			100			ns
Duration of address strobe pulse, $t_{w(AS)}$		200			200			ns
Address setup time, t_{su}		50			50			ns
Address hold time, t_h		50			50			ns
Input voltage hold time		8			8			clock periods
Operating free-air temperature, T_A		−40		85	−40		85	°C

TL522 recommended operating conditions

		MIN	NOM	MAX	UNIT
Supply voltage, V_{CC} (see Note 4)	$T_A = 0$ °C to 85 °C	2.75	3	5.5	V
	$T_A = -40$ °C to 0 °C	3		5.5	
Positive reference voltage, V_{REF+} (see Notes 3 and 4)		2.75		V_{CC}	V
Negative reference voltage, V_{REF-} (see Note 3)		0		0.3	V
Supply voltage relative to V_{REF+}, $V_{CC} - V_{REF+}$		0		1	V
Analog input voltage (see Note 3)		V_{REF-}		V_{REF+}	V
High-level control input voltage, V_{IH}		$0.7V_{CC}$			V
Low-level control input voltage, V_{IL}				$0.3V_{CC}$	V
Clock frequency, f_{clock}	$V_{REF+} = 5$ V		100	260	kHz
	$V_{REF+} = 2.75$ V (see Note 4)		100	130	
Conversion time, t_{conv} (see Note 5)		200			µs
Duration of start pulse, $t_{w(S)}$		600			ns
Duration of address strobe pulse, $t_{w(AS)}$		600			ns
Address setup time, t_{su}		200			ns
Address hold time, t_h		150			ns
Input voltage hold time			8		clock periods
Operating free-air temperature, T_A (see Note 4)		−40		85	°C

NOTES: 3. Analog input voltage greater than V_{REF+} converts as all highs and less than V_{REF-} converts as all lows.
4. For proper operation of TL522 at free-air temperatures below 0 °C, V_{CC} and differential reference voltage ($V_{REF+} - V_{REF-}$) must never be less than 3 volts.
5. Conversion time is a function of clock frequency, with 200 µs corresponding to a maximum clock frequency of 130 kHz.

Data Acquisition

7

183

TEXAS
INSTRUMENTS
POST OFFICE BOX 225012 • DALLAS, TEXAS 75265

TL520, TL521 electrical characteristics over recommended operating free-air temperature range, V_{CC} = 4.5 V to 5.25 V (unless otherwise noted)

	PARAMETER	TEST CONDITIONS	MIN	TYP[†]	MAX	UNIT
V_{OH}	High-level output voltage	I_O = −360 μA	4			V
V_{OL}	Low-level output voltage	I_O = 1.6 mA			0.4	V
I_{OZ}	Off-state (high-impedance state) output current	V_O = 5 V			1	μA
		V_O = 0			−1	
I_{IH}	High-level control input current	V_I = V_{CC} + 0.3 V			1	μA
I_{IL}	Low-level control input current	V_I = 0			−1	μA
$I_{I(op)}$	Peak analog input current (operating) (see Note 6)	V_{CC} = V_{REF+} = 5 V, V_I = 2.5 V f_{clock} = 200 kHz, T_A = 25 °C		−5	−10	μA
$I_{I(stdby)}$	Analog input current (standby) (see Note 7)	V_{CC} = 5 V, V_I = 5 V		10	200	nA
		T_A = 25 °C V_I = 0		−10	−200	
		V_{CC} = 5 V, V_I = 5 V			1	μA
		T_A = 85 °C V_I = 0			−1	
I_{CC}	Supply current (see Note 8)	REF+ and REF− terminals open, f_{clock} = 200 kHz		10	50	μA
I_{CC} + I_{REF+}	Supply current plus reference current (see Note 8)	V_{CC} = V_{REF+} = 5 V, $V_{REF−}$ = 0, f_{clock} = 200 kHz		0.5	1	mA
		V_{CC} = V_{REF+} = 3 V, $V_{REF−}$ = 0, f_{clock} = 100 kHz		0.1		

[†]All typical characteristics are at V_{CC} = 5 V and T_A = 25 °C (unless otherwise specified)

NOTES: 6. $I_{I(op)}$ is measured on a selected channel and decays exponentially during the first clock pulse.
      7. $I_{I(stdby)}$ is measured on a selected channel with the clock input at 0 V.
      8. Current increases linearly with frequency of the clock at the rate of approximately 10% per 100 kHz.

TL520, TL521 operating characteristics, T_A = 25 °C, V_{CC} = V_{REF+} = 5 V, $V_{REF−}$ = 0, f_{clock} = 370 kHz for TL520 and 260 kHz for TL521 (unless otherwise noted)

	PARAMETER	TEST CONDITIONS	TL520			TL521			UNIT
			MIN	TYP[‡]	MAX	MIN	TYP[‡]	MAX	
k_{SVS}	Supply voltage sensitivity	V_{CC} = V_{REF+} = 4.75 V to 5.25 V		0.05			0.05		%/V
	Linearity error (see Note 9)			±0.25			±0.5		LSB
	Origin error (see Note 9)			±0.25			±0.25		LSB
	Total unadjusted error (see Note 9)	T_A = 25 °C		±0.25	±0.5		±0.5		LSB
		T_A = −40 °C to 85 °C			±0.75			±1	
t_{en}	Output enable time	C_L = 50 pF		100	250		100		ns
t_{dis}	Output disable time	C_L = 10 pF, R_L = 10 kΩ		100	250		100		ns
$t_{d(EOC-L)}$	Delay time, end-of-conversion output		0		100	0		100	ns

[‡]Typical values for all except supply voltage sensitivity are at V_{CC} = 5 V.
NOTE 9: All errors are measured with reference to an ideal straight-line transfer curve from 9.8 mV to 4.99 V with REF+ = V_{CC}.

Data Acquisition

7

TEXAS
INSTRUMENTS
POST OFFICE BOX 225012 • DALLAS, TEXAS 75265

TL522 electrical characteristics over recommended operating free-air temperature range,
V_{CC} = 3 V to 5.25 V, f_{clock} = 125 kHz (unless otherwise noted)

PARAMETER		TEST CONDITIONS		MIN	TYP[†]	MAX	UNIT
V_{OH}	High-level output voltage	I_O = −1 μA		V_{CC} − 0.05			V
		V_{CC} = 2.75 V, T_A = 0 °C to 85 °C	I_O = −0.1 mA,	2.35			
		I_O = −0.36 mA,	V_{CC} = 5 V				
V_{OL}	Low-level output voltage	I_O = −1 μA				0.05	V
		V_{CC} = 2.75 V, T_A = 0 °C to 85 °C	I_O = 0.4 mA,			0.4	
		V_{CC} = 5 V,	I_O = 1.6 mA			0.4	
I_{OZ}	Off-state (high-impedance state) output current	V_{CC} = 5.25 V,	V_O = 5.5 V			1	μA
			V_O = 0			−1	
I_{IH}	High-level input current	V_{CC} = 5.25 V,	V_I = 5.5 V			1	μA
I_{IL}	Low-level input current	V_I = 0				−1	μA
$I_{I(op)}$	Peak analog input current (operating) (see Note 6)	V_{CC} = V_{REF+} = 3 V, f_{clock} = 125 kHz,	V_I = 1.5 V, T_A = 25 °C		−5	−10	μA
$I_{I(stdby)}$	Analog input current (see Note 7)	V_{CC} = 3 V, T_A = 25 °C	V_I = 3 V		10	200	nA
			V_I = 0		−10	−200	
		V_{CC} = 3 V, T_A = 85 °C	V_I = 3 V			1	μA
			V_I = 0			−1	
I_{CC}	Supply current from V_{CC1}	REF+ and REF− terminals open			10	50	μA
I_{CC} + I_{REF}	Supply current plus reference current (see Note 8)	V_{CC} = V_{REF+} = 5 V, f_{clock} = 200 kHz	V_{REF-} = 0,		0.5		mA
		V_{CC} = V_{REF+} = 3 V, f_{clock} = 125 kHz	V_{REF-} = 0,		0.1	0.2	
C_i	Input capacitance					10	pF
C_o	Output capacitance					10	pF

NOTES: 6. $I_{I(op)}$ is measured on a selected channel and decays exponentially during the first clock pulse.
      7. $I_{I(stdby)}$ is measured on a selected channel with the clock input at 0 V.
      8. Current increases linearly with frequency of the clock at the rate of approximately 10% per 100 kHz.

TL522 operating characteristics, T_A = 25 °C, V_{REF+} = 3 V to 5.5 V, V_{REF-} = 0, f_{clock} = 130 kHz (unless otherwise noted)

PARAMETER		TEST CONDITIONS	MIN	TYP[†]	MAX	UNIT
k_{SVS}	Supply voltage sensitivity			0.05		%/V
	Linearity error (see Note 9)			±0.25		LSB
	Origin error (see Note 9)			±0.25		LSB
	Total unadjusted error (see Note 9)	V_{CC} = 2.75 V, T_A = 0 °C to 70 °C		±0.25	±0.5	LSB
		T_A = −40 °C to 85 °C		±0.25	±0.5	
t_{en}	Output enable time	C_L = 50 pF, R_L = 10 kΩ		0.7	1	μs
t_{dis}	Output disable time	C_L = 10 pF, R_L = 10 kΩ		0.6	0.8	μs
$t_{d(EOC-L)}$	Delay time, end-of-conversion output		0		100	ns

[†]All typical values are at V_{CC} = 3 V, T_A = 25 °C (unless otherwise noted)
NOTE 9: All errors are measured with reference to an ideal straight-line transfer curve from 9.8 mV to 4.99 V with REF+ = V_{CC}.

Data Acquisition

7

TEXAS
INSTRUMENTS
POST OFFICE BOX 225012 • DALLAS, TEXAS 75265

timing diagram

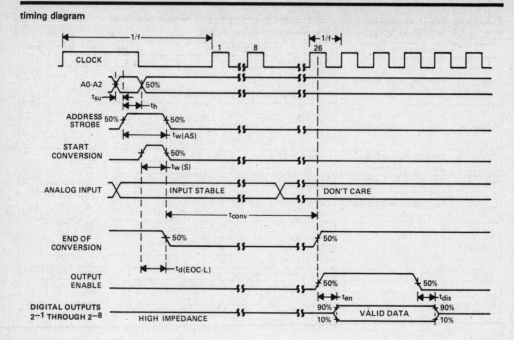

TEXAS
INSTRUMENTS

POST OFFICE BOX 225012 • DALLAS, TEXAS 75265

11

PRINCIPLES OF OPERATION

timing diagram

The analog multiplexer selects 1 of 8 single-ended input channels as determined by the input address code. The address strobe transfers and latches the address into the decoder on the positive-going edge of the signal. The output latch is reset by the postivie-going edge of the start pulse. Sampling also starts with the positive-going edge of the start pulse and lasts for 8 clock periods. The conversion process may be interrupted by a new start pulse before the end of 24 clock periods. The previous data will be lost if a new start of conversio occurs before the 24th clock pulse. Continuous conversion may be accomplished by connecting the end-of-pconversion output to the start input. If used in this mode an external pulse should be applied after power up to assure start up.

converter

The CMOS threshold detector in the successive-approximation conversion system determines each bit by examining the charge on a series of binary-weighted capacitors (Figure 1). In the first phase of the conversion process, the analog input is sampled by closing switch S_C and all S_T switches, and by simultaneously charging all the capacitors to the input voltage.

In the next phase of the conversion process, all S_T and S_C switches are opened and the threshold detector begins identifying bits by identifying the charge (voltage) on each capacitor relative to the reference voltage. In the switching sequence, all eight capacitors are examined separately until all 8 bits are identified, and then the charge-convert sequence is repeated. In the first step of the conversion phase, the threshold detector looks at the first capacitor (weight = 128). Node 128 of this capacitor is switched to the reference voltage, and the equivalent nodes of all the other capacitors on the ladder are switched to REF −. If the voltage at the summing node is greater than the trip-point of the threshold detector (approximately one-half the reference voltage), a bit is placed in the output register, and the 128-weight capacitor is switched to REF −. If the voltage at the summing node is less than the trip point of the threshold detector, this 128-weight capacitor remains connected to REF + through the remainder of the capacitor-sampling (bit-counting) process. The process is repeated for the 64-weight capacitor, the 32-weight capacitor, and so forth down the line, until all bits are counted.

With each step of the capacitor-sampling process, the initial charge is redistributed among the capacitors. The conversion process is succcessive approximation, but relies on charge shifting rather than a successive-approximation register (and reference D/A) to count and weigh the bits from MSB to LSB.

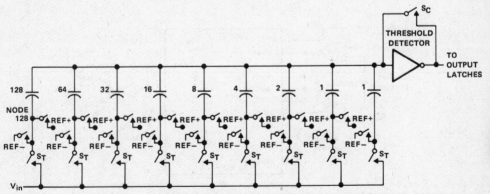

FIGURE 1—SIMPLIFIED MODEL OF THE SUCCESSIVE-APPROXIMATION SYSTEM

TEXAS
INSTRUMENTS
POST OFFICE BOX 225012 • DALLAS, TEXAS 75265

TYPICAL APPLICATION INFORMATION

The TL520, TL521, and TL522 are CMOS devices using charge redistribution to achieve A/D conversion. In typical applications as a ratiometric conversion system for a microprocessor, REF− will be connected to ground and REF+ will be connected to V_{CC}. The output will then be a simple proportional ratio between the analog input voltage and V_{CC} (Figure 3). The general relationship is

$$\frac{D_{out}}{2^8} = \frac{V_{in}}{V_{REF+} - V_{REF-}}$$

where
D_{out} = decimal value of binary output word
V_{in} = analog input voltage
V_{REF+} = positive reference voltage = V_{CC}
V_{REF-} = negative reference voltage = V_{GND}

Latchup may overheat and destroy the device and may occur by either of two kinds of circumstances: out of range reference voltages or by incorrect power-up sequence. V_{REF+} should not be more positive than V_{CC} by more than 300 millivolts or V_{REF-} should not be more negative than GND by more than 300 millivolts. Apply V_{CC} before either of the reference voltages. The advantage of the compressed reference potential is that the full 8-bit resolution applies to be compressed voltage range (Figure 4). However, the cautions mentioned above must be observed. Operation at voltages down to V_{CC} = 3 volts is possible but limits the frequency to 100 kilohertz maximum and thus conversion time to 260 microseconds minimum. Interface for the common microprocessors is shown in Figure 2.

MICROPROCESSOR INTERFACE TABLE

PROCESSOR	READ	WRITE	INTERRUPT (COMMENT)
TMS7000	RD	WR	EINT
TMS9900	MEMEN	WE	INTREQ
8080	MEMR	MEMW	INTR (Thru RST Circuit)
8085	RD	WR	INTR (Thru RST Circuit)
Z-80	RD	WR	INT (Thru RST Circuit, Mode 0)
SC/MP	NRDS	NWDS	SA (Thru Sense A)
6800	VMA +2 R/W	VMA +2 R/W	IRQA or IRQB (Thru PIA)

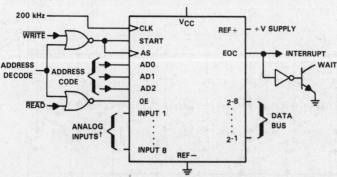

† The full-scale value of the analog input voltage can be shifted between 3 volts and 6.5 volts by varying V_{REF-} and V_{CC}, but only 5 volts guarantees TTL compatibility.

FIGURE 2 — TYPICAL MICROPROCESSOR APPLICATION

TEXAS INSTRUMENTS
POST OFFICE BOX 225012 • DALLAS, TEXAS 75265

Data Acquisition

7

TYPICAL APPLICATION INFORMATION

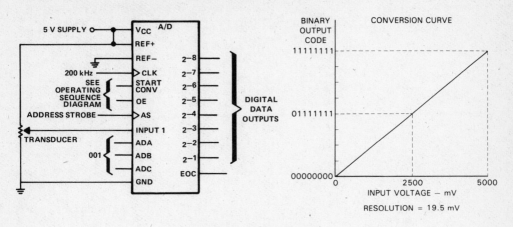

FIGURE 3 — RATIOMETRIC SYSTEM

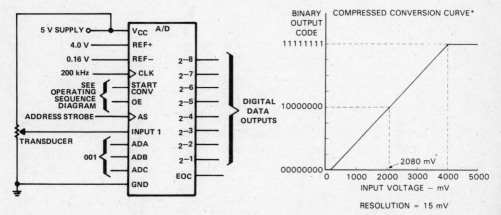

NOTE: Input voltage below V_{REF-} converts as all zeros
Input voltage above V_{REF+} converts as all ones

*Equivalent to 9-bit resolution over a 5-V range

FIGURE 4—COMPRESSED RATIOMETRIC SYSTEM

Data Acquisition

7

TEXAS
INSTRUMENTS
POST OFFICE BOX 225012 • DALLAS, TEXAS 75265

Data Acquisition

7

DATA
ACQUISITION
CIRCUITS

TYPES TL530, TL531
CMOS 8-BIT ANALOG-TO-DIGITAL PERIPHERALS WITH
ADDITIONAL DIGITAL INPUT/OUTPUT CAPABILITY
D2750, NOVEMBER 1983

- 8-Bit Resolution
- Total Unadjusted Error . . . ± 0.5 LSB Max for TL530 and ± 1 LSB Max for TL531
- Ratiometric Conversion
- Conversion Time (Including Access Time) . . . 300 μs (290 Clock Cycles)
- 3-State, Bidirectional I/O Data Bus
- Up to 12 Digital Inputs Including 3 I/O Pins
- 10 Analog and 6 Multipurpose Analog or Digital Inputs
- On-Chip 16-Channel Analog Multiplexer
- Three On-Chip 16-Bit Data Registers
- Polled or Interrupt Driven
- Single 5-V Supply Operation
- Low Power Consumption . . . 15 mW Typ
- Pin-for-Pin Compatible Functional Replacements for Motorola MC14444 and National Semiconductor ADC0830

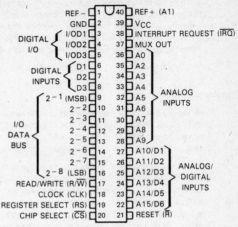

N DUAL-IN-LINE PACKAGE
(TOP VIEW)

REF –	1	40	REF + (A1)
GND	2	39	VCC
I/OD1	3	38	INTERRUPT REQUEST (IRQ)
I/OD2	4	37	MUX OUT
I/OD3	5	36	A0
D1	6	35	A2
D2	7	34	A3
D3	8	33	A4
2⁻¹ (MSB)	9	32	A5
2⁻²	10	31	A6
2⁻³	11	30	A7
2⁻⁴	12	29	A8
2⁻⁵	13	28	A9
2⁻⁶	14	27	A10/D1
2⁻⁷	15	26	A11/D2
2⁻⁸ (LSB)	16	25	A12/D3
READ/WRITE (R/W̄)	17	24	A13/D4
CLOCK (CLK)	18	23	A14/D5
REGISTER SELECT (RS)	19	22	A15/D6
CHIP SELECT (C̄S̄)	20	21	RESET (R̄)

DIGITAL I/O — I/OD1, I/OD2, I/OD3
DIGITAL INPUTS — D1, D2, D3
I/O DATA BUS — 2⁻¹ through 2⁻⁸
ANALOG INPUTS — A2 through A9
ANALOG/DIGITAL INPUTS — A10/D1 through A15/D6

description

The TL530 and TL531 are monolithic CMOS peripheral integrated circuits each designed to interface a microprocessor for analog data acquisition. These devices are complete peripheral data acquisition systems on a single chip and can convert analog signals to digital data from up to 15 external analog terminals. Each device features operation from a single 5-volt supply and additional digital input/output capabilities. Each contains a 16-channel analog multiplexer, an 8-bit ratiometric analog-to-digital (A/D) converter, three 16-bit registers, and microprocessor-compatible control logic circuitry. Additional features include a built-in self-test, six multipurpose (analog or digital) inputs, nine external analog inputs, and an 8-pin input/output (I/O) data port. The three on-chip data registers store the control data, the conversion results, and the input digital data that can be accesssed via the microprocessor data bus in two 8-bit bytes (most-significant byte first). In this manner, a microprocessor can access up to 15 external analog inputs or 6 digital signals and the positive reference voltage that may be used for self-test.

FUNCTION TABLE

ADDRESS/CONTROL					DESCRIPTION
R/W̄	RS	C̄S̄	R̄	CLK	
X	X	X	L†		Reset
L	H	L	H	↓	Write bus data to control register
H	L	L	H	↑	Read data from analog conversion register
H	H	L	H	↑	Read data from ditigal data register
X	X	H	H	X	No response

H = High-level, L = Low-level, X = Irrelevant,
↓ = High-to-low transition, ↑ = Low-to-high transition
†For proper operation, Reset must be low for at least three clock cycles.

Data Acquisition

7

description (continued)

The A/D conversion uses the successive-approximation technique employing a high-impedance chopper-stabilized comparator, a 256R end-compensated voltage divider with analog switch tree, and a successive-approximation register (SAR). This method eliminates the possibility of missing codes, nonmonotonicity, and a need for zero or full-scale adjustment. Positive and negative reference voltage inputs make possible ratiometric conversion and reference isolation from supply noises.

The TL530I and TL531I are characterized for operation from −40°C to 85°C.

functional description

The TL530 and TL531 provide direct interface to a microprocessor-based system. Control of the TL530 and TL531 is handled via the 8-line TTL-compatible 3-state data bus, the three control inputs (Read/Write, Register Select, and Chip Select), and the Clock input. Each device contains three 16-bit internal registers. These registers are the control register, the analog conversion data register, and the digital data register.

A high level at the Read/Write input and a low level at the Chip Select input set the device to output data on the 8-line data bus for the processor to read. A low level at the Read/Write input and a low level at the Chip Select input set the device to receive instructions into the internal control register on the 8-line data bus from the processor. When the device is in the read mode and the Register Select input is low, the processor will read the the data contained in the analog conversion data register. However, when the Register Select input is high, the processor reads the data contained in the digital data register.

The control register is a write-only register into which the microprocessor writes command instructions for the device to start A/D conversion and to select the analog channel to be converted, to select the output logic levels and the direction (input or output) of the 3-bit digital I/O port, and to set interrupt enable for the Interrupt Request output. The analog conversion data register is a read-only register that contains the current converter status and most recent conversion results. The digital data register is also a read-only register that holds the 3-bit I/O port status and digital input logic levels from the six multipurpose and the three digital inputs.

Internally each device contains a byte pointer that selects the appropriate byte during two cycles of the Clock input in a normal 16-bit microprocessor instruction. The internal pointer will automatically point to the most-significant (MS) byte after the first complete clock cycle any time that the Chip Select is at the high level for at least one clock cycle. This causes the device to treat the next signal on the 8-line data bus as the MS byte. A low level at the Chip Select input activates the inputs and outputs and an internal function decoder. However, no data is transferred until the Clock goes high. The internal byte pointer first points to the MS byte of the selected register during the first clock cycle. After the first clock cycle in which the MS byte is accessed, the internal pointer switches to the LS byte and remains there for as long as Chip Select is low. The MS byte of any register may be accessed by either an 8-bit or a 16-bit microprocessor instruction; however, the LS byte may only be accessed by a 16-bit microprocessor instruction.

Normally, a two-byte word is written into or read from the controlling processor, but a single byte can be read by the processor by proper manipulation of the Chip Select input. This can be used to read conversion status from the analog conversion data register or the digital multipurpose input levels from the digital data register. The format and content of each two-byte word is shown in Figures 1 through 3.

A conversion cycle is started after a two-byte instruction is written into the control register and the start conversion (SC) bit is a logic high. This two-byte instruction also selects the input analog channel, configures the 3-bit digital I/O pins, and sets the interrupt enable bit. The status (EOC) bit in the analog conversion data register is reset and remains reset until the conversion is completed, at that time the status bit is then set again. After conversion, the results are loaded into the analog conversion data register. These results remain in the analog conversion data register until the next conversion cylce is completed. If the interrupt enable bit is set to a logic high level in the control register, the Interrupt Request ($\overline{\text{IRQ}}$) output

TEXAS
INSTRUMENTS

POST OFFICE BOX 225012 • DALLAS, TEXAS 75265

functional description (continued)

will go low after the next conversion cycle is completed. If a new conversion command is entered into the control register while the conversion cycle is in progress, the on-going conversion will be aborted and a new channel acquisition cycle will immediately begin.

The Reset input allows the device to be externally forced to a known state. When a low level is applied to the Reset input for a minimum of three clock periods, the start conversion, interrupt enable, and the I/O port data direction bits of the control register are cleared. The A/D converter is then idled and all the outputs including the 3-bit I/O digital port are placed in the high-impedance off-state. However, the content of the analog conversion data register is not affected by the Reset input going to a low level.

functional block diagram

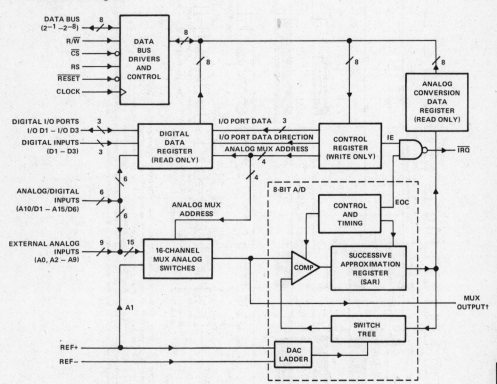

†Loading of the MUX output affects the changing times of the DAC; it is recommended that no connection be made to this pin.

Data Acquisition

7

TEXAS
INSTRUMENTS
POST OFFICE BOX 225012 • DALLAS, TEXAS 75265

typical operating sequence

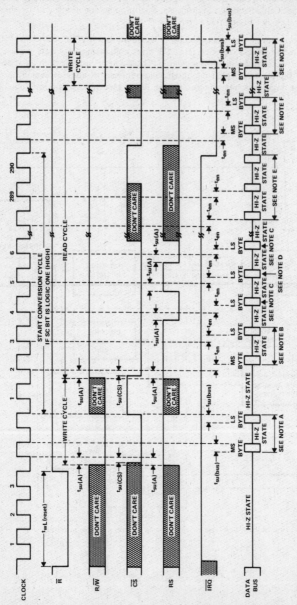

NOTES: A. This is a 16-bit input instruction from the microprocessor being sent to the control data register.
      B. This is the 2-byte (16-bit) content of the digital data register being sent to the microprocessor.
      C. This is the LS byte (8-bit) content of the analog conversion data register being sent to the microprocessor.
      D. This is the LS byte (8-bit) content of the digital data register being sent to the microprocessor.
      E. These are 8-bit or 16-bit output data from either the analog conversion data register or the digital data register being sent to the microprocessor.
      F. This is the 2-byte (16-bit) content of the analog conversion data register being sent to the microprocessor.

TEXAS INSTRUMENTS
POST OFFICE BOX 225012 • DALLAS, TEXAS 75265

Data Acquisition

7

DATA BUS LINES	$2-1$	$2-2$	$2-3$	$2-4$	$2-5$	$2-6$	$2-7$	$2-8$	$2-1$	$2-2$	$2-3$	$2-4$	$2-5$	$2-6$	$2-7$	$2-8$
	X (MSB)	IE	X	X	X	X	X	SC (LSB)	DD (MSB)	I/O D3	I/O D2	I/O D1	A3	A2	A1	A0 (LSB)

← MOST-SIGNIFICANT BYTE → ← LEAST-SIGNIFICANT BYTE →

← 16-BIT WRITE →

Interrupt Enable (IE) — The interrupt enable bit, when set to a logical 1 (high level), allows the $\overline{IRQ}$ pin to be activated at the completion of the next analog-to-digital conversion.

Unused Bits (X) — The MS byte bits $2-1$ through $2-7$ and LS byte bits $2-1$ through $2-4$ of the control register are not used internally.

Start Conversion (SC) — When the SC bit in the MS byte is set to a logical 1 (high level), analog-to-digital conversion on the specified analog channel will begin immediately after the completion of the control register write.

I/O Port Data Direction (DD) — The MSB of the LS byte is the data direction bit for the 3-bit I/O port. A logical 1 configures the port as the output while a logical 0 configures the port as an input.

Digital I/O Output (I/OD1-I/OD3) — When the microcompressor configures the 3-bit I/O port as an output, these are the bit locations into which the output states are written. A logical 1 written by the microprocesssor will cause the output to be high, while a logical 0 will cause the output to be low.

Analog Multiplex Address (A0-A3) — These four address bits are decoded by the analog multiplexer and used to select the appropriate analog channel as shown below:

Hexadecimal Address (A3 = MSB)	Channel Select
0	A0
1	REF+ (A1)
2-9	A2-A9
A-F	A10-A15

FIGURE 1—CONTROL REGISTER TWO-BYTE WRITE WORD FORMAT AND CONTENT

DATA BUS LINES	$2-1$	$2-2$	$2-3$	$2-4$	$2-5$	$2-6$	$2-7$	$2-8$	$2-1$	$2-2$	$2-3$	$2-4$	$2-5$	$2-6$	$2-7$	$2-8$
	EOC (MSB)	0	0	0	0	0	0	0 (LSB)	R7 (MSB)	R6	R5	R4	R3	R2	R1	R0 (LSB)

← MOST-SIGNIFICANT BYTE → ← LEAST-SIGNIFICANT BYTE →

← 8-BIT READ →

← 16-BIT READ →

A/D Status (EOC) — The A/D status-end-of-converesion (EOC) bit is set whenever an analog-to-digital conversion is successfully completed by the A/D converter. The status bit is cleared by a 16-bit write from the microprocessor to the control register. The remainder of the bits in the MS byte of the analog conversion data are always reset to logical 0 to simplify microprocessor interrogation of the A/D converter status.

A/D Result (R0-R7) — The LS byte of the analog conversion data register contains the result of the analog-to-digital conversion. Result bit R7 is the MSB and the converter follows the standard convention of assigning a code of all ones (11111111) to a full-scale analog voltage. There are no special overflow or underflow indications.

FIGURE 2—ANALOG CONVERSION DATA REGISTER ONE-BYTE OR TWO-BYTE READ WORD FORMAT AND CONTENT

DATA BUS LINES	$2-1$	$2-2$	$2-3$	$2-4$	$2-5$	$2-6$	$2-7$	$2-8$	$2-1$	$2-2$	$2-3$	$2-4$	$2-5$	$2-6$	$2-7$	$2-8$
	A15 /D6 (MSB)	A14 /D5	A13 /D4	A12 /D3	A11 /D2	A10 /D1	A3	A2 (LSB)	A1 (MSB)	A0	D3	D2	D1	I/O D3	I/O D2	I/O D1 (LSB)

← MOST-SIGNIFICANT BYTE → ← LEAST-SIGNIFICANT BYTE →

← 8-BIT READ →

← 16-BIT READ →

Shared Digital Port (A10/D1-A15/D6) — The voltage present on these pins is interpreted as a digital signal and the corresponding states are read from these bits. A digital value will be given for each pin even if some or all of these pins are being used as analog inputs.

Analog Multiplexer Address (A0-A3) — The address of the selected analog channel presently addressed is given by these bits.

Digital Inputs (D1-D3) — The states of the three digital inputs are read from these bits.

Digital I/O Port (I/OD1-I/OD3) — The states of the three digital I/O pins are read from these bits regardless of whether the port is configured as input or output.

FIGURE 3—DIGITAL DATA REGISTER ONE-BYTE AND TWO-BYTE READ WORD FORMAT AND CONTENT

Data Acquisition

7

3

absolute maximum ratings over operating free-air temperature range (unless otherwise noted)

Supply voltage, V_{CC} (see Note 1) . . . . −0.3 V to 6.5 V
Input voltage range: Positive reference voltage . . . . V_{REF-} to V_{CC} + 0.3 V
Negative reference voltage . . . . −0.3 V to V_{REF+}
All other inputs . . . . −0.3 V to V_{CC} + 0.3 V
Input current, I_I (any input) . . . . ±10 mA
Total input current (all inputs) . . . . ±20 mA
Continuous total dissipation at (or below) 25°C free-air temperature (see Note 2) . . . . 1250 mW
Storage temperature range . . . . −65°C to 150°C
Operating free-air temperature range . . . . −40°C to 85°C
Lead temperature 1,6 mm (1/16 inch) from case for 10 seconds . . . . 260°C

NOTES: 1. All voltage values are with respect to network ground terminal.
2. For operation above 25°C free-air temperature, refer to Dissipation Derating Curves, Section 2.

recommended operating conditions

		TL530			TL531			UNIT
		MIN	NOM	MAX	MIN	NOM	MAX	
Supply voltage, V_{CC}		4.75	5	5.5	4.75	5	5.5	V
Positive reference voltage, V_{REF+} (see Note 3)		4.6	V_{CC}		4.6	V_{CC}		V
Negative reference voltage, V_{REF-} (see Note 3)		0		0.1	0		0.1	V
Average voltage across ladder		$V_{CC}-0.2$	V_{CC}	$V_{CC}+0.2$	$V_{CC}-0.2$	V_{CC}	$V_{CC}+0.2$	V
High-level input voltage, V_{IH}	Clock input	$V_{CC}-0.8$			$V_{CC}-0.8$			V
	All other digital inputs	2			2			
Low-level input voltage, V_{IL}	Any digital input			0.8			0.8	V
Clock frequency, f_{CLK}		0.1	1.048		0.1	1.048		MHz
Clock period, $1/f_{CLK}$		943			943			ns
$\overline{CS}$ setup time, $t_{su(CS)}$		100			100			ns
Address (R/$\overline{W}$ and RS) setup time, $t_{su(A)}$		145			145			ns
Data bus input setup time, $t_{su(bus)}$		185			185			ns
Control (R/$\overline{W}$, RS, and $\overline{CS}$) hold time, $t_{h(C)}$		20			20			ns
Data bus input hold time, $t_{h(bus)}$		20			25			ns
Pulse duration of control inputs during read cycle, $t_{w(C)}$		575			575			ns
Pulse duration, reset low, $t_{wL(reset)}$		3			3			Clock Cycles
Pulse duration of clock high, $t_{wH(CLK)}$		440			440			ns
Pulse duration of clock low, $t_{wL(CLK)}$		410			410			ns
Clock rise time, $t_{r(CLK)}$				25			25	ns
Clock fall time, $t_{f(CLK)}$				30			30	ns
Operating free-air temperature, T_A		−40		85	−40		85	°C

NOTE 3: Analog input voltages greater than or equal to that applied to the REF+ terminal convert to all ones (11111111), while input voltages equal to or less than that applied to the REF− terminal convert to all zeros (00000000).

Data Acquisition

7

TEXAS
INSTRUMENTS
POST OFFICE BOX 225012 • DALLAS, TEXAS 75265

electrical characteristics over recommended ranges of V_{CC}, V_{REF+}, and operating free-air temperature, $V_{REF-} = 0$, $f_{CLK} = 1.048$ MHz (unless otherwise noted)

PARAMETER			TEST CONDITIONS	MIN	TYP[†]	MAX	UNIT
V_{OH}	High-level	I/O Digital pins 3, 4, and 5	$I_{OH} = -190\ \mu A$	$V_{CC}-0.4$			V
	output voltage	I/O Data Bus	$I_{OH} = -1.6$ mA	2.4			
V_{OL}	Low-level	I/O Digital pins 3, 4, and 5	$I_{OL} = 975\ \mu A$			0.4	V
	output voltage	$\overline{IRQ}$ and I/O Data Bus	$I_{OL} = 1.6$ mA			0.4	
I_{IH}	High-level	Any control input	$V_{IH} = 5.5$ V			1	μA
	input current	Any other digital input				10	
I_{IL}	Low-level	Any control input	$V_{IL} = 0$			-1	μA
	input current	Any other digital input				-10	
I_{OH}	High-level output current	$\overline{IRQ}$ output	$V_{OH} = V_{CC} = 4.75$ V			10	μA
I_{OZ}	Off-state (high-impedance-state) output current		$V_O = V_{CC}$			10	μA
			$V_O = 0$			-10	
I_I	Analog input current (see Note 4)		$V_I = 0$ to V_{CC}			± 500	nA
	Leakage current between selected channel and all other analog channels		$V_I = 0$ to V_{CC}, Clock input at 0 V			± 400	nA
$r_{i(on)}$	Analog channel input on-state resistance					5	kΩ
	Ladder resistance (REF+ to REF−)			1	5	10	kΩ
C_i	Input capacitance	Digital pins 9 thru 16			7	30	pF
		Any other input pin			5	15	
$I_{CC}+I_{REF+}$	Supply current plus reference current		$V_{CC} = V_{REF+} = 5.5$ V, Outputs open		3	16	mA
I_{CC}	Supply current		$V_{CC} = 5.5$ V		2	10	mA

[†]Typical values are at $V_{CC} = 5$ V, $T_A = 25°C$.
NOTE 4: Analog input current is an average of the current flowing into a selected analog channel input during one full conversion cycle.

Data Acquisition

7

TEXAS
INSTRUMENTS
POST OFFICE BOX 225012 • DALLAS, TEXAS 75265

operating characteristics over recommended ranges of V_{CC}, V_{REF+}, and operating free-air temperature, V_{REF-} at ground, f_{CLK} = 1.048 MHz (unless otherwise noted)

PARAMETER		TEST CONDITIONS	MIN	TYP†	MAX	UNIT
	Linearity error	See Note 5			±0.5	LSB
	Zero error	See Note 6			±0.5	LSB
	Full-scale error	See Note 6			±0.5	LSB
	Total unadjusted error	See Note 7			±0.5	LSB
	Absolute accuracy error	See Note 8			±1	LSB
t_{conv}	Conversion time (including channel acquisition time)				290	Clock Cycles
t_{acq}	Channel acquisition time				30	Clock Cycles
t_{en}	Data output enable time (see Note 9)	R_L = 3 kΩ, C_L = 50 pF,			335	ns
t_{dis}	Data output disable time	R_L = 3 kΩ, C_L = 50 pF	10			ns
$t_{r(bus)}$	Data bus output rise time	High-impedance to high-level	R_L = 3 kΩ, C_L = 50 pF			150	ns
		Low to high-level				300	
$t_{r(I/O)}$	Digital I/O output rise time	High-impedance to high-level				0.5	µs
		Low to high-level				1	
$t_{f(bus)}$	Data bus output fall time	High-impedance to low-level	R_L = 3 kΩ, C_L = 50 pF			150	ns
		High to low-level				300	
$t_{f(I/O)}$	Digital I/O output fall time	High-impedance to low-level				0.5	µs
		High to low-level				1	
$t_{f(IRQ)}$	IRQ output fall time	High-impedance or high-level to low-level				0.5	µs

†Typical values are at V_{CC} = 5 V, T_A = 25°C.

NOTES: 5. Linearity error is the deviation from the best straight line through the A/D transfer characteristics.
6. Zero error is the difference between the output of an ideal and an actual A/D for zero input voltage; full-scale error is that same difference for full-scale input voltage.
7. Total unadjusted error is the sum of linearity, zero, and full-scale errors.
8. Absolute accuracy error is the maximum difference between an analog value and the nominal midstep value within any step. This includes all errors including inherent quantization error, which is the ±0.5 LSB uncertainty caused by the A/D converters finite resolution.
9. If chip-select setup time, $t_{su(CS)}$, is less than 0.14 microseconds, the effective data output enable time, t_{en}, may extend such that $t_{su(CS)}$ + t_{en} is equal to a maximum of 0.475 microseconds.

TEXAS
INSTRUMENTS
POST OFFICE BOX 225012 • DALLAS, TEXAS 75265

electrical characteristics over recommended ranges of V_{CC}, V_{REF+}, and operating free-air temperature, $V_{REF-} = 0$, $f_{CLK} = 1.048$ MHz (unless otherwise noted)

PARAMETER			TEST CONDITIONS	MIN	TYP[†]	MAX	UNIT
V_{OH}	High-level output voltage	I/O Digital pins 3, 4, and 5	$I_{OH} = -190$ A	$V_{CC}-0.4$			V
		I/O Data Bus	$I_{OH} = -1.6$ mA	2.4			
V_{OL}	Low-level output voltage	I/O Digital pins 3, 4, and 5	$I_{OL} = 975$ µA			0.4	V
		$\overline{IRQ}$ and I/O Data Bus	$I_{OL} = 1.6$ mA			0.4	
I_{IH}	High-level input current	Any control input	$V_{IH} = 5.5$ V			1	µA
		Any other digital input				10	
I_{IL}	Low-level input current	Any control input	$V_{IL} = 0$			-1	µA
		Any other digital input				-10	
I_{OH}	High-level output current	$\overline{IRQ}$ output	$V_{OH} = V_{CC} = 4.75$ V			10	µA
I_{OZ}	Off-state (high-impedance-state) output current		$V_O = V_{CC}$			10	µA
			$V_O = 0$			-10	
I_I	Analog input current (see Note 4)		$V_I = 0$ to V_{CC}			±700	nA
	Leakage current between selected channel and all other analog channels		$V_I = 0$ to V_{CC}, Clock input at 0 V			±500	nA
$r_{i(on)}$	Analog channel input on-state resistance					5	kΩ
	Ladder resistance (REF+ to REF−)			1	5	10	kΩ
C_i	Input capacitance	Digital pins 9 thru 16			7	30	pF
		Any other input pins			5	15	
$I_{CC}+I_{REF+}$	Supply current plus reference current		$V_{CC} = V_{REF+} = 5.5$ V, Outputs open		3	16	mA
I_{CC}	Supply current		$V_{CC} = 5.5$ V		2	10	mA

[†]Typical values are at $V_{CC} = 5$ V, $T_A = 25°C$.
NOTE 4: Analog input current is an average of the current flowing into a selected analog channel input during one full conversion cycle.

Data Acquisition

7

operating characteristics over recommended operating free-air temperature range, $V_{REF+} = V_{CC}$, V_{REF-} at ground, f_{CLK} = 2 MHz (unless otherwise noted)

PARAMETER		TEST CONDITIONS		MIN	TYP†	MAX	UNIT
	Linearity error	See Note 5			±0.5		LSB
	Zero error	See Note 6			±0.25		LSB
	Full-scale error	See Note 6			±0.25		LSB
	Total unadjusted error	See Note 7	T_A = 25°C		±0.5		LSB
			T_A = −40°C to 85°C			±1	
	Absolute accuracy error	See Note 8				±1.5	LSB
t_{conv}	Conversion time (including channel acquisition time)					290	Clock Cycles
t_{acq}	Channel acquisition time					30	Clock Cycles
t_{en}	Data output enable time (see Note 9)	R_L = 3 kΩ, C_L = 50 pF				335	ns
t_{dis}	Data output disable time	R_L = 3 kΩ, C_L = 50 pF		10			ns
$t_{r(bus)}$	Data bus output rise time	High-impedance to high-level	R_L = 3 kΩ, C_L = 50 pF			150	ns
		Low to high-level				300	
$t_{r(I/O)}$	Digital I/O output rise time	High-impedance to high-level				0.5	μs
		Low to high-level				1	
$t_{f(bus)}$	Data bus output fall time	High-impedance to low-level	R_L = 3 kΩ, C_L = 50 pF			150	ns
		High to low-level				300	
$t_{f(I/O)}$	Digital I/O output fall time	High-impedance to low-level				0.5	μs
		High to low-level				1	
$t_{f(IRQ)}$	IRQ output fall time	High-impedance or high-level to low-level				0.5	μs

NOTES: 5. Linearity error is the deviation from the best straight line through the A/D transfer characteristics.
6. Zero error is the difference between the output of an ideal and an actual A/D for zero input voltage; full-scale error is that same difference for full-scale input voltage.
7. Total unadjusted error is the sum of linearity, zero, and full-scale errors.
8. Absolute accuracy error is the maximum difference between an analog value and the nominal midstep value within any step. This includes all errors including inherent quantization error, which is the ±0.5 LSB uncertainty caused by the A/D converters finite resolution.
9. If chip-select setup time, $t_{su(CS)}$, is less than 0.14 microseconds, the effective data output enable time, t_{en}, may extend such that $t_{su(CS)}$ + t_{en} is equal to a maximum of 0.475 microseconds.

Data Acquisition

7

TEXAS
INSTRUMENTS
POST OFFICE BOX 225012 • DALLAS, TEXAS 75265

DATA
ACQUISITION
CIRCUITS

TYPES TL532, TL533
CMOS 8-BIT ANALOG-TO-DIGITAL PERIPHERALS WITH
ADDITIONAL DIGITAL INPUT/OUTPUT CAPABILITY
D2818, OCTOBER 1983

- **8-Bit Resolution**

- **Total Unadjusted Error . . . ±0.5 LSB Max for TL532 and ±1 LSB Max for TL533**

- **Access Plus Conversion Time . . . 300 μs (290 Clock Cycles)**

- **Ratiometric Conversion**

- **3-State, Bidirectional I/O Data Bus**

- **5 Analog and 6 Multipurpose Inputs**

- **On-Chip 12-Channel Analog Multiplexer**

- **Three On-Chip 16-Bit Data Registers**

- **Software Compatible with Larger TL530 and TL531 (21-Input Versions)**

- **Single 5-V Supply Operation**

- **Low Power Consumption . . . 15 mW Typ**

- **Pin-for-Pin Compatible and Functionally Compatible with Motorola MC14442**

- **Direct Replacement for National Semiconductor ADC0829**

N DUAL-IN-LINE PACKAGE
(TOP VIEW)

```
                  REF –  [1   U 28]  REF + (A1)
                   GND  [2     27]  VCC
           2 – 1 (MSB)  [3     26]  A0
              2 – 2     [4     25]  A2
      I/O     2 – 3     [5     24]  A3     ANALOG
     DATA     2 – 4     [6     23]  A4     INPUTS
      BUS     2 – 5     [7     22]  A5
              2 – 6     [8     21]  A10/D1
              2 – 7     [9     20]  A11/D2
       2 – 8 (LSB)     [10    19]  A12/D3   ANALOG/
    READ/WRITE (R/W)   [11    18]  A13/D4   DIGITAL
        CLOCK (CLK)    [12    17]  A14/D5   INPUTS
  REGISTER SELECT (RS) [13    16]  A15/D6
      CHIP SELECT (CS) [14    15]  RESET (R)
```

FUNCTION TABLE

ADDRESS/CONTROL					DESCRIPTION
R/W	RS	CS	R	CLK	
X	X	X	L†		Reset
L	H	L	H	↓	Write bus data to control register
H	L	L	H	↑	Read data from analog conversion register
H	H	L	H	↑	Read data from digital data register
X	X	H	H	X	No response

H = High-level, L = Low-level, X = Irrelevant,
↓ = High-to-low transition, ↑ = Low-to-high transition
†For proper operation, Reset must be low for at least three clock cycles.

description

The TL532 and TL533 are monolithic CMOS peripheral integrated circuits each designed to interface a microprocessor for analog data acquisition. These devices are complete peripheral data acquisition systems on a single chip that can convert analog signals to digital data from up to 11 external analog terminals. Each device features operation from a single 5-volt supply. Each contains a 12-channel analog multiplexer, an 8-bit ratiometric analog-to-digital (A/D) converter, three 16-bit registers, and microprocessor-compatible control logic circuitry. Additional features include a built-in self-test, six multipurpose (analog or digital) inputs, five external analog inputs, and an 8-pin input/output (I/O) data port. The three on-chip data registers store the control data, the conversion results, and the input digital data that can be accessed via the microprocessor data bus in two 8-bit bytes (most-significant byte first). In this manner, a microprocessor can access up to 11 external analog inputs or 6 digital signals and the positive reference voltage that may be used for self-test.

The A/D conversion uses the successive-approximation technique employing a high-impedance chopper-stabilized comparator, a 256R end-compensated voltage divider with analog switch tree, and a successive-approximation register (SAR). This method eliminates the possibility of missing codes, nonmonotonicity, and a need for zero or full-scale adjustment. Positive and negative reference voltage inputs make possible ratiometric conversion and reference isolation from supply noises.

The TL532I and TL533I are characterized for operation from −40°C to 85°C.

Data Acquisition

7

TEXAS
INSTRUMENTS
POST OFFICE BOX 225012 • DALLAS, TEXAS 75265

functional description

The TL532 and TL533 provide direct interface to a microprocessor-based system. Control of the TL532 and TL533 is handled via the 8-line TTL-compatible 3-state data bus, the three control inputs (Read/Write, Register Select, and Chip Select), and the Clock input. Each device contains three 16-bit internal registers. These registers are the control register, the analog conversion data register, and the digital data register.

A high level at the Read/Write input and a low level at the Chip Select input set the device to output data on the 8-line data bus for the processor to read. A low level at the Read/Write input and a low level at the Chip Select input set the device to receive instructions into the internal control register on the 8-line data bus from the processor. When the device is in the read mode and the Register Select input is low, the processor will read the data contained in the analog conversion data register. However, when the Register Select input is high, the processor reads the data contained in the digital data register.

The control register is a write-only register into which the microprocessor writes command instructions for the device to start A/D conversion and to select the analog channel to be converted. The analog conversion data register is a read-only register that contains the current converter status and most recent conversion results. The digital data register is also a read-only register that holds the digital input logic levels from the six multipurpose inputs.

Internally each device contains a byte pointer that selects the appropriate byte during two cycles of the Clock input in a normal 16-bit microprocessor instruction. The internal pointer will automatically point to the most-significant (MS) byte after the first complete clock cycle any time that the Chip Select is at the high level for at least one clock cycle. This causes the device to treat the next signal on the 8-line data bus as the MS byte. A low level at the Chip Select input activates the inputs and outputs and an internal function decoder. However, no data is transferred until the Clock goes high. The internal byte pointer first points to the MS byte of the selected register during the first clock cycle. After the first clock cycle in which the MS byte is accessed, the internal pointer switches to the LS byte and remains there for as long as Chip Select is low. The MS byte of any register may be accessed by either an 8-bit or a 16-bit microprocessor instruction; however, the LS byte may only be accessed by a 16-bit microprocessor instruction.

Normally, a two-byte word is written into or read from the controlling processor, but a single byte can be read by the processor by proper manipulation of the Chip Select input. This can be used to read conversion status from the analog conversion data register or the digital multipurpose input levels from the digital data register. The format and content of each two-byte word is shown in Figures 1 through 3.

A conversion cycle is started after a two-byte instruction is written into the control register and the start conversion (SC) bit is a logic high. This two-byte instruction also selects the input analog channel to be converted. The status (EOC) bit in the analog conversion data register is reset and remains reset until the conversion is completed, at that time the status bit is then set again. After conversion, the results are loaded into the analog conversion data register. These results remain in the analog conversion data register until the next conversion cylce is completed. If a new conversion command is entered into the control register while the conversion cycle is in progress, the on-going conversion will be aborted and a new channel acquisition cycle will immediately begin.

The Reset input allows the device to be externally forced to a known state. When a low level is applied to the Reset input for a minimum of three clock periods, the start conversion bit of the control register is cleared. The A/D converter is then idled and all the outputs are placed in the high-impedance off-state. However, the content of the analog conversion data register is not affected by the Reset input going to a low level.

128

TEXAS
INSTRUMENTS
POST OFFICE BOX 225012 • DALLAS, TEXAS 75265

functional block diagram

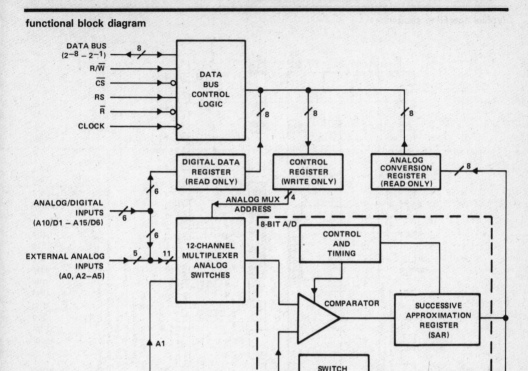

Data Acquisition

7

TEXAS INSTRUMENTS
POST OFFICE BOX 225012 • DALLAS, TEXAS 75265

typical operating sequence

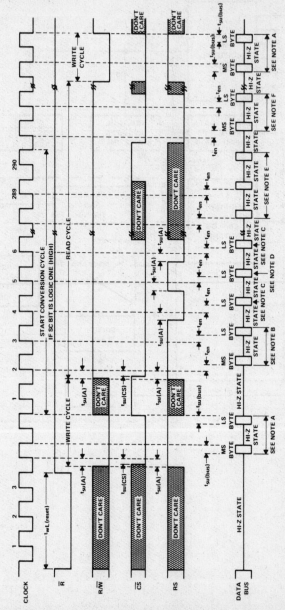

NOTES: A. This is a 16-bit input instruction from the microprocessor being sent to the control data register.
   B. This is the 2-byte (16-bit) content of the digital data register being sent to the microprocessor.
   C. This is the LS byte (8-bit) content of the analog conversion data register being sent to the microprocessor.
   D. This is the LS byte (8-bit) content of the digital data register being sent to the microprocessor.
   E. These are 8-bit or 16-bit output data from either the analog conversion data register or the digital data register being sent to the microprocessor.
   F. This is the 2-byte (16-bit) content of the analog conversion data register being sent to the microprocessor.

TEXAS
INSTRUMENTS
POST OFFICE BOX 225012 • DALLAS, TEXAS 75265

Data Acquisition

7

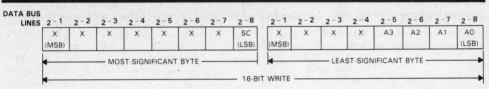

Unused Bits (X) — The MS byte bits 2^{-1} through 2^{-7} and LS byte bits 2^{-1} through 2^{-4} of the control register are not used internally.

Start Conversion (SC) — When the SC bit in the MS byte is set to a logical 1 (high level), analog-to-digital conversion of the specified analog channel will begin immediately after the completion of the control register write.

Analog Multiplex Address (A0-A3) — These four address bits are decoded by the analog multiplexer and used to select the appropriate analog channel as shown below:

Hexadecimal Address (A3 = MSB)	Channel Select
0	A0
1	REF+ (A1)
2-5	A2-A5
6-9 (not used)	
A-F	A10-A15

FIGURE 1—CONTROL REGISTER TWO-BYTE WRITE WORD FORMAT AND CONTENT

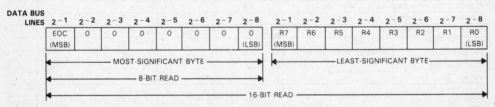

A/D Status (EOC) — The A/D status end-of-conversion (EOC) bit is set whenever an analog-to-digital conversion is successfully completed by the A/D converter. The status bit is cleared by a 16-bit write from the microprocessor to the control register. The remainder of the bits in the MS byte of the analog conversion data register are always reset to logical 0 to simplify microprocessor interrogation of the A/D converter status.

A/D Result (R0-R7) — The LS byte of the analog conversion data register contains the result of the analog-to-digital conversion. Result bit R7 is the MSB and the converter follows the standard convention of assigning a code of all ones (11111111) to a full-scale analog voltage. There are no special overflow or underflow indications.

FIGURE 2—ANALOG CONVERSION DATA REGISTER ONE-BYTE AND TWO-BYTE READ WORD FORMAT AND CONTENT

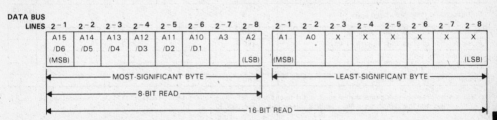

Shared Digital Port (A10/D1-A15/D6) — The voltage present on these pins is interpreted as a digital signal and the corresponding states are read from these bits. A digital value will be given for each pin even if some or all of these pins are being used as analog inputs.

Analog Multiplexer Address (A0-A3) — The address of the selected analog channel presently addressed is given by these bits.

Unused Bits (X) — LS byte bits 2^{-3} through 2^{-8} of the digital data register are not used.

FIGURE 3—DIGITAL DATA REGISTER ONE-BYTE AND TWO-BYTE READ WORD FORMAT AND CONTENT

Data Acquisition

7

TEXAS INSTRUMENTS
POST OFFICE BOX 225012 • DALLAS, TEXAS 75265

absolute maximum ratings over operating free-air temperature range (unless otherwise noted)

Supply voltage, V_{CC} (see Note 1) . . . . −0.3 V to 6.5 V
Input voltage range: Positive reference voltage . . . . $V_{REF}-$ to V_{CC} + 0.3 V
                Negative reference voltage . . . . −0.3 V to $V_{REF}+$
                All other inputs . . . . −0.3 V to V_{CC} + 0.3 V
Input current, I_I (any input) . . . . ± 10 mA
Total input current, (all inputs) . . . . ± 20 mA
Continuous total dissipation at (or below) 25 °C free-air temperature (see Note 2) . . . . 1250 mW
Storage temperature range . . . . −65 °C to 150 °C
Operating free-air temperature range . . . . −40 °C to 85 °C
Lead temperature 1,6 mm (1/16 inch) from case for 10 seconds . . . . 260 °C

NOTES: 1. All voltage values are with respect to network ground terminal.
         2. For operation above 25 °C free-air temperature, refer to Dissipation Derating Curves in Section 2.

recommended operating conditions

		TL532			TL533			UNIT
		MIN	NOM	MAX	MIN	NOM	MAX	
Supply voltage, V_{CC}		4.75	5	5.5	4.75	5	5.5	V
Positive reference voltage, $V_{REF}+$ (see Note 3)		4.6		V_{CC}	4.6		V_{CC}	V
Negative reference voltage, $V_{REF}-$ (see Note 3)		0		0.1	0		0.1	V
Average voltage across ladder		$V_{CC}-0.2$	V_{CC}	$V_{CC}+0.2$	$V_{CC}-0.2$	V_{CC}	$V_{CC}+0.2$	V
High-level input voltage, V_{IH}	Clock input	$V_{CC}-0.8$			$V_{CC}-0.8$			V
	All other digital inputs	2			2			
Low-level input voltage, V_{IL}	Any digital input			0.8			0.8	V
Clock frequency, f_{CLK}		0.1	1.048		0.1	1.048		MHz
Clock period, $1/f_{CLK}$		943			943			ns
$\overline{CS}$ setup time, $t_{su(CS)}$		100			100			ns
Address (R/$\overline{W}$ and RS) setup time, $t_{su(A)}$		145			145			ns
Data bus input setup time, $t_{su(bus)}$		185			185			ns
Control (R/$\overline{W}$, RS, and $\overline{CS}$) hold time, $t_{h(C)}$		20			20			ns
Data bus input hold time, $t_{h(bus)}$		20			25			ns
Pulse duration of control inputs during read cycle, $t_{w(C)}$		575			575			ns
Pulse duration, reset low, $t_{wL(reset)}$		3			3			Clock Periods
Pulse duration of clock high, $t_{wH(CLK)}$		440			440			ns
Pulse duration of clock low, $t_{wL(CLK)}$		410			410			ns
Clock rise time, $t_{r(CLK)}$				25			25	ns
Clock fall time, $t_{f(CLK)}$				30			30	ns
Operating free-air temperature, T_A		−40		85	−40		85	°C

NOTE 3: Analog input voltages greater than or equal to that applied to the REF+ terminal convert to all ones (11111111), while input voltages equal to or less than that applied to the REF− terminal convert to all zeros (00000000).

TEXAS INSTRUMENTS
POST OFFICE BOX 225012 • DALLAS, TEXAS 75265

electrical characteristics over recommended ranges of V_{CC}, V_{REF+}, and operating free-air temperature, V_{REF-} at ground, f_{CLK} = 1.048 MHz (unless otherwise noted)

PARAMETER		TEST CONDITIONS	MIN	TYP†	MAX	UNIT
V_{OH}	High-level output voltage I/O Data bus	I_{OH} = 1.6 mA	2.4			V
V_{OL}	Low-level output voltage I/O Data bus	I_{OL} = 1.6 mA			0.4	V
I_{IH}	High-level input current — Any control input	V_{IH} = 5.5 V			1	µA
	Any other digital input				10	
I_{IL}	Low-level input current — Any control input	V_{IL} = 0			−1	µA
	Any other digital input				−10	
I_{OZ}	Off-state (high-impedance-state) output current	V_O = V_{CC}			10	µA
		V_O = 0			−10	
I_I	Analog input current (see Note 4)	V_I = 0 to V_{CC}			±500	nA
	Leakage current between selected channel and all other analog channels	V_I = 0 to V_{CC} Clock input at 0 V			±400	nA
$r_{i(on)}$	Analog channel input on-state resistance				5	kΩ
	Ladder resistance (REF+ to REF−)		1	5	10	kΩ
C_i	Input capacitance — Digital pins 3 thru 10			7	30	pF
	Any other input pin			5	15	
$I_{CC}+I_{REF+}$	Supply current plus reference current	V_{CC} = V_{REF+} = 5.5 V, Outputs open		3	16	mA
I_{CC}	Supply current	V_{CC} = 5.5 V		2	10	mA

†Typical values are at V_{CC} = 5 V, T_A = 25°C.
NOTE 4:   Analog input current is an average of the current flowing into a selected analog channel input during one full conversion cycle.

operating characteristics over recommended ranges of V_{CC}, V_{REF+}, and operating free-air temperature, V_{REF-} at ground, f_{CLK} = 1.048 MHz (unless otherwise noted)

PARAMETER		TEST CONDITIONS	MIN	TYP†	MAX	UNIT
	Linearity error	See Note 5			±0.5	LSB
	Zero error	See Note 6			±0.5	LSB
	Full-scale error	See Note 6			±0.5	LSB
	Total unadjusted error	See Note 7			±0.5	LSB
	Absolute accuracy error	See Note 8			±1	LSB
t_{conv}	Conversion time (including channel acquisition time)				290	Clock Cycles
t_{acq}	Channel acquisition time				30	Clock Cycles
t_{en}	Data output enable time (See Note 9)	R_L = 3 kΩ, C_L = 50 pF			335	ns
t_{dis}	Data output disable time	R_L = 3 kΩ, C_L = 50 pF	10			ns
$t_{r(bus)}$	Data bus output rise time — High-impedance to high-level	R_L = 3 kΩ, C_L = 50 pF			150	ns
	Low to high-level				300	
$t_{f(bus)}$	Data bus output fall time — High-impedance to low-level	R_L = 3 kΩ, C_L = 50 pF			150	ns
	High to low-level				300	

†Typical values are at V_{CC} = 5 V, T_A = 25°C.
NOTES: 5. Linearity error is the deviation from the best straight line through the A/D transfer characteristics.
   6. Zero error is the difference between the output of an ideal and an actual A/D for zero input voltage; full-scale error is that same difference for full-scale input voltage.
   7. Total unadjusted error is the sum of linearity, zero, and full-scale errors.
   8. Absolute accuracy error is the maximum difference between an analog value and the nominal midstep value within any step. This includes all errors including inherent quantization error, which is the ±0.5 LSB uncertainty caused by the A/D converters finite resolution.
   9. If chip-select setup time, $t_{su(CS)}$, is less than 0.14 microseconds, the effective data output enable time, t_{en}, may extend such that $t_{su(CS)}$ + t_{en} is equal to a maximum of 0.475 microseconds.

Data Acquisition

7

TEXAS
INSTRUMENTS
POST OFFICE BOX 225012 ● DALLAS, TEXAS 75265

electrical characteristics over recommended ranges of V_{CC}, V_{REF+}, and operating free-air temperature, V_{REF-} at ground, $f_{CLK} = 1.048$ MHz (unless otherwise noted)

	PARAMETER		TEST CONDITIONS	MIN	TYP[†]	MAX	UNIT
V_{OH}	High-level output voltage	I/O Data bus	$I_{OH} = -1.6$ mA	2.5			V
V_{OL}	Low-level output voltage	I/O Data bus	$I_{OL} = 1.6$ mA			0.4	V
I_{IH}	High-level input current	Any control input	$V_{IH} = 5.5$ V			1	μA
		Any other digital input				10	
I_{IL}	Low-level input current	Any control input	$V_{IL} = 0$			-1	μA
		Any other digital input				-10	
I_{OZ}	Off-state (high-impedance-state) output current		$V_O = V_{CC}$			10	μA
			$V_O = 0$			-10	
I_I	Analog input current (see Note 4)		$V_I = 0$ to V_{CC}			±700	nA
	Leakage current between selected channel and all other analog channels		$V_I = 0$ to V_{CC}, Clock input at 0 V			±500	nA
$r_{i(on)}$	Analog channel input on-state resistance					5	kΩ
	Ladder resistance (REF+ to REF−)			1	5	10	kΩ
C_i	Input capacitance	Digital pins 3 thru 10			7	30	pF
		Any other input pin			5	15	
$I_{CC}+I_{REF+}$	Supply current plus reference current		$V_{CC} = V_{REF+} = 5.5$ V, Outputs open		3	16	mA
I_{CC}	Supply current		$V_{CC} = 5.5$ V		2	10	mA

[†]Typical values are at $V_{CC} = 5$ V, $T_A = 25$°C.
NOTE 4: Analog input current is an average of the current flowing into a selected analog channel input during one full conversion cycle.

operating characteristics over recommended ranges of V_{CC}, V_{REF+}, and operating free-air temperature, V_{REF-} at ground, $f_{CLK} = 1.048$ MHz (unless otherwise noted)

	PARAMETER		TEST CONDITIONS	MIN	TYP[†]	MAX	UNIT
	Linearity error		See Note 5		±0.5		LSB
	Zero error		See Note 6		±0.25		LSB
	Full-scale error		See Note 6		±0.25		LSB
	Total unadjusted error		See Note 7	$T_A = 25$°C	±0.5		LSB
				$T_A = -0$°C to 85°C	±1		
	Absolute accuracy error		See Note 8			±1.5	LSB
t_{conv}	Conversion time (including channel acquisition time)					290	Clock Cycles
t_{acq}	Channel acquisition time					30	Clock Cycles
t_{en}	Data output enable time (See Note 9)		$R_L = 3$ kΩ, $C_L = 50$ pF			335	ns
t_{dis}	Data output disable time		$R_L = 3$ kΩ, $C_L = 50$ pF	10			ns
$t_{r(bus)}$	Data bus output rise time	High-impedance to high-level	$R_L = 3$ kΩ, $C_L = 50$ pF			150	ns
		Low to high-level				300	
$t_{f(bus)}$	Data bus output fall time	High-impedance to low-level	$R_L = 3$ kΩ, $C_L = 50$ pF			150	ns
		High to low-level				300	

[†]Typical values are at $V_{CC} = 5$ V, $T_A = 25$°C.
NOTES: 5. Linearity error is the deviation from the best straight line through the A/D transfer characteristics.
　　　　6. Zero error is the difference between the output of an ideal and an actual A/D for zero input voltage; full-scale error is that same difference for full-scale input voltage.
　　　　7. Total unadjusted error is the sum of linearity, zero, and full-scale errors.
　　　　8. Absolute accuracy error is the maximum difference between an analog value and the nominal midstep value within any step. This includes all errors including inherent quantization error, which is the ±0.5 LSB uncertainty caused by the A/D converters finite resolution.
　　　　9. If chip-select setup time, $t_{su(CS)}$, is less than 0.14 microseconds, the effective data output enable time, t_{en}, may extend such that $t_{su(CS)} + t_{en}$ is equal to a maximum of 0.475 microseconds.

TEXAS INSTRUMENTS
POST OFFICE BOX 225012 • DALLAS, TEXAS 75265

DATA
ACQUISITION
CIRCUITS

TYPES TL601, TL604, TL607, TL610
P-MOS ANALOG SWITCHES

D2401, JUNE 1976—REVISED OCTOBER 1983

NOT RECOMMENDED FOR NEW DESIGN

For New Design, see TL182 Series

description

The TL601, TL604, TL607, and TL610 are a family of monolithic P-MOS analog switches that provide fast switching speeds with high r_{off}/r_{on} ratio and no offset voltage. The p-channel enhancement-type MOS switches will accept analog signals up to ±10 volts and are controlled by TTL-compatible logic inputs. The monolithic structure is made possible by BI-MOS technology, which combines p-channel MOS with standard bipolar transistors.

These switches are particularly suited for use in military, industrial, and commercial applications such as data acquisition, multiplexers, A/D and D/A converters, MODEMS, sample-and-hold systems, signal multiplexing, integrators, programmable operational amplifiers, programmable voltage regulators, crosspoint switching networks, logic interface, and many other analog systems.

The TL601 is an SPDT switch with two logic control inputs. The TL604 is a dual complementary SPST switch with a single control input. The TL607 is an SPDT switch with one logic control input and one enable input. The TL610 is an SPST switch with three logic control inputs. The TL610 features a higher r_{off}/r_{on} ratio than the other members of the family.

The TL601M, TL604M, TL607M, and TL610M are characterized for operation over the full military temperature range of −55°C to 125°C, the TL601I, TL604I, TL607I, and TL610I are characterized for operation from −25°C to 85°C, and the TL601C, TL604C, TL607C, and TL610C are characterized for operation from 0°C to 70°C.

JG OR P DUAL-IN-LINE PACKAGE
(TOP VIEW)

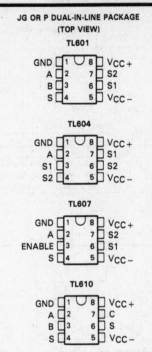

TYPICAL OF
ALL INPUTS

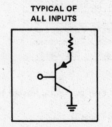

TYPICAL OF
ALL SWITCHES

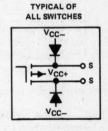

Data Acquisition

7

TEXAS
INSTRUMENTS
POST OFFICE BOX 225012 • DALLAS, TEXAS 75265

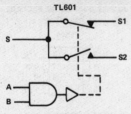

TL601

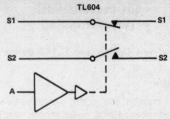

TL604

FUNCTION TABLE

LOGIC INPUTS		ANALOG SWITCH	
A	B	S1	S2
L	X	OFF (OPEN)	ON (CLOSED)
X	L	OFF (OPEN)	ON (CLOSED)
H	H	ON (CLOSED)	OFF (OPEN)

FUNCTION TABLE

LOGIC INPUT	ANALOG SWITCH	
A	S1	S2
H	ON (CLOSED)	OFF (OPEN)
L	OFF (OPEN)	ON (CLOSED)

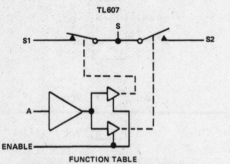

TL607

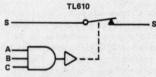

TL610

FUNCTION TABLE

INPUTS		ANALOG SWITCH
A	ENABLE	S1	S2
X	L	OFF (OPEN)	OFF (OPEN)
L	H	OFF (OPEN)	ON (CLOSED)
H	H	ON (CLOSED)	OFF (OPEN)

FUNCTION TABLE

INPUTS			ANALOG SWITCH
A	B	C	S
L	X	X	OFF (OPEN)
X	X	X	OFF (OPEN)
X	X	L	OFF (OPEN)
H	H	H	ON (CLOSED)

H = high logic level
L = low logic level
X = irrelevant

Switch positions shown are for all inputs high.

absolute maximum ratings over operating free-air temperature range (unless otherwise noted)

Supply voltage, V_{CC+} (see Note 1) . . . . 30 V
Supply voltage, V_{CC-} . . . . −30 V
V_{CC+} to V_{CC-} supply voltage differential . . . . 35 V
Control input voltage . . . . V_{CC+}
Switch off-state voltage . . . . 30 V
Switch on-state current . . . . 10 mA
Operating free-air temperature range: TL601M, TL604M, TL607M, TL610M . . . . . . −55°C to 125°C
                       TL601I, TL604I, TL607I, TL610I . . . . −25°C to 85°C
                       TL601C, TL604C, TL607C, TL610C . . . . . . 0°C to 70°C
Storage temperature range . . . . −65°C to 150°C
Lead temperature 1/16 inch (1,6 mm) from case for 60 seconds: JG package . . . . 300°C
Lead temperature 1/16 inch (1,6 mm) from case for 10 seconds: P package . . . . 260°C

NOTE 1: All voltage values are with respect to network ground terminal.

Data Acquisition

7

Texas
Instruments
POST OFFICE BOX 225012 ● DALLAS, TEXAS 75265

recommended operating conditions

	TL601M, TL604M TL607M, TL610M			TL601I, TL604I TL607I, TL610I			TL601C, TL604C TL607C, TL610C			UNIT
	MIN	NOM	MAX	MIN	NOM	MAX	MIN	NOM	MAX	
Supply voltage, V_{CC+} (see Figure 1)	5	10	25	5	10	25	5	10	25	V
Supply voltage, V_{CC-} (see Figure 1)	−5	−20	−25	−5	−20	−25	−5	−20	−25	V
V_{CC+} to V_{CC-} supply voltage differential (see Figure 1)	15		30	15		30	15		30	V
Control input voltage	0		5.5	0		5.5	0		5.5	V
Voltage at any analog switch (S) terminal	$V_{CC-}+8$		V_{CC+}	$V_{CC-}+8$		V_{CC+}	$V_{CC-}+8$		V_{CC+}	V
Switch on-state current			10			10			10	mA
Operating free-air temperature, T_A	−55		125	−25		85	0		70	°C

Figure 1 shows power supply boundary conditions for proper operation of the TL601 Series. The range of operation for supply V_{CC+} from +5 V to +25 V is shown on the vertical axis. The range of V_{CC-} from −5 volts to −25 volts is shown on the horizontal axis. A recommended 30-volt maximum voltage differential from V_{CC+} to V_{CC-} governs the maximum V_{CC+} for a chosen V_{CC-} (or vice versa). A minimum recommended difference of 15 volts from V_{CC+} to V_{CC-} and the boundaries shown in Figure 1 allow the designer to select the proper combinations of the two supplies.

The designer-selected V_{CC+} for a chosen V_{CC-} supply values limit the maximum input voltage that can be applied to either switch terminal; that is, the input voltage should be between V_{CC-} +8 V and V_{CC+} to keep the on-state resistance within specified limits.

RECOMMENDED COMBINATIONS
OF SUPPLY VOLTAGES

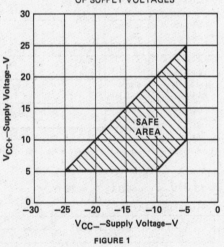

FIGURE 1

Data Acquisition

7

TEXAS
INSTRUMENTS
POST OFFICE BOX 225012 • DALLAS, TEXAS 75265

TYPES TL601, TL604, TL607, TL610
P-MOS ANALOG SWITCHES

electrical characteristics over recommended operating free-air temperature range,
V_{CC+} = 10 V, V_{CC-} = −20 V, analog switch test current = 1 mA (unless otherwise noted)

PARAMETER		TEST CONDITIONS†		TL6__M / TL6__I			TL6__C			UNIT
				MIN	TYP‡	MAX	MIN	TYP‡	MAX	
V_{IH}	High-level input voltage			2			2			V
V_{IL}	Low-level input voltage	Enable input of TL607M				0.6				V
		All other inputs				0.8			0.8	
I_{IH}	High-level input current	V_I = 5.5 V			0.5	10		0.5	10	µA
I_{IL}	Low-level input current	V_I = 0.4 V			−50	−250		−50	−250	µA
I_{off}	Switch off-state current	$V_{I(sw)}$ = −10 V, See Note 2	T_A = 25°C			−400			−500	pA
			T_A = MAX		−50	−100		−10	−20	nA
r_{on}	Switch on-state resistance	$V_{I(sw)}$ = 10 V, $I_{O(sw)}$ = −1 mA	TL601 TL604 TL607		55	100		75	200	Ω
			TL610		40	80		40	100	
		$V_{I(sw)}$ = −10 V, $I_{O(sw)}$ = −1 mA	TL601 TL604 TL607		220	400		220	600	
			TL610		120	300		120	400	
r_{off}	Switch off-state resistance				1×10^{11}			5×10^{10}		Ω
C_{on}	Switch on-state input capacitance	$V_{I(sw)}$ = 0 V, f = 1 MHz			16			16		pF
C_{off}	Switch off-state input capacitance	$V_{I(sw)}$ = 0 V, f = 1 MHz			8			8		pF
I_{CC+}	Supply current from V_{CC+}	Logic input(s) at 5.5 V, All switch terminals open	TL601 TL604		5	10		5	10	mA
			Enable input high — TL607		5	10		5	10	
			Enable input low — TL607		3	5		3	5	
			TL610		5	10		5	10	
I_{CC-}	Supply current from V_{CC-}	Logic input(s) at 5.5 V, All switch terminals open	TL601 TL604		−1.2	−2.5		−1.2	−2.5	mA
			Enable input high — TL607		−2.5	−5		−2.5	−5	
			Enable input low — TL607		−0.05	−0.5		−0.05	−0.5	
			TL610		−1.2	−2.5		−1.2	−2.5	

†For conditions shown as MIN or MAX, use the appropriate value specified under recommended operating conditions.
‡All typical values are at T_A = 25°C.
NOTE 2: The other terminal of the switch under test is at V_{CC+} = 10 V.

switching characteristics, V_{CC} = 10 V, V_{CC-} = −20 V, T_A = 25°C

PARAMETER		TEST CONDITIONS	MIN	TYP	MAX	UNIT
t_{off}	Switch turn-off time	R_L = 1 kΩ, C_L = 35 pF, See Figure 2		400	500	ns
t_{on}	Switch turn-on time			100	150	

TEXAS
INSTRUMENTS
POST OFFICE BOX 225012 • DALLAS, TEXAS 75265

108

PARAMETER MEASUREMENT INFORMATION

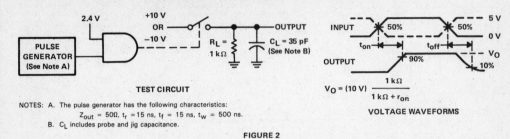

TEST CIRCUIT

$$V_O = (10\ V) \frac{1\ k\Omega}{1\ k\Omega + r_{on}}$$

VOLTAGE WAVEFORMS

NOTES: A. The pulse generator has the following characteristics:
$Z_{out} = 50\Omega$, $t_r = 15$ ns, $t_f = 15$ ns, $t_w = 500$ ns.
B. C_L includes probe and jig capacitance.

FIGURE 2

TYPICAL CHARACTERISTICS

SWITCH ON-STATE RESISTANCE
vs
FREE-AIR TEMPERATURE

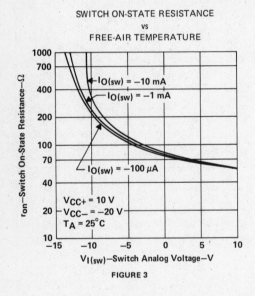

FIGURE 3

SWITCH ON-STATE RESISTANCE
vs
SWITCH ANALOG VOLTAGE

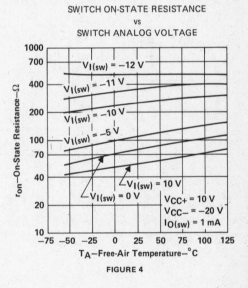

FIGURE 4

Data Acquisition

7

TEXAS
INSTRUMENTS
POST OFFICE BOX 225012 • DALLAS, TEXAS 75265

DATA ACQUISITION CIRCUITS

- LinCMOS™ Technology
- 8-Bit Resolution
- Total Unadjusted Error . . . ±0.5 LSB Max
- Ratiometric Conversion
- Access Plus Conversion Time:
  TLC532A . . . 15 μs Max
  TLC533A . . . 30 μs Max
- 3-State, Bidirectional I/O Data Bus
- 5 Analog and 6 Multipurpose Inputs
- On-Chip 12-Channel Analog Multiplexer
- Three On-Chip 16-Bit Data Registers
- Software Compatible with Larger TL530 and TL531 (21-Input Versions)
- On-Chip Sample-and-Hold Circuit
- Single 5-V Supply Operation
- Low Power Consumption . . . 6.5 mW Typ
- Improved Direct Replacements for Texas Instruments TL532 and TL533, National Semiconductor ADC0829, and Motorola MC14442

description

The TLC532A and TLC533A are monolithic LinCMOS™ peripheral integrated circuits each designed to interface a microprocessor for analog data acquisition. These devices are complete peripheral data acquisition systems on a single chip and can convert analog signals to digital data from up to 11 external analog terminals. Each device features operation from a single 5-volt supply. Each contains a 12-channel analog multiplexer, an 8-bit ratiometric analog-to-digital (A/D) converter, a sample-and-hold, three 16-bit registers, and microprocessor-compatible control logic circuitry. Additional features include a built-in self-test, six multipurpose (analog or digital) inputs, five external analog inputs, and an 8-pin input/output (I/O) data port. The three on-chip data registers store the control data, the conversion results, and the input digital data that can be accesssed via the microprocessor data bus in two 8-bit bytes (most-significant byte first). In this manner, a microprocessor can access up to 11 external analog inputs or 6 digital signals and the positive reference voltage that may be used for self-test.

J OR N DUAL-IN-LINE PACKAGE
(TOP VIEW)

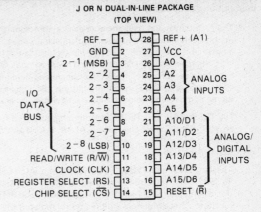

FH CHIP-CARRIER PACKAGE
(TOP VIEW)

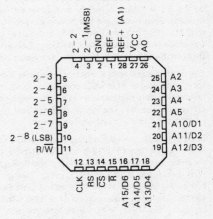

FUNCTION TABLE

ADDRESS/CONTROL					DESCRIPTION
R/W̄	RS	C̄S̄	R̄	CLK	
X	X	X	L†		Reset
L	H	L	H	↓	Write bus data to control register
H	L	L	H	↑	Read data from analog conversion register
H	H	L	H	↑	Read data from ditigal data register
X	X	H	H	X	No response

H = High-level, L = Low-level, X = Irrelevant,
↓ = High-to-low transition, ↑ = Low-to-high transition
†For proper operation, Reset must be low for at least three clock cycles.

Data Acquisition

7

POST OFFICE BOX 225012 • DALLAS, TEXAS 75265

description (continued)

The A/D conversion uses the successive-approximation technique and switched-capacitor circuitry. This method eliminates the possibility of missing codes, nonmonotonicity, and a need for zero or full-scale adjustment. Any one of 11 analog inputs (or self-test) can be converted to an 8-bit digital word and stored in 10 microseconds (TLC532A) or 20 microseconds (TLC533A) after instructions from the microprocessor have been recognized. The on-chip sample-and-hold functions automatically to minimize errors due to noise on the analog inputs. Furthermore, differential high-impedance reference inputs are available to help isolate the analog circuitry from the logic and supply noises while easing ratiometric conversion and scaling.

The TLC532AM and TLC533AM are characterized for operation over the full military temperature range of −55°C to 125°C. The TLC532AI and TLC533AI are characterized for operation from −40°C to 85°C.

functional description

The TLC532A and TLC533A provide direct interface to a microprocessor-based system. Control of the TLC532A and TLC533A is handled via the 8-line TTL-compatible 3-state data bus, the three control inputs (Read/Write, Register Select, and Chip Select), and the Clock input. Each device contains three 16-bit internal registers. These registers are the control register, the analog conversion data register, and the digital data register.

A high level at the Read/Write input and a low level at the Chip Select input set the device to output data on the 8-line data bus for the processor to read. A low level at the Read/Write input and a low level at the Chip Select input set the device to receive instructions into the internal control register on the 8-line data bus from the processor. When the device is in the read mode and the Register Select input is low, the processor will read the data contained in the analog conversion data register. However, when the Register Select input is high, the processor reads the data contained in the digital data register.

The control register is a write-only register into which the microprocessor writes command instructions for the device to start A/D conversion and to select the analog channel to be converted. The analog conversion data register is a read-only register that contains the current converter status and most recent conversion results. The digital data register is also a read-only register that holds the digital input logic levels from the six multipurpose inputs.

Internally each device contains a byte pointer that selects the appropriate byte during two cycles of the Clock input in a normal 16-bit microprocessor instruction. The internal pointer will automatically point to the most-significant (MS) byte after the first complete clock cycle any time that the Chip Select is at the high level for at least one clock cycle. This causes the device to treat the next signal on the 8-line data bus as the MS byte. A low level at the Chip Select input activates the inputs and outputs and an internal function decoder. However, no data is transferred until the Clock goes high. The internal byte pointer first points to the MS byte of the selected register during the first clock cycle. After the first clock cycle in which the MS byte is accessed, the internal pointer switches to the LS byte and remains there for as long as Chip Select is low. The MS byte of any register may be accessed by either an 8-bit or a 16-bit microprocessor instruction; however, the LS byte may only be accessed by a 16-bit microprocessor instruction.

Normally, a two-byte word is written into or read from the controlling processor, but a single byte can be read by the processor by proper manipulation of the Chip Select input. This can be used to read conversion status from the analog conversion data register or the digital multipurpose input levels from the digital data register. The format and content of each two-byte word is shown in Figures 1 through 3.

TEXAS
INSTRUMENTS

POST OFFICE BOX 225012 • DALLAS, TEXAS 75265

functional description (continued)

A conversion cycle is started after a two-byte instruction is written into the control register and the start conversion (SC) bit is a logic high. This two-byte instruction also selects the input analog channel to be converted. The status (EOC) bit in the analog conversion data register is reset and remains reset until the conversion is completed, at that time the status bit is then set again. After conversion, the results are loaded into the analog conversion data register. These results remain in the analog conversion data register until the next conversion cycle is completed. If a new conversion command is entered into the control register while the conversion cycle is in progress, the on-going conversion will be aborted and a new channel acquisition cycle will immediately begin.

The Reset input allows the device to be externally forced to a known state. When a low level is applied to the Reset input for a minimum of three clock periods, the start conversion bit of the control register is cleared. The A/D converter is then idled and all the outputs are placed in the high-impedance off-state. However, the content of the analog conversion data register is not affected by the Reset input going to a low level.

functional block diagram

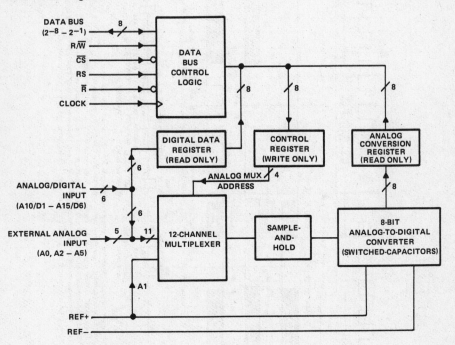

Data Acquisition

7

TEXAS
INSTRUMENTS
POST OFFICE BOX 225012 • DALLAS, TEXAS 75265

typical operating sequence

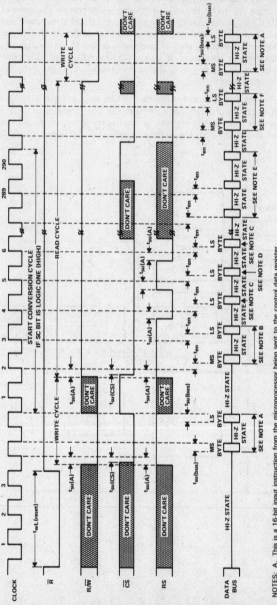

NOTES: A. This is a 16-bit input instruction from the microprocessor being sent to the control data register.
       B. This is the 2-byte (16-bit) content of the digital data register being sent to the microprocessor.
       C. This is the LS byte (8-bit) content of the analog conversion data register being sent to the microprocessor.
       D. This is the LS byte (8-bit) content of the digital data register being sent to the microprocessor.
       E. These are 8-bit or 16-bit output data from either the analog conversion data register or the digital data register being sent to the microprocessor.
       F. This is the 2-byte (16-bit) content of the analog conversion data register being sent to the microprocessor.

Data Acquisition

7

TEXAS
INSTRUMENTS
POST OFFICE BOX 225012 • DALLAS, TEXAS 75265

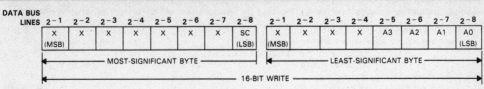

Unused Bits (X)— The MS byte bits 2^{-1} through 2^{-7} and LS byte bits 2^{-1} through 2^{-4} of the control register are not used internally.

Start Conversion (SC)— When the SC bit in the MS byte is set to a logical 1 (high level), analog-to-digital conversion of the specified analog channel will begin immediately after the completion of the control register write.

Analog Multiplex Address (A0-A3)— These four address bits are decoded by the analog multiplexer and used to select the appropriate analog channel as shown below:

Hexadecimal Address (A3 = MSB)	Channel Select
0	A0
1	REF + (A1)
2-5	A2-A5
6-9 (not used)	
A-F	A10-A15

FIGURE 1—CONTROL REGISTER TWO-BYTE WRITE WORD FORMAT AND CONTENT

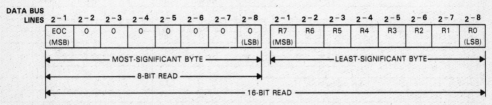

A/D Status (EOC)— The A/D status end-of-conversion (EOC) bit is set whenever an analog-to-digital conversion is successfully completed by the A/D converter. The status bit is cleared by a 16-bit write from the microprocessor to the control register. The remainder of the bits in the MS byte of the analog conversion data register are always reset to logical 0 to simplify microprocessor interrogation of the A/D converter status.

A/D Result (R0-R7)— The LS byte of the analog conversion data register contains the result of the analog-to-digital conversion. Result bit R7 is the MSB and the converter follows the standard convention of assigning a code of all ones (11111111) to a full-scale analog voltage. There are no special overflow or underflow indications.

FIGURE 2—ANALOG CONVERSION DATA REGISTER ONE-BYTE AND TWO-BYTE READ WORD FORMAT AND CONTENT

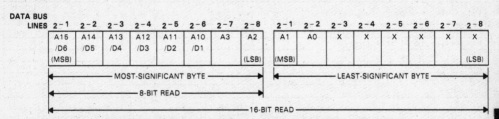

Shared Digital Port (A10/D1-A15/D6)— The voltage present on these pins is interpreted as a digital signal and the corresponding states are read from these bits. A digital value will be given for each pin even if some or all of these pins are being used as analog inputs.

Analog Multiplexer Address (A0-A3)— The address of the selected analog channel presently addressed is given by these bits.

Unused Bits (X)— LS byte bits 2^{-3} through 2^{-8} of the digital data register are not used.

FIGURE 3—DIGITAL DATA REGISTER ONE-BYTE AND TWO-BYTE READ WORD FORMAT AND CONTENT

Data Acquisition

7

TEXAS
INSTRUMENTS
POST OFFICE BOX 225012 • DALLAS, TEXAS 75265

TYPES TLC532AM, TLC532AI, TLC533AM, TLC533AI
LinCMOS™ 8-BIT ANALOG-TO-DIGITAL PERIPHERALS WITH
5 ANALOG AND 6 MULTIPURPOSE INPUTS

absolute maximum ratings over operating free-air temperature range (unless otherwise noted)

Supply voltage, V_{CC} (see Note 1) . . . . −0.3 V to 6.5 V
Input voltage range: Positive reference voltage . . . . V_{REF-} to V_{CC} + 0.3 V
                     Negative reference voltage . . . . −0.3 V to V_{REF+}
                     All other inputs . . . . −0.3 V to V_{CC} + 0.3 V
Input current, I_I (any input) . . . . ±10 mA
Total input current, (all inputs) . . . . ±20 mA
Continuous total dissipation at (or below) 25 °C free-air temperature (see Note 2) . . . . . . 1025 mW
Operating free-air temperature range: TLC532AM, TLC533AM . . . . −55 °C to 125 °C
                            TLC532AI, TLC533AI . . . . −40 °C to 85 °C
Storage temperature range . . . . −65 °C to 150 °C
Lead temperature 1,6 mm (1/16 inch) from case for 60 seconds: FH or J package . . . . 300 °C
Lead temperature 1,6 mm (1/16 inch) from case for 10 seconds: N package . . . . 260 °C

NOTES: 1. All voltage values are with respect to network ground terminal.
         2. For operation above 25 °C free-air temperature, refer to Dissipation Derating Curves, Section 2.

recommended operating conditions

		TLC532A			TLC533A			UNIT
		MIN	NOM	MAX	MIN	NOM	MAX	
Supply voltage, V_{CC}		4.75	5	5.5	4.75	5	5.5	V
Positive reference voltage, V_{REF+} (see Note 3)		2.5	V_{CC}	V_{CC}+0.1	2.5	V_{CC}	V_{CC}+0.1	V
Negative reference voltage, V_{REF-} (see Note 3)		−0.1	0	2.5	−0.1	0	2.5	V
Differential reference voltage, $V_{REF+} - V_{REF-}$		1	V_{CC}	V_{CC}+0.2	1	V_{CC}	V_{CC}+0.2	V
High-level input voltage, V_{IH}	Clock input	V_{CC}−0.8			V_{CC}−0.8			V
	All other digital inputs	2			2			
Low-level input voltage, V_{IL}	Any digital input			0.8			0.8	V
Clock frequency, f_{CLK}		0.1	2	2.048	0.1	1.048	1.06	MHz
$\overline{CS}$ setup time, $t_{su(CS)}$		75			100			ns
Address (R/$\overline{W}$ and RS) setup time, $t_{su(A)}$		100			145			ns
Data bus input setup time, $t_{su(bus)}$		140			185			ns
Control (R/$\overline{W}$, RS, and $\overline{CS}$) hold time, $t_{h(C)}$		10			20			ns
Data bus input hold time, $t_{h(bus)}$		15			20			ns
Pulse duration of control during read, $t_{w(C)}$		305			575			ns
Pulse duration, reset low, $t_{wL(reset)}$		3			3			Clock Cycles
Pulse duration of clock high, $t_{wH(CLK)}$		230			440			ns
Pulse duration of clock low, $t_{wL(CLK)}$		200			410			ns
Clock rise time, $t_{r(CLK)}$				15			25	ns
Clock fall time, $t_{f(CLK)}$				16			30	ns
Operating free-air temperature, T_A	TLC__AM	−55		125	−55		125	°C
	TLC__AI	−40		85	−40		85	

NOTE 3: Analog input voltages greater than or equal to that applied to the REF+ terminal convert to all ones (11111111), while input voltages equal to or less than that applied to the REF− terminal convert to all zeros (00000000). For proper operation, the positive reference voltage, V_{REF+}, must be at least 1-volt greater than the negative reference voltage, V_{REF-}. In addition, unadjusted errors may increase as the differential reference voltage, $V_{REF+} - V_{REF-}$, falls below 4.75 volts.

TEXAS
INSTRUMENTS
POST OFFICE BOX 225012 • DALLAS, TEXAS 75265

12

electrical characteristics over recommended operating free-air temperature range, $V_{REF+} = V_{CC}$, V_{REF-} at ground, $f_{CLK} = 2$ MHz (unless otherwise noted)

PARAMETER		TEST CONDITIONS	MIN	TYP†	MAX	UNIT
V_{OH}	High-level output voltage	$I_{OH} = -1.6$ mA	2.4			V
V_{OL}	low-level output voltage	$I_{OL} = 1.6$ mA			0.4	V
I_{IH}	High-level input current — Any digital or Clock input	$V_{IH} = 5.5$ V			10	µA
	High-level input current — Any control input				1	
I_{IL}	Low-level input current — Any digital or Clock input	$V_{IL} = 0$			-10	µA
	Low-level input current — Any control input				-1	
I_{OZ}	Off-state (high impedance-state) output current	$V_O = V_{CC}$			10	µA
		$V_O = 0$			-10	
I_I	Analog input current (see Note 4)	$V_I = 0$ to V_{CC}			±500	nA
	Leakage current between selected channel and all other analog channels	$V_I = 0$ to V_{CC}, Clock input at 0 V			±400	nA
C_i	Input capacitance — Digital pins 3 thru 10			4	30	pF
	Input capacitance — Any other input pin			2	15	
$I_{CC} + I_{REF+}$	Supply current plus reference current	$V_{CC} = V_{REF+} = 5.5$ V, Outputs open		1.5	3	mA
I_{CC}	Supply current	$V_{CC} = 5.5$ V		1.4	2	mA

NOTE 4: Analog input current is an average of the current flowing into a selected analog channel input during one full conversion cycle.

operating characteristics over recommended operating free-air temperature range, $V_{REF+} = V_{CC}$, V_{REF-} at ground, $f_{CLK} = 2$ MHz (unless otherwise noted)

PARAMETER		TEST CONDITIONS	MIN	TYP†	MAX	UNIT
	Linearity error	See Note 5			±0.5	LSB
	Zero error	See Note 6			±0.5	LSB
	Full-scale error	See Note 6			±0.5	LSB
	Total unadjusted error	See Note 7			±0.5	LSB
	Absolute accuracy error	See Note 8			±1	LSB
t_{conv}	Conversion time (including channel acquisition time)			30		Clock Cycles
t_{acq}	Channel acquisition time			10		Clock Cycles
t_{en}	Data output enable time (see Note 9)	$C_L = 50$ pF, $R_L = 3$ kΩ,			250	ns
t_{dis}	Data output disable time	$C_L = 50$ pF, $R_L = 3$ kΩ	10			ns
$t_{r(bus)}$	Data bus output rise time — High-impedance to high-level	$C_L = 50$ pF, $R_L = 3$ kΩ			150	ns
	Data bus output rise time — Low to high-level				300	
$t_{f(bus)}$	Data bus output fall time — High-impedance to low-level	$C_L = 50$ pF, $R_L = 3$ kΩ			150	ns
	Data bus output fall time — High to low-level				300	

† Typical values are at $V_{CC} = 5$ V, $T_A = 25$°C.
NOTES: 5. Linearity error is the deviation from the best straight line through the A/D transfer characteristics.
6. Zero error is the difference between the output of an ideal and an actual A/D for zero input voltage; full-scale error is that same difference for full-scale input voltage.
7. Total unadjusted error is the sum of linearity, zero, and full-scale errors.
8. Absolute accuracy error is the maximum difference between an analog value and the nominal midstep value within any step. This includes all errors including inherent quantization error, which is the ±0.5 LSB uncertainty caused by the A/D converters finite resolution.
9. If chip-select setup time, $t_{su(CS)}$, is less than 0.14 microseconds, the effective data output enable time, t_{en}, may extend such that $t_{su(CS)} + t_{en}$ is equal to a maximum of 0.475 microseconds.

Data Acquisition

7

TEXAS
INSTRUMENTS
POST OFFICE BOX 225012 • DALLAS, TEXAS 75265

electrical characteristics over recommended ranges of V_{CC}, V_{REF+}, and operating free-air temperature, V_{REF-} at ground, f_{CLK} = 1.048 MHz (unless otherwise noted)

PARAMETER		TEST CONDITIONS	MIN	TYP[†]	MAX	UNIT
V_{OH}	High-level output voltage	I_{OH} = − 1.6 mA	2.4			V
V_{OL}	Low-level output voltage	I_{OL} = 1.6 mA			0.4	V
I_{IH}	High-level Any digital or Clock input	V_{IH} = 5.5 V			10	µA
	input current Any control input				1	
I_{IL}	Low-level Any digital or Clock input	V_{IL} = 0			− 10	µA
	input current Any control input				− 1	
I_{OZ}	Off-state (high impedance-state)	$V_O = V_{CC}$			10	µA
	output current	V_O = 0			− 10	
I_I	Analog input current (see Note 4)	V_I = 0 to V_{CC}			±500	nA
	Leakage current between selected channel and all other analog channels	V_I = 0 to V_{CC}, Clock input at 0 V			±400	nA
C_i	Input capacitance Digital pins 3 thru 10			4	30	pF
	Any other input pin			2	15	
$I_{CC} + I_{REF+}$	Supply current plus reference current	$V_{CC} = V_{REF+}$ = 5.5 V, Outputs open		1.3	3	mA
I_{CC}	Supply current	V_{CC} = 5.5 V		1.2	2	mA

NOTE 4: Analog input current is an average of the current flowing into a selected analog channel input during one full conversion cycle.

operating characteristics over recommended ranges V_{CC}, V_{REF+}, and operating free-air temperature, V_{REF-} at ground, f_{clock} = 1.048 MHz for TLC532A and f_{clock} = 1.048 MHz for TLC533A (unless otherwise noted)

PARAMETER		TEST CONDITIONS	MIN	TYP[†]	MAX	UNIT
Linearity error		See Note 5			±0.5	LSB
Zero error		See Note 6			±0.5	LSB
Full-scale error		See Note 6			±0.5	LSB
Total unadjusted error		See Note 7			±0.5	LSB
Absolute accuracy error		See Note 8			±1	LSB
t_{conv}	Conversion time (including channel acquisition time)			30		Clock Cycles
t_{acq}	Channel acquisition time			10		Clock Cycles
t_{en}	Data output enable time (see Note 9)	C_L = 50 pF, R_L = 3 kΩ,			335	ns
t_{dis}	Data output disable time	C_L = 50 pF, R_L = 3 kΩ	10			ns
$t_{r(bus)}$	Data bus output High-impedance to high-level	C_L = 50 pF, R_L = 3 kΩ			150	ns
	rise time Low to high-level				300	
$t_{f(bus)}$	Data bus output High-impedance to low-level	C_L = 50 pF, R_L = 3 kΩ			150	ns
	fall time High to low-level				300	

[†]Typical values are at V_{CC} = 5 V, T_A = 25°C.
NOTES: 5. Linearity error is the deviation from the best straight line through the A/D transfer characteristics.
     6. Zero error is the difference between the output of an ideal and an actual A/D for zero input voltage; full-scale error is that same difference for full-scale input voltage.
     7. Total unadjusted error is the sum of linearity, zero, and full-scale errors.
     8. Absolute accuracy error is the maximum difference between an analog value and the nominal midstep value within any step. This includes all errors including inherent quantization error, which is the ±0.5 LSB uncertainty caused by the A/D converters finite resolution.
     9. If chip-select setup time, $t_{su(CS)}$, is less than 0.14 microseconds, the effective data output enable time, t_{en}, may extend such that $t_{su(CS)} + t_{en}$ is equal to a maximum of 0.475 microseconds.

Data Acquisition

7

**TEXAS
INSTRUMENTS**

POST OFFICE BOX 225012 • DALLAS, TEXAS 75265

- LinCMOS™ Technology
- 8-Bit Resolution A/D Converter
- On-Chip 12-Channel Analog Multiplexer
- Built-In Self-Test Mode
- Software-Controllable Sample and Hold
- Total Unadjusted Error . . . ±0.5 LSB Max
- Direct Replacement for Motorola MC145040

TYPICAL PERFORMANCE:	TLC540	TLC541
Channel Acquisition Time	2 µs	7 µs
Conversion Time	10 µs	19 µs
Sampling Rate	71×10^3	29×10^3
Power Dissipation	6 mW	6 mW

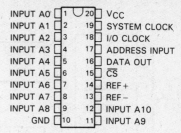

**J OR N DUAL-IN-LINE PACKAGE
(TOP VIEW)**

INPUT A0	1	20 VCC
INPUT A1	2	19 SYSTEM CLOCK
INPUT A2	3	18 I/O CLOCK
INPUT A3	4	17 ADDRESS INPUT
INPUT A4	5	16 DATA OUT
INPUT A5	6	15 $\overline{CS}$
INPUT A6	7	14 REF +
INPUT A7	8	13 REF −
INPUT A8	9	12 INPUT A10
GND	10	11 INPUT A9

**FK OR FN PACKAGE
(TOP VIEW)**

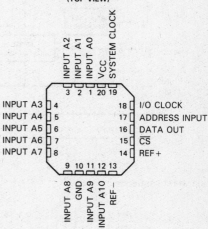

description

The TLC540 and TLC541 are LinCMOS™ A/D peripherals built around an 8-bit switched-capacitor successive-approximation A/D converter. They are designed for serial interface to a microprocessor or peripheral via a three-state output with up to four control inputs [including independent System Clock, I/O Clock, Chip Select ($\overline{CS}$), and Address Input]. A 4-megahertz system clock for the TLC540 and a 2.1-megahertz system clock for the TLC541 with a design that includes simultaneous read/write operation allow high-speed data transfers and sample rates of up to 71,910 samples per second for the TLC540 and 29,144 samples per second for the TLC541. In addition to the high-speed converter and versatile control logic, there is an on-chip 12-channel analog multiplexer that can be used to sample any one of 11 inputs or an internal "self-test" voltage, and a sample-and-hold that can operate automatically or under processor control.

The converters incorporated in the TLC540 and TLC541 feature differential high-impedance reference inputs that facilitate ratiometric conversion, scaling, and analog circuitry isolation from logic and supply noises. A totally switched-capacitor design allows guaranteed low-error (±0.5 LSB) conversion in 10 microseconds for the TLC540 and 19 microseconds for the TLC541 over the full operating temperature range.

The TLC540M and the TLC541M are characterized for operation over the full military temperature range of −55°C to 125°C. The TLC540I and the TLC541I are characterized for operation from −40°C to 85°C.

ADVANCE INFORMATION

This document contains information on a new product.
Specifications are subject to change without notice.

TEXAS INSTRUMENTS

POST OFFICE BOX 225012 • DALLAS, TEXAS 75265

functional block diagram

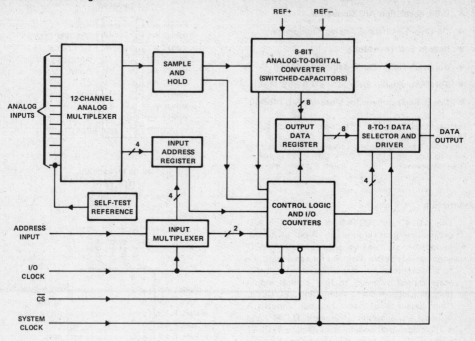

operating sequence

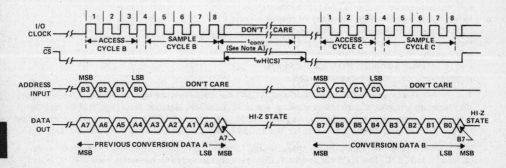

NOTE A: The conversion cycle, which requires 40 system clock periods, is initiated with the 8th I/O clock↓after CS ↓ for the channel whose address exists in memory at that time.

TEXAS
INSTRUMENTS
POST OFFICE BOX 225012 • DALLAS, TEXAS 75265

Data Acquisition

7

absolute maximum ratings over operating free-air temperature range (unless otherwise noted)

Supply voltage, V_{CC} (see Note 1) .... 6.5 V
Input voltage range (any input) .... −0.3 V to V_{CC} + 0.3 V
Output voltage range .... −0.3 V to V_{CC} + 0.3 V
Peak input current range (any input) .... ±10 mA
Peak total input current (all inputs) .... ±30 mA
Continuous total dissipation at (or below) 25 °C free-air temperature (see Note 2) .... 875 mW
Operating free-air temperature range: TLC540I, TLC541I .... −40 °C to 85 °C
                           TLC540M, TLC541M .... −55 °C to 125 °C
Storage temperature range .... −65 °C to 150 °C
Lead temperature 1,6 mm (1/16 inch) from case for 60 seconds: FK or J package .... 300 °C
Lead temperature 1,6 mm (1/16 inch) from case for 10 seconds: FN or N package........ 260 °C

NOTES: 1. All voltage values are with respect to digital ground with REF− and GND wired together (unless otherwise noted).
2. For operation above 25 °C free-air temperature, see Dissipation Derating Curves, Section 2. In the J package, TLC540M and TLC541M chips are alloy mounted, TLC540I and TLC541I chips are glass mounted.

recommended operating conditions

			TLC540			TLC541			UNIT
			MIN	NOM	MAX	MIN	NOM	MAX	
Supply voltage, V_{CC}			4.75	5	5.5	4.75	5	5.5	V
Positive reference voltage, V_{REF+} (see Note 3)			1.25	V_{CC}	$V_{CC}+0.1$	1.25	V_{CC}	$V_{CC}+0.1$	V
Negative reference voltage, V_{REF-} (see Note 3)			−0.1	0	$V_{CC}-1.25$	0.1	0	$V_{CC}-1.25$	V
Differential reference voltage, $V_{REF+}-V_{REF-}$ (see Note 3)			1	V_{CC}	$V_{CC}+0.2$	1	V_{CC}	$V_{CC}+0.2$	V
Analog input voltage (see Note 3)			0		V_{CC}	0		V_{CC}	V
High-level control input voltage, V_{IH}			2			2			V
Low-level control input voltage, V_{IL}					0.8			0.8	V
Setup time, address bits at data input, before I/O CLK↑, $t_{su(A)}$			200			400			ns
Setup time, $\overline{CS}$ low before clocking in first address bit, $t_{su(CS)}$ (see Note 4)			2			2			System clock cycles
Input/Output clock frequency, $f_{CLK(I/O)}$			0.005		2.048	0		0.525	MHz
System clock frequency, $f_{CLK(SYS)}$			$f_{CLK(I/O)}$			$f_{CLK(I/O)}$			MHz
System clock high, $t_{wH(SYS)}$			110			210			ns
System clock low, $t_{wL(SYS)}$			100			190			ns
Input/Output clock high, $t_{wH(I/O)}$			200			808			ns
Input/Output clock low, $t_{wL(I/O)}$			200			808			ns
Clock transition time (see Note 5)	System	$f_{CLK(SYS)} \leq 1048$ kHz			30			30	ns
		$f_{CLK(SYS)} > 1048$ kHz			20			20	
	I/O	$f_{CLK(I/O)} \leq 525$ kHz			100			100	ns
		$f_{CLK(I/O)} > 525$ kHz			40			40	
Operating free-air temperature, T_A	TLC540M, TLC541M		−55		125	−55		125	°C
	TLC540I, TLC541I		−40		85	−40		85	

NOTES: 3. Analog input voltages greater than that applied to REF+ convert as all "1"s (11111111), while input voltages less than that applied to REF− convert as all "0"s (00000000). For proper operation, REF+ voltage must be at least 1 volt higher than REF− voltage. Also, adjusted errors may increase as this differential reference voltage falls below 4.75 volts.
4. To minimize errors caused by noise at the Chip Select input, the internal circuitry waits for two system clock cycles (or less) after a chip select falling edge is detected before responding to control input signals. Therefore, no attempt should be made to clock-in address data until the chip select setup time has elapsed.
5. This is the time required for the clock input signal to fall from V_{IH} min to V_{IL} max or to rise from V_{IL} max to V_{IH} min.

Data Acquisition

7

TEXAS
INSTRUMENTS
POST OFFICE BOX 225012 • DALLAS, TEXAS 75265

electrical characteristics over recommended operating temperature range,
$V_{CC} = V_{REF+} = 4.75$ V to 5.5 V (unless otherwise noted), $f_{CLK(I/O)} = 2.028$ MHz for
TLC540 or $f_{CLK(I/O)} = 0.525$ MHz for TLC541

	PARAMETER	TEST CONDITIONS		MIN	TYP[†]	MAX	UNIT
V_{OH}	High-level output voltage (pin 16)	$V_{CC} = 4.75$ V,	$I_{OH} = 360\ \mu A$	2.4			V
V_{OL}	Low-level output voltage	$V_{CC} = 4.75$ V,	$I_O = 3.2$ mA			0.4	V
I_{OZ}	Off-state (high-impedance state)	$V_O = V_{CC}$,	$\overline{CS}$ at V_{CC}			10	μA
	output current	$V_O = 0$,	$\overline{CS}$ at V_{CC}			−10	
I_{IH}	High-level input current	$V_I = V_{CC} + 0.3$ V			0.005	2.5	μA
I_{IL}	Low-level input current	$V_I = 0$			−0.005	−2.5	μA
I_{CC}	Operating supply current	$\overline{CS}$ at 0 V			1.2	2	mA
	Selected channel leakage current	Selected channel at V_{CC}, Unselected channel at 0 V			0.4	1	μA
		Selected channel at 0 V, Unselected channel at V_{CC}			−0.4	−1	
$I_{CC} + I_{REF}$	Supply and reference current	$V_{REF+} = V_{CC}$, $\overline{CS}$ at 0 V			1.3	3	mA
C_i	Input capacitance	Analog inputs			7	55	pF
		Control inputs			5	15	

[†]All typical values are at $T_A = 25\,°C$.

operating characteristics over recommended operating free-air temperature range,
$V_{CC} = V_{REF+} = 4.75$ V to 5.5 V, $f_{CLK(I/O)} = 2.048$ MHz for TLC540 or
0.525 MHz for TLC541, $f_{CLK(SYS)} = 4$ MHz for TLC540 or 2.097 MHz for TLC541.

	PARAMETER	TEST CONDITIONS	TLC540 MIN	TYP	MAX	TLC541 MIN	TYP	MAX	UNIT
	Linearity error	See Note 6			±0.5			±0.5	LSB
	Zero error	See Note 7			±0.5			±0.5	LSB
	Full-scale error	See Note 7			±0.5			±0.5	LSB
	Total unadjusted error	See Note 8			±0.5			±0.5	LSB
	Self-test output code	Input address = 1011 (A11) (See Note 9)	01111101 (125)		10000011 (131)	01111101 (125)		10000011 (131)	
t_{conv}	Conversion time				10			19	μs
t_{acq}	Channel acquisition time				4			4	I/O clock cycles
t_v	Time output data remains valid after I/O clock↓		10			10			ns
t_d	Delay time, I/O clock↓ to data output valid				200			400	ns
t_{acc}	Output access time (delay to valid output after chip select↓)	See Parameter Measurement Information	1		3	1		3	System clock cycles
t_{en}	Output enable time				150			150	ns
t_{dis}	Output disable time				150			150	ns
$t_{r(bus)}$	Data bus rise time				300			300	ns
$t_{f(bus)}$	Data bus fall time				300			300	ns

NOTES: 6. Linearity error is the maximum deviation from the best straight line through the A/D transfer characteristics.
7. Zero Error is the difference between the output of an ideal and an actual A/D for zero input voltage; full-scale error is that same difference for full-scale input voltage.
8. Total Unadjusted Error is the sum of linearity, zero, and full-scale errors.
9. Both the input address and the output codes are expressed in positive logic.

Data Acquisition

7

PARAMETER MEASUREMENT INFORMATION

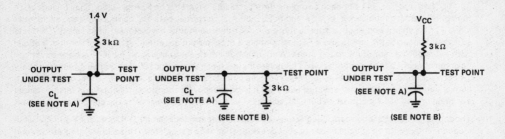

LOAD CIRCUIT FOR
t_d, t_{acc}, t_r, t_f

LOAD CIRCUIT FOR
t_{PZH} AND t_{PHZ}

LOAD CIRCUIT FOR
t_{PZL} AND t_{PLZ}

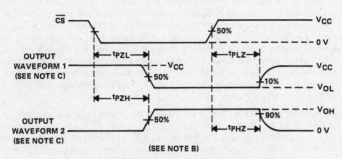

VOLTAGE WAVEFORMS FOR ENABLE AND DISABLE TIMES

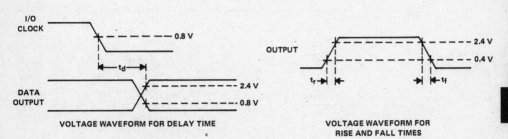

VOLTAGE WAVEFORM FOR DELAY TIME

VOLTAGE WAVEFORM FOR
RISE AND FALL TIMES

NOTES: A. C_L = 50 pF for TLC540 and 100 pF for TLC541

B. t_{en} = t_{PZH} or t_{PZL}, t_{dis} = t_{PHZ} or t_{PLZ}

C. Waveform 1 is for an output with internal conditions such that the output is low except when disabled by the output control. Waveform 2 is
for an output with internal conditions such that the output is high except when disabled by the output control.

Data Acquisition

7

principles of operation

The TLC540 and TLC541 are each complete data acquisition systems on a single chip. They include such functions as analog multiplexer, sample-and-hold, 8-bit A/D converter, data and control registers, and control logic. For flexibility and access speed, there are four control inputs [two clocks, chip select ($\overline{CS}$), and address]. These control inputs and a TTL-compatible 3-state output are intended for serial communications with a microprocessor or microcomputer. With judicious interface timing, with TLC540 a conversion can be completed in 10 microseconds, while complete input-conversion-output cycles are being repeated every 14 microseconds. With TLC541 a conversion can be completed in 19 microseconds, while complete input-conversion-output cycles are repeated every 35 microseconds. Furthermore, this fast conversion can be executed on any of 11 inputs or its built-in "self-test," and in any order desired by the controlling processor.

Though they can be operated "tied" together, the System Clock and I/O Clock are normally used independently, with no special phase or speed relationship to be considered. This allows integrated circuit operation to continue independent of serial Input/Output timing, permitting manipulation of the I/O Clock as desired for a wide range of software and hardware needs.

The I/O Clock, Data Input, and Data Output are controlled by $\overline{CS}$. It floats the 3-state output and shuts off signals to other control inputs while it is high. This allows any pins except pin 15 to share lines with other integrated circuits. A normal control sequence is as follows: (1) $\overline{CS}$ goes low; (2) a new positive-logic multiplexer address is clocked in through the address input on the first four I/O Clock rising edges while previous conversion results are brought out on the first seven I/O Clock falling edges. Input and output most-significant bits (MSB) are first, with the output MSB available at the start of the cycle; (3) the on-chip sample-and-hold begins sampling a newly addressed input after the 4th falling edge, and goes into the hold mode on the 8th falling I/O Clock edge just before conversion; (4) $\overline{CS}$ must then go high or the I/O Clock must remain low for at least 40 system clock cycles to allow conversion. A new address may then be loaded or the previous conversion results read any time $\overline{CS}$ is brought low, but it should be noted that any pending conversion may stop.

The instant that the TLC540 or TLC541 holds a sample of the analog input, conversion can be determined under software control (or by external logic), by keeping the 8th I/O Clock cycle high. Any output data will have already been shifted out, and TLC540 or TLC541 will continue sampling a new analog input. At the desired time, the I/O Clock signal can then be lowered freezing the voltage and turning off all analog inputs. In this manner, signals can be sampled at precise intervals for a wide range of comparison or processing applications, in much the same manner as a strobe light is used to determine engine speed.

TEXAS
INSTRUMENTS
POST OFFICE BOX 225012 ● DALLAS, TEXAS 75265

TYPE TLC549
LinCMOS™ 8-BIT ANALOG-TO-DIGITAL
PERIPHERAL WITH SERIAL CONTROL
D2816, NOVEMBER 1983

- LinCMOS™ Technology
- 8-Bit Resolution A/D Converter
- Differential Reference Input Voltages
- Conversion Time . . . 19 µs Max
- Total Access and Conversion
  Cycles . . . 29,144 cps
- On-Chip Software-Controllable
  Sample-and-Hold
- Total Unadjusted Error . . . ±0.5 LSB Max
- 4-MHz Internal System Clock
- Single 5-V Supply Operation
- Low Power Consumption . . . 6 mW Typ
- Dual-In-Line 8-pin Package

P DUAL-IN-LINE PACKAGE
(TOP VIEW)

```
        ┌───┬─┐
REF +  │1  ∪ 8│ VCC
ANALOG IN │2    7│ INPUT/OUTPUT CLOCK
REF −  │3    6│ DATA OUT
GND    │4    5│ CS
        └─────┘
```

description

The TLC549 is a LinCMOS™ A/D Peripheral integrated circuit built around an 8-bit switched-capacitor successive-approximation A/D converter. It is designed for serial interface with a microprocessor or peripheral through a 3-state data output and an analog input. The TLC549 uses only the Input/Output Clock (I/O Clock) input along with the Chip Select ($\overline{CS}$) input for data control. The I/O Clock input frequency of the TLC549 is guaranteed up to 525 kilohertz.

Operation of the TLC549 is very similar to that of the more complex TLC540 and TLC541 devices; however, unlike the TLC540 and TLC541, the TLC549 provides an on-chip system clock that operates typically at 4 megahertz and requires no external components. The on-chip system clock allows internal device operation to proceed independently of serial input/output data timing, permitting manipulation of the TLC549 as desired for a wide range of software and hardware requirements. The I/O Clock together with the internal system clock allow high-speed data transfer and sample rates of up to 29,144 cycles per second.

Additional TLC549 features include versatile control logic, an on-chip sample-and-hold circuit that can operate automatically or under processor control, and a high-speed converter with differential high-impedance reference voltage inputs that ease ratiometric conversion, scaling, and analog circuit isolation from logic and supply noises. Design of the totally switched-capacitor successive-approximation converter circuit allows guaranteed low-error conversion of ±0.5 least-significant bit (LSB) in less than 19 microseconds.

The TLC549M is characterized for operation over the full military temperature range of −55°C to 125°C. The TLC549I is characterized for operation from −40°C to 85°C.

Data Acquisition

7

Copyright © 1983 by Texas Instruments Incorporated

**TEXAS
INSTRUMENTS**
POST OFFICE BOX 225012 • DALLAS, TEXAS 75265

TYPE TLC549
LinCMOS™ 8-BIT ANALOG-TO-DIGITAL
PERIPHERAL WITH SERIAL CONTROL

functional block diagram

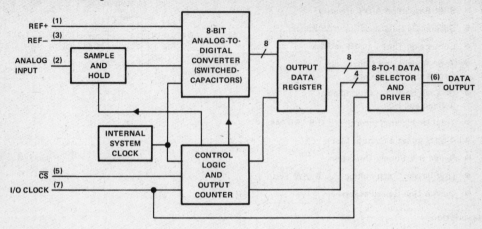

operating sequence

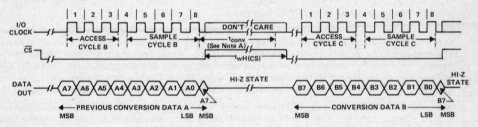

NOTE A: The conversion cycle, which requires 40 system clock periods, is initiated with the 8th I/O clock ↓ after $\overline{CS}$ ↓ for the channel whose address exists in memory at that time.

absolute maximum ratings over recommended operating free-air temperature range (unless otherwise noted)

Supply voltage, V_{CC} (see Note 1) . . . . 6.5 V
Input voltage range at any input . . . . −0.3 V to V_{CC}+ 0.3 V
Output voltage range . . . . −0.3 V to V_{CC}+ 0.3 V
Continuous total dissipation at (or below) 25 °C free-air temperature (see Note 2) . . . . 725 mW
Storage temperature range . . . . −65 °C to 150 °C
Operating free-air temperature range: TLC549M . . . . −55 °C to 125 °C
                                                TLC549I . . . . −40 °C to 85 °C
Lead temperature 1,6 mm (1/16 inch) from case for 60 seconds . . . . 260 °C

NOTES: 1. All voltage values are with respect to the network ground terminal with the REF − and GND terminal pins connected together, unless otherwise noted.
        2. For operation above 25 °C, refer to the Dissipation Derating Tables, Section 2.

Data Acquisition

7

TEXAS
INSTRUMENTS
POST OFFICE BOX 225012 • DALLAS, TEXAS 75265

recommended operating conditions

		MIN	NOM	MAX	UNIT
Supply voltage, V_{CC}		3	5	6	V
Positive reference voltage, V_{REF+} (see Note 3)		1.25	V_{CC}	$V_{CC}+0.1$	V
Negative reference voltage, V_{REF-} (see Note 3)		-0.1	0	$V_{CC}-1.25$	V
Differential reference voltage, $V_{REF+} - V_{REF-}$ (see Note 3)		1	V_{CC}	$V_{CC}+0.2$	V
Analog input voltage (see Note 3)		0		V_{CC}	V
High-level control input voltage, V_{IH}		2			V
Low-level control input voltage, V_{IL}				0.8	V
Peak input current, I_I				10	mA
Input/Output clock frequency, $f_{CLK(I/O)}$		0		525	kHz
Input/Output clock high, $t_{wH(I/O)}$		808			ns
Input/Output clock low, $t_{wL(I/O)}$		808			ns
Input/Output clock transition time, $t_{t(I/O)}$ (see Note 4)				100	ns
Duration of $\overline{CS}$ input high state during conversion, $t_{wH(CS)}$		19			μs
Operating free-air temperature, T_A	TLC549M	-55		125	°C
	TLC549I	-40		85	

NOTES: 3. Analog input voltages greater than that applied to REF+ convert as all ones (11111111), while input voltages less than that applied to REF− convert to all zeros (00000000). For proper operation, the positive reference voltage V_{REF+}, must be at least 1-volt greater than the negative reference voltage V_{REF-}. In additon, unadjusted errors may increase as the differential reference voltage $V_{REF+} - V_{REF-}$, falls below 4.75 volts.
          4. This is the time required for the input/output clock input signal to fall from V_{IH} min to V_{IL} max or to rise from V_{IL} max to V_{IH} min.

electrical characteristics over recommended operating free-air temperature range, $V_{CC} = V_{REF+} = 4.75$ V to 5.5 V, $f_{CLK(I/O)} = 525$ kHz (unless otherwise noted)

PARAMETER			TEST CONDITIONS		MIN	TYP[†]	MAX	UNIT
V_{OH}	High-level output voltage		$V_{CC} = 4.75$ V,	$I_{OH} = -360$ μA	2.4			V
V_{OL}	Low-level output voltage		$V_{CC} = 4.75$ V,	$I_{OL} = 3.2$ mA			0.4	V
I_{OZ}	Off-state (high-impedance		$V_O = V_{CC}$,	$\overline{CS}$ at V_{CC}			10	V
	state) output current		$V_O = 0$,	$\overline{CS}$ at V_{CC}			-10	
I_{IH}	High-level input current	Control inputs	$V_I = V_{CC}+0.3$ V			0.005	2.5	μA
I_{IL}	Low-level input current	Control inputs	$V_I = 0$			-0.005	-2.5	μA
$I_{I(on)}$	Analog channel on-state input current, during sample cycle		Analog input at V_{CC}			0.4	1	μA
			Analog input at 0 V			-0.4	-1	
I_{CC}	Operating supply current		$\overline{CS}$ at 0 V			1.2	2	mA
$I_{CC}+I_{REF}$	Supply and reference current		$V_{REF+} = V_{CC}$			1.3	3	mA
C_i	Input capacitance	Analog inputs				7	55	pF
		Control inputs				5	15	

[†] All typicals are at $T_A = 25$ °C.

Data Acquisition

7

operating characteristics over recommended operating free-air temperature range, $V_{CC} = V_{REF+} = 4.75$ V to 5.5 V, $f_{CLK(I/O)} = 525$ kHz (unless otherwise noted)

	PARAMETER	TEST CONDITIONS	MIN	TYP	MAX	UNIT
	Linearity error	See Note 5			±0.5	LSB
	Zero error	See Note 6			±0.5	LSB
	Full-scale error	See Note 6			±0.5	LSB
	Total unadjusted error	See Note 7			±0.5	LSB
t_{conv}	Conversion time				19	μs
t_{acq}	Channel acquisition time				4	I/O Clock Cycles
t_v	Time output data remains valid after I/O clock↓		10			ns
t_d	Delay time, internal system clock to data output valid				400	ns
t_{acc}	Output access time (Delay to valid output after $\overline{CS}$↓)				975	ns
t_{en}	Output enable time				150	ns
t_{dis}	Output disable time				150	ns
$t_{r(bus)}$	Data bus rise time				300	ns
$t_{f(bus)}$	Data bus fall time				300	ns

NOTES: 5. Linearity error is the deviation from the best straight line through the A/D transfer characteristics.
6. Zero error is the difference between the output of an ideal and an actual A/D converter for zero input voltage; full-scale error is that same difference for full-scale input voltage.
7. Total unadjusted error is the sum of linearity, zero, and full-scale errors.

TEXAS
INSTRUMENTS
POST OFFICE BOX 225012 • DALLAS, TEXAS 75265

- LinCMOS™ Technology
- Zero Reading With 0-V Input on All Scales
- Precision Null Detection With True Polarity at Zero
- 1 pA Typical Input Leakage Current
- True Differential Input and Reference
- Direct LCD Display Drive, No External Components Required
- Low Noise: Less Than 15 μV Peak-to-Peak
- On-Chip Clock and Reference
- No Additional Active Circuits Required
- Convenient 9-V Battery Operation With Low Power Dissipation, Less than 1 mW
- Pin Compatible With the Intersil ICL7106 and Teledyne TSC7106, Direct Replacement for ICL7126 and TSC7126A

N DUAL-IN-LINE PACKAGE
(TOP VIEW)

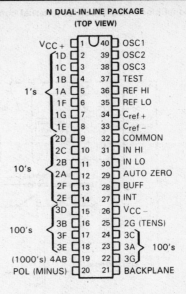

V_{CC}+	1	40 OSC1
1D	2	39 OSC2
1C	3	38 OSC3
1B	4	37 TEST
1A (1's)	5	36 REF HI
1F	6	35 REF LO
1G	7	34 C_{ref}+
1E	8	33 C_{ref}−
2D	9	32 COMMON
2C	10	31 IN HI
2B (10's)	11	30 IN LO
2A	12	29 AUTO ZERO
2F	13	28 BUFF
2E	14	27 INT
3D	15	26 V_{CC}−
3B	16	25 2G (TENS)
3F (100's)	17	24 3C
3E	18	23 3A (100's)
(1000's) 4AB	19	22 3G
POL (MINUS)	20	21 BACKPLANE

description

The TLC7126, operating with externally connected passive elements, is a versatile high-performance low-power 3 1/2-digit dual-slope-integrating analog-to-digital (A/D) converter. All the necessary active devices are contained on a single LinCMOS™ integrated circuit including seven-segment decoders, display drivers, backplane driver, reference, and clock. The device is designed to interface with a liquid-crystal display (LCD). The supply current is 100 microamperes maximum and is suited for 9-volt battery operation.

The TLC7126 is characterized by accuracy, versatility, and economy. Accuracy is obtained by an automatic compensation for zero offset to less than 10 microvolts, a zero-reading temperature coefficient less than 1 microvolt/°C, an input bias current of 10 picoamperes maximum, and a rollover error of less than one count. Applications include the conversion of analog data from high-impedance sensors of pressure, temperature, light, moisture, position, and many others. Analog-to-digital conversion logic provides display signals for weight scales, thermometers, light-level indicators, and many other applications. By changing the passive components, an ICL7106 socket can be upgraded to TLC7126.

The TLC7126 will be characterized for operation from 0°C to 70°C.

Data Acquisition

7

Copyright © 1983 by Texas Instruments Incorporated

TEXAS INSTRUMENTS
POST OFFICE BOX 225012 • DALLAS, TEXAS 75265

functional block diagram (with external components)

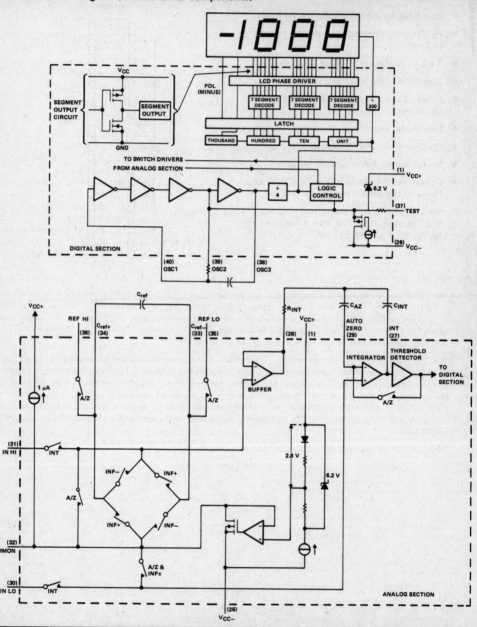

TEXAS
INSTRUMENTS

POST OFFICE BOX 225012 • DALLAS, TEXAS 75265

108

absolute maximum ratings over operating free-air temperature range (unless otherwise noted)

Supply voltage ($V_{CC}+$ with respect to $V_{CC}-$), V_{CC} . . . . 15 V
Voltage range for any input except clock (see Note 1) . . . . $V_{CC}-$ to $V_{CC}+$
Clock input voltage range . . . . TEST to $V_{CC}+$
Continuous total power dissipation . . . . 1000 mW
Operating free-air temperature range . . . . 0°C to 70°C
Storage temperature range . . . . −65°C to 150°C
Lead temperature 1,6 mm (1/16 inch) from case for 60 seconds . . . . 260°C

NOTE 1: Input voltages may exceed the supply voltages provided the input current is limited to ± 100 μA.

recommended operating conditions

			MIN	NOM	MAX	UNIT
V_{CC}	Supply voltage			9		V
V_{ref}	Reference input voltage	FS (full scale) = 200 mV		100		mV
		FS = 2 V		1		V
	Full-scale input voltage				2 V_{ref}	V
V_I	Input voltage, differential input		$V_{CC}-+1$		$V_{CC}+-0.5$	V
C_{ref}	Reference capacitor		0.1		1	μF
C_z	Auto-zero Capacitor		0.033		0.32	μF
C_x	Integrator Capacitor		0.047		0.15	μF
R_s	Integrator Resistor	FS = 200 mV		180		kΩ
		FS = 2 V		1.8		MΩ
T_A	Operating free-air temperature		0		70	°C

electrical characteristics, V_{CC} = 9 V, f_{clock} = 16 kHz, T_A = 25°C (unless otherwise noted)

PARAMETER	TEST CONDITIONS	MIN	TYP†	MAX	UNIT
Common-mode rejection ratio	$V_{IC} = \pm 1$ V, $V_{ID} = 0$, FS = 200 mV		50		μV/V
Peak-to-peak output noise voltage	$V_I = 0$, FS = 200 mV		15		μV
Input leakage current	$V_I = 0$		1	10	pA
Scale factor temperature coefficient	$V_I = 199$ mV, $T_A = 0$ to 70°, See Note 2		1	5	ppm/°C
Analog common voltage (with respect to $V_{CC}+$)	250 kΩ between COMMON and V_{CC}	2.4	2.8	3.2	V
Temperature coefficient of analog common voltage	250 kΩ between COMMON and V_{CC}		35	75	ppm/°C
Peak-to-peak segment drive voltage (see Note 3)		4	5	6	V
Peak-to-peak backplane drive voltage (see Note 3)		4	5	6	V
Supply current (see Note 4)	$V_I = 0$		50	100	μA
Power dissipation capacitance	See Note 5		40		pF

†This is the value not exceeded 95% of the time.
NOTES: 2. This is measured using a fixed external reference voltage.
3. Backplane drive is in phase with segment drive for a turned-off segment, 180° out of phase for a turned-on segment. Backplane frequency is 20 times the conversion rate. The average dc component is less than 50 mV.
4. This does not include current through the common terminal. During the auto-zero phase, current is 10 to 20 μA higher. Use of a 48-kHz oscillator increases current by typically 8 μA.
5. This can be used to determine the no-load dynamic power dissipation. $P_D = C_{pd} \cdot V_{CC}^2 \cdot f + I_{CC} \cdot V_{CC}$.

Data Acquisition

7

1083

TEXAS
INSTRUMENTS
POST OFFICE BOX 225012 • DALLAS, TEXAS 75265

TYPE TLC7126
LinCMOS™ SINGLE-CHIP 3 1/2-DIGIT LOW-POWER
A/D CONVERTER AND LCD DRIVER

operating characteristics over recommended operating free-air temperature range, $V_{CC} = 9$ V

PARAMETER	TEST CONDITIONS	MIN	TYP†	MAX	UNIT
Zero-input digital reading	$V_I = 0$, FS = 200 mV	−0.000	±0.000	+0.000	
Ratiometric digital reading	$V_I = V_{ref} = 100$ mV	999	999/1000	1000	
Rollover error (see Note 6)	$V_{I-} = V_{I+} \approx 200$ mV		±0.2	±1	Count
Linearity error	FS = 200 mV or 2 V		±0.2	±1	Count
Zero-reading temperature coefficient	$V_I = 0$, $T_A = 0°$ to 70°C		0.2	1	µV/°C

NOTE 6: Difference in reading for equal positive and negative reading near full scale.

PARAMETER MEASUREMENT INFORMATION

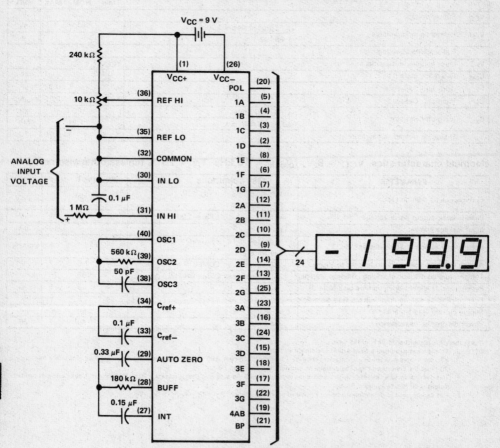

FIGURE 2—TEST CIRCUIT (CLOCK = 16 kHz, 1 READING PER SECOND)

Data Acquisition

7

TEXAS
INSTRUMENTS
POST OFFICE BOX 225012 • DALLAS, TEXAS 75265

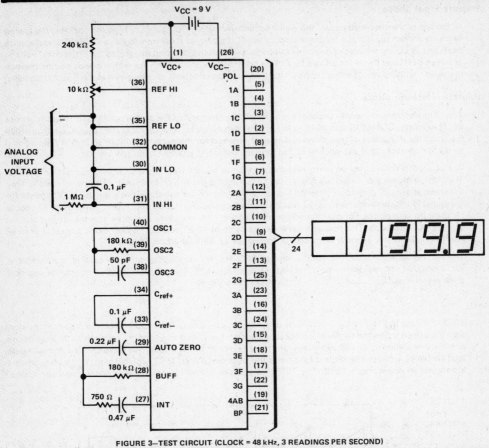

FIGURE 3—TEST CIRCUIT (CLOCK = 48 kHz, 3 READINGS PER SECOND)

PRINCIPLES OF OPERATION

The principles of operation of the TLC7126 are similar to the combined operation of the TL500C and TL502C. The TLC7126 has an analog section similar to TL500C and a digital section similar to TL502C (see functional block diagram). Each measurement cycle is divided into three phases. The phases are auto-zero, integrate input, and integrate reference.

auto-zero phase

The cycle begins at the end of the integrate-reference phase when the digital section control logic applies a low-level signal to the analog section. This action initiates three switching operations: IN HI and IN LO signals are disconnected from the input pins and internally shorted to analog COMMON; the reference capacitor C_{ref} charges to the reference voltage; and a feedback loop is connected from the comparator output to the integrator input. The feedback loop charges the auto-zero capacitor, C_Z, to compensate for offset voltages in the buffer amplifier, integrator, and comparator. Since the comparator is included in the loop, the auto-zero accuracy is limited only by the system noise.

Data Acquisition

7

integrate input phase

At the end of the auto-zero phase, the digital section control logic output goes high and operates the analog section switches to open the feedback loop, removes the internal short from input to COMMON, and applies IN HI and IN LO signals to the analog section input. The converter then integrates the differential voltage between IN HI and IN LO for a fixed time. This differential voltage can vary within a wide common-mode range (i.e., within a volt of either supply). The integrated signal polarity is determined at the end of this phase.

integrate reference phase

After a predetermined number of counts, the digital section operates the analog section switches to connect IN LO to analog COMMON and IN HI to the previously charged reference capacitor, C_{ref}. C_{ref} discharges returning the integrator output to zero. The time required for the output to return to zero is proportional to the differential input signal. The digital reading displayed will be 1000 V_I/V_{REF}.

Digital-section timing may be controlled by an external RC network, crystal, or oscillator (see Figure 4). The clock frequency is divided by four prior to clocking the decade counters and is further divided to time the three phases of operation (refer to Figure 1). The timing for the three phases uses 1000 counts for the signal integrate phase, 0 to 2000 counts for the integrate reference phase, and 1000 to 3000 counts for the auto-zero phase. For digital readings of less than full count, auto-zero time gets the unused portion of integrate reference. This makes a full measure cycle of 4000 counts (16,000 clock pulses). For three readout updates per second, an oscillator frequency of 48 kilohertz is required.

The backplane frequency is the clock divided by 800. For three readout updates per second, the backplane signal is a 60-hertz squarewave with a nominal amplitude of 5 volts. To turn on a segment, the segment drive signal must be out of phase with the backplane signal.

test

The test output has two possible functions. It is connected to an internal supply through a 500-ohm resistor, thus it can be used as the negative supply for externally generated segment drivers such as decimal points or any other presentation the user may want to include on the LCD display. No more than a 1-microampere load should be applied. The second function is the lamp test. When TEST is pulled high (V_{CC}), all segments are turned on displaying −1888. The TEST pin will sink about 10 milliamperes under these conditions.

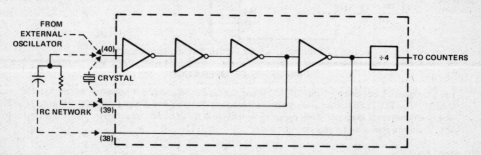

NOTE: This figure shows all three external control circuits connected; however, only one external circuit (crystal, RC network or external oscillator) is connected for proper operation.

FIGURE 4—CLOCK CIRCUITS

external component selection guide

The integrating resistor, R_X, should be large enough to maintain linear drive current from the buffer amplifier and integrator, but small enough to minimize leakage currents. For a 2-volt scale, 1.8 megohm is near optimum and for the 200-millivolt full scale, a 180-kilohm resistor is correct.

The integrating capacitor, C_X, should be selected to give the maximum voltage swing without saturation. For three readout updates per second (48-kilohertz clock), the nominal value for C_X is 0.047 microfarad; for one readout update per second (16-kilohertz clock), the nominal value is 0.15 microfarads. Capacitor values should be changed in inverse proportion to changes in oscillator frequency. The integrating capacitor should have low dielectric absorption to prevent roll-over errors. Polypropylene capacitors give undetectable errors at reasonable cost. At three readout updates per second, a 750-ohm resistor should be placed in series with C_X to compensate for comparator delay.

The size of the auto-zero capacitor, C_Z, affects the system noise. On the 200-millivolt scale, a 0.32-microfarad capacitor is recommended. On the 2-volt scale, a 0.033-microfarad capacitor decreases the recovery time from overload and is adequate for noise on this scale.

A reference capacitor, C_{ref}, of 0.1 microfarad is suitable for most applications. If a large common-mode voltage exists (reference-low pin is not at the same potential as analog COMMON) and a 200-millivolt scale is used, a larger C_{ref} is required to prevent roll-over error. A 1-microfarad capacitor will usually hold the roll-over error to 0.5 count in this case.

For all ranges of oscillator frequency, a 50-picofarad capacitor is recommended and the resistor is selected using the relationship $f \approx 0.45/RC$.

R = 180 kilohm for a 48-kilohertz clock. The analog input voltage for a full-scale output (2000 counts) is $V_I = 2\ V_{ref}$.

Data Acquisition

7

Linear Circuits

General Information   1

Thermal Information   2

Operational Amplifiers   3

Voltage Comparators   4

Special Functions   5

Voltage Regulators   6

Data Acquisition   7

Appendix   A

ORDERING INSTRUCTIONS

Electrical characteristics presented in this data book, unless otherwise noted, apply for the circuit type(s) listed in the page heading regardless of package. The availability of a circuit function in a particular package is denoted by an alphabetical reference above the pin-connection diagram(s). These alphabetical references refer to mechanical outline drawings shown in this section.

Factory orders for circuits described in this data book should include a four-part type number as explained in the following example.

EXAMPLE: TL 062M JG /883B

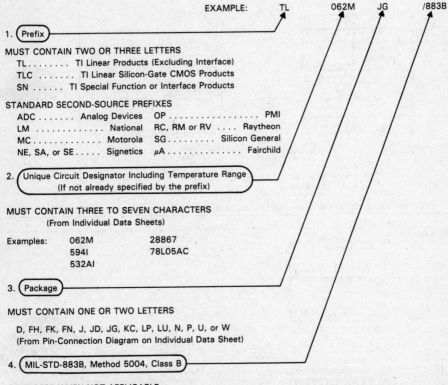

1. Prefix

MUST CONTAIN TWO OR THREE LETTERS

TL . . . . TI Linear Products (Excluding Interface)
TLC . . . . TI Linear Silicon-Gate CMOS Products
SN . . . . . . TI Special Function or Interface Products

STANDARD SECOND-SOURCE PREFIXES

ADC . . . . Analog Devices   OP . . . . PMI
LM . . . . National   RC, RM or RV . . . . Raytheon
MC . . . . Motorola   SG . . . . Silicon General
NE, SA, or SE . . . . . Signetics   μA . . . . Fairchild

2. Unique Circuit Designator Including Temperature Range
(If not already specified by the prefix)

MUST CONTAIN THREE TO SEVEN CHARACTERS
(From Individual Data Sheets)

Examples: 062M   28867
           594I   78L05AC
           532AI

3. Package

MUST CONTAIN ONE OR TWO LETTERS

D, FH, FK, FN, J, JD, JG, KC, LP, LU, N, P, U, or W
(From Pin-Connection Diagram on Individual Data Sheet)

4. MIL-STD-883B, Method 5004, Class B

OMIT/883B WHEN NOT APPLICABLE

Circuits are shipped in one of the carriers below. Unless a specific method of shipment is specifed by the customer (with possible additional costs), circuits will be shipped on the most practical carrier.

Dual-In-Line (D, J, JD, JG, N, P)   Plug-In (LP, LU)   Flat (U, W)

   —Slide Magazines   —Barnes Carrier   —Barnes Carrier
   —A-Channel Plastic Tubing   —Sectional Cardboard Box   —Milton Ross Carrier
   —Barnes Carrier   —Individual Cardboard Box
   —Sectioned Cardboard Box
   —Individual Cardboard Box   Chip Carriers (FH, FK, FN)   TO-220AB (KC)
                              —Anti-Static Plastic Tubing   —Sleeves

Appendix

A

MECHANICAL DATA

D plastic dual-in-line packages

Each of these dual-in-line packages consists of a circuit mounted on a lead frame and encapsulated within a plastic compound. The compound will withstand soldering temperature with no deformation, and circuit performance characteristics will remain stable when operated in high-humidity conditions. Leads require no additional cleaning or processing when used in soldered assembly.

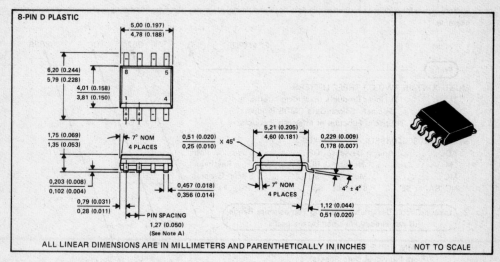

8-PIN D PLASTIC

ALL LINEAR DIMENSIONS ARE IN MILLIMETERS AND PARENTHETICALLY IN INCHES — NOT TO SCALE

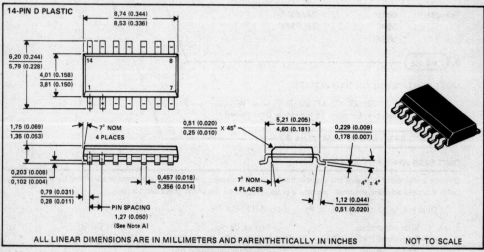

14-PIN D PLASTIC

ALL LINEAR DIMENSIONS ARE IN MILLIMETERS AND PARENTHETICALLY IN INCHES — NOT TO SCALE

NOTE A: Each pin centerline is located within 0,25 (0.010) of its true longitudinal position.

Appendix

A

TEXAS
INSTRUMENTS
POST OFFICE BOX 225012 • DALLAS, TEXAS 75265

FH and FK ceramic chip carrier packages

Both versions of these hermetically sealed chip carrier packages have ceramic bases. The FH package is an all-ceramic package with a glass seal. The FK package has a three-layer base with a metal lid and braze seal.

The packages are intended for surface mounting on solder lands on 1,27 (0.050) centers. Terminals require no additional cleaning or processing when used in soldered assembly.

FH and FK package terminal assignments conform to JEDEC Standards 1 and 2.

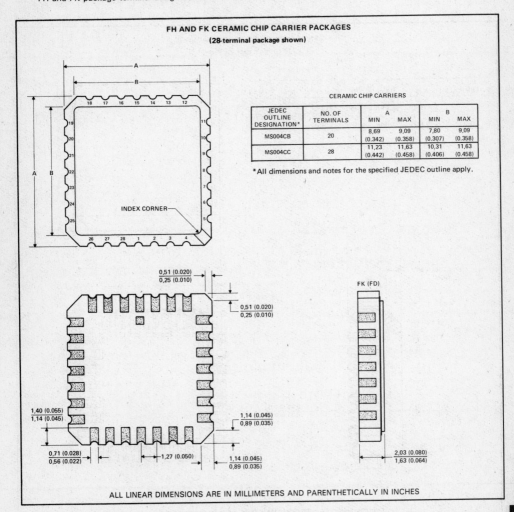

FH AND FK CERAMIC CHIP CARRIER PACKAGES
(28-terminal package shown)

CERAMIC CHIP CARRIERS

JEDEC OUTLINE DESIGNATION*	NO. OF TERMINALS	A		B	
		MIN	MAX	MIN	MAX
MS004CB	20	8,69 (0.342)	9,09 (0.358)	7,80 (0.307)	9,09 (0.358)
MS004CC	28	11,23 (0.442)	11,63 (0.458)	10,31 (0.406)	11,63 (0.458)

*All dimensions and notes for the specified JEDEC outline apply.

ALL LINEAR DIMENSIONS ARE IN MILLIMETERS AND PARENTHETICALLY IN INCHES

Appendix

A

MECHANICAL DATA

FN plastic chip carrier package

Each of these chip carrier packages consists of a circuit mounted on a lead frame and encapsulated within an electrically nonconductive plastic compound. The compound withstands soldering temperatures with no deformation, and circuit performance characteristics remain stable when the devices are operated in high-humidity conditions. The packages are intended for surface mounting on solder lands on 1,27 (0.050) centers. Leads require no additional cleaning or processing when used in soldered assembly.

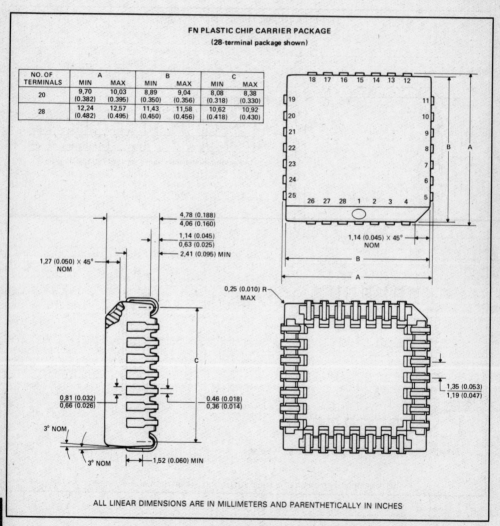

FN PLASTIC CHIP CARRIER PACKAGE
(28-terminal package shown)

NO. OF TERMINALS	A		B		C	
	MIN	MAX	MIN	MAX	MIN	MAX
20	9,70 (0.382)	10,03 (0.395)	8,89 (0.350)	9,04 (0.356)	8,08 (0.318)	8,38 (0.330)
28	12,24 (0.482)	12,57 (0.495)	11,43 (0.450)	11,58 (0.456)	10,62 (0.418)	10,92 (0.430)

ALL LINEAR DIMENSIONS ARE IN MILLIMETERS AND PARENTHETICALLY IN INCHES

TEXAS INSTRUMENTS
POST OFFICE BOX 225012 • DALLAS, TEXAS 75265

J ceramic dual-in-line packages

Each of these hermetically sealed dual-in-line packages consists of a ceramic base, ceramic cap, and a lead frame. Hermetic sealing is accomplished with glass. The packages are intended for insertion in mounting-hole rows on 7,62 (0.300) centers (see Note A). Once the leads are compressed and inserted, sufficient tension is provided to secure the package in the board during soldering. Leads require no additional cleaning or processing when used in soldering assembly.

14-PIN J CERAMIC

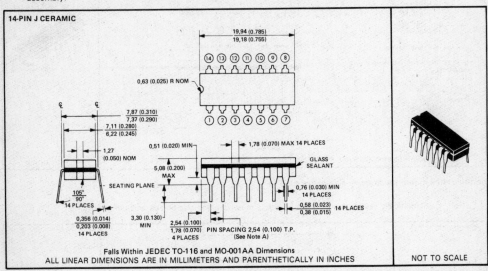

Falls Within JEDEC TO-116 and MO-001AA Dimensions
ALL LINEAR DIMENSIONS ARE IN MILLIMETERS AND PARENTHETICALLY IN INCHES

NOT TO SCALE

16-PIN J CERAMIC

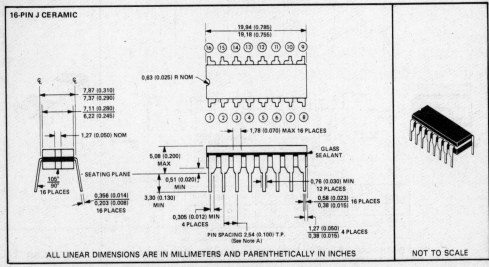

ALL LINEAR DIMENSIONS ARE IN MILLIMETERS AND PARENTHETICALLY IN INCHES

NOT TO SCALE

Appendix

A

NOTE A: Each pin centerline is located within 0,25 (0.010) of its true longitudinal position.

MECHANICAL DATA

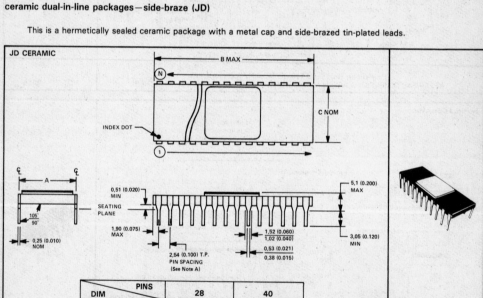

20-PIN J CERAMIC

ALL LINEAR DIMENSIONS ARE IN MILLIMETERS AND PARENTHETICALLY IN INCHES

NOT TO SCALE

ceramic dual-in-line packages—side-braze (JD)

This is a hermetically sealed ceramic package with a metal cap and side-brazed tin-plated leads.

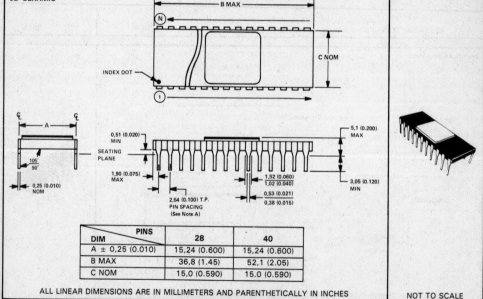

JD CERAMIC

DIM	PINS	28	40
A ± 0,25 (0.010)		15,24 (0.600)	15,24 (0.600)
B MAX		36,8 (1.45)	52,1 (2.05)
C NOM		15,0 (0.590)	15,0 (0.590)

ALL LINEAR DIMENSIONS ARE IN MILLIMETERS AND PARENTHETICALLY IN INCHES

NOT TO SCALE

NOTE A: Each pin centerline is located within 0,25 (0.010) of its true longitudinal position.

Texas Instruments

POST OFFICE BOX 225012 • DALLAS, TEXAS 75265

JG ceramic dual-in-line package

This hermetically sealed dual-in-line package consists of a ceramic base, ceramic cap, and a lead frame. The package is intended for insertion in mounting-hole rows 7,62 (0.300) centers (see Note A). Once the leads are compressed and inserted, sufficient tension is provided to secure the package in the board during soldering.

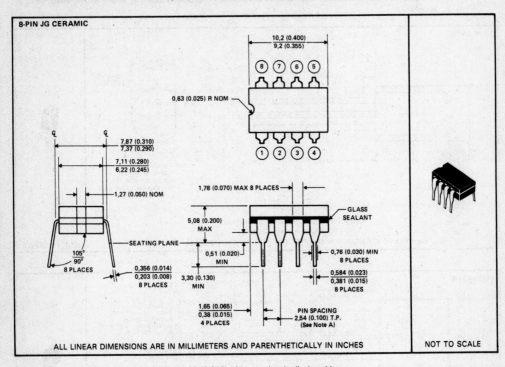

8-PIN JG CERAMIC

ALL LINEAR DIMENSIONS ARE IN MILLIMETERS AND PARENTHETICALLY IN INCHES

NOT TO SCALE

NOTE A: Each pin centerline is located within 0,25 (0.010) of its true longitudinal position.

Appendix

A

MECHANICAL DATA

LP and LU plastic packages

These packages each consists of a circuit mounted on a lead frame and encapsulated within a plastic compound. The compound will withstand soldering temperature with no deformation and circuit performance characteristics remain stable when operated in high-humidity conditions. Leads require no additional cleaning or processing when used in soldered assembly.

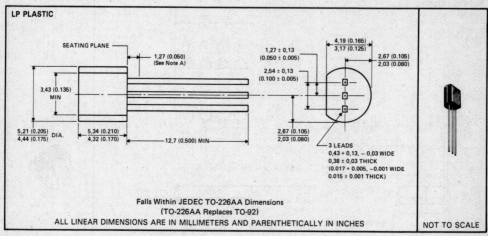

LP PLASTIC

Falls Within JEDEC TO-226AA Dimensions
(TO-226AA Replaces TO-92)
ALL LINEAR DIMENSIONS ARE IN MILLIMETERS AND PARENTHETICALLY IN INCHES

NOT TO SCALE

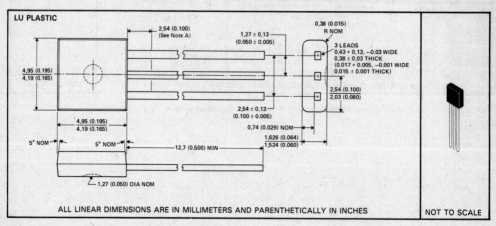

LU PLASTIC

ALL LINEAR DIMENSIONS ARE IN MILLIMETERS AND PARENTHETICALLY IN INCHES

NOT TO SCALE

NOTE A: Lead dimensions are not controlled within this area.

Appendix

A

TEXAS
INSTRUMENTS
POST OFFICE BOX 225012 • DALLAS, TEXAS 75265

N plastic dual-in-line package

Each of these dual-in-line packages consists of a circuit mounted on a lead frame and encapsulated within a plastic compound. The compound will withstand soldering temperature with no deformation, and circuit performance characteristics will remain stable when operated in high-humidity conditions. The packages are intended for insertion in mounting-hole rows on 7,62 (0.300) or 15,24 (0.600) centers (see Note A). Once the leads are compressed and inserted, sufficient tension is provided to secure the package in the board during soldering. Leads require no additional cleaning or processing when used in soldered assembly.

14-PIN N PLASTIC

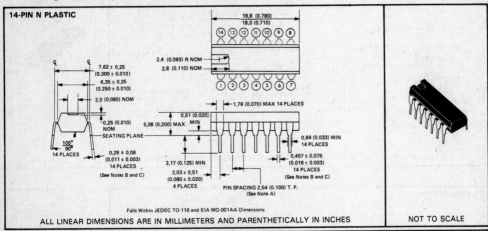

Falls Within JEDEC TO-116 and EIA MO-001AA Dimensions

ALL LINEAR DIMENSIONS ARE IN MILLIMETERS AND PARENTHETICALLY IN INCHES

NOT TO SCALE

16-PIN N PLASTIC

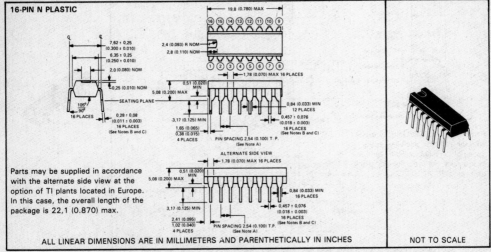

Parts may be supplied in accordance with the alternate side view at the option of TI plants located in Europe. In this case, the overall length of the package is 22,1 (0.870) max.

ALL LINEAR DIMENSIONS ARE IN MILLIMETERS AND PARENTHETICALLY IN INCHES

NOT TO SCALE

NOTES: A. Each pin centerline is located within 0,25 (0.010) of its true longitudinal position.
  B. This dimension does not apply for solder-dipped leads.
  C. When solder-dipped leads are specified, dipped area of the lead extends from the lead tip to at least 0,51 (0.020) above seating plane.

Appendix

A

MECHANICAL DATA

N plastic dual-in-line packages (continued)

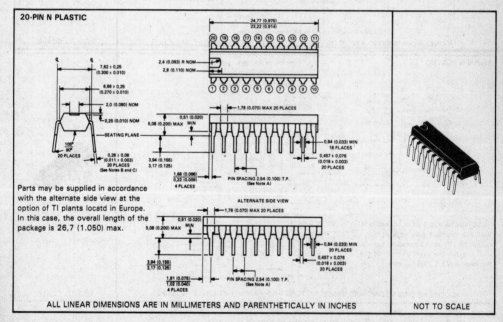

18-PIN N PLASTIC

ALL LINEAR DIMENSIONS ARE IN MILLIMETERS AND PARENTHETICALLY IN INCHES

NOT TO SCALE

20-PIN N PLASTIC

Parts may be supplied in accordance with the alternate side view at the option of TI plants located in Europe. In this case, the overall length of the package is 26,7 (1.050) max.

ALTERNATE SIDE VIEW

ALL LINEAR DIMENSIONS ARE IN MILLIMETERS AND PARENTHETICALLY IN INCHES

NOT TO SCALE

NOTES: A. Each pin centerline is located within 0,25 (0.010) of its true longitudinal position.
        B. This dimension does not apply for solder-dipped leads.
        C. When solder-dipped leads are specified, dipped area of the lead extends from the lead tip to at least 0,51 (0.020) above seating plane.

Appendix

A

TEXAS
INSTRUMENTS
POST OFFICE BOX 225012 • DALLAS, TEXAS 75265

N plastic packages (continued)

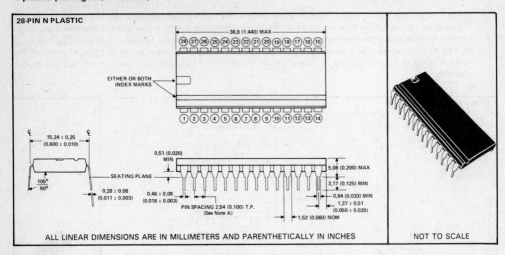

28-PIN N PLASTIC

ALL LINEAR DIMENSIONS ARE IN MILLIMETERS AND PARENTHETICALLY IN INCHES

NOT TO SCALE

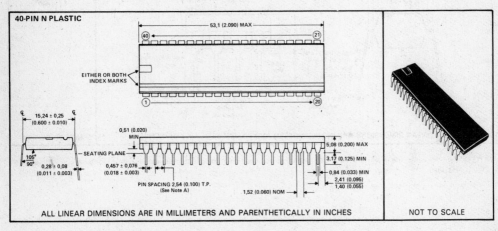

40-PIN N PLASTIC

ALL LINEAR DIMENSIONS ARE IN MILLIMETERS AND PARENTHETICALLY IN INCHES

NOT TO SCALE

NOTE A: Each pin centerline is located within 0,25 (0.010) of its true longitudinal position.

TEXAS INSTRUMENTS
POST OFFICE BOX 225012 ● DALLAS, TEXAS 75265

Appendix

A

MECHANICAL DATA

P dual-in-line plastic package

This dual-in-line package consists of a circuit mounted on a lead frame and encapsulated within a plastic compound. The compound will withstand soldering temperature with no deformation, and circuit performance characteristics remain stable when operated under high-humidity conditions. The package is intended for insertion in mounting hole rows on 7,62 (0.300) centers. Once the leads are compressed and inserted, sufficient tension is provided to secure the package in the board during soldering. Leads require no additional cleaning or processing when used in soldering assembly.

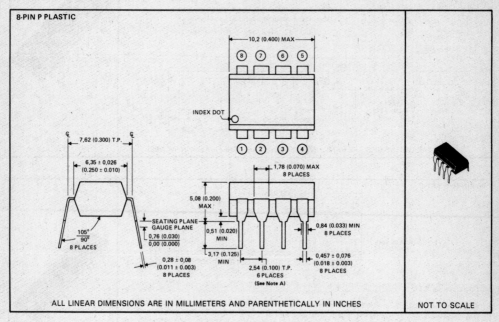

8-PIN P PLASTIC

ALL LINEAR DIMENSIONS ARE IN MILLIMETERS AND PARENTHETICALLY IN INCHES

NOT TO SCALE

NOTE A: Each pin centerline is within 0,13 (0.005) radius of true position at the gauge plane with maximum material condition and unit installed.

Appendix

A

TEXAS
INSTRUMENTS
POST OFFICE BOX 225012 ● DALLAS, TEXAS 75265

U ceramic flat packages

This flat package consists of a ceramic base, ceramic cap, and lead frame. Circuit bars are alloy mounted. Hermetic sealing is accomplished with glass. Leads require no additional cleaning or processing when used in soldered assembly.

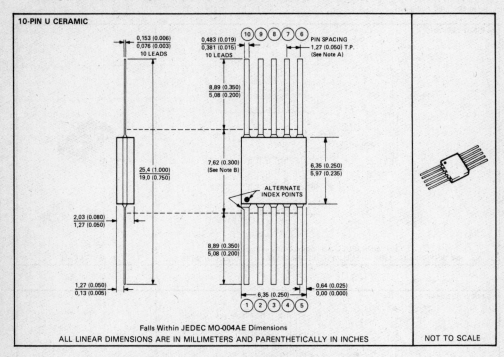

10-PIN U CERAMIC

Falls Within JEDEC MO-004AE Dimensions

ALL LINEAR DIMENSIONS ARE IN MILLIMETERS AND PARENTHETICALLY IN INCHES

NOT TO SCALE

NOTES: A. Leads are within 0.005 radius of true position (TP) at maximum material condition.
B. This dimension determines a zone within which all body and lead irregularities lie.

MECHANICAL DATA

W ceramic flat packages

These hermetically sealed flat packages consist of an electrically nonconductive ceramic base and cap and a lead frame. Hermetic sealing is accomplished with glass. Leads require no additional cleaning or processing when used in soldered assembly.

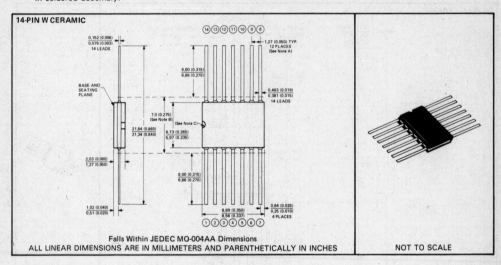

14-PIN W CERAMIC

Falls Within JEDEC MO-004AA Dimensions
ALL LINEAR DIMENSIONS ARE IN MILLIMETERS AND PARENTHETICALLY IN INCHES

NOT TO SCALE

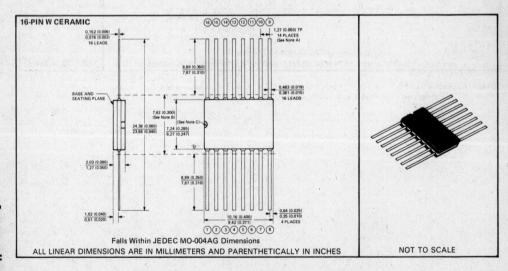

16-PIN W CERAMIC

Falls Within JEDEC MO-004AG Dimensions
ALL LINEAR DIMENSIONS ARE IN MILLIMETERS AND PARENTHETICALLY IN INCHES

NOT TO SCALE

NOTES: A. Leads are within 0,13 (0.005) radius of true position (TP) at maximum material condition.
   B. This dimension determines a zone within which all body and lead irregularities lie.
   C. Index point is provided on cap for terminal identification only.

Appendix

A

TEXAS INSTRUMENTS
POST OFFICE BOX 225012 ● DALLAS, TEXAS 75265

NOTES

NOTES

Texas Instruments
Semiconductor Technical Literature

TTL Data Book, Vol. 1, 1984, 336 pages.
Product guide for all TI TTL devices, functional indexes, alphanumeric index, and general information.

TTL Data Book, Vol. 2, 1984, 1,000 pages.
Detailed specifications and application information on the TI family of Low-power Schottky (LS), Schottky (S), and standard TTL logic devices.

TTL Data Book, Vol. 3, 1984, 792 pages.
Detailed specifications and application information on the TI family of Advanced Low-power Schottky (ALS) and Advanced Schottky (AS) logic devices.

TTL Data Book, Vol. 4, 1984, 416 pages.
Detailed specifications and application information on the TI family of bipolar field-programmable logic (FPL), programmable read-only memories (PROM), random-access memories (RAM), microprocessors, and support circuits.

High-speed CMOS Logic Data Book, 1984, 580 pages.
Detailed specifications and application information on the TI family of High-speed CMOS logic devices. Includes product selection guide, glossary, and alphanumeric index.

Linear Circuits Data Book, 1984, 820 pages.
Detailed specifications on operational amplifiers, voltage comparators, voltage regulators, data-acquisition devices, a/d converters, timers, switches, amplifiers, and special functions. Includes LinCMOS™ functions. Contains product guide, interchangeability guide, glossary, and alphanumeric index.

Interface Circuits Data Book, 1981, 700 pages.
Includes specifications and applications information on TTL logic interface circuits, as well as product profiles on the line drivers/receivers and peripheral drivers.

Optoelectronics Data Book, 1983, 480 pages.
Contains more than 300 device types representing traditional optoelectronics (IREDs, LEDs, detectors, couplers, and displays), special components (avalanche, photodiodes, and transimpedance amplifiers), fiber optic components (sources, detectors, and interconnecting cables), and new image sensors (linear and arrays).

MOS Memory Data Book, 1984, 456 pages.
Detailed specifications on dynamic RAMs, static RAMs, EPROMs, ROMs, cache address comparators, and memory controllers. Contains product guide, interchangeability guide, glossary, and alphanumeric index. Also, chapters on testing and reliability.

TMS7000 Family Data Manual, 1983, 350 pages.
Detailed specifications and application information on TI's family of microprogrammable 8-bit microcomputers. Includes architecture description, device operation, instruction set, electrical characteristics, and mechanical data. TMS7000 microcomputers include versions in CMOS and SMOS and with on-board UART.

TMSxxxxx Microcomputer Data Manuals
These manuals contain detailed specifications and application information on specific TMSxxxxx microcomputers and peripherals. Include architecture description, device operation, instruction set, electrical characteristics, and mechanical data.

Assembly Language Programmer's Guides.
TMS32010, 1983, 160 pages.
TMS99000, 1983, 322 pages.
TMS7000, 1983, 160 pages.
Include general programming information, assembly instructions, assembler directives, assembler output, and application notes.

TMS32010 User's Guide, 1984, 400 pages.
Detailed application information on the TMS32010 Digital Signal Processor. Detailed reference manual on use of the TMS320 instruction set. Data sheets included.

Fundamentals of Microcomputer Design, 1982, 584 pages.
University textbook. Subjects include microprocessors, software, instruction sets, microcomputer programming, high-level languages, hardware features, microcomputer memory, and I/O design. A design example is included.

Understanding Series™ Books
The Understanding Series books form a library written for anyone who wants to learn quickly and easily about today's technology, its impact on our world, and its application in our lives. Each book is written in bright, clear, down-to-earth language and focuses on one aspect of what's new in today's electronics. Engineering concepts and theory are explained using simple arithmetic. Technical terms are explained in layman's language. Ideal for self-paced, individualized instruction. Currently 13 different titles in the series.

Write for current availability and prices to:
Texas Instruments
Information Publishing Center
P.O. Box 225012 MS-54
Dallas, TX 75265

April 1984

NOTES

TI Sales Offices

ALABAMA: Huntsville (205) 837-7530.

ARIZONA: Phoenix (602) 995-1007.

CALIFORNIA: Irvine (714) 660-8187;
Sacramento (916) 929-1521;
San Diego (619) 278-9601;
Santa Clara (408) 980-9000;
Torrance (213) 217-7010;
Woodland Hills (818) 704-7759.

COLORADO: Aurora (303) 368-8000.

CONNECTICUT: Wallingford (203) 269-0074.

FLORIDA: Ft. Lauderdale (305) 973-8502;
Maitland (305) 660-4600; **Tampa** (813) 870-6420.

GEORGIA: Norcross (404) 662-7900.

ILLINOIS: Arlington Heights (312) 640-2925.

INDIANA: Ft. Wayne (219) 424-5174;
Indianapolis (317) 248-8555.

IOWA: Cedar Rapids (319) 395-9550.

MARYLAND: Baltimore (301) 944-8600.

MASSACHUSETTS: Waltham (617) 895-9100.

MICHIGAN: Farmington Hills (313) 553-1500.

MINNESOTA: Eden Prairie (612) 828-9300.

MISSOURI: Kansas City (816) 523-2500;
St. Louis (314) 569-7600.

NEW JERSEY: Iselin (201) 750-1050.

NEW MEXICO: Albuquerque (505) 345-2555.

NEW YORK: East Syracuse (315) 463-9291;
Endicott (607) 754-3900; **Melville** (516) 454-6600;
Pittsford (716) 385-6770;
Poughkeepsie (914) 473-2900.

NORTH CAROLINA: Charlotte (704) 527-0930;
Raleigh (919) 876-2725.

OHIO: Beachwood (216) 464-6100;
Dayton (513) 258-3877.

OKLAHOMA: Tulsa (918) 250-0633.

OREGON: Beaverton (503) 643-6758.

PENNSYLVANIA: Ft. Washington (215) 643-6450;
Coraopolis (412) 771-8550.

PUERTO RICO: Hato Rey (809) 753-8700

TEXAS: Austin (512) 250-7655;
Houston (713) 778-6592; **Richardson** (214) 680-5082;
San Antonio (512) 496-1779.

UTAH: Murray (801) 266-8972.

VIRGINIA: Fairfax (703) 849-1400.

WASHINGTON: Redmond (206) 881-3080.

WISCONSIN: Brookfield (414) 785-7140.

CANADA: Nepean, Ontario (613) 726-1970;
Richmond Hill, Ontario (416) 884-9181;
St. Laurent, Quebec (514) 334-3635.

TI Regional Technology Centers

CALIFORNIA: Irvine (714) 660-8140.
Santa Clara (408) 748-2220.

GEORGIA: Norcross (404) 662-7945.

ILLINOIS: Arlington Heights (312) 640-2909.

MASSACHUSETTS: Waltham (617) 890-6671.

TEXAS: Richardson (214) 680-5066.

CANADA: Nepean, Ontario (613) 726-1970

Technical Support Center

TOLL FREE: (800) 232-3200

TI Distributors

TI AUTHORIZED DISTRIBUTORS IN USA

Arrow Electronics
Diplomat Electronics
General Radio Supply Company
Graham Electronics
Harrison Equipment Co.
International Electronics
JACO Electronics
Kierulff Electronics
LCOMP, Incorporated
Marshall Industries
Milgray Electronics
Newark Electronics
Rochester Radio Supply
Time Electronics
R.V. Weatherford Co.
Wyle Laboratories

TI AUTHORIZED DISTRIBUTORS IN CANADA

Arrow/CESCO Electronics, Inc.
Future Electronics
ITT Components
L.A. Varah, Ltd.

ALABAMA: Arrow (205) 882-2730;
Kierulff (205) 883-6070; Marshall (205) 881-9235.

ARIZONA: Arrow (602) 968-4800;
Kierulff (602) 243-4101; Marshall (602) 968-6181;
Wyle (602) 866-2888.

CALIFORNIA: Los Angeles/Orange County:
Arrow (818) 701-7500, (714) 838-5422;
Kierulff (213) 725-0325, (714) 731-5711, (714) 220-6300;
Marshall (818) 999-5001, (818) 442-7204,
(714) 660-0951; R.V. Weatherford (714) 634-9600,
(213) 849-3451, (714) 623-1261; Wyle (213) 322-8100,
(818) 880-9001, (714) 863-9953; **Sacramento:** Arrow
(916) 925-7456; Wyle (916) 638-5282; **San Diego:**
Arrow (619) 565-4800; Kierulff (619) 278-2112;
Marshall (619) 578-9600; Wyle (619) 565-9171;
San Francisco Bay Area: Arrow (408) 745-6600;
(415) 487-4600; Kierulff (408) 971-2600;
Marshall (408) 732-1100; Wyle (408) 727-2500;
Santa Barbara: R.V. Weatherford (805) 965-8551.

COLORADO: Arrow (303) 696-1111;
Kierulff (303) 790-4444; Wyle (303) 457-9953.

CONNECTICUT: Arrow (203) 265-7741;
Diplomat (203) 797-9674; Kierulff (203) 265-1115;
Marshall (203) 265-3822; Milgray (203) 795-0714.

FLORIDA: Ft. Lauderdale: Arrow (305) 429-8200;
Diplomat (305) 974-8700; Kierulff (305) 486-4004;
Orlando: Arrow (305) 725-1480;
Milgray (305) 647-5747; **Tampa:**
Arrow (813) 576-8995; Diplomat (813) 443-4514;
Kierulff (813) 576-1966.

GEORGIA: Arrow (404) 449-8252;
Kierulff (404) 447-5252; Marshall (404) 923-5750.

TEXAS INSTRUMENTS
Creating useful products
and services for you.

ILLINOIS: Arrow (312) 397-3440;
Diplomat (312) 595-1000; Kierulff (312) 250-0500;
Marshall (312) 490-0155; Newark (312) 784-5100.

INDIANA: Indianapolis: Arrow (317) 243-9353;
Graham (317) 634-8202; Marshall (317) 297-0483;
Ft. Wayne: Graham (219) 423-3422.

IOWA: Arrow (319) 395-7230.

KANSAS: Kansas City: Marshall (913) 492-3121;
Wichita: LCOMP (316) 265-9507.

MARYLAND: Arrow (301) 995-0003;
Diplomat (301) 995-1226; Kierulff (301) 636-5800;
Milgray (301) 793-3993.

MASSACHUSETTS: Arrow (617) 933-8130;
Diplomat (617) 935-6611; Kierulff (617) 667-8331;
Marshall (617) 272-8200; Time (617) 935-8080.

MICHIGAN: Detroit: Arrow (313) 971-8220;
Marshall (313) 525-5850; Newark (313) 967-0600;
Grand Rapids: Arrow (616) 243-0912.

MINNESOTA: Arrow (612) 830-1800;
Kierulff (612) 941-7500; Marshall (612) 559-2211.

MISSOURI: Kansas City: LCOMP (816) 221-2400;
St. Louis: Arrow (314) 567-6888;
Kierulff (314) 739-0855.

NEW HAMPSHIRE: Arrow (603) 668-6968.

NEW JERSEY: Arrow (201) 575-5300, (609) 596-8000;
Diplomat (201) 785-1830;
General Radio (609) 964-8560; Kierulff (201) 575-6750;
(609) 235-1444; Marshall (201) 882-0320,
(609) 234-9100; Milgray (609) 983-5010.

NEW MEXICO: Arrow (505) 243-4566;
International Electronics (505) 345-8127.

NEW YORK: Long Island: Arrow (516) 231-1000;
Diplomat (516) 454-6400; JACO (516) 273-5500;
Marshall (516) 273-2053; Milgray (516) 420-9800;
Rochester: Arrow (716) 427-0300;
Marshall (716) 235-7620;
Rochester Radio Supply (716) 454-7800;
Syracuse: Arrow (315) 652-1000;
Diplomat (315) 652-5000; Marshall (607) 798-1611.

NORTH CAROLINA: Arrow (919) 876-3132.
(919) 725-8711; Kierulff (919) 872-8410.

OHIO: Cincinnati: Graham (513) 772-1661;
Cleveland: Arrow (216) 248-3990;
Kierulff (216) 587-6558; Marshall (216) 248-1788.
Columbus: Graham (614) 895-1590;
Dayton: Arrow (513) 435-5563; Kierulff (513) 439-0045;
Marshall (513) 236-8088.

OKLAHOMA: Kierulff (918) 252-7537.

OREGON: Arrow (503) 684-1690; Kierulff
(503) 641-9153; Wyle (503) 640-6000; Marshall
(503) 644-5050.

PENNSYLVANIA: Arrow (412) 856-7000.
(215) 928-1800; General Radio (215) 922-7037.

RHODE ISLAND: Arrow (401) 431-0980

TEXAS: Austin: Arrow (512) 835-4180;
Kierulff (512) 835-2090; Marshall (512) 837-1991;
Wyle (512) 834-9957; **Dallas:** Arrow (214) 380-6464;
International Electronics (214) 233-9323;
Kierulff (214) 343-2400; Marshall (214) 233-5200;
Wyle (214) 235-9953;
El Paso: International Electronics (915) 598-3406;
Houston: Arrow (713) 530-4700;
Marshall (713) 789-6600;
Harrison Equipment (713) 879-2600;
Kierulff (713) 530-7030; Wyle (713) 879-9953.

UTAH: Diplomat (801) 486-4134;
Kierulff (801) 973-6913; Wyle (801) 974-9953.

VIRGINIA: Arrow (804) 282-0413.

WASHINGTON: Arrow (206) 643-4800;
Kierulff (206) 575-4420; Wyle (206) 453-8300; Marshall
(206) 747-9100.

WISCONSIN: Arrow (414) 764-6600; Kierulff
(414) 784-8160.

CANADA: Calgary: Future (403) 235-5325; Varah
(403) 255-9550; **Edmonton:** Future (403) 486-0974;
Varah (403) 437-2755; **Montreal:** Arrow/CESCO
(514) 735-5511; Future (514) 694-7710; ITT
Components (514) 735-1177; **Ottawa:** Arrow/CESCO
(613) 226-6903; Future (613) 820-8313; ITT
Components (613) 226-7406; Varah (613) 726-8884;
Quebec City: Arrow/CESCO (418) 687-4231; **Toronto:**
CESCO (416) 661-0220;
Future (416) 638-4771; ITT Components
(416) 736-1144; Varah (416) 842-8484;
Vancouver: Future (604) 438-5545; Varah
(604) 873-3211; **Winnipeg:** Varah (204) 633-6190   BL

TI Worldwide Sales Offices

TEXAS INSTRUMENTS
Creating useful products
and services for you